D0844193

NELSON

Physics 11

NELSON

Physics 11

Authors

Alan Hirsch
Formerly of Peel District School Board

David Martindale
Formerly of Waterloo Region District School Board

Steve Bibla
Toronto District School Board

Charles Stewart
Toronto District School Board

Program Consultant

Maurice Di Giuseppe
Toronto Catholic District School Board

Australia • Canada • Mexico • Singapore • Spain • United Kingdom • United States

Nelson Physics 11

Authors
Alan Hirsch
David Martindale
Steve Bibla
Charles Stewart

Contributing Authors
Robert Heath
William Konrad

Contributing Writers
Alison Armstrong
Lloyd Gill
Roche Kelly
Barry LeDrew
Edward Somerton

Director of Publishing
David Steele

Publisher
Kevin Martindale

Program Manager
John Yip-Chuck

Developmental Editors
Vicki Austin, Julie Bedford, Betty Robinson, Winnie Siu

Researcher
David Frair

Editorial Assistant
Matthew Roberts

Senior Managing Editor
Nicola Balfour

Senior Production Editor
Linh Vu

Production Coordinator
Sharon Latta Paterson

Creative Director
Angela Cluer

Art Management
Suzanne Peden
Peggy Rhodes

Design and Composition Team
Marnie Benedict
Anne Bradley
Susan Calverley
Angela Cluer
Zenaida Diores
Krista Donnelly
Erich Falkenberg
Tammy Gay
Julie Greener
Alicja Jamorski
Linda Neale
Peter Papayanakis
Suzanne Peden
Ken Phipps
June Reynolds
Peggy Rhodes
Katherine Strain
Janet Zanette

Cover Design
Katherine Strain

Cover Image
Bruce Coleman Collection/
First Light

Photo Research and Permissions
Vicki Gould

Printer
Transcontinental Printing Inc.

Printed and bound in Canada
1 2 3 4 04 03 02 01

For more information contact Nelson Thomson Learning,
1120 Birchmount Road,
Toronto, Ontario, M1K 5G4.
Or you can visit our Internet site at
http://www.nelson.com

National Library of Canada Cataloguing in Publication Data

Main entry under title:
Nelson physics 11

(Nelson science)
For use in grade 11 Ontario curriculum.
Includes index.
ISBN 0-17-612102-1

1. Physics–Study and teaching (Secondary).
I. Hirsch, Alan J. II. Title: Physics eleven. III. Title: Physics 11.
IV. Title: Nelson physics eleven.
V. Series.

QC47.C32O57 2001 530
C2001-930535-4

Reviewers

NELSON SENIOR SCIENCE ADVISORY PANEL

Oksana Baczynsky
Toronto District School Board, ON

Terry Brown
Thames Valley District School Board, ON

Doug De La Matter
Renfrew District School Board, ON

Xavier Fazio
Halton Catholic District School Board, ON

Ted Gibb
Thames Valley District School Board, ON

Diana Hall
Ottawa-Carleton District School Board, ON

William Konrad
Formerly of Lambton Kent District School Board, ON

Clifford Sampson
Appleby College, ON

ASSESSMENT CONSULTANT

Damian Cooper
Halton District School Board, ON

SKILLS HANDBOOK CONSULTANT

Barry LeDrew
Formerly of Newfoundland Department of Education

ACCURACY REVIEWERS

Professor Ernie McFarland
Department of Physics, University of Guelph

Dr. Stefan Zukotynski
Department of Electrical and Computer Engineering, University of Toronto

SAFETY REVIEWER

Stella Heenan
STAO Safety Committee

TEACHER REVIEWERS

Carolyn Black
Western School Board, PEI

Jason Braun
St. James-Assiniboine School Division #2, MB

Elizabeth Dunning
Toronto Catholic District School Board, ON

Bonnie Jean Edwards
Wellington Catholic District School Board, ON

Duncan Foster
York Region District School Board, ON

Martin Furtado
Thunder Bay Catholic District School Board, ON

Lloyd Gill
Formerly of Avalon East School Board, NF

Christopher Howes
Durham District School Board, ON

David Jensen
Bluewater District School Board, ON

Mark Kinoshita
Toronto District School Board, ON

Scott Leedham
Grand Erie District School Board, ON

Geary MacMillan
Halifax Regional School Board, NS

Michael McArdle
Dufferin-Peel Catholic District School Board, ON

Tim Murray
Upper Grand District School Board, ON

Kristen Niemi
Near North District School Board, ON

Jeannette Rensink
Durham District School Board, ON

Ron Ricci
Greater Essex District School Board, ON

Brian Schroeder
Dufferin-Peel Catholic District School Board, ON

Julian Smith
Dufferin-Peel Catholic District School Board, ON

Shelly Stazyk
Thames Valley District School Board, ON

Jim Young
Limestone District School Board, ON

Contents

Unit 1

Forces and Motion

Unit 2

Energy, Work, and Power

Unit 3

Waves and Sound

Unit 4

Light and Geometric Optics

Unit 5

Electricity and Magnetism

Unit 1

Forces and Motion

When Julie Payette first watched the Apollo moon missions on television she knew she wanted to be an astronaut. In 1999, she became a crew member of the space shuttle *Discovery*. During the ten-day mission, Ms Payette had to account for the reduced force of gravity and its effect on force and motion.

Forces affect the way that everything is done in space, from docking with other spacecraft to the design of tools. Tools must be specially designed so they will not impart a force on the user. Using an ordinary screwdriver could cause a person to fly off in the opposite direction.

Julie Payette, Canadian astronaut

Ms Payette hopes that her involvement in space exploration inspires others. "Students who want to be astronauts should be driven to succeed. You should develop your talent and go for it — get out and see as much as you can, learn as much as you can, and accomplish as much as you can!"

Overall Expectations

In this unit, you will be able to

- demonstrate an understanding of the relationship between forces and the acceleration of an object in linear motion
- investigate, through experimentation, the effect of a net force on the linear motion of an object, and analyze the effect in quantitative terms, using graphs, free-body diagrams, and vector diagrams
- describe the contributions of Galileo and Newton to the understanding of dynamics; evaluate and describe technological advances related to motion; and identify the effects of societal influences on transportation and safety issues

Unit 1

Forces and Motion

Are You Ready?

Knowledge and Understanding

1. **Figure 1** shows the motion of a car along a straight road. The images are taken at time intervals of 1.0 s. Describe the motion of the car using your vocabulary of motion.

Figure 1

2. A playful dog runs along the path shown in **Figure 2**, starting at A and following through all arrows. The distance from A to B is 16 m, and the distance from B to C is 12 m. The total time the dog takes to go from A along the path back to A again is 16 s.
   (a) State the compass direction the dog is moving in during each part of the run.
   (b) Determine the total distance travelled by the dog.
   (c) What is the net displacement of the dog over the entire path?
   (d) Calculate the dog's average speed of motion.
   (e) What is the dog's average velocity for the entire trip?

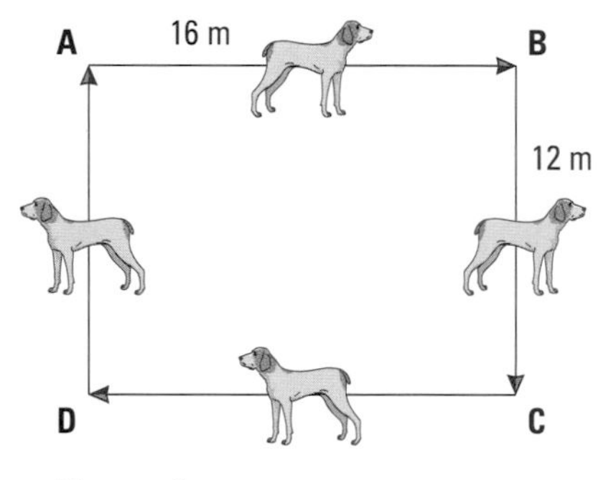

Figure 2

3. A "physics" golf ball, attached with a light that flashes regularly with time, is dropped in a dark room from shoulder height to the floor. Which set of dots representing the flashing of light in **Figure 3** would you observe in a photograph of the golf ball's downward motion? Explain your choice.

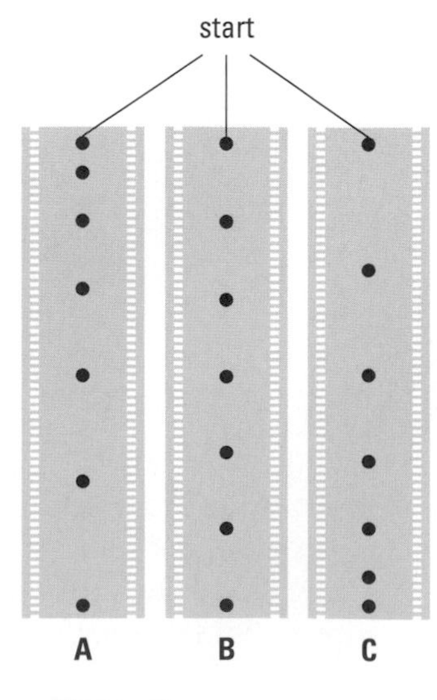

Figure 3

4. Refer to **Figure 4**.
   (a) Name all the forces on the diver at position A; at position B.
   (b) Draw a sketch of the diving board with the diver on it, and label the following in your diagram: tension in the board, compression in the board, force of gravity on the diver, and force of the board acting on the diver.
   (c) Which of the forces you labelled in (b) is a non-contact force? Explain how you can tell.
5. The scale used to draw the two geometric shapes in **Figure 5** is 1.0 cm = 1.0 m. Determine the surface area of each shape.
6. The motions of three cars, L, M, and N, are illustrated by the graphs in **Figure 6**. Compare the times of travel and average speeds of the three cars.
7. Calculate the slope of the line and the area under the line on the graph in **Figure 7**. State what the slope and area represent.

Figure 4
For question 4

Inquiry and Communication

8. Four groups of students perform an experiment to determine the density of a liquid taken from the same container. **Table 1** gives the calculated densities from the four groups.
   (a) Which result is the most reasonable for the density of a liquid determined by such an experiment?
   (b) Describe why the other three results are unacceptable in this case.

Table 1

Group	A	B	C	D
Density	0.76 g/mL	0.757 786 g/mL	1.3 g/mL	0.76 g/L

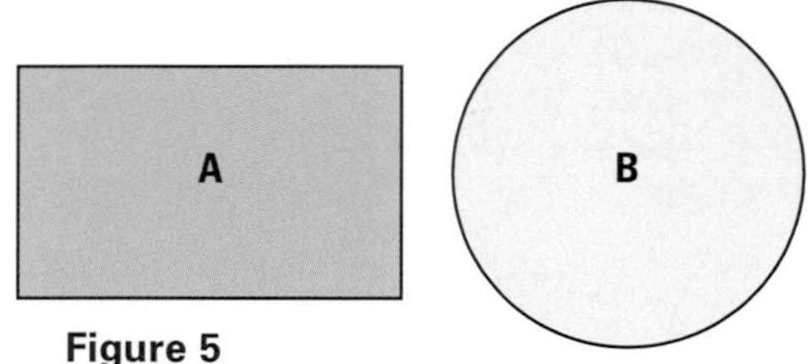

Figure 5
For question 5

9. You are asked to calculate the speed of a jogger who runs beside you along a straight track. Describe how you would perform an experiment to calculate the required quantity.
10. The results of a motion experiment can be communicated in various forms. Given one form, as shown in **Table 2** below, communicate the same results using the following other forms:
    (a) a description
    (b) a graph
    (c) a calculation

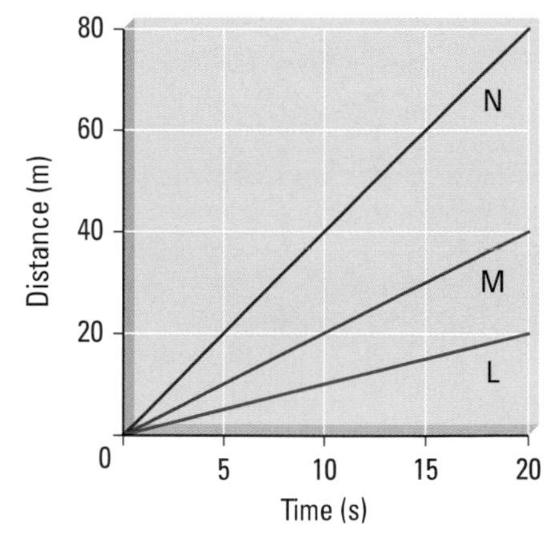

Figure 6
For question 6

Table 2

Time (s)	0	1.0	2.0	3.0	4.0	5.0
Distance (m)	0	4.0	8.0	12.0	16.0	20.0

Making Connections

11. Draw three dollar-size circles so they are spread out on a piece of paper. Within each circle, write one of these topics: transportation safety, space exploration, and sports applications. For each topic, write around the outside of the circle the physics concepts and other facts that you already know about the topic.

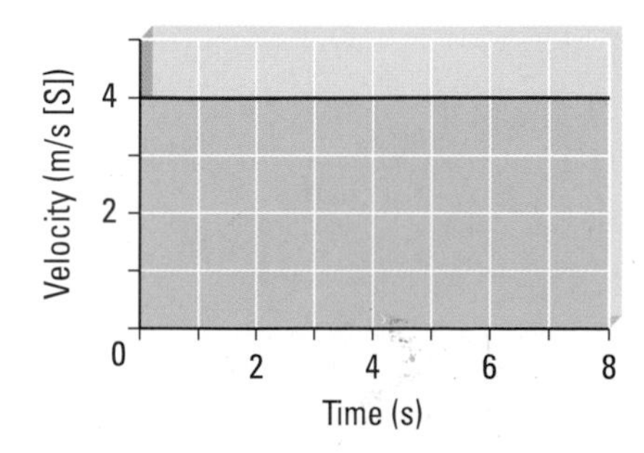

Figure 7
For question 7

Chapter

1

Motion

In this chapter, you will be able to

- define and describe motion concepts, such as speed, velocity, acceleration, scalar quantities, and vector quantities
- use motion-related units, such as metres per second and kilometres per hour for speed, in proper context
- describe and explain different kinds of motion, including uniform motion and uniform acceleration for both horizontal and vertical motion, and apply equations relating the variables displacement, velocity, acceleration, and time interval
- use vector diagrams to analyze motion in two dimensions
- interpret patterns observed on motion graphs drawn by hand or by computer, and calculate or infer relationships among the variables plotted on the graphs
- evaluate the design of technological solutions to transportation needs

Humans have climbed mountains, trekked to the North and South Poles, explored the ocean floor, and walked on the Moon. The next frontier people may explore is Mars, a planet whose properties resemble Earth's most closely. If you were part of a team planning a trip to Mars, what would you need to know in order to determine how long the trip would take? How does the average speed of the spacecraft you would use compare with the speeds of the spacecrafts in science fiction movies that can travel faster than light (**Figure 1**)? The skill of analyzing the motion of these spacecraft will help you determine how realistic some of the distances, velocities, and accelerations described in movies are. As you study this chapter, your ability to estimate values such as time and average speeds of moving objects will improve.

Only recently has our ability to measure time accurately been possible. About 400 years ago, the first pendulum clock was invented. You can imagine how difficult it was to determine seconds or fractions of a second using a swinging pendulum. The accuracy of time measurement has improved tremendously over the years. Many of today's watches can measure time to the nearest hundredth of a second (centisecond). This advancement in time measurement in turn leads to more questions, experiments, and applications. For example, in a mission to Mars, exact times of engine propulsions would be needed to ensure that the spacecraft could land safely on Mars' surface.

Reflect on your Learning

1. List all the terms regarding motion that you can recall from your earlier studies. Write the metric units for the terms that are quantitative measures.
2. Describe how you would calculate the time required for a specific motion. For example, calculate the time required for a spacecraft to travel from Earth to Mars with this information: Earth's distance from the Sun is 1.50×10^8 km, Mars' average distance from the Sun is 2.28×10^8 km, and a typical spacecraft that orbits Earth travels at an average speed of 3.00×10^4 km/h.
3. Suggest several reasons why the measurement of time is important in the study of motion.
4. Reflect on your understanding of "uniform motion." List five examples of motion that you think demonstrate uniform motion.

Throughout this chapter, note any changes in your ideas as you learn new concepts and develop your skills.

Comparing Speeds

Estimating quantities is a valuable skill, not only in physics, but also in everyday life. It will help you to become a more aware consumer who is able to analyze the truthfulness of numeric claims in advertisements. In this activity, you can practise estimating the average speeds of various interesting examples of motion.

(a) In your notebook, arrange the following examples of motion from the slowest average speed to the fastest. Beside each description, write your best estimate of the average speed of the moving object in kilometres per hour (km/h).
Motion examples:
A A migrating whale
B A passenger aircraft flying from Vancouver to Toronto
C Your index and middle fingers walking across the desk
D Mars travelling around the Sun

(b) Use the information below to determine the average speed of each motion in kilometres per hour (km/h). Remember that the average speed equals total distance travelled divided by time of travel.
A A migrating blue whale can travel from Baja California, Mexico, to the Bering Sea near Alaska in about 2.5 months.
B Prevailing winds may increase the speed of aircraft flying eastward.
C Devise and carry out a simple experiment to determine the speed.
D Assume Mars orbits the Sun once each 687 (Earth) days.

(c) Compare your estimates to your calculations. Comment on how accurate your original estimates of speeds were. What could you do to improve your skills of estimating average speeds?

Figure 1
Spacecraft in science fiction movies, such as the *Enterprise D*, can travel at warp speeds. How much faster is warp 9.6 than the speed of light?

1.1 Motion in Our Lives

Everything in our universe is in a state of motion. Our solar system moves through space in the Milky Way Galaxy. Earth revolves around the Sun while rotating about its own axis. People, animals, air, and countless other objects move about on Earth's surface. The elementary particles that make up all matter, too, are constantly in motion.

kinematics: the study of motion

uniform motion: movement at a constant speed in a straight line

nonuniform motion: movement that involves change in speed or direction or both

scalar quantity: quantity that has magnitude, but no direction

base unit: unit from which other units are derived or made up

Scientists call the study of motion **kinematics**, a word that stems from the Greek word for motion, *kinema*. (A "cinema" is a place where people watch motion pictures.) **Uniform motion** is a movement at a constant speed in a straight line. (It is presented in section 1.2.) However, most motions in our lives are classified as **nonuniform**, which means the movement involves changes in speed or direction or both. A roller coaster is an obvious example of such motion—it speeds up, slows down, rises, falls, and travels around corners.

Practice

Understanding Concepts

1. Which of the motions described below are nonuniform? Explain your choices.
   (a) A rubber stopper is dropped from your raised hand to the floor.
   (b) A car is travelling at a steady rate of 85 km/h due west.
   (c) A rocket begins rising from the launch pad.
   (d) A motorcycle rider applies the brakes to come to a stop.

Scalar Quantities

Speeds we encounter in our daily lives are usually given in kilometers per hour (km/h) or metres per second (m/s). Thus, speed involves both distance and time. Speed, distance, and time are examples of a **scalar quantity**, a quantity that has magnitude (or size) only, but no direction. The magnitude is made up of a number and often an appropriate unit. Specific examples of scalar quantities are a distance of 2.5 m, a time interval of 15 s, a mass of 2.2 kg, and the grade of a mountain highway of 0.11 or 11%. (Vectors, which have both magnitude and direction, are described later in the chapter.)

Practice

Understanding Concepts

2. State which measurements are scalar quantities:
   (a) 12 ms (c) 3.2 m [up] (e) 15 cm^2
   (b) 500 MHz (d) 100 km/h [west] (f) 50 mL
3. (a) Name eight scalar quantities presented so far.
   (b) What other scalar quantities can you think of?

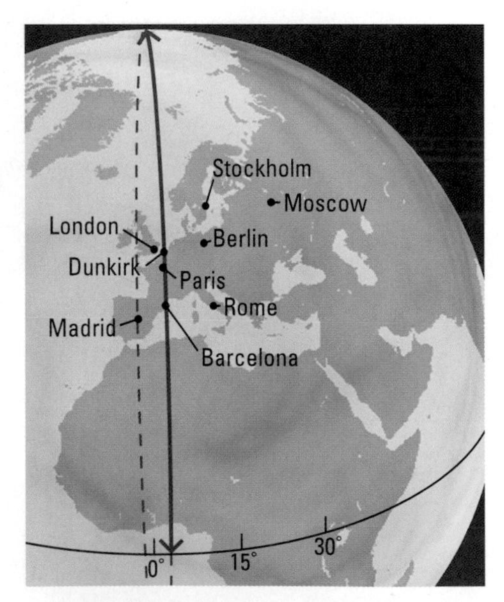

Figure 1
The original metre was defined in terms of the "assumed to be constant" distance from the equator to the geographic North Pole. The distance between two European cities, Dunkirk and Barcelona, was measured by surveyors. Calculations were then made to determine the distance from the equator to the North Pole. The resulting distance was divided by 10^7 to obtain the length of one metre.

Base Units and Derived Units

Every measurement system, including the SI (Système International), consists of base units and derived units. A **base unit** is a unit from which other units are derived or made up. In the metric system, the base unit of length is the metre (m). The metre was originally defined as one ten-millionth of the distance from the equator to the geographic North Pole (**Figure 1**). Then, in 1889, the metre was redefined as the distance between two fine marks on a metal bar now kept in Paris, France. Today, the length of one metre is defined as the distance that light travels in $\frac{1}{299\,792\,458}$ of a second in a vacuum. This quantity does not change and is reproducible anywhere in the world, so it is an excellent standard.

The base unit of time is the second (s). It was previously defined as $\frac{1}{86\,400}$ of the time it takes Earth to rotate once about its own axis. Now, it is defined as the time for 9 192 631 770 cycles of a microwave radiation emitted by a cesium-133 atom, another unchanging quantity.

The kilogram (kg) is the base unit of mass. It has not yet been defined based on any naturally occurring quantity. Currently the one-kilogram standard is a block of iridium alloy kept in France. Copies of this kilogram standard are kept in major cities around the world (**Figure 2**).

In addition to the metre, the second, and the kilogram, there are four other base units in the metric system. All units besides these seven are called **derived units** because they can be stated in terms of the seven base units. One example of a derived unit is the common unit for speed, metres per second, or m/s; it is expressed in terms of two SI base units, the metre and the second.

Figure 2
The kilogram standard kept in France was used to make duplicate standards for other countries. Each standard is well protected from the atmosphere. The one shown is the Canadian standard kept in Ottawa, Ontario.

derived unit: unit that can be stated in terms of the seven base units

Practice

Understanding Concepts

4. Describe possible reasons why the original definitions of the metre and the second were not precise standards.
5. Express the derived units for surface area and volume in terms of SI base units.

Average Speed

Although everyone entering a race (**Figure 3**) must run the same distance, the winner is the person finishing with the fastest time. During some parts of the race, other runners may have achieved a greater **instantaneous speed**, the speed at a particular instant. However, the winner has the greatest average speed. **Average speed** is the total distance travelled divided by the total time of travel. (The symbol for average speed, v_{av}, is taken from the word "velocity.") The equation for average speed is

$$v_{av} = \frac{d}{t} \quad \text{where } d \text{ is the total distance travelled in a total time } t.$$

Figure 3
Every runner covers the same distance, so the person with the least time has the greatest average speed.

instantaneous speed: speed at a particular instant

average speed: total distance of travel divided by total time of travel

Sample Problem

A track star, aiming for a world outdoor record, runs four laps of a circular track that has a radius of 15.9 m in 47.8 s. What is the runner's average speed for this motion?

Solution

The total distance run is four times the track circumference, C.

$r = 15.9$ m

$t = 47.8$ s

$v_{av} = ?$

$$\begin{aligned} d &= 4C \\ &= 4(2\pi r) \\ &= 8\pi\,(15.9 \text{ m}) \\ d &= 4.00 \times 10^2 \text{ m} \end{aligned}$$

The average speed is

$$v_{av} = \frac{d}{t}$$

$$= \frac{4.00 \times 10^2 \text{ m}}{47.8 \text{ s}}$$

$$v_{av} = 8.36 \text{ m/s}$$

The runner's average speed is 8.36 m/s.

Practice

Understanding Concepts

6. Assume that the backwards running marathon record is 3 h 53 min 17 s. Determine the average speed of this 42.2 km race. Express your answer in both metres per second and kilometres per hour.
7. Electrons in a television tube travel 38 cm from their source to the screen in 1.9×10^{-7} s. Calculate the average speed of the electrons in metres per second.
8. Write an equation for each of the following:
   (a) total distance in terms of average speed and total time
   (b) total time in terms of average speed and total distance
9. Copy **Table 1** into your notebook and calculate the unknown values.

Table 1

Total distance (m)	Total time (s)	Average speed (m/s)
3.8×10^5	?	9.5×10^{-3}
?	2.5	480
1800	?	24

10. In the human body, blood travels faster in the aorta, the largest blood vessel, than in any other blood vessel. Given an average speed of 28 cm/s, how far does blood travel in the aorta in 0.20 s?
11. A supersonic jet travels once around Earth at an average speed of 1.6×10^3 km/h. The average radius of its orbit is 6.5×10^3 km. How many hours does the trip take?

Answers

6. 3.01 m/s; 10.9 km/h
7. 2.0×10^6 m/s
9. 4.0×10^7 s; 1.2×10^3 m; 75 s
10. 5.6 cm
11. 26 h

Measuring Time

Time is an important quantity in the study of motion. The techniques used today are much more advanced than those used by early experimenters such as Galileo Galilei, a famous Italian scientist from the 17th century (**Figure 4**). Galileo had to use his own pulse as a time-measuring device in experiments.

In physics classrooms today, various tools are used to measure time. A stopwatch is a simple device that gives acceptable values of time intervals whose duration is more than 2 s. However, for more accurate results, especially for very short time intervals, elaborate equipment must be used.

Figure 4
Galileo Galilei (1564–1642), considered by many to be the originator of modern science, was the first to develop his theories using the results of experiments he devised to test the hypotheses.

Most physics classrooms have instruments that measure time accurately for demonstration purposes. A digital timer is an electronic device that measures time intervals to a fraction of a second. An electronic stroboscope has a light, controlled by adjusting a dial, that flashes on and off at regular intervals. The stroboscope illuminates a moving object in a dark room while a camera records the object's motion on film. The motion is analyzed by using the known time between flashes of the strobe (**Figure 5**). A computer with an appropriate sensor can be used to measure time intervals. A video camera can record motion and have it played back on a monitor; the motion can also be frozen on screen at specific times for analyzing the movement.

Figure 5
This photograph of a golf swing was taken with a stroboscopic light. At which part of the swing is the club moving the fastest?

Two other devices, a spark timer and a ticker-tape timer, are excellent for student experimentation. These devices produce dots electrically on paper at a set frequency. A ticker-tape timer, shown in **Figure 6**, has a metal arm that vibrates at constant time intervals. A needle on the arm strikes a piece of carbon paper and records dots on a paper tape pulled through the timer. The dots give a record of how fast the paper tape is pulled. The faster the motion, the greater the spaces between the dots.

Some spark timers and ticker-tape timers make 60 dots each second. They are said to have a frequency of 60 Hz or 60 vibrations per second. (The SI unit hertz (Hz) is named after German physicist Heinrich Hertz, 1857–1894, and is discussed in greater detail in Chapter 6, section 6.1.) **Figure 7** illustrates why an interval of six spaces produced by a spark timer represents a time of 0.10 s. The period of vibration, which is the time between successive dots, is the reciprocal of the frequency.

Figure 6
A ticker-tape timer

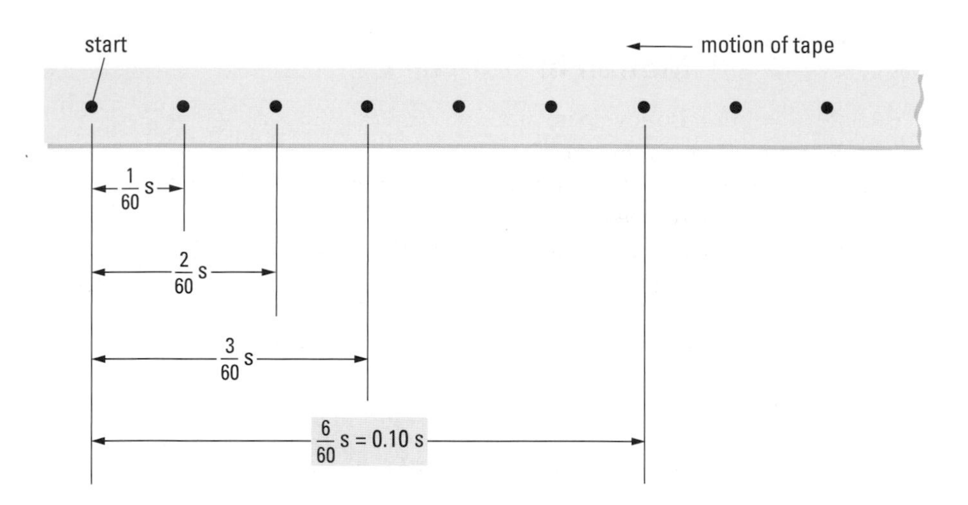

Figure 7
Measuring time with a spark timer

Activity 1.1.1

Calibrating a Ticker-Tape Timer

This activity will introduce you to the use of a ticker-tape timer. You will need a stopwatch as well as the timer and related apparatus.

Before the activity, familiarize yourself with the operation of the ticker-tape timer available.

Procedure

1. Obtain a piece of ticker tape about 200 cm long and position it in the timer. With the timer off and held firmly in place on the lab bench, practise pulling the tape through it so the motion takes about 3 s. Repeat until you can judge what speed of motion works well.
2. Connect the timer to an electrical source, remembering safety guidelines. As you begin pulling the tape through the timer at a steady rate, have your partner turn on the timer and start the stopwatch at the same instant. Just before the tape leaves the timer, have your partner simultaneously turn off both the timer and the stopwatch.

Analysis

(a) Calibrate the timer by determining its frequency (dots per second) and period.
(b) Calculate the percent error of your measurement of the period of the timer. Your teacher will tell you what the "accepted" value is. (To review percentage error, refer to Appendix A.)
(c) What are the major sources of error that could affect your measurements and calculation of the period? If you were to perform this activity again, what would you do to improve the accuracy?

Practice

Understanding Concepts

12. Calculate the period of vibration of a spark timer set at (a) 60.0 Hz and (b) 30.0 Hz.
13. Determine the frequency of a spark timer set at a period of 0.10 s.

Answers

12. (a) 0.0167 s
    (b) 0.0333 s
13. 1.0×10^1 Hz

SUMMARY Motion in Our Lives

- Uniform motion is movement at a constant speed in a straight line. Most motions are nonuniform.
- A scalar quantity has magnitude but no direction. Examples include distance, time, and speed.
- The Système International (SI) base units, the metre (m), the kilogram (kg), and the second (s), can be used to derive other more complex units, such as metres per second (m/s).
- Average speed is the ratio of the total distance travelled to the total time, or $v_{av} = \frac{d}{t}$.
- A ticker-tape timer is just one of several devices used to measure time intervals in a school laboratory.

Section 1.1 Questions

Understanding Concepts

1. In Hawaii's 1999 Ironman Triathlon, the winning athlete swam 3.9 km, biked 180.2 km, and then ran 42.2 km, all in an astonishing 8 h 17 min 17 s. Determine the winner's average speed, in kilometres per hour and also in metres per second.
2. Calculate how far light can travel in a vacuum in (a) 1.00 s and (b) 1.00 ms.
3. Estimate, in days, how long it would take you to walk nonstop at your average walking speed from one mainland coast of Canada to the other. Show your reasoning.

Applying Inquiry Skills

4. Refer to the photograph taken with the stroboscopic light in **Figure 5**.
   (a) Describe how you could estimate the average speed of the tip of the golf club.
   (b) How would you determine the slowest and fastest instantaneous speeds of the tip of the club during the swing?
5. A student, using a stopwatch, determines that a ticker-tape timer produces 138 dots in 2.50 s.
   (a) Determine the frequency of vibration according to these results.
   (b) Calculate the percent error of the frequency, assuming that the true frequency is 60.0 Hz. (To review percentage error, refer to Appendix A.)

Making Connections

6. What scalar quantities are measured by a car's odometer and speedometer?
7. **Figure 8** shows four possible ways of indicating speed limits on roads. Which one communicates the information best? Why?
8. Find out what timers are available in your classroom and describe their features. If possible, compare the features of old and new technologies.

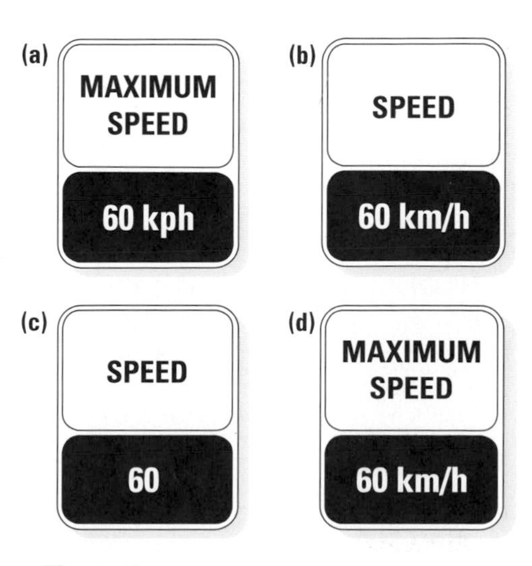

Figure 8
For question 7

Figure 1
When traffic becomes this heavy, the vehicles in any single lane tend to move at approximately the same speed.

1.2 Uniform Motion

On major urban highways, slow-moving traffic is common (**Figure 1**). One experimental method to keep traffic moving is a totally automatic guidance system. Using this technology, cars cruise along at the same speed with computers controlling the steering and the speed. Sensors on the road and on all cars work with video cameras to ensure that cars are at a safe distance apart. Magnetic strips on the road keep the cars in the correct lanes. Could this system be a feature of driving in the future?

The motion shown in **Figure 1** and the motion controlled on a straight section of an automated highway are examples of uniform motion, which is movement at a constant speed in a straight line. Motion in a straight line is also called *linear motion*. Learning to analyze uniform motion helps to understand more complex motions.

Practice

Understanding Concepts

1. Give an example in which linear motion is not uniform motion.

Vector Quantities

vector quantity: quantity that has both magnitude and direction

position: the distance and direction of an object from a reference point

displacement: change in position of an object in a given direction

In studying motion, directions are often considered. A **vector quantity** is one that has both magnitude and direction. In this text, a vector quantity is indicated by a symbol with an arrow above it and the direction is stated in square brackets after the unit. A common vector quantity is **position**, which is the distance and direction of an object from a reference point. For example, the position of a friend in your class could be at a distance of 2.2 m and in the west direction relative to your desk. The symbol for this position is $\vec{d}$ = 2.2 m [W].

Another vector quantity is **displacement**, which is the change in position of an object in a given direction. The symbol for displacement is $\Delta\vec{d}$, where the Greek letter delta "Δ" indicates change. **Figure 2** illustrates a displacement that occurs in moving from one position, $\vec{d}_1$, to another, $\vec{d}_2$, relative to an observer.

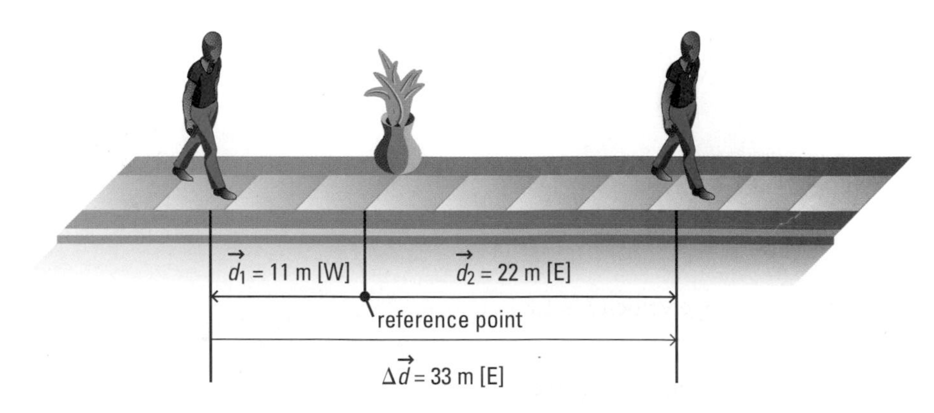

Figure 2
A person walks from a position $\vec{d}_1$ = 11 m [W] to another position $\vec{d}_2$ = 22 m [E]. The displacement for this motion is $\Delta\vec{d} = \vec{d}_2 - \vec{d}_1$ = 22 m [E] – 11 m [W] = 22 m [E] + 11 m [E] = 33 m [E]. Thus, the person's displacement, or change of position, for this motion is 33 m [E].

Practice

Understanding Concepts

2. A curling rock leaves a curler's hand at a point 2.1 m from the end of the ice and travels southward [S]. What is its displacement from its point of release after it has slid to a point 9.7 m from the same edge?
3. A dog, initially at a position 2.8 m west of its owner, runs to retrieve a stick that is 12.6 m east of its owner. What displacement does the dog need in order to reach the stick?
4. **Table 1** gives the position-time data of a ball that has left a bowler's hand and is rolling at a constant speed forward. Determine the displacement between the times:
   (a) $t = 0$ s and $t = 1.0$ s
   (b) $t = 1.0$ s and $t = 2.0$ s
   (c) $t = 1.0$ s and $t = 3.0$ s

Answers

2. 7.6 m [S]
3. 15.4 m [E]
4. (a) 4.4 m [fwd]
   (b) 4.4 m [fwd]
   (c) 8.8 m [fwd]

Table 1

Time (s)	Position (m [fwd])
0.0	0.0
1.0	4.4
2.0	8.8
3.0	13.2

Average Velocity

Most people consider that speed, which is a scalar quantity, and velocity are the same. However, physicists make an important distinction. **Velocity**, a vector quantity, is the rate of change of position. The **average velocity** of a motion is the change of position divided by the time interval for that change. The equation for average velocity is

$$\vec{v}_{av} = \frac{\Delta\vec{d}}{\Delta t}$$

where $\Delta\vec{d}$ is the displacement (or change of position)
Δt is the time interval

velocity: the rate of change of position

average velocity: change of position divided by the time interval for that change

Sample Problem 1

The world's fastest coconut tree climber takes only 4.88 s to climb barefoot 8.99 m up a coconut tree. Calculate the climber's average velocity for this motion, assuming that the climb was vertically upward.

Solution

$\Delta\vec{d} = 8.99$ m [up]

$\Delta t = 4.88$ s

$\vec{v}_{av} = ?$

$$\vec{v}_{av} = \frac{\Delta\vec{d}}{\Delta t} = \frac{8.99 \text{ m [up]}}{4.88 \text{ s}}$$

$$\vec{v}_{av} = 1.84 \text{ m/s [up]}$$

The climber's average velocity is 1.84 m/s [up].

In situations when the average velocity of an object is given, but a different quantity such as displacement or time interval is unknown, you will have to use the equation for average velocity to find the unknown. It is left as an exercise to write equations for displacement and time interval in terms of average velocity.

Practice

Understanding Concepts

5. For objects moving with uniform motion, compare the average speed with the magnitude of the average velocity.
6. While running on his hands, an athlete sprinted 50.0 m [fwd] in a record 16.9 s. Determine the average velocity for this feat.
7. Write an equation for each of the following:
   (a) displacement in terms of average velocity and time interval
   (b) time interval in terms of average velocity and displacement
8. At the snail racing championship in England, the winner moved at an average velocity of 2.4 mm/s [fwd] for 140 s. Determine the winning snail's displacement during this time interval.
9. The women's record for the top windsurfing speed is 20.8 m/s. Assuming that this speed remains constant, how long would it take the record holder to move 178 m [fwd]?

Answers

6. 2.96 m/s [fwd]
8. 34 cm [fwd]
9. 8.56 s

Try This **Activity**

Attempting Uniform Motion

How difficult is it to move at constant velocity? You can find out in this activity.

- Use a motion sensor connected to a graphics program to determine how close to uniform motion your walking can be. Try more than one constant speed, and try moving toward and away from the sensor.
  (a) How can you judge from the graph how uniform your motion was?
  (b) What difficulties occur when trying to create uniform motion?
- Repeat the procedure using a different moving object, such as a glider on an air track or a battery-powered toy vehicle.

Graphing Uniform Motion

In experiments involving motion, the variables that can be measured directly are usually time and either position or displacement. The third variable, velocity, is often obtained by calculation.

In uniform motion, the velocity is constant, so the displacement is the same during equal time intervals. For instance, assume that an ostrich, the world's fastest bird on land, runs 18 m straight west each second for 8.0 s. The bird's velocity is steady at 18 m/s [W]. **Table 2** shows a position-time table describing this motion, starting at 0.0 s.

Table 2

Time (s)	Position (m [W])
0.0	0
2.0	36
4.0	72
6.0	108
8.0	144

Figure 3 shows a graph of this motion, with position plotted as the dependent variable. Notice that, for uniform motion, a position-time graph yields a straight line, which represents a direct variation.

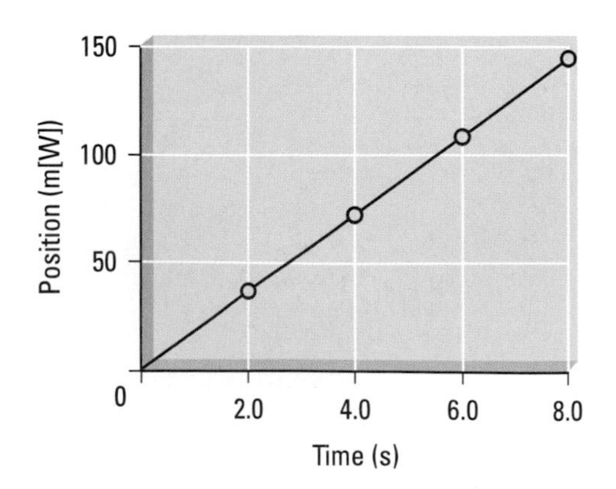

Figure 3
A graph of uniform motion

Sample Problem 2

Calculate the slope of the line in **Figure 3** and state what the slope represents.

Solution

$$m = \frac{\Delta\vec{d}}{\Delta t}$$

$$= \frac{144\text{ m [W]} - 0\text{ m [W]}}{8.0\text{ s} - 0.0\text{ s}}$$

$$m = 18\text{ m/s [W]}$$

The slope of the line is 18 m/s [W]. Judging from the unit of the slope, the slope represents the ostrich's average velocity.

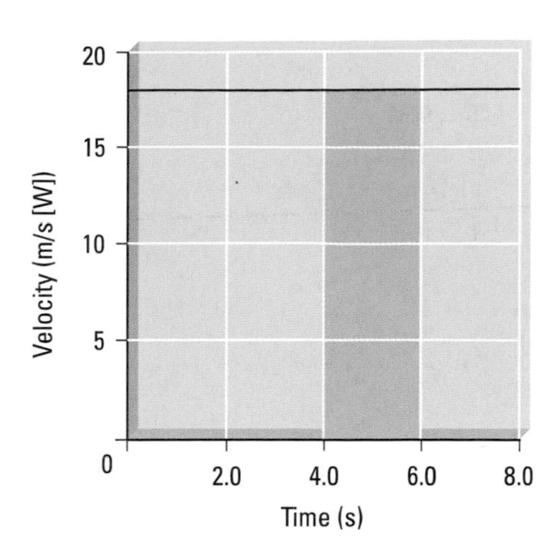

Figure 4
A velocity-time graph of uniform motion
(The shaded region is for Sample Problem 3.)

If the line on a position-time graph such as **Figure 3** has a negative slope, then the slope calculation yields a negative value, –18.0 m/s [W], for example. Since "negative west" is equivalent to "positive east," the average velocity in this case would be 18 m/s [E].

The slope calculation in Sample Problem 2 is used to plot a velocity-time graph of the motion. Because the slope of the line is constant, the velocity is constant from $t = 0.0$ s to $t = 8.0$ s. **Figure 4** gives the resulting velocity-time graph.

A velocity-time graph can be used to find the displacement during various time intervals. This is accomplished by finding the area under the line on the velocity-time graph (Sample Problem 3).

Sample Problem 3

Find the area of the shaded region in **Figure 4**. State what that area represents.

Solution

For a rectangular shape,

$$A = lw$$

$$= \vec{v}(\Delta t)$$

$$= (18\text{ m/s [W]})(2.0\text{ s})$$

$$A = 36\text{ m [W]}$$

The area of the shaded region is 36 m [W]. This quantity represents the ostrich's displacement from $t = 4.0$ s to $t = 6.0$ s. In other words, $\Delta\vec{d} = \vec{v}_{av}(\Delta t)$.

Practice

Understanding Concepts

10. Refer to the graph in **Figure 3**. Show that the slope of the line from $t = 4.0$ s to $t = 6.0$ s is the same as the slope of the entire line found in Sample Problem 2.
11. A military jet is flying with uniform motion at 9.3×10^2 m/s [S], the magnitude of which is approximately Mach 2.7. At time zero, it passes a mountain top, which is used as the reference point for this question.
    (a) Construct a table showing the plane's position relative to the mountain top at the end of each second for a 12 s period.
    (b) Use the data from the table to plot a position-time graph.
    (c) Find the slope of two different line segments on the position-time graph. Is the slope constant? What does it represent?
    (d) Plot a velocity-time graph of the plane's motion.
    (e) Calculate the total area under the line on the velocity-time graph. What does this area represent?

Answers

11. (c) 9.3×10^2 m/s [S]
    (e) 1.1×10^4 m [S]

Answers

12. (a) 1.5×10^2 m/s [E]
    (b) 5.0×10^1 m/s [E]
    (c) 5.0×10^1 m/s [W]
13. (a) 1.2×10^2 m [N]
    (b) 1.2×10^2 m [N]
    (c) 1.2×10^2 m [S]

DECISION MAKING SKILLS

- Define the Issue
- Identify Alternatives
- Research
- Analyze the Issue
- Defend the Proposition
- Evaluate

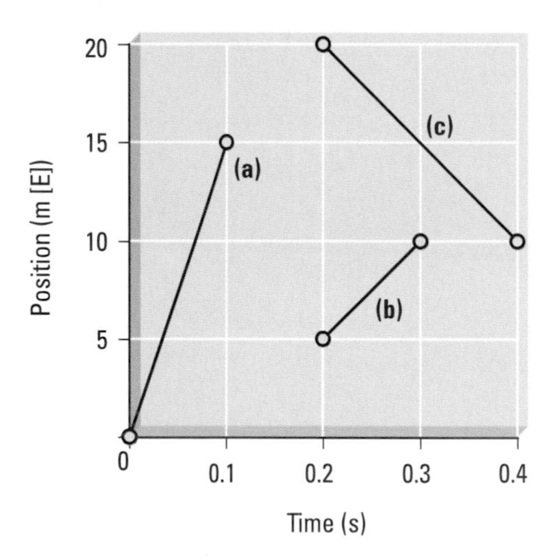

Figure 5
For question 12

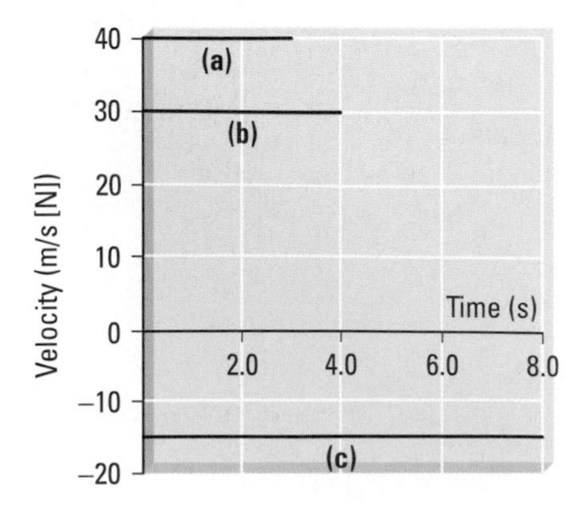

Figure 6
For question 13

12. Determine the average velocities of the three motions depicted in the graph in **Figure 5**.
13. Determine the displacement for each motion shown in the graph in **Figure 6**.

Explore an Issue

Tailgating on Highways

The *Official Driver's Handbook* states that the minimum safe following distance is the distance a vehicle can travel in 2.0 s at a constant speed. People who fail to follow this basic rule, called "tailgaters," greatly increase their chances of an accident if an emergency occurs (**Figure 7**).

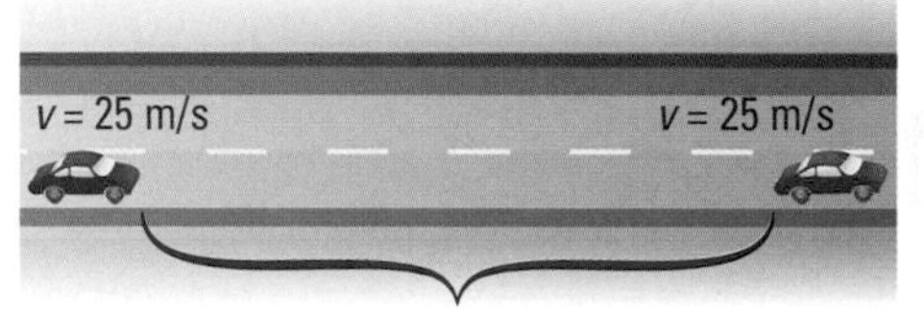

Figure 7
How can you determine the safe following distances knowing the highway speed limit and applying the two-second rule?

Take a Stand

Should tailgaters be fined for dangerous driving?

Proposition

People who drive behind another vehicle too closely should be considered dangerous drivers and fined accordingly.

There are arguments for fining tailgaters:

- Following another vehicle too closely is dangerous because there is little or no time to react to sudden changes in speed of the vehicle ahead.
- If an accident occurs, it is more likely to involve several other vehicles if they are all close together.
- Large vehicles, especially transport trucks, need longer stopping distances, so driving too closely enhances the chance of a collision.

There are arguments against fining tailgaters:

- It is difficult to judge how close is "too close." It could mean two car lengths for a new car equipped with antilock brakes, or it could mean five car lengths for an older car with weak or faulty brakes.
- Tailgaters should not be fined unless other drivers with unsafe driving practices, such as hogging the passing lane, are also fined.

Forming an Opinion

- Read the arguments above and add your own ideas to the list.
- Find more information about the issue to help you form opinions to support your argument. Follow the links for Nelson Physics 11, 1.2.

GO TO **www.science.nelson.com**

- In a group, discuss the ideas.
- Create a position paper in which you state your opinions and present arguments based on these opinions. The "paper" can be a Web page, a video, a scientific report, or some other creative way of communicating.

SUMMARY Uniform Motion

- A vector quantity has both magnitude and direction. Examples are position, displacement, and velocity.
- Position, $\vec{d}$, is the distance and direction of an object from a reference point. Displacement, $\Delta\vec{d}$, is the change in position of an object from a reference point.
- Average velocity is the ratio of the displacement to the time interval, or $\vec{v}_{av} = \dfrac{\Delta\vec{d}}{\Delta t}$.
- The straight line on a position-time graph indicates uniform motion and the slope of the line represents the average velocity between any two times.
- The area under a line on a velocity-time graph represents the displacement between any two times.

Section 1.2 Questions

Understanding Concepts

1. State what each of the following represents:
   (a) the slope of a line on a position-time graph
   (b) the area under the line on a velocity-time graph
2. What is the relationship between the magnitude of the slope of the line on a position-time graph and the magnitude of the velocity of the motion?
3. A runner holds the indoor track record for the 50.0 m and 60.0 m sprints.
   (a) How do you think this runner's average velocities in the two events compare?
   (b) To check your prediction, calculate the average velocities, assuming that the direction of both races is eastward and the record times are 5.96 s and 6.92 s, respectively.
4. To prove that ancient mariners could have crossed the oceans in a small craft, in 1947 a Norwegian explorer named Thor Heyerdahl and his crew of five sailed a wooden raft named the Kon-Tiki westward from South America across the Pacific Ocean to Polynesia. At an average velocity of 3.30 km/h [W], how long did this journey of 8.00×10^3 km [W] take? Express your answer in hours and days.
5. Determine the time to complete a hurdle race in which the displacement is 110.0 m [fwd] and the average velocity is 8.50 m/s [fwd]. (This time is close to the men's record for the 110 m hurdle.)
6. To maintain a safe driving distance between two vehicles, the "two-second" rule for cars and single motorcycles is altered for motorcycle group riding. As shown in **Figure 8**, the leading rider is moving along the left side of the lane, and is "two seconds" ahead of the third rider. At a uniform velocity of 90.0 km/h [E], what is the position of the second rider relative to the leading rider? (Express your answer in kilometres and metres, with a direction.)

(continued)

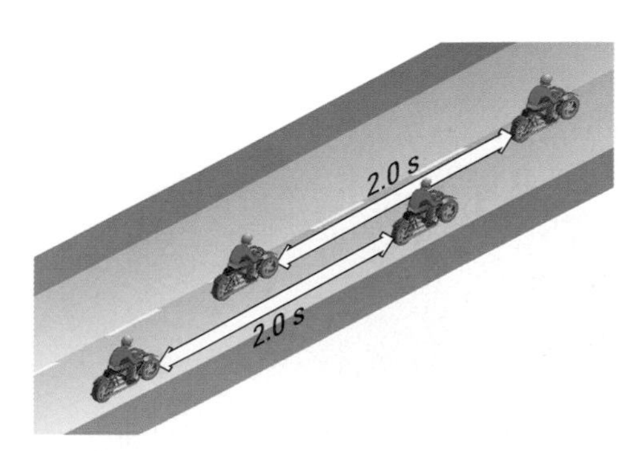

Figure 8
The *Motorcycle Handbook* suggests a staggered format for group motorcycle riding.

Applying Inquiry Skills

7. With the period of the spark timer on a horizontal air table set at 0.10 s, students set two pucks, A and B, moving in the same direction. The resulting dots are shown in **Figure 9.**
   (a) Which puck has a higher average velocity for the entire time interval? How can you tell?
   (b) Use a ruler to determine the data you will need to plot a position-time graph of each motion. Enter your data in a table. Plot both graphs on the same set of axes.
   (c) Are the two motions over their entire time intervals examples of uniform motion? How can you tell? What may account for part of the motion that is not uniform?
   (d) Use the information on the graph to determine the average velocity of puck A for the entire time interval. Then plot a velocity-time graph of that motion.
   (e) Determine the area under the line for the entire time interval on the velocity-time graph for puck A. What does this area represent?
   (f) Describe sources of error in this activity.

puck A • • • • • • •
start

puck B • • • • • • •
start

Figure 9
The motions of pucks A and B

1.3 Two-Dimensional Motion

Suppose you are responsible for designing an electronic map that uses the Global Positioning System to show a rescue worker the best route to travel from the ambulance station to the site of an emergency (**Figure 1**). How would your knowledge about motion in two dimensions help?

Although uniform motion, as discussed in the previous section, is the simplest motion to analyze, it is not as common as nonuniform motion. A simple change of direction renders a motion nonuniform, even if the speed remains constant. In this section, you will explore motion in the horizontal plane, which, like all planes, is two-dimensional.

Figure 1
Any location on Earth's surface can be determined using a global positioning receiver that links to a minimum of three satellites making up the Global Positioning System (GPS). The GPS can also indicate the displacement to some other position, such as the location of an emergency.

Communicating Directions

Vector quantities can have such directions as up, down, forward, and backward. In the horizontal plane, the four compass points, north, east, south, and west, can be used to communicate directions. However, if a displacement or velocity is at some angle between any two compass points, a convenient and consistent method of communicating the direction must be used. In this text, the direction of a vector will be indicated using the smaller angle measured from one of the compass points. **Figure 2** shows how a protractor can be used to determine a vector's direction.

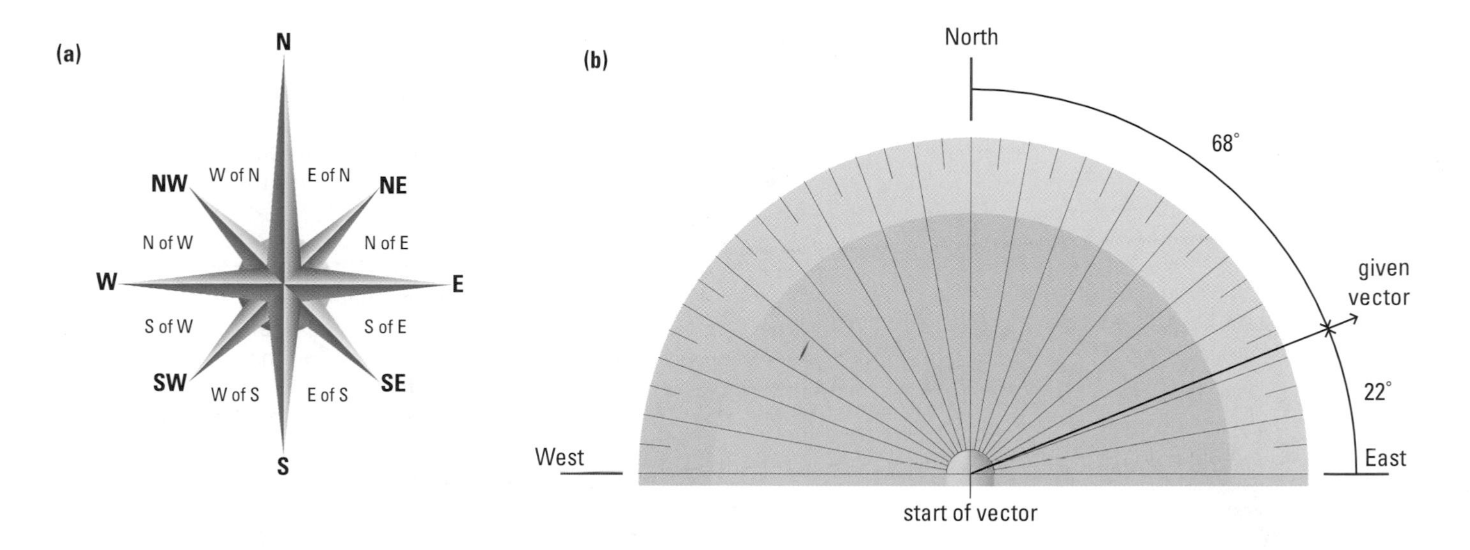

Figure 2
(a) Directions can be labelled from either side of the compass points N, E, S, and W. Notice that NE means exactly 45° N of E or 45° E of N.
(b) To find the direction of a given vector, place the base of the protractor along the east-west (or north-south) direction with the origin of the protractor at the starting position of the vector. Measure the angle to the closest compass point (N, E, S, or W) and write the direction using that angle. In this case, the direction is 22° N of E.

Practice

Understanding Concepts

1. Use a ruler and a protractor to draw these vectors. For (c), make up a convenient scale.
   (a) $\Delta\vec{d}_1 = 3.7$ cm [25° S of E]
   (b) $\Delta\vec{d}_2 = 41$ mm [12° W of N]
   (c) $\Delta\vec{d}_3 = 4.9$ km [18° S of W]

Resultant Displacement in Two Dimensions

On a rainy day a boy walks from his home 1.7 km [E], and then 1.2 km [S] to get to a community skating arena. On a clear, dry day, however, he can walk straight across a vacant field to get to the same arena. As shown in **Figure 3**, the resultant displacement is the same in either case. The **resultant displacement**, $\Delta\vec{d}_R$, is the vector sum of the individual displacements ($\Delta\vec{d}_1 + \Delta\vec{d}_2 + ...$). Notice in **Figure 3** that in order to add individual vectors, the tail of one vector must touch the head of the previous vector. Sometimes, vectors on a horizontal plane have to be moved in order to be added. A vector can be moved anywhere on the plane as long as its magnitude and direction remain the same. That is, the vector in the new position is parallel and equal in length to the original vector.

resultant displacement: vector sum of the individual displacements

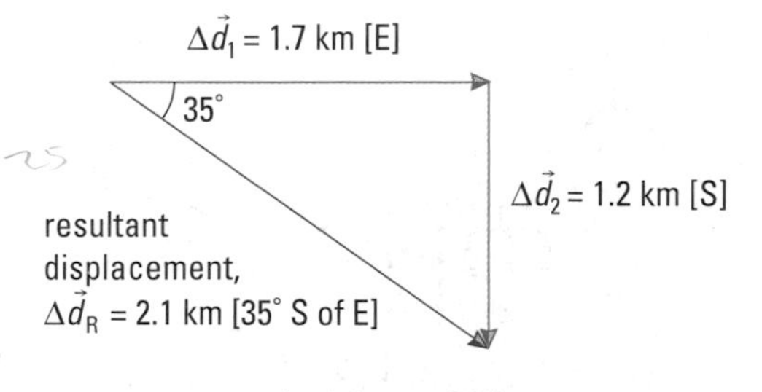

Figure 3
The resultant displacement of 1.7 km [E] + 1.2 km [S] is 2.1 km [35° S of E].

Sample Problem 1

A cyclist travels 5.0 km [E], then 4.0 km [S], and then 8.0 km [W]. Use a scale diagram to determine the resultant displacement and a protractor to measure the angle of displacement.

Solution

A convenient scale in this case is 1.0 cm = 1.0 km. **Figure 4** shows the required vector diagram using this scale. Since the resultant displacement in the diagram, going from the initial position to the final position, indicates a length of 5.0 cm and an angle of 37° west of the south direction, the actual resultant displacement is 5.0 km [37° W of S].

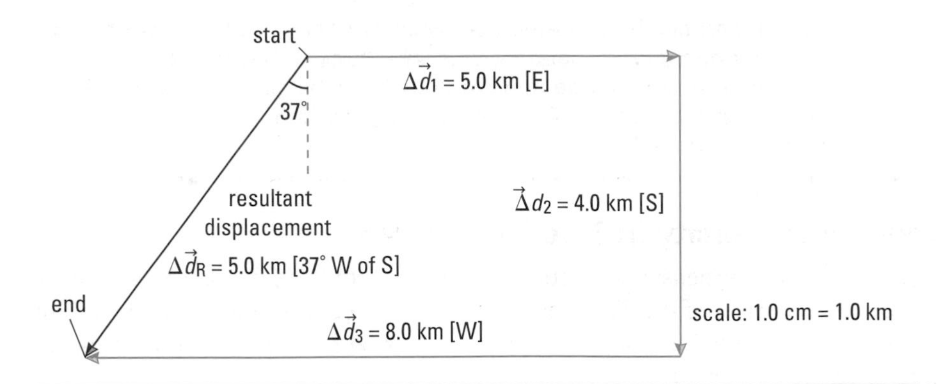

Figure 4
To determine the resultant displacement on a vector diagram, the individual displacements are added together, with the head of one vector attached to the tail of the previous one. A protractor can be used to measure the angles in the diagram.

DID YOU KNOW ?

Indicating Directions

Various ways can be used to communicate directions between compass points. For example, the direction 22° N of E can be written as E22°N or N68°E. Another convention uses the north direction as the reference with the angle measured clockwise from north. In this case, 22° N of E is simply written as 68°.

In situations where finding the resultant displacement involves solving a right-angled triangle, the Pythagorean theorem and simple trigonometric ratios (sine, cosine, and tangent) can be used.

Sample Problem 2

Determine the resultant displacement in **Figure 3** by applying the Pythagorean theorem and trigonometric ratios.

Solution

The symbol Δd is used to represent the magnitude of a displacement.
The Pythagorean theorem can be used to determine the magnitude of the resultant displacement.

$\Delta d_1 = 1.7$ km
$\Delta d_2 = 1.2$ km
$\Delta d_R = ?$

$$(\Delta d_R)^2 = (\Delta d_1)^2 + (\Delta d_2)^2$$
$$\Delta d_R = \sqrt{(\Delta d_1)^2 + (\Delta d_2)^2}$$
$$= \sqrt{(1.7 \text{ km})^2 + (1.2 \text{ km})^2}$$
$$\Delta d_R = 2.1 \text{ km} \quad \text{(Use the positive root to two significant digits.)}$$

$$\theta = \tan^{-1} \frac{\Delta d_2}{\Delta d_1}$$
$$= \tan^{-1} \frac{1.2 \text{ km}}{1.7 \text{ km}}$$
$$\theta = 35° \quad \text{(also to two significant digits)}$$

The resultant displacement is 2.1 km [35° S of E].

Practice

Understanding Concepts

2. Show that the resultant displacement in Sample Problem 1 remains the same when the vectors are added in a different order.
3. An outdoor enthusiast aims a kayak northward and paddles 26 m [N] across a swift river that carries the kayak 36 m [E] downstream.
   (a) Use a scale diagram to determine the resultant displacement of the kayak relative to its initial position.

Answers

3. (a) close to 44 m [36° N of E]

(b) Use an algebraic method (such as the Pythagorean theorem and trigonometry) to determine the resultant displacement.
(c) Find the percentage difference between the angles you found in (a) and (b) above. (To review percentage difference, refer to Appendix A.)

Answers

3. (b) 44 m [36° N of E]

Average Velocity in Two Dimensions

Just as for one-dimensional motion, the average velocity for two-dimensional motion is the ratio of the displacement to the elapsed time. Since more than one displacement may be involved, the average velocity is described using the resultant displacement.

Thus, $\vec{v}_{av} = \dfrac{\Delta \vec{d}_R}{\Delta t}$.

Sample Problem 3

After leaving the huddle, a receiver on a football team runs 8.5 m [E] waiting for the ball to be snapped, then he turns abruptly and runs 12.0 m [S], suddenly changes directions, catches a pass, and runs 13.5 m [W] before being tackled. If the entire motion takes 7.0 s, determine the receiver's (a) average speed and (b) average velocity.

Solution

(a) $d = 8.5 \text{ m} + 12.0 \text{ m} + 13.5 \text{ m} = 34.0 \text{ m}$

$t = 7.0 \text{ s}$

$v_{av} = ?$

$$v_{av} = \frac{d}{t}$$

$$= \frac{34.0 \text{ m}}{7.0 \text{ s}}$$

$$v_{av} = 4.9 \text{ m/s}$$

The receiver's average speed is 4.9 m/s.

(b) As shown in the scale diagram in **Figure 5**, $\Delta \vec{d}_R = 13.0 \text{ m [23° W of S]}$.

$$\vec{v}_{av} = \frac{\Delta \vec{d}_R}{\Delta t}$$

$$= \frac{13.0 \text{ m [23° W of S]}}{7.0 \text{ s}}$$

$$\vec{v}_{av} = 1.9 \text{ m/s [23° W of S]}$$

The receiver's average velocity is 1.9 m/s [23° W of S].

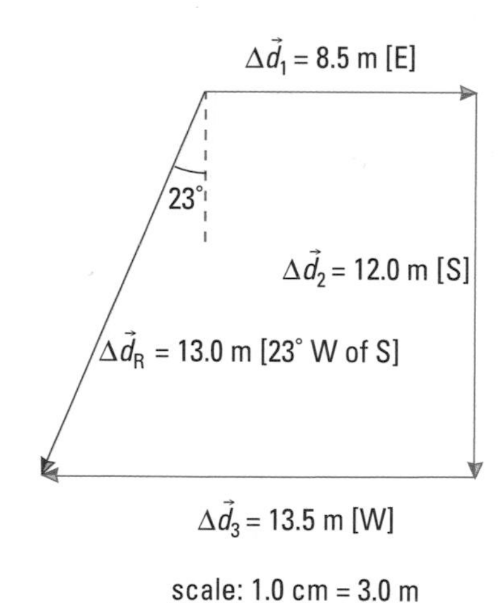

Figure 5
For Sample Problem 3

Practice

Understanding Concepts

4. To get to the cafeteria entrance, a teacher walks 34 m [N] in one hallway, and then 46 m [W] in another hallway. The entire motion takes 1.5 min. Determine the teacher's
   (a) resultant displacement (using trigonometry or a scale diagram)
   (b) average speed
   (c) average velocity

Answers

4. (a) 57 m [36° N of W]
   (b) 53 m/min, or 0.89 m/s
   (c) 38 m/min [36° N of W], or 0.64 m/s [36° N of W]

Answers

5. (a) 7.7 m/s
   (b) 4.9 m/s [E]

5. A student starts at the westernmost position of a circular track of circumference 200 m and runs halfway around the track in 13 s. Determine the student's (a) average speed and (b) average velocity. (Assume two significant digits.)

Relative Motion

frame of reference: coordinate system relative to which a motion can be observed

relative velocity: velocity of a body relative to a particular frame of reference

Suppose a large cruise boat is moving at a velocity of 5.0 m/s [S] relative to the shore and a passenger is jogging at a velocity of 3.0 m/s [S] relative to the boat. Relative to the shore, the passenger's velocity is the addition of the two velocities, 5.0 m/s [S] and 3.0 m/s [S], or 8.0 m/s [S]. The shore is one frame of reference, and the boat is another. More mathematically, a **frame of reference** is a coordinate system "attached" to an object, such as the boat, relative to which a motion can be observed. Any motion observed depends on the frame of reference chosen.

The velocity of a body relative to a particular frame of reference is called **relative velocity.** In all previous velocity discussions, we have assumed that Earth or the ground is the frame of reference, even though it has not been stated. To analyze motion with more than one frame of reference, we introduce the symbol for relative velocity, $\vec{v}$ with two subscripts. In the cruise boat example above, if E represents Earth's frame of reference (the shore), B represents the boat, and P represents the passenger, then

$\vec{v}_{BE}$ = the velocity of the boat B relative to Earth E (or relative to the shore)
$\vec{v}_{PB}$ = the velocity of the passenger P relative to the boat B
$\vec{v}_{PE}$ = the velocity of the passenger P relative to Earth E (or relative to the shore)

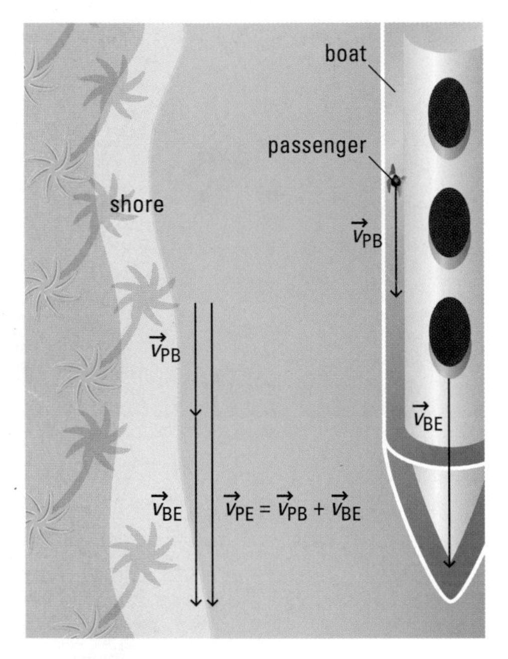

Figure 6
The velocity of the passenger is 3.0 m/s [S] relative to the boat, but is 8.0 m/s [S] relative to the shore.

Notice that the first subscript represents the object whose velocity is stated relative to the object represented by the second subscript. In other words, the second subscript is the frame of reference.

To relate the above velocities, we use a relative velocity equation. For this example, it is

$$\vec{v}_{PE} = \vec{v}_{PB} + \vec{v}_{BE} \quad \text{(where "+" represents a vector addition)}$$
$$= 3.0 \text{ m/s [S]} + 5.0 \text{ m/s [S]}$$
$$\vec{v}_{PE} = 8.0 \text{ m/s [S]}$$

The velocity of the passenger relative to the shore (Earth) is 8.0 m/s [S], as illustrated in **Figure 6**.

The relative velocity equation also applies to motion in two dimensions, in which case it is important to remember the vector nature of velocity. Before looking at examples of relative velocity in two dimensions, be sure you see the pattern of the subscripts in the symbols in any relative velocity equation. In the equation, the first subscript of the vector on the left side is the same as the first subscript of the first vector on the right side, and the second subscript of the vector on the left side is the same as the second subscript of the second vector on the right side. This pattern is illustrated for the boat example as well as other examples in **Figure 7**.

DID YOU KNOW ?

Alternative Communication

An alternative way of communicating the relative velocity equation is to place the symbol for the observed object before the v and the symbol for the frame of reference after the v. In this way, the equation for the boat example is written:

$${}_{P}\vec{v}_{E} = {}_{P}\vec{v}_{B} + {}_{B}\vec{v}_{E}$$

(a) $\vec{v}_{XZ} = \vec{v}_{XY} + \vec{v}_{YZ}$

(b) $\vec{v}_{AC} = \vec{v}_{AB} + \vec{v}_{BC}$

(c) $\vec{v}_{PE} = \vec{v}_{PB} + \vec{v}_{BE}$

Figure 7
The pattern in a relative velocity equation

Sample Problem 4

Suppose the passenger in the boat example above is jogging at a velocity of 3.0 m/s [E] relative to the boat as the boat is travelling at a velocity of 5.0 m/s [S] relative to the shore. Determine the jogger's velocity relative to the shore.

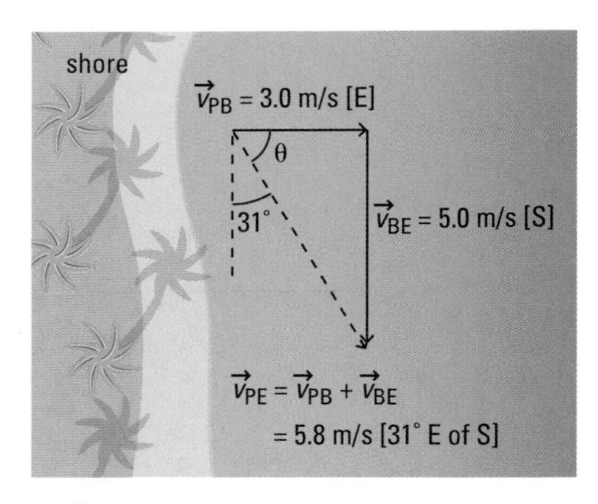

Figure 8
The scale used to draw this vector diagram is 1.0 cm = 2.0 m/s.

Solution

$\vec{v}_{PB}$ = 3.0 m/s [E]
$\vec{v}_{BE}$ = 5.0 m/s [S]
$\vec{v}_{PE}$ = ?

$\vec{v}_{PE} = \vec{v}_{PB} + \vec{v}_{BE}$ (This is a vector addition.)

Figure 8 shows this vector addition. Using trigonometry, the Pythagorean theorem, or a scale diagram, we find that the magnitude of $\vec{v}_{PE}$ is 5.8 m/s. To find the direction, we first find the angle θ.

$$\theta = \tan^{-1} \frac{5.0 \text{ m/s}}{3.0 \text{ m/s}}$$

$$\theta = 59^\circ$$

$\vec{v}_{PE}$ = 5.8 m/s [31° E of S]

The jogger's velocity relative to the shore is 5.8 m/s [31° E of S].

Practice

Understanding Concepts

6. Determine the velocity of a canoe relative to the shore of a river if the velocity of the canoe relative to the water is 3.2 m/s [N] and the velocity of the water relative to the shore is 2.3 m/s [E].
7. A blimp is travelling at a velocity of 22 km/h [E] relative to the air. A wind is blowing from north at an average speed of 15 km/h relative to the ground. Determine the velocity of the blimp relative to the ground.

Making Connections

8. Is a passenger in an airplane more concerned about the plane's "air speed" (velocity relative to the air) or "ground speed" (velocity relative to the ground)? Explain.

Answers

6. 3.9 m/s [36° E of N]
7. 27 km/h [34° S of E]

SUMMARY Two-Dimensional Motion

- In two-dimensional motion, the resultant displacement is the vector sum of the individual displacements, $\Delta\vec{d}_R = \Delta\vec{d}_1 + \Delta\vec{d}_2$.
- The average velocity in two-dimensional motion is the ratio of the resultant displacement to the time interval, $\vec{v}_{av} = \frac{\Delta\vec{d}_R}{\Delta t}$.
- All motion is relative to a frame of reference. We usually use Earth as our frame of reference. For example, the velocity of a train relative to Earth or the ground can be written $\vec{v}_{TG}$.
- When two motions are involved, the relative velocity equation is $\vec{v}_{AC} = \vec{v}_{AB} + \vec{v}_{BC}$, which is a vector addition.

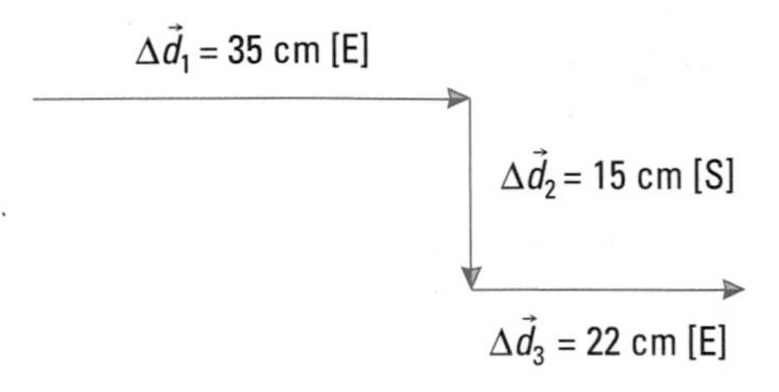

Figure 9
For question 3

Section 1.3 Questions

Understanding Concepts

1. (a) Can the magnitude of the displacement of an object from its original position ever exceed the total distance moved? Explain.
   (b) Can the total distance moved ever exceed the magnitude of an object's displacement from its original position? Explain.
2. Cheetahs, the world's fastest land animals, can run up to about 125 km/h. A cheetah chasing an impala runs 32 m [N], then suddenly turns and runs 46 m [W] before lunging at the impala. The entire motion takes only 2.7 s.
   (a) Determine the cheetah's average speed for this motion.
   (b) Determine the cheetah's average velocity.
3. Air molecules travel at high speeds as they bounce off each other and their surroundings. In 1.50 ms, an air molecule experiences the motion shown in **Figure 9**. For this motion, determine the molecule's (a) average speed, and (b) average velocity.
4. How do $\vec{v}_{AB}$ and $\vec{v}_{BA}$ compare?
5. An airplane pilot checks the instruments and finds that the velocity of the plane relative to the air is 320 km/h [35° S of E]. A radio report indicates that the wind velocity relative to the ground is 75 km/h [E]. What is the velocity of the plane relative to the ground as recorded by an air traffic controller in a nearby airport?

Making Connections

6. Highway accidents often occur when drivers are distracted by non-driving activities such as talking on a hand-held phone, listening to loud music, and reading maps. Some experts fear that the installation of new technology in cars, such as electronic maps created by signals from the Global Positioning System (GPS), will cause even more distraction to drivers. Research more on GPS and other new technology. Follow the links for Nelson Physics 11, 1.3. Assuming that money is not an obstacle, how would you design a way of communicating location, map information, driving times, road conditions, and other details provided by technological advances to the driver in the safest way possible?

GO TO www.science.nelson.com

7. A wind is blowing from the west at an airport with an east-west runway. Should airplanes be travelling east or west as they approach the runway for landing? Why?

DID YOU KNOW ?

Particle Accelerators

You have heard of particles such as protons, neutrons, and electrons. Have you also heard of quarks, muons, neutrinos, pions, and kaons? They are examples of tiny elementary particles that are found in nature. Physicists have discovered hundreds of these types of particles and are researching to find out more about them. To do so, they study the properties of matter in particle accelerators. These high-tech, expensive machines use strong electric fields to cause the particles to reach extremely high speeds and then collide with other particles. Analyzing the resulting collisions helps to unlock the mysteries of the universe.

1.4 Uniform Acceleration

On the navy aircraft carrier shown in **Figure 1**, a steam-powered catapult system can cause an aircraft to accelerate from speed zero to 265 km/h in only 2.0 s! Stopping a plane also requires a high magnitude of acceleration, although in this case, the plane is slowing down. From a speed of about 240 km/h, a hook extended from the tail section of the plane grabs onto one of the steel cables stretched across the deck, causing the plane to stop in about 100 m. Basic motion equations can be used to analyze these motions and compare the accelerations of the aircraft with what you experience in cars and on rides at amusement parks.

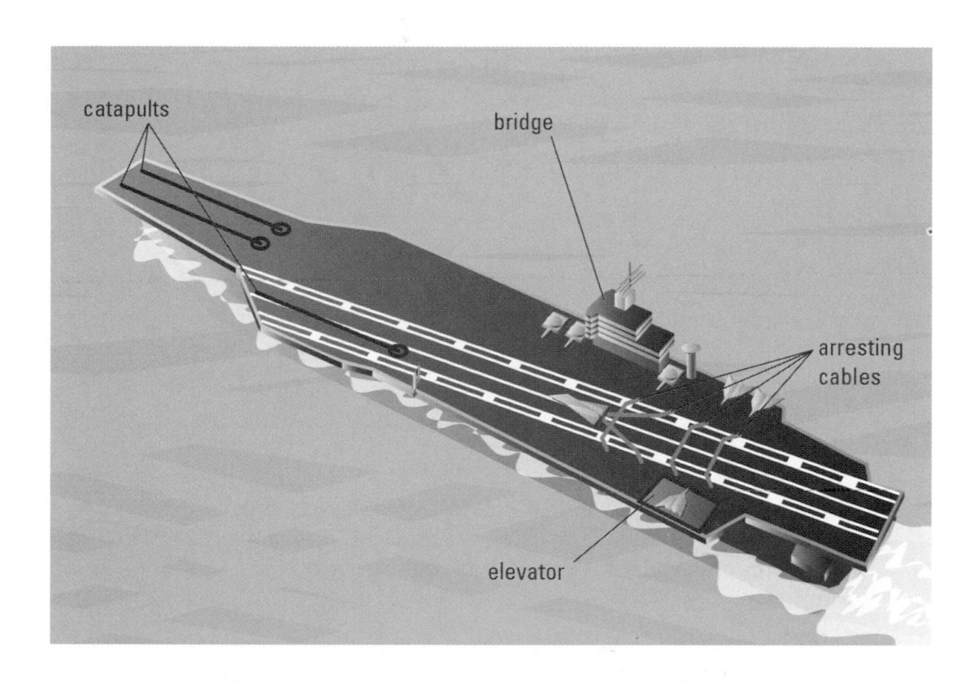

Figure 1
Pilots of planes that take off from and land on an aircraft carrier experience high magnitudes of acceleration.

accelerated motion: nonuniform motion that involves change in an object's speed or direction or both

uniformly accelerated motion: motion that occurs when an object travelling in a straight line changes its speed uniformly with time

Comparing Uniform Motion and Uniformly Accelerated Motion

You have learned that uniform motion occurs when an object moves at a steady speed in a straight line. For uniform motion the velocity is constant; a velocity-time graph yields a horizontal, straight line.

Most moving objects, however, do not display uniform motion. Any change in an object's speed or direction or both means that its motion is not uniform. This nonuniform motion, or changing velocity, is called **accelerated motion**. Since the direction of the motion is involved, acceleration is a vector quantity. A car ride in a city at rush hour during which the car must speed up, slow down, and turn corners is an obvious example of accelerated motion.

One type of accelerated motion, called **uniformly accelerated motion**, occurs when an object travelling in a straight line changes its speed uniformly with time. **Figure 2** shows a velocity-time graph for a motorcycle whose motion is given in **Table 1**. (In real life, the acceleration is unlikely to be so uniform, but it can be close.) The motorcycle starts from rest and increases its speed by 6.0 m/s every second in a westerly direction.

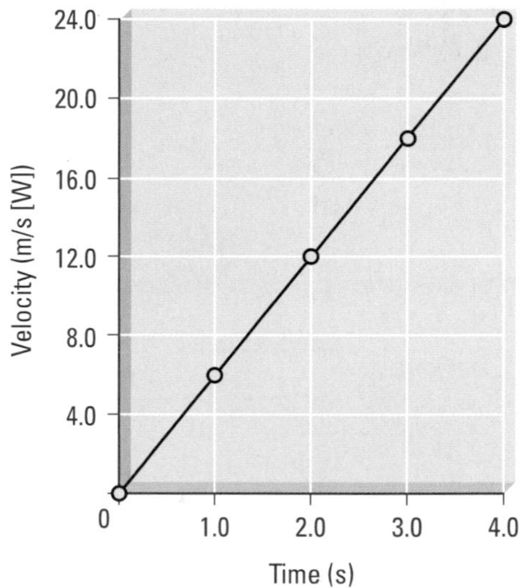

Figure 2
Uniform acceleration

Table 1

Time (s)	0.0	1.0	2.0	3.0	4.0
Velocity (m/s [W])	0.0	6.0	12.0	18.0	24.0

Uniform acceleration also occurs when an object travelling in a straight line slows down uniformly. (In this case, the object is sometimes said to be *decelerating*.) Refer to **Table 2** below and **Figure 3**, which give an example of uniform acceleration in which an object, such as a car, slows down uniformly from 24.0 m/s [E] to 0.0 m/s in 4.0 s.

Table 2

Time (s)	0.0	1.0	2.0	3.0	4.0
Velocity (m/s [E])	24.0	18.0	12.0	6.0	0.0

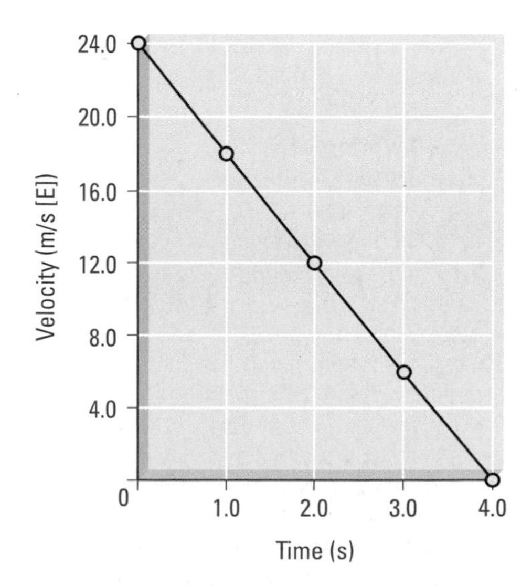

Figure 3
Uniformly accelerated motion for a car slowing down

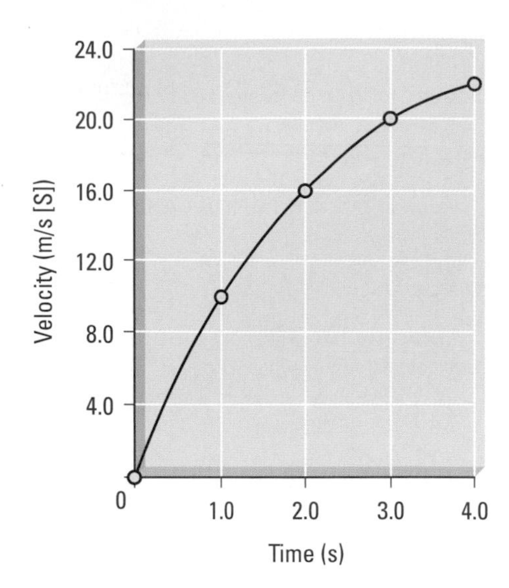

Figure 4
Nonuniform acceleration

If an object is changing its speed in a nonuniform fashion, its acceleration is nonuniform. Such motion is more difficult to analyze than motion with uniform acceleration, but an example of the possible acceleration of a sports car is given in **Table 3** below and **Figure 4** for comparison purposes.

Table 3

Time (s)	0.0	1.0	2.0	3.0	4.0
Velocity (m/s [S])	0.0	10.0	16.0	20.0	22.0

Practice

Understanding Concepts

1. **Table 4** shows five different sets of velocities at times of 0.0 s, 1.0 s, 2.0 s, and 3.0 s. Which of them involve uniform acceleration with an increasing velocity for the entire time? Describe the motion of the other sets.

Table 4

	Time (s)	0.0	1.0	2.0	3.0
(a)	Velocity (m/s [E])	0.0	8.0	16.0	24.0
(b)	Velocity (cm/s [W])	0.0	4.0	8.0	8.0
(c)	Velocity (km/h [N])	58	58	58	58
(d)	Velocity (m/s [W])	15	16	17	18
(e)	Velocity (km/h [S])	99	66	33	0

2. Describe the motion illustrated in each velocity-time graph shown in **Figure 5**. Where possible, use terms such as uniform motion, uniform acceleration, and increasing or decreasing velocity. In (c), you can compare the magnitudes.

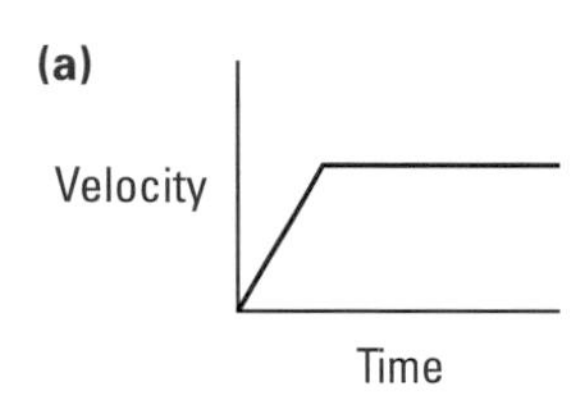

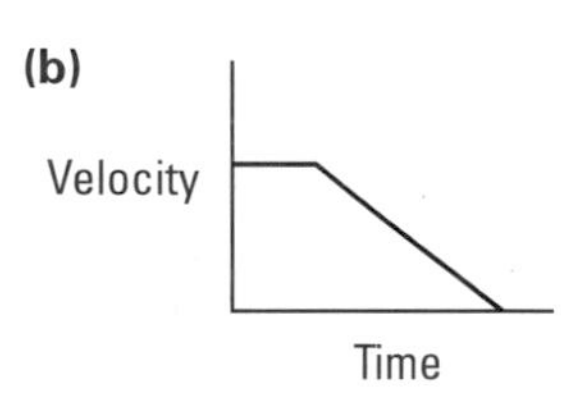

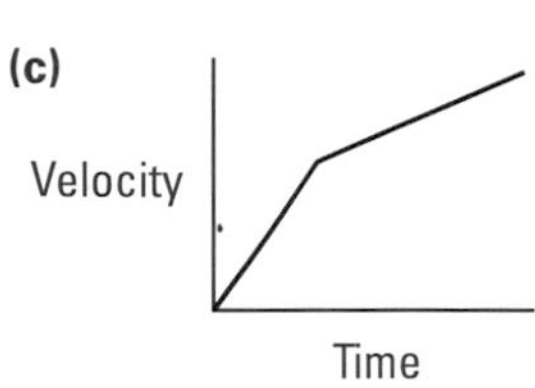

Figure 5
For question 2

Analyzing Motion Graphs

A cart is pushed so it travels up a straight ramp, stops for an instant, and then travels back down to the point from which it was first pushed. Consider the motion just after the force pushing the cart upward is removed until the cart is caught on its way down. For this activity, assume that "up the ramp" is the positive direction.

(a) Sketch what you think the $\vec{d}$-t, $\vec{v}$-t, and $\vec{a}$-t graphs would be for the motion of the cart up the ramp.
(b) Repeat (a) for the motion of the cart down the ramp.
(c) Your teacher will set up a motion sensor or a "smart pulley" to generate the graphs by computer as the cart undergoes the motion described. Compare your predicted graphs with those generated by computer.

Be sure the moving cart is caught safely as it completes its downward motion.

Calculating Acceleration

Acceleration is defined as the rate of change of velocity. Since velocity is a vector quantity, acceleration is also a vector quantity. The **instantaneous acceleration** is the acceleration at a particular instant. For uniformly accelerated motion, the instantaneous acceleration has the same value as the average acceleration. The **average acceleration** of an object is found using the equation

$$\text{average acceleration} = \frac{\text{change of velocity}}{\text{time interval}}$$

$$\vec{a}_{av} = \frac{\Delta\vec{v}}{\Delta t}$$

Since the change of velocity ($\Delta\vec{v}$) of a moving object is the final velocity ($\vec{v}_f$) minus the initial velocity ($\vec{v}_i$), the equation for acceleration can be written

$$\vec{a}_{av} = \frac{\vec{v}_f - \vec{v}_i}{\Delta t}$$

acceleration: rate of change of velocity

instantaneous acceleration: acceleration at a particular instant

average acceleration: change of velocity divided by the time interval for that change

Sample Problem 1

A motorbike starting from rest and undergoing uniform acceleration reaches a velocity of 21.0 m/s [N] in 8.4 s. Find its average acceleration.

Solution

$\vec{v}_f = 21.0$ m/s [N]

$\vec{v}_i = 0.0$ m/s [N]

$\Delta t = 8.4$ s

$\vec{a}_{av} = ?$

$$\vec{a}_{av} = \frac{\vec{v}_f - \vec{v}_i}{\Delta t}$$

$$= \frac{21.0 \text{ m/s [N]} - 0.0 \text{ m/s [N]}}{8.4 \text{ s}}$$

$$\vec{a}_{av} = 2.5 \text{ m/s}^2 \text{ [N]}$$

The bike's average acceleration is 2.5 m/s^2 [N], or 2.5 (m/s)/s [N].

In Sample Problem 1, the uniform acceleration of 2.5 m/s^2 [N] means that the velocity of the motorbike increases by 2.5 m/s [N] every second. Thus, the bike's velocity is 2.5 m/s [N] after 1.0 s, 5.0 m/s [N] after 2.0 s, and so on.

If an object is slowing down, its acceleration is opposite in direction to the velocity, which means that if the velocity is positive, the acceleration is negative. This is illustrated in the sample problem that follows.

Sample Problem 2

A cyclist, travelling initially at 14 m/s [S], brakes smoothly and stops in 4.0 s. What is the cyclist's average acceleration?

Solution

$\vec{v}_f = 0$ m/s [S]

$\vec{v}_i = 14$ m/s [S]

$\Delta t = 4.0$ s

$\vec{a}_{av} = ?$

$$\vec{a}_{av} = \frac{\vec{v}_f - \vec{v}_i}{\Delta t}$$

$$= \frac{0 \text{ m/s [S]} - 14 \text{ m/s [S]}}{4.0 \text{ s}}$$

$$= -3.5 \text{ m/s}^2 \text{ [S]}$$

$$\vec{a}_{av} = 3.5 \text{ m/s}^2 \text{ [N]}$$

The cyclist's average acceleration is 3.5 m/s^2 [N]. Notice that the direction "negative south" is the same as the direction "positive north."

The equation for average acceleration can be rearranged to solve for final velocity, initial velocity, or time interval, provided that other variables are known. Some of the following questions will allow you to practise this skill.

Practice

Answers

3. 1.3×10^2 (km/h)/s
5. (a) 59 m/s [E]
   (b) 75 m/s [E]
   (c) 6.0×10^1 s
6. 2.9×10^2 s
8. 22 m/s [up]

Understanding Concepts

3. Determine the magnitude of the average acceleration of the aircraft that takes off from the aircraft carrier described in the first paragraph of this section.
4. Rewrite the equation $\vec{a}_{av} = \frac{\vec{v}_f - \vec{v}_i}{\Delta t}$ to solve for the following:
   (a) final velocity (b) initial velocity (c) time interval
5. Calculate the unknown quantities in **Table 5**.

Table 5

	Acceleration (m/s^2 [E])	Initial velocity (m/s [E])	Final velocity (m/s [E])	Time interval (s)
(a)	8.5	?	93	4.0
(b)	0.50	15	?	120
(c)	–0.20	24	12	?

6. In the second stage of a rocket launch, the rocket's upward velocity increased from 1.0×10^3 m/s to 1.0×10^4 m/s, with an average acceleration of magnitude 31 m/s^2. How long did the acceleration last?
7. A truck driver travelling at 90.0 km/h [W] applies the brakes to prevent hitting a stalled car. In order to avoid a collision, the truck would have to be stopped in 20.0 s. At an average acceleration of −4.00 (km/h)/s [W], will a collision occur? Try to solve this problem using two or three different techniques.
8. When a ball is thrown upward, it experiences a downward acceleration of magnitude 9.8 m/s^2, neglecting air resistance. With what velocity must a ball leave a thrower's hand in order to climb for 2.2 s before stopping?

Applying Inquiry Skills

9. (a) Estimate your maximum running velocity, and estimate the average acceleration you undergo from rest to reach that velocity.
   (b) Design an experiment to check your estimates in (a). Include the equations you would use.
   (c) Get your design approved by your teacher, and then carry it out. Compare your results to your estimates.

Try This Activity

Student Accelerometers

Figure 6 illustrates various designs of an accelerometer, a device used to determine horizontal acceleration.

(a) Based on the diagrams, what do you think would happen in each accelerometer if the object to which it is attached accelerated to the right? Explain why in each case.

(b) If you have access to a horizontal accelerometer, discuss its safe use with your teacher. Then use it to test your answers in (a). Describe what you discover.

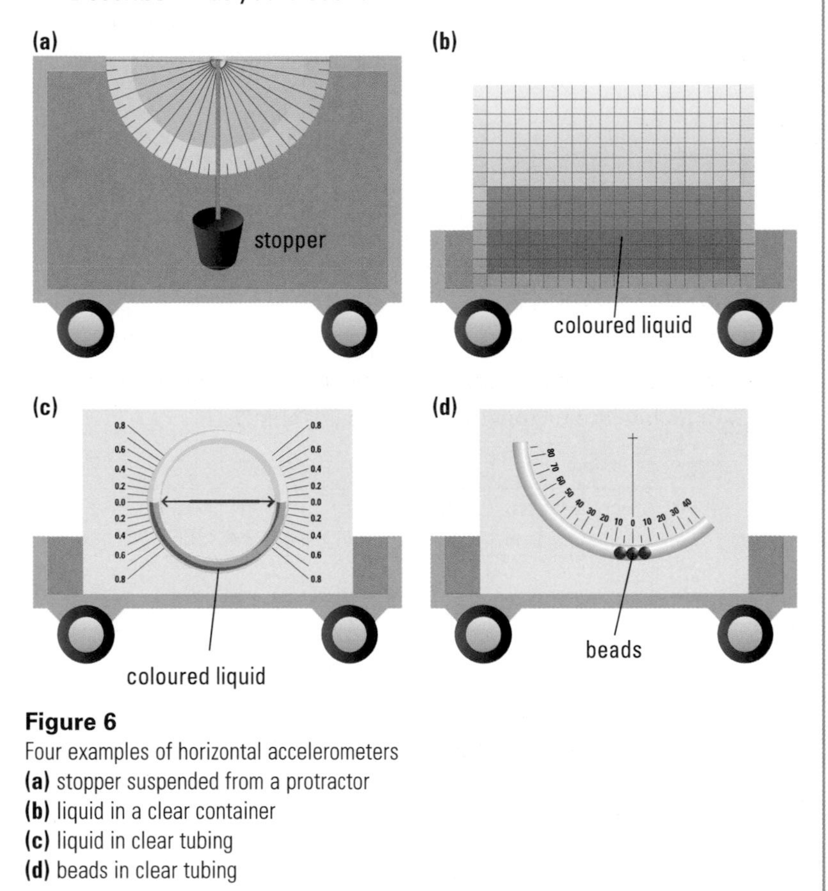

Figure 6
Four examples of horizontal accelerometers
(a) stopper suspended from a protractor
(b) liquid in a clear container
(c) liquid in clear tubing
(d) beads in clear tubing

DID YOU KNOW ?

Electronic Accelerometers

Engineers who want to test their roller coaster design attach an electronic accelerometer to one of the coaster cars. This type of accelerometer measures the acceleration in three mutually perpendicular directions. Electronic accelerometers with computer interfaces are also available for student use.

DID YOU KNOW ?

Accelerometers in Nature

Fish use hairs that are sensitive to pressure change as accelerometers. In an attempt to copy nature's adaptations, scientists are experimenting with hairs on the surfaces of robots to detect changing conditions surrounding them. Eventually, robotic vehicles and even astronauts may apply the resulting technology on Mars or the Moon.

Using Velocity-Time Graphs to Find Acceleration

You have learned that the slope of a line on a position-time graph indicates the velocity. We use an equation to determine the slope of a line on a velocity-time graph.

Consider the graph in **Figure 7**. The slope of the line is constant and is

$$m = \frac{\Delta \vec{v}}{\Delta t}$$

$$= \frac{24.0 \text{ m/s [E]} - 0.0 \text{ m/s [E]}}{8.0 \text{ s} - 0.0 \text{ s}}$$

$$m = 3.0 \text{ m/s}^2 \text{ [E]}$$

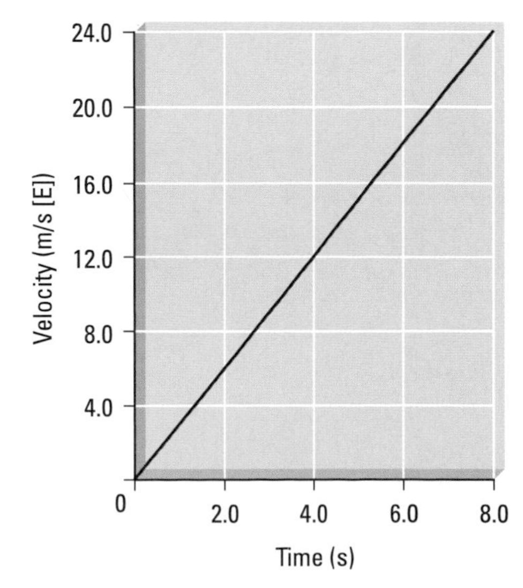

Figure 7
Velocity-time graph

In uniformly accelerated motion, the acceleration is constant, so the slope of the entire line represents the average acceleration. In equation form,

average acceleration = slope of velocity-time graph (for uniform acceleration)

$$\vec{a}_{av} = \frac{\Delta \vec{v}}{\Delta t}$$

This equation is equivalent to the one used previously for average acceleration:

$$\vec{a}_{av} = \frac{\vec{v}_f - \vec{v}_i}{\Delta t}$$

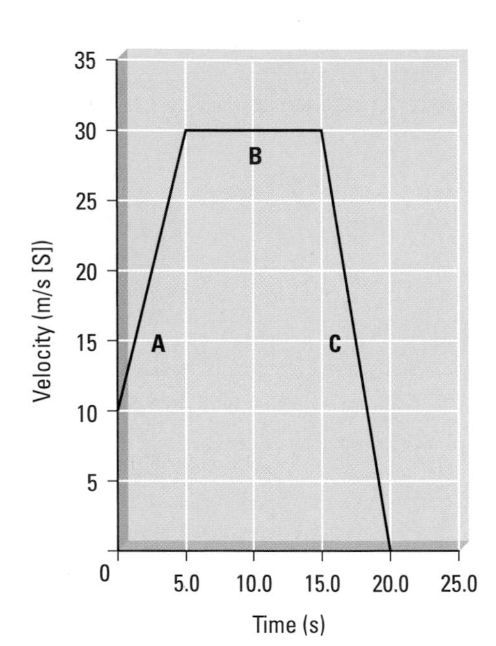

Figure 8
For Sample Problem 3

Sample Problem 3

For the motion shown in **Figure 8**, determine the average acceleration in segments A, B, and C.

Solution

(a) **Segment A:**

$$\vec{a}_{av} = \frac{\vec{v}_f - \vec{v}_i}{\Delta t}$$

$$= \frac{30 \text{ m/s [S]} - 10 \text{ m/s [S]}}{5.0 \text{ s} - 0.0 \text{ s}}$$

$$\vec{a}_{av} = 4.0 \text{ m/s}^2 \text{ [S]}$$

The average acceleration is 4.0 m/s^2 [S].

(b) **Segment B:**

$$\vec{a}_{av} = \frac{\vec{v}_f - \vec{v}_i}{\Delta t}$$

$$= \frac{30 \text{ m/s [S]} - 30 \text{ m/s [S]}}{15.0 \text{ s} - 5.0 \text{ s}}$$

$$\vec{a}_{av} = 0.0 \text{ m/s}^2 \text{ [S]}$$

The average acceleration from 5.0 s to 15.0 s is zero.

(c) **Segment C:**

$$\vec{a}_{av} = \frac{\vec{v}_f - \vec{v}_i}{\Delta t}$$

$$= \frac{0 \text{ m/s [S]} - 30 \text{ m/s [S]}}{20.0 \text{ s} - 15.0 \text{ s}}$$

$$\vec{a}_{av} = -6.0 \text{ m/s}^2 \text{ [S], or } 6.0 \text{ m/s}^2 \text{ [N]}$$

The average acceleration is –6.0 m/s^2 [S], or 6.0 m/s^2 [N].

For nonuniform accelerated motion, the velocity-time graph for the motion is not a straight line and the slope changes. In this case, the slope of the graph at a particular instant represents the instantaneous acceleration.

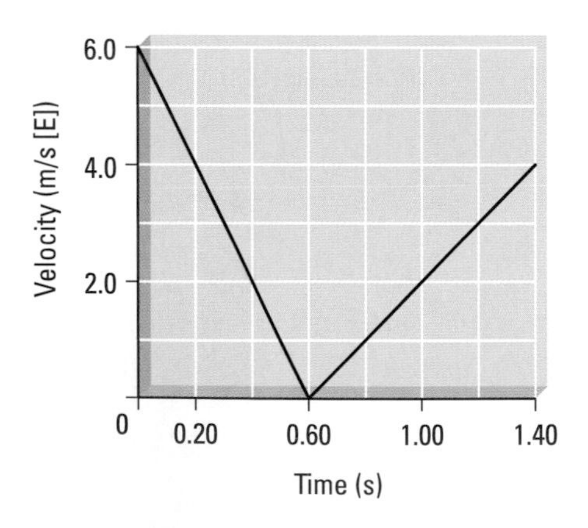

Figure 9
For question 10

Practice

Understanding Concepts

10. Use the velocity-time graph in **Figure 9** to determine the following:
   (a) the velocity at 0.40 s and 0.80 s
   (b) the average acceleration between 0.0 s and 0.60 s
   (c) the average acceleration between 0.60 s and 1.40 s
   (d) the average acceleration between 0.80 s and 1.20 s

Answers

10. (a) 2.0 m/s [E]; 1.0 m/s [E]
   (b) 1.0×10^1 m/s^2 [W]
   (c) 5.0 m/s^2 [E]
   (d) 5.0 m/s^2 [E]

11. Sketch a velocity-time graph for the motion of a car travelling south along a straight road with a posted speed limit of 60 km/h, except in a school zone where the speed limit is 40 km/h. The only traffic lights are found at either end of the school zone, and the car must stop at both sets of lights. (Assume that when $t = 0.0$ s, the velocity is 60 km/h [S].)
12. The data in **Table 6** represent test results on a recently built, standard transmission automobile. Use the data to plot a fully labelled, accurate velocity-time graph of the motion from $t = 0.0$ s to $t = 35.2$ s. Assume that the acceleration is constant in each time segment. Then use the graph to determine the average acceleration in each gear and during braking.

Answers

12. 12 (km/h)/s [fwd]; 5.4 (km/h)/s [fwd]; 1.8 (km/h)/s [fwd]; −27 (km/h)/s [fwd]

Table 6

Acceleration mode	Change in velocity (km/h [fwd])	Time taken for the change (s)
first gear	0.0 to 48	4.0
second gear	48 to 96	8.9
third gear	96 to 128	17.6
braking	128 to 0.0	4.7

Using Position-Time Graphs to Find Acceleration

Assume that you are asked to calculate the acceleration of a car as it goes from 0.0 km/h to 100.0 km/h [W]. The car has a speedometer that can be read directly, so the only instrument you need is a watch. The average acceleration can be calculated from knowing the time it takes to reach maximum velocity. For instance, if the time taken is 12 s and if we assume two significant digits, the average acceleration is

$$\vec{a}_{av} = \frac{\Delta \vec{v}}{\Delta t}$$

$$= \frac{100.0 \text{ km/h [W]} - 0.0 \text{ km/h [W]}}{12 \text{ s}}$$

$$\vec{a}_{av} = 8.3 \text{ (km/h)/s [W]}$$

In a science laboratory, however, an acceleration experiment is not so simple. Objects that move (for example, a cart, a ball, or a metal mass) do not come equipped with speedometers, so their speeds cannot be found directly. (Motion sensors can indicate speeds, but they are not always available or convenient.)

One way to solve this problem is to measure the position of the object from its starting position at specific times. Then a position-time graph can be plotted and a mathematical technique used to calculate the average acceleration. To begin, consider **Figure 10**, which shows a typical position-time graph for a skier

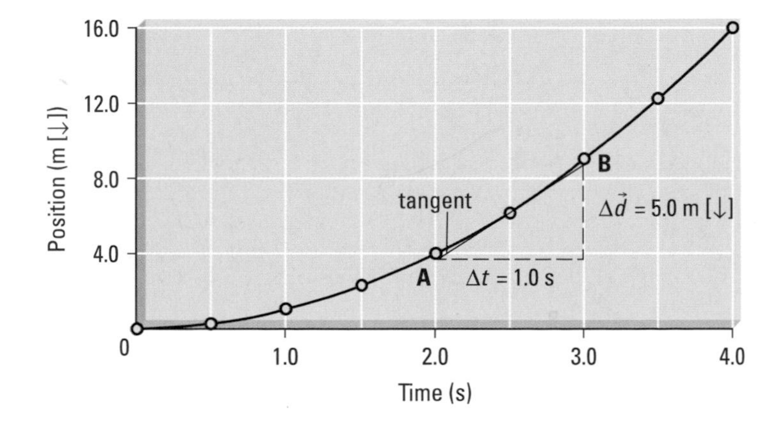

Figure 10
Position-time graph of uniform acceleration

starting from rest and accelerating downhill [↓]. Notice that the skier's displacement in each time interval increases as time increases.

Since the slope of a line on a position-time graph indicates the velocity, we perform slope calculations first. Because the line is curved, however, its slope keeps changing. Thus, we must find the slope of the curved line at various times. The technique we use is called the **tangent technique.** A **tangent** is a straight line that touches a curve at a single point and has the same slope as the curve at that point. To find the velocity at 2.5 s, for example, we draw a tangent to the curve at that time. For convenience, the tangent in our example is drawn so that its Δt value is 1.0 s. Then the slope of the tangent is

tangent technique: a method of determining velocity on a position-time graph by drawing a line tangent to the curve and calculating the slope

tangent: a straight line that touches a curve at a single point and has the same slope as the curve at that point

$$m = \frac{\Delta \vec{d}}{\Delta t}$$

$$= \frac{5.0 \text{ m } [\downarrow]}{1.0 \text{ s}}$$

$$m = 5.0 \text{ m/s } [\downarrow]$$

That is, the skier's velocity at 2.5 s is 5.0 m/s [↓].

This velocity is known as an **instantaneous velocity,** one that occurs at a particular instant. Refer to **Figure 11(a)**, which shows the same position-time graph with some tangents drawn and velocities shown.

instantaneous velocity: velocity that occurs at a particular instant

In **Figure 11(b)**, the instantaneous velocities calculated from the position-time graph are plotted on a velocity-time graph. Notice that the line is extended to 4.0 s, the same final time as that found on the position-time graph. The slope of the line on the velocity graph is then calculated and used to plot the acceleration-time graph, shown in **Figure 11(c)**.

(a)

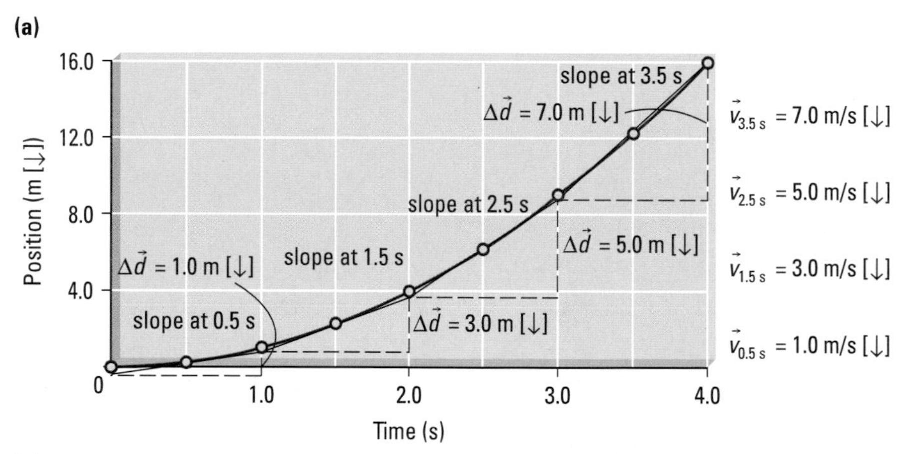

(b)

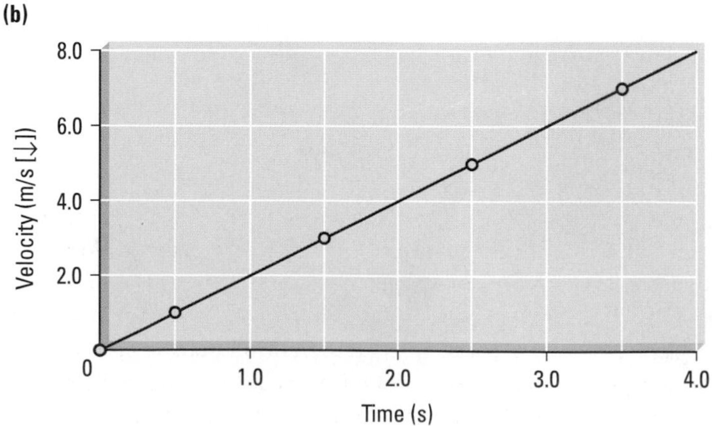

Figure 11
Graphing uniform acceleration
(a) Position-time graph
(b) Velocity-time graph
(c) Acceleration-time graph

(c)

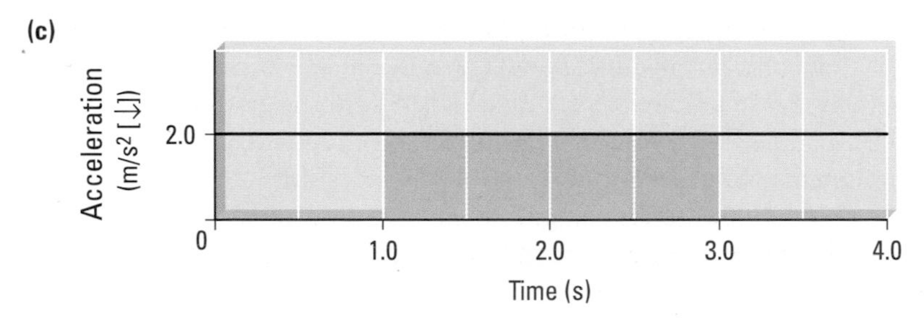

An acceleration-time graph can be used to find the change of velocity during various time intervals. This is accomplished by determining the area under the line on the acceleration-time graph.

$$\begin{aligned} A &= lw \\ &= \vec{a}_{av}(\Delta t) \\ &= (2.0 \text{ m/s}^2 \ [\downarrow])(2.0 \text{ s}) \\ A &= 4.0 \text{ m/s} \ [\downarrow] \end{aligned}$$

The area of the shaded region in **Figure 11(c)** is 4.0 m/s [↓]. This quantity represents the skier's change of velocity from $t = 1.0$ s to $t = 3.0$ s. In other words,

$$\Delta \vec{v} = \vec{a}_{av}(\Delta t)$$

In motion experiments involving uniform acceleration, the velocity-time graph should yield a straight line. However, because of experimental error, this might not occur. If the points on a velocity-time graph of uniform acceleration do not lie on a straight line, draw a straight line of best fit, and then calculate its slope to find the acceleration.

Table 7

Time (s)	Position (m [N])
0.0	0.0
2.0	8.0
4.0	32.0
6.0	72.0
8.0	128.0

Practice

Understanding Concepts

13. **Table 7** shows a set of position-time data for uniformly accelerated motion.
    (a) Plot a position-time graph.
    (b) Find the slopes of tangents at appropriate times.
    (c) Plot a velocity-time graph.
    (d) Plot an acceleration-time graph.
    (e) Determine the area under the line on the velocity-time graph and then on the acceleration-time graph. State what these two areas represent.

Answers

13. (e) 1.3×10^2 m [N]; 32 m/s [N]

Investigation 1.4.1

Attempting Uniform Acceleration

INQUIRY SKILLS

- ○ Questioning
- ○ Hypothesizing
- ● Predicting
- ○ Planning
- ● Conducting
- ● Recording
- ● Analyzing
- ● Evaluating
- ● Communicating

Various methods are available to gather data to analyze the position-time relationship of an object in the laboratory. For example, you can analyze the dots created as a puck slides down an air table raised slightly on one side, or the dots created by a ticker-tape timer on a paper strip attached to a cart rolling down an inclined plane (**Figure 12**). A computer-interfaced motion sensor can also be used to determine the positions at pre-set time intervals as a cart accelerating down an inclined plane moves away from the sensor. Or a picket fence mounted on the cart can be used with a photogate system. Yet another method involves the frame-by-frame analysis of a videotape of the motion. The following investigation shows analysis using a ticker-tape timer, but this investigation can be easily modified for other methods of data collection. If you choose to use a picket fence with the photogate method, you may want to look ahead to Investigation 1.5.1, where the procedure for analysis using this equipment is given.

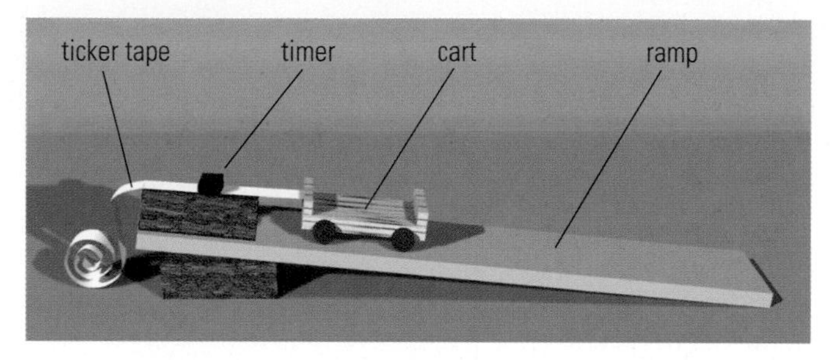

Figure 12
Using a ticker-tape timer to determine acceleration in the laboratory

Question

What type of motion (uniform motion, uniform acceleration, or nonuniform acceleration) is experienced by an object moving down an inclined plane?

Hypothesis/Prediction

(a) Make a prediction to answer the Question.

Try to provide some reasoning for your prediction. Ask yourself questions such as: What causes acceleration? How can you make something accelerate at a greater rate? at a lower rate? What effect will friction have? Relate these ideas to the inclined plane example. Be sure to do this step before starting so that you will have something to compare your results to.

Materials

ticker-tape timer and ticker tape
dynamics cart or smooth-rolling toy truck
one 2-m board
bricks or books
clamp
masking tape

Procedure

1. Set up an inclined plane as shown in **Figure 12**. Clamp the timer near the top of the board. Set up a fixed stop at the bottom of the board.
2. Obtain a length of ticker tape slightly shorter than the incline. Feed the tape through the timer and attach the end of the tape to the back of the cart.
3. Turn on the timer and allow the cart to roll down the incline for at least 1.0 s. Have your partner stop the cart.
4. Record the time intervals for the dots on the tape.
5. Choose a clear dot at the beginning to be the reference point at a time of 0.00 s. The timer likely makes dots at a frequency of 60 Hz. The time interval between two points is therefore $\frac{1}{60}$ s. For convenience, mark off intervals every six dots along the tape so that each interval represents 0.10 s.
6. Measure the position of the cart with respect to the reference point after each 0.10 s, and record the data in a position-time table.

Analysis

(b) Plot the data on a position-time graph. Draw a smooth curve that fits the data closely.

(c) Use the tangent technique to determine the instantaneous velocity of the cart every 0.20 s at five specific times. Use these results to plot a velocity-time graph of the motion.

(d) If the velocity-time graph is a curve, use the tangent method to determine acceleration values every 0.20 s and plot an acceleration-time graph. If the velocity-time graph is a straight line, what does this reveal about the motion? How can you determine the acceleration of the cart?

(e) Calculate the area under the line on the velocity-time graph and state what that area represents. Compare this value to the actual measured value.

(f) Use your results to plot an acceleration-time graph and draw a line of best fit.

(g) What type of motion do your results show? If any aspect of the motion is constant, state the corresponding numerical value.

Evaluation

(h) What evidence is there to support your answer to the Question? Refer to the shapes of your three graphs.

(i) Do your results reflect an object experiencing uniform acceleration? How can you tell?

(j) Describe any sources of error in this experiment. (To review errors, refer to Appendix A.)

(k) If you were to perform this experiment again, what would you do to improve the accuracy of your results?

Synthesis

In this investigation, graphing is suggested for determining the acceleration. However, it is possible to apply the defining equation for average acceleration to determine the acceleration of the moving object.

(l) Describe how you would do this, including the assumption(s) that you must make to solve the problem. (*Hint:* How would you obtain a fairly accurate value of v_f?)

Analysis for Other Methods of Data Gathering

Motion Sensor

If a motion sensor is used, the position-time graph will be constructed using the computer. Tangent tools can be used to determine the instantaneous velocities. Depending on the experiment template used, interfacing software can be used to plot all or part of the data automatically.

Video Analysis

If a videotape of the motion is used, it can be analyzed either using a software program designed for this purpose or manually by moving frame by frame through the videotape and making measurements on a projection. Depending on the instructions, data obtained manually from the projected images can be analyzed as above or using a computer graphing program. For either method, be sure to accurately scale the measurements in order to get correct results.

SUMMARY Uniform Acceleration

- Uniformly accelerated motion occurs when an object, travelling in a straight line, changes its speed uniformly with time. The acceleration can be one that is speeding up or slowing down.
- The equation for average acceleration is $\vec{a}_{av} = \frac{\Delta \vec{v}}{\Delta t}$, or $\vec{a}_{av} = \frac{\vec{v}_f - \vec{v}_i}{\Delta t}$.
- On a velocity-time graph of uniform acceleration, the slope of the line represents the average acceleration between any two times.
- A position-time graph of uniform acceleration is a curve whose slope continually changes. Tangents to the curve at specific times indicate the instantaneous velocities at these times, which can be used to determine the acceleration.
- You should be able to determine the acceleration of an accelerating object using at least one experimental method.

Section 1.4 Questions

Understanding Concepts

1. Describe the motion in each graph in **Figure 13**.
2. State what each of the following represents:
   (a) the slope of a tangent on a position-time graph of nonuniform motion
   (b) the slope of a line on a velocity-time graph
   (c) the area under the line on an acceleration-time graph
3. Under what condition can an object have an eastward velocity and a westward acceleration at the same instant?
4. The world record for motorcycle acceleration occurred when it took a motorcycle only 6.0 s to go from rest to 281 km/h [fwd]. Calculate the average acceleration in
   (a) kilometres per hour per second
   (b) metres per second squared
5. One of the world's fastest roller coasters has a velocity of 8.0 km/h [fwd] as it starts its descent on the first hill. Determine the coaster's maximum velocity at the base of the hill, assuming the average acceleration of 35.3 (km/h)/s [fwd] lasts for 4.3 s.
6. With what initial velocity must a badminton birdie be travelling if in the next 0.80 s its velocity is reduced to 37 m/s [fwd], assuming that air resistance causes an average acceleration of -46 m/s^2 [fwd]?
7. **Table 8** is a set of position-time data for uniform acceleration.
   (a) Plot a position-time graph.
   (b) Determine the instantaneous velocity at several different times by finding the slopes of the tangents at these points on the graph.
   (c) Plot a velocity-time graph.
   (d) Plot an acceleration-time graph.
   (e) Determine the area under the line on the velocity-time graph and then on the acceleration-time graph.

(a)

Position
0
Time

(b)

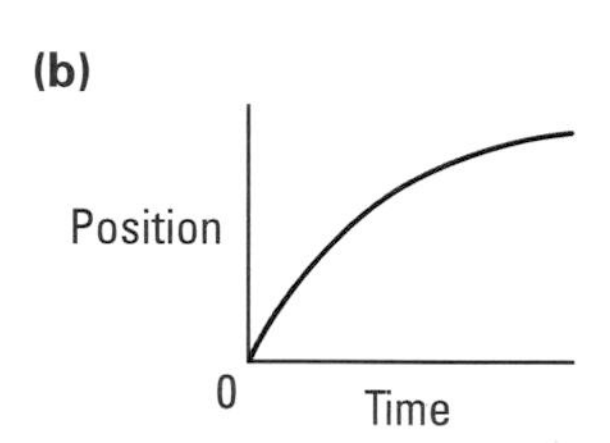

(c)

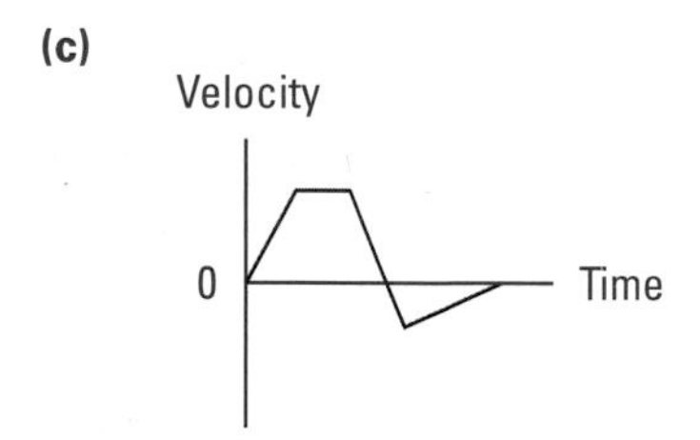

Figure 13
For question 1

Table 8

Time (s)	Position (mm [W])
0.0	0.0
0.10	3.0
0.20	12.0
0.30	27.0
0.40	48.0

8. **Figure 14** shows three accelerometers attached to carts that are in motion. In each case, describe two possible motions that would create the condition shown.

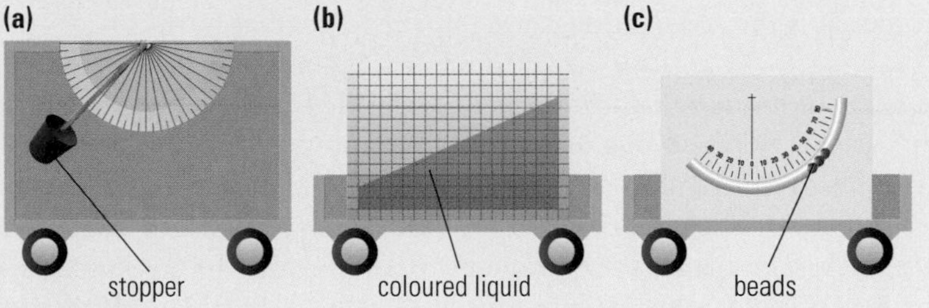

Figure 14

Making Connections

9. During part of the blastoff of a space shuttle, the velocity of the shuttle changes from 125 m/s [up] to 344 m/s [up] in 2.30 s.
   (a) Determine the average acceleration experienced by the astronauts on board during this time interval.
   (b) This rate of acceleration would be dangerous if the astronauts were standing or even sitting vertically in the shuttle. What is the danger? Research the type of training that astronauts are given to avoid the danger.

Reflecting

10. Describe the most common difficulties you have in applying the tangent technique on position-time graphs. What do you do to reduce these difficulties?
11. Think about the greatest accelerations you have experienced. Where did they occur? Did they involve speeding up or slowing down? What effects did they have on you?

1.5 Acceleration Near Earth's Surface

Amusement park rides that allow passengers to drop freely toward the ground attract long lineups (**Figure 1**). Riders are accelerated toward the ground until a braking system causes the cars to slow down over a small distance.

If two solid metal objects of different masses, 20 g and 1000 g, for example, are dropped from the same height above the floor, they land at the same time. This fact proves that the acceleration of falling objects near the surface of Earth does not depend on mass.

It was Galileo Galilei who first proved that, if we ignore the effect of air resistance, the acceleration of falling objects is constant. He proved this experimentally by measuring the acceleration of metal balls rolling down a ramp. Galileo found that, for a constant slope of the ramp, the acceleration was constant—it did not depend on the mass of the metal ball. The reason he could not measure vertical acceleration was that he had no way of measuring short periods of time accurately. You will appreciate the difficulty of measuring time when you perform the next experiment.

Figure 1
The Drop Zone at Paramount Canada's Wonderland, north of Toronto, allows the riders to accelerate toward the ground freely for approximately 3 s before the braking system causes an extreme slowing down.

Had Galileo been able to evaluate the acceleration of freely falling objects near Earth's surface, he would have measured it to be approximately 9.8 m/s^2 [down]. This value does not apply to objects influenced by air resistance. It is an average value that changes slightly from one location on Earth's surface to another. It is the acceleration caused by the force of gravity.

acceleration due to gravity: the vector quantity 9.8 m/s^2 [down], represented by the symbol $\vec{g}$

The vector quantity 9.8 m/s^2 [down], or 9.8 m/s^2 [↓], occurs so frequently in the study of motion that from now on, we will give it the symbol $\vec{g}$, which represents the **acceleration due to gravity**. (Do not confuse this $\vec{g}$ with the g used as the symbol for "gram.") More precise magnitudes of $\vec{g}$ are determined by scientists throughout the world. For example, at the International Bureau of Weights and Measures in France, experiments are performed in a vacuum chamber in which an object is launched upwards by using an elastic. The object has a system of mirrors at its top and bottom that reflect laser beams used to measure time of flight. The magnitude of $\vec{g}$ obtained using this technique is 9.809 260 m/s^2. Galileo would have been pleased with the precision!

In solving problems involving the acceleration due to gravity, $\vec{a}_{av}$ = 9.8 m/s^2 [down] can be used if the effect of air resistance is assumed to be negligible. When air resistance on an object is negligible, we say the object is "falling freely."

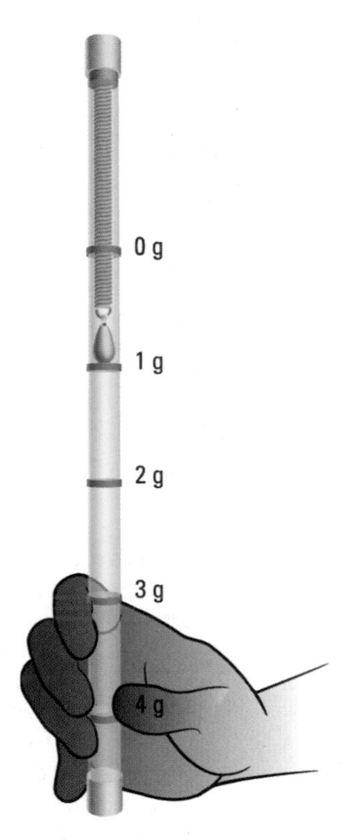

Figure 2
A typical vertical accelerometer for student use

A Vertical Accelerometer

Vertical accelerometers, available commercially in kit form, can be used to measure acceleration in the vertical direction (**Figure 2**).

(a) Predict the reading on the accelerometer if you held it and
- kept it still
- moved it vertically upward at a constant speed
- moved it vertically downward at a constant speed

(b) Predict what happens to the accelerometer bob if you
- thrust the accelerometer upward
- dropped the accelerometer downward

(c) Use an accelerometer to test your predictions in (a) and (b). Describe what you discover.

(d) How do you think this device could be used on amusement park rides?

Practice

Understanding Concepts

1. In a 1979 movie, a stuntman leaped from a ledge on Toronto's CN Tower and experienced free fall for 6.0 s before opening the safety parachute. Assuming negligible air resistance, determine the stuntman's velocity after falling for (a) 3.0 s and (b) 6.0 s.
2. A stone is thrown from a bridge with an initial vertical velocity of magnitude 4.0 m/s. Determine the stone's velocity after 2.2 s if the direction of the initial velocity is (a) upward and (b) downward. Neglect air resistance.

Answers

1. (a) 29 m/s [↓]
   (b) 59 m/s [↓]
2. (a) 18 m/s [↓]
   (b) 26 m/s [↓]

Investigation 1.5.1

Acceleration Due to Gravity

INQUIRY SKILLS

- ○ Questioning
- ○ Hypothesizing
- ● Predicting
- ○ Planning
- ● Conducting
- ○ Recording
- ● Analyzing
- ● Evaluating
- ● Communicating

As with Investigation 1.4.1, there are several possible methods for obtaining position-time data of a falling object in the laboratory. The ticker-tape timer, the motion sensor, and the videotape were suggested before. In this investigation, a picket fence and photogate can also be used to get very reliable results. If possible, try to use a different method from that used in the previous investigation. Here, the analysis will be shown for the picket fence and photogate method. If other methods of data collection are used, refer back to Investigation 1.4.1 for analysis.

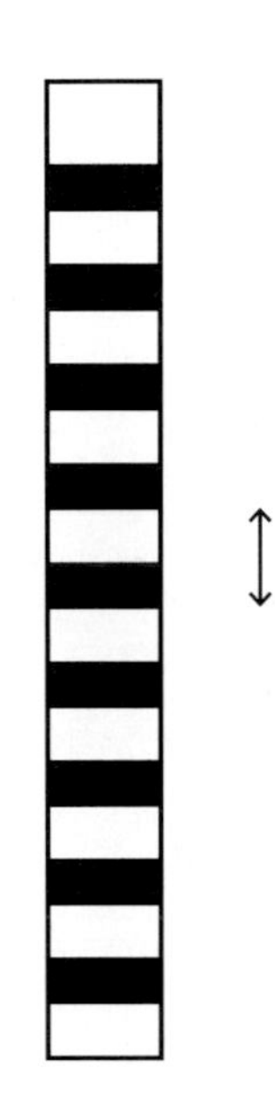

Figure 3
A picket fence is a clear strip of plastic with several black wide bands marked at regular intervals along the length. The black bands interrupt the beam of the photogate. As each band interrupts the beam, it triggers a clock to measure the time required for the picket fence to travel a distance equal to the spacing between the leading edges of two successive bands. Picket fences can be used with computer software applications or with stand-alone timing devices.

Question

What type of motion is experienced by a free-falling object?

Hypothesis/Prediction

(a) How will this motion compare with that on the inclined plane studied in Investigation 1.4.1? Make a prediction with respect to the general type of motion and the quantitative results.

Also, think about how the motion will differ if the mass of the object is altered.

Materials

picket fence with photogate
computer interfacing software
light masses to add to the picket fence
masking tape

Procedure

1. Open the interface software template designed for use with a picket fence.
2. Obtain a picket fence and measure the distance between the leading edges of two bands as shown in **Figure 3**. Enter this information into the appropriate place in the experimental set-up window.
3. Before performing the experiment, become familiar with the picket fence and the software to find out how the computer obtains the values shown.
4. Enable the interface and get a pad ready for the picket fence to land on.
5. Hold the picket fence vertically just above the photogate. Drop the picket fence straight through the photogate and have your partner catch it.
6. After analyzing this trial, tape some added mass to the bottom of the picket fence and repeat the experiment.

Analysis

(b) The position-time data should appear automatically on the computer screen. Look at the position-time graph of the data collected. What type of motion is represented by the graph?
(c) Look at the velocity-time graph. What type of motion does it describe?
(d) Determine the average acceleration from the velocity-time graph.
(e) What type of motion is experienced by a free-falling object? State the average acceleration of the picket fence. How did the acceleration of the heavier object compare with that of the lighter one?

DID YOU KNOW ?

Escape Systems

One area of research into the effect of acceleration on the human body deals with the design of emergency escape systems from high-performance aircraft. In an emergency, the pilot would be shot upward away from the damaged plane from a sitting position through an escape hatch. The escape system would have to be designed to produce a high enough acceleration to quickly remove a pilot from danger, but not too high that the acceleration would cause injury to the pilot.

Evaluation

(f) Explain how the computer calculates the velocity values. Are these average or instantaneous velocities?

(g) What evidence is there to support your answer to the Question? Refer to shapes of three graphs.

(h) Look back in this text for the type of motion that a free-falling object should experience and the accepted value for the acceleration due to gravity on Earth's surface. How do your results compare with the accepted value? Determine the percentage error between the experimental value for the acceleration due to gravity and the accepted value.

(i) Are your results the same as what you predicted? If not, what incorrect assumption did you make?

(j) Identify any sources of error in this investigation. Do they reasonably account for the percentage error for your results?

(k) How does the mass of an object affect its acceleration in a free-fall situation?

(l) If you were to repeat the investigation, what improvements could you make in order to increase the accuracy of the results?

Applications of Acceleration

Galileo Galilei began the mathematical analysis of acceleration, and the topic has been studied by physicists ever since. However, only during the past century has acceleration become a topic that relates closely to our everyday lives.

The study of acceleration is important in the field of transportation. Humans undergo acceleration in automobiles, airplanes, rockets, amusement park rides, and other vehicles. The acceleration in cars and passenger airplanes is usually small, but in a military airplane or a rocket, it can be great enough to cause damage to the human body. A person can faint when blood drains from the head and goes to the lower part of the body. In 1941, a Canadian pilot and inventor named W.R. Franks designed an "anti-gravity" suit to prevent pilot blackouts in military planes undergoing high-speed turns and dives. The suit had water encased in the inner lining to prevent the blood vessels from expanding outwards (**Figure 4**).

Figure 4
This 1941 photograph shows W.R. Franks in the "anti-gravity" suit he designed.

Modern experiments have shown that the maximum acceleration a human being can withstand for more than about 0.5 s is approximately $|30\vec{g}|$ (the vertical bars represent the magnitude of the vector, in this case, 294 m/s^2). Astronauts experience up to $|10\vec{g}|$ (98 m/s^2) for several seconds during a rocket launch. At this acceleration, if the astronauts were standing, they would faint from loss of blood to the head. To prevent this problem, astronauts must sit horizontally during blastoff (**Figure 5**).

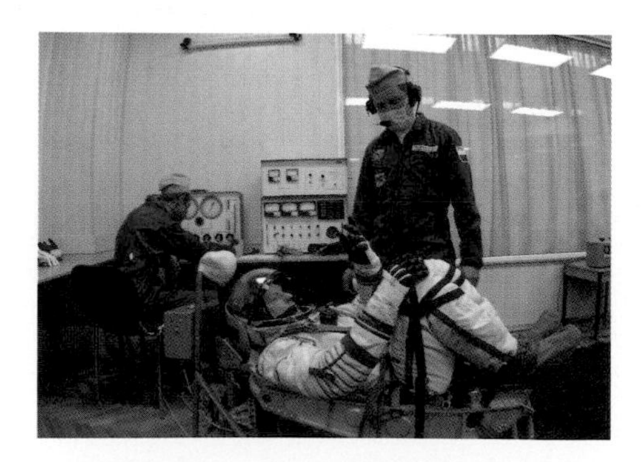

Figure 5
An astronaut participates in a launch simulation exercise as two crew members assist.

In our day-to-day lives, we are more concerned with braking in cars and other vehicles than with blasting off in rockets. Studies are continually being done to determine the effect on the human body when a car has a collision or must stop quickly. Seatbelts, headrests, and airbags help prevent many injuries caused by rapid braking (**Figure 6**).

In the exciting sport of skydiving, the diver jumps from an airplane and accelerates toward the ground, experiencing free fall for the first while (**Figure 7**). While falling, the skydiver's speed will increase to a maximum amount called

Figure 6
As the test vehicle shown crashes into a barrier, the airbag being researched expands rapidly and prevents the dummy's head from striking the windshield or steering wheel. After the crash, the airbag deflates quickly so that, in a real situation, the driver can breathe.

terminal speed. Air resistance prevents a higher speed. At terminal speed, the diver's acceleration is zero; in other words, the speed remains constant. For humans, terminal speed in air is about 53 m/s or 190 km/h. After the parachute opens, the terminal speed is reduced to between 5 m/s and 10 m/s.

terminal speed: maximum speed of a falling object at which point the speed remains constant and there is no further acceleration

Terminal speed is also important in other situations. Certain plant seeds, such as dandelions, act like parachutes and have a terminal speed of about 0.5 m/s. Some industries take advantage of the different terminal speeds of various particles in water when they use sedimentation to separate particles of rock, clay, or sand from one another. Volcanic eruptions produce dust particles of different sizes. The larger dust particles settle more rapidly than the smaller ones. Thus, very tiny particles with low terminal speeds travel great horizontal distances around the world before they settle. This phenomenon can have a serious effect on Earth's climate.

Practice

Understanding Concepts

3. Sketch the general shape of a velocity-time graph for a skydiver who accelerates, then reaches terminal velocity, then opens the parachute and reaches a different terminal velocity. Assume that downward is positive.

Figure 7
This skydiver experiences "free fall" immediately upon leaving the aircraft, but reaches terminal speed later.

SUMMARY Acceleration Near Earth's Surface

- On average, the acceleration due to gravity on Earth's surface is $\vec{g} = 9.8 \text{ m/s}^2$ [↓]. This means that in the absence of air resistance, an object falling freely toward Earth accelerates at 9.8 m/s^2 [↓].
- Various experimental ways can be used to determine the local value of $\vec{g}$.
- The topic of accelerated motion is applied in various fields, including transportation and the sport of skydiving.

Section 1.5 Questions

Understanding Concepts

1. An apple drops from a tree and falls freely toward the ground. Sketch the position-time, velocity-time, and acceleration-time graphs of the apple's motion, assuming that (a) downward is positive, and (b) upward is positive.
2. An astronaut standing on the Moon drops a feather, initially at rest, from a height of over 2.0 m above the Moon's surface. The feather accelerates downward, just as a ball or any other object would on Earth. In using frame-by-frame analysis of a videotape of the falling feather, the data in **Table 1** are recorded.

Table 1

Time (s)	0.000	0.400	0.800	1.200	1.600
Position (m [down])	0.0	0.128	0.512	1.512	2.050

   (a) Use the data to determine the acceleration due to gravity on the Moon.
   (b) Why can a feather accelerate at the same rate as all other objects on the Moon?

3. Give examples to verify the following statement: "In general, humans tend to experience greater magnitudes of acceleration when slowing down than when speeding up."
4. Sketch an acceleration-time graph of the motion toward the ground experienced by a skydiver from the time the diver leaves the plane and reaches terminal speed. Assume downward is positive.
5. During a head-on collision, the airbag in a car increases the time for a body to stop from 0.10 s to 0.30 s. How will the airbag change the magnitude of acceleration of a person travelling initially at 28 m/s?

Applying Inquiry Skills

6. Two student groups choose different ways of performing an experiment to measure the acceleration due to gravity. Group A chooses to use a ticker-tape timer with a mass falling toward the ground. Group B chooses to use a motion sensor that records the motion of a falling steel ball. If both experiments are done well, how will the results compare? Why?
7. Describe how you would design and build an accelerometer that measures vertical acceleration directly using everyday materials.

Making Connections

8. Today's astronauts wear an updated version of the anti-gravity suit invented by W.R. Franks. Research and describe why these suits are required and how they were developed. Follow the links for Nelson Physics 11, 1.5.

GO TO www.science.nelson.com

1.6 Solving Uniform Acceleration Problems

Now that you have learned the definitions and basic equations associated with uniform acceleration, it is possible to extend your knowledge so that you can solve more complex problems. In this section, you will learn how to derive and use some important equations involving the following variables: initial velocity, final velocity, displacement, time interval, and average acceleration. Each equation derived will involve four of these five variables and thus will have a different purpose. It is important to remember that these equations only apply to uniformly accelerated motion.

The process of deriving equations involves three main stages:

1. State the given facts and equations.
2. Substitute for the variable to be eliminated.
3. Simplify the equation to a convenient form.

The derivations involve two given equations. The first is the equation that defines average acceleration, $\vec{a}_{av} = \frac{\vec{v}_f - \vec{v}_i}{\Delta t}$. A second equation can be found by applying the fact that the area under the line on a velocity-time graph indicates the displacement. **Figure 1** shows a typical velocity-time graph for an object that undergoes uniform acceleration from an initial velocity ($\vec{v}_i$) to a final velocity ($\vec{v}_f$) during a time (Δt). The shape of the area under the line is a trapezoid, so the area is $\Delta\vec{d} = \frac{1}{2}(\vec{v}_i + \vec{v}_f)\Delta t$. (The area of a trapezoid is the product of the average length of the two parallel sides and the perpendicular distance between them.)

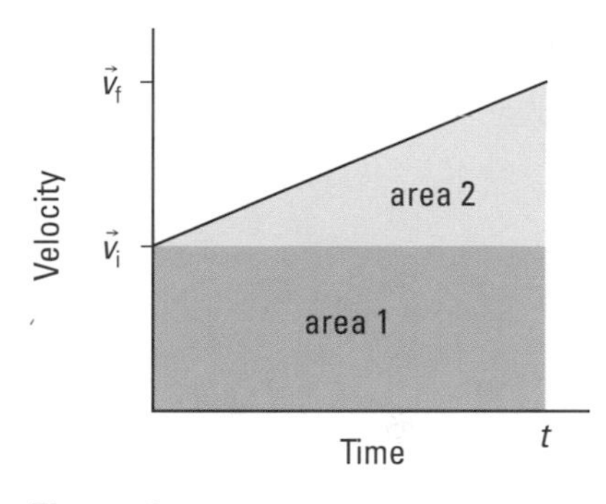

Figure 1
A velocity-time graph of uniform acceleration

Notice that the defining equation for the average acceleration has four of the five possible variables ($\Delta\vec{d}$ is missing), and the equation for displacement also has four variables ($\vec{a}_{av}$ is missing). These two equations can be combined to derive three other uniform acceleration equations, each of which involves four variables. (Two such derivations are shown next, and the third one is required in a section question.)

To derive the equation in which $\vec{v}_f$ is eliminated, we rearrange the defining equation of acceleration to get

$$\vec{v}_f = \vec{v}_i + \vec{a}_{av}\Delta t$$

Substituting this equation into the equation for displacement eliminates $\vec{v}_f$.

$$\Delta\vec{d} = \frac{1}{2}(\vec{v}_i + \vec{v}_f)\Delta t$$

$$= \frac{1}{2}(\vec{v}_i + \vec{v}_i + \vec{a}_{av}\Delta t)\Delta t$$

$$= \frac{1}{2}(2\vec{v}_i + \vec{a}_{av}\Delta t)\Delta t$$

$$\Delta\vec{d} = \vec{v}_i\Delta t + \frac{\vec{a}_{av}(\Delta t)^2}{2}$$

Next, we derive the equation in which Δt is eliminated. This derivation is more complex because using vector notation would render the results invalid. (We would encounter mathematical problems if we tried to multiply two vectors.) To overcome this problem, only magnitudes of vector quantities are used,

DID YOU KNOW ?

An Alternative Calculation

In **Figure 1**, the area can also be found by adding the area of the triangle to the area of the rectangle beneath it.

$$\Delta\vec{d} = \text{area 1} + \text{area 2}$$

$$= lw + \tfrac{1}{2}bh$$

$$= \vec{v}_i\Delta t + \tfrac{1}{2}(\vec{v}_f - \vec{v}_i)\Delta t$$

$$= \vec{v}_i\Delta t + \tfrac{1}{2}\vec{v}_f\Delta t - \tfrac{1}{2}\vec{v}_i\Delta t$$

$$\Delta\vec{d} = \tfrac{1}{2}(\vec{v}_i + \vec{v}_f)\Delta t$$

and directions of vectors involved will be decided based on the context of the situation. This resulting equation is valid only for one-dimensional, uniform acceleration. From the defining equation for average acceleration,

$$\Delta t = \frac{v_f - v_i}{a_{av}} \quad \text{which can be substituted into the equation for displacement}$$

$$\Delta d = \frac{1}{2}(v_f + v_i)\Delta t$$

$$= \frac{1}{2}(v_f + v_i)\left(\frac{v_f - v_i}{a_{av}}\right) \quad (v_f + v_i) \text{ and } (v_f - v_i) \text{ are factors of the difference of two squares}$$

$$\Delta d = \frac{v_f^2 - v_i^2}{2a_{av}}$$

$$\text{or} \quad 2a_{av}\Delta d = v_f^{\,2} - v_i^2$$

Therefore, $v_f^{\,2} = v_i^2 + 2a_{av}\Delta d$

The equations for uniform acceleration are summarized for your convenience in **Table 1**. Applying the skill of unit analysis to the equations will help you check to see if your derivations are appropriate.

Table 1 Equations for Uniformly Accelerated Motion

Variables involved	General equation	Variable eliminated
$\vec{a}_{av}$, $\vec{v}_f$, $\vec{v}_i$, Δt	$\vec{a}_{av} = \frac{\vec{v}_f - \vec{v}_i}{\Delta t}$	$\Delta\vec{d}$
$\Delta\vec{d}$, $\vec{v}_i$, $\vec{a}_{av}$, Δt	$\Delta\vec{d} = \vec{v}_i\Delta t + \frac{\vec{a}_{av}(\Delta t)^2}{2}$	$\vec{v}_f$
$\Delta\vec{d}$, $\vec{v}_i$, $\vec{v}_f$, Δt	$\Delta\vec{d} = \vec{v}_{av}\Delta t$ or $\Delta\vec{d} = \frac{1}{2}(\vec{v}_i + \vec{v}_f)\Delta t$	$\vec{a}_{av}$
$\vec{v}_f$, $\vec{v}_i$, $\vec{a}_{av}$, $\Delta\vec{d}$	$v_f^2 = v_i^2 + 2a_{av}\Delta d$	Δt
$\Delta\vec{d}$, $\vec{v}_f$, Δt, $\vec{a}_{av}$	$\Delta\vec{d} = \vec{v}_f\Delta t - \frac{\vec{a}_{av}(\Delta t)^2}{2}$	$\vec{v}_i$

Sample Problem 1

Starting from rest at $t = 0.0$ s, a car accelerates uniformly at 4.1 m/s² [S]. What is the car's displacement from its initial position at 5.0 s?

Solution

$\vec{v}_i = 0.0$ m/s

$\vec{a}_{av} = 4.1$ m/s² [S]

$\Delta t = 5.0$ s

$\Delta\vec{d} = ?$

$$\Delta\vec{d} = \vec{v}_i\Delta t + \frac{1}{2}\vec{a}_{av}\Delta t^2$$
$$= (0.0 \text{ m/s})(5.0 \text{ s}) + \frac{1}{2}(4.1 \text{ m/s}^2 [\text{S}])(5.0 \text{ s})^2$$
$$\Delta\vec{d} = 51 \text{ m [S]}$$

The car's displacement at 5.0 s is 51 m [S].

In Sample Problem 1, as in many motion problems, there is likely more than one method for finding the solution. Practice is necessary to help you develop skill in solving this type of problem efficiently.

Sample Problem 2

An Olympic diver falls from rest from the high platform. Assume that the fall is the same as the official height of the platform above water, 10.0 m. At what velocity does the diver strike the water?

Solution

Since no time interval is given or required, the equation to be used involves v_f^2, so we will not use vector notation for our calculation. Both the acceleration and the displacement are downward, so we choose downward to be positive.

$v_i = 0.0$ m/s
$\Delta d = +10.0$ m
$a_{av} = +9.8$ m/s^2
$v_f = ?$

$$v_f^2 = v_i^2 + 2a_{av}\Delta d$$
$$= (0.0 \text{ m/s})^2 + 2(9.8 \text{ m/s}^2)(10.0 \text{ m})$$
$$v_f^2 = 196 \text{ m}^2/\text{s}^2$$
$$v_f = \pm\ 14 \text{ m/s}$$

The diver strikes the water at a velocity of 14 m/s [↓].

Activity 1.6.1

Human Reaction Time

Earth's acceleration due to gravity can be used to determine human reaction time (the time it takes a person to react to an event that the person sees). Determine your own reaction time by performing the following activity. Your partner will hold a 30 cm wooden ruler or a metre stick at a certain position, say the 25 cm mark, in such a way that the ruler is vertically in line with your thumb and index finger (**Figure 2**). Now, as you look at the ruler, your partner will drop the ruler without warning. Grasp it as quickly as possible. Repeat this several times for accuracy, and then find the average of the displacements the ruler falls before you catch it.

Figure 2
Determining reaction time

(a) Knowing that the ruler accelerates from rest at 9.8 m/s^2 [↓] and the displacement it falls before it is caught, calculate your reaction time using the appropriate uniform acceleration equation.

(b) Compare your reaction time to that of other students.

(c) Assuming that your leg reaction time is the same as your hand reaction time, use your calculated value to determine how far a car you are driving at 100 km/h would travel between the time you see an emergency and the time you slam on the brakes. Express the answer in metres.

(d) Repeat the procedure while talking to a friend this time. This is to simulate distraction by an activity, such as talking on a hand-held phone, while driving.

SUMMARY Solving Uniform Acceleration Problems

- Starting with the defining equation of average acceleration, $\vec{a}_{av} = \frac{\vec{v}_f - \vec{v}_i}{\Delta t}$, and a velocity-time graph of uniform acceleration, equations involving uniform acceleration can be derived. The resulting equations, shown in **Table 1**, can be applied to find solutions to a variety of motion problems.

Section 1.6 Questions

Understanding Concepts

1. In an acceleration test for a sports car, two markers 0.30 km apart were set up along a road. The car passed the first marker with a velocity of 5.0 m/s [E] and passed the second marker with a velocity of 33.0 m/s [E]. Calculate the car's average acceleration between the markers.

2. A baseball travelling at 26 m/s [fwd] strikes a catcher's mitt and comes to a stop while moving 9.0 cm [fwd] with the mitt. Calculate the average acceleration of the ball as it is stopping.

3. A plane travelling at 52 m/s [W] down a runway begins accelerating uniformly at 2.8 m/s^2 [W].
   (a) What is the plane's velocity after 5.0 s?
   (b) How far has it travelled during this 5.0 s interval?

4. A skier starting from rest accelerates uniformly downhill at 1.8 m/s^2 [fwd]. How long will it take the skier to reach a point 95 m [fwd] from the starting position?

5. For a certain motorcycle, the magnitude of the braking acceleration is $|4\vec{g}|$. If the bike is travelling at 32 m/s [S],
   (a) how long does it take to stop?
   (b) how far does the bike travel during the stopping time?

6. (a) Use the process of substitution to derive the uniform acceleration equation in which the initial velocity has been omitted.
   (b) An alternative way to derive the equation in which the initial velocity, $\vec{v}_i$, is eliminated is to apply the fact that the area on a velocity-time graph indicates displacement. Sketch a velocity-time graph (like the one in **Figure 1** of section 1.6) and use it to derive the equation required. (*Hint:* Find the area of the large rectangle on the graph, subtract the area of the top triangle, and apply the fact that $\vec{v}_f - \vec{v}_i = \vec{a}\Delta t$.)

7. A car travelling along a highway must uniformly reduce its velocity to 12 m/s [N] in 3.0 s. If the displacement travelled during that time interval is 58 m [N], what is the car's average acceleration? What is its initial velocity?

Applying Inquiry Skills

8. Make up a card or a piece of other material such that, when it is dropped in the same way as the ruler in Activity 1.6.1, the calibrations on it indicate the human reaction times. Try your calibrated device.
9. Design an experiment to determine the maximum height you can throw a ball vertically upward. This is an outdoor activity, requiring the use of a stopwatch and an appropriate ball, such as a baseball. Assume that the time for the ball to rise (or fall) is half the total time. Your design should include the equations you will use and any safety considerations. Get your teacher's approval and then perform the activity. After you have calculated the height, calculate how high you could throw a ball on Mars. The magnitude of the acceleration due to gravity on the surface of Mars is 3.7 m/s^2.

Making Connections

10. How could you use the device suggested in question 8 as a way of determining the effect of taking a cold medication that causes drowsiness on human reaction time?

Reflecting

11. This chapter involves many equations, probably more than any other chapter in this text. Describe the ways that you and others in your class learn how to apply these equations to solve problems.
12. Visual learners tend to like the graphing technique for deriving acceleration equations, and abstract learners tend to like the substitution technique. Which technique do you prefer? Why?

Chapter 1 Summary

Key Expectations

Throughout this chapter, you have had opportunities to

- define and describe concepts and units related to motion (e.g., vector quantities, scalar quantities, displacement, uniform motion, instantaneous and average velocity, uniform acceleration, and instantaneous and average acceleration); (1.1, 1.2, 1.3, 1.4, 1.5, 1.6)
- describe and explain different kinds of motion, and apply quantitatively the relationships among displacement, velocity, and acceleration in specific contexts, including vertical acceleration; (1.2, 1.3, 1.4, 1.5, 1.6)
- analyze motion in the horizontal plane in a variety of situations, using vector diagrams; (1.2, 1.3)
- interpret patterns and trends in data by means of graphs drawn by hand or by computer, and infer or calculate linear and non-linear relationships among variables (e.g., analyze and explain the motion of objects, using position-time graphs, velocity-time graphs, and acceleration-time graphs); (1.2, 1.4, 1.5, 1.6)
- evaluate the design of technological solutions to transportation needs (e.g., the safe following distances of vehicles); (1.2, 1.3, 1.5)

Key Terms

kinematics
uniform motion
nonuniform motion
scalar quantity
base unit
derived unit
instantaneous speed
average speed
vector quantity
position
displacement
velocity
average velocity
resultant displacement
frame of reference
relative velocity
accelerated motion
uniformly accelerated motion
acceleration
instantaneous acceleration
average acceleration
tangent technique
tangent
instantaneous velocity
acceleration due to gravity
terminal speed

Make a Summary

Almost all the concepts in this chapter can be represented on graphs or by scale diagrams. On a single piece of paper, draw several graphs and scale diagrams to summarize as many of the key words and concepts in this chapter as possible. Where appropriate, include related equations on the graphs and diagrams.

Reflect on your Learning

Revisit your answers to the Reflect on Your Learning questions at the beginning of this chapter.

- How has your thinking changed?
- What new questions do you have?

Chapter 1 Review

Understanding Concepts

1. Describe the differences between uniform and non-uniform motion. Give a specific example of each type of motion.
2. Laser light, which travels in a vacuum at 3.00×10^8 m/s, is used to measure the distance from Earth to the Moon with great accuracy. On a clear day, an experimenter sends a laser signal toward a small reflector on the Moon. Then, 2.51 s after the signal is sent, the reflected signal is received back on Earth. What is the distance between Earth and the Moon at the time of the experiment?
3. The record lap speed for car racing is about 112 m/s (or 402 km/h). The record was set on a track 12.5 km in circumference. How long did it take the driver to complete one lap?
4. A fishing boat leaves port at 04:30 A.M. in search of the day's catch. The boat travels 4.5 km [E], then 2.5 km [S], and finally 1.5 km [W] before discovering a large school of fish on the sonar screen at 06:30 A.M.
   (a) Draw a vector scale diagram of the boat's motion.
   (b) Calculate the boat's average speed.
   (c) Determine the boat's average velocity.
5. State what is represented by each of the following calculations:
   (a) the slope on a position-time graph
   (b) the area on a velocity-time graph
   (c) the area on an acceleration-time graph
6. **Table 1** shows data recorded in an experiment involving motion.

Table 1

Time (s)	0.00	0.10	0.20	0.30	0.40	0.50	0.60
Position (cm [W])	0	25	50	75	75	75	0

   (a) Use the data to plot a position-time graph of the motion. Assume that the lines between the points are straight.
   (b) Use the graph from (a) to find the instantaneous velocity at times 0.10 s, 0.40 s, and 0.55 s.
   (c) Plot a velocity-time graph of the motion.
   (d) Find the total area between the lines and the time-axis on the velocity-time graph. Does it make sense that this area indicates the resultant displacement?
7. A ferry boat is crossing a river that is 8.5×10^2 m wide. The average velocity of the water relative to the shore is 3.8 m/s [E] and the average velocity of the boat relative to the water is 4.9 m/s [S].
   (a) Determine the velocity of the ferry boat relative to the shore.
   (b) How long does the crossing take?
   (c) Determine the displacement of the boat as it crosses from the north shore to the south shore.
8. (a) Is it possible to have zero velocity but non-zero acceleration at some instant in a motion? Explain.
   (b) Is it possible to have zero acceleration but non-zero velocity at some instant in a motion? Explain.
9. A ball is thrown vertically upward. What is its acceleration
   (a) after it has left the thrower's hand and is travelling upward?
   (b) at the instant it reaches the top of its flight?
   (c) on its way down?
10. (a) If the instantaneous speed of an object remains constant, can its instantaneous velocity change? Explain.
    (b) If the instantaneous velocity of an object remains constant, can its instantaneous speed change? Explain.
    (c) Can an object have a northward velocity while experiencing a southward acceleration? Explain.
11. Show that (cm/s)/s is mathematically equivalent to cm/s^2.
12. A cyclist on a ten-speed bicycle accelerates from rest to 2.2 m/s in 5.0 s in third gear, then changes into fifth gear. After 10.0 s in fifth gear, the cyclist reaches 5.2 m/s. Assuming that the direction of travel remains the same, calculate the magnitude of the average acceleration in the third and fifth gears.
13. For the graph shown in **Figure 1**, determine the following:
    (a) velocity at 1.0 s, 3.0 s, and 5.5 s
    (b) acceleration at 1.0 s, 3.0 s, and 5.5 s

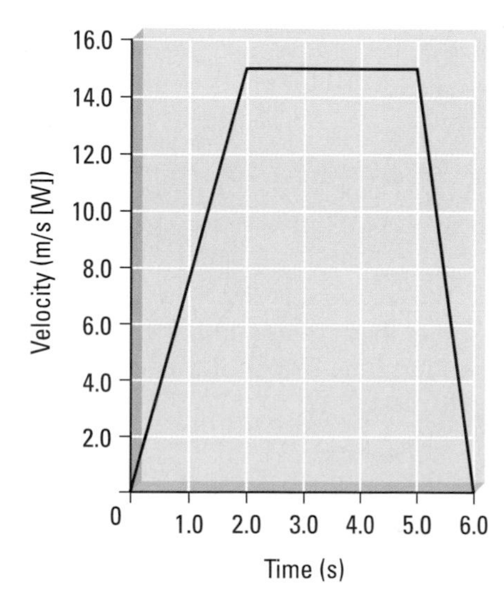

Figure 1

14. A student throws a baseball vertically upward, and 2.8 s later catches it at the same level. Neglecting air resistance, calculate the following:
    (a) the velocity at which the ball left the student's hand (*Hint:* Assume that, when air resistance is ignored, the time it takes to rise equals the time it takes to fall for an object thrown upward.)
    (b) the height to which the ball climbed above the student's hand.
15. An arrow is accelerated for a displacement of 75 cm [fwd] while it is on the bow. If the arrow leaves the bow at a velocity of 75 m/s [fwd], what is its average acceleration while on the bow?
16. An athlete in good physical condition can land on the ground at a speed of up to 12 m/s without injury. Calculate the maximum height from which the athlete can jump without injury. Assume that the takeoff speed is zero.
17. Two cars at the same stoplight accelerate from rest when the light turns green. Their motions are shown by the velocity-time graph in **Figure 2**.
    (a) After the motion has begun, at what time do the cars have the same velocity?
    (b) When does the car with the higher final velocity overtake the other car?
    (c) How far from the starting position are they when one car overtakes the other?

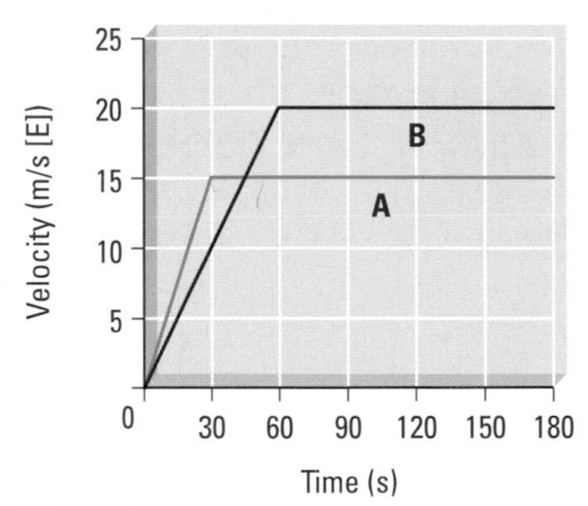

Figure 2

18. (a) Discuss the factors that likely affect the terminal speed of an object falling in Earth's atmosphere.
    (b) Is there a terminal speed for objects falling on the Moon? Explain.
19. Describe the motion represented by each graph in **Figure 3**.

(a)

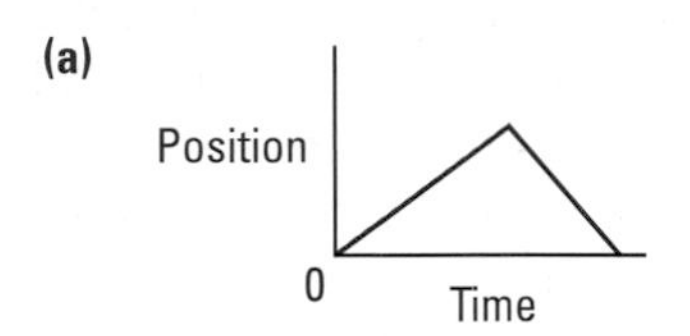

(b)

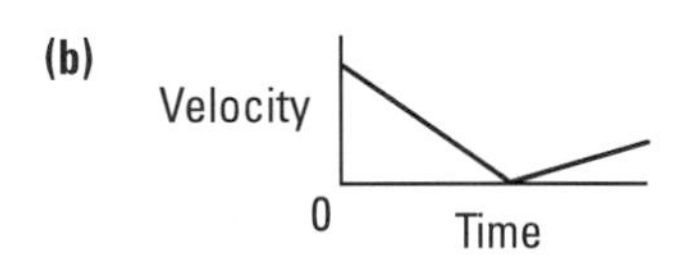

(c)

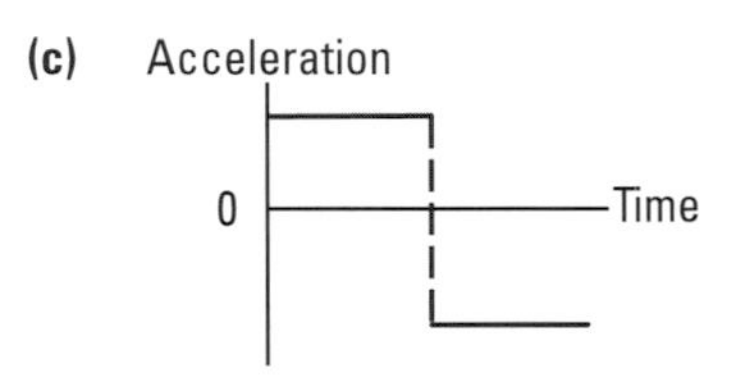

Figure 3
(a) Position-time graph
(b) Velocity-time graph
(c) Acceleration-time graph

20. At a certain location the acceleration due to gravity is 9.82 m/s^2 [down]. Calculate the percentage error of the following experimental values of $\vec{g}$ at that location:
    (a) 9.74 m/s^2 [down]
    (b) 9.95 m/s^2 [down]
21. You can learn to estimate how far light travels in your classroom in small time intervals, such as 1.0 ns, 4.5 ns, etc.
    (a) Verify the following statement: "Light travels the length of a 30.0 cm ruler (about one foot in the Imperial system) in 1.00 ns."
    (b) Estimate how long it takes light to travel from the nearest light source in your classroom to your eyes.
22. The results of an experiment involving motion are summarized in **Table 2**. Apply your graphing skills to generate the velocity-time and acceleration-time graphs of the motion.

Table 2

Time (s)	0.0	1.0	2.0	3.0	4.0
Position (m [S])	0	19	78	176	315

23. In a certain acceleration experiment, the initial velocity is zero and the initial position is zero. The acceleration is shown in **Figure 4**. From the graph, determine the information needed to plot a velocity-time graph. Then, from the velocity-time graph, find the information needed to plot a position-time graph. (*Hint:* You should make at least four calculations on the velocity-time graph to be sure you obtain a smooth curve on the position-time graph.)

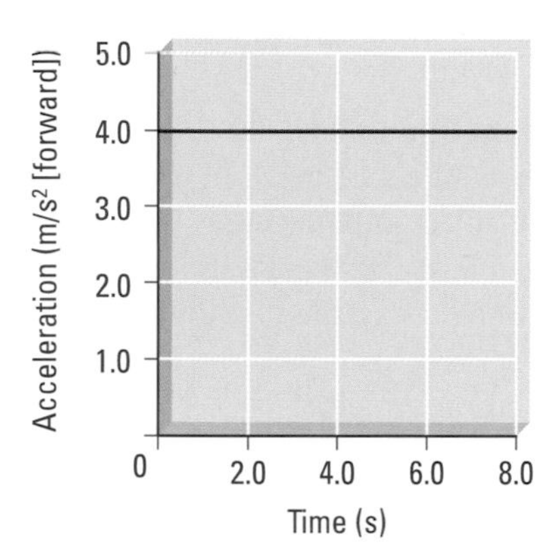

Figure 4

Applying Inquiry Skills

24. For experiments involving motion, state
    (a) examples of random error
    (b) examples of systematic error
    (c) an example of a measurement that has a high degree of precision but a low degree of accuracy (Make up a specific example.)
    (d) an example of when you would calculate the percentage error of a measurement
    (e) an example of when you would find the percentage difference between two measurements

25. Small distances in the lab, such as the thickness of a piece of paper, can be measured using instruments with higher precision than a millimetre ruler. Obtain a vernier caliper and a micrometer caliper. Learn how to use them to measure small distances. Then compare them to a millimetre ruler, indicating the advantages and disadvantages.

Making Connections

26. Car drivers and motorcycle riders can follow the two-second rule for following other vehicles at a safe distance. But truck drivers have a different rule. They must maintain a distance of at least 60.0 m between their truck and other vehicles while on a highway at any speed above 60.0 km/h (unless they are overtaking and passing another vehicle). In this question, assume two significant digits.
    (a) At 60.0 km/h, how far can a vehicle travel in 2.0 s?
    (b) Repeat (a) for a speed of 100.0 km/h.
    (c) Compare the two-second rule values in (a) and (b) to the 60-m rule for trucks. Do you think the 60-m rule is appropriate? Justify your answer.
    (d) Big, heavy trucks need a long space to slow down or stop. One of the dangerous practices of aggressive car drivers is cutting in front of a truck, right into the supposed 60.0-m gap. Suggest how to educate the public about this danger.

27. Describe why the topic of acceleration has more applications now than in previous centuries.

Exploring

28. In this chapter, several world records of sporting and other events are featured in questions and sample problems. But records only stand until somebody breaks them. Research the record times for various events of interest to you. Follow the links for Nelson Physics 11, Chapter 1 Review. Calculate and compare the average speeds or accelerations of various events.

GO TO www.science.nelson.com

29. Critically analyze the physics of motion in your favourite science fiction movie. Describe examples in which the velocities or accelerations are exaggerated.

Chapter

2

Forces and Newton's Laws of Motion

In this chapter, you will be able to

- define and describe concepts related to force, such as types of forces and their vector nature
- develop skill in drawing diagrams of single objects to analyze the forces and determine the acceleration of the objects
- state Newton's three laws of motion, and apply them to explain the observed motion of objects
- analyze the mathematical relationship among the net force acting on an object, its acceleration, and its mass
- interpret patterns observed on force and acceleration graphs and calculate or infer relationships among the variables plotted on the graphs
- apply scientific principles to explain the design of technological solutions to transportation needs, and evaluate the design
- explain and analyze how an understanding of forces and motion combined with an understanding of societal issues (political, economic, environmental, and safety) leads to the development and use of transportation technologies and recreation and sports equipment

We constantly experience forces, such as the force of gravity pulling us, the force of the wind pushing on us, and many other pulls and pushes. The paraglider in **Figure 1** must learn how to pull on the cords attached to the sail in order to control the direction that the wind pushes on the sail. The downward force on the glider is a result of Earth's force of gravity. If suddenly the glider became detached from the cords, he would accelerate downward due to that force. It is evident that forces influence motion.

You studied and analyzed accelerated motion in Chapter 1. In this chapter, you will explore the forces that cause objects to accelerate. You will see how the study of forces, analyzed scientifically by Galileo and another great scientist, Sir Isaac Newton, has evolved to a sophisticated level. Data analysis and simulations using computers help to apply the study of forces to the development of better materials and equipment such as airbags in cars, tread patterns in car tires, and braking systems for the safety of transportation vehicles.

Figure 1
In the exciting sport of paragliding, the motion is controlled by such forces as the force of gravity, tension in the cords attached to the sail, and the wind against the sail.

Reflect on your Learning

1. Forces are everywhere. List several forces that you observe or experience, indicating the direction of each force and naming the object on which the force is acting.
2. When your family is on a car trip, why is it unsafe to pile small bags and cases inside the car on the rear window shelf?
3. When each of three boats of different masses is driven by the same-size motor at its maximum power, do you think the difference in mass has an effect on the acceleration of the boats? Explain.
4. A skater on an ice rink is pushed by the coach with a force in the west direction. How does the force of the skater on the coach, with respect to magnitude and direction, compare to the pushing force?

Throughout this chapter, note any changes in your ideas as you learn new concepts and develop your skills.

Measuring and Estimating Forces

For this activity, you will need either a force sensor or a spring scale calibrated in the SI unit of force, the newton (N), shown in **Figure 2**. You will also need various common objects that can be hung from the sensor or the scale, graph paper, and the following masses: one 100-g mass, two 200-g masses, one 500-g mass. Be sure that the objects are not too heavy for the force scale available.

- Carefully pull on the sensor or spring so that you can "feel" forces of 1 N, 2 N, and so on. Pick up various objects, one at a time, and estimate the magnitude of the force in newtons required to suspend the object vertically from the sensor or the scale. After each trial, test your estimation so as to improve your skill at estimating the next item.
  - (a) Pick up a 100-g mass and estimate the magnitude of the force needed to hold it. Hang it from the sensor or the scale and record the value.
  - (b) Measure and record the magnitude of the force needed to hold each of these masses: 200 g, 300 g, 400 g, . . . , up to the maximum suggested by your teacher.
  - (c) Plot a graph of the magnitude of the force (on the vertical axis) as a function of mass. Determine the slope of the line of best fit on the graph.
  - (d) Based on the results you have found, apply your skills of interpolation and extrapolation to calculate the magnitude of the force required to hold a mass of
    (i) 350 g (ii) 1200 g (iii) 2.0 kg
- Use one or two more objects not used previously to determine whether your skill at estimating force has improved.

Protect the masses from dropping or recoiling violently.

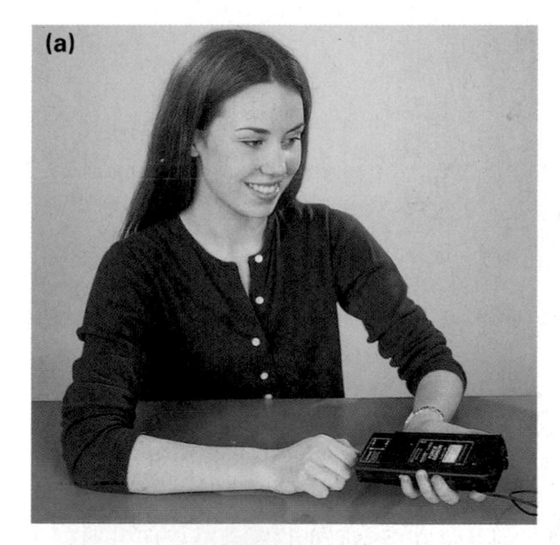
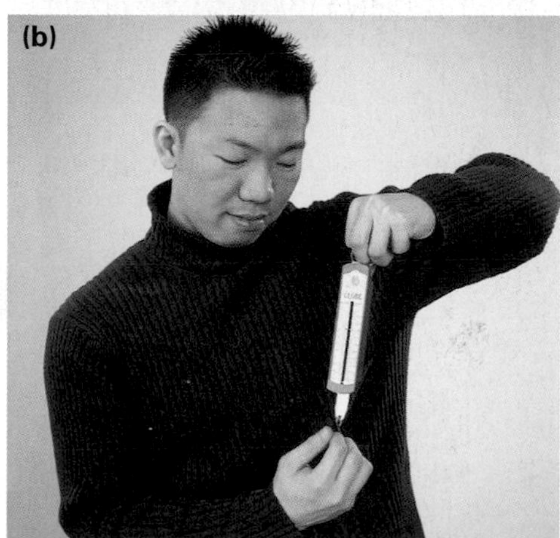

Figure 2
Measuring force in newtons
(a) Using a force sensor
(b) Using a spring scale

2.1 Forces in Nature

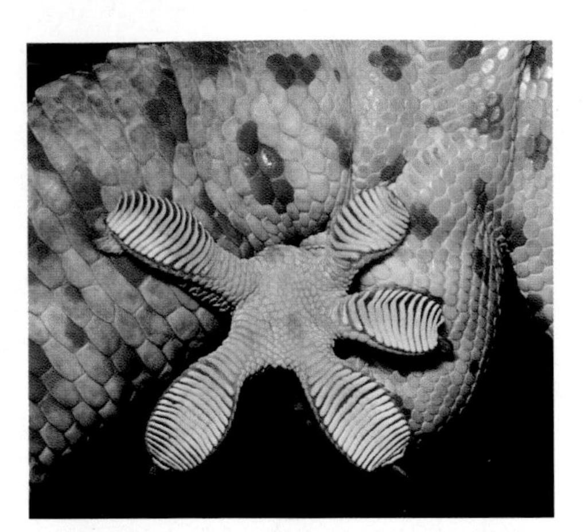

Figure 1
Scientists study examples of force in nature, such as this gecko's ability to cling to surfaces, in the hope of finding yet unknown applications.

Forces are everywhere. If there were no forces in our universe, our Earth would not be trapped in its orbit around the Sun; the uses of electricity would never have been discovered; we would not be able to operate automobiles or even walk — in fact, we would not exist, because objects need forces to keep their shape. **Figure 1** shows the underside of the foot of an animal called a gecko. Geckos have the ability to climb up a glass wall and hang from one toe. Each gecko foot has about half a million setae, or hairs, each split into hundreds of ends. These setae exert a force of adhesion on materials such as glass, but they let go if they are tipped at an angle of 30° or more from the surface. Researchers estimate that this small creature could lift 40 kg of mass if all the setae operated together.

In simple terms, a **force** is a push or a pull. Forces act on almost anything. They speed things up, slow them down, push them around corners and up hills. Forces can also distort matter by compressing, stretching, or twisting.

Force is a vector quantity. Like other vector quantities, its direction can be stated in various ways, such as forward, up, down, east, northeast, and so on.

force: a push or a pull

fundamental forces: forces are classified into four categories—gravitational, electromagnetic, strong nuclear, and weak nuclear

Practice

Understanding Concepts

1. State an everyday life example in which a force causes an object to (a) decrease its speed; (b) become compressed; (c) become stretched.
2. You are facing eastward, standing in front of a gate that can swing. In what direction is your force if you pull on the gate? push on the gate?

Applying Inquiry Skills

3. Assume you are given an empty matchbox, a magnet, a metal paperclip, an elastic, and a balloon. Make a list of ways you could make the matchbox move with or without touching the matchbox with the given materials. (*Hint:* You may place items in the box.)

The Four Fundamental Forces

There may seem to be many different types of forces around us, but physicists have found that they are able to understand how objects interact with one another by classifying forces into only four categories. The four **fundamental forces** of nature are the gravitational force, the electromagnetic force, the strong nuclear force, and the weak nuclear force. A comparison of these forces is shown in **Table 1**.

Table 1 Comparing the Fundamental Forces

Force	Relative strength (approx.)	Range	Effect
gravitational	1	∞	attraction only
electromagnetic	10^{36}	∞	attraction and repulsion
weak nuclear	10^{25}	less than 10^{-18} m	attraction and repulsion
strong nuclear	10^{38}	less than 10^{-15} m	attraction and repulsion

DID YOU KNOW ?

Combining Fundamental Forces

Researchers have discovered that two of the fundamental forces, the electromagnetic force and the weak nuclear force, have a common origin. The combined force is called the *electroweak* force. Its effects can be verified in experiments performed using high-energy collisions in particle accelerators. As these accelerators become stronger, researchers are hoping to observe evidence of even more unification of forces, to verify the Grand Unified Theory (GUT) and, eventually, the Theory of Everything (TOE). This will help scientists understand more about what occurred shortly after the "big bang" start of the universe. At first, an unimaginably huge force existed and particles were indistinguishable (TOE: all four forces were unified). Then, the force of gravity separated from the other three forces (GUT). Much later, as the universe expanded and cooled, the forces all separated, leaving matter as we know it.

The **gravitational force**, or the **force of gravity**, is the force of attraction between all objects in the universe. It is important for large objects such as stars, planets, and moons. It holds them together and controls their motions in the same way that it controls the motion of falling objects here on Earth. You can see in **Table 1** that the gravitational force is tiny compared with the other fundamental forces. However, it has an important role in the universe because it exerts attraction only. The gravitational force is an example of an "action-at-a-distance" force, in other words, a force that acts even if the objects involved are not touching. The force of gravity between two objects is noticed only if at least one of the objects has a large mass.

gravitational force or **force of gravity**: force of attraction between all objects

electromagnetic force: force caused by electric charges

The **electromagnetic force** is the force caused by electric charges. It includes both electric forces (such as static electricity) and magnetic forces (such as the force that affects a magnetic compass). The electromagnetic force can exert either an attraction or a repulsion (**Figure 2**), so on average, the forces tend to cancel each other out. If this were not the case, then electromagnetic forces would completely overwhelm the force of gravity. It is the electromagnetic force that holds atoms and molecules together, making diamonds hard and cotton weak. It tenses muscles and explodes sticks of dynamite. In fact, most common forces that we experience are electromagnetic in origin. Sometimes it is convenient to treat electric and magnetic forces separately, even though they are both caused by electric charges.

Figure 2
The magnetic force of repulsion keeps this magnetic levitation train separated from the track. With a maximum speed of 450 km/h, the train runs smoothly and quietly.

There are strong and weak nuclear forces acting between the particles within the nucleus of an atom. The nucleus contains positively charged particles and neutral particles called protons and neutrons, respectively. The **strong nuclear force** holds the protons and neutrons together, even though the protons are influenced by the electric force of repulsion. This nuclear force is a short-range force but is much stronger than the electromagnetic force. It is significant only when the particles are close together.

strong nuclear force: force that holds protons and neutrons together in the nucleus of an atom

weak nuclear force: force responsible for interactions involving elementary particles such as protons and neutrons

Besides the proton and the neutron, there are many more "elementary" particles. The electron is but one of the others. Many of these particles, including the neutron, are unstable and break up. The **weak nuclear force** is responsible for the interactions involved. This type of force is noticed only at extremely small distances.

Practice

Understanding Concepts

4. List the fundamental forces in order from the strongest to the weakest.
5. In what way is gravitational force unique among the fundamental forces?
6. Which of the fundamental forces do you notice most often in your everyday activities? Give some examples to illustrate your answer.

Forces We Experience Daily

We experience several types of forces daily. The most obvious one is the force of gravity between Earth and objects at or near its surface. The direction of this force is toward Earth's centre, a direction referred to as vertically downward (**Figure 3**).

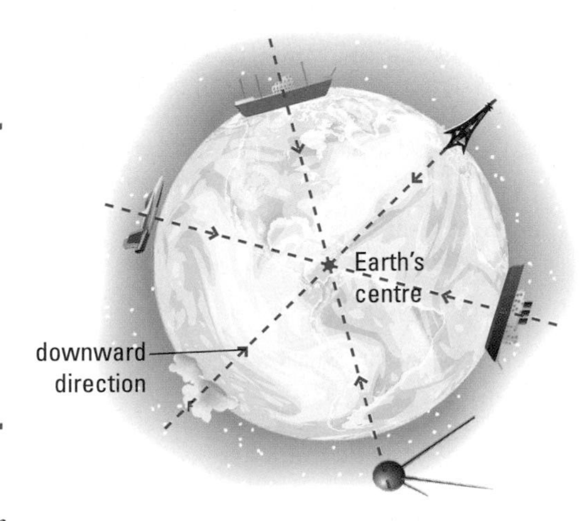

Figure 3
The force of gravity between Earth and objects at or near its surface is directed in a line to Earth's centre. This direction defines what we mean by vertical at any location on Earth.

normal force: force perpendicular to the surfaces of the objects in contact

friction: force between objects in contact and parallel to contact surfaces

tension: force exerted by string, ropes, fibres, and cables

If your pen is resting on your desk, the force of gravity is pulling down on it, but since it is at rest, it must also be experiencing a force of the desk pushing upward. This force is called the **normal force**, which is a force that is perpendicular to the surfaces of the objects in contact. In other words, normal means perpendicular in this context. (See **Figure 4**.)

Another force that we cannot live without is **friction**, which is a force between objects in contact and parallel to the contact surfaces. If you push your pen across your desk, it soon comes to rest as the force of friction between the desktop and the pen causes the pen to slow down. Friction on an object acts in a direction opposite to the direction of the object's motion or attempted motion.

Another common force is **tension**, which is a force exerted by string, ropes, fibres, and cables. A spider web (**Figure 5**) consists of numerous fine strands that pull on each other. We say that the strands are under tension. Muscle fibres of our body experience tension when they contract.

The normal force, friction, and tension are all contact forces, in other words, forces that exist when objects are in direct contact with each other. They are caused by the interaction of particles on the contact surfaces and are thus a result of the electromagnetic force. More details about these forces as well as the force of gravity are presented throughout the rest of this unit.

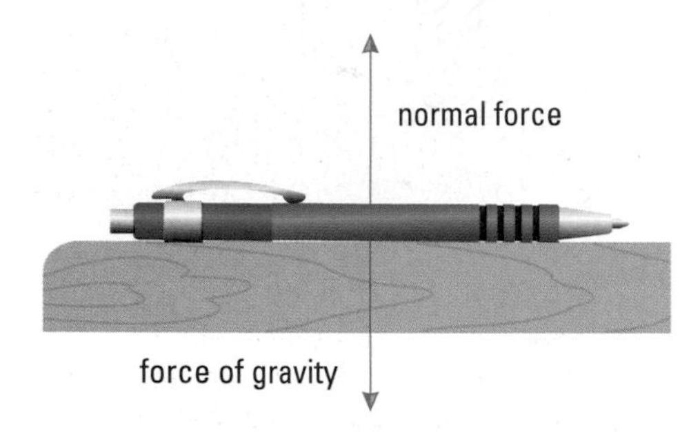

Figure 4
The normal force is perpendicular to both surfaces in contact.

Practice

Understanding Concepts

7. Draw a sketch to show the force of gravity, the normal force, tension, and friction in each case.
   (a) A toboggan is on a horizontal surface being pulled by a rope that is also horizontal.
   (b) A toboggan is being pulled by a rope up a hill with the rope parallel to the hillside.
8. Describe, with examples, the difference between a "contact" force and an "action-at-a-distance" force.

Figure 5
Tension keeps this spider web together.

Measuring Force

newton: (N) the SI unit of force

The title of this chapter refers to one of the greatest scientists in history, Sir Isaac Newton (**Figure 6**). Many of Newton's brilliant ideas resulted from Galileo's discoveries. Since Newton developed important ideas about force, it is fitting that the unit of force is called the **newton** (N).

The newton is a derived SI unit, which means that it can be expressed as a combination of the base units of metres, kilograms, and seconds. For the remainder of your study of forces and motion, it is important that you perform calculations to express distance in metres, mass in kilograms, and time in seconds. In other words, all units must conform to the preferred SI units of metres, kilograms, and seconds.

Two devices used to measure force in the laboratory are the spring scale and the force sensor, both introduced in the chapter-opening activity. The spring scale has a spring that extends under tension. Attached to the spring is a needle that points to a graduated scale to indicate the force. The force sensor uses an electronic gauge to measure force, with a digital readout or a graph of the forces when interfaced with a computer. Force sensors, especially the high quality ones, measure force with a high degree of accuracy.

Drawing Force Diagrams

In order to analyze forces and the effects they have on objects, the skill of drawing force diagrams is essential. Two types of force diagrams, "system diagrams" and "free-body diagrams," are useful.

A **system diagram** is a sketch of all the objects involved in a situation. A **free-body diagram** (FBD) is a drawing in which only the object being analyzed is drawn, with arrows showing all the forces acting on the object. **Figure 7** shows three examples of each type of diagram. Notice the following features of the FBDs: the force vectors are drawn with their lengths proportional to the magnitudes of the forces; each force vector is labelled with the symbol $\vec{F}$, with a subscript (for example, $\vec{F}_g$ is the force of gravity, $\vec{F}_N$ is the normal force, $\vec{F}_f$ is friction, $\vec{F}_T$ is tension, and $\vec{F}_A$ is the applied force); and positive directions are indicated, such as $+y$ for the vertical direction. (The positive direction in any FBD is chosen for convenience and must be adhered to carefully in a specific situation of problem solving.)

Understanding Concepts

9. For each situation described below, draw a system diagram and an FBD. Be careful when deciding what forces are acting on each object. If you cannot think of a cause for the force, the force may not even exist.
   (a) Your *textbook* is resting on your desk.
   (b) A *tennis ball* is falling through the air from the server's hand. Neglect air resistance.
   (c) A fully loaded *dog sled*, moving slowly along a flat, snowy trail, is being pushed horizontally by the sled owner while being pulled horizontally by dogs attached to it by rope.

Figure 6
Sir Isaac Newton (1642–1727) was both an eccentric and a genius. He had a difficult childhood and in his adult life cared little for his personal appearance and social life. By the age of 26, he had made profound discoveries in mathematics, mechanics, and optics. His classic book, *Principia Mathematica*, written in Latin, appeared in 1687. It laid the foundations of physics that still apply today. In 1705, he became the first scientist to be knighted. In his later years, he turned his attention to politics and theology and became the Official Master of England's mint.

system diagram: sketch of all the objects involved in a situation

free-body diagram: (FBD) drawing in which only the object being analyzed is drawn, with arrows showing all the forces acting on the object

(a) the system diagram **the FBD**

$\vec{F}_T$ $+y$ $\vec{F}_g$

(b) the system diagram **the FBD**

$\vec{F}_N$ $+y$ $\vec{F}_g$

(c) the system diagram

the FBD

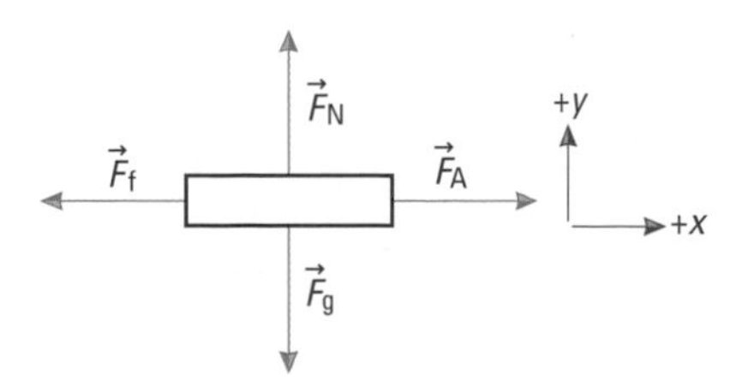

Figure 7
Examples of system diagrams and free-body diagrams.
(a) A fish is held by a fishing line. (The only direction labelled is $+y$ because there are no horizontal forces.)
(b) A volleyball is held by a hand.
(c) A book is pushed across a desk by a horizontal applied force.

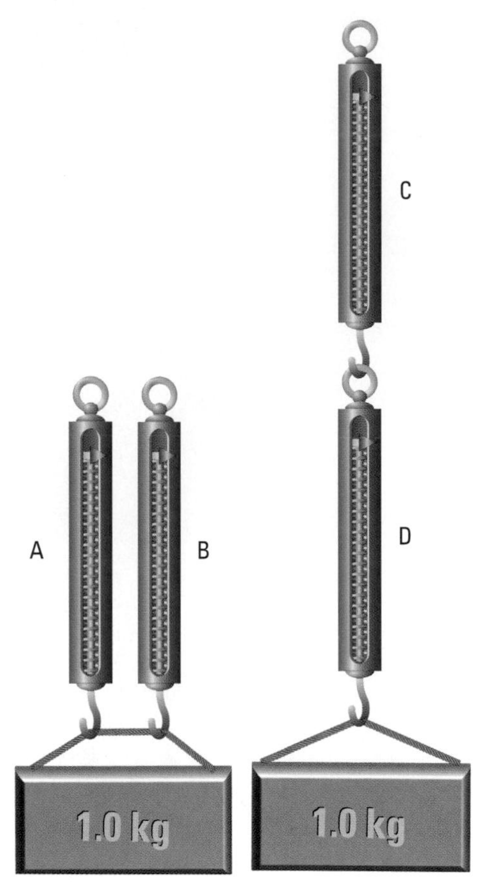

Figure 8
Predicting scale readings on four spring scales

Activity 2.1.1

Forces on Springs

In this activity, you will predict what you will observe in a demonstration of forces. Then, you will check your predictions.

Figure 8 shows four identical springs, A, B, C, and D, arranged in two different ways to hold a 1.0-kg mass. The readings are not shown on the spring scales.

(a) Name the force that is pulling downward on the 1.0-kg mass and the force exerted by the spring.

(b) Assume that each spring scale has a negligible mass. Predict the readings on all four spring scales.

(c) Now assume that the spring scales have mass that cannot be ignored. Predict the readings on all four spring scales. Explain your reasoning.

(d) Set up the demonstration to check your predictions. Describe what you discover.

(e) Draw a system diagram and an FBD of each of the following items:
 (i) the 1.0-kg mass held by springs A and B (Show three forces.)
 (ii) the 1.0-kg mass held by springs C and D (Show two forces.)
 (iii) spring D (Show two forces.)
 (iv) spring C (Show three forces.)

SUMMARY Forces in Nature

- A force, which is a push or a pull, is a vector quantity; its SI unit is the newton (N).
- The fundamental forces responsible for all forces are gravitational, electromagnetic, strong nuclear, and weak nuclear.
- Some common forces we experience are friction, tension, and the normal force.
- Drawing free-body diagrams is an important skill that helps in solving problems involving forces.

Section 2.1 Questions

Understanding Concepts

1. State the direction of the named force acting on the object in *italics*.
   (a) A *puck* experiences friction on rough ice as it slides northward.
   (b) The force of gravity exerts a force on a falling *leaf*.
   (c) The gravitational force acts on the *Moon*, keeping it in orbit around Earth.
   (d) The force of the wind is pushing against a *cyclist* who is cycling westward.
2. Name the fundamental force that is responsible for
   (a) preventing protons from flying apart in the nucleus of an atom
   (b) keeping stars in huge groups called galaxies
   (c) the function of a magnetic compass

3. Explain why friction and tension are not classified as fundamental forces.
4. Assuming that you are not allowed to jump, estimate the maximum force you think you could exert on a strong rope attached firmly to
   (a) the ceiling
   (b) the floor

Applying Inquiry Skills

5. (a) Estimate the force in newtons required to hold up this physics book.
   (b) How would you test your estimation?
6. Draw a system diagram and an FBD for each object named in *italics*.
   (a) A golfer is holding a *pail* of golf balls.
   (b) A *cue ball* is slowing down as it rolls in a straight line on a billiard table.
   (c) A batter tips a baseball and the *ball* is rising vertically in the air. (Neglect air resistance.)

2.2 Newton's First Law of Motion

Each year in Canada there are more than 150 000 traffic collisions resulting in over 200 000 injuries and almost 3000 deaths. Of these deaths, the greatest number in any single age category is the 15-to-25-year-old group. Although the average number of deaths on Canada's highways has decreased in the past 20 years, there is still a great demand for improvement on safety.

Sadly, many of the deaths and injuries could have been prevented so easily, simply by wearing seatbelts or by using airbags. Understanding Newton's first law of motion will help you appreciate how wearing a seatbelt could improve safety.

You learned about kinematics, the study of motion, in Chapter 1. Here you will study **dynamics**, the branch of physics that deals with the causes of motion. The word "dynamics" stems from the Greek word *dynamis*, which means power.

dynamics: the study of the causes of motion

Galileo's View of Force and Motion

Prior to the 1600s, most natural philosophers (early scientists) held a simplistic view of motion. They believed that a constant force was needed to produce constant velocity. Increasing the force caused increased velocity. Decreasing the force caused decreased velocity. Removing the force caused the velocity to become zero. Galileo, the Renaissance physicist, questioned this view. Galileo performed real experiments with a ball rolling down and up sloped ramps. He also performed virtual experiments (thought experiments) to try to explain his ideas. Galileo's virtual experiments are illustrated in **Figure 1**.

In **Figure 1**(a), he reasoned that a ball speeds up as it rolls down a slope, then moves with constant velocity along the horizontal surface, and finally rolls up the far slope to the same level it started from. Of course, when he performed the real experiment, he observed that the ball rolling along the horizontal surface stopped without rolling up the far slope. However, he was able to assume correctly that the ball slowed down to zero velocity because of friction.

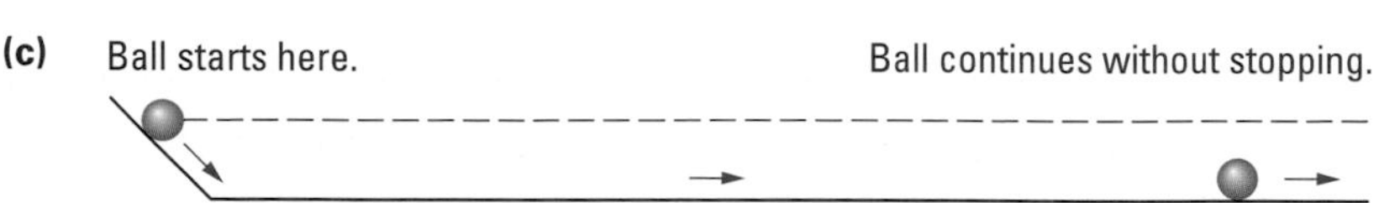

Figure 1
Galileo's virtual experiments
(a) The ascending ramp has a steep slope.
(b) The ascending ramp has a low slope.
(c) The ascending ramp is removed.

In the second virtual experiment, shown in **Figures 1(b)** and **1(c)**, Galileo reasoned that if the slope of the rising plane was decreased to make it less steep than the descending slope, the ball would roll farther along this slope, stopping when it reached the height from which it was released on the first slope. The fact that the ball did not quite reach that height he attributed, once again, to friction. He concluded, logically, that if the slope of the second plane was zero, the ball, once rolling, would continue forever with no loss of speed, in an attempt to reach its original height.

We now have the technology to illustrate what Galileo could only visualize. There are devices available that can slide along a smooth surface with almost no friction. For example, you may have used an air table or an air track in previous investigations. Both devices have air pumped between the moving object and the surface to virtually eliminate the effect of friction.

Once an object, such as an almost frictionless air puck, is moving, it needs almost no force to keep it moving, as long as no other force acts on it. Galileo explained his theory of motion by stating that every object possesses a property that Newton later called inertia. **Inertia** is the property of matter that causes a body to resist changes in its state of motion. Suppose, for example, you are standing in the aisle of a stationary bus. As the bus starts to move, your body wants to stay at rest because of its inertia. As a result you may fall toward the rear of the bus unless you brace yourself. When the bus reaches a uniform velocity, you will have no trouble standing in the aisle because you are also moving with uniform velocity. Suppose the bus driver suddenly applies the brakes. The bus will slow down. Your body, however, has a tendency to keep moving because of its inertia. Again, you will need to brace yourself or you will fall toward the front of the bus. The amount of inertia an object possesses depends directly on its mass: *the greater the mass, the greater the inertia the body possesses.*

inertia: the property of matter that causes a body to resist changes in its state of motion

Practice

Understanding Concepts

1. What is a virtual experiment?
2. (a) Assuming the following objects are at rest, rank their inertia in order from the least to the greatest: a school bus, a small child, a compact car, and yourself.
   (b) Repeat (a) assuming all the objects are moving at the same velocity.

DID YOU KNOW ?

Changing Influences

A major shift in the relationship between science and society occurred during and after the Renaissance. Up to the end of Galileo's time, society influenced scientific thought. Galileo had to overcome the prejudices of society as he performed experiments to test hypotheses of scientific principles. Newton was born in the same year that Galileo died, and during his time, reasoning and logic grew in importance as scientific endeavours gained credibility. Thus, by the end of the 17th century, science influenced society rather than the other way around.

Free-Body Diagrams and Resultant Forces

To study the effects of forces acting on any object, we can apply the skill of drawing force diagrams. Since force is a vector quantity, the vector sum of all the forces acting on an object is the **resultant force.** The resultant force can also be called the **net force**. These two terms can be used interchangeably. They will be represented by the same symbol, $\vec{F}_{net}$, in this text.

resultant force or **net force:** the vector sum of all the forces acting on an object

Sample Problem 1

A weight lifter holds a weight above his head by exerting a force of 1.6 kN [↑]. The force of gravity acting on the weight is 1.6 kN [↓]. Draw a system diagram and an FBD of the weight, and state the net force on the weight at that instant.

Solution

The required diagrams are shown in **Figure 2**. The upward force exerted by the weight lifter can be called a normal force or an applied force. The net force at the instant shown is zero.

(a) the system diagram

(b) the FBD

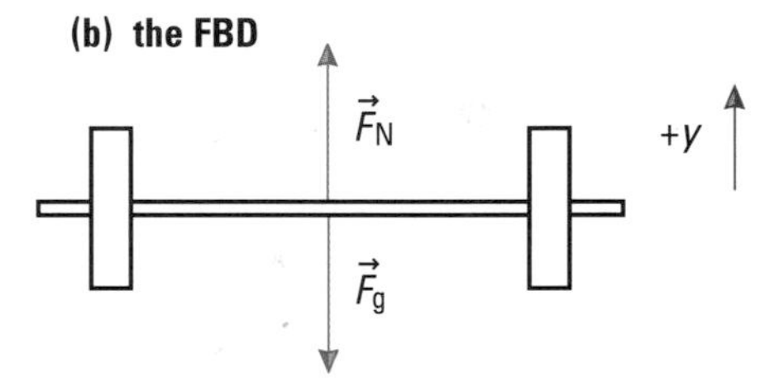

$$\vec{F}_{net} = \vec{F}_N + \vec{F}_g$$
$$= 1.6 \text{ kN } [\uparrow] + 1.6 \text{ kN } [\downarrow]$$
$$\vec{F}_{net} = 0$$

Figure 2
Force diagrams for Sample Problem 1

Sample Problem 2

As an arrow leaves a bow, the string exerts a force of 180 N [fwd] on the arrow. Draw a system diagram and an FBD of the arrow as the string is applying the force to it. (Neglect any vertical forces because the force of gravity on the arrow is so small that it will not show up in the diagram.)

Solution

Figure 3 shows the required diagrams. At the instant shown, the net force on the arrow is 180 N [fwd].

(a) the system diagram

(b) the FBD

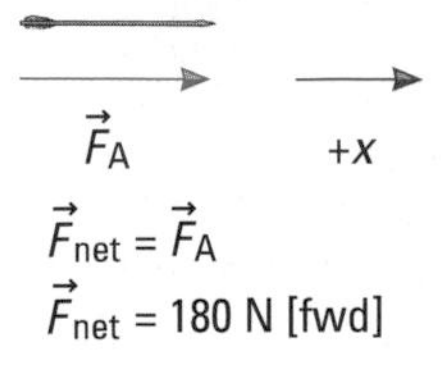

$$\vec{F}_{net} = \vec{F}_A$$
$$\vec{F}_{net} = 180 \text{ N [fwd]}$$

Figure 3
Force diagrams for Sample Problem 2

Try This Activity

Observing Objects at Rest and in Motion

Your teacher will set up demonstrations of objects initially at rest and others initially in motion. In each case, predict what you think will happen. Then, observe carefully the result of each motion and summarize your observations in tabular form using these headings:

Object observed	Initial state	Predicted result	Observed result	Tendency of object

In the last column, indicate whether the object observed tends to remain at rest or moves at a constant velocity.

Practice

Understanding Concepts

3. Calculate the net force when each of the following sets of forces acts on the same object:
   (a) 2.4 N [N], 1.8 N [N], and 8.6 N [S]
   (b) 65 N [down], 92 N [up], and 54 N [up]

Answers

3. (a) 4.4 N [S]
   (b) 81 N [up]
4. 1.1 N [W]

4. A store clerk pushes a parcel on the counter toward a customer with a force of 7.6 N [W]. The frictional resistance on the parcel is 6.5 N [E], and both the force of gravity and the normal force have a magnitude of 9.9 N. Draw an FBD of the parcel and determine the net force acting on it.

first law of motion: if the net force acting on an object is zero, the object will maintain its state of rest or constant velocity

Newton's First Law of Motion: The Law of Inertia

Galileo published his ideas about motion early in the 17th century. Newton summarized his own work and that of Galileo in his book *Principia Mathematica.* In the book, Newton described what Galileo had discovered about inertia much the same way Galileo had described it. Because it was included with Newton's other laws of motion, it is often referred to as Newton's **first law of motion.**

First Law of Motion

If the net force acting on an object is zero, the object will maintain its state of rest or constant velocity.

This law has several important implications.

- An external force is required to change the velocity of an object. Internal forces have no effect on an object's motion. For example, a driver pushing on the car's dashboard does not cause the car's velocity to change.
- To cause a change in velocity, the net force on an object must not be zero; that is, two equal opposing forces acting on an object will not change its velocity.
- Objects at rest remain at rest unless acted upon by a net external force greater than zero.
- Moving objects continue to move in a straight line at a constant speed unless acted upon by a net external force greater than zero.

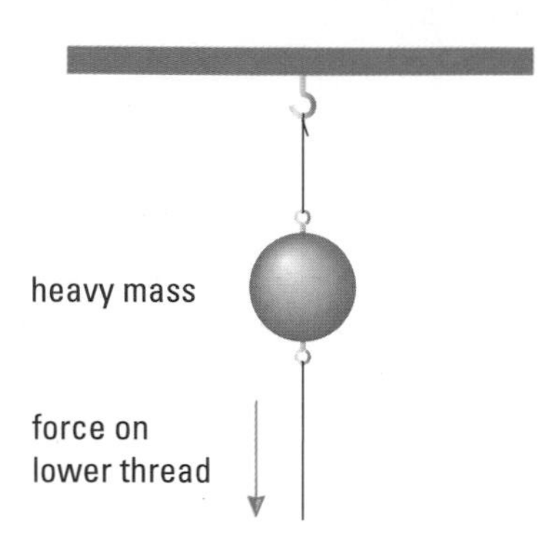

Figure 4
For question 5

Practice

Understanding Concepts

5. A thread supports a mass hung from the ceiling; another thread of equal dimension is suspended from the mass (**Figure 4**). Which thread is likely to break if the bottom thread is pulled slowly? quickly? Use the concept of inertia to explain your answer.

Examples of the First Law

Examples of Newton's first law of motion will help clarify its meaning. Consider this example of objects initially at rest. A carefully constructed vintage car, without many of the safety features of today's cars, is stopped at an intersection when a truck crashes into the rear of the car. Both the driver's head and the passenger's head are initially at rest, and tend to remain there as the seats push their bodies forward. Severe whiplash occurs as the head jerks backward relative to the body. Modern car seats have headrests designed to reduce this type of injury. **Figure 5** shows the two different types of seat designs.

(a)

(b)

Figure 5
Comparing seat designs
(a) Seats in older car designs had no support for the head, so neck injuries were difficult to avoid.
(b) Current seat designs include a headrest that helps to keep the head, neck, and body moving together, thus reducing the chance of whiplash.

Next consider an object in uniform motion. It will continue with constant speed and direction unless a net external force acts on it.

A car approaching a curve on an icy highway has the tendency to continue in a straight line, thus failing to follow the curve. This explains why drivers must pay special attention when driving on slippery roads. A stapler being pushed across a table at a constant velocity has four forces acting on it, as illustrated in **Figure 6**. In the vertical direction, the force of gravity downward is equal in magnitude to the normal force exerted upward by the desk. In the horizontal direction, the backward force of moving friction is equal in magnitude to the forward applied force on the stapler. Thus, the net external force on the stapler is zero, and the stapler does not accelerate.

Let us summarize Newton's first law of motion by stating four important results of it:

(a) Objects at rest tend to remain at rest.
(b) Objects in motion tend to remain in motion.
(c) If the velocity of an object is constant, the net external force acting on it must be zero.
(d) If the velocity of an object is changing in either magnitude or direction or both, the change must be caused by a net external force acting on the object. This fact sets the stage for experimentation in dynamics (section 2.3).

(a) the system diagram

(b) the FBD

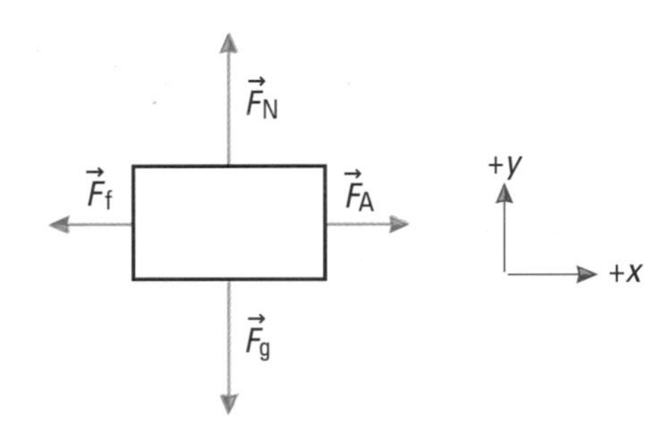

Figure 6
The net external force on the stapler is zero; since it is already moving, it remains moving with a constant velocity.

Practice

Making Connections

6. Astronauts are placed horizontally in their space capsule during blastoff from the launching pad. Explain why this is a good example of Newton's first law of motion.
7. Explain how it would be possible to apply the first law of motion when trying to get a heap of snow off a shovel. Try to give more than one solution.

Explore an Issue

Transportation Safety Technologies

As our population grows and the number of vehicles on the roads increases, people are becoming more aware of how important transportation safety is. Using research, automobile manufacturers know that car buyers place a high priority on safety systems. Governments, together with institutions such as the police force, create and enforce transportation safety laws. Insurance companies also influence transportation safety by charging rates based on the safety designs of vehicles.

Despite new technologies and safety laws, each year thousands of people are killed and injured in traffic mishaps in Canada. In one crash at a low speed, a driver was killed by a rapidly expanding airbag while sitting close to the steering wheel without a seatbelt on. In some other cases, toddlers have been injured because parents do not follow closely the instructions for how to place infants in the child restraint system (**Figure 7**).

Improving Technologies

Automobile companies perform thousands of tests each year in an effort to improve the technologies of seatbelts and airbags (**Figure 8**). Many other safety features are available and others are being researched.

(continued)

DECISION MAKING SKILLS

- ○ Define the Issue
- ● Identify Alternatives
- ● Research
- ● Analyze the Issue
- ● Defend the Proposition
- ○ Evaluate

Figure 7
Studies reveal that many parents do not secure their infants in their car seats according to instructions. What is the advantage of having the child face the rear of the car?

Figure 8
Front-end collisions using crash-test dummies are among the many tests performed by automobile manufacturers to try to improve safety features. In such a test, computers analyze the effectiveness of seatbelts and airbags.

Side airbags and safety screens that drop from above the windows help prevent injuries during side collisions or roll-overs. Shatter-resistant glass has a clear laminated layer that prevents the glass from breaking, at least on first impact. Steering control systems and antilock brakes help keep the vehicle going in the right direction, even under extreme conditions of reacting to an emergency. Night-vision systems allow drivers to see from three to five times farther ahead than their regular headlights allow. And on-board communication systems allow drivers to contact emergency and information services quickly.

The Issue

Since transportation safety affects everyone, a fair question we can explore is "To what extent should government agencies be involved in transportation safety?" Should they bring in more legislation to force auto companies to create newer and better technologies? Should they offer more financial aid to scientific researchers? Should they enforce the safety laws, such as the use of seatbelts, more rigidly? Should they create new laws regarding seatbelts for public transportation vehicles, such as school buses? Should they do nothing more than try to educate the public about safety and leave the innovations to private industry?

As you can see, there are several aspects of the safety issue that can be discussed.

Understanding the Issue

1. Even if every vehicle were 100% safe and if the weather conditions were always perfect, there would still be many deaths caused by collisions. Explain why.
2. What role does physics play in researching transportation safety technologies?
3. Some car manufacturers are installing passenger airbags that can be switched on or off. Describe situations in which switching the airbag mechanism off would be (a) advantageous and (b) unwise.

Take a Stand

How involved should government agencies be in transportation safety?

Proposition

Governments should do more to promote safety in transportation.

There are points in favour of more government intervention:

- There have been some serious cases in which the irresponsible action of a private company has resulted in many deaths and injuries. Without government checks, this might happen more often.
- Injuries to people who do not obey the law (for example, by not wearing their seatbelts) cost the entire society money for health care, rehabilitation, and lost productivity.

There are points in favour of less government intervention:

- The automobile industry is very competitive and will create new safety technologies without government intervention.
- Every new technology costs money to develop and manufacture, so if governments force more features, costs to the consumer will rise.

Forming an Opinion

- Read the points above and add your own ideas to the list.
- Find information to help you understand more about the issue. Follow the links for Nelson Physics 11, 2.2.

GO TO www.science.nelson.com

- In a group, discuss the ideas.
- Communicate your opinions and arguments in a creative way.

DID YOU KNOW?

The Interaction of Science and Technology

The invention and use of airbags provides a typical example of the interaction of science and technology in our society. A problem in society, in this case injuries and deaths in traffic mishaps, leads to research by scientists, which in turn leads to technological development. But then, more problems are identified that are a result of the initial solution to the problem — the use of airbags causes injuries and deaths in certain instances. Thus, the process of scientific research starts again, followed by more technological development.

SUMMARY Newton's First Law of Motion

- Galileo's real and virtual experiments led the way for Newton to formulate his three laws of motion.
- The net force acting on an object, $\vec{F}_{net}$, is the vector sum of all the forces acting on the object.
- Newton's first law of motion, often called the law of inertia, states that if the net force acting on an object is zero, the object will maintain its state of rest or uniform velocity.
- The first law of motion is observed and applied in many situations, including the need for restraint systems in automobiles.

Section 2.2 Questions

Understanding Concepts

1. State the net force acting on an object that is (a) at rest and (b) moving with constant velocity.
2. You are helping a friend move furniture. The friend asks you to stand inside the back of a pickup truck to hold a piano because there is no rope available to tie it to the truck. Explain why you should refuse this request. Your answer should show that you understand the reasoning used by Galileo and Newton.
3. A badminton birdie is struck by the racket so that the birdie goes vertically upward, stops, and falls downward. Draw a system diagram and an FBD for each case below.
   (a) The racket is in contact with the birdie.
   (b) The birdie is rising in the air just above the racket.
   (c) The birdie is stopped for a brief instant at the top of its flight.
   (d) The birdie is beginning to fall, so air resistance exists but has not reached maximum value.

Applying Inquiry Skills

4. Describe how you would demonstrate both parts of the first law of motion (objects initially at rest; objects initially in motion) to a class of elementary school students using your favourite toys or recreation device (scooter, skateboard, in-line skates, ice skates, etc.). Include safety considerations.
5. Describe how you would set up a safe classroom demonstration to show that an object moving in a circle travels with uniform velocity when the force keeping it moving in a circle is removed.

Making Connections

6. Explain how modern technology makes it easier to demonstrate Galileo's virtual experiments than in his time.
7. Describe why the application of Newton's first law of motion is important in transportation safety.
8. Some thrill rides at exhibitions or fairs create sensations that may be explained using the first law of motion. Describe two such rides and how the law applies to them.

Reflecting

9. If you were designing the "ultimate" safe automobile, what features would it have? Consider the safety of the occupants, pedestrians, and other vehicles.

DID YOU KNOW ?

Pedestrian Airbags

Airbags have greatly reduced deaths and injuries for people inside cars. Now, a European manufacturer is testing airbags that reduce deaths and injuries of pedestrians struck by cars. Similar to interior airbags, the exterior ones are computer-controlled. As the car approaches a pedestrian, an infrared detector senses the heat radiating from the human body. A front-end airbag deploys at a speed that depends on the car's speed relative to the pedestrian. If needed, a second airbag on the car's hood deploys, softening the pedestrian's landing.

Figure 1
Launching a space vehicle, such as this shuttle, requires a huge upward thrust larger in magnitude than the downward force of gravity.

2.3 Investigating Force, Mass, and Acceleration

During the launch of a space shuttle (**Figure 1**), a huge net force is needed to cause the entire mass to accelerate upward. How does the acceleration of the shuttle relate to the net or resultant force provided by the rocket engines and the shuttle's entire mass being blasted off the ground?

Knowing the relationship among acceleration, net force, and mass helps in the design and use of numerous devices. Engineers decide how powerful they should make the rocket engines needed to lift a space vehicle off the launch pad. Other engineers design tires, motors, and many other components of vehicles, as well as the highways, to maximize the acceleration potential of a variety of vehicles, from motorcycles to cars, trucks, airplanes, ships, and spacecraft. They also design brakes, bumpers, and restraint systems to improve the chances of people surviving extreme braking and collisions.

Practice

Understanding Concepts

1. Draw an FBD of the rocket and shuttle system shown in **Figure 1** just after it lifts off the pad in a successful launch.

Making Connections

2. The text describes examples of applying the knowledge of acceleration, force, and mass to transportation. How do you think this knowledge can be applied to sports activities?

Acceleration and Net Force

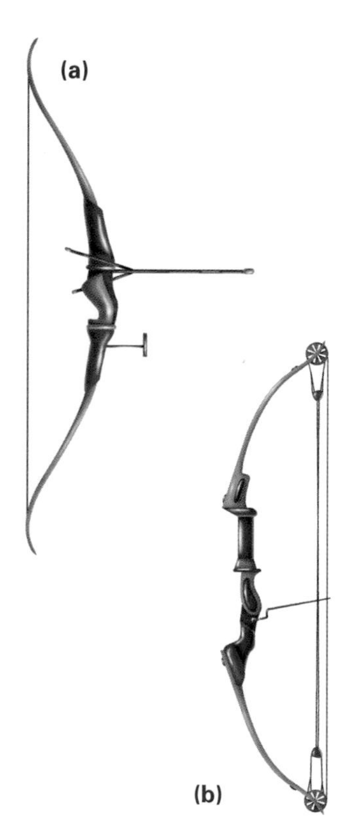

Figure 2
The recurve bow can apply only about half the force on an arrow that a compound bow can.
(a) the recurve bow
(b) the compound bow.

In the sport of archery (**Figure 2**), there are two types of bows used. One is the simple recurve bow that can typically apply a force of magnitude 150 N to the arrow. The other is the compound bow that can typically apply a force of magnitude 300 N. Assuming the arrows shot from these bows have the same mass, how will their accelerations compare? It is logical that the arrow shot from the compound bow will undergo greater acceleration than the one shot off the simple bow. A similar argument can be given for the acceleration of objects of the same mass with different net forces applied.

Practice

Understanding Concepts

3. In order to compare the effects of a changing net force on acceleration, what must be kept constant?
4. Sketch a graph in which you show how you think the acceleration of a particular object (vertical axis) depends on the net force applied to the object.

Acceleration and Mass

In the discus throw (**Figure 3**), the minimum discus mass for the female competition is 1.0 kg and and that for the male competition is 2.0 kg. Assuming that a person is able to exert the same net force in throwing both discuses, how will their accelerations compare? It is logical that the acceleration during the throw would be greater for the discus of lower mass. The same argument can be given for the acceleration of objects of different masses with the same net force applied.

Figure 3
A discus throw

Practice

Understanding Concepts

5. In order to compare the effect of changing the mass on acceleration, what must be kept constant?
6. Sketch a graph in which you show how you think the acceleration of an object (vertical axis) depends on the mass of the object, with a constant resultant force applied to the object.

Designing a Controlled Experiment

To discover the relationship among three variables, a controlled experiment must be carried out. This means that there is only one independent variable at a time. For example, to discover how the acceleration of an object depends on the net force on the object, the mass of the object must remain constant. Then, to discover how the acceleration of an object depends on the mass of the object, the net force must be kept constant. (To review the steps of a controlled experiment, refer to Appendix A.)

Imagine that the experiment is performed using racing cars. The mass of one of the cars, with the driver included, must remain constant while determining the acceleration of the car using three different engines with different powers, such that three different net forces can be applied to the car: a small force, a medium force, and a large force. In the second part of the experiment, you can use the same engine in three different race cars of different masses, say small, medium, and large mass. Again, you will determine the acceleration of each car. After finding out how the acceleration depends on the net force and mass, you try to express the relationships in a single statement, both in written words and in mathematical form.

Practice

Understanding Concepts

7. Name the dependent and independent variables in the virtual investigation involving the racing cars.

Applying Inquiry Skills

8. (a) How could you determine the acceleration of a cart in a physics laboratory using position-time data? (Assume there is uniform acceleration.)
   (b) How could you apply a uniform acceleration equation involving average acceleration, position, and time interval to determine the acceleration of a cart undergoing uniform acceleration in the laboratory? (Assume the initial velocity is zero.)
   (c) Which method would you prefer to use? Why?

INQUIRY SKILLS

- ○ Questioning
- ○ Hypothesizing
- ● Predicting
- ● Planning
- ● Conducting
- ● Recording
- ● Analyzing
- ● Evaluating
- ● Communicating

Investigation 2.3.1

The Relationship Involving Acceleration, Net Force, and Mass

The instructions for this experiment are based on the assumption that you have performed motion experiments previously using sensors, a spark timer, a ticker-tape timer, or some other common device. In whichever method you choose to use, you must determine the acceleration of the object tested under the conditions of changing the net force, and then changing the mass. Since force is expressed in the SI unit of newtons, you must express mass in kilograms and acceleration in metres per second squared.

(a)

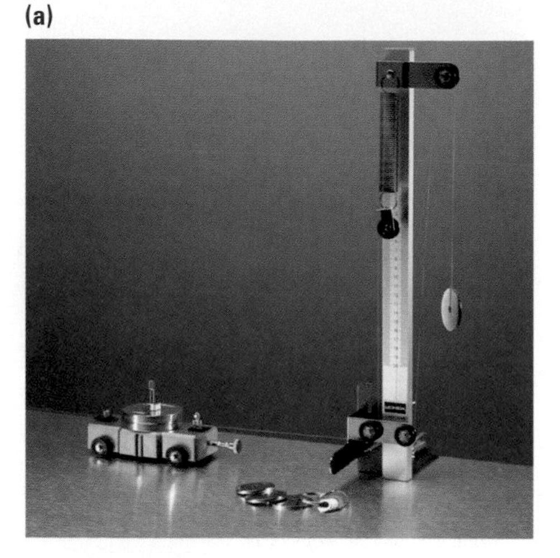

(b)

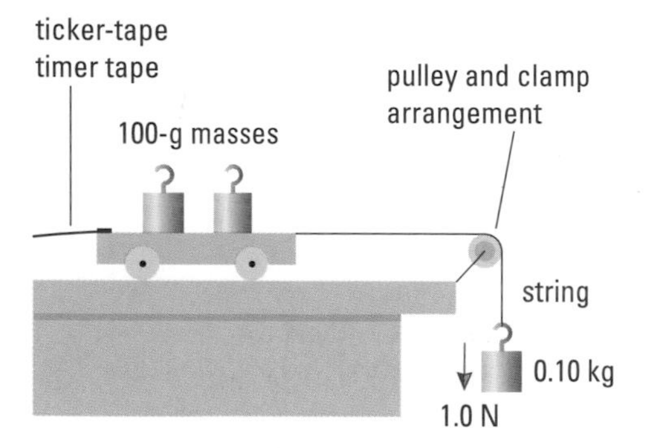

Figure 4
Two possible designs of the investigation involving acceleration, net force, and mass
(a) Using a specially designed cart acceleration apparatus
(b) Using a system with a cart, pulley, and clamp

Question

How does the acceleration of a cart depend on the resultant force acting on the cart and the mass of the cart, and how can this relationship be expressed in a single equation?

Hypothesis/Prediction

(a) Communicate in various ways (such as words, graphs, and mathematical variation statements) your answer to the first part of the Question. Then provide reasons for your prediction.

Design

As in Investigation 1.4.1 in Chapter 1, there are various ways to perform this investigation. **Figure 4** shows two of the ways in which a constant force can be applied to a cart. In both cases, the force of gravity pulls on a falling mass, and the mass is connected by a low-friction string to the cart. With a known mass hung vertically, the magnitude of the force of gravity on the mass is known, and thus the magnitude of the force causing the acceleration of the entire system is known. For example, if the mass is 100 g, the magnitude of the force of gravity is 0.98 N, or about 1.0 N. However, the mass of the system has to be carefully controlled: the total mass being accelerated includes the cart, the mass on the cart, and the suspended mass. When you want to keep the force constant and change the mass being accelerated, you add other masses to the cart.

Decide with your group how you will determine the acceleration of each trial (using a computer, photogates, a uniform acceleration equation, position-time and velocity-time graphs, or some other appropriate technique). Also discuss safety issues, such as how you will safely stop any moving object. Then design and set up an appropriate data table to record the observations and calculations for the trials you perform.

Materials

dynamics cart
three 100-g masses (or other suitable sizes)
two 1.0-kg masses
string
pulley
clamp
ticker-tape timer and related apparatus
beam balance, spring balance, or heavy-duty electronic balance

Procedure

Part A: Acceleration and Net Force

1. Verify that the equipment you intend to use is functioning properly.
2. Measure the mass of the cart.
3. Set up the apparatus so that the least net force will act on the cart (based on **Figure 4** or another method of your choice). Allow the motion to occur and obtain the data required to find the acceleration $\vec{a}_1$.
4. Repeat the procedure with an increased net force. For example, you can transfer one of the 100-g masses from the cart to the string hanging over the pulley. This allows the mass of the system to remain constant, as shown in **Figure 5**(a). Determine the data for $\vec{a}_2$.
5. Repeat the procedure with the highest net force to determine the data for $\vec{a}_3$ as in **Figure 5**(b).

Part B: Acceleration and Mass

6. Use the data for $\vec{a}_3$ as the first set of data in this part of the experiment. Call the acceleration $\vec{a}_4$.
7. Keep the net force constant at the highest value (such as 3.0 N), but add a 1.0-kg mass to the cart, as illustrated in **Figure 6**(a). Perform the trial to obtain the data for $\vec{a}_5$.
8. Add another 1.0-kg mass, as in **Figure 6**(b). Repeat the step to determine the data for $\vec{a}_6$.

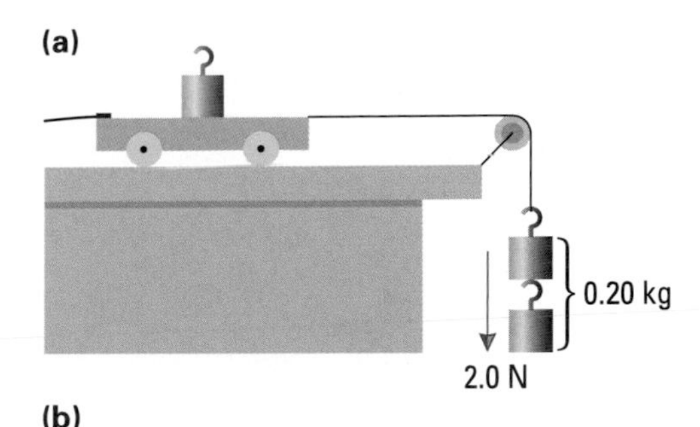

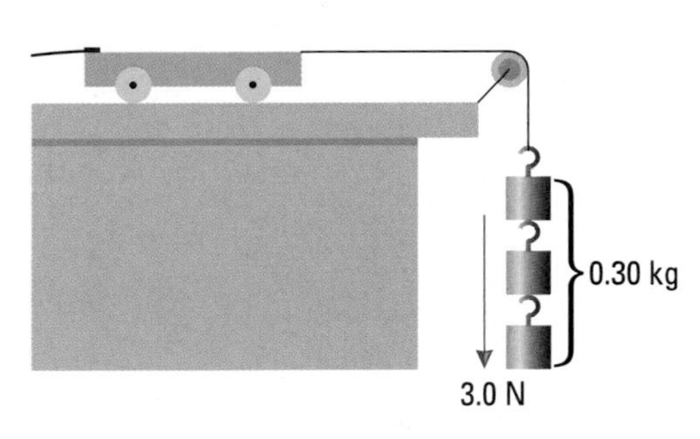

Figure 5
Changing the net force (The total mass must remain constant.)
(a) A net force of medium magnitude
(b) The maximum net force

Be sure to set up the apparatus such that the cart and the moving masses can be stopped safely.

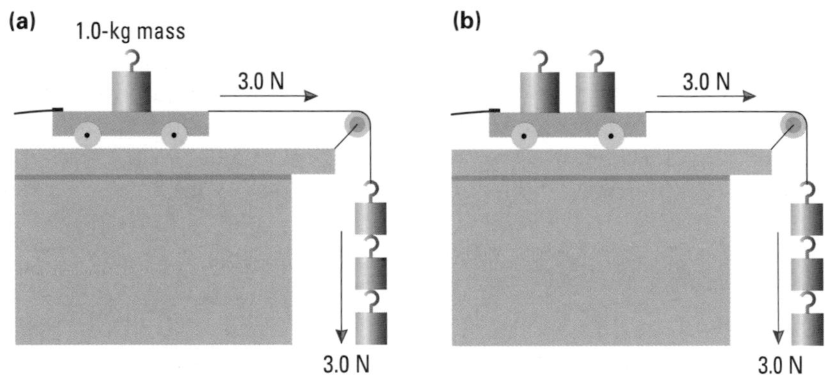

Figure 6
Changing the mass (The net force must be kept constant.)
(a) A total mass of medium value
(b) Maximum mass

Observations

(b) For each trial, use an appropriate technique you have chosen to determine the acceleration of the cart. Summarize all your data in your table.

Analysis

(c) For each trial, calculate the ratio of the net force to the total mass of the system, expressing your answer to the appropriate number of significant digits in N/kg, and enter the calculations in your table.

(d) Plot a graph of the acceleration (vertical axis) as a function of the net force for the trials in which the mass remained constant. Draw a line of best fit and calculate its slope. What does the graph indicate about the relationship between acceleration and net force?

(e) Plot a graph of the acceleration (vertical axis) as a function of the mass of the system for the trials in which the net force was kept constant. (Theoretically, the graph is a smooth curve.) What does the graph indicate about the relationship between acceleration and mass?

(f) Plot a graph of acceleration (vertical axis) as a function of the reciprocal of the mass ($\frac{1}{m}$) for a constant net force. Draw a line of best fit and calculate its slope. What does the slope represent?

(g) From the observation table, compare the calculated acceleration in each case to the ratio $\frac{\vec{F}_{net}}{m}$. Derive an equation relating all three variables in this experiment.

(h) Plot a graph of acceleration as a function of the ratio $\frac{\vec{F}_{net}}{m}$. What do you discover?

(i) Answer the Question posed at the beginning of this investigation.

Evaluation

(j) Comment on the accuracy of your prediction.

(k) How could you determine whether friction had an effect on the results of this investigation?

(l) Identify systematic, random, and human errors in this investigation. (To review errors refer to Appendix A.)

Synthesis

(m) Do you think the equation you derived in question (g) of Analysis applies to Newton's first law of motion? Explain your answer.

SUMMARY Force, Mass, and Acceleration

- To determine the relationship among three variables, a controlled investigation can be performed in which one independent variable is kept constant while the other independent variable is varied, and vice versa.
- A controlled investigation can be performed to determine how the acceleration of an object depends on the mass of the object and the net force acting on the object.

Section 2.3 Questions

Understanding Concepts

1. Compare the direction of the acceleration of an object to the direction of the resultant force acting on it.
2. Determine the false statements below and rewrite them to make them true.
   (a) The acceleration of an object is inversely proportional to the resultant force applied to the object.
   (b) The acceleration of an object is proportional to the mass of the object.
   (c) The acceleration of an object is proportional to the ratio of the resultant force to the mass.

Applying Inquiry Skills

3. When an acceleration experiment is performed using a cart, friction occurs and can affect the results. Describe what you can do so that the effect of friction is eliminated or at least is accounted for. (Try to describe more than one way.)
4. How many significant digits are appropriate in the calculation of forces and accelerations in investigations? Justify your answer.

2.4 Newton's Second Law of Motion

In recent tests of different vehicles, the minimum time for each one to accelerate from zero to 96.5 km/h [fwd] was measured. The results are shown in **Table 1**. It is evident from the table that the acceleration of a vehicle depends greatly on the power of the engine causing the acceleration and the mass of the vehicle. Logically, the magnitude of acceleration increases as the engine power increases and decreases as the mass increases.

Table 1 Data for Vehicles Accelerating from Zero to 96.5 km/h (60.0 mi/h)

Vehicle	Mass (kg)	Engine power (kW)	Time to accelerate (s)
Chevrolet Camaro	1500	239	5.5
Rolls-Royce	2300	240	8.2
Cadillac Seville	1800	224	6.8
Lamborghini Diablo	1800	410	3.6
Volkswagen Beetle	1260	86	10.6
Suzuki Bandit Bike	245	74	2.8

You observed a similar relationship in Investigation 2.3.1, although the variables you tested were net force and mass. These observations led to Newton's second law of motion.

Newton's first law of motion deals with situations in which the net external force acting on an object is zero, so no acceleration occurs. His second law deals with situations in which the net external force acting on an object is not zero, so acceleration occurs in the direction of the net force. The acceleration increases as the net force increases, but decreases as the mass of the object increases.

Newton's **second law of motion** states:

second law of motion: if the net external force on an object is not zero, the object accelerates in the direction of the net force, with magnitude of acceleration proportional to the magnitude of the net force and inversely proportional to the object's mass

Second Law of Motion

If the net external force on an object is not zero, the object accelerates in the direction of the net force. The magnitude of the acceleration is proportional to the magnitude of the net force and is inversely proportional to the object's mass.

Using mathematical notation, we can derive an equation for the second law of motion.

$\vec{a} \propto \vec{F}_{\text{net}}$ when m is constant

$\vec{a} \propto \frac{1}{m}$ when $\vec{F}_{\text{net}}$ is constant

Thus, $\vec{a} \propto \frac{\vec{F}_{\text{net}}}{m}$.

Now, we insert a proportionality constant, k, to create the equation relating all three variables:

$$\vec{a} = k\frac{\vec{F}_{\text{net}}}{m}$$

If the units on both sides of the equation are consistent, with newtons for force and the preferred SI units of metres, kilograms, and seconds for the acceleration and mass, the value of k is 1. Thus, the final equation for Newton's second law of motion is

$\vec{a} = \frac{\vec{F}_{net}}{m}$ Acceleration equals net force divided by mass.

This equation is often written in rearranged form:

$\vec{F}_{net} = m\vec{a}$ Net force equals mass times acceleration.

$$\vec{F}_{net} = m\vec{a}$$

As stated earlier, the unit for force is the newton. We can use the second law equation to define the newton in terms of the SI base units.

Therefore, 1 N = 1 kg•m/s^2.

Thus, we can define *one newton* as the magnitude of the net force required to give a 1.0 kg object an acceleration of magnitude 1.0 m/s^2.

Newton's second law of motion is important in physics. It affects all particles and objects in the universe. Because of its mathematical nature, the law is applied in finding the solutions to many problems and questions.

Sample Problem 1

A net force of 58 N [W] is applied to a water polo ball of mass 0.45 kg. Calculate the ball's acceleration.

Solution

$\vec{F}_{net} = 58\text{ N [W]}$

$m = 0.45\text{ kg}$

$\vec{a} = ?$

$$\begin{aligned}\vec{a} &= \frac{\vec{F}_{net}}{m} \\ &= \frac{58\text{ N [W]}}{0.45\text{ kg}} \\ &= \frac{58\text{ kg}\bullet\text{m/s}^2\text{ [W]}}{0.45\text{ kg}} \\ \vec{a} &= 1.3 \times 10^2\text{ m/s}^2\text{ [W]}\end{aligned}$$

The ball's acceleration is 1.3×10^2 m/s^2 [W].

Sample Problem 2

In an extreme test of its braking system under ideal road conditions, a Toyota Celica, travelling initially at 26.9 m/s [S], comes to a stop in 2.61 s. The mass of the car with the driver is 1.18×10^3 kg. Calculate (a) the car's acceleration and (b) the net force required to cause that acceleration.

Solution

(a) $\vec{v}_f = 0.0$ m/s [S]

$\vec{v}_i = 26.9$ m/s [S]

$\Delta t = 2.61$ s

$\vec{a} = ?$

$$\vec{a} = \frac{\vec{v}_f - \vec{v}_i}{\Delta t}$$

$$= \frac{0.0 \text{ m/s [S]} - 26.9 \text{ m/s [S]}}{2.61 \text{ s}}$$

$$\vec{a} = -10.3 \text{ m/s}^2 \text{ [S]}$$

The car's acceleration is -10.3 m/s^2 [S], or 10.3 m/s^2 [N].

(b) $m = 1.18 \times 10^3$ kg

$\vec{F}_{net} = ?$

$$\vec{F}_{net} = m\vec{a}$$

$$= (1.18 \times 10^3 \text{ kg})(10.3 \text{ m/s}^2 \text{ [N]})$$

$$\vec{F}_{net} = 1.22 \times 10^4 \text{ N [N]}$$

The net force is 1.22×10^4 N [N].

Does Newton's second law agree with his first law of motion? According to the second law, $\vec{a} = \frac{\vec{F}_{net}}{m}$, so the acceleration is zero when the net force is zero. This is in exact agreement with the first law. In fact, the first law is simply a special case ($\vec{F}_{net} = 0$) of the second law of motion.

Practice

Understanding Concepts

1. Calculate the acceleration in each situation.
   (a) A net force of 27 N [W] is applied to a cyclist and bicycle having a total mass of 63 kg.
   (b) A bowler exerts a net force of 18 N [fwd] on a 7.5-kg bowling ball.
   (c) A net force of 32 N [up] is applied to a 95-g model rocket.
2. Find the magnitude and direction of the net force in each situation.
   (a) A cannon gives a 5.0-kg shell a forward acceleration of 5.0×10^3 m/s^2 before it leaves the muzzle.
   (b) A 28-g arrow is given an acceleration of 2.5×10^3 m/s^2 [E].
   (c) A 500-passenger Boeing 747 jet (with a mass of 1.6×10^5 kg) undergoes an acceleration of 1.2 m/s^2 [S] along a runway.
3. Write an equation expressing the mass of an accelerated object in terms of its acceleration and the net force causing that acceleration.
4. Determine the mass of a regulation shot in the women's shot-put event (**Figure 1**) if a net force of 7.2×10^2 N [fwd] is acting on the shot, giving the shot an average acceleration of 1.8×10^2 m/s^2 [fwd].
5. Derive an equation for net force in terms of mass, final velocity, initial velocity, and time.
6. Assume that during each pulse a mammalian heart accelerates 21 g of blood from 18 cm/s to 28 cm/s during a time interval of 0.10 s. Calculate the magnitude of the force (in newtons) exerted by the heart muscle on the blood.

Answers

1. (a) 0.43 m/s^2 [W]
   (b) 2.4 m/s^2 [fwd]
   (c) 3.4×10^2 m/s^2 [up]
2. (a) 2.5×10^4 N [fwd]
   (b) 7.0×10^1 N [E]
   (c) 1.9×10^5 N [S]

4. 4.0 kg

6. 2.1×10^{-2} N

Figure 1
For question 4

SUMMARY Newton's Second Law of Motion

- Newton's second law of motion relates the acceleration of an object to the mass of the object and the net force acting on it. The equation is

$$\vec{a} = \frac{\vec{F}_{net}}{m} \quad \text{or} \quad \vec{F}_{net} = m\vec{a}.$$

- Newton's second law is applied in many problem-solving situations.

Section 2.4 Questions

Understanding Concepts

1. When the Crampton coal-fired train engine was built in 1852, its mass was 48.3 t (1.0 t = 1.0×10^3 kg) and its force capability was rated at 22.4 kN. Assuming it was pulling train cars whose total mass doubled its own mass and the total friction on the engine and cars was 10.1 kN, what was the magnitude of the acceleration of the train?
2. Determine the net force needed to cause a 1.31×10^3-kg sports car to accelerate from zero to 28.6 m/s [fwd] in 5.60 s.
3. As you have learned from Chapter 1, the minimum safe distance between vehicles on a highway is the distance a vehicle can travel in 2.0 s at a constant speed. Assume that a 1.2×10^3-kg car is travelling 72 km/h [S] when the truck ahead crashes into a northbound truck and stops suddenly.
   (a) If the car is at the required safe distance behind the truck, what is the separation distance?
   (b) If the average net braking force exerted by the car is 6.4×10^3 N [N], how long would it take the car to stop?
   (c) Determine whether a collision would occur. Assume that the driver's reaction time is an excellent 0.09 s.

2.5 Newton's Third Law of Motion

Figure 1
The astronaut is wearing a manoeuvring unit (the backpack) that illustrates an application of the third law of motion. Expanding gases expelled from the unit propel the astronaut in a direction opposite to the direction of the expelled gases.

When astronauts go for a "space walk" outside the International Space Station (the ISS), they travel along with the station at a speed of about 30 000 km/h relative to Earth's surface. (You should be able to use the first law of motion to explain why: since the station and the astronaut are both in motion, they remain in motion together.) However, to move around outside the station to make repairs, the astronaut must be able to manoeuvre in different directions relative to the station. To do so, the astronaut wears a special backpack called a mobile manoeuvring unit, or MMU (**Figure 1**), a device that applies another important principle named after Sir Isaac Newton.

Newton's first law of motion is descriptive and his second law is mathematical. In both cases, we consider the forces acting on only one object. However, when your hand pushes on the desk in one direction, you feel a force of the desk pushing back on your hand in the opposite direction. This brings us to the third law, which considers forces acting in pairs on two objects.

Newton's **third law of motion**, often called the action-reaction law, states:

Third Law of Motion

For every action force, there is a reaction force equal in magnitude, but opposite in direction.

third law of motion: for every action force, there is a reaction force equal in magnitude, but opposite in direction

To illustrate the third law, imagine a ball shot horizontally out of the tube of a toy cart on wheels, as shown in **Figure 2**. When the ball is pushed into the cart, a spring becomes compressed. Then, when the spring is released, the spring (and thus the cart) pushes forward on the ball. We call this force the *action force*. At the same instant, the ball pushes backward on the spring (and thus the cart). We call this the *reaction force* of the ball on the cart. The action and reaction forces are equal in size but opposite in direction, and act on different objects. (Do not worry if you have difficulty deciding which of an action-reaction force pair is the action force and which is the reaction force. Both forces occur at the same time, so either way works.)

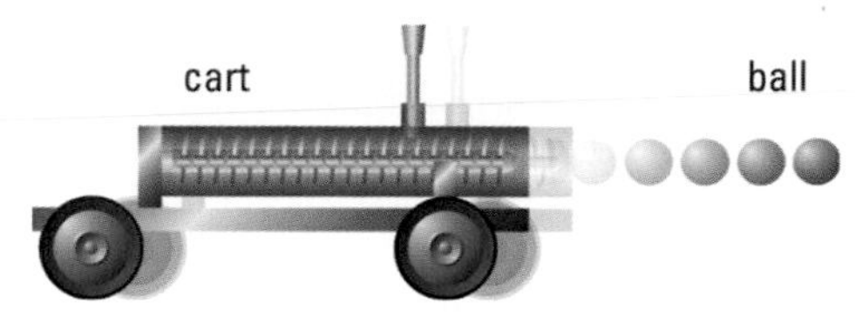

Figure 2
When the spring is released, it exerts a forward force on the ball, and simultaneously the ball exerts a force on the spring (and cart) in the opposite direction. The ball moves forward and the cart moves backward.

Sample Problem

Draw an FBD of the ball shown in **Figure 2** while it is still pushed by the spring.

Solution

The FBD is shown in **Figure 3**. The force of the spring on the ball is to the right. While the ball is still in the tube, the normal force exerted by the tube is equal in magnitude to the force of gravity on the ball. Notice that there are no action-reaction pairs of forces in the diagram because an FBD is a drawing of a single object, not two objects, and an FBD shows only the forces exerted on the object, not any forces that the object might exert on something else.

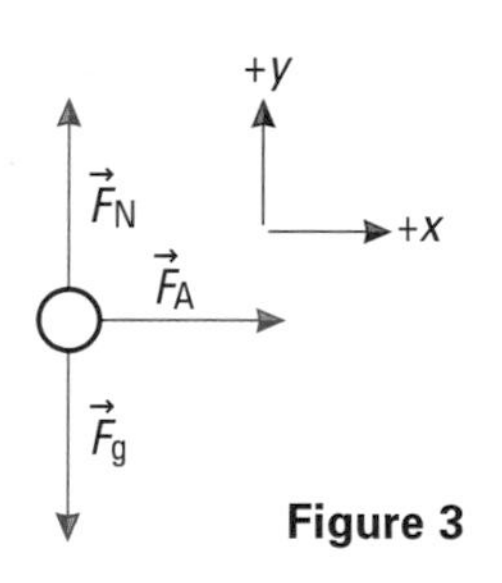

Figure 3

Try This **Activity**

Demonstrating Newton's Third Law

Your teacher will set up demonstrations of the third law of motion. In each case, predict what you think will occur, then observe what happens, and finally summarize your observations in tabular form using these headings:

Object observed	Predicted result	Observed result	Description of the action force(s)	Description of the reaction force(s)	Diagram of the action-reaction pair(s)

Do action-reaction forces exist on stationary objects? Yes, they do, but they might not seem as obvious as the example of the ball being shot out of the toy cart. Consider, for instance, an apple hanging in a tree, as in **Figure 4**. The force of gravity on the apple pulling downward is balanced by the upward force of the stem holding the apple. However, these two forces act on the same object (the apple), so they are not an action-reaction pair. In fact, there are two action-

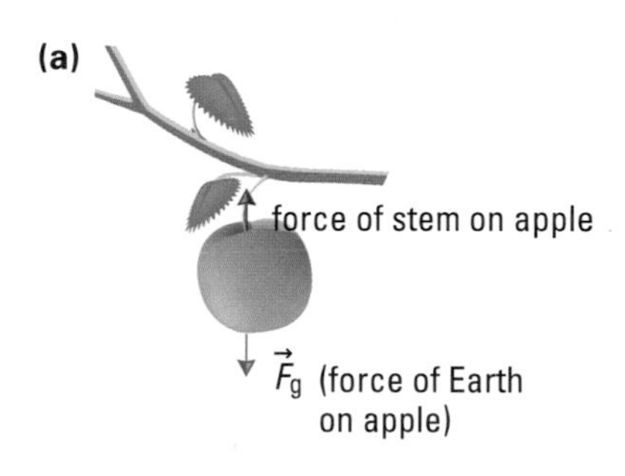

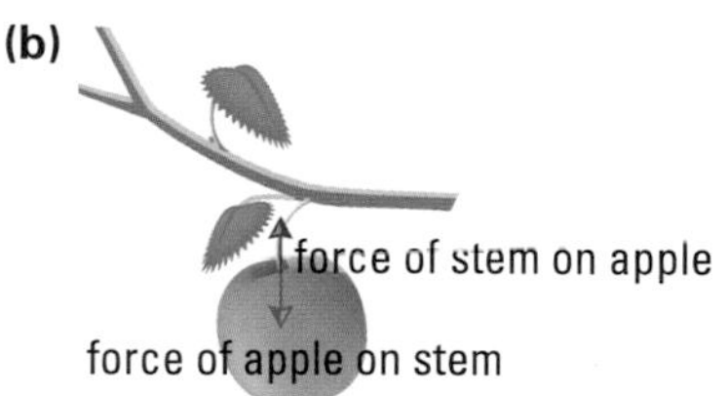

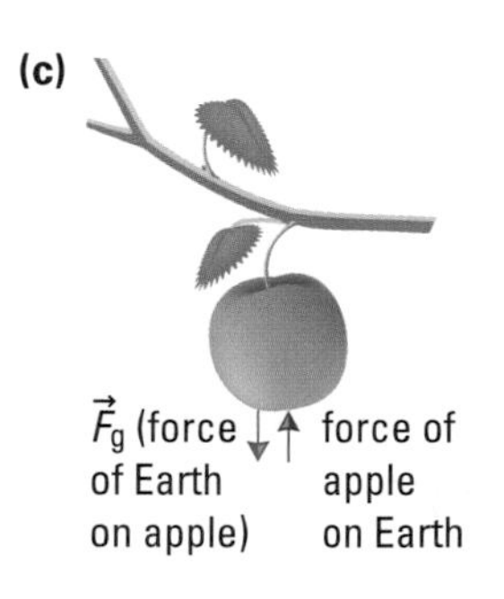

Figure 4
(a) The downward force of gravity on the apple is balanced by the upward force exerted by the stem. (This is not an action-reaction pair of forces.)
(b) One action-reaction pair of forces exists where the stem and apple are attached.
(c) A second action-reaction pair exists between Earth and the apple.

DID YOU KNOW?

Observing Newton's Third Law Electronically

You can observe verification of Newton's third law of motion by using simulation software or by accessing the Nelson science Web site. Follow the links for Nelson Physics 11, 2.5. Alternatively, if the interfacing technology is available, your teacher can demonstrate the third law using force sensors mounted on two dynamics carts. When the carts collide, the magnitude of the force each cart exerts on the other can be measured. The law for various mass combinations and initial velocities can be observed.

GO TO www.science.nelson.com

reaction pairs in this example. One is the downward force of the apple on the stem, equal in size but opposite in direction to the force of the stem on the apple. The other is the downward force of Earth's gravity, equal in size but opposite in direction to the upward force of the apple on Earth. Of course, if the stem breaks, the apple accelerates toward Earth, and Earth also accelerates toward the apple. However, because Earth has such a large mass, and acceleration is inversely proportional to mass, Earth's acceleration is extremely small.

Practice

Understanding Concepts

1. Draw an FBD of the cart in **Figure 2** when the spring is released.
2. Draw an FBD of the apple hanging from the stem in **Figure 4**. Are there any action-reaction pairs of forces in your diagram?

Applying the Third Law of Motion

The third law of motion has many interesting applications. As you read the following descriptions of some of them, remember there are always two objects to consider. One object exerts the action force while simultaneously the other exerts the reaction force. In certain cases, one of the "objects" may be a gas such as air.

(a) When someone is swimming, the person's arms and legs exert an action force backward against the water. The water exerts a reaction force forward against the person's arms and legs, pushing his or her body forward.

(b) A jet engine on an aircraft allows air to enter a large opening at the front of the engine. The engine compresses the air, heats it, then expels it rapidly out the rear (**Figure 5**). The action force is exerted by the engine backward on the expelled air. The reaction force is exerted by the expelled air forward on the engine, pushing the engine and, thus, the entire airplane in the opposite direction.

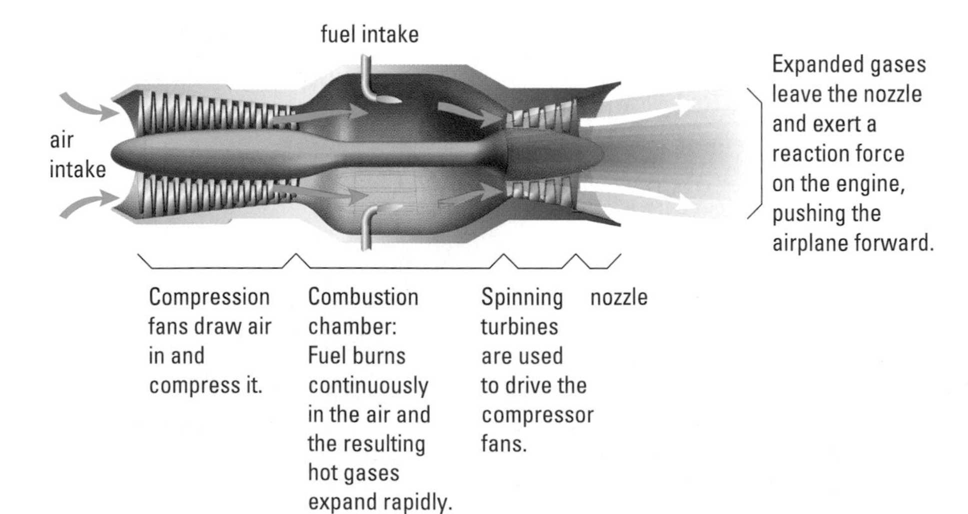

Figure 5
The design of a turbo jet engine

(c) A squid is a marine animal with a body size ranging from about 3 cm to 6 m. It propels itself by taking in water and expelling it in sudden spurts.The action force is applied by the squid backward on the discharged water. The reaction force of the expelled water pushes the squid in the opposite direction.

Practice

Understanding Concepts

3. Explain each event described below in terms of Newton's third law of motion.
   (a) A space shuttle vehicle, like the one shown in the photograph in **Figure 1** of section 2.3, is launched.
   (b) When a toy balloon is blown up and released, it flies erratically around the room.
4. You are a passenger on a small rowboat. You are about to step from the boat onto a nearby dock. Explain why you may end up in the water instead.
5. According to Newton's third law, when a horse pulls on a cart, the cart pulls back with an equal force on the horse. If, in fact, the cart pulls back on the horse as hard as the horse pulls forward on the cart, how is it possible for the horse to move the cart?

SUMMARY Newton's Third Law of Motion

- Newton's third law of motion, which always involves two objects, states that for every action force, there is a reaction force equal in magnitude, but opposite in direction.
- Action-reaction pairs of forces are applied in many situations, such as a person walking, a car accelerating, and a rocket blasting off into space.

Section 2.5 Questions

Understanding Concepts

1. Use the third law of motion together with a diagram of the action-reaction pair(s) to explain each situation.
   (a) A person with ordinary shoes is able to walk on a sidewalk.
   (b) A rocket accelerates in the vacuum of outer space.
2. (a) A certain string breaks when a force of 225 N is exerted on it. If two people pull on opposite ends of the string, each with a force of 175 N, will the string break? Explain.
   (b) Draw a diagram of the situation in (a) showing all the action-reaction pairs of forces.

Making Connections

3. What is meant by the term "whiplash" in an automobile collision? Explain how and why whiplash occurs by applying Newton's laws of motion.
4. An "ion propulsion system" is a proposed method of space travel using ejected charged particles (**Figure 6**). Locate information on this system and analyze how it relates to the third law of motion. Follow the links for Nelson Physics 11, 2.5.

www.science.nelson.com

Reflecting

5. How important are diagrams in helping you solve problems involving forces and motion?

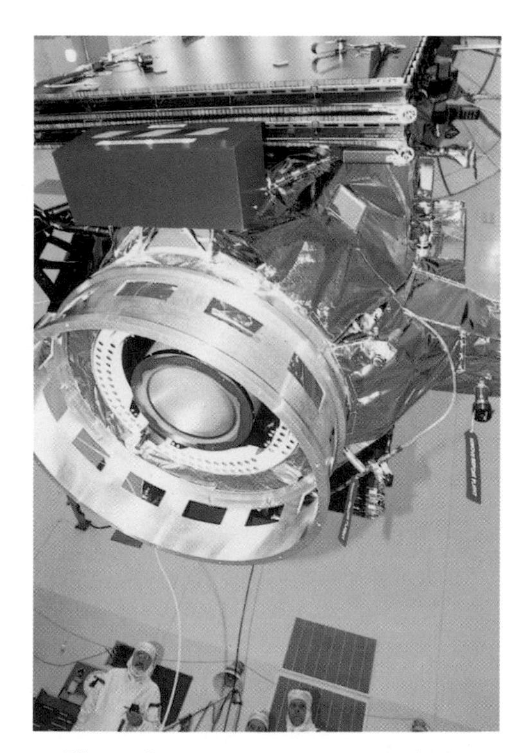

Figure 6
The space probe, *Deep Space 1*, was launched in 1998 to observe the asteroid Braille. The probe used ions ejected from the exhaust grid (the grey disk) to propel itself forward. This successful mission will help scientists develop ion propulsion systems for the future.

Careers in Forces and Motion

There are many different types of careers that involve the study of forces and motion in one form or another. Have a look at the careers described on this page and find out more about one of them or another career in forces and motion that interests you.

Civil Engineer

It takes about four years at university to complete a civil engineering degree. Qualified civil engineers help to plan and design buildings, foundations, roads, sewage systems, and bridges. These engineers work with architects and use computer programs to design structures. Some work for private construction companies and government agencies, while others work as private consultants to businesses or the government. All civil engineers must be members of the Association of Professional Engineers.

Mechanical Engineer

Mechanical engineering is a four-year university degree course. Upon completion of their studies, many mechanical engineers obtain work in the manufacturing sector, in the areas of safety and quality control. The work done by mechanical engineers can vary from designing heating and cooling systems to overseeing assembly line work in car manufacturing. This line of work often requires extensive use of computers. A mechanical engineer must be a member of the Association of Professional Engineers.

Kinesiologist

Kinesiologists must study for about four years to obtain a bachelor's degree in science, majoring in kinesiology. To obtain entry into this course, good marks in high school mathematics, physics, and biology are essential. Most of the time, kinesiologists work with their hands and with computers. They also use electromyography instruments that measure the electrical activity in muscles to treat and diagnose physical problems. These professionals often work in clinics and hospitals. Increasingly, kinesiologists may be found working as ergonomic consultants, in offices and factories, and teaching people how to work safely with various forms of technology.

Practice

Making Connections

1. Identify several careers that require knowledge about forces and motion. Select a career you are interested in from the list you made or from the careers described above. Imagine that you have been employed in your chosen career for five years and that you are applying to work on a new project of interest.
   (a) Describe the project. It should be related to some of the new things you learned in this unit. Explain how the concepts from this unit are applied in the project.
   (b) Create a résumé listing your credentials and explaining why you are qualified to work on the project. Include in your résumé
   - your educational background — what university degree or diploma program you graduated with, which educational institute you attended, post-graduate training (if any);
   - your duties and skills in previous jobs; and
   - your salary expectations.

   Follow the links for Nelson Physics 11, 2.5.

   GO TO www.science.nelson.com

Chapter 2 Summary

Key Expectations

Throughout this chapter, you have had opportunities to

- define and describe concepts and units related to force (e.g., applied force and net force); (2.1, 2.4, 2.5)
- identify and describe the fundamental forces of nature; (2.1)
- analyze and describe the forces acting on an object, using free-body diagrams, and determine the acceleration of the object; (2.4)
- state Newton's laws, and apply them to explain the motion of objects in a variety of contexts; (2.2, 2.4, 2.5)
- analyze in quantitative terms, using Newton's laws, the relationships among the net force acting on an object, its mass, and its acceleration; (2.4)
- carry out experiments to verify Newton's second law of motion; (2.3)
- interpret patterns and trends in data by means of graphs drawn by hand or by computer, and infer or calculate linear and non-linear relationships among variables (e.g., analyze data related to the second law experiments); (2.3)
- analyze the motion of objects, using vector diagrams, free-body diagrams, uniform acceleration equations, and Newton's laws of motion; (2.4)
- explain how the contributions of Galileo and Newton revolutionized the scientific thinking of their time and provided the foundation for understanding the relationship between motion and force; (2.1, 2.2)
- evaluate the design of technological solutions to transportation needs and, using scientific principles, explain the way they function (e.g., evaluate the design and explain the operation of airbags in cars); (2.2)
- analyze and explain the relationship between an understanding of forces and motion and an understanding of political, economic, environmental, and safety issues in the development and use of transportation technologies (including terrestrial and space vehicles) and recreation and sports equipment; (2.2, 2.4, 2.5)
- identify and describe science and technology-based careers related to forces and motion; (career feature)

Key Terms

force
fundamental forces
gravitational force
force of gravity
electromagnetic force
strong nuclear force
weak nuclear force
normal force
friction
tension
newton
system diagram
free-body diagram (FBD)
dynamics
inertia
resultant force
net force
first law of motion
second law of motion
third law of motion

Make a Summary

Imagine you have been chosen to join a team of explorers to travel from Earth to the Moon. Make up a "physics" story of your trip to the Moon, starting from the launching pad and ending when you land on the Moon's surface. You can use a written description as well as diagrams, graphs, and equations. In your story, use as many of the key words and concepts from this chapter as you can.

Reflect on your Learning

Revisit your answers on the Reflect on Your Learning questions at the beginning of this chapter.

- How has your thinking changed?
- What new questions do you have?

Chapter 2 Review

Understanding Concepts

1. Describe briefly the four fundamental forces that can account for all known forces.
2. Name the type(s) of force responsible for each of the following:
   (a) A nickel is attracted to a special steel bar.
   (b) A coasting cyclist gradually comes to a stop.
   (c) An electron in a hydrogen atom travels in an orbit around a proton.
   (d) A meteor accelerates toward Earth.
   (e) The meteor in (d) begins to burn and give off light.
3. An airplane is travelling with constant speed, heading east at a certain altitude. What forces are acting on the airplane? What is the net force on the airplane?
4. The scale used to draw the forces in **Figure 1** is 1.0 cm = 2.0 N.
   (a) Find the net force acting on the cart.
   (b) If the cart's mass is 1.2 kg, what is its acceleration?

Figure 1

5. Apply Newton's first law of motion to explain the danger in travelling too quickly on a curve of an icy highway.
6. A tractor pulls forward on a moving plow with a force of 2.5×10^4 N, which is just large enough to overcome friction.
   (a) What are the action and reaction forces between the tractor and the plow?
   (b) Are these action and reaction forces equal in magnitude but opposite in direction? Can there be any acceleration? Explain.
7. A football player kicks a 410-g football, giving it an acceleration of magnitude $|25\vec{g}|$ for 0.10 s.
   (a) What net force is imparted to the ball?
   (b) Name and state the magnitude of the reaction force.
8. One of the world's greatest jumpers is the flea. For a brief instant a flea is estimated to accelerate with a magnitude of 1.0×10^3 m/s^2. What is the magnitude of the net force a 6.0×10^{-7}-kg flea would need to produce this acceleration?
9. A net force of magnitude 36 N gives a mass m_1 an acceleration of magnitude 4.0 m/s^2. The same net force gives another mass m_2 an acceleration of magnitude 12 m/s^2. What magnitude of acceleration will this net force give to the entire mass if m_1 and m_2 are fastened together?
10. In an electronic tube, an electron of mass 9.1×10^{-31} kg experiences a net force of magnitude 8.0×10^{-15} N over a distance of 2.0 cm.
    (a) Calculate the magnitude of the electron's acceleration.
    (b) Assuming it started from rest, how fast would the electron be travelling at the end of the 2.0-cm motion?
11. A 1.2×10^4-kg truck is travelling south at 22 m/s.
    (a) What net force is required to bring the truck to a stop in 330 m?
    (b) What is the cause of this net force?
12. Calculate the acceleration of the cart shown in **Figure 2**, given the following assumptions:
    (a) No friction is acting on the cart.
    (b) A frictional resistance of magnitude 2.0 N is acting on the cart.

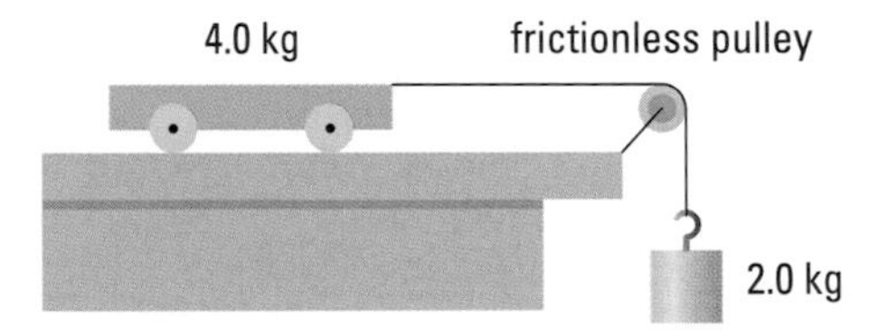

Figure 2

13. You are standing on the edge of a frozen pond where friction is negligible. In the centre of the ice is a red circle 1.0 m in diameter. A prize of a megadollar will be offered if you can apply all three of Newton's laws of motion to get to the red circle and stop there. Describe what you would do to win the prize.
14. A shuffleboard disk of mass 0.50 kg accelerates under an applied force of 12.0 N [forward].
    (a) If the magnitude of the frictional resistance is 8.0 N, find the magnitude of the disk's acceleration.
    (b) If the disk moves from rest for 0.20 s, how far does it travel while accelerating?
15. Each of the four wheels of a car pushes on the road with a force of 4.0×10^3 N [down]. The driving force on the car is 8.0×10^3 N [W]. The frictional resistance on the car is 6.0×10^3 N [E]. Calculate the following:
    (a) the mass of the car
    (b) the net force on the car
    (c) the car's acceleration

16. Draw a free-body diagram to determine the net force acting on the object in *italics* in each of the following situations:
    (a) Two teenagers are pushing a *dirt bike* through a freshly plowed field. One exerts a force of 390 N [W] on the bike while the other exerts a force of 430 N [W]. Frictional resistance amounts to 810 N.
    (b) A *water-skier* is being pulled directly behind a motorboat at a constant speed of 20.0 m/s. The tension in the horizontal rope is 520 N.
    (c) An *elevator*, including passengers, has a mass of 1.0×10^3 kg. The cable attached to the elevator exerts an upward force of 1.2×10^4 N. Friction opposing the motion of the elevator is 1.5×10^3 N.
17. Use unit analysis to check the validity of this equation:

$$\vec{F}_{\text{net}} = \frac{m(\vec{v}_{\text{f}} - \vec{v}_{\text{i}})}{\Delta t}$$

Applying Inquiry Skills

18. A group performing an investigation uses a 100.0-g mass to cause the acceleration of a cart. They record the force of gravity on the mass as 1.0 N. Assuming the true value of the force is 0.98 N, what is the percentage error of their value?
19. (a) State at least one possible source of error when using a spring scale to determine the force needed to support an object.
    (b) What would you do to account for that source of error in an experiment?
20. Design a virtual experiment in which you test the advantages of wearing a safety helmet while mountain biking.
21. A student performs an experiment to study the relationship between applied force and acceleration on a dynamics cart. After applying five different forces and determining the resulting acceleration, the student plots a graph of acceleration versus applied force and obtains the result shown in **Figure 3**.
    (a) Explain why the graph does not pass through the origin.
    (b) How would the graph change if the experiment were repeated using a surface where friction between the cart and the surface was greater?
    (c) What was the value of the frictional force (assumed constant) that acted upon the cart when the student carried out the experiment?

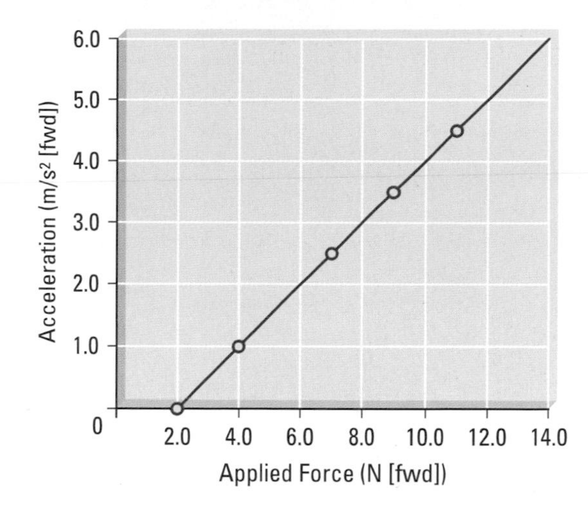

Figure 3

Making Connections

22. A car has a fuel consumption of 7.2 L/100.0 km on an expressway and 9.5 L/100.0 km on city roads. Explain why there is a difference.
23. List possible careers that apply the principles presented in this chapter.
24. Explain these warnings found on the visor of a new automobile:

    "Children 12 and under should be seated in the rear seat."

    "Never seat a rearward-facing child in the front."

Exploring

25. Research the current progress toward a "unified field theory" of forces. Follow the links for Nelson Physics 11, Chapter 2 Review. Use a concept map to summarize your findings about the electroweak force, the Grand Unified Theory (GUT), and the Theory of Everything (TOE).

GO TO www.science.nelson.com

26. Many resources are available that depict the life and contributions of Galileo and Newton, the two scientists whose work has been featured frequently in this chapter. Research some interesting aspects of either of these scientists. Write a summary of what you discover.

Chapter

3

Gravitational Force and Friction

In this chapter, you will be able to

- define and describe concepts related to the forces of friction and gravity
- describe and analyze the gravitational force acting on an object either at Earth's surface or at a known distance above Earth's surface
- design and carry out experiments to identify how friction affects the motion of a sliding object
- use scientific principles to explain the design of technological solutions to transportation needs and evaluate the design
- explain and analyze how an understanding of forces and motion combined with an understanding of societal issues leads to the development and use of transportation technologies and recreation and sports equipment

Which task would require you to apply a greater force, pushing a single textbook across your desk, or pushing a pile of 10 of the same books across the desk? Of course, you would have to apply a greater force to push the 10 books. The force of gravity is greater on the 10 books and so is the force of friction that you must overcome to move the books. Evidently, in this situation, friction is somehow related to the force of gravity. Both forces were introduced in Chapter 2, and we will discuss their properties in greater detail.

Without gravitational forces, Earth would not be trapped in its orbit around the Sun. In fact, Earth would not even exist because gravitational forces between particles were needed to form our planet. What causes the gravitational force, and what properties of this force must be understood in order to control the orbit of the International Space Station (**Figure 1**) or plan space travel to other parts of the solar system? On Earth's surface, do these forces have an effect on the records set at sports events? Is there a mathematical relationship between gravitational force and friction? These are some of the questions we will explore in this chapter.

Friction is important in transportation. When friction between the tires and the road becomes too small, such as in the case of a snowstorm or an ice storm, vehicles can be out of the drivers' control and crash. Scientists and engineers are continually researching better materials for manufacturing tires, constructing roads, and building bridges to improve their properties in an effort to control friction. Such research helps to improve transportation safety.

However, drivers, too, have great responsibility for transportation safety. Those who know how to apply what they have learned about the physics of motion and forces, whether or not they learned it in school, tend to be better equipped to drive safely.

Reflect on your Learning

1. When you drop a book in your classroom, the book falls on the floor. Will the book drop faster if it is heavier? Why?
2. Would your weight be different on the top of a mountain than it is at sea level? Why?
3. Imagine that you are on Mars. Would you walk faster than you do on Earth? How would this be related to the force of gravity?
4. List three ways in which friction is useful to you and three ways in which friction is unwanted or not helpful. Explain.
5. When brakes are applied to stop a bicycle, what factors affect the stopping distance?

Throughout this chapter, note any changes in your ideas as you learn new concepts and develop your skills.

Try This **Activity**

Modelling an Important Relationship

When you apply a spray, such as window cleaner, spray paint, or hair spray, you know that the farther the target is from the spray source, the lower the concentration of the spray. This principle can be used to help visualize the inverse square relationship, an important relationship in nature. This relationship applies to the force of gravity (presented in this chapter) and other physical situations.

To simplify the calculations in this activity, assume that the spray nozzle is square rather than round, and it sprays 25 g of paint in each of five trials, as depicted in **Figure 2**. In the first trial, the target is 10.0 cm from the source, in the second trial, the target is 20.0 cm away, and so on.

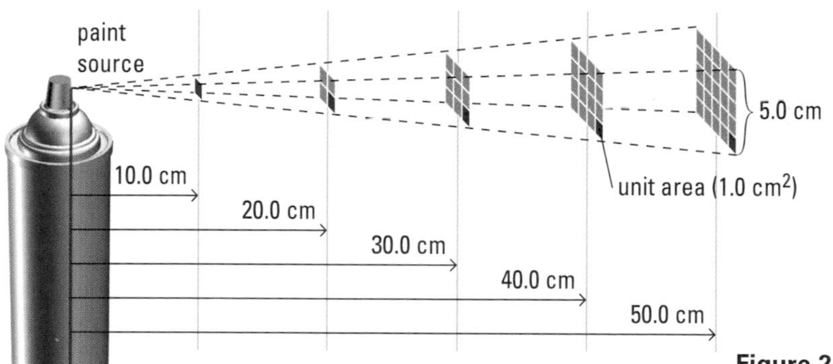

Figure 2

For this activity, you will need graph paper and a calculator.

(a) Sketch a graph of inverse square relationship $y \propto \frac{1}{x^2}$.

(b) Copy **Table 1** into your notebook, and complete it by referring to **Figure 2**.

(c) Describe any pattern you observe in your completed table of data and calculations.

(d) Plot a graph of the mass of paint per unit area as a function of the distance from the source. Compare this graph to the sketch you drew in (a) above.

(e) Describe in your own words the meaning of "inverse square relationship."

(f) How do you think the inverse square relationship applies to the gravitational forces surrounding Earth?

Figure 1
The International Space Station (ISS) is an orbiting science laboratory with scientists living aboard for several months at a time. To build and maintain this technological marvel above Earth's atmosphere requires a thorough understanding of the gravitational forces that keep the ISS in orbit. Experiments conducted on the ISS are unique because everything on board is under free fall conditions.

Table 1

Trial	1	2	3	4	5
Distance from source (cm)	10.0				
Side length of target (cm)	1.0				
Surface area of target (cm²)	1.0				
Mass of paint per unit area (g/cm²)	25				

Figure 1
Usually, a new Olympic record replaces the old one by a small difference. In the 1968 Olympics, however, some records were shattered by a large amount. The previous pole vault record, for example, was exceeded by 30 cm in that year.

3.1 Gravitational Force on Earth's Surface

The Summer Olympic Games are held every four years in different cities around the world. During each Olympic Games, many of the Olympic records, as well as several world records, are broken. When the Olympic Games were held in Mexico City in 1968, a higher than expected number of Olympic and world records were broken, especially for track and field events (**Figure 1**). At the same time, the results of certain other events were very poor. Which types of events were easier in Mexico City, and which were more difficult? What factors do you think contributed to these differences?

Mexico City is located in a mountainous region of central Mexico, at a latitude of about 20° north of the equator and at an elevation of 2300 m above sea level. These two factors—the proximity to the equator and the high elevation—contributed to the results of the 1968 Olympics. Both factors affect the force of gravity at a location, while the latter, the elevation above sea level, also affects the density of the air. Thus, with a smaller force of gravity, track and field events, such as short races, jumping events, and throwing events, are all easier to perform. However, an approximate 30% reduction in oxygen in the low-density air makes events that require great stamina, such as long-distance running, much more difficult.

The force of gravity affects much more than just sports events. The properties of this ever-present force have been applied in many ways, from designing and making simple devices, such as a weigh scale or a plumb bob (**Figure 2**), to developing and building more complex devices, such as rockets and space probes.

Figure 2
A *plumb bob* is a relatively large mass hung on a cord. To be sure that fence posts, walls, door and window frames, and strips of wallpaper are vertical, a construction expert or an interior decorator suspends a bob to check whether the vertical lines are parallel to its cord. To obtain a horizontal line, a 90° angle can be drawn from the vertical.

Earth's Gravitational Field

A **force field** is a space surrounding an object in which the object exerts a force on other objects placed in the space. Earth is surrounded by a gravitational force field (**Figure 3**). This means that every mass, no matter how large or small, and whether it is in the space on or above Earth's surface, feels a force pulling it toward Earth. The **gravitational field strength** is the amount of force per unit mass acting on objects in the gravitational field. It is a vector quantity that has the direction downward or toward Earth's centre. Using SI units, gravitational field strength is measured in newtons per kilogram (N/kg). The gravitational field strength is not the same everywhere. It depends on how close the object is to the centre of Earth—it is greater in valleys and smaller on mountaintops.

A simple way to measure the force of gravity acting on a mass is to suspend the mass from a force sensor or a spring balance calibrated in newtons, as you did

force field: space surrounding an object in which the object exerts a force on other objects placed in the space

gravitational field strength: the amount of force per unit mass acting on objects in the gravitational field

Figure 3
This model illustrates the gravitational force field surrounding Earth. All the vectors point toward Earth's centre, and their magnitudes indicate that the field becomes weaker as the distance from Earth's centre increases. As seen in Chapter 2, it is the "direction toward the centre" that defines what we mean by "downward" on Earth's surface.

in the Chapter 2 opener activity. **Table 1** shows data that were collected in this way, and **Figure 4** shows that the graph of these data yields a straight line.

The slope of the line in **Figure 4** is

$$\text{slope} = \frac{\Delta\vec{F}}{\Delta m} = \frac{490 \text{ N } [\downarrow]}{50.0 \text{ kg}} = 9.8 \text{ N/kg } [\downarrow]$$

The slope gives the gravitational field strength on Earth's surface in newtons per kilogram. Where have you seen the number 9.8 before? This is the magnitude of the acceleration due to gravity (9.8 m/s^2 [↓]). It is left as an exercise to show that N/kg and m/s^2 are equivalent.

Since the gravitational field strength and the acceleration due to gravity are numerically equal, the same symbol, $\vec{g}$, is used for both. Therefore, on Earth's surface, $\vec{g} = 9.8$ N/kg [↓], or $\vec{g} = 9.8$ m/s^2 [↓].

The gravitational field strength can be applied using the equation for Newton's second law of motion, $\vec{F}_g = m\vec{g}$, to determine the force of gravity acting on an object at Earth's surface.

Table 1 The Force of Gravity on Masses

Mass (kg)	Force of gravity (N [↓])
0.0	0
10.0	98
20.0	196
30.0	294
40.0	392
50.0	490

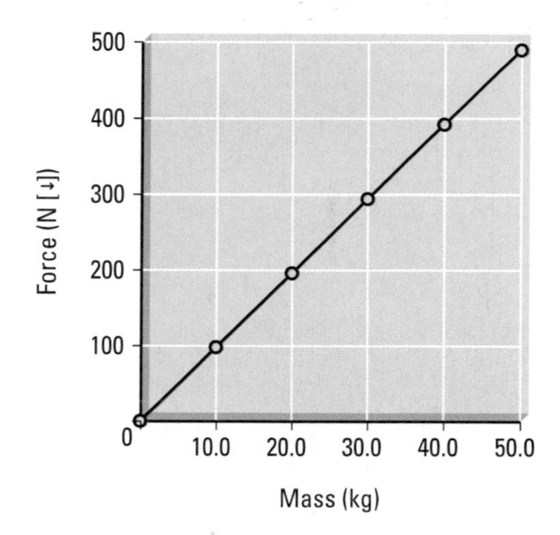

Figure 4
The slope of the line on this force-mass graph indicates the gravitational field strength.

Sample Problem 1

The maximum train load pulled through the Chunnel, the train tunnel under the English Channel that links England and France, is 2434 t. Determine the force of gravity on this huge mass.

Solution

$m = 2434$ t

$\vec{g} = 9.8$ N/kg [↓]

$\vec{F}_g = ?$

We must first convert the mass in tonnes to kilograms:

$$2434 \text{ t} = 2434 \text{ t} \times \frac{1000 \text{ kg}}{1 \text{ t}} = 2.434 \times 10^6 \text{ kg}$$

$$\vec{F}_g = m\vec{g}$$
$$= (2.434 \times 10^6 \text{ kg})(9.8 \text{ N/kg } [\downarrow])$$
$$\vec{F}_g = 2.4 \times 10^7 \text{ N } [\downarrow]$$

The force on the load is 2.4×10^7 N [↓].

Practice

Understanding Concepts

1. Show that N/kg is equivalent to m/s^2.
2. The average mass of a basketball is 0.63 kg. What is the force of gravity acting on the ball?
3. The force of gravity on the heaviest person in history is about 6.2 kN [↓]. Determine the mass of this record-holder in kilograms.
4. The force of gravity on a 251-kg spacecraft on the Moon's surface is 408 N [↓].
   (a) What is the gravitational field strength there?
   (b) What is the acceleration of a free-falling object on the surface of the Moon?
5. Assume you are in a space colony on Mars, where the gravitational field strength is 3.7 N/kg [↓]. What is the force of gravity on you?

Answers

2. 6.2 N [↓]
3. 6.3×10^2 kg
4. (a) 1.63 N/kg [↓]
   (b) 1.63 m/s^2 [↓]

mass: the quantity of matter in an object

weight: the force of gravity on an object

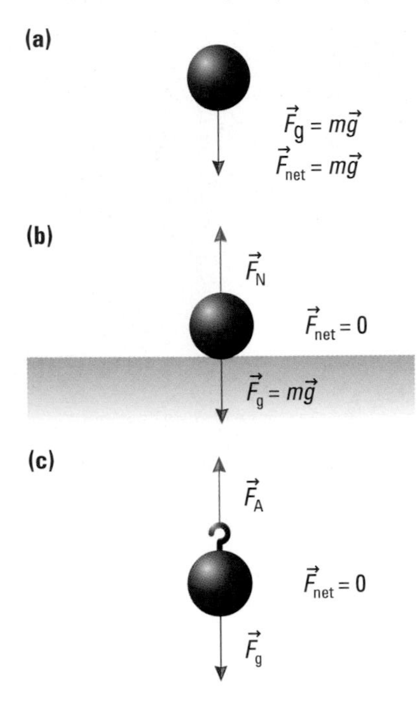

Figure 5
(a) For an object undergoing free fall, the weight, $\vec{F}_g$, is the force which causes the object to accelerate downward.
(b) When an object rests on a horizontal surface, the object's weight causes a downward force on the surface. This force is balanced by an upward force of the surface, the normal force, on the object.
(c) The upward force required to raise an object at a constant speed is equal in magnitude to the object's weight.

Mass and Weight

Weight is a commonly used term that is often confused with mass. In daily transactions, weight and mass may be used interchangeably, but they are quite different things.

Mass is the quantity of matter in an object. As long as the amount of matter in an object remains the same, its mass stays the same. Mass is measured using a balance, a device that compares an unknown mass to a standard (the kilogram, for example). The label on a kilogram of ground meat should say "mass: 1 kg."

Weight is the force of gravity on an object. Thus, we will use the same symbol for weight as for the force of gravity, $\vec{F}_g$. Being a force, weight is measured in newtons, not in kilograms. In the laboratory, a spring scale or a force sensor can be used to measure weight. Since the force of gravity can vary, the weight of an object will also vary according to its location. The magnitude of an object's weight is equal in magnitude to the force required to hold the object or to raise or lower the object without acceleration (**Figure 5**).

If you were to travel to the Moon, you would find that the force of gravity there is only about $\frac{1}{6}$ of that on Earth. That is, the gravitational field strength on the Moon is only about 1.6 N/kg [↓].

Sample Problem 2

Calculate the weight of a fully outfitted astronaut who has a mass of 150 kg on the Moon.

Solution

$m = 150$ kg
$\vec{g} = 1.6$ N/kg [↓]
$\vec{F}_g = ?$

$$\vec{F}_g = m\vec{g}$$
$$= (150 \text{ kg})(1.6 \text{ N/kg } [\downarrow])$$
$$\vec{F}_g = 2.4 \times 10^2 \text{ N } [\downarrow]$$

The astronaut's weight on the Moon is 2.4×10^2 N [↓].

Practice

Understanding Concepts

6. To summarize the differences between mass and weight, set up and complete a table using these titles:

Quantity	Definition	Symbol	SI unit	Method of measuring	Variation with location

7. (a) What is the weight of a 19-kg curling stone?
   (b) What force is required to raise the curling stone without acceleration?
8. Calculate the weight of a 54-kg robot on the surface of Venus where the gravitational field strength is 8.9 N/kg [↓].

Answers

7. (a) 1.9×10^2 N [↓]
   (b) 1.9×10^2 N [↑]
8. 4.8×10^2 N [↓]

Variation in Earth's Gravitational Field Strength

From the equation $\vec{F}_g = m\vec{g}$, it is evident that the weight of an object of known mass changes if $\vec{g}$ changes. Although the average value of $\vec{g}$ on Earth, to two significant digits, is 9.8 m/s^2 [↓], this quantity may change slightly depending on the object's location.

Earth is not a sphere. Since it rotates on an axis through the North and South Poles, it bulges out slightly at the equator. A sled located at the North Pole is about 21 km closer to the centre of Earth than a boat floating in the ocean at the equator. As a result, the gravitational field strength at the North Pole is slightly greater than that at the equator, as shown in **Table 2**.

Table 2 Variation in $\vec{g}$ with Latitude (at sea level)

Latitude (°)	$\vec{g}$ (N/kg [↓])	Distance from Earth's centre (km)
0 (equator)	9.7805	6378
15	9.7839	6377
30	9.7934	6373
45	9.8063	6367
60	9.8192	6362
75	9.8287	6358
90 (North Pole)	9.8322	6357

The altitude, or height above sea level, also affects the gravitational field strength, as shown in **Table 3**. Again, the greater the distance from Earth's centre, the smaller the gravitational field strength.

Table 3 Variation in $\vec{g}$ with Altitude (at similar latitudes)

Location	Latitude (°)	$\vec{g}$ at sea level (N/kg [↓])	Altitude (m)	$\vec{g}$ (N/kg [↓])
Toronto	44	9.8054	162	9.8049
Mount Everest	28	9.7919	8848	9.7647
Dead Sea	32	9.7950	–397	9.7962

DID YOU KNOW ?

Gravitational Attraction of Polar Ice Caps

The huge masses of major polar ice caps, such as the one covering much of Greenland, have a large gravitational attraction on nearby ocean waters. As polar ice caps melt due to global warming, their gravitational attraction decreases, so they are unable to keep as much water near them. Thus, sea levels farther from the melting ice caps will rise more than sea levels near the ice caps.

Practice

Understanding Concepts

9. (a) Calculate Earth's force of gravity on each of two steel balls of masses 6.0 kg and 12.0 kg.
   (b) If the force of gravity on the 12-kg ball is greater than that on the other ball, why do the two balls accelerate at the same rate when dropped?
10. (a) What is your mass in kilograms?
    (b) Calculate your own weight at sea level at
        (i) the equator
        (ii) the North Pole

Answers

9. (a) 59 N [↓]; 1.2×10^2 N [↓]

11. A vegetable vendor sets up a stall in an elevator of a tall building. The vendor uses a spring scale to measure the weight of vegetables in newtons.
    (a) Under what conditions of changing velocity would it be advantageous for you to buy from the vendor? for the vendor to sell to you? Explain your answers.
    (b) Under what condition would neither of you have an advantage?

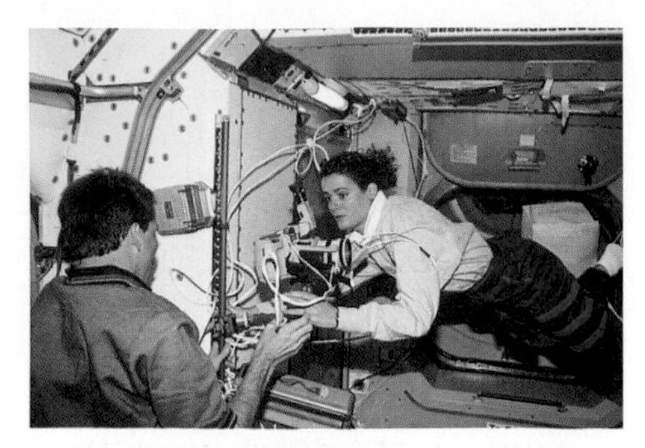

Figure 6
Astronauts aboard the International Space Station appear to be floating as they move along with the station. Although this sensation is commonly called weightlessness, a better term is "constant free fall."

Free Fall, Weightlessness, and Microgravity

Julie Payette, Canada's first astronaut aboard the International Space Station in May 1999, experienced a sensation often referred to as "weightlessness" or "microgravity" while on the station orbiting Earth (**Figure 6**). During this motion, astronauts appear to be floating in space, so they appear weightless. However, the terms *weightless* and *microgravity* are misleading because they do not explain what is really happening.

The weight of an object is defined as the force of gravity acting on it. At the altitude where the space station orbits Earth, the force of gravity acting on the astronauts and the station is about 90% of what it is on Earth's surface. With such a large force, "microgravity" is not a good description, since micro refers to a very small amount of gravity. Without this large force of gravity, the space station would not have stayed in its orbit around Earth.

Figure 7
Newton's reasoning to explain constant free fall: Only an object moving at a high enough speed will fall at the same rate as the surface of Earth curves.

It was Newton who first saw the connection between falling objects, projectiles, and satellites in orbit. The virtual experiment he conducted is still as good an explanation of this relationship as any others. Imagine a large cannon on the top of a high mountain firing cannon balls horizontally at greater and greater speeds, as illustrated in **Figure 7**. At first the cannon balls fall quickly to the ground. As their initial speeds increase, the cannon balls travel farther and farther. At very high speeds, a new factor affects the distance. Since Earth is round, the surface of landing curves downwards. The cannon balls must travel down and around before landing. When a certain critical speed is reached, the cannon ball's path curves downward at the same rate as Earth's curvature. The orbiting cannon ball is then in constant free fall, always falling toward Earth, but never landing. The space station and everything inside undergo the same type of accelerated motion as the imaginary cannon ball does.

When you jump from a height to the ground, you momentarily experience free fall. Virtually no force is acting upward on you as the force of gravity pulls you down toward Earth. However, the time interval is so short that the sensation does not really have time to take effect. The interval of free fall or "weightlessness" is extended for astronauts during training. The astronauts are placed in the cargo hold of a large military transport plane. First, the plane climbs to a high altitude. Then, it dives to gain speed. Next, it pulls into a large parabolic path and into a dive. It is during the upper part of the parabolic path (as in **Figure 8**) that the plane is in free fall for about 30 s, and the astronauts can "float" around inside the cargo hold, experiencing the sensation of "weightlessness."

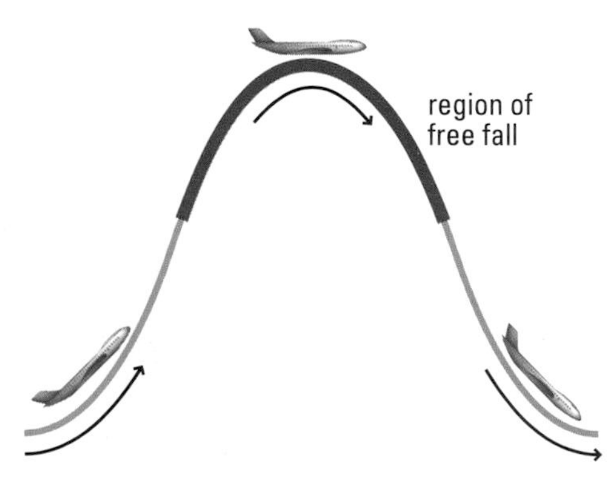

Figure 8
As the airplane repeats the curved path several times, the astronauts in training experience short periods of free fall when they float around inside the airplane.

Practice

Understanding Concepts

12. (a) In the SI metric system, what does the prefix "micro" mean?
    (b) Use your answer in (a) to explain why it is misleading to call the forces acting on an astronaut in orbit microgravity.

13. In **Figure 7,** what would happen to a cannon ball travelling at a speed higher than any of those shown?
14. If you wanted to reduce the number of forces acting on your body to a minimum, what would you have to do?

SUMMARY Gravitational Force on Earth's Surface

- Earth's gravitational field is the space surrounding Earth in which the force of gravity has an effect.
- On Earth's surface, the average gravitational field strength is $\vec{g} = 9.8$ N/kg [↓].
- Mass is the quantity of matter (measured in kilograms) and weight is the force of gravity (measured in newtons and determined using the equation $\vec{F}_g = m\vec{g}$).
- The magnitude of $\vec{g}$ decreases as the distance from Earth's centre increases, so this magnitude is smaller at higher altitudes, and it is greater at the poles than at the equator because Earth is slightly flattened at the poles.
- An object orbiting another object maintains its orbit by constantly free falling toward the central body. For example, the International Space Station and its contents constantly free fall toward Earth.

DID YOU KNOW ?

Weight in Orbit

If an astronaut, while in orbit, stood on a bathroom scale to determine his or her weight, the scale would read zero—the scale would be accelerating toward Earth at the same rate as the astronaut. To help understand this, imagine what a bathroom scale would read if you were standing on it while you were on a vertical-drop ride at an amusement park and the ride was undergoing temporary free fall. The result would be the same. Any time a person is in free fall, the sensation that the person experiences is similar to what the astronauts experience while their spacecraft is in orbit.

Section 3.1 Questions

Understanding Concepts

1. In a videotape of the Apollo astronauts on the Moon, it seems that the astronauts are moving about in slow motion. Explain why this is the case.
2. Why is the gravitational field strength halfway up Mount Everest the same as at sea level at the equator?
3. Why is the gravitational field strength at the South Pole less than that at the North Pole? (*Hint:* Look at the globe to see which pole is at sea level and which is on a thick ice shelf.)
4. Suppose you wanted to make some money by purchasing precious materials such as gold at one altitude and selling them for the same price in dollars per newton at another altitude. Describe the conditions that would favour your "buy high and sell low" strategy.
5. What is the force of gravity at Earth's surface on each of the following masses?
   (a) 75.0 kg (b) 454 g (c) 2.00 t
6. What is the weight of each of the following masses at Earth's surface?
   (a) 25 g (b) 102 kg (c) 12 mg
7. Use these magnitudes of the forces of gravity at Earth's surface to determine the masses of the objects on which they act.
   (a) 0.98 N (b) 62 N (c) 44.5 MN
8. Copy **Table 4** into your notebook and complete it for a 57-kg instrument on each planet.

(continued)

DID YOU KNOW ?

Students on a Training Flight

High school students are sometimes allowed to travel aboard NASA's astronaut-training airplane (the KC-135 aircraft). Follow the links for Nelson Physics 11, 3.1 to find information about a student flight conducted on April 13, 1999.

GO TO **www.science.nelson.com**

Table 4

Planet	$\vec{F}_g$ (N [↓])	$\vec{g}$ (N/kg [↓])
(a) Mercury	188	?
(b) Venus	462	?
(c) Jupiter	?	26

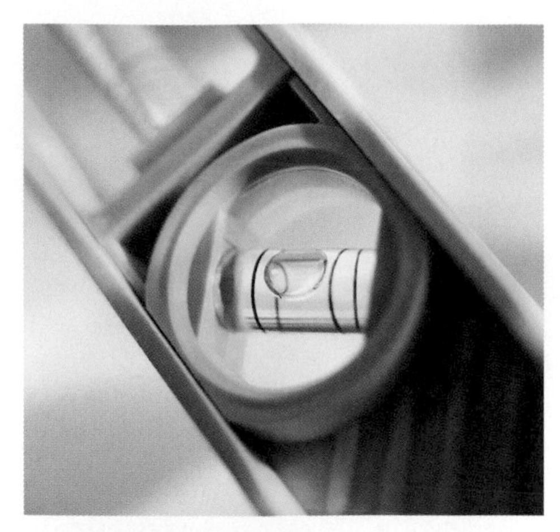

Figure 9
A level

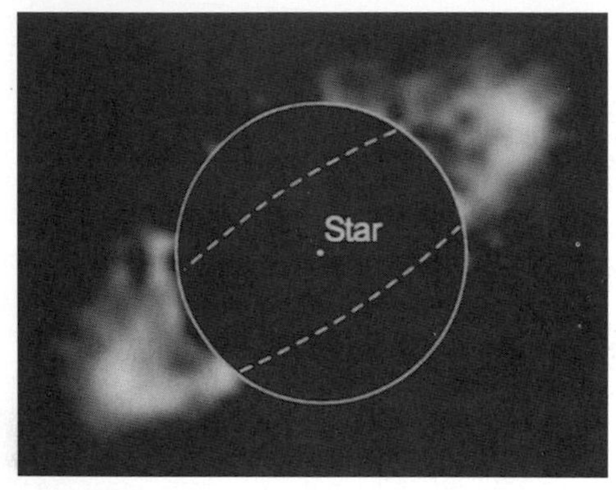

Figure 1
Astronomers are constantly searching for evidence of planets orbiting stars. This photograph shows a planet likely forming around a star in our galaxy.

DID YOU KNOW ?

Universal Truths

Newton's law of universal gravitation was a great breakthrough in scientific thinking. Just as a cannon ball in orbit around Earth is constantly free falling, so is the Moon in its orbit around Earth and all the planets in their orbits around the Sun. Not long before Galileo and Newton, people did not even realize that planets travelled around the Sun. Newton's ideas explained what was actually observed.

Newton's law of universal gravitation, formulated in about 1670, was among the first indications that there existed "universal" truths. The laws of nature could be applied everywhere, which was a concept not welcomed by some of the authorities in those days.

9. An astronaut on the surface of Mars finds that a rock accelerates at a magnitude of 3.7 m/s^2 when it is dropped. The astronaut also finds that a force scale reads 180 N when the astronaut steps on it.
   (a) What is the astronaut's mass as determined on the surface of Mars?
   (b) What would the force scale read if the astronaut stepped on it on Earth?

Applying Inquiry Skills

10. A 500.0-g mass is hung on a force scale at a location where the magnitude of the gravitational field strength is 9.8 N/kg, but the scale reads 4.7 N.
    (a) What should the scale reading be?
    (b) What could have accounted for the error in the measurement?
    (c) Determine the percent error of the measurement.

Making Connections

11. The photograph in **Figure 9** shows a device called a level, which is a common alternative to a plumb bob. Describe how the level can be used to determine vertical and horizontal lines.
12. Some science fiction movies and television programs show people walking around in a spacecraft the way we walk around a room. How scientifically accurate is this representation?

3.2 Universal Gravitation

Are we alone in the universe? How do astronomers search for intelligent life elsewhere? First, they look for evidence of planets travelling around stars that appear to be relatively close (**Figure 1**). They search for these planets by applying what they know about universal gravitation. Perhaps, one day, you, or a physics student in your class who has become an astronomer, will announce a discovery of life elsewhere in the universe.

Newton's Law of Universal Gravitation

One day, so the story goes, while sitting under an apple tree, Isaac Newton observed a falling apple. According to science historians, this seemingly unimportant event led to Newton's formation of a fundamental law of nature involving gravitational forces. The apple fell because it was pulled toward Earth by the force of gravity. Why then, Newton hypothesized, could it not be concluded that the Moon stays in its orbit around Earth because the Moon, too, is "pulled" toward Earth by the force of gravity?

After consulting known data about the Moon's orbit around Earth and determining how the force of gravity depends on other variables, Newton devised the law of universal gravitation. Considering only the magnitudes of the forces, the relationships he discovered can be stated using these symbols:

F_G is the force of gravitational attraction between any two objects.
m_1 is the mass of one object.
m_2 is the mass of a second object.
d is the distance between the centres of the two objects.
(Objects are assumed to be spherical.)

If m_2 and d are constant, $F_G \propto m_1$ (direct variation).
If m_1 and d are constant, $F_G \propto m_2$ (direct variation).

If m_1 and m_2 are constant, $F_G \propto \frac{1}{d^2}$ (inverse square variation).

Thus, $F_G \propto \frac{m_1 m_2}{d^2}$ (joint variation).

Therefore, $F_G = \frac{Gm_1 m_2}{d^2}$, where G is a constant other than zero (**Figure 2**).

Newton's **law of universal gravitation** is therefore stated as follows:

law of universal gravitation: The force of gravitational attraction between any two objects is directly proportional to the product of the masses of the objects, and inversely proportional to the square of the distance between their centres.

Law of Universal Gravitation

The force of gravitational attraction between any two objects is directly proportional to the product of the masses of the objects, and inversely proportional to the square of the distance between their centres.

In considering the law of universal gravitation, it is important to notice that there are two equal but opposite forces present. For example, Earth pulls on the Moon and the Moon pulls on Earth with a force of equal magnitude. At Earth's surface, Earth pulls down on a 1.0-kg mass with a force of magnitude 9.8 N, and the 1.0-kg mass pulls upward on Earth with a force of magnitude 9.8 N. (Recall Newton's third law of motion.)

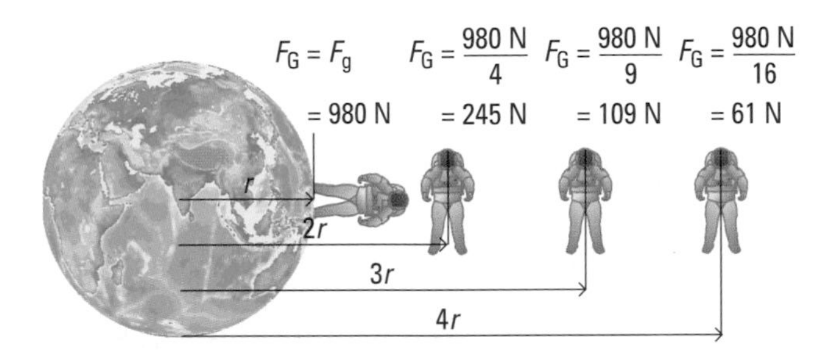

Figure 2
The force of gravitational attraction between two objects is inversely proportional to the square of the distance between their centres. In this case, the distance is measured between the centre of Earth and a fully outfitted astronaut whose mass is 1.0×10^2 kg.

Notice also the implications of the inverse square relationship. If an object is moved seven times as far from the centre of Earth as Earth's surface, the force of gravity will be reduced to $\frac{1}{7^2}$, or $\frac{1}{49}$ of what it is at Earth's surface. The forces of attraction decrease rapidly as objects move apart. On the other hand, there is no value of d, no matter how large it can be, that would reduce the forces of attraction to zero. Every object in the universe exerts a force of attraction on all other objects, near or far.

Finally, it is important to realize that the equation for Newton's law of universal gravitation cannot be applied to objects of all shapes. The equation is valid only for two spheres (such as Earth and the Sun), for two "particles" whose sizes are much smaller than the distance separating them (for example, two people 1.0 km apart), or for a "particle" and a sphere (such as you standing on Earth).

DID YOU KNOW ?

The Inverse Square Relationship

The inverse square relationship applies to many phenomena, including electric forces and light intensity. For example, the intensity of light from a point source is inversely proportional to the square of the distance from the source.

Practice

Understanding Concepts

1. Plot a graph of the force of gravity as a function of the distance from the centre of Earth, using the data in **Figure 2**. Compare this graph to the one you drew in the chapter opener activity.
2. Explain why "universal" is a good word to describe gravitation.

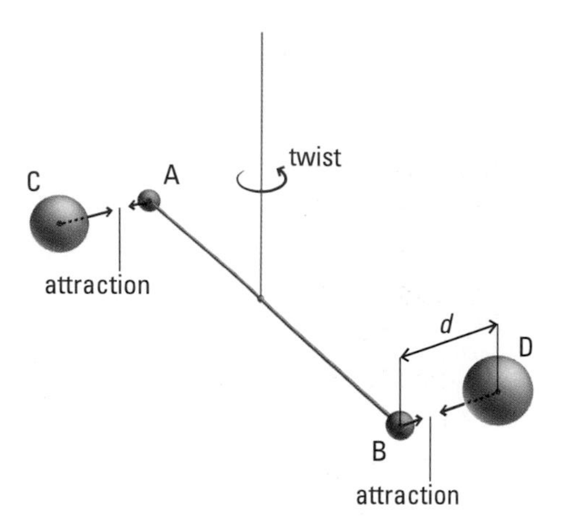

Figure 3
In Cavendish's experiment, the gravitational attraction between the sets of balls caused the fine wire to twist a measurable amount.

Determining the Universal Gravitational Constant

The value of the constant, G, in Newton's gravitational equation is extremely small; Newton determined it by calculation. Verification of this value by experiment did not occur until more than a century after the law was formulated. Then, Henry Cavendish (1731–1810), an English physicist, performed an experiment using a setup similar to that in **Figure 3.** When balls C and D were brought close to balls A and B, the force of gravitational attraction caused the double ball system (A and B) to rotate a measured amount. Since Cavendish had previously determined the force needed for a certain twist, he could now find the force of attraction between A and C and between B and D. Using this technique, he calculated the value of the constant in Newton's equation, which now has the symbol G, to be 6.67×10^{-11} N•m^2/kg^2.

Thus, Newton's law of universal gravitation in equation form is

$$F_G = \frac{Gm_1m_2}{d^2}$$

where G is 6.67×10^{-11} N•m^2/kg^2
m_1 and m_2 are the masses of the objects in kilograms
d is the distance between the centres of the objects in metres

DID YOU KNOW ?

Another Great Triumph

Cavendish's experimental determination of G was a great triumph for science. It indicates how strong gravitational attraction is as a fundamental force of nature, and its magnitude influences the fate of the entire universe. Knowing G allowed scientists to calculate the mass of Earth and other heavenly bodies (see Practice question 4 later in this section) and helped scientists make judgements about Earth's interior.

Sample Problem

Calculate the magnitude of the force of gravitational attraction between two metal balls used in the men's shot put competitions, each of mass 7.26 kg and whose centres are separated by a distance of 24 cm.

Solution

$G = 6.67 \times 10^{-11}$ N•m^2/kg^2

$m_1 = 7.26$ kg
$m_2 = 7.26$ kg
$d = 0.24$ m
$F_G = ?$

$$F_G = \frac{Gm_1m_2}{d^2}$$

$$= \frac{(6.67 \times 10^{-11} \text{ N•m}^2/\text{kg}^2)(7.26 \text{ kg})(7.26 \text{ kg})}{(0.24 \text{ m})^2}$$

$$F_G = 6.1 \times 10^{-8} \text{ N}$$

The magnitude of the force of attraction is 6.1×10^{-8} N, which is an extremely small force.

Obviously, the force of attraction is insignificant if both of the objects have a low mass. It takes the whole Earth, with a mass of 5.98×10^{24} kg, to exert a force of magnitude 9.8 N on a 1.0-kg mass. The gravitational force operating between relatively small objects, such as two cars on the road, is not worth considering.

Practice

Understanding Concepts

Note: "Force" or "weight" in the following questions refers to the magnitude of the force or the magnitude of the weight.

3. Calculate the force of gravitational attraction between a concrete ball ($m_1 = 2.0 \times 10^3$ kg) and a steel ball ($m_2 = 2.0 \times 10^4$ kg) whose centres are separated by a distance of 4.0 m.
4. The mass of Earth can be calculated using the fact that the weight of an object (in newtons) is equal to the force of gravity between the object and Earth. Given that the radius of Earth is 6.4×10^6 m, determine its mass.
5. Find the force of attraction between Earth and the Moon, using the following data:

   mass of the Moon = 7.35×10^{22} kg

   mass of Earth = 5.98×10^{24} kg

   average distance between centres = 3.84×10^8 m
6. The force of gravity at Earth's surface on an astronaut is 634 N. What is the force of gravity on the same person at each of the following distances, in multiples of Earth's radius, from the centre of Earth?
   (a) 2 (b) 5 (c) 10 (d) 17.2
7. As a rocket carrying a space probe accelerates away from Earth, the fuel is being used up and the rocket's mass becomes less. When the mass of a rocket (and its fuel) is M and the distance of the rocket from Earth's centre is $1.5r_E$, the force of gravitational attraction between Earth and the rocket is F_1. When some fuel is consumed causing the mass to become $0.5M$ and the distance from Earth's centre is $2.5r_E$, the new gravitational attraction is F_2. Determine the ratio of F_2 to F_1. The symbol r_E is Earth's radius.

Answers

3. 1.7×10^{-4} N
4. 6.0×10^{24} kg
5. 1.99×10^{20} N
6. (a) 158 N
   (b) 25.4 N
   (c) 6.34 N
   (d) 2.14 N
7. 0.18 : 1

Gravitational Attraction in the Solar System and Beyond

Learning how the gravitational attraction forces influence objects in space, both in the solar system and beyond it, allows us to answer many questions not answered by scientists in the past. What causes the ocean tides? How do scientists decide where satellites should orbit Earth? What will happen to the universe in the future? Let's look at these questions in more detail.

The force that keeps the Moon in its orbit around Earth is the force of gravity. The reaction force of the Moon pulling on Earth is evident in the formation of ocean tides on Earth. At any given time, ocean waters on two opposite sides of Earth are at high tide while the waters on the other sides are at low tide (**Figure 4**). Since Earth rotates about its own axis once every 24 h and since the

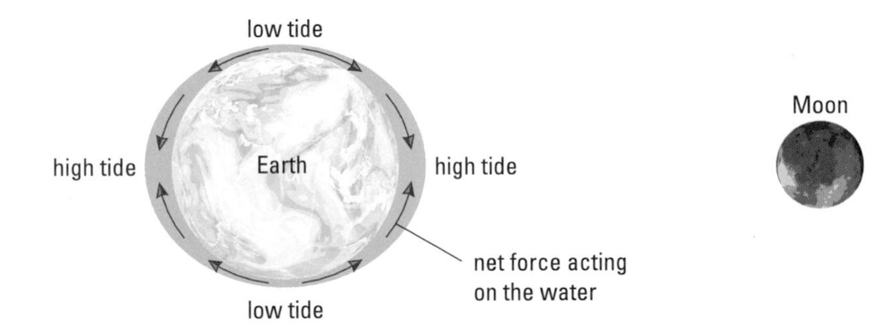

Figure 4
Ocean tides are caused by a complex set of forces involving the Moon, Earth, the Sun, and the motions of these bodies. The diagram shows the result of these forces on an exaggerated layer of water. High tides result when water on Earth is made to bulge outward on the sides of Earth in line with the Moon. The other sides of Earth experience low tide at this time.

Moon rises in the sky about 50 min later each day, the change of tides from one high tide to the next high tide occurs approximately every 12 h 25 min. The tides in the Bay of Fundy in Nova Scotia are among the highest in the world. **Figure 5** is a photograph taken during low tide at the Bay of Fundy.

Figure 5
The difference between low and high tides along the coast of the Bay of Fundy is very evident in this photograph.

Like the Moon, communications satellites orbit Earth. Canada is among the world's leaders in placing such satellites in a 24-h orbit directly above the equator. This type of orbit is described as *geosynchronous* because the period of revolution coincides with Earth's rotation. Each satellite must remain within a predetermined space so that messages can be sent from and received by communication dishes on Earth. **Figure 6** illustrates the principle of satellite communication.

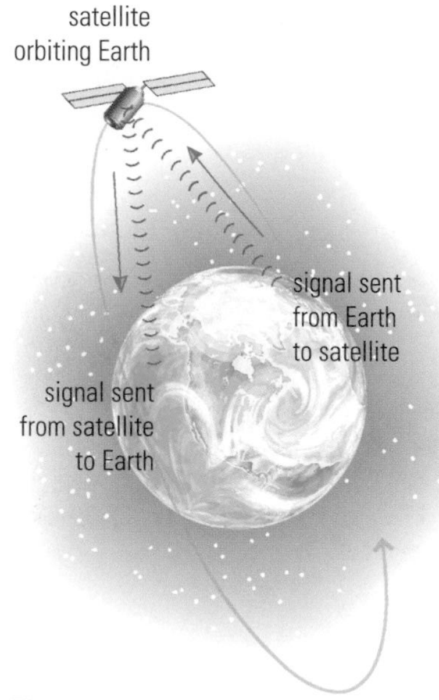

Figure 6
The force of gravity keeps the communications satellite in a stable orbit around Earth.

Of course, Earth and the other eight planets of the solar system are all satellites of the Sun, held in orbit by the Sun's force of gravity. So are the objects called comets. A comet is a dense heavenly body that has an elongated orbit around the Sun. As a comet is attracted closer to the Sun, a huge current of charged particles called the solar wind causes the comet's head to become brilliant (**Figure 7**). Probably the most famous of all comets is Halley's comet, which has a period of revolution around the Sun of about 76 years. It is named after Edmund Halley (1656–1742). Halley, who lived in England, sometimes worked with Sir Isaac Newton. He charted the comet's path in 1682. The comet's appearance for several months during 1985 and 1986 was predicted and anticipated by people all over the world.

Figure 7
This photograph of Comet Hale-Bopp, was taken in 1997. The comet's tail can stretch for millions of kilometres. (The word "comet" is derived from the Greek word *kometes*, which means "wearing long hair.")

Between Mars and Jupiter, there are countless chunks of material called *asteroids*. Like planets and comets, these bodies orbit the Sun. However, as their motion is influenced by the force of gravity of larger bodies, especially Jupiter, their orbits can change. From time to time, an asteroid passes closer to Earth than in previous orbits.

The force of gravity exists throughout the universe. Our Sun is but one ordinary star in our galaxy, the Milky Way Galaxy, which consists of hundreds of billions of stars orbiting the galaxy's centre. Our galaxy is only one of the millions of galaxies that make up the known universe and influence each other through the force of gravity. Studying the properties of stars, galaxies, and other components of the universe helps scientists discover more about the origin and possible future of our Earth.

Practice

Understanding Concepts

8. Assume that in a coastal village a low tide occurs at 04:20. Predict the times for the next two high tides and two low tides. (Your answer may not necessarily coincide with the real-life situation because tides are also affected by local conditions.)
9. Determine the number of times Halley's comet has come near Earth since Edmund Halley made his discovery. Predict the year of its next passing.

Answers

8. High tides.
   Low tides:
9. 4 times; 2062

SUMMARY Universal Gravitation

- Newton's law of universal gravitation applies to all objects in the universe. It is an example of an inverse square relationship in which the force of gravitational attraction between any two objects in the universe is inversely proportional to the square of the distance between the objects.
- For particles or spherical objects, the equation for the law of universal gravitation is $F_G = \frac{Gm_1m_2}{d^2}$, where G is 6.67×10^{-11} N•m²/kg².
- The concept of universal gravitation is applied in astronomy, space science, and satellite technology.

DID YOU KNOW ?

The Ultimate Gravitational Force

A "black hole" is an extremely dense object that exerts a huge gravitational force on relatively close bodies. The force is so powerful that nothing, not even light, can escape from the object, which explains why the object is called black. Evidence of black holes in the universe comes as material surrounding a black hole gets sucked into it, emitting radiation as it does so.

Section 3.2 Questions

Understanding Concepts

1. What is the magnitude of the force of gravitational attraction between a 55-kg student and a 65-kg student, whose centres are 1.0 km apart?
2. At a certain instant, a 255-kg meteoroid moving toward Earth is located 6.75 Mm from Earth's centre. What is the magnitude of the force of gravitational attraction between the two bodies? (Earth's mass is 5.98×10^{24} kg.)
3. Is it possible for a body to exist somewhere in the universe that has no forces whatsoever acting on it? Explain your answer.

Applying Inquiry Skills

4. Describe how you would use a globe to illustrate the orbit of a geosynchronous communications satellite.

Making Connections

5. Research the meanings and causes of "spring tides" and "neap tides." Draw diagrams to show what you discover.
6. Astronomers have used equations derived by Sir Isaac Newton to estimate that the Milky Way Galaxy has a mass of approximately 4×10^{41} kg. If the mass of our Sun is 2×10^{30} kg, how many stars are there in our galaxy? Assume for this question that all stars have the same average mass and that the masses of planets are negligible.

(continued)

7. A communication satellite has small engines aboard that are used to keep the satellite within its proper space. Use at least one of Newton's laws to explain the use of the engines.
8. Find out about the asteroid impact in the Yucatan Peninsula in Mexico about 65 million years ago, which is believed to have caused the extinction of dinosaurs and numerous other species. Also find out about the chances that an asteroid or a huge meteor could crash into Earth in the future. Follow the links for Nelson Physics 11, 3.2.
   (a) How does an asteroid impact relate to gravitational forces?
   (b) What might possibly be done to prevent an asteroid impact in the future?
   (c) How do the research scientists apply their knowledge of local gravitational fields to determine the shape of the impact crater?

GO TO www.science.nelson.com

3.3 The Effects of Friction

Figure 1
To reduce the friction between skis and the snow, skiers choose a wax that is designed for use within a specific temperature range. At temperatures of, say, −10°C, the ski wax is more slippery than the wax used at higher temperatures.

Friction is part of everyday life. Removing a fried egg from a frying pan is easier if the pan has a non-stick surface. You find running easier on a dry sidewalk than on a skating rink. Skiers can choose different waxes for their skis, depending on the air temperature and characteristics of the snow (**Figure 1**). Cars need friction to speed up, slow down, and go around corners.

Static and Kinetic Friction

Friction resists motion and acts in a direction opposite to the direction of motion. It occurs because of the electrical forces between the surfaces where two objects are in contact. No one would put on a pair of ice skates to try to glide along a concrete sidewalk. The friction between the sidewalk and the skate blades would prevent any skating.

One type of friction, called **static friction**, is the force that tends to prevent a stationary object from starting to move. ("Static" comes from the Greek word *statikos*, which means "causing to stand.") The maximum static friction is called the **starting friction.** It is the amount of force that must be overcome to start a stationary object moving. See **Figure 2(a)**.

static friction: the force that tends to prevent a stationary object from starting to move

starting friction: the amount of force that must be overcome to start a stationary object moving

kinetic friction: the force that acts against an object's motion in a direction opposite to the direction of motion

In certain circumstances static friction is useful; in others, it is not. A person trying to turn a stubborn lid on a jam jar appreciates the extra friction that comes with using a rubberized cloth between the lid and the hand. However, someone attempting to move a heavy filing cabinet across a floor does not appreciate static friction.

Once the force applied to an object overcomes the starting friction, the object begins moving. Then, moving or kinetic friction replaces static friction. **Kinetic friction** is the force that acts against an object's motion in a direction opposite to the direction of motion. For horizontal motion, if the applied force has the same magnitude as the kinetic friction, the moving object will maintain uniform velocity. See **Figure 2(b)**.

Different types of kinetic friction have different names, depending on the situation. *Sliding friction* affects a toboggan; *rolling friction* affects a bicycle; and *fluid friction* affects a boat moving through water and an airplane flying through air.

Practice

Understanding Concepts

1. Compare and contrast starting friction and kinetic friction, giving an example of each.
2. Give examples of friction you have experienced, besides the ones already given, that are
   (a) sliding (b) rolling (c) fluid
3. What type of friction is air resistance? Give two examples of it.

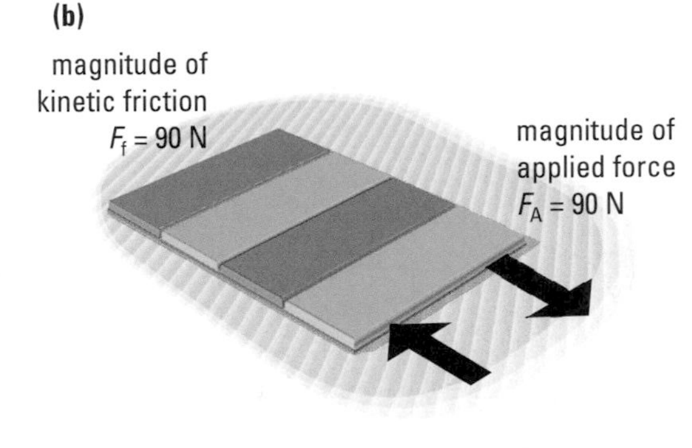

Figure 2
(a) Starting friction must be overcome before an object begins moving.
(b) Kinetic friction occurs with a moving object. In general, kinetic friction between two surfaces is less than starting friction between the same surfaces.

Controlling Friction

About 4500 years ago, the Egyptians built enormous pyramids using huge stone blocks that were difficult to move by sliding. The Egyptians placed logs beneath the blocks to push them and move them. By doing this, people were taking advantage of the fact that rolling friction is much less than sliding friction. Modern technology uses the same principles as the Egyptians did, though in a more sophisticated way. We try to reduce undesirable friction for many reasons. For instance, all machines have moving parts that experience friction during operation. Friction can wear out the machines, reduce efficiency, and cause unwanted heat. (If you rub your hands together vigorously, you can feel the heat produced by friction.) Excess friction in machines can be overcome by making surfaces smooth, using materials with little friction, lubricating with grease or oil, and using bearings.

Bearings function on the principle of the rolling logs used by the Egyptians to move stones. A *bearing* is a device containing many rollers or balls that reduce friction while supporting a load (**Figure 3**). Bearings change sliding friction into rolling friction, reducing friction by up to 100 times.

Ways of reducing undesirable friction in other situations are also common. The wax applied to skis mentioned earlier reduces sliding friction. A layer of air between a hovercraft and the water reduces fluid friction in a manner similar to the use of air pucks and the linear air track in a physics laboratory (**Figure 4(a)**). A human joint is lubricated by *synovial fluid* between the layers of cartilage lining the joint. The amount of lubrication provided by synovial fluid increases when a person moves, giving an excellent example of the efficiency of the human body (**Figure 4(b)**). In fact, our lubrication systems work so well that it is difficult for technologists to design artificial joints that function to the same standard.

Figure 3
Ball bearings are used to reduce friction in a wheel.

(a)

(b)

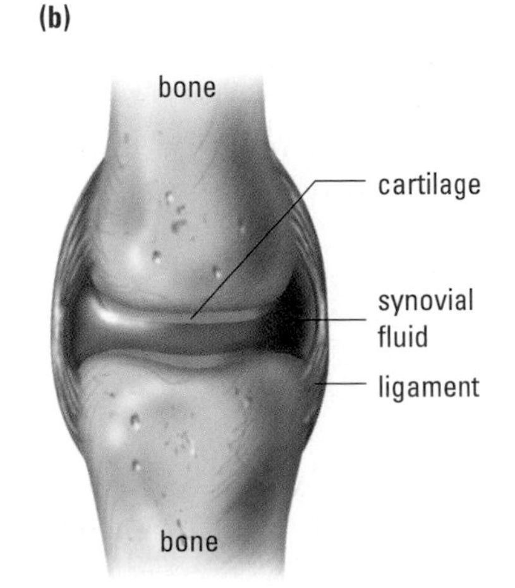

Figure 4
Reducing undesirable friction
(a) This hovercraft carries cars across the English Channel.
(b) A typical joint in the human body

Figure 5
Tight-fitting stones of a large Inca wall near Cuzco, Peru

It has been pointed out that, although friction is often undesirable, it can be useful. Consider, for example, a problem encountered by the Incas, who dominated a large portion of South America before the Europeans arrived in the 16th century. South America has many earthquake zones and, of course, buildings have a tendency to crumble during an earthquake. To help overcome this difficulty, the Inca stonemasons developed great skill in fitting building stones together very tightly so that a great deal of sliding friction would help hold their buildings together, even during an earthquake. **Figure 5** shows an example of the skill of the Incas. Technological applications of useful friction are presented in the next section.

Practice

Making Connections

4. Explain each of the following statements, taking into consideration the force of friction.
   (a) Streamlining is important in the transportation industry.
   (b) Friction is necessary to open a closed door that has a doorknob.
   (c) A highway sign reads, "Reduce speed on wet pavement."
   (d) Screwnails are useful for holding pieces of wood tightly together.

INQUIRY SKILLS

- ○ Questioning
- ● Hypothesizing
- ● Predicting
- ● Planning
- ● Conducting
- ● Recording
- ● Analyzing
- ● Evaluating
- ● Communicating

Investigation 3.3.1

Factors That Affect Friction

Since this is the last investigation in Unit 1 and you have had lots of day-to-day experience with friction, you are expected to design, perform, and report on your own friction investigation. To determine the magnitude of the kinetic friction, you can measure the horizontal push or pull required to keep an object moving with uniform velocity on a horizontal surface. (You should be able to explain, using Newton's laws, why a force greater than this push or pull would cause accelerated motion.)

If possible, use a force sensor connected to a computer interface to measure force rather than a spring scale. The force sensor obtains several readings per second and the accompanying software yields an average force, which smooths out the "peaks" and "valleys" of the actual force.

Question

What are the factors that affect the force of friction between two objects or materials?

Hypothesis/Prediction

(a) In your group, discuss your hypothesis and prediction. Then, write them out.

Design

There are several factors to test to determine how they affect friction, and by how much. Use the following questions to decide what variables you want to test.

- To what extent does the type of friction (starting, sliding, rolling) affect its magnitude?
- Does an object's mass influence friction?
- How does the contact area affect sliding friction (**Figure 6**)?
- How do the types of surfaces in contact affect friction?

(b) In your group, decide how you will test, analyze, and report on your variables. Discuss safety precautions, such as how you will control moving objects. Write out your procedural steps and prepare any data tables you will need.

Materials

(c) Create a list of materials available that you intend to use, with quantities.

Procedure

1. When your teacher has approved your procedure and the materials that you will use, carry out the procedure.

Analysis

(d) Answer the Question.

(e) Plot a graph to show the relationship between mass and friction for an object moving horizontally under the influence of an applied horizontal force.

(f) Make a general statement that relates friction to the types of materials in contact. Are there any apparent exceptions? Explain your answer.

Evaluation

Evaluate your investigation by considering the following:

(g) the design of your experiment

(h) the equipment you used to measure force and any other quantities

(i) sources of error in your investigation (other than those already mentioned before)

(j) If you were to perform the same investigation again, would you do it differently? Explain.

(a) $\vec{F}_f$ $\vec{F}_A$ (horizontal)

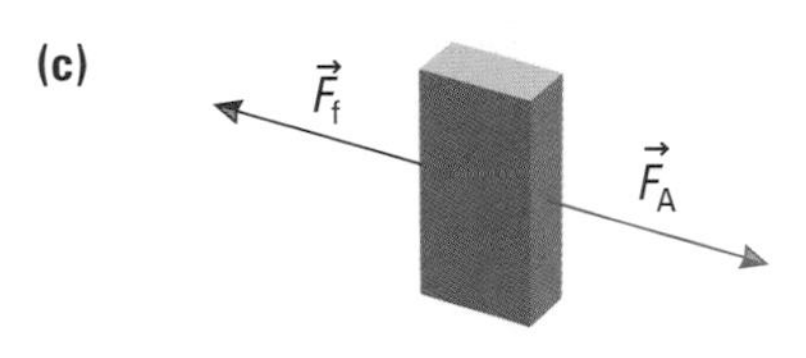

(c) $\vec{F}_f$ $\vec{F}_A$

Figure 6
Determining the force needed to cause uniform velocity
(a) Block lying flat
(b) Same block on its edge
(c) Same block on its end

SUMMARY The Effects of Friction

- Friction acts parallel to two surfaces in contact in a direction opposite to the motion or attempted motion of an object.
- Static friction tends to prevent a stationary object from starting to move. Kinetic friction acts against an object's motion; it is usually less than static friction.
- For an object to maintain uniform velocity, the net force acting on it must be zero, so for an object moving at uniform velocity on a horizontal surface, the applied horizontal force must be equal in magnitude to the kinetic friction.
- Unwanted friction can be reduced by changing sliding to rolling, by using bearings, and by using lubrication.
- The extent to which certain factors affect friction, such as the types of surfaces in contact or the mass of the object, can be determined using a controlled experiment.

DID YOU KNOW?

Teflon: An Amazing Material

The material called Teflon has many uses when low friction is desired, such as in non-stick frying pans. Although two research scientists created this chemical by chance in 1938, its usefulness was not realized until 20 years later. Since Teflon does not stick to any materials, the process used to make it stick onto a frying pan surface is unique: the Teflon is blasted into tiny holes in the surface of the pan where the material sticks well enough for use.

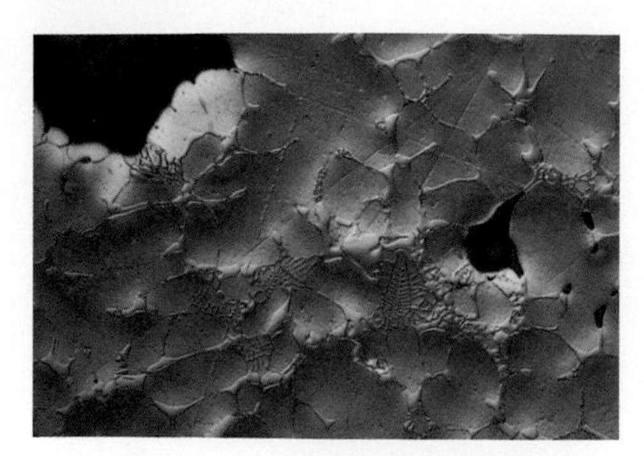

Figure 7
This micrograph of a polished aluminum surface was taken at a magnification of 50×.

Section 3.3 Questions

Understanding Concepts

1. The photograph in **Figure 7** shows a microscopic view of an aluminum surface that appears smooth to the unaided eye. Describe what the photograph reveals about friction.
2. Is friction desirable or undesirable when you tie a knot in a string? Explain.

Applying Inquiry Skills

3. Exercise bikes have a control that allows the rider to adjust the amount of friction in the wheel. Describe how you would perform an experiment to determine the relationship between the setting on the control and the minimum force needed to move the bicycle's pedal.
4. (a) At speeds typical of transportation vehicles (trucks, bikes, boats, airplanes, etc.), how do you think fluid friction depends on the speed?
   (b) Describe a controlled experiment that could be conducted to check your answer in (a).

Making Connections

5. Friction may be a help in some situations and a hindrance in others. Describe two examples for each situation. For those situations where friction is undesirable, what efforts are made to reduce it as much as possible? For those situations where friction is desirable, how is it increased?
6. Explain how you would solve each of the following problems. Relate your answer to friction.
   (a) A refrigerator door squeaks when opened or closed.
   (b) A small throw rug at the front entrance of a home slips easily on the hard floor.
   (c) A picture frame hung on a wall falls down because the nail holding it slips out of its hole.

3.4 Analyzing Motion with Friction

Most frictional forces are complex because they are affected by a number of factors, such as the nature of the materials involved and the size, shape, and speed of the moving object. In this section, we shall focus on the force of friction acting on an object on a horizontal surface experiencing only horizontal forces.

Solving problems involving friction brings together many concepts presented in this unit, including velocity, acceleration, forces, free-body diagrams, Newton's laws of motion, gravitational attraction, and weight. It also relies on different skills you have applied, such as the use of a calculator, graphing, and analyzing experimental results. Thus, it is fitting that you complete this unit by combining the concepts and skills from all three chapters to solve motion problems involving friction.

Coefficient of Friction

coefficient of friction: ratio of the magnitude of friction to the magnitude of the normal force

The **coefficient of friction** is a number that indicates the ratio of the magnitude of the force of friction, F_f, between two surfaces to the magnitude of the force perpendicular to these surfaces. Recall that the magnitude of the force perpendicular to the surface is called the normal force, F_N, so the coefficient of friction is the ratio of F_f to F_N. We will use the Greek letter μ to represent the coefficient of friction.

Thus, $\mu = \frac{F_f}{F_N}$

where F_f is the magnitude of the force of friction, in newtons;
F_N is the magnitude of the normal force pushing the surfaces together, in newtons (it acts at right angles to the surfaces); and
μ is the coefficient of friction (it has no units because it is a ratio of forces).

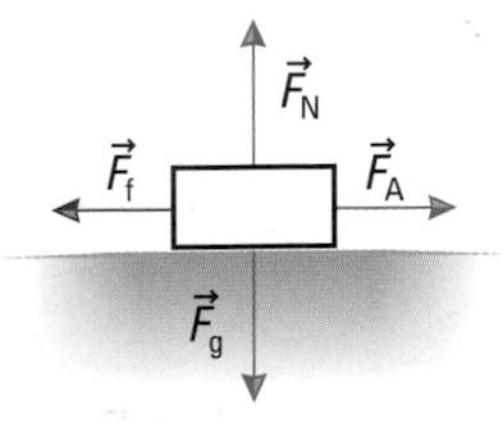

Figure 1
For an object experiencing an applied force on a horizontal surface, four forces are involved. The magnitudes of two of these forces, F_f and F_N, are used to calculate the coefficient of friction.

A rearrangement of this equation will give the equation for calculating the force of friction: $F_f = \mu F_N$ (see **Figure 1**).

In almost all situations, the force needed to start the motion of an object initially at rest is greater than the force needed to keep it going at a constant velocity. This means that the maximum static friction is slightly greater than the kinetic friction, and the coefficients of friction for these situations are different. To account for the difference, two coefficients of friction can be determined. The **coefficient of kinetic friction** is the ratio of the magnitude of the kinetic friction to the magnitude of the normal force. The **coefficient of static friction** is the ratio of the magnitude of the maximum static friction to the magnitude of the normal force. This "maximum" occurs just when the stationary object begins to move. Using "K" for kinetic and "S" for static, the corresponding equations are:

coefficient of kinetic friction: ratio of the magnitude of kinetic friction to the magnitude of the normal force

coefficient of static friction: ratio of the magnitude of the maximum static friction to the magnitude of the normal force

$$\mu_K = \frac{F_K}{F_N} \quad \text{and} \quad \mu_S = \frac{F_S}{F_N}$$

Determining the coefficients of friction for various surfaces can be done only experimentally. Even with careful control of other variables, results obtained are often inconsistent. Consider, for example, performing an experiment to determine the coefficient of kinetic friction of steel on ice using skates at a hockey arena. Using a skate with a sharp blade on very clean smooth ice, the coefficient of friction may be, say, 0.010. However, using a different skate on a slightly rougher ice surface, the coefficient of friction will be higher, 0.014, for example. **Table 1** indicates typical coefficients of kinetic friction and static friction for sets of common materials in contact based on empirical observations.

Table 1 Approximate Coefficients of Kinetic Friction and Static Friction

Materials in contact	μ_K	μ_S
oak on oak, dry	0.30	0.40
waxed hickory on dry snow	0.18	0.22
steel on steel, dry	0.41	0.60
steel on steel, greasy	0.12	
steel on ice	0.010	
rubber on asphalt, dry	1.07	
rubber on asphalt, wet	0.95	
rubber on concrete, dry	1.02	
rubber on concrete, wet	0.97	
rubber on ice	0.005	
leather on oak, dry	0.50	

Sample Problem 1

In the horizontal starting area of a four-person bobsled race, the four athletes, with a combined mass including outfits of 295 kg, exert a minimum horizontal force of 41 N [fwd] to get the 315-kg sled to begin moving. After the sled has travelled for almost 15 m, all four people jump into the sled, and the sled then experiences a kinetic friction of magnitude 66 N. Determine the coefficient of (a) static friction and (b) kinetic friction.

DID YOU KNOW ?

Determining the Normal Force

In the situations described in Sample Problems 1 and 2, we apply the fact that the normal force is equal in magnitude to the weight of the object. You can show this in an FBD of the object. However, it is only true for objects on a horizontal surface with horizontal applied forces acting on them. If the applied force is at an angle to the horizontal, the normal force either increases or decreases, as shown in **Figure 2**.

Solution

(a) The normal force is equal in magnitude to the weight of the sled. Thus, omitting directions,

$$|\vec{F}_N| = |\vec{F}_g| = m|\vec{g}|$$
$$= (315 \text{ kg})(9.8 \text{ N/kg})$$
$$F_N = 3.1 \times 10^3 \text{ N}$$

$$\text{Now, } \mu_S = \frac{F_S}{F_N}$$
$$= \frac{41 \text{ N}}{3.1 \times 10^3 \text{ N}}$$
$$\mu_S = 0.013$$

The coefficient of static friction is 0.013.

(b) The normal force is much greater when the people are in the sled as the total mass now is 315 kg + 295 kg = 610 kg.

$$|\vec{F}_N| = |\vec{F}_g| = m|\vec{g}|$$
$$= (610 \text{ kg})(9.8 \text{ N/kg})$$
$$F_N = 6.0 \times 10^3 \text{ N}$$

$$\text{Now, } \mu_K = \frac{F_K}{F_N}$$
$$= \frac{66 \text{ N}}{6.0 \times 10^3 \text{ N}}$$
$$\mu_K = 0.011$$

The coefficient of kinetic friction is 0.011.

(a) The normal force increases.

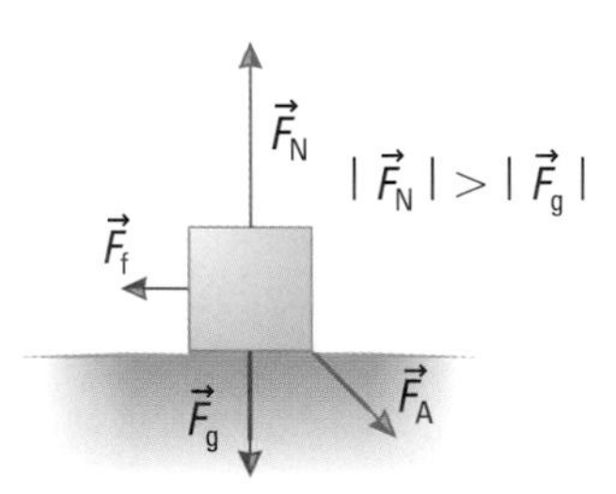

(b) The normal force decreases.

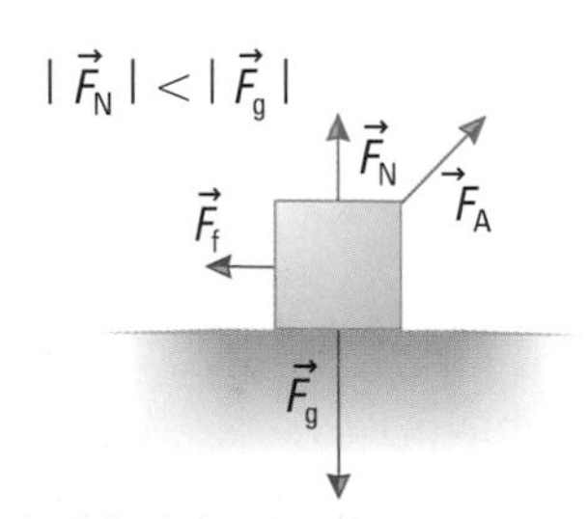

Figure 2
Influencing the normal force

Sample Problem 2

A truck's brakes are applied so hard that the truck goes into a skid on the dry asphalt road. If the truck and its contents have a mass of 4.2×10^3 kg, determine the force of kinetic friction on the truck.

Solution

From **Table 1**, μ_K for rubber on dry asphalt is 1.07. Thus,

$$F_K = \mu_K F_N, \text{ where } |\vec{F}_N| = |\vec{F}_g| = m|\vec{g}|$$
$$= (1.07)(4.2 \times 10^3 \text{ kg})(9.8 \text{ N/kg})$$
$$F_K = 4.4 \times 10^4 \text{ N}$$

The kinetic friction is 4.4×10^4 N (in the direction opposite to the truck's initial motion).

Practice

Understanding Concepts

Note: "Force" in the following questions refers only to the magnitude of the force.

1. Based on **Table 1**, which type of road, asphalt or concrete, provides better traction (friction of a tire on a road) for rubber tires under
   (a) dry conditions?
   (b) wet conditions?
2. Use the data in **Table 1** to verify that driving on an icy highway is much more dangerous than on a wet one.
3. Determine the appropriate coefficient of friction in each case.
   (a) It takes 59 N of horizontal force to get a 22-kg leather suitcase just starting to move across a floor.
   (b) A horizontal force of 54 N keeps the suitcase in (a) moving at a constant velocity.
4. A 73-kg hockey player glides across the ice on skates with steel blades. What is the force of friction acting on the skater?
5. A 1.5-Mg car moving along a concrete road has its brakes locked and skids to a smooth stop. Calculate the force of friction (a) on a dry road and (b) on a wet road.
6. A worker of a moving company places a 252-kg trunk on a piece of carpeting and slides it across the floor at constant velocity by exerting a horizontal force of 425 N on the trunk.
   (a) What is the coefficient of kinetic friction?
   (b) What happens to the coefficient of kinetic friction if another 56-kg trunk is placed on top of the 252-kg trunk?
   (c) What horizontal force must the mover apply to move the combination of the two trunks at constant velocity?

Answers

3. (a) $\mu_S = 0.27$
   (b) $\mu_K = 0.25$
4. 7.2 N
5. (a) 1.5×10^4 N
   (b) 1.4×10^4 N
6. (a) 0.17
   (c) 5.2×10^2 N

Lab Exercise 3.4.1

Determining Coefficients of Friction

Determining the coefficient of friction for any set of materials is empirical; in other words, the coefficient of friction must be found experimentally. In this lab exercise, you will apply the equations for static friction and kinetic friction to determine the coefficients of friction for several sets of materials. Once you have completed this exercise, your teacher may arrange for you to perform similar measurements in the classroom using a spring scale or a force sensor to pull (or push) objects across various flat surfaces.

In the experiment, the applied forces are all horizontal, and the objects move on a horizontal surface. Thus, the magnitude of the normal force is equal to the magnitude of the weight of each object. The data in **Table 2** refer only to the magnitudes of the forces. Once you calculate the coefficients of friction, you can refer back to **Table 1** to identify the materials in contact.

Observations

See **Table 2**.

Table 2

Surface	Object	Mass of object (kg)	F_S (N)	F_K (N)
1	A	0.24	0.96	0.72
1	B	0.36	1.5	1.1
1	C	0.48	1.9	1.4
1	D	0.60	2.3	1.7
2	E	1.2	1.8	1.4
3	F	1.2	7.1	4.8
4	G	86	9.3	8.4
5	H	0.86	5.4	4.2

Analysis

(a) Copy **Table 2** into your notebook and add three columns, two to record the calculated coefficients of static friction and kinetic friction, and a third to identify the likely materials in contact. Perform the calculations and enter the data in your table.

(b) List all patterns you can find in your data table.

(c) Assuming objects A, B, C, and D are made of the same material, plot a graph of the kinetic friction as a function of the normal force. Calculate the slope of the line. What do you conclude?

(d) Find the percentage error of the measurements involving objects A, B, C, and D, assuming that the values given in **Table 1** in this section are the accepted values. (To review percentage error, refer to Appendix A.)

Evaluation

(e) List random and systematic errors likely to occur in an experiment of this nature. In each case, state what you would do to minimize the error. (To review random and systematic errors, refer to Appendix A.)

Solving More Complex Friction Problems

A single sample problem will convince you that there are many concepts and skills you can apply when dealing with motion and forces, including friction. The important thing is to practise your problem-solving skills in a variety of contexts.

Sample Problem 3

A 0.17 kg hockey puck, sliding on an outdoor rink, has a velocity of 19 m/s [fwd] when it suddenly hits a rough patch of ice that is 5.1 m across. Assume that the coefficient of kinetic friction between the puck and the rough ice is 0.47.

(a) Draw a system diagram and an FBD of the puck moving on the rough ice.
(b) Calculate the kinetic friction acting on the puck.
(c) Determine the puck's average acceleration while on the rough ice.
(d) Calculate the puck's velocity as it leaves the rough ice and returns to the smooth ice.

Solution

(a) The required diagrams are shown in **Figure 3**. The magnitudes of the vertical forces are equal, and there is only one horizontal force, the kinetic friction, acting in a direction opposite to the puck's velocity.

(b) As the puck is sliding along the ice, there is no vertical acceleration, so the magnitude of $\vec{F}_N$ must be equal to the magnitude of $\vec{F}_g$. Thus,

$$|\vec{F}_N| = |\vec{F}_g| = m|\vec{g}|$$
$$= (0.17 \text{ kg})(9.8 \text{ N/kg})$$
$$F_N = 1.666 \text{ N}$$

Now,
$$F_K = \mu_K F_N$$
$$= (0.47)(1.666 \text{ N})$$
$$F_K = 0.78 \text{ N (to two significant digits)}$$

The magnitude of kinetic friction acting on the puck is 0.78 N. The direction of this force is to the left as in the diagram.

(a)

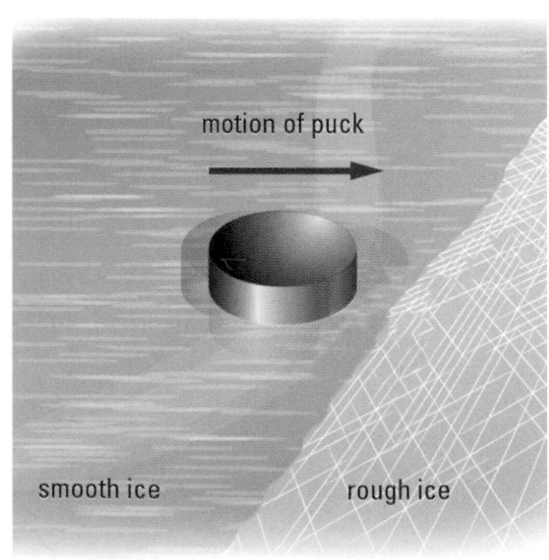

(b)

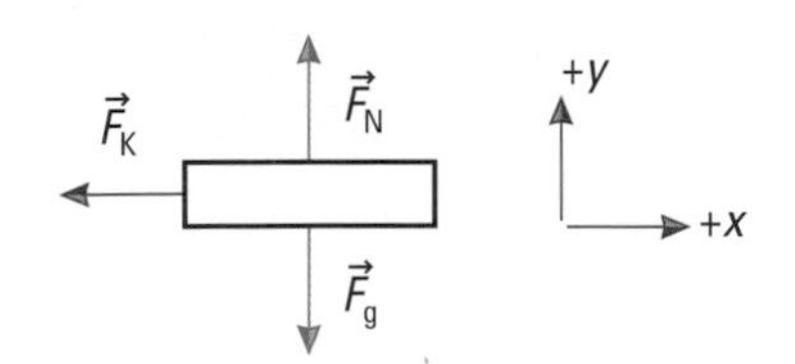

Figure 3
Diagrams of a puck sliding on rough ice
(a) The system diagram
(b) The FBD

(c) Considering the horizontal forces,

$$\vec{F}_{net} = m\vec{a} = \vec{F}_K$$

Therefore, $\vec{a} = \frac{\vec{F}_K}{m}$

$$= \frac{-0.78 \text{ N [fwd]}}{0.17 \text{ kg}}$$

$$\vec{a} = -4.6 \text{ m/s}^2 \text{ [fwd]}$$

The puck's acceleration is -4.6 m/s^2 [fwd].

(d) Since we have known values for the variables v_i, Δd, and a, to find v_f we can use the following equation, which considers only magnitudes.

$$v_f^2 = v_i^2 + 2a\Delta d$$

$$v_f^2 = (19 \text{ m/s})^2 + 2(-4.6 \text{ m/s}^2)(5.1 \text{ m})$$

$$v_f = 18 \text{ m/s}$$

Since the puck must still be moving forward, the velocity of the puck, to two significant digits, is 18 m/s [fwd].

Practice

Understanding Concepts

7. In the last few seconds of a hockey game, a player aims a slap shot at the opponent's empty net from a distance of 32.5 m. The speed of the puck upon leaving the stick is 41.5 m/s. The coefficient of kinetic friction of rubber on ice is given in **Table 1** in this section. Use the steps shown in Sample Problem 3 to verify that the puck's speed hardly changes when friction is so low. (Use 9.80 N/kg for $|\vec{g}|$ and a puck of mass 0.170 kg to keep to the accuracy of three significant digits.)
8. In a brake test on dry asphalt, a Chevrolet Camaro, travelling with an initial speed of 26.8 m/s, can stop without skidding after moving 39.3 m. The mass of the Camaro, including the driver, is 1580 kg. Refer to **Table 1** in this section for friction data of rubber on various surfaces.
   (a) Determine the magnitude of the average acceleration of the car during the non-skidding braking.
   (b) Calculate the magnitude of the average stopping friction force.
   (c) Assume that the test is now done with skidding on dry asphalt. Determine the magnitude of the kinetic friction, the magnitude of the average acceleration, and stopping distance during the skid. Compare this situation with the non-skid test.
   (d) Repeat (c) for the car skidding on ice.
   (e) In the skidding tests, does the mass of the car have an effect on the average acceleration? Explain, using examples.

Making Connections

9. When a driver faces an emergency and must brake to a stop as quickly as possible, do you think it is better to lock the brakes and skid to a stop, or apply as large a force as possible to the brakes to avoid skidding? In the latter case, the friction of the brake pads on the rotors is responsible for stopping, not the friction of the road on the tires. Explain your choice. (Your answers to question 8 above will help you understand this question.)

Answers

8. (a) 9.14 m/s^2
   (b) 1.44×10^4 N
   (c) 1.66×10^4 N; 10.5 m/s^2; 34.2 m
   (d) 8×10^1 N; 5×10^{-2} m/s^2; 7×10^3 m

Friction and Technology

Friction plays a major role in technologies involved in designing roads, bridges, automobile tires, athletic shoes, and surfaces of playing fields. Without friction, driving on highways and running on playing field surfaces would not only be dangerous, but simply impossible!

Car tires need friction to provide traction for steering, speeding up, and stopping. You know from Newton's third law of motion that the force of the road on the tires pushing the car forward is equal in magnitude and opposite in direction to the force of the tires pushing backward on the road. Each of these forces is a friction force. The tires must be able to grip the road in all conditions, including rain and snow. If a film of water develops between the tires and the road, friction is reduced. Tire treads are designed to disperse the water so that friction will be well maintained (**Figure 4**).

Figure 4
Tire treads have thin cuts, called sipes, to gather water on a wet road. This water is then pushed by the zigzag channels out behind the moving car. Tires are rated according to what degree of traction they provide.

Cars rely on friction for stopping. Most new cars have disk brakes, especially on the front wheels, which operate in a way that resembles brakes on bicycles. **Figure 5** shows that when the brakes are applied, a piston pushes the brake pads toward a rotating disk. The disk is attached to the hub of the wheel, so exerting friction on the disk causes the wheel to slow down.

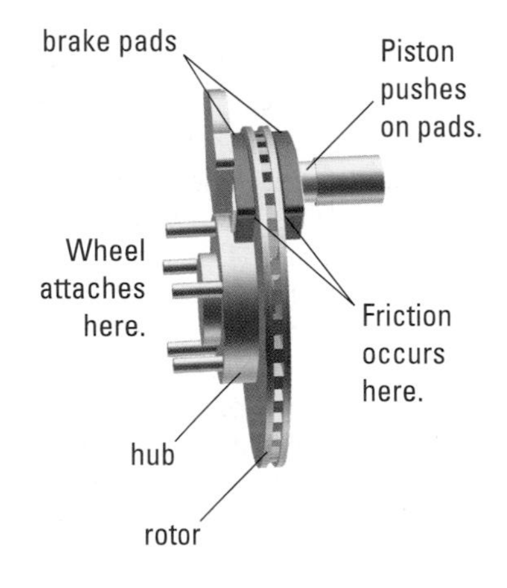

Figure 5
Basic operation of disk brakes

Antilock brakes, too, are features on most new cars. They help to control stopping under extreme conditions. Without such brakes, slamming the brakes on hard often results in skidding and losing control of the car's direction. With a computer-controlled antilock system, the friction on the brakes of individual wheels is adjusted 20 times or more per second to prevent wheels from locking.

The Bicycle Braking System

Look carefully at the braking system of a bicycle with hand brakes and obtain the Owner's Manual for the same bike, or find another resource that features the technology of bicycle brakes.

(a) Describe in words and/or diagrams how the brakes operate. Include any discussion of friction.
(b) Describe how to adjust the brakes to increase their ability to help the rider stop. Why are such adjustments needed?
(c) Under what conditions does maximum braking occur?
(d) List suggestions to maintain the safety of brakes used on a bicycle.
(e) Use at least one of Newton's laws of motion to describe why you think you should transfer your weight backward when you are braking hard.

Practice

Making Connections

10. Complete this statement: "Friction is important to transportation engineers because . . . "

SUMMARY Analyzing Motion with Friction

- The coefficient of friction, μ, is the ratio of the magnitude of the force of friction to the magnitude of the normal force between two surfaces in contact.
- The coefficient of static friction is $\mu_S = \frac{F_S}{F_N}$ and the coefficient of kinetic friction is $\mu_K = \frac{F_K}{F_N}$. These coefficients can be determined using a controlled experiment in which a horizontal applied force is used to move an object with constant velocity across a horizontal surface.
- Problem-solving skills developed throughout this unit can be synthesized in order to solve problems that include friction and coefficients of friction.

DID YOU KNOW ?

Using Physics to Make Roads Safer

A very dangerous situation arises when ice forms on the surfaces of roads, leaving hardly any friction between car tires and the road surface. (Look at the coefficient of kinetic friction between rubber and ice in **Table 1** in section 3.4.) Probably the most dangerous form of ice is *black ice*, a name that means the drivers cannot see the ice. Engineers are researching several possible ways of warning drivers of icy conditions. In one design, ice sensors on roadside posts emit a flashing blue light to warn drivers of possible black ice. In another design, devices mounted on highway maintenance trucks use infrared radiation (part of the electromagnetic spectrum) to keep track of moisture and road temperatures. When the temperature drops below the freezing point of water, the trucks disperse salt to melt the ice while travelling at highway speeds. This system reduces the chance of wasting salt on roads that do not need it. A third design uses radio waves (also part of the electromagnetic spectrum) to communicate messages of dangerous conditions from roadside detectors or emergency vehicles to any other vehicle that has an appropriate receiver. Advanced systems display the information on a liquid crystal display (LCD) screen in the car.

Section 3.4 Questions

Understanding Concepts

1. Assume you are on a luge toboggan that has a regulation mass of 22 kg and no brakes. The luge relies partly on friction to slow it down. If the coefficient of kinetic friction between the luge and the horizontal icy surface is 0.012, what is the kinetic friction acting on the luge?
2. A 12-kg toboggan is pulled along at a constant velocity on a horizontal surface by a horizontal force of 11 N.
   (a) What is the force of gravity on the toboggan?
   (b) What is the coefficient of friction?
   (c) How much horizontal force is needed to pull the toboggan at a constant velocity if two 57-kg girls are sitting in it?
3. In a tug-of-war contest on a firm, horizontal sandy beach, Team A consists of six players with an average mass of 65 kg and Team B consists of five players with an average mass of 84 kg. Team B, pulling with a force of 3.2 kN, dislodges Team A and then applies good physics principles and pulls on Team A with a force of 2.9 kN, just enough to keep Team A moving at a low constant velocity. Determine Team A's coefficient of
   (a) static friction
   (b) kinetic friction on sand
4. If the coefficient of kinetic friction is 0.25, how much horizontal force is needed to pull each of the following masses along a rough desk at a constant speed?
   (a) 25 kg (b) 15 kg (c) 250 g
5. For each situation given below, draw an FBD of the object and then answer the question. The forces described are acting horizontally; the net vertical force is zero.
   (a) A butcher pulls on a freshly cleaned side of beef with a force of 2.2×10^2 N. The frictional resistance between the beef and the countertop is 2.1×10^2 N. What is the net force exerted on the beef?
   (b) A net force of 12 N [S] results when a force of 51 N [S] is applied to a box filled with books. What is the frictional resistance on the box?

(continued)

(c) Two students exert a horizontal force on a piano. The frictional resistance on the piano is 92 N [E] and the net force on it is 4.0 N [E]. What is the force on the piano applied by the students? Describe and explain what is happening to the piano.

6. A 63-kg sprinter accelerates from rest toward a strong wind that exerts an average frictional resistance of magnitude 63 N. If the ground applies a forward force of magnitude 240 N on the sprinter's body, calculate the following:
   (a) the net force on the sprinter
   (b) the sprinter's acceleration
   (c) the distance travelled in the first 2.0 s
   (d) the coefficient of friction between the sprinter's shoes and the track (Explain whether this friction is static or kinetic.)

Applying Inquiry Skills

7. Describe how to determine the coefficients of static and kinetic friction experimentally.
8. Values of coefficients of friction are less exact than almost all other measurements made in physics. Why?

Making Connections

9. An average car tire in good condition throws out about 5 L of water per second on a wet road. At a speed of 100 km/h, approximately how much water is dispersed by a single tire in a distance of 75 km?
10. (a) Why do you think race cars use smooth tires during dry track conditions?
    (b) What do you think is meant by the term *aquaplaning*? Research to find out if you are right. Is it the same as hydroplaning? Explain.
11. Tire tread ratings are indicated on the side walls of tires. Find out about traction ratings (AA, A, B, and C), treadwear ratings, and temperature ratings (A, B, and C). Check out the labelling on tires of several cars and describe how safe you think these tires are. Follow the links for Nelson Physics 11, 3.4.

GO TO www.science.nelson.com

12. Cars use hydraulic brakes and big trucks use air brakes. Research these two systems, and write a short paragraph to compare and contrast them.

Chapter 3 Summary

Key Expectations

Throughout this chapter, you have had opportunities to

- define and describe concepts and units related to force (e.g., universal gravitation, static friction, kinetic friction, coefficients of friction); (3.1, 3.2, 3.3, 3.4)
- analyze and describe the gravitational force acting on an object near, and at a distance from, the surface of Earth; (3.1, 3.2)
- design and carry out an experiment to identify specific variables that affect motion (e.g., conduct an experiment to determine the factors that affect the motion of an object sliding along a surface); (3.3, 3.4)
- evaluate the design of technological solutions to transportation needs and, using scientific principles, explain the way they function (e.g., evaluate the design, and explain the operation of, tread patterns on car tires, braking systems, and devices to warn of black ice); (3.4)
- analyze and explain the relationship between an understanding of forces and motion and an understanding of political, economic, environmental, and safety issues in the development and use of transportation technologies (including terrestrial and space vehicles) and recreation and sports equipment; (3.2, 3.4)

Key Terms

force field
gravitational field strength
mass
weight
law of universal gravitation
static friction
starting friction
kinetic friction
coefficient of friction
coefficient of kinetic friction
coefficient of static friction

Make a Summary

The Canadian Space Agency (CSA) has asked students to create a set of diagrams or cartoons to illustrate their concept understanding and skills development in the topics of gravitational force and friction. The context of the design is a mission to send astronauts to Mars to explore the possibility of living there. A large rocket and the space vehicle it will carry into space must first be moved from the storage building to the launch pad on a flat car that rolls on wide railway tracks. The rocket is then launched into space, sending the vehicle to Mars, where it lands. People on the Mars mission will stay there for a year and then use a small rocket to blast off the surface of Mars and return to Earth. Draw the following diagrams or cartoons, and describe what is happening in each one, using as many of the key words, concepts, equations, and force diagrams as you can from this chapter.

(a) The loaded flatbed railway car is being pulled along the tracks.
(b) The rocket and its payload are launched from the launch pad.
(c) The space vehicle is about halfway to Mars.
(d) The vehicle is landing on Mars.
(e) The vehicle is blasting off from Mars.

Reflect on your Learning

Revisit your answers to the Reflect on Your Learning questions at the beginning of this chapter.

- How has your thinking changed?
- What new questions do you have?

Chapter 3 Review

Understanding Concepts

1. Sketch a graph to show how the first variable in the list below depends on the second one, and write the corresponding proportionality statement.
   (a) the gravitational field strength surrounding an object; the force the object exerts on surrounding objects
   (b) the gravitational field strength of a body; the mass of the body
   (c) the force of gravity on an object; the distance of the object from the body exerting the force of gravity on it
   (d) the force of friction an object experiences; the coefficient of friction between the object and the surface it is on
   (e) your weight; the altitude of your location
   (f) your weight; the latitude of your location
2. (a) What quantity has the same numerical value as the gravitational field strength, but is expressed in a different unit?
   (b) What is the gravitational field strength in your classroom?
3. A student hangs a 500.0-g mass on a spring scale. Correct to two significant digits, state the reading on the scale when the mass and the scale are
   (a) stationary
   (b) being lifted upward at a constant velocity
   (c) released, so they fall downward freely
4. A 65-kg student can safely lift 73% of the student's own mass off the floor.
   (a) Determine the mass the student can lift (on Earth).
   (b) How would the mass the student could lift on the Moon compare with your answer in (a)?
5. If you were to assign a direction to the force of gravity both at and high above Earth's surface, what single direction would apply in all cases?
6. If Earth's radius is 6.4×10^3 km, calculate the force of gravity on a 1.0×10^5 kg space station situated
   (a) on Earth's surface
   (b) 1.28×10^5 km from the centre of Earth
   (c) 3.84×10^5 km from the centre of Earth (about the distance to the Moon)
   (d) 1.5×10^8 km from Earth's centre (about the distance to the Sun)
7. The magnitude of the force of gravity on a meteoroid approaching Earth is 5.0×10^4 N at a certain point. Calculate the magnitude of the force of gravity on the meteoroid when it reaches a point one-quarter of this distance from the centre of Earth.
8. What is the gravitational field strength at a place 220 km above Earth's surface, the altitude of many piloted space flights?
9. Sirius is the brightest star in the night sky. It has a radius of 2.5×10^9 m and a mass of 5.0×10^{31} kg. What is the gravitational force on a 1.0-kg mass at its surface?
10. Sirius B is a white dwarf star, in orbit around Sirius, with a mass of 2.0×10^{30} kg (approximately the mass of the Sun), and a radius of 2.4×10^7 m (approximately one-thirtieth of the radius of the Sun).
    (a) What is the force on a 1.0-kg mass on the surface of Sirius B?
    (b) What is the acceleration due to gravity on the surface of Sirius B?
11. Determine the distance from Earth's centre where the force of gravity acting on a space probe is only 11% of the force acting on the same probe at Earth's surface. Express your answer in terms of Earth's radius, r_E.
12. As a space vehicle travels toward the Moon, it will eventually reach a location where Earth's pull of gravity is balanced by the Moon's pull.
    (a) Draw a sketch of this situation. Label distances and forces using appropriate symbols.
    (b) Determine the location. (*Hints:* You will need to know the average distance separating Earth and the Moon as well as the mass of both bodies. Also, after you set up a single equation involving the magnitude of the forces caused by both bodies, you can apply the quadratic formula to solve the equation.)
    (c) One of the roots of the equation used in (b) is greater than the Earth–Moon distance, so it does not answer the original question. However, the root does have a meaning. What is it?
13. Describe two specific ways in which people try to
    (a) increase friction between surfaces
    (b) decrease friction between surfaces
14. Relate the difference in the magnitudes of starting friction and kinetic friction to Newton's first law of motion.
15. Six dogs, each having a mass of 32 kg, pull a 325-kg sled horizontally across the snow ($\mu_S = 0.16$ and $\mu_K = 0.14$).
    (a) How much force must each dog exert in order to move the sled at a constant velocity?
    (b) If each dog applies a force of 86 N [fwd], what average acceleration does the sled experience? (Draw an FBD of this situation.)
    (c) One dog becomes lame and must be transported on the sled. With what force must each of the remaining dogs pull to achieve an average acceleration of 0.75 m/s^2 [fwd]?

16. A 650-N student is Rollerblading at a velocity of 7.8 m/s [W] when the student trips and slides horizontally along the trail, coming to a stop in 0.95 s.
    (a) Determine the student's average acceleration.
    (b) Draw an FBD of the student sliding on the trail.
    (c) What is the kinetic friction as the student slides?
    (d) Determine the coefficient of kinetic friction between the student and the trail.

Applying Inquiry Skills

17. In writing a test on this chapter, a student could not recall the unit of the universal gravitation constant, *G*. Help the student determine the unit by rearranging the equation of the law of universal gravitation to isolate *G* and then perform unit analysis on the other variables.
18. (a) If you were in a physics classroom on a Mars colony, how would you determine the gravitational field strength there experimentally?
    (b) If you measured the kinetic friction between two surfaces on Mars and between the same surfaces on Earth, how would you expect the results to compare?
19. Design a controlled experiment to answer one of the following questions. Your design should include all the features of a formal experimental design.
    (a) What factors affect the air resistance acting on a moving car?
    (b) What factors affect the rolling friction of a moving object of your choice? (See **Figure 1**.)

Figure 1
For question 19, choose some device with wheels.

Making Connections

20. (a) Explain why it will likely be many years before humans attempt to colonize Mars.
    (b) Why would intergalactic travel be impossible for any humans?
21. Using concepts from this chapter, explain why signs like the one in **Figure 2** have proven to reduce injury.

Figure 2
For question 21

22. (a) Why do weight lifters and gymnasts put powder on their hands before performing their exercises?
    (b) What other athletes use the same type of powder?
23. What tire tread design is appropriate for mountain bikes? Explain why, using principles you have learned in this chapter.

Exploring

24. Research more about the historical developments of the study of gravitational forces, such as more details about Newton's contributions or Cavendish's determination of *G*. Write a brief report on what you discover.
25. Find out what gravitometry is and how it is applied. Follow the links for Nelson Physics 11, Chapter 3 Review.

GO TO www.science.nelson.com

26. With your teacher's guidance, use a force sensor or an acceleration sensor to demonstrate free fall or "weightlessness" using a computer interface. Describe what you observe.
27. Research what is meant by a "libration point." In the 1970s, a group of interested people, including many scientists, proposed a space colony to be situated at a libration point between Earth and the Moon. This group, called the "L5 Society," suggested such a colony would have distinct advantages. Suggest what these advantages would be, and use research to check your answers. Follow the links for Nelson Physics 11, Chapter 3 Review.

GO TO www.science.nelson.com

Motion and Space Exploration

Galileo and Newton, who led the way in understanding the relationship between forces and motion, would be impressed by recent achievements in space exploration. Humans have landed on the Moon, launched communication and telescope satellites, and sent space probes to explore other planets and the Sun. In the future, humans may colonize other bodies in the solar system. Mars will likely be one of them. Robots and rovers have been sent to Mars to discover features that must be fully understood before humans can land on Mars. However, the technological challenges of sending probes and robots to Mars are immense. In the spring of 2000, a disaster occurred when the Mars Polar Lander (**Figure 1**) went missing for reasons still not fully known. One assumption of the disappearance is that the Lander crashed into a sloped surface and tumbled onto its side, making the communications system useless.

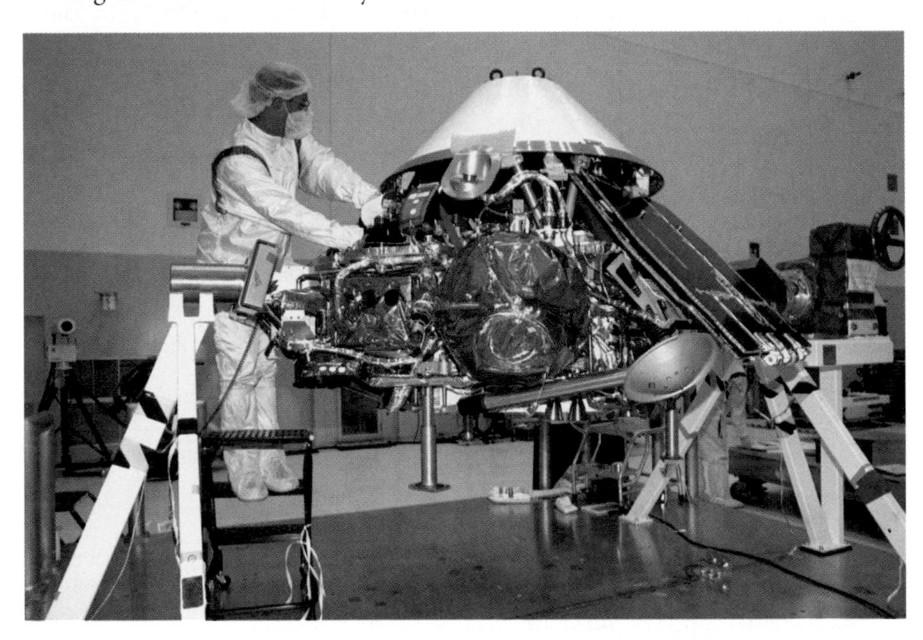

Figure 1
The rocket carrying the Mars Polar Lander was launched from Earth in January 1999. The mission of the Polar Lander was to search for water and possible evidence of some form of life that might have existed on Mars. As the 290-kg Polar Lander was descending to Mars' surface 15 months later, something went wrong, ending the Polar Lander's communication with Earth.

Perhaps the Polar Lander was designed to land only on a relatively flat surface. But whatever the reason for the failure, the costs were enormous and the need for a better design and more tests became obvious. These procedures are just part of the process of technological development: a need is identified; design, construction, and tests are carried out; a new problem arises, which identifies a new need; design changes are made to accommodate the new need.

In this Unit Task, you are expected to demonstrate an understanding of the relationship between forces and motion and apply the skills of inquiry and communication that you have developed both in this unit and in other science courses.

Spinoffs

The space exploration programs of the past half century have resulted in many designs, discoveries, and innovations that have been applied to everyday life. These spinoffs include freeze-dried foods, robotic tools used in industry, hard plastics used in safety helmets and in-line skates, and Velcro. Your design can also have spinoffs. For example, your ideas may be applied to create safer helmets for sports activities or restraint systems for automobiles.

The Task

Your task will be to design a landing device of relatively low mass that can make a safe landing when dropped to a hard surface from a height of at least 1.0 m. The "instrumentation" carried by the device will be a raw egg (encased in a small plastic bag in case the landing isn't "safe") to simulate an actual instrument sent to explore Mars. To simulate realistic landing conditions, the Lander should be tested using three different drops, one to land upright on a hard, flat surface, one to land sideways on a hard, flat surface, and one to strike a ramp that simulates a rigid hillside.

You will be expected to analyze the design, make modifications to improve its performance, describe practical spinoffs, and explain how the design can be applied to other situations.

Basic safety and design rules should be discussed before brainstorming begins. For example, any tools used in constructing the device must be used safely, and the tests of the device should be carried out in a way that ensures absolutely no damage is done to people or furniture. Considering the Lander design, no parachute should be allowed, and the ratio of the mass of the Lander to the mass of the raw egg can be used to compare different landers.

Analysis

Your analysis should include answers to the questions below.

(a) When dropped from rest, with what maximum speed will the Lander crash in your tests?
(b) From what height could the Lander be dropped from rest to land at the same maximum speed on Mars? on the Moon?
(c) What effect would different surfaces likely found on Mars or the Moon have on the safe landing speed of your device?
(d) What is the ratio of the mass of your Lander to the mass of its cargo (the egg)?
(e) Why is this ratio a useful criterion?
(f) What other criteria can be used to evaluate the Lander design?
(g) How can your design be applied to the design of safety helmets for sports and restraint systems for automobiles?
(h) Create your own analysis questions and answer them.

Evaluation

(i) How does your design compare to the design of other students or groups?
(j) Draw a flow chart of the process you used in working on this task. How do the steps in your flow chart compare to the steps you would follow in a typical motion investigation in this unit?
(k) If you were to begin this task again, what would you modify to ensure a better process and a better final product?

Assessment

Your completed task will be assessed according to the following criteria:

Process

- Draw up detailed plans of the technological design, tests, and modifications.
- Choose appropriate materials for the Lander.
- Carry out the construction, tests, and modification of the Lander.
- Analyze the process (as described in Analysis).
- Evaluate the task (as described in Evaluation).

Product

- Submit a report containing your design plans, the flow chart, test results, analysis, and evaluation.
- Demonstrate an understanding of the process of technological design and related physics concepts, principles, laws, and theories.
- Use terms, symbols, equations, and SI metric units correctly.
- Prepare a final product of the Lander to be tested.

Understanding Concepts

1. Set up a five-column table with these headings:

Quantity	Symbol	Type of quantity (scalar or vector)	SI unit	Typical value in daily life

Complete the table for the following quantities: distance, time interval, average speed, position, displacement, average velocity, average acceleration, net force, mass, and coefficient of kinetic friction.

2. The Yukon Quest, an annual dog-sled race between Whitehorse in the Yukon and Fairbanks in Alaska, covers a gruelling distance of 1.61×10^3 km. Assuming that the average speed of the dog team that set a course record is 6.28 km/h, determine the best time for this race. Express your answer in days, hours, and minutes.

3. The highest average lap speed on a closed circuit in motorcycling is about 72 m/s or 258 km/h. If a cyclist takes 56 s to complete one lap of the circular track, what is the track's circumference?

4. Under what condition(s) will the magnitude of the displacement of a moving object be the same as the distance it travels?

5. The distance between bases on a baseball diamond is 27.4 m. Use a scale diagram, the Pythagorean theorem, or trigonometry to find the magnitude of a runner's displacement from home plate to second base.

6. Toronto is about 50 km from Hamilton. A freight train travels from Hamilton to Toronto at an average speed of 50 km/h. At the same time, a passenger train travels from Toronto to Hamilton at an average speed of 75 km/h. Find the time passed, in minutes, before the trains meet each other. (Assume two significant digits.)

7. The motion depicted in **Figure 1** shows the path of a team in a car rally held in a rural area where the speed limit is 80 km/h. (Assume two significant digits.) In the car rally, points are deducted for anyone determined to be breaking the speed limit.

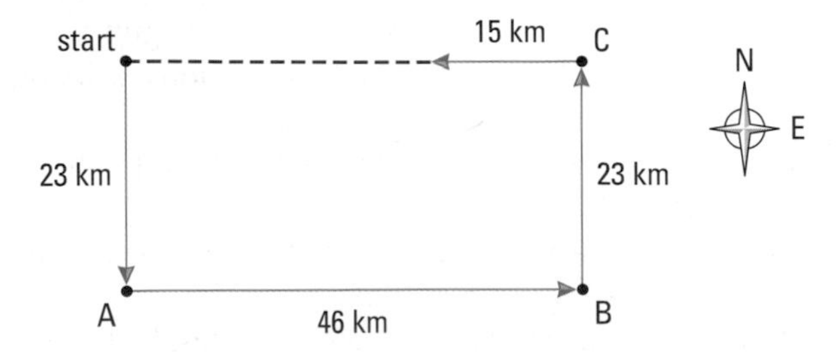

Figure 1

   (a) If the car took 1 h 15 min for the trip, will the team get points deducted?
   (b) Determine the displacement from the starting position when the car arrives at Checkpoint A, then at Checkpoint B.
   (c) Determine the average velocity (magnitude and direction) of the car's motion for the entire trip.

8. Communicate the meaning of the following equation: $\vec{v}_{AD} = \vec{v}_{AB} + \vec{v}_{BC} + \vec{v}_{CD}$

9. Satellites are often launched so that they orbit Earth from west to east. What is the advantage of selecting this direction instead of east to west?

10. Sometimes, when your car stops at a traffic light and you look at the car beside you, you feel that your car is moving backward. In fact, your car is still at rest on the road. How is this possible?

11. An airplane pilot points the plane due east. A wind is blowing from the north. Assuming the destination is due east of the starting position, explain why the plane will not reach it.

12. As a GO Train (T) is leaving a station, a passenger (P), walking toward the front of the car, waves at a friend (F) standing on the platform. At the instant of the wave, the train is travelling at 1.3 m/s [W] relative to the platform and the passenger is walking at 1.1 m/s [W] relative to the train.
   (a) Determine the velocity of the passenger relative to the friend.
   (b) What is the velocity of the friend relative to the passenger?

13. For which of the following situations does the net force have a magnitude of zero?
   (a) A ball rolls down a hill at an ever-increasing speed.
   (b) A car travels along a straight, level expressway with the cruise control set at 100 km/h (as read from the speedometer).
   (c) A boulder falls from the edge of a cliff to the lake below.
   (d) A communications satellite is located in an orbit such that it remains in the same spot above Earth's surface at all times.

14. A 45-t electric train accelerates uniformly from a station with the train's motion depicted in the graph in **Figure 2**. (Recall that $1 \text{ t} = 1.0 \times 10^3$ kg.)
   (a) Use the information in the velocity-time graph to generate the position-time and acceleration-time graphs of the motion.
   (b) Determine the net force acting on the train during the uniform acceleration portion of the motion.

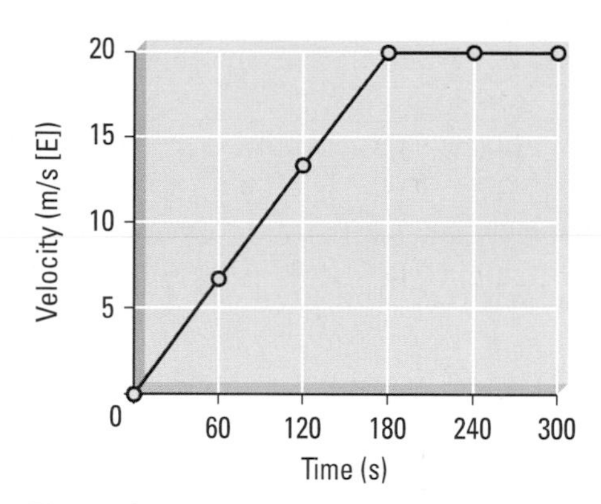

Figure 2

15. Assuming the best high jump record from a standing position in a school was 1.52 m, what initial velocity was needed for this jump? Neglect any sideways motion.
16. When a world-class pole-vaulter is gliding over the pole, the jumper is approximately 5.5 m above the top of the mat where the jumper will land. Assuming that the jumper's initial vertical velocity is zero at the top of the jump, with what vertical velocity does the jumper strike the matt? Neglect any sideways motion.
17. A hockey referee drops the puck from rest for a face-off, and the puck reaches a speed of 5.6 m/s just before striking the ice.
    (a) Assuming free fall, how long did it take the puck to fall to the ice?
    (b) From what height above the ice did the referee drop the puck?
18. A sports car takes 2.8 s to accelerate from 17 m/s [fwd] to 28 m/s [fwd].
    (a) Determine the average acceleration of the car during this time interval.
    (b) What is the car's displacement during the acceleration?
19. A student suggests that a wet dog shaking water off its back is a good example of Newton's first law of motion. Do you agree? Explain your answer.
20. Use Newton's laws to explain why each of the following statements is correct. In each case, indicate which of the three laws best explains the situation.
    (a) It takes longer for a car to accelerate from 0 km/h to 100 km/h if it has five passengers in it than when it has only one.
    (b) Many a novice hunter has experienced a sore shoulder after firing a shotgun.
    (c) Subway cars provide posts and overhead rails for standing passengers to hold.
21. (a) Calculate the magnitude of the force of gravity acting on you, in other words your weight, using the equation of Newton's second law.
    (b) Calculate the magnitude of the force of gravity acting on you by applying the law of universal gravitation.
    (c) Compare your answers to (a) and (b).
    (d) Based on the results, derive an equation that could be used to determine Earth's mass.
22. **Figure 3** shows a cart with a battery-operated fan on it. The cart has a slot into which a card can be inserted. Predict and explain what will happen if the fan is turned on when the card is
    (a) absent
    (b) present

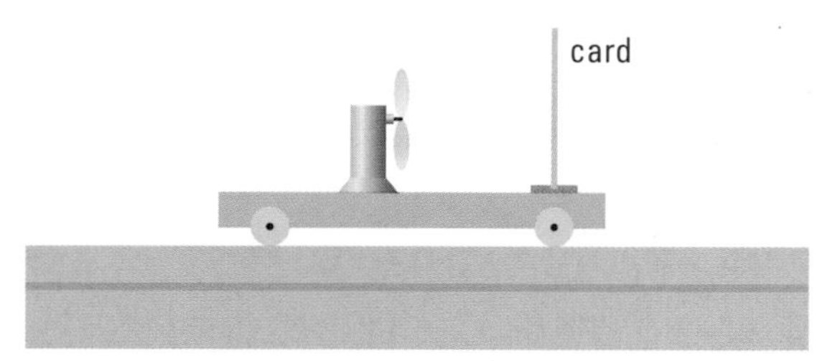

Figure 3

23. Determine the magnitude of the force of gravity acting on a 340-kg satellite, 850 km above Earth's surface.
24. An applied horizontal force of magnitude 9.1 N is needed to push a 2.8-kg lamp across a table at a constant velocity.
    (a) Determine the coefficient of kinetic friction between the lamp and the table.
    (b) How would the coefficient of static friction compare with your answer in (a)? Why?
25. A 5.0-kg object made of one material is being pulled at a constant velocity along a table made of another material. The coefficient of sliding friction between these two materials is 0.35. What is the magnitude of the force of friction?
26. A student pushes horizontally eastward on a 1.2-kg textbook initially resting 1.3 m from the edge of a long table. The coefficient of static friction is 0.24 and the coefficient of kinetic friction is 0.18. The force is just large enough to get the book to start moving, and is maintained for 1.0 s after the motion has begun.
    (a) Draw an FBD of the book while the force is being applied and the book is moving.
    (b) Determine the force applied to the book by the student.
    (c) Determine the acceleration of the book while the student is applying the push.
    (d) Calculate the maximum velocity reached by the book.
    (e) Draw an FBD of the book after the push is removed and the book is sliding on its own.

(f) Determine the book's acceleration while it is sliding on its own.
(g) Will the book slide off the end of the table? Show your calculations.

27. A 950-kg communications satellite in synchronous orbit 42 400 km from Earth's centre has a period of 24 h. Placed in orbit above the equator, and moving in the same direction as Earth is rotating, it stays above the same point on Earth at all times. This makes it useful for the reception and retransmission of telephone, radio, and TV signals.
(a) Calculate the magnitude of the force of gravity acting on the satellite.
(b) What is the magnitude of the gravitational field strength at this altitude?

Applying Inquiry Skills

28. To check your skill in estimating speeds in sports events, estimate in kilometres per hour and metres per second the highest speed in each of these situations: a baseball pitch, a hockey slap shot, a female tennis serve, a golf drive, a badminton serve, and a cricket bowl. Rank the estimated speeds from lowest to highest. Then check your estimations using the following information:
(a) After leaving the pitcher's hand, a fastball can travel 9.0 m in the first 0.20 s.
(b) A slap shot drives the puck 18 m from just inside the blue line to the goal in 0.34 s.
(c) The fastest female tennis serve takes 0.21 s to travel 12 m from the baseline to the net.
(d) After being driven off the tee, a golf ball travels the first 5.0 m in 91 ms.
(e) A well-hit birdie travels the first 0.98 m off the serve in 11 ms, although air resistance slows it down quickly.
(f) The fastest cricket bowl takes 0.25 s to travel the first 11 m toward the hitter.

29. Two students perform an activity to determine the frequency of a ticker-tape timer that is labelled 60.0 Hz. One student gets a frequency of 55.5 Hz, and the other gets a frequency of 65.5 Hz. Determine the percentage error of each of the results.

30. Sketch a graph of time as a function of speed for a uniform motion experiment in which the distance travelled is always the same.

31. Several students set up an investigation to obtain data about a cart moving at a constant speed in a straight line. The data for the distance the cart travels from start after each time interval of 0.10 s are shown in **Table 1**.

Table 1

Time (s)	0.0	0.10	0.20	0.30	0.40
Distance (m)	0.0	2.1	4.2	6.1	8.3

(a) Which variable is dependent? independent?
(b) Plot a graph of the data and determine the slope of the line of best fit on the graph. State what the slope represents.
(c) Give an example of interpolation and extrapolation, using the graph.

32. Sketch a position-time graph for a sprinter who runs as fast as possible until she or he crosses the finish line, and then slows down and jogs at a constant speed for a while longer to cool down.

33. None of the three graphs shown in **Figure 4** could be a position-time graph for a moving object. Explain why in each case.

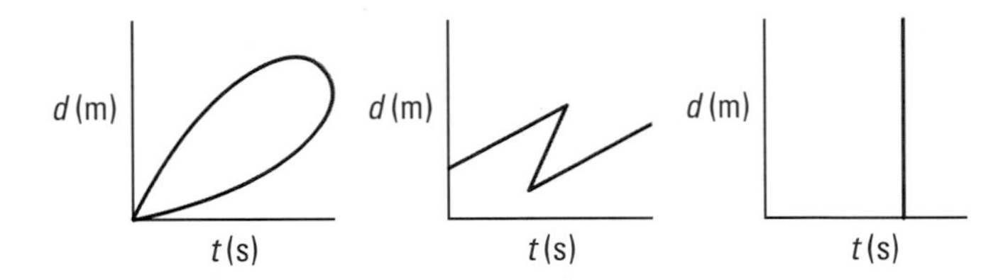

Figure 4

34. In an investigation involving the horizontal motion of a cart with a horizontal force acting on it, a computer program generates the acceleration-time graph shown in **Figure 5**. Assume that the initial velocity of the cart is 0 and its initial mass is 1.2 kg.
(a) Generate the velocity-time graph of the motion.
(b) What is the net force acting on the accelerating cart?
(c) If the mass of the cart is doubled, what happens to the acceleration graph?

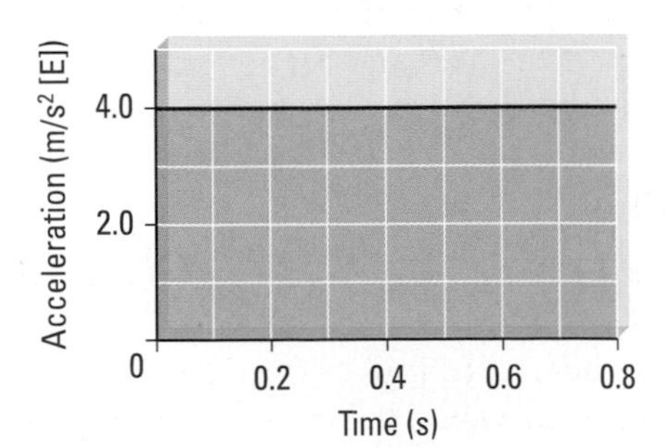

Figure 5

35. **Figure 6** shows a device that can be used as an accelerometer. Predict how the bob in the water will move (if at all) in each of the following situations:

Figure 6

(a) the accelerometer is moving to the right at a constant velocity
(b) the accelerometer is accelerating uniformly to the left
(c) the accelerometer is moving to the right but is slowing down uniformly
(d) Suggest a similar setup that can act as the accelerometer without the use of water.

36. A student is designing simulation software to demonstrate an experiment that tests the acceleration due to gravity on a planetary moon. The magnitude of the acceleration due to gravity on the moon is 1.6 m/s^2.
(a) Determine the height from which a ball can be dropped from rest on this moon in order for it to take 2.0 s to land on the surface.
(b) Set up a table of data of position and time (for time intervals of 0.4 s) for the situation in (a).
(c) Plot a graph of the data in (b). From the graph, generate the velocity-time graph by applying the tangent technique.
(d) From the velocity-time graph, calculate the average acceleration, and determine the percentage error of your graphing technique.

Making Connections

37. Two towns, 100 km apart, are linked by a highway with a speed limit of 100 km/h.
(a) Determine the time in minutes a trip from one town to the other would take at an average speed of 100 km/h, 110 km/h, 120 km/h, . . . , 200 km/h.
(b) Use the data to plot a graph of *time saved* as a function of *speed above the limit.* What can you conclude about the shape of the graph?
(c) Describe the disadvantages of exceeding a speed limit by a large amount.

38. **Figure 7** shows three possible arrangements of an eavestrough installed on a shed. Assuming the down spout is on the right side in each design, which design would you recommend? Use physics principles to explain why.

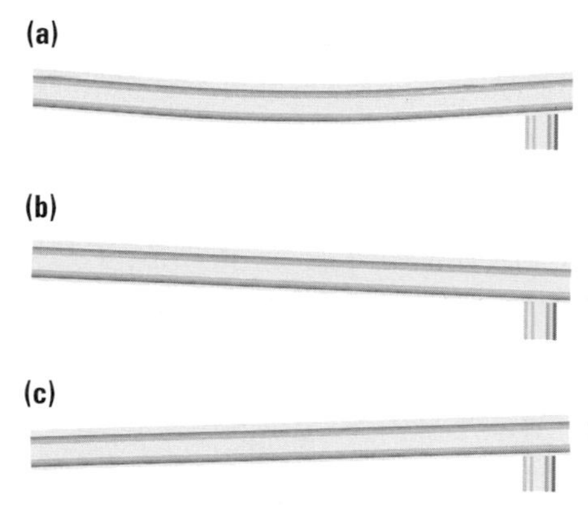

Figure 7

39. Some large pendulum rides at amusement parks, such as the one shown in **Figure 8**, use large rotating wheels to cause the large mass to accelerate, first as the mass is speeding up and then as it is slowing down.
(a) Describe the role of friction in this type of ride.
(b) What effect would a heavy rain have on the operation of this ride? Explain your answer.

Figure 8

Unit 2

Energy, Work, and Power

Companies need energy to maintain comfortable working conditions for employees and to provide lighting systems and security and fire alarms. With increasing energy costs, companies must use energy effectively and efficiently to be productive. Reducing energy use and minimizing energy waste are also important issues. Edwin Lim, an electrical engineer, develops energy management technologies for companies. Mr Lim designs individual control systems and integrated systems that enhance comfort, save energy, and increase security. New technology makes it possible to determine the number of people in a building and adjust energy use accordingly; wireless technology can monitor energy use and carry out maintenance management. Mr. Lim uses this leading-edge technology to solve energy problems faced by companies today.

Edwin Lim, engineer at Honeywell Limited

Overall Expectations

In this unit, you will be able to

- demonstrate an understanding (qualitative and quantitative) of the concepts of work, energy (kinetic energy, gravitational potential energy, and thermal energy and its transfer [heat]), energy transformations, efficiency, and power
- design and carry out experiments and solve problems involving energy transformations and the law of conservation of energy
- analyze the costs and benefits of various energy sources and energy-transformation technologies that are used around the world, and explain how the application of scientific principles related to mechanical energy has led to the enhancement of sports and recreational activities

Unit

2

Energy, Work, and Power

Are You Ready?

Knowledge and Understanding

1. Refer to **Figure 1**.
   (a) List all the forms of energy you can find in the drawing.
   (b) List several examples of energy transformations you can tell are happening in the drawing. (For example, the TV is changing electrical energy into sound energy and light energy.)

Figure 1

2. The electromagnetic spectrum, shown in **Figure 2**, consists of many different components.
   (a) What characteristics do the components of the electromagnetic spectrum have in common?
   (b) Describe how visible light differs from the other components.

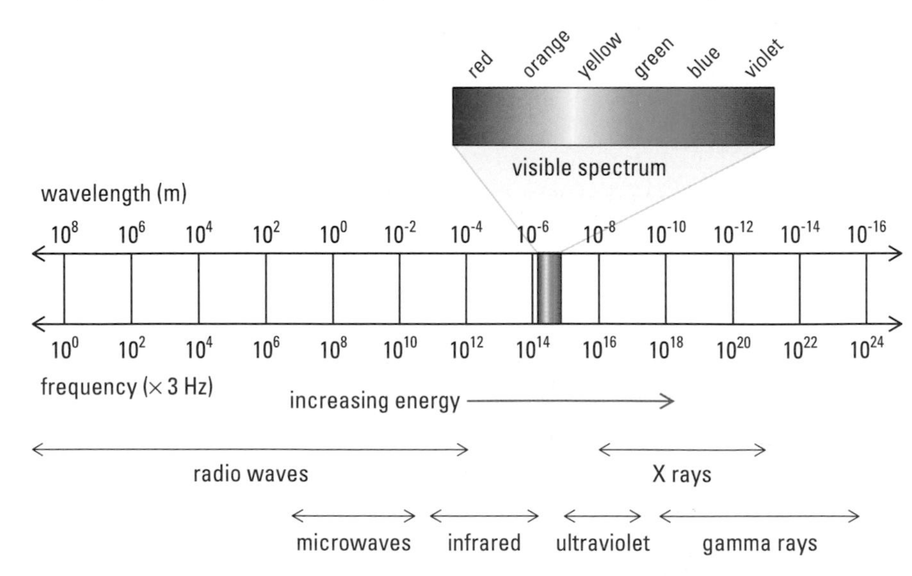

Figure 2

3. Energy can be transferred by conduction, by convection, and by radiation. Sketch a diagram based on **Figure 3** into your notebook and use symbols (for example, particles and waves) to show how these three methods relate to what is happening in the drawing.
4. To check your understanding of efficiency, copy **Table 1** into your notebook and complete the missing data.

Table 1

Device	lever	electric heater	incandescent light bulb	car
Input energy (J)	90	100	100	?
Output energy (J)	100	100	?	250
Efficiency (%)	?	?	5	25

Figure 3
For question 3

Inquiry and Communication

5. **Figure 4** shows water being heated in an investigation in the laboratory. What safety precautions should be followed in this type of investigation?
6. You are asked to do a cost-benefit analysis of operating a radio using a battery as compared to using an electric outlet. In the analysis, you are to include economic, social, and environmental impacts. Describe what is meant by a "cost-benefit analysis" using the radio example in your answer. Show that you understand the meaning of costs, benefits, economic impact, social impact, and environmental impact.

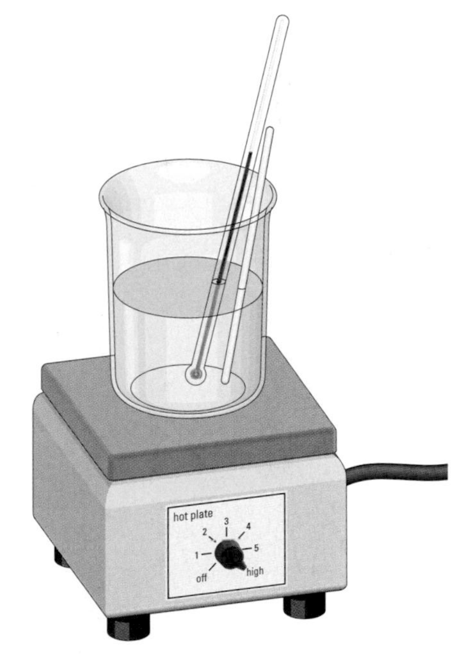

Figure 4
For question 5

Making Connections

7. Set up a table using the headings shown to classify at least eight different energy sources as renewable or non-renewable (for example, wind and natural gas). Describe a major advantage and a major disadvantage of each source.

Energy source	**Classification**	**Major advantage**	**Major disadvantage**

8. Draw a simple sketch of a small house, like the one in **Figure 5**. On the sketch, use a variety of labels and symbols, such as coloured arrows, to show
   (a) ways in which energy can enter the house
   (b) ways in which energy can be used wisely in the house
   (c) ways in which energy can be wasted in the house

Figure 5

Chapter

4

Energy, Work, Heat, and Power

In this chapter, you will be able to

- define and describe the concepts of work, energy, heat, and power
- apply the equation that relates work, force, and displacement in situations where the force and displacement are parallel
- use the law of conservation of energy to analyze, both qualitatively and quantitatively, situations involving energy, work, and heat
- apply the equation that relates power, energy, and time interval in a variety of contexts
- apply the equation that relates percent efficiency, input energy, and useful output energy to several energy transformations
- design and carry out experiments related to energy transformations
- analyze and interpret experimental data or computer simulations involving work, energy, and heat

When you hear screams coming from passengers on a roller coaster (**Figure 1**), you want to ask why people seem to love to be frightened. Roller coasters have become so big and so fast that their cars can reach speeds exceeding 160 km/h!

A roller coaster combines the concepts presented in Chapters 1 to 3 with the concepts you will explore in this chapter. The roller coaster experiences changing speeds and directions, requires a force to go up against the force of gravity, and is subject to frictional forces. These principles are extended to analyze the *power* needed by the motor to do the *work* to pull a coaster up the first big hill. Once at the top of the hill, the ever-present force of gravity is responsible for changing the speed of the roller coaster from slow to extremely fast, increasing the *energy of motion*. As the coaster goes up and down the hills, the energy of motion changes mostly into (and out of) another form of energy, *energy of position*, with some into the form of *thermal energy* due to friction.

It is these concepts (power, work, energy of motion, energy of position, and thermal energy) that you will study in detail in this chapter. You will apply these principles to many examples, including sports, transportation, and, of course, roller coasters.

Reflect on your Learning

1. What kind of relationship do you think exists between the following variables?
   (a) the *work* you do in pushing a piano across the floor and the *distance* the piano moves
   (b) the *energy of position* of a roller coaster as it goes up the first hill and the *height* of the roller coaster
   (c) the *energy of motion* of the coaster and the *speed* at which it travels
   (d) the *power* of a volunteer shifting 100 sandbags to control flooding and the *time interval* used to shift the 100 bags
2. To get to a building entrance that is above street level, you can go straight up a set of stairs, or along a ramp. What is the advantage of either way with regard to the force applied and to the energy consumed?
3. What are the methods in which heat can be transferred from one place to another? Give an example of each of these methods.
4. What is meant by the phrase "the law of conservation of energy?"

Throughout this chapter, note any changes in your ideas as you learn new concepts and develop your skills.

Analyzing a Toy's Action

The toy in **Figure 2** may look much simpler than a roller coaster, but it operates on similar principles, including work, energy of position, energy of motion, and friction. (Of course, it operates on these principles only if you discover how to make it work. Good luck!)

- Look carefully at the design of the toy. Try to predict what you will have to do to get it to move in its most interesting way. Test your prediction using trial and error until you think you have mastered the operation of the toy.
- Have your teacher inspect your attempted motion and tell you if you are ready to do the following.
  (a) Explain the operation of the toy using as many of the physics words as possible presented in this introduction.
  (b) Explain this statement: "The toy provides a good example of the law of conservation of energy."

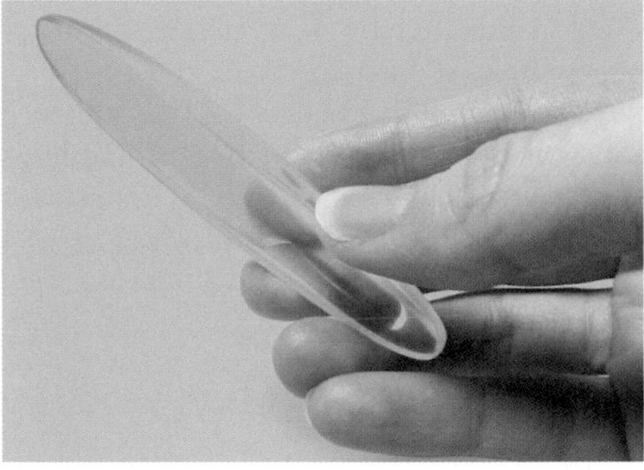

Figure 2
Can you figure out how to make this toy operate?

Figure 1
What physics principles help to explain the operation of roller coasters?

4.1 Energy and Energy Transformations

energy: the capacity to do work

Without light and other radiant energies that come to us from the Sun, life on Earth would not exist. With these energies, plants can grow and the oceans and atmosphere can maintain temperature ranges that support life. Although energy is difficult to define comprehensively, a simple definition is that **energy** is the capacity to do work. The word energy comes from the Greek prefix *en*, which means "in," and *ergos*, a Greek word which means "work." Thus, when you think of energy, think of what work is involved. For example, energy must be supplied to a car's engine in order for the engine to do work in moving the car. In this case, the energy may come from burning gasoline.

Forms of Energy

It took many years for scientists to develop a system of classifying the various forms of energy. **Table 1** lists nine forms of energy and describes them briefly, and **Figure 1** shows examples of these forms.

Notice in **Table 1** and **Figure 1** that eight of the energy forms listed involve particles (atoms, molecules, or objects). The other form is radiant energy. Visible light and other types of radiant energy belong to the electromagnetic (e.m.) spectrum. Components of the electromagnetic spectrum have characteristics of waves, such as wavelength, frequency, and energy, and they all can travel in a vacuum at the speed of light, 3.00×10^8 m/s. (Refer to section 9.1 to review the e.m. spectrum.)

heat: the transfer of energy from a warmer body or region to a cooler one

Notice also that heat is not listed as a "form" of energy. **Heat** is the transfer of energy from a warmer body or region to a cooler one. Various methods of heat transfer are described later.

In this chapter, we focus on gravitational potential energy, kinetic energy, and thermal energy, which will be defined formally in the appropriate sections. Later in the text, you will study more about the other forms listed.

Table 1 Several Forms of Energy

Form of energy	Description	Example
thermal	The atoms and molecules of a substance possess thermal energy. The more rapid the motion of the atoms and molecules, the greater the total thermal energy.	Figure 1(a)
electrical	This form of energy is possessed by charged particles. The charges can transfer energy as they move through an electric circuit.	Figure 1(a)
radiant	Radiant energy travels by means of waves without requiring particles of matter.	Figure 1(b)
nuclear potential	The nucleus of every atom has stored energy. This energy can be released by nuclear reactions such as nuclear fission and nuclear fusion.	Figure 1(b)
gravitational potential	A raised object has stored energy due to its position above some reference level.	Figure 1(c)
kinetic	Every moving object has energy of motion, or kinetic energy.	Figure 1(c)
elastic potential	This potential energy is stored in objects that are stretched or compressed.	Figure 1(c)
sound	This form of energy, produced by vibrations, travels by waves through a material to the receiver.	Figure 1(d)
chemical potential	Atoms join together in various combinations to form many different kinds of molecules, involving various amounts of energy. In chemical reactions, new molecules are formed and energy is released or absorbed.	Figure 1(d)

Figure 1
Examples of forms of energy
(a) Electrical energy delivered to the stove heats the water in the pot. Thermal energy in the boiling water transfers to the pasta to cook it.
(b) The Sun emits radiant energies, such as infrared radiation, visible light, and ultraviolet radiation. The Sun's energy comes from nuclear fusion reactions in its core.
(c) At the highest position above the trampoline, this athlete has the greatest amount of gravitational potential energy. The energy changes to kinetic energy as her downward velocity increases. The kinetic energy then changes into elastic potential energy in the trampoline to help her bounce back up.
(d) Chemical potential energy is released when fireworks explode. Some of that energy is changed into sound energy. What other forms of energy are involved in this example?

Practice

Understanding Concepts

1. Give three examples of energy use available today that your grandparents would not have had when they were your age.
2. Name at least one form of energy associated with each object in *italics*.
   (a) A *bonfire* roasts a marshmallow.
   (b) A *baseball* smashes a window.
   (c) A *solar collector* heats water for a swimming pool.
   (d) A *stretched rubber band* is used to launch a rolled-up T-shirt toward the audience during intermission at a hockey game.
   (e) The *siren* of an ambulance warns of an emergency.

Energy Transformations

The forms of energy listed in **Table 1** are able to change from one to another; such a change is called an **energy transformation**. For example, in a microwave oven, electrical energy transforms into radiant energy (microwaves), which is then transformed into thermal energy in the food being cooked. Undoubtedly you can give many other examples of energy transformations.

energy transformation: the change from one form of energy to another

We can summarize these changes using an *energy transformation equation*. For the microwave oven example described above, the equation is

electrical energy → radiant energy → thermal energy

Practice

Understanding Concepts

3. Write the energy transformation equation for each example below.
   (a) Fireworks explode.
   (b) An arrow is shot off a bow and flies through the air.
   (c) A paved driveway feels hot on a clear, sunny day.
   (d) A camper raises an axe to chop a chunk of wood.
   (e) A lawn mower with a gasoline engine cuts a lawn.
4. Make up an example of an energy transformation involving the creation of your favourite sounds. Then, write the corresponding energy transformation equation.

SUMMARY Energy and Energy Transformations

- Energy is the capacity to do work.
- Energy exists in many forms, such as thermal energy and kinetic energy.
- In an energy transformation, energy changes from one form into another. The transformation can be described using an equation with arrows.

Section 4.1 Questions

Understanding Concepts

1. Show how energy is transformed for each situation by using an energy transformation equation.
   (a) A hotdog is being cooked at an outdoor concession stand.
   (b) A truck is accelerating along a level highway.
   (c) A child jumps on a trampoline.
   (d) A tree is knocked over by a strong wind.
   (e) An incandescent light bulb is switched on.
2. Provide an example (not yet given in this text) of a situation that involves the following energy transformations.
   (a) electrical energy → thermal energy
   (b) kinetic energy → sound energy
   (c) chemical potential energy → thermal energy
   (d) electrical energy → kinetic energy

work: the energy transferred to an object by an applied force over a measured distance

4.2 Work

The term *work* has a specific meaning in physics. **Work** is the energy transferred to an object by an applied force over a measured distance. For example, work is done when a crane lifts a steel beam for a new building, when a truck's engine makes the truck accelerate, when an archer bends a bow as the arrow is pulled back, and when the bow releases the arrow. However, when holding a heavy box on your shoulder, you may feel pain and even break into a sweat, but you are not doing any work on the box because you are not moving it.

To determine what factors affect the amount of work done in moving an object, consider a situation in which an employee at a grocery store pulls a long string of empty carts at a constant velocity with a horizontal force (**Figure 1**).

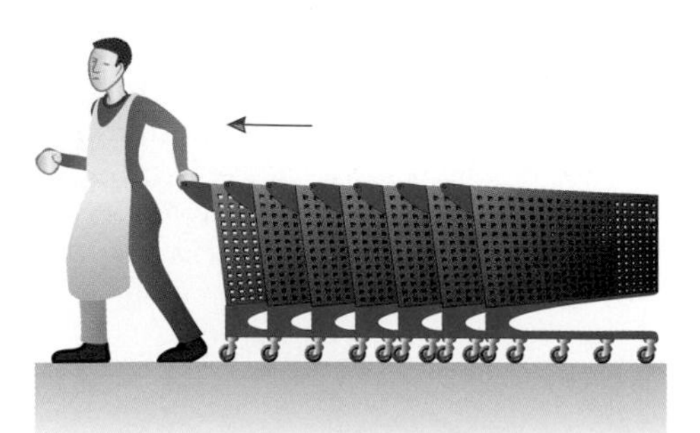

Figure 1
The work done in pulling the carts, which are locked together, depends on the force applied to the carts and the distance the carts move.

By fixing the distance that the carts are moved and doubling the applied force required to pull the carts (by adding twice as many carts), the amount of work done is doubled. Similarly, with a constant applied force, if the distance the carts are pulled is doubled, then again the work done is doubled. Thus, the work done by the applied force is directly proportional to the magnitude of the force and directly proportional to the magnitude of the displacement (distance) over which the force acts.

Using the symbols W for work, F for the magnitude of the applied force, and Δd for the magnitude of the displacement, the relationships among these variables are

$$W \propto F \quad \text{and} \quad W \propto \Delta d$$

Combining these proportionalities,

$$W \propto F{\cdot}\Delta d$$

$$W = kF{\cdot}\Delta d \quad \text{where } k \text{ is the proportionality constant}$$

Choosing the value of k to be 1, we obtain the equation for work:

$$W = F\Delta d$$

joule: (J) the SI unit for work

Notice that work is a scalar quantity; it has magnitude but no direction. Therefore, vector notations for F and Δd are omitted.

The equation $W = F\Delta d$ has important limitations. It applies only when the applied force and the displacement are in the same direction. (In more complex situations where two-dimensional motion is analyzed, the equation that is used is $W = F\Delta d(\cos\theta)$, where θ is the angle between the applied force and the displacement.)

Since force is measured in newtons and displacement is measured in metres, work is measured in newton metres (N•m). The newton metre is called the **joule** (J) in honour of James Prescott Joule, an English physicist who studied heat and electrical energy (**Figure 2**). Since the joule is a derived SI unit, it can be expressed in terms of metres, kilograms, and seconds. (Recall that the newton, the unit of force, can also be expressed in these base units: 1 N = 1 kg (m/s^2).) You will gain a lot of practice with the joule, kilojoule, megajoule, etc., for the rest of this chapter and in other parts of this course.

Figure 2
The joule is named after James Prescott Joule (1818 – 1889), owner of a Manchester brewery, who showed that heat was not a substance but, instead, the transfer of energy. He found that thermal energy produced by stirring water or mercury is proportional to the amount of energy transferred in the stirring.

Sample Problem 1

An airport terminal employee is pushing a line of carts at a constant velocity with a horizontal force of magnitude 95 N. How much work is done in pushing the carts 16 m in the direction of the applied force? Express the answer in kilojoules.

Solution

$F = 95$ N
$\Delta d = 16$ m
$W = ?$

$$W = F\Delta d$$
$$= (95 \text{ N})(16 \text{ m})$$
$$W = 1.5 \times 10^3 \text{ J}$$

The work done in pushing the carts is 1.5 kJ.

DID YOU KNOW?

Joules and Calories

Although the joule is the SI unit of energy and work, we still hear of the heat calorie (cal), a former unit of heat, and the food calorie (Cal), a former unit of food energy. These units are related in the following ways:

$$1.000 \text{ Cal} = 1.000 \times 10^3 \text{ cal} = 1.000 \text{ kcal}$$
$$1.000 \text{ cal} = 4.184 \text{ J}$$
$$1.000 \text{ Cal} = 4.184 \text{ kJ}$$

Thus, a piece of apple pie with 395.0 Cal contains 1.652×10^6 J, or 1.652 MJ, of chemical potential energy.

Answers

1. 67 J
4. 1.8×10^3 N
5. 0.73 m
6. 12 J

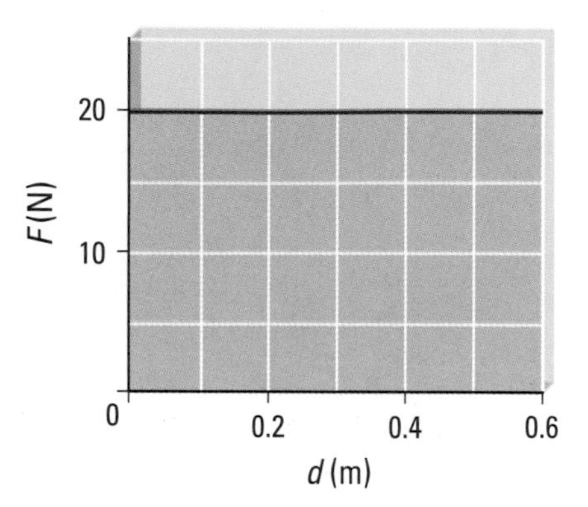

Figure 3
You can analyze the units on this force-displacement graph to determine what the area calculation represents. (Only magnitudes are considered.)

Practice

Understanding Concepts

1. A farmer uses a constant horizontal force of magnitude 21 N on a wagon and moves it a horizontal displacement of magnitude 3.2 m. How much work has the farmer done on the wagon?
2. Express joules in the base units of metres, kilograms, and seconds.
3. Rearrange the equation $W = F\Delta d$ to express (a) F by itself and (b) Δd by itself.
4. A tow truck does 3.2 kJ of work in pulling horizontally on a stalled car to move it 1.8 m horizontally in the direction of the force. What is the magnitude of the force?
5. A store clerk moved a 4.4-kg box of soap without acceleration along a shelf by pushing it with a horizontal force of magnitude 8.1 N. If the employee did 5.9 J of work on the box, how far did the box move?
6. Determine the area under the line on the graph shown in **Figure 3**. What does that area represent?

Applying Inquiry Skills

7. (a) Consider a constant force applied to an object moving with uniform velocity. Sketch a graph of the work (done on the object) as a function of the magnitude of the object's displacement.
   (b) What does the slope of the line on the graph represent?

Positive and Negative Work

In the examples presented so far, work has been positive, which is the case when the force is in the same direction as the displacement. Positive work indicates that the force tends to increase the speed of the object. However, if the force is opposite to the direction of the displacement, negative work is done. Negative work means that the force tends to decrease the speed of the object. For example, a force of kinetic friction does negative work on an object.

Thus, $W = F\Delta d$ yields a positive value when the force and displacement are in the same direction, and yields a negative value when the force and displacement are in opposite directions.

Sample Problem 2

A toboggan carrying two children (total mass = 85 kg) reaches its maximum speed at the bottom of a hill, and then glides to a stop in 21 m along a horizontal surface (see **Figure 4(a)**). The coefficient of kinetic friction between the toboggan and the snowy surface is 0.11.

(a) Draw an FBD of the toboggan when it is moving on the horizontal surface.
(b) Determine the magnitude of kinetic friction acting on the toboggan.
(c) Calculate the work done by the kinetic friction.

(a) the system diagram

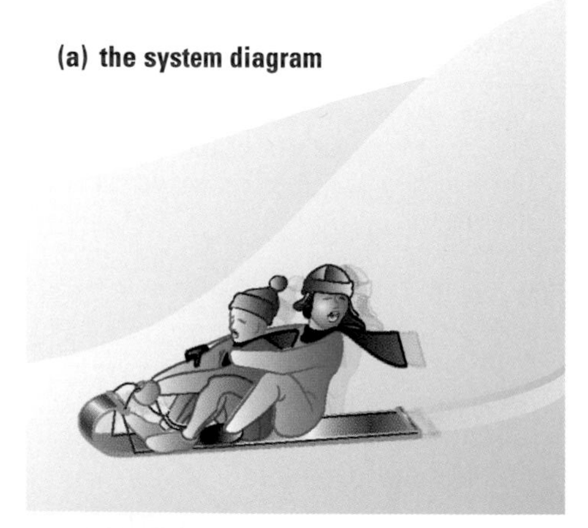

(b) the FBD

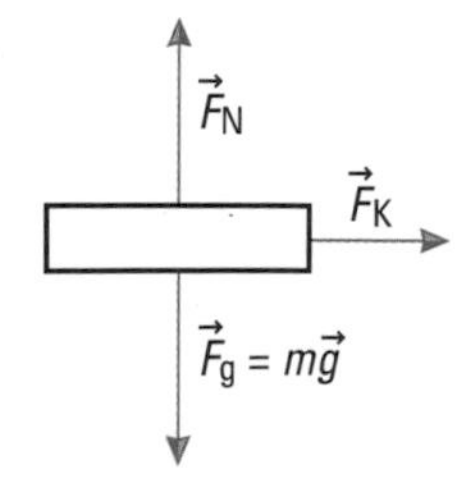

Figure 4
For Sample Problem 2

Solution

(a) The required FBD is shown in **Figure 4(b)**.

(b) $m = 85$ kg

$|\vec{g}| = 9.8$ N/kg

$\mu_K = 0.11$

$F_K = ?$

$$F_K = \mu_K F_N$$
$$= \mu_K |\vec{F}_g|$$
$$= \mu_K m |\vec{g}|$$
$$= (0.11)(85 \text{ kg})(9.8 \text{ N/kg})$$
$$F_K = 92 \text{ N}$$

The kinetic friction has a magnitude of 92 N.

(c)
$$W = F\Delta d$$
$$= (92 \text{ N})(21 \text{ m})$$
$$W = 1.9 \times 10^3 \text{ J}$$

The work done by the kinetic friction is -1.9×10^3 J because the force of friction is opposite in direction to the displacement.

Practice

Understanding Concepts

8. A student pushes a 0.85-kg textbook across a cafeteria table toward a friend. As soon as the student withdraws the hand (the force is removed), the book starts slowing down, coming to a stop after moving 65 cm horizontally. The coefficient of kinetic friction between the surfaces in contact is 0.38.
   (a) Draw a system diagram and an FBD of the book as it slows down, and calculate the magnitudes of all the forces in the diagram.
   (b) Calculate the work done on the book by the friction of the table.

Answers

8. (a) $F_g = 8.3$ N; $F_N = 8.3$ N; $F_K = 3.2$ N
   (b) −2.1 J

Work Done Against Gravity

In order to lift an object to a higher position, a force must be applied upward against the downward force of gravity on the object. If the force applied and the displacement are both vertically upward and no acceleration occurs, the work done by the force against gravity is positive, and is $W = F\Delta d$. The force in this case is equal in magnitude to the weight of the object or the force of gravity on the object, $F = |\vec{F}_g| = m|\vec{g}|$.

Sample Problem 3

A bag of groceries of mass 8.1 kg is raised vertically without acceleration from the floor to a counter top, over a distance of 92 cm. Determine:

(a) the force needed to raise the bag without acceleration
(b) the work done on the bag of groceries against the force of gravity

Solution

(a) $m = 8.1$ kg
$|\vec{g}| = 9.8$ N/kg
$F = ?$

$$F = |\vec{F}_g| = m|\vec{g}|$$
$$= (8.1 \text{ kg})(9.8 \text{ N/kg})$$
$$F = 79 \text{ N}$$

The force needed is 79 N.

(b) $\Delta d = 0.92\ \text{m}$
$W = ?$

$$W = F\Delta d$$
$$= (79\ \text{N})(0.92\ \text{m})$$
$$W = 73\ \text{J}$$

The work done against the force of gravity is 73 J.

Practice

Understanding Concepts

9. A 150-g book is lifted from the floor to a shelf 2.0 m above. Calculate the following:
   (a) the force needed to lift the book without acceleration
   (b) the work done by this force on the book to lift it up to the shelf
10. A world-champion weight lifter does 5.0×10^3 J of work in jerking a weight from the floor to a height of 2.0 m. Calculate the following:
    (a) the average force exerted to lift the weight
    (b) the mass of the weight
11. An electric forklift truck is capable of doing 4.0×10^5 J of work on a 4.5×10^3 kg load. To what height can the truck lift the load?

Answers

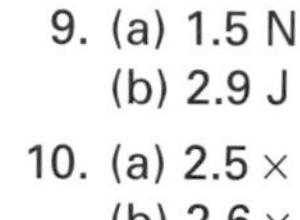

9. (a) 1.5 N
   (b) 2.9 J
10. (a) 2.5×10^3 N
    (b) 2.6×10^2 kg
11. 9.1 m

Zero Work

Situations exist in which an object experiences a force, or a displacement, or both, yet no work is done on the object. If you are holding a box on your shoulder, you may be exerting an upward force on the box, but the box is not moving, so the displacement is zero, and the work done on the box, $W = F\Delta d$, is also zero.

In another example, if a puck on an air table is moving, it experiences negligible friction while moving for a certain displacement. The force in the direction of the displacement is zero, so the work done on the puck is also zero.

In a third example, consider the force exerted by the figure skater who glides along the ice while holding his partner above his head (**Figure 5**). There is both a force on the partner and a horizontal displacement. However, the displacement is perpendicular (not parallel) to the force, so no work is done on the woman. Of course, work was done in lifting the woman vertically to the height shown.

Figure 5
If the applied force and the displacement are perpendicular, no work is done by the applied force.

Practice

Understanding Concepts

12. A student pushes against a large maple tree with a force of magnitude 250 N. How much work does the student do on the tree?
13. A 500-kg meteoroid is travelling through space far from any measurable force of gravity. If it travels at 100 m/s for 100 years, how much work is done on the meteoroid?
14. A nurse holding a newborn 3.0-kg baby at a height of 1.2 m off the floor carries the baby 15 m at constant velocity along a hospital corridor. How much work has the force of gravity done on the baby?
15. Based on questions 12, 13, and 14, write general conclusions regarding when work is or is not done on an object.

SUMMARY Work

- Work is the energy transferred to an object by an applied force over a distance.
- If the force and displacement are in the same direction, the work done by the force, $W = F\Delta d$, is a positive value. If the force and displacement are in opposite directions, the work done is a negative value.
- Work is a scalar quantity measured in joules (J).

Figure 6
An off-road dump truck

Section 4.2 Questions

Understanding Concepts

1. An average horizontal force of magnitude 32 N is exerted on a box on a horizontal floor. If the box moves 7.8 m along the floor, how much work does the force do on the box?
2. An elevator lifts you upward without acceleration a distance of 36 m. How much work does the elevator do against the force of gravity to move you this far?
3. An off-road dump truck can hold 325 t of gravel (**Figure 6**). How much work must be done on the gravel to raise it an average of 9.2 m to get it into the truck?
4. A camper does 7.4×10^2 J of work in lifting a pail filled with water 3.4 m vertically up a well at a constant speed.
   (a) What force is exerted by the camper on the pail of water?
   (b) What is the mass of water in the pail?
5. In an emergency, the driver of a 1.3×10^3-kg car slams on the brakes, causing the car to skid forward on the road. The coefficient of kinetic friction between the tires and the road is 0.97, and the car comes to a stop after travelling 27 m horizontally. Determine the work done by the force of friction during the skidding.
6. For the equation $W = F\Delta d$, describe
   (a) when the equation applies
   (b) when the equation yields a nil or zero value of work

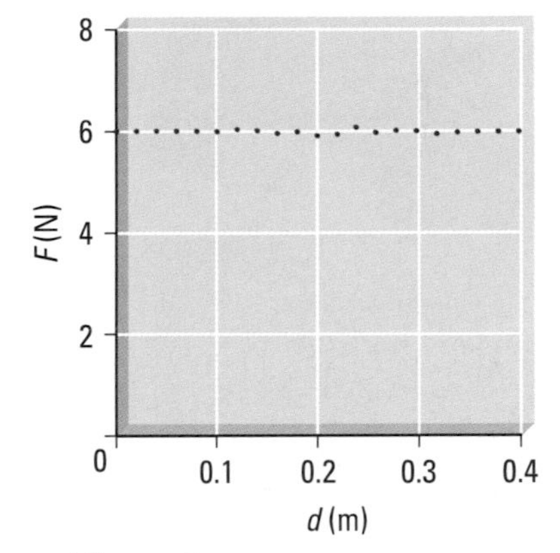

Figure 7
For question 7

Applying Inquiry Skills

7. The graph shown in **Figure 7** was generated by a computer interfaced to a force sensor that collected data several times per second as a block of wood was pulled with a horizontal force across a desk.
   (a) Estimate the work done by the force applied to the block. Show your calculations.
   (b) Describe sources of systematic error when using a force sensor in this type of investigation.

4.3 Mechanical Energy

A constant change our society experiences is tearing down old buildings to make way for new ones. One way to do this is by chemical explosions. However, if that is considered to be too dangerous, a much slower way is to use a wrecking ball (**Figure 1**). What energy transformations allow such a ball to destroy a building?

Figure 1
Several principles of mechanics are applied in the demolition of this large structure. Can you write the energy transformation equation for this situation?

As work is done by a machine on the wrecking ball to raise it to a high level, the ball gains gravitational potential energy. This potential energy arises from the fact that the force of gravity is pulling down on the ball. The type of energy possessed by an object because of its position relative to a lower position is called **gravitational potential energy**, E_g. This potential energy can be used to do work on some object at a lower level.

gravitational potential energy: the energy possessed by an object because of its position relative to a lower position

When the ball is released, it falls. As the ball falls, it loses gravitational potential energy and gains kinetic energy as its speed increases. Energy due to the motion of an object is called **kinetic energy**, E_k. ("Kinetic," like the word "kinematics," comes from the Greek word *kinema*, which means "motion.")

kinetic energy: the energy possessed by an object due to its motion

The sum of the gravitational potential energy and kinetic energy is called **mechanical energy**. When the wrecking ball strikes the wall, its mechanical energy allows it to do work on the wall.

mechanical energy: the sum of the gravitational potential energy and the kinetic energy

Determining Gravitational Potential Energy

Suppose you are erecting a tent and using a hammer to pound the tent pegs into the ground, as in **Figure 2**.

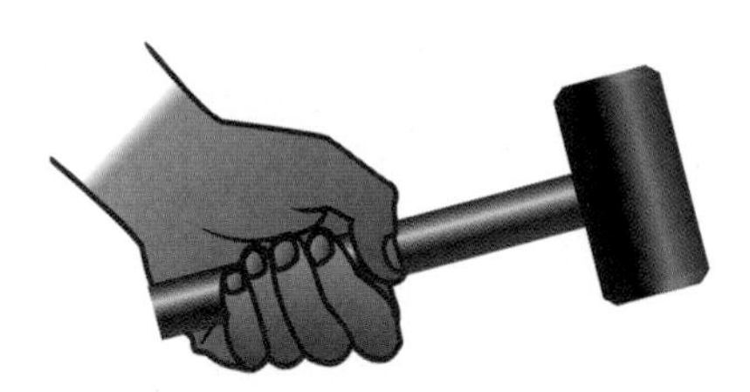

Figure 2
Raising a hammer gives it gravitational potential energy.

In order to lift the hammer a height Δh, you would have to transfer energy $W = F\Delta h$ to it, based on the equation for work, $W = F\Delta d$. Here, F is the magnitude of the force required to lift the hammer from the ground without acceleration; it is equal to the magnitude of the hammer's weight, which is $m|\vec{g}|$. From here on, $|\vec{g}|$ will simply be represented by g. The transferred energy, or work, equals the hammer's gravitational potential energy above a reference level such as the ground. That is, $E_g = F\Delta h$, where E_g is the gravitational potential energy of the hammer raised a height Δh above the original level. Since $F = mg$, we can now write the common equation for gravitational potential energy:

$$E_g = F\Delta h = mg\Delta h \quad \text{where } g = 9.8 \text{ N/kg}$$

In SI, energy is measured in joules, mass in kilograms, and height (or displacement) in metres.

Often we are concerned about the potential energy relative to a particular **reference level**, the level to which an object may fall. Then Δh in the gravitational potential energy equation is equivalent to the height h of the object above the reference level. Thus, the equation for the potential energy of an object relative to a reference level is

reference level: the level to which an object may fall

$$E_g = mgh$$

When answering questions on relative potential energy, it is important for you to indicate the reference level. In the hammer and tent peg example, the hammer has a greater potential energy relative to the ground than it has relative to the top of the peg.

Sample Problem 1

In the sport of pole vaulting, the jumper's centre of mass must clear the pole. Assume that a 59-kg jumper must raise the centre of mass from 1.1 m off the ground to 4.6 m off the ground. What is the jumper's gravitational potential energy at the top of the bar relative to where the jumper started to jump?

Solution

The height of the jumper's centre of mass above the reference level indicated is $h = 4.6 \text{ m} - 1.1 \text{ m} = 3.5 \text{ m}$.

$m = 59 \text{ kg}$
$g = 9.8 \text{ N/kg}$
$E_g = ?$

$$E_g = mgh$$
$$= (59 \text{ kg})(9.8 \text{ N/kg})(3.5 \text{ m})$$
$$E_g = 2.0 \times 10^3 \text{ J}$$

The jumper's gravitational potential energy relative to the lower position is 2.0×10^3 J.

DID YOU KNOW ?

Gravitational Potential Energy in Nature

Some animals are known to take advantage of gravitational potential energy. One example is the bearded vulture (or lammergeyer), the largest of all vultures. This bird, found in South Africa, is able to digest bones. Often its food consists of bones picked clean by other animals. When confronted with a bone too large to digest, the vulture carries the bone to a great height and drops it onto a rock. Then it circles down to scoop out the marrow with its tongue.

Practice

Understanding Concepts

1. A 485-g book is resting on a desk 62 cm high. Calculate the book's gravitational potential energy relative to (a) the desktop and (b) the floor.
2. Estimate your own gravitational potential energy relative to the lowest floor in your school when you are standing at the top of the stairs of the highest floor.
3. Rearrange the equation $E_g = mgh$ to obtain an equation for
   (a) m (b) g (c) h
4. The elevation at the base of a ski hill is 350 m above sea level. A ski lift raises a skier (total mass = 72 kg, including equipment) to the top of the hill. If the skier's gravitational potential energy relative to the base of the hill is now 9.2×10^5 J, what is the elevation at the top of the hill?
5. The spiral shaft in a grain auger raises grain from a farmer's truck into a storage bin (**Figure 3**). Assume that the auger does 6.2×10^5 J of work on a certain amount of grain to raise it 4.2 m from the truck to the top of the bin. What is the total mass of the grain moved? Ignore friction.
6. A fully dressed astronaut, weighing 1.2×10^3 N on Earth, is about to jump down from a space capsule that has just landed safely on Planet X. The drop to the surface of X is 2.8 m, and the astronaut's gravitational potential energy relative to the surface is 1.1×10^3 J.
   (a) What is the magnitude of the gravitational field strength on Planet X?
   (b) How long does the jump take?
   (c) What is the astronaut's maximum speed?

Answers

1. (a) 0.0 J
   (b) 2.9 J
4. 1.7×10^3 m
5. 1.5×10^4 kg
6. (a) 3.2 N/kg
   (b) 1.3 s
   (c) 4.2 m/s

Figure 3
A grain auger

Applications of Mechanical Energy

The example in **Figure 1** from earlier in this section illustrates a useful application of gravitational potential energy. An object is raised to a position above a reference level. Then the force of gravity causes the object to accelerate. The object gains speed and, thus, kinetic energy, allowing it to crash into the wall and do work in demolishing the wall. Another example is a roller coaster at a high position where its gravitational potential energy is maximum. The force of gravity causes the roller coaster to accelerate downward, giving it enough speed and, thus, kinetic energy, to travel around the track.

(a)

(b)

Figure 4
Applications of gravitational potential energy
(a) Pile driver
(b) Damming a river to produce hydroelectricity

Figure 4 shows two more applications of gravitational potential energy. In **Figure 4(a)**, a pile driver is about to be lifted by a motor high above the pile. It will then have the gravitational potential energy to do the work of driving the pile into the ground. The pile will act as a support for a high-rise building. In **Figure 4(b)**, water stored on a dammed river has gravitational potential energy relative to the base of the dam. At hydroelectric generating stations, this gravitational potential energy is transformed into electrical energy.

Practice

Understanding Concepts

7. Write the energy transformation equation for each example below:
   (a) the wrecking ball (Assume the machine has a diesel engine.)
   (b) the hammer used to pound in the tent peg (Start with the energy stored in the food eaten by the camper.)

Determining Kinetic Energy

A dart held stationary in your hand has no kinetic energy; once you have thrown it at the dartboard and it is moving, it has kinetic energy (**Figure 5**).

To determine an equation for kinetic energy, we will use concepts from this chapter and Chapter 1. Assume that an object of mass m, travelling at a speed v_i, has a net force of magnitude F exerted on it over a displacement of magnitude Δd. The object will undergo an acceleration, with magnitude a, to reach a speed v_f. The work done on the object is

$$W = F\Delta d = ma\Delta d$$

From Chapter 1, we know that

$$v_f^2 = v_i^2 + 2a\Delta d$$

$$a\Delta d = \frac{v_f^2 - v_i^2}{2}$$

Substituting this equation into the equation for work we have

$$W = m\left(\frac{v_f^2 - v_i^2}{2}\right)$$

or $$W = \frac{mv_f^2}{2} - \frac{mv_i^2}{2}$$

To simplify this equation, let us assume that the object starts from rest, so $v_i = 0$, and the last term in the equation is zero. Then the work done by the force to cause the object to reach a speed v is

$$W = \frac{mv^2}{2}$$

This quantity is equal to the object's kinetic energy. Thus, an object of mass m, travelling at a speed v, has a kinetic energy of

$$E_k = \frac{mv^2}{2}$$

Again, in SI, energy is measured in joules, mass in kilograms, and speed in metres per second.

Figure 5
A moving object has kinetic energy.

Sample Problem 2

Find the kinetic energy of a 48-g dart travelling at a speed of 3.4 m/s.

Solution

$m = 0.048$ kg
$v = 3.4$ m/s
$E_k = ?$

$$E_k = \frac{mv^2}{2}$$
$$= \frac{(0.048 \text{ kg})(3.4 \text{ m/s})^2}{2}$$
$$E_k = 0.28 \text{ J}$$

The kinetic energy of the dart is 0.28 J.

Figure 6
Powerful legs carry an ostrich at high speeds.

Practice

Understanding Concepts

8. Calculate the kinetic energy of the item in *italics* in each case.
   (a) A 7.2-kg *shot* leaves an athlete's hand during the shot put at a speed of 12 m/s.
   (b) A 140-kg *ostrich* is running at 14 m/s. (The ostrich, **Figure 6**, is the fastest two-legged animal on Earth.)
9. Prove that the unit for kinetic energy is equivalent to the unit for work.
10. Starting with the equation $E_k = \frac{mv^2}{2}$, find an equation for (a) m and (b) v.
11. A softball travelling at a speed of 34 m/s has a kinetic energy of 98 J. What is its mass?
12. A 97-g cup falls from a kitchen shelf and shatters on the ceramic tile floor. Assume that the maximum kinetic energy obtained by the cup is 2.6 J and that air resistance is negligible.
    (a) What is the cup's maximum speed?
    (b) What do you suppose happened to the 2.6 J of kinetic energy after the crash?
13. There are other ways of deriving the equation for kinetic energy using the uniform acceleration equations from Chapter 1. Show an alternative derivation.

Answers

8. (a) 5.2×10^2 J
   (b) 1.4×10^4 J
11. 0.17 kg
12. (a) 7.3 m/s

DID YOU KNOW ?

Positive Speeds

In Practice questions 10 and 12, when the equation for kinetic energy is rewritten to solve for the speed, the solution is a square root. Only the positive square root applies in all cases, so we need to omit the "±" sign in front of the square root symbol.

SUMMARY Mechanical Energy

- Gravitational potential energy, which is energy possessed by an object due to its position above a reference level, is given by the equation $E_g = mgh$, where m is the mass of the object, g is the magnitude of the gravitational field, and h is the height above the reference level.
- Kinetic energy, which is energy of motion, is found using the equation $E_k = \frac{mv^2}{2}$.

Section 4.3 Questions

Understanding Concepts

1. Explain why a roller coaster is called a "gravity ride."
2. In April 1981, Arnold Boldt of Saskatchewan set a world high-jump record for disabled athletes in Rome, Italy, jumping to a height of 2.04 m. Calculate Arnold's gravitational potential energy relative to the ground. (Assume that his mass was 68 kg at the time of the jump.)
3. A hockey puck has a gravitational potential energy of 2.3 J when it is held by a referee at a height of 1.4 m above the rink surface. What is the mass of the puck?
4. A 636-g basketball has a gravitational potential energy of 19 J near the basket. How high is the ball off the floor?
5. Determine your own kinetic energy when you are running at a speed of 5.5 m/s.
6. At what speed must a 1200-kg car be moving to have a kinetic energy of (a) 2.0×10^3 J and (b) 2.0×10^5 J?
7. How high would a 1200-kg car have to be raised above a reference level to give it a gravitational potential energy of (a) 2.0×10^3 J and (b) 2.0×10^5 J?

Applying Inquiry Skills

8. Use your graphing skills to show the relationship between each set of variables listed.
   (a) gravitational potential energy; gravitational field strength (with a constant mass and height)
   (b) kinetic energy; mass of the object (at a constant speed)
   (c) kinetic energy; speed of the object (with a constant mass)

Making Connections

9. (a) By what factor does the kinetic energy of a car increase when its speed doubles? triples?
   (b) What happens to this kinetic energy in a car crash?
   (c) Determine the speed a car is travelling if its kinetic energy is double what it would be at a highway speed limit of 100 km/h.
   (d) Make up a cartoon to educate drivers about the relationships among speed, higher energy, and damage done in collisions.

4.4 The Law of Conservation of Energy and Efficiency

As you have learned, energy can change from one form to another. Scientists say that when any such change occurs, energy is conserved. This is expressed as the **law of conservation of energy**.

law of conservation of energy: When energy changes from one form to another, no energy is lost.

Law of Conservation of Energy

When energy changes from one form to another, no energy is lost.

This law applies to all the forms of energy listed in **Table 1** of section 4.1.

Let us consider a practical application of the law of conservation of energy. Our example deals with the use of a pile driver, shown in **Figure 4** of section 4.3.

Figure 1(a) illustrates the design of a pile driver. A hammer is lifted by an electric or gasoline engine (not shown) to a position above the pile. From there, the shaft guides the hammer as it falls and strikes the pile to knock it further into the ground to act as a structural support for a building. We will analyze the energy changes occurring in these events, shown in **Figure 1**(b).

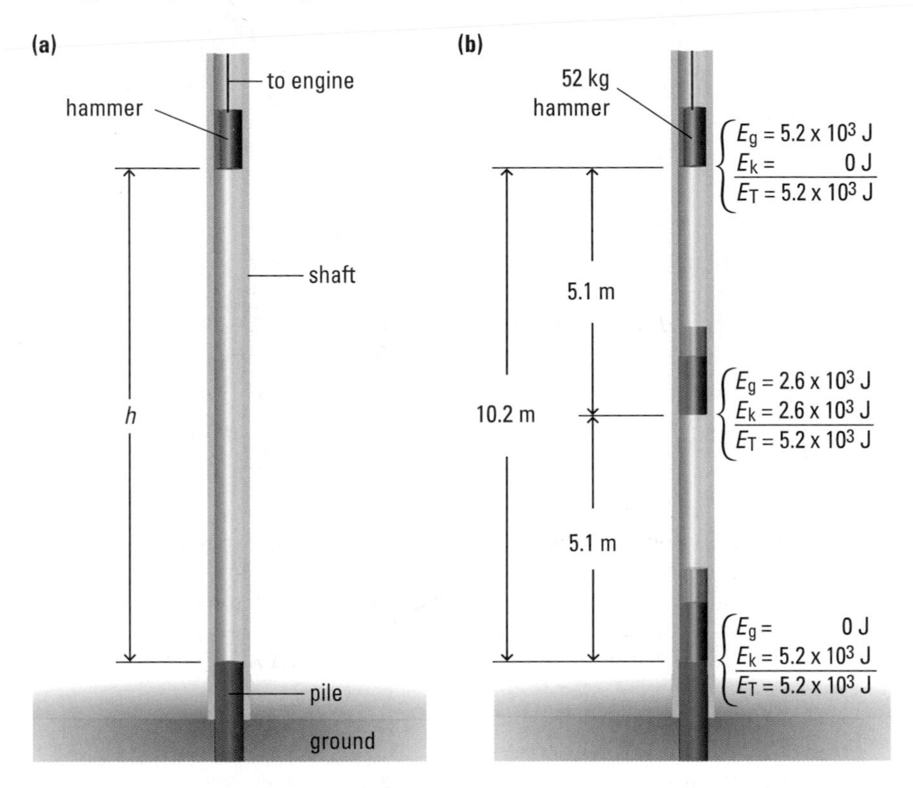

Figure 1
The design and energy changes of a pile driver
(a) Design features
(b) Energy changes

In hoisting the hammer from the level of the pile to the top position, the engine does the following work on the hammer:

$$
\begin{aligned}
W &= F\Delta d \\
&= mgh \\
&= (52\text{ kg})(9.8\text{ N/kg})(10.2\text{ m}) \\
W &= 5.2 \times 10^3\text{ J}
\end{aligned}
$$

Notice that we are concerned here with the work done *on the hammer*. We will not consider the work required because of friction in the shaft or in the engine itself. Including these factors would involve an extra series of calculations, though the final conclusions would be more or less the same.

At the top position, then, the hammer has a gravitational potential energy ($E_g = mgh$) of 5.2×10^3 J. Its kinetic energy is zero because its speed is zero. The hammer's total mechanical energy, $E_T = E_g + E_k$, is thus 5.2×10^3 J.

When the hammer is released, it accelerates down the shaft at 9.8 m/s². (We assume that friction is negligible when the shaft is vertical.) At any position, the gravitational potential energy and kinetic energy can be calculated. We will choose the halfway point, where the hammer is 5.1 m above the pile. Its gravitational potential energy there is

$$
\begin{aligned}
E_g &= mgh \\
&= (52\text{ kg})(9.8\text{ N/kg})(5.1\text{ m}) \\
E_g &= 2.6 \times 10^3\text{ J}
\end{aligned}
$$

DID YOU KNOW?

Comparing Conservation and Conserving

The word "conservation" means that the total amount remains constant. This differs from the everyday usage of the expression "energy conservation," which refers to not wasting energy. Thus, you should distinguish between *conservation of energy* (a scientific phenomenon) and *conserving energy* (a wise thing to do).

To find the kinetic energy, we must first find the hammer's speed after it has fallen 5.1 m. From Chapter 1, we know that

$$v_f^2 = v_i^2 + 2a\Delta d$$

$$v = \sqrt{v_i^2 + 2a\Delta d} \quad (v_i = 0)$$

$$= \sqrt{2(9.8 \text{ m/s}^2)(5.1 \text{ m})}$$

$$v = 1.0 \times 10^1 \text{ m/s}$$

Thus,

$$E_k = \frac{mv^2}{2}$$

$$= \frac{(52 \text{ kg})(1.0 \times 10^1 \text{ m/s})^2}{2}$$

$$E_k = 2.6 \times 10^3 \text{ J}$$

Again, the total mechanical energy, $E_g + E_k$, is 5.2×10^3 J.

Next, at the instant just prior to striking the pile, the hammer has zero gravitational potential energy ($h = 0$ in $E_g = mgh$) but a kinetic energy based on these calculations:

$$v = \sqrt{v_i^2 + 2a\Delta d} \quad (v_i = 0)$$

$$= \sqrt{2(9.8 \text{ m/s}^2)(10.2 \text{ m})}$$

$$v = 14.1 \text{ m/s}$$

Thus,

$$E_k = \frac{mv^2}{2}$$

$$= \frac{(52 \text{ kg})(14.1 \text{ m/s})^2}{2}$$

$$E_k = 5.2 \times 10^3 \text{ J}$$

Once again, the total mechanical energy is 5.2×10^3 J.

Finally, the hammer strikes the pile. Its kinetic energy changes into other forms of energy such as sound, thermal energy (due to increased motion of the molecules), and the kinetic energy of the pile as it is driven further into the ground.

This series of energy changes illustrates the law of conservation of energy. The work done on the object in raising it gives the object gravitational potential energy. This energy changes into kinetic energy and other forms of energy. Energy is not lost; it simply changes into other forms.

The law of conservation of energy can be applied to describe the energy transformations mentioned earlier in this chapter, namely using a wrecking ball to tear down old buildings, using a hammer to pound in a tent peg, operating a thrill ride, and producing hydroelectric power at a dam. The law can also be used to solve various types of problems, as Sample Problem 1 and the Practice questions illustrate.

Sample Problem 1

As the water in a river approaches a 5.7-m vertical drop, its average speed is 5.1 m/s. For each kilogram of water in the river, determine the following:

(a) the kinetic energy at the top of the waterfall
(b) the gravitational potential energy at the top of the falls relative to the bottom
(c) the total mechanical energy at the bottom of the falls, not considering friction (use the law of conservation of energy)
(d) the speed at the bottom of the falls (use the law of conservation of energy)

Solution

(a) $m = 1.0\ \text{kg}$
$v = 5.1\ \text{m/s}$
$E_k = ?$

$$E_k = \frac{mv^2}{2}$$
$$= \frac{(1.0\ \text{kg})(5.1\ \text{m/s})^2}{2}$$
$$E_k = 13\ \text{J}$$

The kinetic energy of each kilogram of water at the top of the falls is 13 J.

(b) $m = 1.0\ \text{kg}$
$g = 9.8\ \text{N/kg}$
$h = 5.7\ \text{m}$
$E_g = ?$

$$E_g = mgh$$
$$= (1.0\ \text{kg})(9.8\ \text{N/kg})(5.7\ \text{m})$$
$$E_g = 56\ \text{J}$$

The gravitational potential energy relative to the bottom of the falls for each kilogram of water is 56 J.

(c) As the water falls, the gravitational potential energy changes to kinetic energy. By the law of conservation of energy, the total energy at the bottom of the falls is the sum of the initial kinetic energy and the kinetic energy gained due to the energy change. Thus, the total energy for each kilogram of water at the bottom of the falls is 13 J + 56 J = 69 J.

(d) $m = 1.0\ \text{kg}$
$E_k = 69\ \text{J}$
$v = ?$

From $E_k = \frac{mv^2}{2}$,

$$v = \sqrt{\frac{2E_k}{m}}$$
$$= \sqrt{\frac{2(69\ \text{J})}{1.0\ \text{kg}}}$$
$$v = 12\ \text{m/s}$$

The speed of water at the bottom of the falls is approximately 12 m/s.

Practice

Understanding Concepts

1. Use the law of conservation of energy to describe the energy changes that occur in the operation of a roller coaster at an amusement park.
2. A 91-kg kangaroo exerts enough force to acquire 2.7 kJ of kinetic energy in jumping straight upward.
   (a) Apply the law of conservation of energy to determine how high this agile marsupial jumps.
   (b) What is the magnitude of the kangaroo's maximum velocity?
3. A ball is dropped vertically from a height of 1.5 m; it bounces back to a height of 1.3 m. Does this violate the law of conservation of energy? Explain.

Answers

2. (a) 3.0 m
   (b) 7.7 m/s

Answers

4. (a) 4.5×10^3 J
   (b) 1.0×10^1 m/s
   (c) 13 m/s

4. A 56-kg diver jumps off the end of a 7.5-m platform with an initial horizontal speed of 3.6 m/s.
   (a) Determine the diver's total mechanical energy at the end of the platform relative to the surface of the water in the pool below.
   (b) Apply the law of conservation of energy to determine the diver's speed at a height of 2.8 m above the water.
   (c) Repeat (b) to find the maximum speed of the diver upon reaching the water.

INQUIRY SKILLS

- ● Questioning
- ● Hypothesizing
- ● Predicting
- ○ Planning
- ● Conducting
- ● Recording
- ● Analyzing
- ● Evaluating
- ● Communicating

Investigation 4.4.1

Testing the Law of Conservation of Energy

The energy transformations involved in the motion of a pendulum bob will be analyzed to test the law of conservation of energy.

Question

(a) Make up a question for this investigation about the law of conservation of mechanical energy for a mechanical system, such as a swinging pendulum.

Hypothesis/Prediction

(b) Write your own hypothesis and prediction.

Design

When a raised pendulum bob (**Figure 2**) is released, the bob gains speed and kinetic energy, both of which reach a maximum at the bottom of the swing. You can use a simple method to determine the gravitational potential energy of the bob relative to its lowest position. Then, you can determine the maximum speed of the bob, and thus the maximum kinetic energy. The method suggested here uses a photocell that determines the time a light beam, aimed at it, is blocked by the moving pendulum bob. Finally, you can compare the maximum gravitational potential energy and the maximum kinetic energy.

To achieve the highest accuracy, the distance measurements related to the pendulum bob should be taken from the centre of the bob.

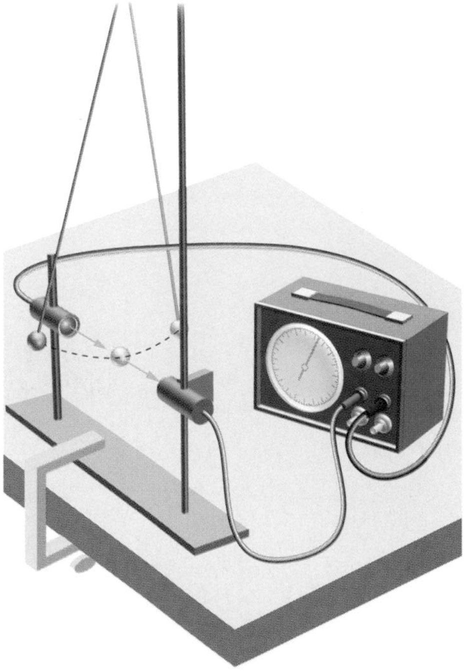

Figure 2
This is one way to determine the maximum speed of the pendulum bob.

Materials

50-g (or smaller) mass
strong string or wire
tall stand (more than 1 m) and clamp
millisecond timer (or computer)
light source and photocell

Procedure

1. Construct a pendulum by attaching a 50-g mass to a strong string about 1 m long. Clamp the stand to the table and tie the string to the clamp so that the pendulum bob can swing freely between the light source and photocell.
2. Measure the distance from the table to the bottom of the pendulum bob. The table is the reference level for determining the gravitational potential energy of the pendulum bob.

3. Set up the timer using a photocell and light beam so that the beam will break when the pendulum bob is at the lowest position of its swing. The timer will then record to the nearest millisecond the time interval during which the beam is broken.
4. Draw the pendulum aside and set the timer to zero.
5. While the pendulum bob is held aside, have someone measure the height of the bob from the reference level, the table in this case.
6. Release the pendulum bob and allow it to swing through the light beam. Have someone catch it on the other side of its swing.
7. Note the time interval for which the light beam was broken by the pendulum bob.
8. Measure the diameter of the pendulum bob, making sure to measure the bob in the area where it broke the beam.

Place the photocell and light source far enough apart so the pendulum bob does not come near them.

Analysis

(c) Determine the average speed of the pendulum bob at the lowest point in its swing by using the time interval for which the beam was broken and the diameter of the bob. Use this speed to calculate the maximum kinetic energy of the mass.

(d) Calculate the gravitational potential energy of the pendulum bob when it is held aside, and when it swings through the lowest point in its cycle.

(e) Is mechanical energy conserved during one swing of the pendulum? Justify your answer.

(f) We know that any pendulum will eventually stop swinging. Explain where the mechanical energy would have gone.

Evaluation

(g) Was your original question appropriate?

(h) How accurate were your hypothesis and prediction?

(i) Describe the sources of error in this investigation. (To review errors in experiments, refer to Appendix A.)

(j) If you were to perform this experiment again, what improvements would you suggest?

Efficiency

An incandescent light bulb is designed to provide light energy. Unfortunately, it also produces a lot of thermal energy while in use. In fact, only about 5% of the electrical energy delivered to the bulb transforms to light energy; the rest becomes waste thermal energy. We say that the incandescent light bulb is only 5% efficient (**Figure 3**).

Efficiency, expressed as a percentage, is the ratio of the useful energy provided by a device to the energy required to operate the device. The efficiency of an energy transformation is calculated as follows:

$$\text{efficiency} = \frac{\text{useful energy output}}{\text{energy input}} \times 100\%$$

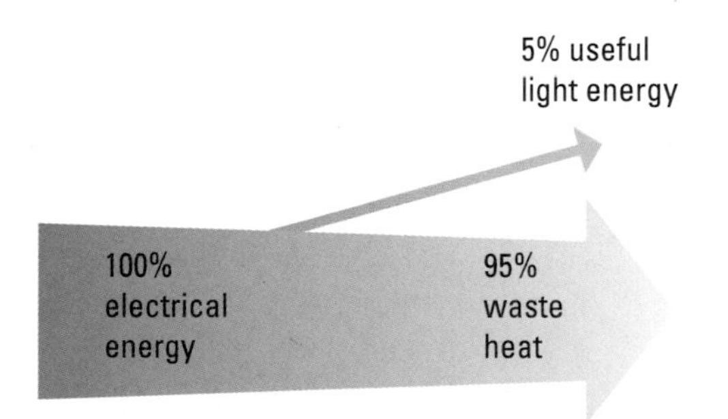

Figure 3
The efficiency in this case is 5%. What would the efficiency of the light bulb be if its purpose were to provide thermal energy, or heat?

efficiency: the ratio of the useful energy provided by a device to the energy required to operate the device

Using E_{out} for useful energy output and E_{in} for energy input, the efficiency equation is

$$\text{efficiency} = \frac{E_{out}}{E_{in}} \times 100\%$$

Sample Problem 2

A family uses several planks to slide a 350-kg piano onto the back of a pickup truck (**Figure 4**). The box on the back of the truck is 81 cm above the ground and the planks are 3.0 m long. An average force of magnitude 1500 N is required to slide the piano up the planks.

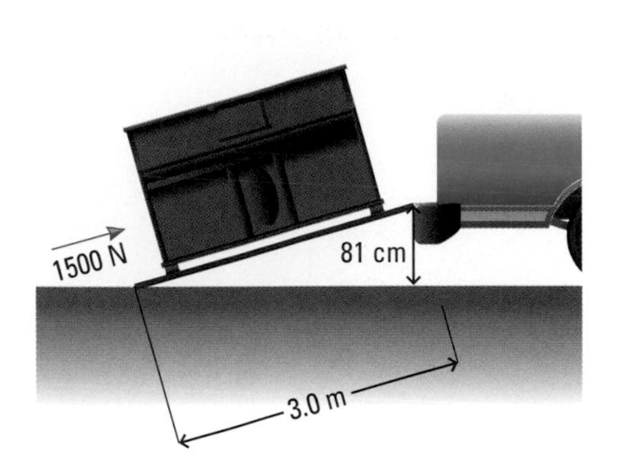

Figure 4
For Sample Problem 2

(a) Determine the work done in loading the piano.
(b) Calculate the efficiency of the planks as a simple machine to load the piano.
(c) Where does the rest of the energy go?

Solution

(a) $F = 1500$ N
$\Delta d = 3.0$ m
$E_{in} = ?$

$$\begin{aligned} E_{in} &= W \\ &= F\Delta d \\ &= (1500 \text{ N})(3.0 \text{ m}) \\ E_{in} &= 4.5 \times 10^3 \text{ J} \end{aligned}$$

The work done, or the energy input in loading the piano, is 4.5×10^3 J.

(b) The useful energy output in this case is the increase in gravitational potential energy of the piano going from ground level onto the back of the truck.

$m = 350$ kg
$g = 9.8$ N/kg
$h = 0.81$ m
$E_{out} = ?$
efficiency = ?

The useful work done, or the energy output, is

$$\begin{aligned} E_{out} &= W \\ &= mgh \\ &= (350 \text{ kg})(9.8 \text{ N/kg})(0.81 \text{ m}) \\ E_{out} &= 2.8 \times 10^3 \text{ J} \end{aligned}$$

The efficiency of the planks is found by using the expression

$$\begin{aligned} \text{efficiency} &= \frac{E_{out}}{E_{in}} \times 100\% \\ &= \frac{2.8 \times 10^3 \text{ J}}{4.5 \times 10^3 \text{ J}} \times 100\% \\ \text{efficiency} &= 62\% \end{aligned}$$

The planks are 62% efficient when used as a machine to load the piano.

(c) Much of the remaining 38% of the energy input is wasted as thermal energy that results as friction causes the molecules to become more agitated. Sound energy is also produced by friction.

Practice

Understanding Concepts

5. A construction worker uses a rope and pulley system to raise a 27-kg can of paint 3.1 m to the top of a scaffold (**Figure 5**). The downward force on the rope is 3.1×10^2 N as the rope is pulled 3.1 m.
   (a) Find the work done in raising the can of paint.
   (b) How much "useful work" is done?
   (c) What is the efficiency of the rope and pulley in raising the can of paint?
   (d) Suggest why the efficiency of this simple machine is not 100%.

Answers

5. (a) 9.6×10^2 J
   (b) 8.2×10^2 J
   (c) 85%

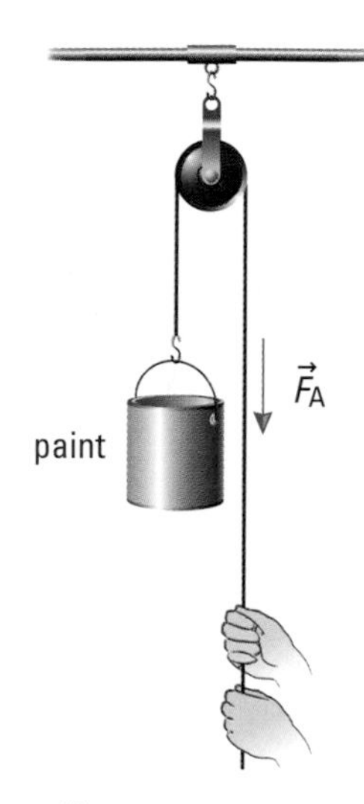

Figure 5
A pulley is one type of simple machine. Here it is used to change the direction of a force.

Investigation 4.4.2

The Efficiency of a Ramp

In Sample Problem 2, the efficiency of the ramp was 62%. Do all ramps have the same efficiency? You can explore this question in a controlled experiment. Three factors will be tested: the mass of the cart, the angle of the ramp, and the type of friction between the cart and the ramp. (Rolling friction occurs when the cart is wheeled up the ramp; sliding friction occurs when the cart is upside down and pulled up the ramp.)

INQUIRY SKILLS

- ○ Questioning
- ● Hypothesizing
- ● Predicting
- ○ Planning
- ● Conducting
- ● Recording
- ● Analyzing
- ● Evaluating
- ● Communicating
- ● Synthesizing

Question

How do the factors affect the efficiency of a ramp used to raise a cart?

Hypothesis/Prediction

(a) Read the procedure given and then write your hypothesis and prediction.

Design

The procedure given suggests using a spring scale to pull a cart up a ramp. Other tools can be used, such as a force meter or simulation software. If you are using these tools, revise your procedural steps.

(b) To record the data for your experiment, copy **Table 1** into your notebook with enough columns for six to nine trials. Then you will have data from two or three trials for each of the variables tested: cart mass, ramp angle, and length and height of the ramp.

Table 1 Data for Ramp Investigation

Variables	Trial 1 ...
mass of cart, m (kg)	
weight of cart, F_g (N)	
length of ramp, Δd (m)	
height of ramp, h (m)	
angle of ramp (°)	
force parallel to ramp, F (N)	
work input, E_{in} (J)	
useful energy output, E_{out} (J)	
efficiency (%)	

Materials

dynamics carts
ramp
beam balance
spring scale calibrated in newtons
several bricks or books
metre stick
extra masses

Make sure the ramp support is stable: clamp if necessary.

Make sure the cart can be stopped safely at the bottom of the ramp.

Make sure the additional masses are mounted in a stable way on the cart.

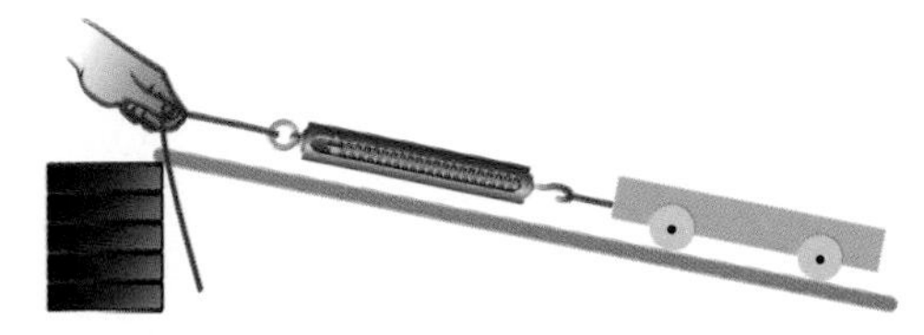

Figure 6
Setup for Investigation 4.4.2

Procedure

1. Measure the mass of the cart.
2. Set up the apparatus, as shown in **Figure 6**, so that the ramp is inclined at an angle of about 20° to the horizontal. Measure the length and height of the ramp.
3. Attach the force scale to the cart and determine the average force needed to pull the cart up the ramp at a constant speed.
4. Repeat steps 1 to 3, using a constant slope but varying the cart mass (by stacking extra masses onto the cart).
5. Repeat steps 2 and 3, using different slopes of the ramp and keeping the cart mass constant.
6. Repeat steps 2 and 3, using a constant cart mass and a constant angle, but changing the friction between the cart and the ramp to sliding friction by turning the cart upside down.

Observations

(c) Complete the data in **Table 1**. Use trigonometry to calculate the angle of the ramp.

Analysis

(d) Describe the effect that each of the following had on the efficiency:
  (i) changing the mass of the cart
  (ii) changing the slope of the ramp
  (iii) changing the friction between the cart and the ramp

(e) What could you do to increase the efficiency of the inclined plane?

(f) Describe how the results of this experiment could be used to determine the frictional resistance acting on the cart in each trial.

Evaluation

(g) Comment on the accuracy of your hypothesis and prediction.

(h) Describe sources of error in this investigation. (To review errors, refer to Appendix A.)

(i) If you were to perform this investigation again, what improvements would you suggest?

Synthesis

(j) Name at least one advantage of moving an object up a ramp instead of lifting it vertically.

(k) Since nature is not in the habit of giving us something for nothing, name at least one disadvantage of moving an object up a ramp instead of lifting it.

Case Study: Physics and Sports Activities

When sports records are broken, do you think it is because the athletes are stronger and faster, and are being trained for longer, or it is because the equipment the athletes use, such as their footwear, is enhancing their performance? In this case study, we will analyze and explain improvements in sports performance using energy concepts such as the law of conservation of energy. After studying pole vaulting examples, you can research other examples in which science and technology have played an important role in sports.

Physics principles have been applied to many sports, whether or not the athlete is aware of them. For example, over the years, footwear designs have evolved, resulting in specialized gear that helps the athletes perform to their maximum potential. Physics research has helped to change the rackets, bats, clubs, and poles for sports such as tennis, baseball, golf, and pole vaulting, all of which involve specialized devices. The designs of the devices and the materials of which they are made have contributed to the enhanced performance of the athletes.

A century ago, the materials used to manufacture sports equipment, such as wood, rubber, twine, and animal skins, were all natural materials. More recently, high-tech materials, such as polymers, composites, and low-density metals, have been used in the designs. **Table 2** lists the properties of some materials that designers consider desirable for a specific application.

Table 2 Properties of Some Materials Used in Sports Equipment

Material	Property
aluminum	stiffness
carbon-fibre composites	low density
magnesium	toughness
metal-matrix composites	corrosion resistance
titanium	strength

A Close Look at Pole Vaulting

Scientists continue to research to find stronger, lighter materials that will aid the athletes. Consider, as an example, the evolution of pole vaulting (**Figure 7**). In this sport, the athlete gains kinetic energy by accelerating to the highest possible speed while carrying a pole that is planted in a box to stop the bottom of the pole from moving while the athlete moves up and over the crossbar. At the maximum height above the ground, the athlete has maximum gravitational potential energy. The athlete then falls safely to the mat below. Considering the law of conservation of energy, it is evident that a greater kinetic energy translates into a greater gravitational potential energy, and thus, a greater maximum height.

In the 1896 Olympics, the record in pole vaulting was 3.2 m, set using a pole made of bamboo, a natural plant product. In the 1960s, aluminum poles were introduced. Aluminum is a low density metal with relatively high strength, so it helped vaulters set new records. By the 1990s, however, scientists had developed artificial materials and used them to make composite poles, that is, poles with layers of fibres that increased stiffness and strength while minimizing twisting.

Figure 7
A pole vaulter reaches the maximum height by first gaining maximum kinetic energy, and then raising the body to that height using a pole that absorbs the least amount of energy.

Let us consider some physics principles underlying the newer composite poles. These poles are lighter in weight than the aluminum or bamboo poles, and they are a definite advantage to the athlete who wants to gain maximum speed in the approach to the jump. (It is the kinetic energy of the jumper that must be maximized, not the kinetic energy of the pole.) When the bottom end of the pole comes to a stop in the box, the pole begins to bend, thus gaining elastic potential energy; this helps the athlete gain gravitational potential energy. The composite pole can return more of the elastic potential energy gained to the jumper than other types of poles because it absorbs less energy. In other words, very little of the input energy is wasted as thermal energy due to internal friction caused by bending and twisting. These changes have helped increase the (current) pole-vaulting records to 4.6 m for women and over 6.1 m for men.

When pole vaulters choose a particular pole, the pole material is only one of the many factors to consider. Length is another important consideration; athletes must experiment to find out the length of the pole that maximizes their ability to convert kinetic energy into gravitational potential energy. Pole stiffness is also important; a stiffer pole is harder to bend, but it can help raise the jumper higher because of the increased ability to return the elastic potential energy to the jumper. In general, a heavier jumper requires a stiffer pole. Other considerations relate to the properties of the poles from different manufacturers. For example, some poles have uniform construction throughout, while others are designed so that the top is different from the bottom.

Practice

Understanding Concepts

Answers

8. (a) 9.1 m/s

6. Write the energy transformation equation for a pole vault, starting with the pole vaulter gaining the maximum kinetic energy while approaching the box and ending when the pole vaulter lands on the safety mat.
7. Explain why a composite vaulting pole is more efficient than a bamboo pole.
8. Assume that at the top of the pole vault, the athlete's speed is essentially zero, and that the safety mat is 0.40 m thick.
   (a) Apply the law of conservation of energy to determine the maximum speed with which the athlete lands after clearing the crossbar set at 4.6 m.
   (b) When comparing your answer in (a) to an estimated value of the athlete's maximum speed prior to jumping, does the answer make sense? Explain.
   (c) Does the mass of the athlete affect the calculations you made in (a)? Explain.

Making Connections

9. Describe the relationship between sports records and advancements in science and technology.
10. Although the principles of physics were discussed in this Case Study, the issue of costs was not discussed. In your opinion, what impact does the cost of new equipment have on sports activities? (You can describe your opinions for sports in general or for one of your favourite sports.)
11. You may argue that new designs and materials give some athletes an advantage over the others. For example, a disabled athlete wearing artificial limbs with highly elastic springs designed by experimental scientists and engineers will have a distinct advantage in a race over another disabled runner wearing less sophisticated artificial limbs.
    (a) Describe, with reasons, both advantages and disadvantages of restricting the use of new designs and materials.
    (b) Find out more about this issue using the Internet or other resources. Follow the links for Nelson Physics 11, 4.4. Sum up what you discover, and then revisit your answer in (a) above.

GO TO www.science.nelson.com

SUMMARY Conservation of Energy and Efficiency

- The law of conservation of energy states that when energy changes from one form into another, no energy is lost.
- The law of conservation of energy is applied in many situations, including playing sports, operating hydroelectric generating stations, and problem solving in physics.
- The efficiency of an energy transformation is found using the equation $\text{efficiency} = \frac{E_{out}}{E_{in}} \times 100\%$. Almost all waste energy goes to heat or thermal energy.

Section 4.4 Questions

Understanding Concepts

1. A 60.0-kg teacher and a 40.0-kg student sit on identical swings. They are each given a push so that both swings move through the same angle from the vertical. How will their speeds compare as they swing through the bottom of the cycle? Explain your answer.
2. At the moment when a shot-putter releases a 7.26-kg shot, the shot is 2.0 m above the ground and travelling at 15.0 m/s. It reaches a maximum height of 8.0 m above the ground and then falls to the ground. Assume that air resistance is negligible.
   (a) What is the gravitational potential energy of the shot as it leaves the hand, relative to the ground?
   (b) What is the kinetic energy of the shot as it leaves the hand?
   (c) What is the total mechanical energy of the shot as it leaves the hand?
   (d) What is the total mechanical energy of the shot as it reaches its maximum height?
   (e) What is the gravitational potential energy of the shot at its maximum height?
   (f) What is the kinetic energy of the shot at its maximum height?
   (g) What is the kinetic energy of the shot just as it strikes the ground?
   (h) Apply the law of conservation of energy to determine the final speed of the shot.
3. A 2.0×10^2-g pendulum bob is raised 22 cm above its rest position. The bob is released, and it reaches its maximum speed as it passes the rest position.
   (a) Calculate its maximum speed at that point by applying the law of conservation of energy and assuming that the efficiency is 100%.
   (b) Repeat (a) if the efficiency is 94%.
4. For an object accelerating uniformly from rest, the speed attained after travelling a certain distance is $v = \sqrt{2ad}$. Substitute this equation into the equation for kinetic energy, $E_k = \frac{mv^2}{2}$. Explain the result.
5. A high-rise window washer (mass 72 kg) is standing on a platform (mass 178 kg) suspended on the side of a building. An electric motor does 1.5×10^3 J of work on the platform and the worker to raise them to the top floor. The electrical energy required to operate the motor for this task is 1.6×10^5 J.
   (a) How high is the platform lifted?
   (b) Calculate the efficiency of the electric motor.

Applying Inquiry Skills

6. (a) Describe how you would perform an investigation to determine the efficiency of a pulley system, such as the one illustrated in **Figure 8**.
   (b) What factor(s) do you think would influence the efficiency of the pulley?

(continued)

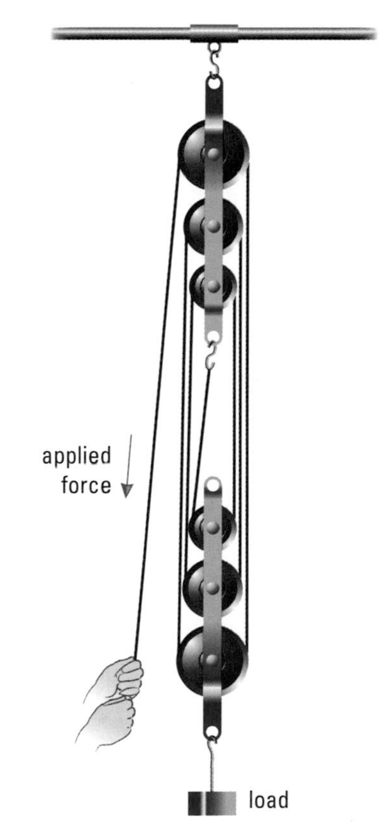

Figure 8
For question 6

Figure 9
A tall pendulum clock is also called a "grandfather clock."

7. Among automobile manufacturers, Volvo has one of the highest safety standards, aided by extensive research into developing safety systems. In one test of the ability of a Volvo sedan to absorb energy in a front-end collision, a stunt driver drives the car off the edge of a cliff, allowing the car to crash straight into the ground below. Assuming the speed of the car was 12 m/s at the top of the 5.4-m cliff, apply the law of conservation of energy to determine the car's impact speed. (Notice that the car's mass is not given. Can you explain why?)

Making Connections

8. A pendulum clock (**Figure 9**) requires a periodic energy input to keep working. Relate the operation of this clock to the law of conservation of energy. If possible, inspect one and describe how the input energy is achieved.

Reflecting

9. One name of the toy suggested in the chapter opener activity is the "switchback." Why is this a good name?

4.5 Thermal Energy and Heat

thermal energy: the total kinetic energy and potential energy of the atoms or molecules of a substance

temperature: a measure of the average kinetic energy of the atoms or molecules of a substance

We are surrounded by the use and effects of heat and thermal energy — thermostats control furnaces, large bodies of water help moderate the climate of certain regions, winds are generated by the uneven heating of Earth's surface and atmosphere, and the weather influences the clothes we wear. Furthermore, much of the energy we consume is eventually transformed into thermal energy. Thus, thermal energy and heat play a significant role in our lives.

Thermal energy and heat are not exactly the same, and temperature is different from both of them. **Thermal energy** is the total kinetic energy and potential energy (caused by electric forces) of the atoms or molecules of a substance. It depends on the mass, temperature, nature, and state of the substance. As stated earlier, heat is the transfer of energy from a hot body to a colder one. **Temperature** is a measure of the average kinetic energy of the atoms or molecules of a substance, which increases if the motion of the particles increases.

Consider, for example, 100 g of water at 50°C and 500 g of water at 50°C. The samples have the same temperature, but the bigger 500-g sample contains more thermal energy. If these samples were mixed, no heat would transfer between them because they are at the same temperature (**Figure 1(a)**).

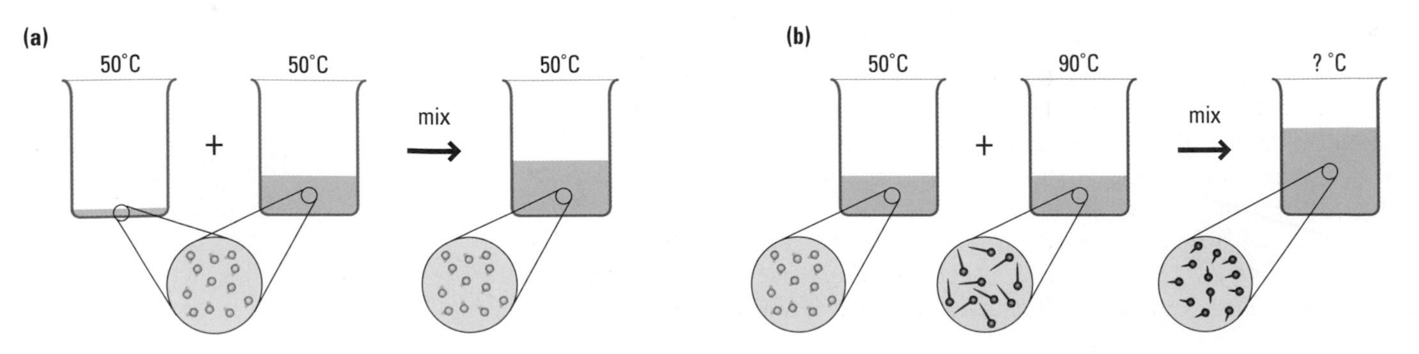

Figure 1
Comparing thermal energy, heat, and temperature
(a) When the average kinetic energy in one sample is the same as in the other, the temperatures are the same and no heat would flow if the samples are mixed.
(b) When the samples are of the same mass, the one with the higher temperature has both higher average kinetic energy and higher thermal energy.

Next, consider 500 g of water at 50°C and 500 g of water at 90°C. The warmer sample has more thermal energy because the motion—in other words, the average kinetic energy—of the molecules is greater at a higher temperature. If the two samples were mixed, heat would transfer from the 90°C sample to the 50°C sample (**Figure 1(b)**).

(a)

(b)

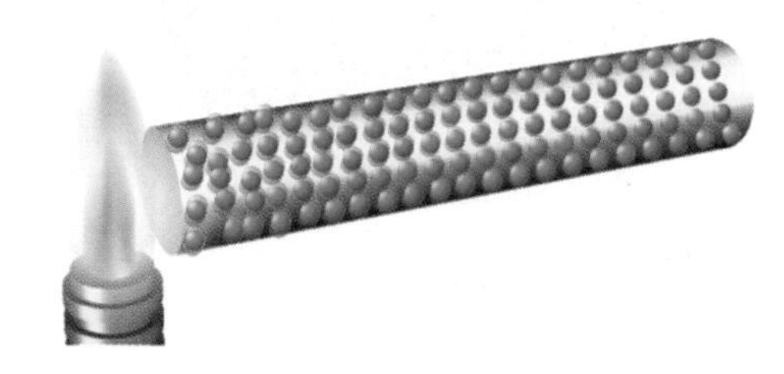

Figure 2
Conduction
(a) The metal rod must be hot before it can be bent into the desired shape. Heat from the fire is transferred through the metal by conduction.
(b) Heat conduction occurs by the collision of atoms.

Practice

Understanding Concepts

1. Explain the difference between the thermal energy and the temperature of a metal coin.
2. A parent places a baby bottle containing 150 mL of milk at 7°C into a pot containing 550 mL of water at 85°C.
   (a) Compare the average kinetic energy of the milk molecules and that of the water molecules.
   (b) Compare the thermal energy of the milk and the water.
   (c) Will the heat stop transferring from the water to the milk at some stage? Explain your answer.

Reflecting

3. Word association often helps in understanding science terminology. To relate thermal energy to various contexts, list as many words as you can that start with the prefix *therm* or *thermo*. Make a list of terms and their meanings for reference.

Methods of Heat Transfer

The definition of heat suggests that energy is transferred from a warmer body to a cooler body. This transfer occurs in three possible ways, which you have studied in previous science classes. These ways are conduction, convection, and radiation.

The skill of bending metal into different shapes, shown in **Figure 2(a)**, relies on heat transfer. The process of heat transferring through a material by the collision of atoms is called **conduction**. A metal rod is composed of billions of vibrating atoms and electrons. When one end of the rod is heated, the atoms there gain kinetic energy and vibrate more quickly. They collide with adjacent atoms, also causing them to vibrate more quickly. This action continues along the rod from the hotter end toward the colder end, as illustrated in **Figure 2(b)**. Conduction occurs best in metals, which have electrons that move much more freely than in other substances. (Metals are good electrical conductors for the same reason.) Conduction occurs much less in solids such as concrete, brick, and glass, and only slightly in liquids and gases.

conduction: the process of transferring heat through a material by the collision of atoms

convection: the process of transferring heat by a circulating path of fluid particles

The process of transferring heat by a circulating path of fluid particles is called **convection**. The circulating path is called a *convection current*. The particles of the fluid actually move, carrying energy with them. Consider, for example, a room in which an electric heater (without a fan) is located along one wall (**Figure 3**). The air particles near the heater gain t hermal energy and move faster. As they collide more, they move farther apart. As they spread out, the heated air becomes less dense than the surrounding cooler air. The warmer air then rises and is replaced with the denser, cooler air. A convection current forms and distributes energy throughout the room.

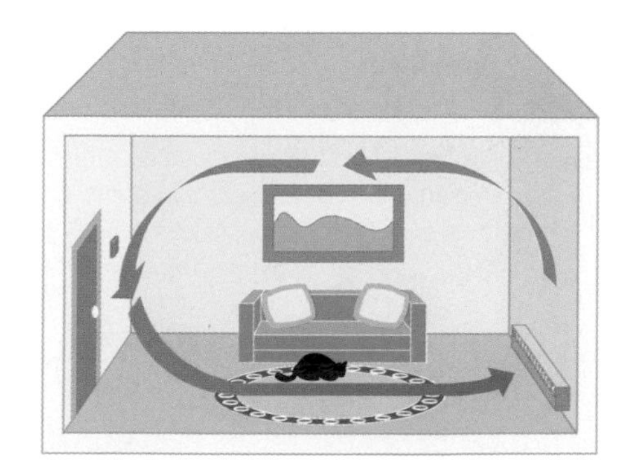

Figure 3
A convection current is set up in a room with an electric heater along one wall.

Both conduction and convection involve particles. However, heat can also transfer through a vacuum, a space with no particles. Evidence of this occurs as energy from the Sun reaches us after travelling through empty space. Thus, there

radiation: the process in which energy is transferred by means of electromagnetic waves

is a third method of heat transfer, one that requires no particles. **Radiation** is the process in which energy is transferred by means of electromagnetic waves. Examples of these waves are visible light, microwaves, radio waves, radar, X rays, and infrared rays. Infrared rays are also called heat radiation because the dominant form of radiation emitted from objects at everyday temperatures is infrared radiation. (See the drawing of the electromagnetic spectrum in section 9.1.)

Heat emitted from an object in the form of infrared rays can be detected by an infrared photograph called a *thermograph*. For example, a cancerous tumour is slightly warmer than its surroundings, so it is detected as a shaded region in a thermograph. Some new cars are equipped with infrared detectors that allow drivers to "see" objects such as a deer or a jogger about four times as far away as the headlights of their cars allow. Another example is shown in **Figure 4**.

Figure 4
This is an infrared photograph of a farm house in Ireland. The darkest colours indicate the highest temperature.

Practice

Understanding Concepts

4. Explain the following:
   (a) Curling irons and clothes irons have plastic handles.
   (b) High-quality cooking pots are often made with copper bottoms.
   (c) Inserting a metal skewer into a potato will decrease the amount of time required to bake the potato in an oven.
   (d) Smoke in a fireplace rises up the chimney.
5. Discuss whether this statement is true or false: In heat conduction, energy is transferred but the particles themselves are not transferred.
6. If air were a good conductor, you would feel cool even on a day when the air temperature is 25°C. Explain why.
7. Would it be better to place an electric room heater near the floor or the ceiling of a room? Explain your answer.
8. What happens to the density of a substance when it is heated?
9. Why is heat radiation vastly different from conduction and convection?

Calculating Heat Transfer

The transfer of heat from one body to another causes either a temperature change, or a change of state, or both. Here we will consider temperature changes.

Different substances require different amounts of energy to increase the temperature of a given mass of the substance. This occurs because different substances have different capacities to hold heat. For example, water holds heat better than steel. Therefore, water is said to have a higher specific heat capacity than steel. The word "specific" indicates that we are considering an equal mass of each substance. In SI units, the mass is 1.0 kg. Thus, **specific heat capacity** (c) is a measure of the amount of energy needed to raise the temperature of 1.0 kg of a substance by 1.0°C. It is measured in joules per kilogram degree Celsius, J/(kg•°C).

specific heat capacity: (c) a measure of the amount of energy needed to raise the temperature of 1.0 kg of a substance by 1.0°C

The English scientist James Prescott Joule performed original investigations to determine the specific heat capacities of various substances. He discovered, for instance, that 4.18×10^3 J of energy is required to raise the temperature of 1.0 kg of water by 1.0°C:

$$c_w = 4.18 \times 10^3 \frac{\text{J}}{\text{kg•°C}}$$ where c_w is the specific heat capacity of water

This value also means that 1.0 kg of water releases 4.18×10^3 J of energy when its temperature drops by 1.0°C.

The quantity of heat gained or lost by a body, Q, is directly proportional to the mass, m, of the body, its specific heat capacity, c, and the change in the body's temperature, Δt. The equation relating these factors is

$$Q = mc\Delta t$$

Sample Problem 1

How much heat is needed to raise the temperature of 2.2 kg of water from 20°C to the boiling point? (Assume two significant digits.)

Solution

$m = 2.2 \text{ kg}$

$c = 4.18 \times 10^3 \dfrac{\text{J}}{\text{kg}\cdot{}^\circ\text{C}}$

$\Delta t = 100^\circ\text{C} - 20^\circ\text{C} = 80^\circ\text{C}$

$Q = \ ?$

$$\begin{aligned} Q &= mc\Delta t \\ &= (2.2 \text{ kg})\left(4.18 \times 10^3 \frac{\text{J}}{\text{kg}\cdot{}^\circ\text{C}}\right)(80^\circ\text{C}) \\ Q &= 7.4 \times 10^5 \text{ J} \end{aligned}$$

The heat required is 7.4×10^5 J, or 0.74 MJ.

The specific heat capacities of different substances are shown in **Table 1**.

Table 1 Specific Heat Capacities of Common Substances

Substance	Specific heat capacity (J/(kg•°C))	Substance	Specific heat capacity (J/(kg•°C))
glass	8.4×10^2	water	4.18×10^3
iron	4.5×10^2	alcohol	2.5×10^3
brass	3.8×10^2	ice	2.1×10^3
silver	2.4×10^2	steam	2.1×10^3
lead	1.3×10^2	aluminum	9.2×10^2

When heat is transferred from one body to another, it normally flows from the hotter body to the colder one. The amount of heat transferred obeys the **principle of heat exchange**, which is stated as follows:

principle of heat exchange: When heat is transferred from one body to another, the amount of heat lost by the hot body equals the amount of heat gained by the cold body.

Principle of Heat Exchange

When heat is transferred from one body to another, the amount of heat lost by the hot body equals the amount of heat gained by the cold body.

Since this is another version of the law of conservation of energy, it can be written using the following equations:

$$Q_{\text{lost}} + Q_{\text{gained}} = 0$$

or

$$\underbrace{m_1c_1\Delta t_1}_{\text{heat lost}} + \underbrace{m_2c_2\Delta t_2}_{\text{heat gained}} = 0$$

Sample Problem 2

A 200-g piece of iron at 350°C is submerged in 300 g of water at 10°C to be cooled quickly. Determine the final temperature of the iron and the water. (Assume two significant digits.)

Solution

$m_i = 0.20$ kg
$m_w = 0.30$ kg
$c_i = 4.5 \times 10^2 \dfrac{\text{J}}{\text{kg•°C}}$
$c_w = 4.18 \times 10^3 \dfrac{\text{J}}{\text{kg•°C}}$

Let the final temperature be t_f.

$\Delta t_i = t_f - 350°\text{C}$
$\Delta t_w = t_f - 10°\text{C}$

$$\underset{\text{iron}}{m_i c_i \Delta t_i} + \underset{\text{water}}{m_w c_w \Delta t_w} = 0$$

$$(0.20\ \text{kg})\left(4.5 \times 10^2 \frac{\text{J}}{\text{kg•°C}}\right)(t_f - 350°\text{C}) + (0.30\ \text{kg})\left(4.18 \times 10^3 \frac{\text{J}}{\text{kg•°C}}\right)(t_f - 10°\text{C}) = 0$$

$$90t_f - 3.15 \times 10^4 °\text{C} + 1.25 \times 10^3 t_f - 1.25 \times 10^4 °\text{C} = 0$$

$$1.34 \times 10^3 t_f = 4.40 \times 10^4 °\text{C}$$

$$t_f = 33°\text{C}$$

The final temperature of the iron and water is 33°C.

Practice

Understanding Concepts

10. Calculate the amount of heat needed to raise the temperature of the following:
    (a) 8.4 kg of water by 6.0°C
    (b) 2.1 kg of alcohol by 32°C
11. Determine the heat lost when
    (a) 3.7 kg of water cools from 31°C to 24°C
    (b) a 540-g piece of silver cools from 78°C to 14°C
12. Rearrange the equation $Q = mc\Delta t$ to obtain an equation for
    (a) c (b) m (c) Δt
13. An electric immersion heater delivers 0.050 MJ of energy to 5.0 kg of a liquid, changing its temperature from 32°C to 42°C. Find the specific heat capacity of the liquid.
14. Determine how much brass can be heated from 20°C to 32°C using 1.0 MJ of energy.
15. A 2.5-kg pane of glass, initially at 41°C, loses 4.2×10^4 J of heat. What is the new temperature of the glass?
16. A 120-g mug at 21°C is filled with 210 g of coffee at 91°C. Assuming all of the heat lost by the coffee is transferred to the mug, what is the final temperature of the coffee? The specific heat capacity of the mug is 7.8×10^2 J/(kg•°C).

Answers

10. (a) 2.1×10^5 J
    (b) 1.7×10^5 J
11. (a) 1.1×10^5 J
    (b) 8.3×10^3 J
13. 1.0×10^3 J/(kg•°C)
14. 2.2×10^2 kg
15. 21°C
16. 84°C

SUMMARY Thermal Energy and Heat

- It is important to distinguish between thermal energy, heat, and temperature.
- Heat transfer can occur by means of conduction, convection, and radiation.
- The quantity of heat, Q, transferred to an object of mass m and specific heat capacity c in raising its temperature by Δt is found using the equation $Q = mc\Delta t$.

Section 4.5 Questions

Understanding Concepts

1. Distinguish between heat and thermal energy.
2. One morning you walk barefoot across a rug onto a tiled floor. The rug and the floor are at the same temperature, yet the tiled floor feels much colder. Explain why.
3. What is the most likely method of heat transfer through
   (a) a metal? (b) a vacuum? (c) a liquid?
4. Water from a tap at 11°C sits in a watering can where it eventually reaches 21°C.
   (a) Where did the energy that warms up the water come from?
   (b) Determine the mass of the water sample if it has absorbed 21 kJ of energy during the temperature change.
5. Hang gliders and birds of prey ride convection currents called thermals. Describe the conditions that cause thermals.
6. Calculate the heat transferred in each case.
   (a) The temperature of a 6.4-kg piece of lead changes from 12°C to 39°C.
   (b) A 2.4-kg chunk of ice cools from −13°C to −19°C.

Applying Inquiry Skills

7. Describe how you would set up a demonstration to show
   (a) convection in water
   (b) convection in air
   (c) conduction in a solid
8. In an experiment to determine the specific heat capacity of a metal sample, a student quickly transfers a 0.70-kg bar of metal M from boiling water into 0.45 kg of water at 16°C. The highest temperature reached by the metal and water together is 28°C.
   (a) Determine the specific heat capacity of metal M.
   (b) What is the possible identity of metal M?
   (c) Describe sources of error in this type of experiment.

Making Connections

9. Police discovered a car that slid off the road and down a cliff. Research to find how a forensic scientist could use infrared photography to determine approximately how long ago the mishap occurred. Follow the links for Nelson Physics 11, 4.5.

GO TO www.science.nelson.com

4.6 Power

power: (P) the rate of doing work or transforming energy

Two students of equal mass, keen on helping a charity drive, run up the stairs of Toronto's CN Tower on the day set aside for the event. The students climb the same vertical displacement, 342 m up the stairs, one in 24 min and the other in 36 min. The work done by each student against the force due to gravity is the same, but the times are different. Thus, some other factor must be influencing the two students. This other factor is the student's "power." **Power** (P) is the rate of doing work or transforming energy. Thus,

$$P = \frac{W}{\Delta t} \quad \text{or} \quad P = \frac{\Delta E}{\Delta t}$$

watt: (W) the SI unit for power

Like work, energy, and time, power is a scalar quantity. Since work and energy are measured in joules and time is measured in seconds, power is measured in joules per second (J/s). This SI unit has the name **watt** (W), in honour of James Watt, a Scottish physicist who invented the first practical steam engine (**Figure 1**). Watts and kilowatts are commonly used to indicate the power of electrical appliances, while megawatts are often used to indicate the power of electric generating stations.

Figure 1
James Watt (1736–1819) introduced a new, improved version of the steam engine that changed its status from that of a minor gadget to that of a great working machine. First used to pump water from coal mines, it soon powered steamships, locomotives, shovels, tractors, cars, and many other mechanical devices.

Sample Problem 1

What is the power of a cyclist who transforms 2.7×10^4 J of energy in 3.0 min?

Solution

$\Delta E = 2.7 \times 10^2$ J
$\Delta t = 30$ min $= 1.8 \times 10^2$ s
$P = ?$

$$\begin{aligned} P &= \frac{\Delta E}{\Delta t} \\ &= \frac{2.7 \times 10^4 \text{ J}}{1.8 \times 10^2 \text{ s}} \\ P &= 1.5 \times 10^2 \text{ W} \end{aligned}$$

The cyclist's power is 1.5×10^2 W.

Sample Problem 2

A 51-kg student climbs 3.0 m up a ladder in 4.7 s. Calculate the student's

(a) gravitational potential energy at the top of the climb
(b) power for the climb

Solution

(a) $m = 51$ kg
$g = 9.8$ N/kg
$h = 3.0$ m
$E_g = ?$

$$\begin{aligned} E_g &= mgh \\ &= (51 \text{ kg})(9.8 \text{ N/kg})(3.0 \text{ m}) \\ E_g &= 1.5 \times 10^3 \text{ J} \end{aligned}$$

The student's gravitational potential energy at the top of the climb is 1.5×10^3 J.

(b) $\Delta E = 1.5 \times 10^3$ J
$\Delta t = 4.7$ s
$P = ?$

$$P = \frac{\Delta E}{\Delta t}$$
$$= \frac{1.5 \times 10^3 \text{ J}}{4.7 \text{ s}}$$
$$P = 3.2 \times 10^2 \text{ W}$$

The student's power for the climb is 3.2×10^2 W.

Practice

Understanding Concepts

1. Express watts in the base SI units of metres, kilograms, and seconds.
2. A fully outfitted mountain climber, complete with camping equipment, has a mass of 85 kg. If the climber climbs from an elevation of 2900 m to 3640 m in exactly one hour, what is the climber's average power?
3. Rearrange the equation $P = \frac{\Delta E}{\Delta t}$ to obtain an equation for (a) ΔE and (b) Δt.
4. The power rating of the world's largest wind generator is 3.0 MW. How long would it take such a generator to produce 1.0×10^{12} J, the amount of energy needed to launch a rocket?
5. An elevator motor provides 32 kW of power while it lifts the elevator 24 m at a constant speed. If the elevator's mass is 2200 kg, including the passengers, how long does the motion take?
6. The nuclear generating station located at Pickering, Ontario, one of the largest in the world, is rated at 2160 MW of electrical power output. How much electrical energy, in megajoules, can this station produce in one day?

Answers

2. 1.7×10^2 W
4. 3.3×10^5 s, or 3.9 d
5. 16 s
6. 1.9×10^8 MJ

Activity 4.6.1

Student Power

You can carry out various activities to determine the power a student can achieve. You might think of ideas other than those suggested here. When you report on this activity, include your own analysis and evaluation.

1. Determine the power of a student, such as yourself, walking up a set of stairs. Safety considerations are important here. Only students wearing running shoes should try this activity and, of course, they should be careful not to trip or pull their arm muscles while pulling on the rail. It may be interesting to compare the student power with that of an average horse, which can exert about 750 W of power for an entire working day. (This quantity bears the old-fashioned name "horsepower.")
2. Determine the power of a student performing a variety of activities. Examples include climbing a rope in the gymnasium, lifting books, doing push-ups, digging in the garden, and shovelling snow.

Students with health problems should not participate in this activity. Those who wear slippery footwear should not participate in option 1.

SUMMARY Power

- Power is the rate of doing work or consuming energy, found using the equation $P = \frac{W}{\Delta t}$, or $P = \frac{\Delta E}{\Delta t}$.
- Power is a scalar quantity measured in watts (W).

Figure 2
A motor grader

Section 4.6 Questions

Understanding Concepts

1. A 60-kg student does 60 push-ups in 40 s. With each push-up, the student must lift an average of 70% of the body mass a height of 40 cm off the floor. Assuming two significant digits, calculate the following:
   (a) the work the student does against the force of gravity for each push-up, assuming work is done only when the student pushes up
   (b) the total work done against the force of gravity in 40 s
   (c) the power achieved for this period
2. A water pump rated at 2.0 kW can raise 55 kg of water per minute at a constant speed from a lake to the top of a storage tank. How high is the tank above the lake? Assume that all the energy from the pump goes into raising the height of the water.
3. (a) Determine how long it would take a hair dryer rated at 1.5×10^3 W to use 5.0 MJ of energy.
   (b) How many times could you dry your hair using the 5.0 MJ of energy described in (a)?
4. The largest motor grader (**Figure 2**) ever built had a mass of about 9.1×10^4 kg and was over 11 m wide. Its two engines had a total maximum power output of about 1.3 MW.
   (a) How much work (in megajoules) could this machine do each hour?
   (b) The grader was used to recondition beaches along a sea coast. What other work required a lot of the energy provided by the engines?

Applying Inquiry Skills

5. (a) Describe how you would conduct an experiment to determine the power output of a battery-powered or wind-up toy car that travels up a ramp inclined at a small angle to the horizontal. Consider only the power output required to overcome the force due to gravity.
   (b) Use actual numbers to estimate the power output of a typical toy car.

Making Connections

6. Each Canadian uses energy at an average rate of about 2 kW per day. (This figure includes energy used outside the home, but not energy used to manufacture products.) Assume that on a bright sunny day, the solar energy striking a horizontal surface provides power at a rate of 7.0×10^2 W/m^2. If a solar collector can capture 20% of the energy striking it, how large a collector in square metres is required to supply the energy requirements of a family of five during the daylight hours of a sunny day?

Chapter 4 Summary

Key Expectations

Throughout this chapter, you have had opportunities to

- define and describe the concepts and units related to energy, work, power, gravitational potential energy, kinetic energy, thermal energy, and heat; (4.1, 4.2, 4.3, 4.5, 4.6)
- identify conditions required for work to be done in one-dimensional motion, and apply quantitatively the relationships among work, force, and displacement along the line of the force; (4.2)
- analyze, in qualitative and quantitative terms, simple situations involving work, gravitational potential energy, kinetic energy, thermal energy, and heat, using the law of conservation of energy; (4.4, 4.5)
- apply quantitatively the relationships among power, energy, and time in a variety of contexts; (4.6)
- analyze, in quantitative terms, the relationships among percent efficiency, input energy, and useful output energy for several energy transformations; (4.4)
- design and carry out experiments related to energy transformations, identifying and controlling major variables (e.g., design and carry out an experiment to identify the energy transformations of a swinging pendulum, and to verify the law of conservation of energy; design and carry out an experiment to determine the power produced by a student); (4.4, 4.6)
- analyze and interpret experimental data or computer simulations involving work, gravitational potential energy, and kinetic energy; (4.4)
- analyze and explain improvements in sports performance, using principles and concepts related to work, kinetic and potential energy, and the law of conservation of energy (e.g., explain the importance of the initial kinetic energy of a pole vaulter or high jumper); (4.4)

Key Terms

energy
heat
energy transformation
work
joule
gravitational potential energy
kinetic energy
mechanical energy
reference level
law of conservation of energy
efficiency
thermal energy
temperature
conduction
convection
radiation
specific heat capacity
principle of heat exchange
power
watt

Reflect on your Learning

Revisit your answers to the Reflect on Your Learning questions at the beginning of this chapter.

- How has your thinking changed?
- What new questions do you have?

Figure 1 shows a horizontal profile of the first part of a typical roller coaster, along with some of the features needed to operate the roller coaster. Draw a larger version of the profile and on your diagram, label and describe as many ideas as you can that relate to what you have learned in this chapter. Include concepts, key terms, equations, and estimations. To help make estimations, assume that the mass of each car is 750 kg and each car holds six passengers.

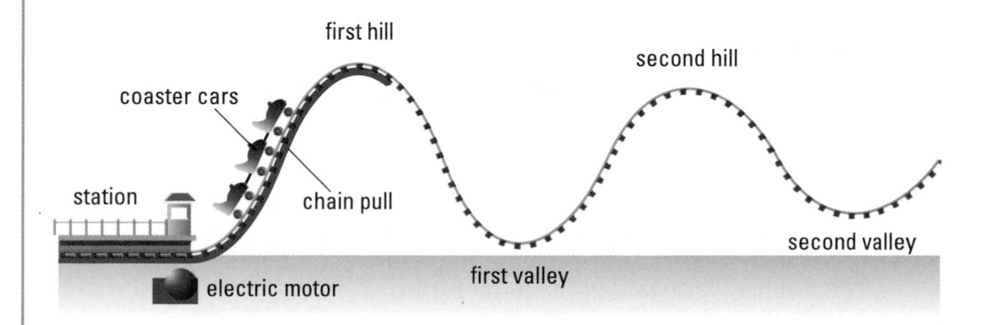

Figure 1
A roller coaster profile

Chapter 4 Review

Understanding Concepts

1. Make up an example of an energy transformation that involves several forms of energy. Write out the energy transformation equation for your example.
2. State the SI unit used to measure
   (a) work (b) kinetic energy (c) power
3. A black bear's greatest enemy is the grizzly bear. To escape a grizzly attack, a black bear does what its enemy cannot do—it climbs a tree whose trunk has a small diameter. Calculate the work done by a 140-kg black bear in climbing 18 m up a tree.
4. Compare the amount of work you would do in climbing a vertical rope with the work done in climbing a stairway inclined at 45°, if both activities get you 6.0 m higher.
5. A golf ball is given 115 J of energy by a club that exerts a force over a distance of 4.5 cm while the club and the ball are in contact.
   (a) Calculate the magnitude of the average force exerted by the club on the ball.
   (b) If the ball's mass is 47 g, find the magnitude of its average acceleration.
   (c) What speed does the club impart to the ball?
6. A roast of beef waiting to be taken out of a refrigerator's freezer compartment has a potential energy of 35 J relative to the floor. If the roast is 1.7 m above the floor, what is the mass of the roast?
7. A 55-kg diver has 1.62 kJ of gravitational potential energy relative to the water when standing on the edge of a diving board. How high is the board above the water?
8. A group of winter enthusiasts returning from the ski slopes are travelling at 95 km/h along a highway. A pair of ski boots having a total mass of 2.8 kg has been placed on the shelf of the rear window.
   (a) What is the kinetic energy of the pair of boots?
   (b) What happens to that energy if the driver must suddenly stop the car?
9. What happens to an object's kinetic energy when its speed doubles? triples?
10. A 50.0-kg cyclist on a 10.0-kg bicycle speeds up from 5.0 m/s to 10.0 m/s.
    (a) What is the total kinetic energy before accelerating?
    (b) What is the total kinetic energy after accelerating?
    (c) How much work is done to increase the kinetic energy of the cyclist and bicycle?
    (d) Is it more work to speed up from 0 to 5.0 m/s than from 5.0 m/s to 10.0 m/s? Explain.
11. A discus travelling at 20.0 m/s has 330 J of kinetic energy. Find the mass of the discus.
12. An archer nocks a 0.20 kg arrow on a bowstring. Then the archer exerts an average force of 110 N to draw the string back 0.60 m. Assume that friction is negligible.
    (a) What speed does the bow give to the arrow?
    (b) If the arrow is shot vertically upward, how high will it rise?
13. It is possible to heat a cold kitchen by opening the oven door, but it is not possible to cool the kitchen by opening the refrigerator door. Why?
14. Some people perform difficult tasks to raise money for charity. For example, walking up the stairs in Toronto's CN Tower helps both charity and personal fitness. Assume that the efficiency of the human body is 25%. If a 70.0-kg participant climbs the 342-m height of the tower 10 times in 4.0 h, calculate the following:
    (a) the work the participant does against the force of gravity on each trip up the stairs
    (b) the energy the participant's body requires for each trip up the stairs (including wasted energy)
    (c) the total energy required for the 10 upward trips
    (d) the power of the participant's body for the upward trips
15. Explain why it is impossible to have a motor that is 100% efficient.
16. State the method of heat transfer
    (a) that does not require particles
    (b) that works because particles collide with their neighbours
    (c) in which thermal energy travels at the speed of light
    (d) that works when particles circulate in a path
17. Given an equal mass of aluminum and brass, which mass would require more heat if the temperature of both were raised the same number of degrees?
18. How much heat is required to raise the temperature of 2.0 kg of water from 25°C to 83°C?
19. What will be the temperature change in each of the following?
    (a) 10.0 kg of water loses 456 kJ
    (b) 4.80 kg of alcohol gains 12.6 kJ
20. A 6.0-g pellet of lead at 32°C gains 36.8 J of heat. What will be its final temperature?
21. When 2.1×10^3 J of heat is added to 0.10 kg of a substance, its temperature increases from 19°C to 44°C. What is the specific heat capacity of the substance?

22. How much water at 82°C must be added to 0.20 kg of water at 14°C to give a final temperature of 36°C?
23. When 0.500 kg of water at 90°C is added to 1.00 kg of water at 10°C, what is the final temperature?
24. A waterfall is 55 m high. If all the gravitational potential energy of the water at the top of the falls were converted to thermal energy at the bottom of the falls, what would be the increase in the temperature of the water at the bottom? (*Hint:* Consider one kilogram of water going over the waterfall.)
25. Calculate the power of a light bulb that transforms 1.5×10^4 J of energy per minute.
26. How much energy is transformed by a 1200-W electric kettle during 5.0 min of operation?
27. An alternative unit to the joule or megajoule is the kilowatt hour (kW•h), which is used in many parts of Canada to measure electrical energy. One kilowatt hour is equivalent to one kilowatt of power used for one hour. Prove that 1.0 kW•h = 3.6 MJ.
28. Use the law of conservation of energy to derive an expression for the speed v acquired by an object allowed to fall freely from rest through a height h at a location where the gravitational field strength is of magnitude g. Assume that air resistance can be ignored.
29. A child of mass m slides down a slide 5.0 m high. The child's speed at the bottom of the slide is 3.0 m/s.
    (a) What percent of the mechanical energy that the child has at the top of the slide is not converted to kinetic energy?
    (b) What feature of the slide determines the percentage of mechanical energy that is converted to other forms of energy?
30. A chair lift takes skiers to the top of a mountain that is 320 m high. The average mass of a skier complete with equipment is 85 kg. The chair lift can deliver three skiers to the top of the mountain every 35 s.
    (a) Determine the power required to carry out this task. (Assume the skiers join the lift at full speed.)
    (b) If friction increases the power required by 25%, what power must the motors running the lift be able to deliver?

Applying Inquiry Skills

31. Suppose you perform an activity in your class to see who can develop the most power in climbing a flight of stairs. Describe the physical characteristics of the person who would have the best chance of developing the most power.

Making Connections

32. In winter the ground may be frozen, but large bodies of water such as the Great Lakes usually are not. Why?
33. An interesting and practical feature of the Montreal subway system is that, in some cases, the level of the station is higher than the level of the adjacent tunnel, as **Figure 1** demonstrates. Explain the advantages of this design. Take into consideration such concepts as force, acceleration, work, potential energy, and kinetic energy.

Figure 1
The Montreal subway system

34. Throughout the year, the average power received from the Sun per unit area in the densely populated regions of Canada is about 150 W/m^2 averaged over a 24-h day.
    (a) Estimate, and then calculate, the average yearly amount of energy received by a roof with a surface area of 210 m^2.
    (b) At a cost of 2.9¢/MJ (the average cost of electrical energy), how much is the energy in (a) worth?
    (c) Estimate the average yearly amount of energy received from the Sun by your province.

Exploring

35. The topic of physics in sports has many areas for further exploration.
    (a) Choose a sport to research. Analyze the improvements in the performance of the athletes of that sport due to the application of physics principles and concepts and present your findings. Consider ideas related to work, kinetic energy, potential energy, and the law of conservation of energy. Follow the links for Nelson Physics 11, Chapter 4 Review.

GO TO www.science.nelson.com

    (b) Describe issues other than the physics principles that you find in your research.

Chapter

5

Using Energy in Our Society

In this chapter, you will be able to

- use the law of conservation of energy to analyze situations involving energy use and energy transformations
- apply the equation for power in terms of energy and time interval to situations involving the production and use of energy, especially electrical energy
- analyze the efficiencies of several energy transformations
- communicate the procedures and results of investigations involving energy and power
- analyze the economic, social, and environmental impact of various energy sources and energy-transformation technologies around the world

Our standard of living is better now than at any time in the past. We have advanced medical facilities, the comfort of temperature-controlled buildings, ample fresh or frozen food, clothing to suit our variable climate, complex transportation systems, the assistance of numerous applications of electricity, and many interesting sports and leisure activities.

For all these advantages to exist, energy is required. Where does this energy come from? Traditionally, we have relied on non-renewable resources, mainly fossil fuels such as coal, oil, and natural gas. Technologies have been developed to transform the energy stored in these resources into other forms of energy, including electrical energy and the kinetic energy of moving vehicles. These technologies are influenced not just by scientific discoveries and inventions, but also by economic and social pressures.

Unfortunately, other than the declining energy resources, there are problems we must overcome if we wish to continue using energy to maintain or improve our standard of living. Burning fossil fuels creates carbon dioxide and other gases that harm the environment. Since the supplies of these fuels are limited, the vast variation in their costs could influence economic decisions worldwide. What have we learned about energy resources from the past that can be applied in order to ensure adequate supplies of energy for future generations? One small part of the solution is shown in **Figure 1**. You will explore many other aspects of this important question as you study this chapter. By the time you have completed the chapter, it is hoped that you can be part of the solution to the problems of energy supply and use.

Reflect on your Learning

1. How does the the consumption of energy per person in Canada compare to that in
   (a) other developed countries?
   (b) developing countries?
   Give reasons for your answers.
2. (a) What forms of energy have you consumed either directly or indirectly in the past week?
   (b) Where did the energy you consumed come from?
3. Explain the difference between renewable and non-renewable energy resources. List as many of each as you can.
4. Describe several ways in which energy resources are currently being wasted in Canada.
5. To ensure that there will be an adequate supply of energy for future generations, decribe what can be done by scientists and engineers.

Throughout this chapter, note any changes in your ideas as you learn new concepts and develop your skills.

Try This Activity

A Job Offer

Both energy consumption and the use of money involve a lot of numbers. This activity uses money, with which you are quite familiar, to get you to think about energy use.

This "Wanted" notice appears on a Web site. Read the notice and answer the questions that follow.

> ***Wanted:*** *A high school student willing to work hard for 10 h each day for 21 days. The Day 1 salary of one cent is doubled on Day 2, and then doubled again each day thereafter.*

Without doing any calculations, guess what the following numbers would be for the job offer:

(a) the salary on Day 21
(b) the total earnings for all 21 days
(c) the average hourly wage
(d) Based on your guesses in (a) to (c), would you consider applying for the job?
(e) Set up a table with these headings:

Day	Salary for the day	Accumulated salary

(f) Perform the calculations needed to fill in the data from Day 1 to Day 21.
(g) Determine the average hourly wage and compare it to your guess in (c) above.
(h) Plot a graph of salary (vertical axis) as a function of the day number.
(i) Comment on the growth rate in this example.
(j) The salary in this activity doubled each day. Other similar growths double in different time periods; for example, the energy consumption in a certain country may double every 12 years, and the world population doubles every 40 to 50 years. Describe why such growth should be of concern to members of our society.

Figure 1
Steady winds turn the test windmills at the Alberta Renewable Energy Test Site at Crowsnest Pass, in the Canadian Rockies.

5.1 The Consumption of Energy

We will begin this chapter by looking at the amount of energy we consume and the effects of the growth in energy demands.

Our Daily Energy Consumption

How much energy (in joules) do you think you have consumed from your physical activities in the past 24 hours? From your study of energy in Chapter 4, you know that it takes 9.8 J of energy (in the form of work) to raise a 1-kg object a vertical distance of 1 m. As your body is not very efficient, for every joule of energy you transform, your body may need to consume 4 J, 5 J, or even 10 J. It usually takes about 200 J to raise your body from a sitting to standing position and about 2×10^3 (2 kJ) to climb a set of stairs. Before you begin reading the next paragraph, estimate an answer to the question above.

If your estimate is less than 10^4 J (10 kJ), you will need to review your work on energy. Your food intake in a day provides about 10 MJ of energy for your activities.

When you compare the energy consumed from your activities to the daily energy consumption of the average Canadian, which is about 10^9 J (1 GJ, or 1000 MJ), you will notice that there is a large amount of other energy you are responsible for consuming. This includes the energy you use to cook your food, the energy that lights and heats your home and school, and perhaps the energy that provides transportation. More indirectly, you are also responsible for the energy used to make your clothes, books, furniture, CD player, and other appliances, as well as the energy to build and light your streets, to remove your sewage and garbage — the list is ever increasing.

Figure 1
The skyline of Toronto, Ontario, after business hours

The energy we consume (about 1000 MJ per person per day) is about 100 times as much as the amount we need to survive (about 10 MJ). Of course, much of the consumed energy is beyond our control, but we should be aware of the entire energy situation. Consider, for example, **Figure 1**. Are we, as individuals, responsible for the energy consumed to illuminate vacant offices at night?

It is interesting to compare our current consumption of energy (1000 MJ per person per day) with that of people in different eras. As civilization progressed, the amount of energy consumed per capita, that is, per person, increased remarkably (**Table 1**). Simultaneously, the world's population has grown, so the net effect is that we are now consuming a vast amount of energy.

Table 1 Daily Average Energy Consumption per Person

Lifestyle	Era	Type of energy use	Energy consumption per day (MJ)
primitive	pre-Stone Age	energy from food only	10
seasonally nomadic	Stone Age	energy from wood fires for cooking and heating	22
agricultural	medieval	energy from domesticated animals, water, wind, and coal	100
industrial	19th century	energy mainly from coal to run industries and steam engines	300
technological	present	energy from fossil fuels and nuclear sources used for electricity, transportation, industry, agriculture, and so on	1000 (in Canada)

DID YOU KNOW?

Comparing Developing and Industrialized Countries

A large portion of the world's population today consumes energy at a rate less than 100 MJ per person per day. Thus, developing countries are not the culprits of excess energy consumption. It is the industrialized, highly technological nations that consume the greatest amount of energy. In fact, about 80% of the energy consumed in the world each year is used by the industrialized countries. North Americans are among the people with the greatest demand for energy. Both the United States and Canada consume much more than their share of the world's energy.

Practice

Understanding Concepts

1. For this question, assume one significant digit.
   (a) What is the daily energy consumption of the average Canadian?
   (b) Estimate Canada's current population.
   (c) Use your answers in (a) and (b) to estimate Canada's total energy consumption per day, and per year.
2. (a) What fuel source replaced wood as the human lifestyle changed from nomadic to agricultural?
   (b) What fuel replaced the one you named in (a)?
   (c) What fuels replaced the one you named in (b)?
3. Refer to the data in **Table 1**, starting with the Stone Age.
   (a) Plot a graph of daily average energy consumption per capita (in megajoules) as a function of year, starting at 6000 B.C. (Assume that the Stone Age value cited in the table applies to 6000 B.C.)
   (b) Compare the shape of this graph to the shape of the graph you drew in question (h) of the chapter opener activity.
4. Why might two countries of similar technological levels have very different per capita energy consumptions? List several reasons.
5. It is known that population growth rates are lowest among the most technologically advanced countries. Suggest reasons why this is so. What consequences does this have in terms of energy consumption?

Making Connections

6. Describe reasons why Canadians, on average, consume much more energy per capita than people living in Mexico.

Answers

1. (a) 1×10^9 J
   (b) 3×10^7
   (c) daily: 3×10^{16} J, yearly: 1×10^{19} J

The Effects of Growth

Let us look at one of the most enlightening and frightening aspects of energy consumption. Frequently we hear news reports indicating the rate of growth *per annum* of some factors in our society. Using "a" to represent *annum* or "year," we have these examples:

- The population increase in Latin America is 2.3%/a.
- The total energy consumption in Africa is increasing at 4.2%/a.
- Consumption of oil for heating is decreasing at 3%/a.
- Consumption of natural gas in Canada is increasing at 6.7%/a.

To analyze the impact of growth rates, we will use the following example. Suppose that a town's annual budget for public transportation is $100 000. However, with the projected increase in salaries, cost of energy, and population growth, an average growth in expenses of 8%/a is expected. The effect of this growth after several years is evident in **Table 2**.

Table 2 shows that at the seemingly low growth rate of 8%/a, the original value has approximately doubled in 9 years, and after 45 years the value is greater by a factor of 32 times! This rapid growth may appear marvellous in the case of salaries. But what about energy consumption? If our energy consumption were to increase at 8%/a, after the average person's period of work expectancy (45 years), our energy use would be 32 times greater! You can now appreciate why scientists are, and politicians should be, concerned about growth in energy consumption.

Another interesting fact emerges from **Table 2**. At a growth rate of 8%/a, the time required for an amount to double, called the **doubling time**, is 9 years. The product of the two numbers is $8\%/\text{a} \times 9\ \text{a} = 72\%$. This percentage can be used

Table 2 Effects upon Budget of a Growth of 8%/a

Year	Budget	
0	$100 000.00	
1	$108 000.00	($100 000 × 1.08^1)
2	$116 640.00	($100 000 × 1.08^2)
3	$125 971.20	($100 000 × 1.08^3)
9	$199 900.46	($100 000 × 1.08^9)
18	$399 601.95	($100 000 × 1.08^{18})
27	$798 806.15	($100 000 × 1.08^{27})
36	$1 596 817.18	($100 000 × 1.08^{36})
45	$3 192 044.94	($100 000 × 1.08^{45})

doubling time: the time required for an amount to double

to estimate the doubling time for particular growth rates. For example, you have seen that at 8%/a, the doubling time is 72% ÷ 8%/a = 9 a. At a growth rate of 10%/a, the doubling time is approximately 72% ÷ 10%/a = 7.2 a. Thus,

$$\text{doubling time} = \frac{72\%}{\text{growth rate}} \quad \text{(an approximation)}$$

This equation gives a relatively accurate result at low rates of growth, but becomes inaccurate at higher rates.

Sample Problem

By what factor will Africa's energy consumption increase in the next 32 a at an average growth rate of 4.5%/a?

Solution

$$\text{doubling time} = \frac{72\%}{\text{growth rate}}$$

$$= \frac{72\%}{4.5\%/\text{a}}$$

$$\text{doubling time} = 16 \text{ a}$$

After 32 years, Africa's energy consumption will be four times its present value.

As you can see from the Sample Problem, the growth rate of energy consumption should be of major concern to everyone. Unfortunately, it is not the only energy-related problem, as you will find out in the rest of the chapter.

Practice

Understanding Concepts

7. Between 1950 and 1989, the world's population grew by 1.85%/a.
   (a) Determine the doubling time for this example.
   (b) If the population in 1950 was 2.5 billion, what was it in 1989?
8. The growth of energy consumption is 3.0%/a in Latin America and 6.0%/a in South Asia (currently the highest in the world). At these rates, by what factor will energy consumption increase after 24 a?
9. (a) With "Budget" as the dependent variable, plot the data in **Table 2** on a graph.
   (b) Discuss the general shape of growth graphs.
   (c) Describe how you can use the graph to estimate the doubling time of the growth.

Answers

7. (a) 39 a
   (b) 5.0 billion
8. 2; 4

SUMMARY The Consumption of Energy

- People in Canada and the United States consume more energy per person than in any other major country in the world.
- Even at low growth rates, both population and energy consumption increase dramatically over many years.
- The approximate doubling time of a growing variable can be determined using the equation

$$\text{doubling time} = \frac{72\%}{\text{growth rate}}.$$

Section 5.1 Questions

Understanding Concepts

1. If our oil reserves are dwindling at a rate of 6%/a, what percent of our current supply will remain after 24 a?
2. (a) What is the approximate current cost per litre of regular gasoline?
   (b) At an estimated average rate of increase of 4.5%/a, what is the doubling time of the cost of gasoline?
   (c) What will be the cost per litre of gasoline when you reach retirement age?
   (d) What factors make it almost impossible to judge what the rate will be in the future?
3. Between 1990 and 2000, the growth of the world's population was 1.5%/a, and in 2000, the population reached 6.0 billion people.
   (a) If this growth continues, when will the population reach 12 billion people?
   (b) Show that at this growth rate, after only 2100 years (which is less time than has elapsed since Aristotle lived), the total mass of all the people on Earth would exceed Earth's total mass.
   (c) Based on this example, explain why continuous growth is not always a good thing.

Applying Inquiry Skills

4. Make up three or four pertinent survey questions that would help you judge how much the average citizen knows and cares about the issue of energy supply and use in Canada.

Making Connections

5. What is meant by a population growth rate of zero? Do you believe that all countries in the world should aim for this rate? Explain.
6. Energy is not only important in physics; it is significant in all walks of life. Follow the links for Nelson Physics 11, 5.1 to discover more. Describe how energy relates to each area of interest listed: industry, economics, technology, communication, travel, agriculture, leisure, medicine, politics, and scientific research.

GO TO www.science.nelson.com

5.2 Energy Transformation Technologies

Imagine if you had to pedal an exercise bike to create the electricity needed to operate the lights, computer, radio, TV, or any other electrical device you use. What a difference there is between the amount of pedalling you would have to do and the simple job of plugging the device into the electrical outlet!

As you know, electrical energy is a very convenient form of energy. However, it never originates in such a convenient form. It must be transformed from some other form of energy before it is delivered to where it will be used. A system that converts energy from some source into a usable form is called an

energy transformation technology or **energy converter:** a system that converts energy from some source into a usable form

energy transformation technology. (Another name for this type of system is an **energy converter.**) Energy transformation technologies are specific examples of energy transformations, presented in Chapter 4.

An electrical generating station, such as the one shown in **Figure 1**, is an important example of an energy transformation technology. It converts energy, such as chemical potential energy stored in coal, oil, or natural gas, into electrical energy. It would be very inconvenient for you to try to generate electricity at your school by burning coal!

Figure 1
The Maritime Electric Generating Station—Borden converts the chemical potential energy in natural gas into electrical energy.

Unfortunately, energy transformation technologies are not 100% efficient. According to the law of conservation of energy, the amount of energy present before an energy transformation is equal to the amount of energy present after. However, very often, some of the energy is not converted into a useful form and is wasted. As from Chapter 4,

$$\text{efficiency} = \frac{E_{\text{out}}}{E_{\text{in}}} \times 100\%$$

Some transformations are more efficient than others, as shown in **Table 1**.

Practice

Understanding Concepts

1. Not many generations ago, homes, schoolhouses, and other buildings were heated by wood stoves or coal furnaces. Describe reasons why today's methods of providing heat are safer and more convenient.
2. Refer to the data in **Table 1**.
   (a) Both fossil-fuel power plants and hydroelectric power plants produce electrical energy. Why is one so much more efficient than the other?
   (b) Both incandescent lights and electric heaters produce heat. Why do they have such a big difference in efficiency?
   (c) Which of the two light-producing devices must operate at a higher temperature? How can you judge?

Making Connections

3. Our society has come to rely heavily on energy transformation technologies, which we often take for granted. However, we soon realize how dependent we are on these technologies when there is an electrical blackout. Describe the effects that would occur in your area as a result of an electrical blackout that lasts from several hours to several days.

DID YOU KNOW ?

Generating Electricity

After several days without electricity and heat during the ice storm in Eastern Canada in 1998, an Ontario family pedalled a bicycle connected to their gas furnace in order to create the electricity needed to operate the furnace. One person in the family was an engineer who knew how to connect the cycle to the furnace safely.

Table 1 Typical Efficiencies of Energy Transformation Technologies

Device	Efficiency (%)
electric heater	100
electric generator	98
hydroelectric power plant	95
large electric motor	95
home gas furnace	85
wind generator	55
fossil fuel power plant	40
automobile engine	25
fluorescent light	20
incandescent light	5

Automobile Efficiency

Cars, trucks, boats, airplanes, and other vehicles that burn fuel to operate are common energy transformation technologies. To learn about the efficiency of these technologies, we will focus on the automobile.

Automobiles are highly inefficient. Suppose that an amount of fuel containing 1000 J of chemical potential energy is used by an automobile's engine. **Figure 2** shows what happens to this energy in a car with an internal combustion engine. The efficiency of the car is

$$\text{efficiency} = \frac{E_{\text{out}}}{E_{\text{in}}} \times 100\%$$

$$= \frac{100\ \text{J}}{1000\ \text{J}} \times 100\%$$

$$\text{efficiency} = 10\%$$

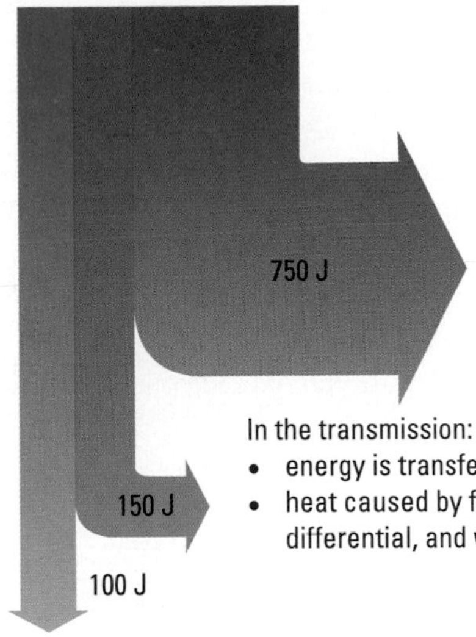

1000 J of energy in the fuel

In the engine:
- fuel is vaporized and mixed with air
- the fuel-air mixture is drawn into the cylinder, which contains a piston
- a spark from the spark plug ignites the mixture, producing a high temperature and pressure
- pressure pushes on the piston, which turns the crankshaft, which turns the transmission
- waste heat is carried away by exhaust gases pushed out through the cylinder's exhaust valve and by a water-antifreeze mixture circulating from the car's radiator

750 J

In the transmission:
- energy is transferred through the differential to the wheels
- heat caused by friction is generated (and lost) in the transmission, differential, and wheels

150 J

100 J

Useful energy:
- only 100 J of useful work is done in producing the kinetic energy of the moving car

Figure 2
Heat loss in a typical car

Engineers around the world are always seeking ways to reduce energy consumption. Making machines more efficient is one way of doing this.

Practice

Understanding Concepts

4. (a) Calculate the efficiency of the car engine described in **Figure 2** up to the point where the energy reaches the transmission.
   (b) Why is the efficiency of the entire car, shown to be only 10%, less than the value you determined in (a)?
5. Assume that a fossil fuel power plant has an output of 2500 MW and an efficiency of 38%.
   (a) Determine its output energy in one day.
   (b) Calculate the input energy required to produce this output energy.

Answers

4. (a) 25%
5. (a) 2.2×10^{14} J
   (b) 5.7×10^{14} J

Lab Exercise 5.2.1

Determining Waste Energy

In Chapter 4, you studied the principle of heat exchange, efficiency, and power. You can combine these concepts to carry out calculations of the results of a laboratory investigation to compare the efficiencies of different ways of transforming electrical energy into thermal energy.

In this lab exercise, you will analyze a controlled experiment in which a 1.1-kg sample of pure water with an initial temperature of 12°C was heated until it reached a temperature of 58°C. The temperatures are measured by a thermometer supported in such a way that the bulb does not touch the container. A stopwatch was used to determine the total time interval needed to cause this temperature change. The sample was slowly and constantly stirred during the heating process. This procedure was repeated using other sources of heat, as summarized in **Table 2** under Evidence.

Prediction

(a) By comparing the five setups shown in the illustration, predict which heat source and corresponding setup will waste the least amount of energy in heating the water. Rank the setups, from least to greatest energy wasted, according to your predictions.

Evidence

Table 2 shows the sources of heat and the observational data for the experiment.

Table 2

Source	Electric kettle	Electric stove	Hot plate	Hot plate	Hot plate
Power (W)	1500	750	600	600	600
Setup		metal container	metal container	metal container	beaker
Δt (s)	173	401	611	787	668

Analysis

(b) Copy **Table 2** into your notebook, and add the following rows: Output energy (J), Input energy (J), Waste energy (J), Efficiency (%).

(c) Apply the Chapter 4 equations to determine the values needed to complete the Evidence table. (The specific heat capacity of water is 4.18×10^3 J/(kg•°C).)

(d) Describe patterns you observe in the calculations of energy wasted and efficiency.

(e) Describe factors that influence the efficiency of the heat sources in this lab exercise. (In your answer, be sure to discuss where the wasted energy has gone.)

(f) If you were to perform a similar experiment, what safety precautions would you implement?

Evaluation

(g) Evaluate your predictions in (a).

(h) Describe sources of random error and systematic error in the experiment. (To review random and systematic errors, refer to Appendix A.)

Synthesis

(i) Assume you were testing an energy transformation technology to determine the efficiency with which it cooks noodles or potatoes.

  (i) What factors would you control?

  (ii) How do you think the efficiency would compare if you were to place a cover on the container? Why?

SUMMARY Energy Transformation Technology

- An energy transformation technology converts energy from some source into a usable form; for example, a fossil-fuel generating station converts chemical potential energy into electrical energy.
- Almost all energy transformation technologies operate at efficiencies less than 100%. A lot of the wasted energy becomes thermal energy.
- The efficiency of an energy transformation technology can be found using the equation

$$\text{efficiency} = \frac{E_{\text{out}}}{E_{\text{in}}} \times 100\%.$$

Section 5.2 Questions

Understanding Concepts

1. State whether each statement below is true or false, and justify your answer.
   (a) Energy transformation technologies are more important to individual Canadians now than they were when Canada first became a country.
   (b) Energy transformation technologies are more important on a daily basis to people living on the Caribbean Islands than to the average Canadian.
2. Write the energy transformation equation for an automobile.
3. An electric kettle, rated at 1.5 kW, heats 1.1 kg of water from 14°C to 99°C in 4 min 45 s. Determine the efficiency of this application of an energy transformation technology.

5.3 Energy Resources

energy resource: raw material obtained from nature that can be used to do work

renewable energy resource: an energy resource that renews itself in the normal human lifespan

non-renewable energy resource: an energy resource that does not renew itself in the normal human lifespan

You have seen that energy transformation technologies convert energy from some source into a useful form of energy. The original source of the energy, called an **energy resource**, is a raw material obtained from nature that can be used to do work. A resource is considered **renewable** if it renews itself in the normal human lifespan. All other resources are considered **non-renewable**. Both types will be explored in this section.

Figure 1 illustrates Canada's main sources of energy. Approximately 11% of Canada's energy consumption originates from water power (at waterfalls, for example). This resource is renewable. Almost all the remaining energy consumption comes from non-renewable resources — crude oil, natural gas, and coal, which are fossil fuels, and uranium.

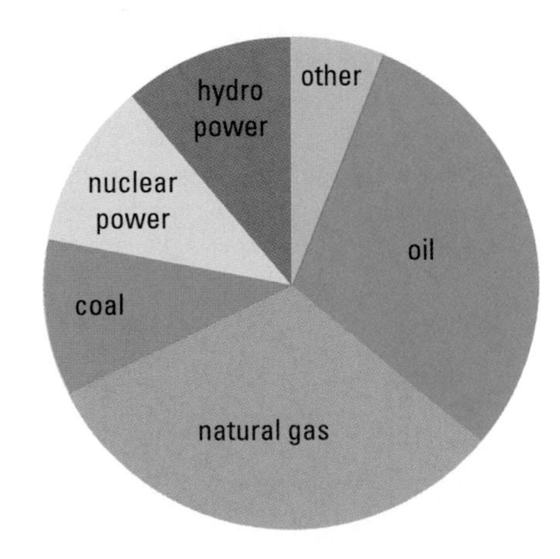

Figure 1
Canada's sources of energy

Non-Renewable Energy Resources

Fossil fuels make up the largest portion of non-renewable energy resources. Energy from fossil fuel begins as radiant energy from the Sun that is absorbed by plants. The plants use the energy to manufacture carbohydrates, which store energy. Most of the stored energy is used during the lifetime of the plants, but some remains after their death. If the plants are buried, they do not disappear;

rather, they are compressed into various new forms. Their energy (chemical potential energy) can be extracted later. Since it takes millions of years for plant life to become useful fuel, once we have consumed the fossil fuels currently available, they are gone forever.

hydrocarbons: compounds that contain only carbon and hydrogen

Fossil fuels are composed mainly of carbons and hydrocarbons. **Hydrocarbons** are compounds containing only carbon and hydrogen. (They differ from carbohydrates such as starch and sugar, which contain oxygen as well as carbon and hydrogen.) Hydrocarbons are found in the solid, liquid, or gaseous state. **Table 1** lists the main categories of fossil fuels currently mined in Canada and their energy content per kilogram.

Table 1 Major Fossil Fuels

Fuel	State	Composition	Energy content per kilogram (MJ)
coal	solid	70% C	28
lignite	solid	30% C	12
gasoline	liquid	varies	44
kerosene	liquid	varies	43
methane	gas	CH_4	49
ethane	gas	C_2H_6	44
propane	gas	C_3H_8	43

(Note: All values are approximate.)

Canada is fortunate in that it happens to be one of the most resource-rich nations in the world. But just how much of Earth's non-renewable resources do we have, and how long will our supplies last? **Table 2** answers these questions for oil, natural gas, coal, and uranium.

Table 2 Canada's Non-Renewable Resources on a World-Wide Scale

Energy resource	Approximate portion of world's supply (%)	Estimated time remaining before depletion of world supply
oil	1	less than 100 a
natural gas	3	less than 200 a
coal	1	more than 1000 a
uranium	20	more than 1000 a

DID YOU KNOW?

Oil Shale Kerogen

A very thick, almost solid, fossil fuel called kerogen is found in oil shale rock. This fuel is mined by underground explosions followed by an injection of steam or hot air. The process heats the shale to at least 500°C to vaporize the oil for recovery. Kerogen was mined at several locations in the world, including the shores of Lake Huron in Ontario, until much cheaper oil became available from oil wells. Canada has significant quantities of kerogen, much of it in New Brunswick. As energy prices rise and supplies of other fossil fuels diminish, this resource may become important.

Fossil fuels are recovered from the ground in raw form. They then undergo a conversion by some form of energy transformation technology to make them useful. When burned directly, fuel products can be used to operate engines of cars and other vehicles and operate furnaces to heat buildings. The fuel products can also be used to generate electricity. In this process, the fossil fuel is used to produce steam, which in turn drives huge electric generators. **Figure 2** shows the basic method of using fuel to generate electricity.

The fossil fuels already mentioned are relatively easy to recover from the ground. However, as the supply of oil and natural gas diminishes, less conventional fossil fuels, which are more difficult to recover, will become important. One source of such fuels is in Western Canada's tar sands, located predominantly in Northern Alberta, which may become extremely valuable in the future. The

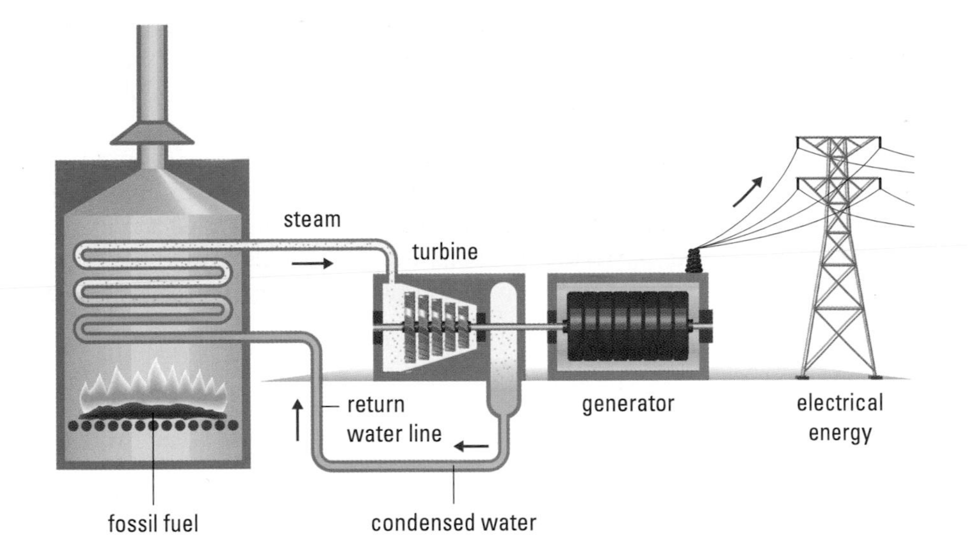

Figure 2
This drawing shows how fossil fuel is used to generate electricity. Chemical potential energy stored in the fuel changes to thermal energy. The thermal energy boils water, which changes to steam. The steam, under pressure, forces the turbine to spin. The generator, connected to the turbine, changes the mechanical energy of spinning into electrical energy.

largest known petroleum accumulation in the world is the deposit called the Athabasca Tar Sands. In fact, the Athabasca Tar Sands are the site of the largest mining operation in the world. This all sounds impressive, but there are numerous problems to overcome before the tar sands can be mined efficiently.

Tar sands are composed of a mixture of sand grains, water, and a thick tar called **bitumen**. Only about 10% of the bitumen can be surface-mined. The remainder lies beneath the surface, down to a depth as far as 600 m. Currently, surface mining is carried out using huge, expensive machines, shown in **Figure 3**. Methods of extracting tar from below the surface use up to 50% of the energy recovered. This is obviously a waste of energy, so much research and development is being carried out to improve the techniques.

A source of energy for the generation of electricity that is not a fossil fuel is uranium. Uranium undergoes **nuclear fission**, in which the nucleus (core) of each atom splits and, in doing so, releases a relatively large amount of energy, which heats water. Thus, for electrical energy production, uranium serves the same function as fossil fuels. The basic operation of Canada's nuclear generating station, called the CANDU reactor, is shown in **Figure 4**.

bitumen: a mixture of hydrocarbons and other substances that occurs naturally or is obtained by distillation from coal or petroleum

Figure 3
Surface mining in the Athabasca Tar Sands is performed by large machines like the one seen in this photograph. The bitumen extracted is refined in the operation nearby.

nuclear fission: process in which the nucleus of each atom splits and releases a relatively large amount of energy

boiler (ordinary water boils, producing steam)
electrical transmission lines
steam drives turbine
generator
pump
pump
control rods
coolant
reactor core
fuel bundles (heat generated during fissioning of uranium)
pump
cooling water from river or lake

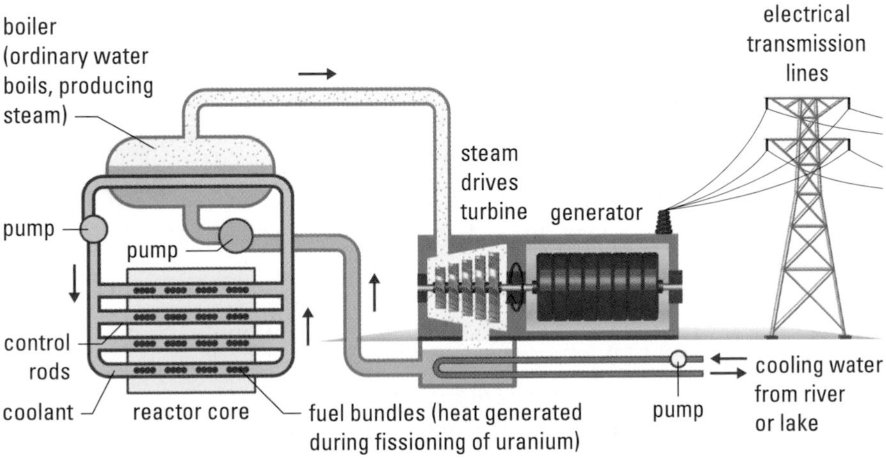

Figure 4
The basic operation of a CANDU generating station. The name CANDU indicates that this fission reactor is **CAN**adian in design, uses **D**euterium oxide (heavy water) to control the rate of the nuclear reaction, and uses **U**ranium as its fuel.

DID YOU KNOW ?

Refuelling a CANDU Reactor

The horizontal design of the calandria in a CANDU reactor allows the fuel bundles to be replaced one at a time without shutting down the reactor to refuel. Most other reactor designs in the world require a shutdown of approximately one week for refuelling, which disrupts not only the supply of electrical energy, but also the local ecology in lakes or rivers where water temperatures are affected.

The fuel used in a CANDU reactor is natural uranium in the form of uranium oxide (UO_2). The fuel is pressed into pellets and placed into long metal tubes sealed at the ends. Several such tubes are assembled into bundles, each bundle having a mass of about 22 kg. Each reactor contains about 5000 fuel bundles placed horizontally in an assembly called a calandria. As the uranium undergoes nuclear fission, the energy released heats the liquid coolant surrounding the bundles. The coolant, a form of water called heavy water, in turn circulates under pressure to heat the ordinary water in a boiler. Steam from the boiler water circulates through the turbines, which in turn drive the generators that produce electrical energy.

Try This Activity

CANDU Reactors

- In a group, research Canada's CANDU reactors. To share the responsibilities, different students can focus on different components of these reactors. Some choices of research are the historical development, physics aspects, cost/benefit analysis, career risks of nuclear industry workers, risks of accidents (such as meltdown), and radioactive wastes. Follow the links for Nelson Physics 11, 5.3.

GO TO www.science.nelson.com

- Coordinate your group's findings to create a summary of what the members found.

Practice

Understanding Concepts

1. Assume that Canadians consume about 1.0×10^{11} kg of methane *per annum*. Use the information in **Table 1** of this section to determine how much energy is available in that amount of methane.
2. Assume Canadians consume 2.0×10^{8} kg of crude oil per day. If the energy content per kilogram of oil is 43 MJ, how much energy do we obtain from crude oil each day?
3. The giant tar sands excavator can move 4.5×10^{7} kg of oil sand per day. How much work is done just in lifting that amount of oil sand a vertical height of 11 m?
4. Write the energy transformation equations for the following energy transformation technologies:
   (a) a fossil-fuel electrical generating station
   (b) a CANDU electrical generating station
   (c) a propane-driven car

Answers

1. 4.9×10^{18} J
2. 8.6×10^{15} J
3. 4.9×10^{9} J

Renewable Energy Resources

Fossil fuels and fissionable materials, such as uranium, will not last many more centuries. But renewable energy resources are alternatives which can supply a seemingly endless amount of energy. Our challenge is to develop means of converting the available renewable energy into usable energy.

Many of the world's renewable energy resources currently being used or researched are briefly described here. As you read each description, try to categorize the resource as being available either locally or non-locally.

Solar energy, radiant energy from the Sun, can be used to produce small amounts of electrical energy when it strikes *photovoltaic cells*. These cells are used in satellites and such instruments as calculators. Solar energy can also be used to heat buildings and swimming pools directly. The expression **passive solar heating** refers to designing and building a structure to take best advantage of the Sun's energy at all times of the year. The Sun's rays enter such a building in winter but not in summer, as illustrated in **Figure 5**. Passive solar heating is much less expensive to install than active solar heating.

solar energy: radiant energy from the Sun

passive solar heating: the process of designing and building a structure to take best advantage of the Sun's energy at all times of the year

Figure 5
The diagram illustrates the basic features of a home with passive solar heating. Other features may include carpets that absorb light energy in winter, and window shutters that are closed at night to prevent heat loss.

An **active solar heating** system (**Figure 6**) absorbs the Sun's energy and converts it into other forms of energy, such as electricity. For instance, an array of solar cells placed on a slanted south-facing wall or roof can convert light energy into electrical energy.

active solar heating: the process of absorbing the Sun's energy and converting it into other forms of energy

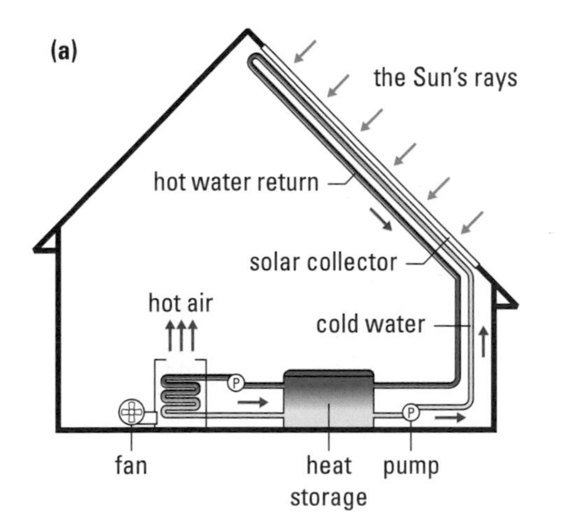

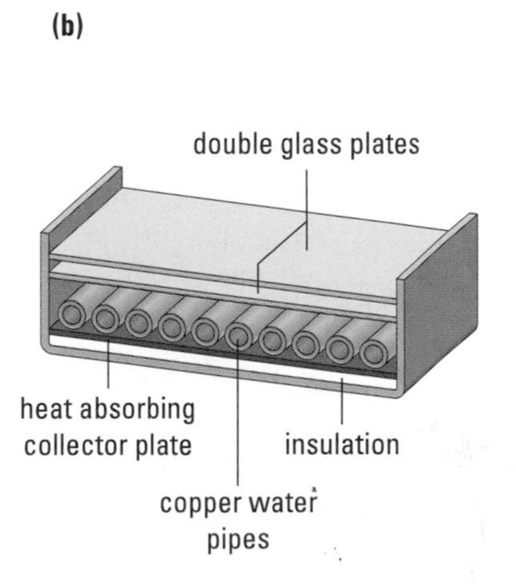

Figure 6
(a) Basic operation of an active solar heating system
(b) An example of solar collector design
(c) A house with an active solar heating system

hydraulic energy: energy generated by harnessing the potential energy of water

Hydraulic energy comes indirectly from solar energy. The Sun's radiant energy strikes water on Earth. The water evaporates, rises, condenses into clouds, and then falls as rain. The rain gathers in rivers and lakes and has gravitational potential energy at the top of a dam or waterfall. This energy can then be changed into another form, such as electricity, which is useful.

wind energy: energy generated by harnessing the kinetic energy of wind

Wind energy, again caused indirectly by solar energy, is a distinct possibility as an energy source in areas of Canada where wind is common throughout the year. Wind generators can change the kinetic energy of the wind into clean, non-polluting electrical energy, or into energy for pumping water. **Figure 7(a)** shows how wind energy is transformed into electrical energy in a wind generator, and **Figure 7(b)** shows the estimated wind energy available in Canada.

(a)

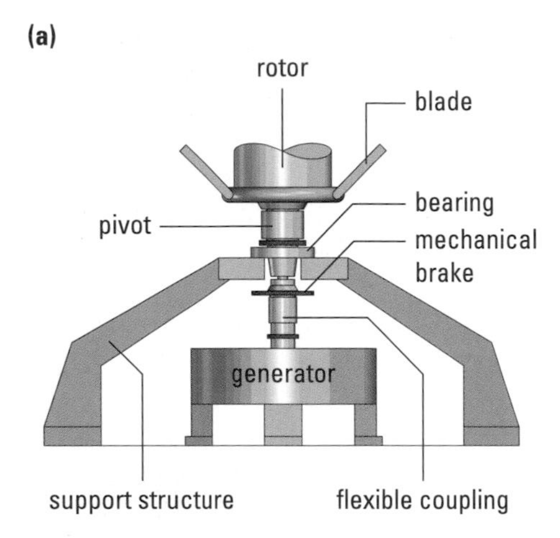

(b)

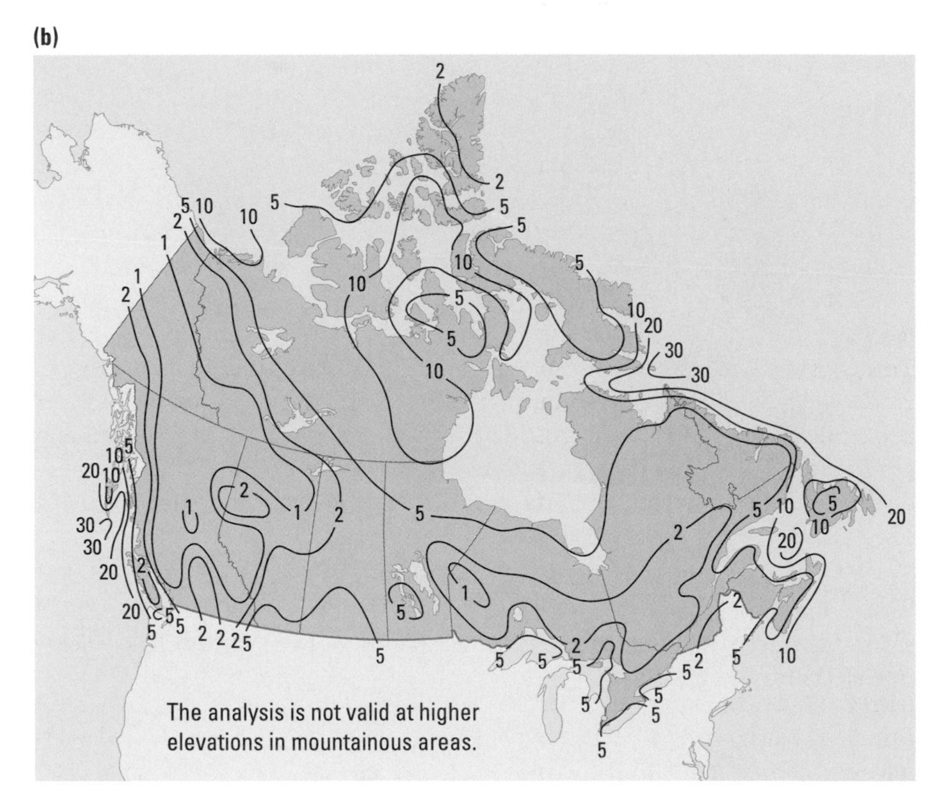

Figure 7
Wind energy
(a) This diagram shows the structure of a wind generator. (The complete vertical-axis generators are shown in the chapter opener.)
(b) This map shows the mean annual wind energy density available in Canada during the period 1967–1976. The units of measurement are GJ/m^2/a at an elevation of 10 m.

tidal energy: energy generated by harnessing the gravitational forces of the Moon and the Sun that act on Earth

Tidal energy is a possible energy resource in regions where ocean tides are large. It is one of the few resources not resulting from the Sun's radiation. It occurs because of gravitational forces of the Moon and the Sun on Earth. To obtain electrical energy from tidal action, a dam must first be built across the mouth of a river that empties into the ocean. The gates of the dam are opened when the tide rolls in. The moving water spins turbines which produce electricity. When the tide stops rising, the gates are closed until low tide approaches. Then the gates are opened and the trapped water rushes out past the turbines, once again producing electricity. A major advantage of this system is that it does not produce either air pollution or thermal pollution. Its disadvantages are that it is difficult to produce a supply of electricity whenever it is most needed; as well, it is hard to judge how the construction of the dam will affect the tides and the local ecology.

North America's first tidal-energy generating station began operation in the mid-1980s. Built on an existing causeway, this station is located at Annapolis Royal, Nova Scotia; it is linked to the Bay of Fundy system, which has among the highest tides (over 15 m) in the world. (Refer to **Figure 8**.) Other possible sites in

Canada are at Ungava Bay in Northern Quebec, Frobisher Bay and Cumberland Sound on Baffin Island, and Jervis and Sechelt Inlets near Vancouver.

Biomass energy is the chemical potential energy stored in plants and animal wastes. Again, this energy comes indirectly from the Sun. Burning wood is a common source of such energy. Wood is used both in home fireplaces and wood stoves, and in large industries that burn the leftover products of the forestry industry. Of course, good planning must be carried out to ensure that the trees are replanted — otherwise, this resource cannot be called renewable.

Numerous schemes are being developed to use forms of biomass other than wood. One proposal is to burn trash to produce heat. Another is the fermentation of sugar molecules in grain by bacteria to produce methane and ethanol (grain alcohol). A mixture of one part of this alcohol in nine parts of gasoline can be used to run automobile engines. This mixture, called gasohol, is being used in various parts of Canada and to a large extent in Brazil, where it is sugar cane, not grain, that is fermented.

One further biomass scheme has interesting possibilities. Certain plants produce not only carbohydrates but also hydrocarbons. An example of such a plant is the rubber tree, which produces latex. Research is underway to develop the use of fast-growing trees and shrubs that produce hydrocarbons directly, and thus require much less processing than carbohydrates before being used as fuel.

Geothermal energy is thermal energy or heat taken from beneath Earth's surface. It results from radioactive decay (the nuclear fission of elements in rocks). This enormous resource increases Earth's subsurface temperature an average of 25°C with each kilometre of depth. Hot springs and geysers spew forth hot water and steam from within Earth's crust. They can be used directly to heat homes and generate electricity. However, most of the thermal energy contained underground does not find its way to the surface, so methods for its extraction are being researched. For example, if the rocks are hot and dry, certainly no water or steam will come to the surface. Still, there is a technique for utilizing this heat. First, two holes are drilled deep into the ground a set distance apart. Water is poured down one hole and it gains energy as it seeps through the hot, porous rocks. Then the water rises up the other hole. The circulating water runs turbines to produce electricity. In Canada, geothermal energy is plentiful in the former volcanic regions of British Columbia and the Yukon Territories, as well as in the Western Canada sedimentary basin in the Prairie provinces. Other places in the world with geothermal activity are Iceland, California, and New Zealand (**Figure 9**).

Nuclear fusion is the process in which the nuclei of the atoms of light elements join together at extremely high temperatures to become larger nuclei. (Notice that this process differs from nuclear fission, in which the nuclei of heavy elements split apart.) With each fusion reaction, some mass is lost, changing into a relatively large amount of energy. Fusion is the energy source for the Sun and stars.

Hydrogen, one of the most abundant substances on Earth, is used, in certain forms, as a common fuel to operate fusion reactors. Nuclear fusion has certain advantages, one of them being a potentially limitless supply of fuel from the world's oceans. Another is that it produces much less radioactive waste than nuclear fission, so it is more desirable from the environmental point of view.

Two main problems must be overcome in order to use fusion to generate electricity. The first is producing temperatures as high as hundreds of millions of degrees, which are needed to begin the fusion reaction. The second is confining the reacting materials, so that fusion may continue. Research is currently progressing in the use of magnetic fields and lasers to solve both problems. (A more detailed explanation of fusion is left for more advanced physics texts.)

Figure 8
The Annapolis Tidal Generating Station, Canada's first tidal generating station, takes advantage of the high tides in the Bay of Fundy.

biomass energy: the chemical potential energy stored in plants and animal wastes

geothermal energy: thermal energy or heat taken from beneath Earth's surface

Figure 9
New Zealand has several active geothermal areas, some of which are used to generate electrical energy. The steam field shown here provides energy for the Wairakei Geothermal Power Station. This station and the nearby Ohaaki Station together provide about 8% of New Zealand's electrical energy needs.

nuclear fusion: the process in which the nuclei of the atoms of light elements join together at extremely high temperatures to become larger nuclei

fuel cell: device that changes chemical potential energy directly into electrical energy

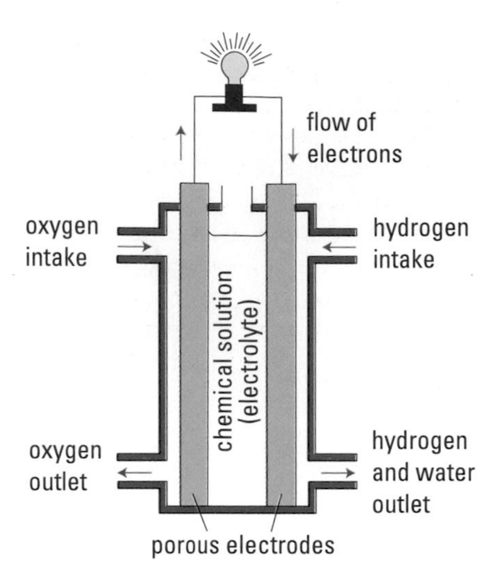

Figure 10
A fuel cell

atmosphere: the air in a specific place that can be used as a source of heat

heat pump: a device that uses evaporation and condensation to heat a home in winter and cool it in summer

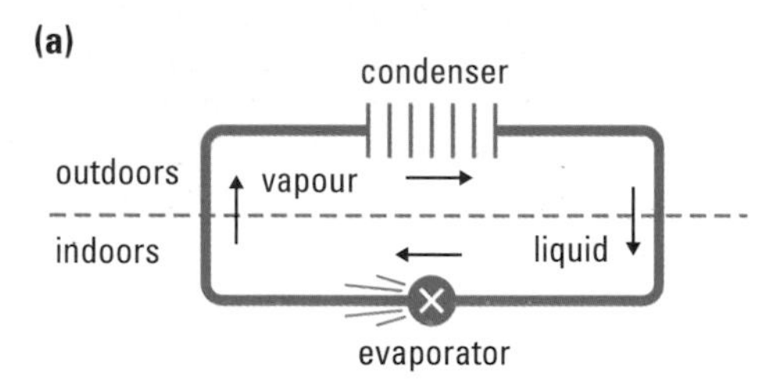

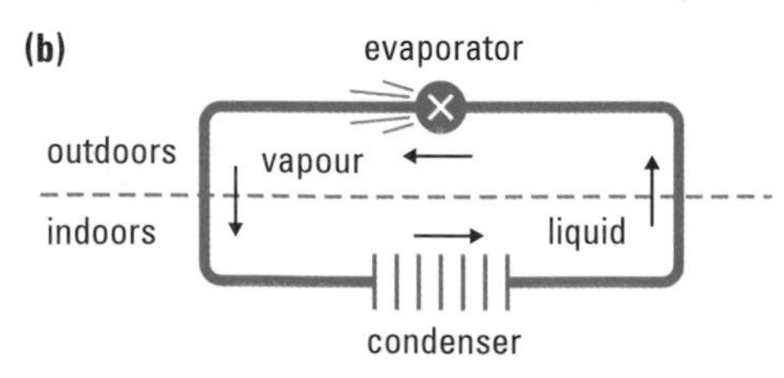

Figure 11
The basic operation of a heat pump
(a) In summer, evaporation occurs indoors to absorb heat; condensation occurs outdoors to give away the heat.
(b) In winter, evaporation occurs outdoors to absorb heat; condensation occurs indoors to give away the heat.

Hydrogen has another important use in an energy transformation technology called a **fuel cell.** This device has attracted much attention after its use in the space program. In a fuel cell, chemical potential energy is changed directly into electrical energy. The chemical fuel used in a fuel cell is usually hydrogen gas. The hydrogen combines with oxygen chemically in the presence of a third chemical called a *catalyst.* The result is the production of water and an electric current. (Refer to **Figure 10.**) Because the energy turns directly from chemical potential into electrical energy, the fuel cell is much more efficient than electrical generating stations. It can operate at a relatively low temperature, so it emits fewer pollutants. Furthermore, it has few moving parts, so it is quiet and easy to maintain. Electricity is taken from the fuel cell in a manner similar to the process for a dry cell.

The **atmosphere** can also be used as a source of heat. An electric **heat pump**, for instance, operates to heat a home in winter and cool it in summer. It works on the principles that when a substance changes from a liquid to a vapour, energy is absorbed (that is, evaporation requires heat), and when a substance changes from a vapour to a liquid, energy is given off (that is, condensation releases heat).

The substance that circulates in a heat pump system is called a *refrigerant.* It flows in one direction in summer and the opposite direction in winter. The refrigerant evaporates inside the home in summer, absorbing heat. It evaporates outside in winter, again absorbing heat. **Figure 11** illustrates the basic operation of a heat pump in both seasons.

Practice

Understanding Concepts

5. (a) List as many renewable energy resources originating from the Sun's radiant energy as you can.
   (b) Which renewable energy resources do not originate from the Sun's radiant energy?
   (c) Classify each of the resources you named in (a) and (b) as either locally or non-locally available.
6. Starting with the Sun, trace the energy transformations that occur in order to cook a roast in an electric oven and write the energy transformation equation. The electricity comes from a hydraulic generating system.

Making Connections

7. Think about the problems related to biomass energy production. Give reasons why this energy source is not used to a great extent in Canada.

SUMMARY Energy Resources

- Non-renewable energy resources include uranium and all the fossil fuels, such as coal, oil, and natural gas.
- Renewable energy resources include solar, hydraulic, wind, tidal, and geothermal energies, as well as biomass, nuclear fusion, and the atmosphere.

Section 5.3 Questions

Understanding Concepts

1. List four non-renewable energy resources and five renewable energy resources.
2. Write the energy transformation equation for each of the following resources used to produce electrical energy:
   (a) hydraulic energy
   (b) the Sun
   (c) biomass
   (d) nuclear fusion
3. List some harmful effects to the environment resulting from each of these types of electrical generating stations:
   (a) coal-fired
   (b) hydraulic
   (c) tidal
   (d) nuclear fission

Making Connections

4. Research the means of converting renewable energy to useful energy, then list some reasons why renewable energy resources are not emphasized more throughout Canada. Follow the links for Nelson Physics 11, 5.3.

GO TO **www.science.nelson.com**

5. Which alternative energy resource described in this section is most likely to be developed in your area? Explain why.
6. Most of Canada's electrical energy is generated at enormous centralized generating stations which use a variety of fuels, mostly non-renewable ones. The generated electricity then enters a grid and spreads out to the consumers, who are often at great distances from the stations. Suggest an alternative generation and supply system for your area which may make more sense considering the variety of renewable energy resources now available.

5.4 Using Energy Wisely

This topic, like the others in this chapter, could fill an entire book. In this section, we will explore what we, as a society and as individuals, can do to ensure that our energy supplies will be available for future generations.

Society's Responsibilities in Using Energy

We all rely heavily on electricity. The production of most of our electricity creates much unwanted pollution and requires vast amounts of non-renewable resources. To improve the efficiency of energy production, some of the thermal energy created during electricity generation could be used as an alternative source of energy. Doing this would reduce not only thermal pollution, but also the need for other resources for heating purposes. The process of producing electricity and using the resulting thermal energy for heating is called **cogeneration.** Cogeneration is used mainly in industrial plants located near generating stations. It is likely to become more important in the future.

Another way to improve the efficiency of generating electricity is to learn how to produce electricity for **local consumption**, so that the energy does not have to be transferred long distances using huge transmission lines. Sources of energy for localized power include renewable resources such as solar, hydraulic, and wind energy.

DID YOU KNOW?

Fuel Cell Applications

Fuel cell technology is being used and developed for many applications. Phosphoric fuel cells, which operate at efficiencies between 40% and 85%, are used in buildings such as hospitals and schools, and in buses and other large vehicles. Solid oxide fuel cells are applied in industry and electrical generating stations. This method of generating electrical energy is cheaper to set up and more efficient than using fossil fuels. You can find information about these and other fuel cell applications on the Internet.

DID YOU KNOW?

Thermoacoustic Refrigeration

Like a heat pump, a refrigerator uses a refrigerant. Old refrigerants pose many difficulties, so scientists try to find ways of producing better refrigerants or not using them at all. The latter case is possible with a new energy transformation technology called a thermoacoustic refrigerator. This device has a driver that is basically a high-powered loudspeaker that sends sound waves vibrating back and forth through gases in a resonating tube. The vibrating gases carry heat away from the food being cooled in the refrigerator to a radiator that emits radiant energy to the air outside the refrigerator.

cogeneration: the process of producing electricity and using the resulting thermal energy for heat

local consumption: generating energy locally to avoid the transfer of energy over long distances using transmission lines

Conserving energy is another vital goal for all members of our society. Governments have provided incentives for people to improve home insulation and to replace old, inefficient furnaces with new ones that use cleaner, more plentiful sources of energy. They also promote the use of active and passive solar heating systems.

In transportation, governments support reduced speed limits, car pools, public transportation, and lanes on city roads restricted to bus use only. **Table 1** makes it clear that a person driving alone in a car consumes a relatively large amount of energy for the distance travelled.

Table 1 A Comparison of Forms of Surface Transportation

Passenger kilometres per litre (0 10 20 30 40 50)	Average number of passengers
highway bus	22
inter-city train	400
compact car	1.5
urban bus	12
large car in city	1.5

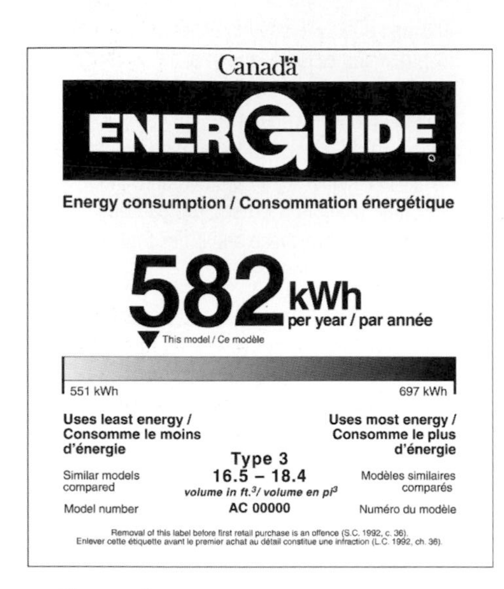

Figure 1
EnerGuide label

Canada's Energy Efficiency Act and Regulations require that most household appliances be tested to a prescribed standard, with the rating printed on the EnerGuide label for each appliance (see **Figure 1**). A directory that lists the ratings is published every year.

Saving Energy

In a group, brainstorm ways in which energy savings could be realized within your school. Consider lighting, heating, air conditioning, the use of water, reusing, recycling, and any other factors you can think of. Describe your ideas, and write a proposal in which your school gains half of the savings generated by the ways you describe. The proposal can be aimed at the school board or your local school council.

Practice

Making Connections

1. Use at least one of the following questions to initiate a project, debate, or class discussion.
   (a) Is cogeneration of electricity possible in your region?
   (b) Is electricity generation for local consumption possible in your area?
   (c) Should highway express lanes in and near large cities be reserved during rush hours for cars with three or more people?
   (d) How could school officials improve energy consumption in your school?
   (e) If you were in control of time zones in Canada, would you advise the use of standard time, daylight savings time, or a combination? Explain your reasons.

Our Personal Responsibilities in Using Energy

Ideally, we all would like our lives to be care-free and happy. To achieve this state, we must all share responsibility for conserving energy. There are many ways to conserve energy at home. Consider the following questions:

- Do you use more hot water than necessary to take a bath or shower?
- Do you leave the refrigerator door open while deciding what to eat?
- Do you keep your home quite cool in the winter and wear a sweater?
- If you have a fireplace, does most of the heat it produces go up the chimney? See **Figure 2**.
- Is your home properly insulated?
- Do you leave lights and electric appliances on when they are not in use?
- Are you aware of which types of lights and appliances are most efficient?
- When you use an appliance such as a toaster, clothes washer, or clothes dryer, do you make maximum use of its energy?

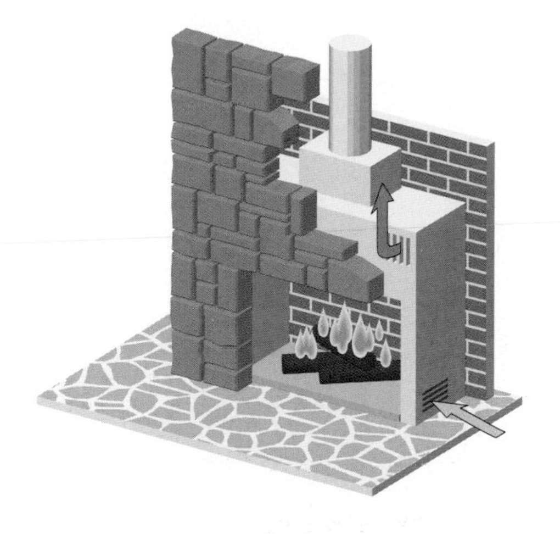

Figure 2
In a fireplace heat circulator, cool air enters through the lower vent and passes through a duct adjacent to the fire. The hot air rises and is discharged through the upper vent or passes through ducts to other rooms.

Besides conserving energy in our homes, we should conserve it outside as well. Consider these questions:

- When you travel short distances, do you usually walk or take a car?
- Do you take part in entertainment and sports activities which are large energy consumers, such as water-skiing behind a motorboat? Can you think of fun activities that would consume less energy, for example, wind surfing?

These are just a few of the many important questions about energy we can ask ourselves. The answers will help determine the fate of future generations who will, no doubt, wish to be able to continue to consume energy.

Practice

Understanding Concepts

2. Describe ways in which you can conserve energy in your own home.
3. Ice cubes that remain in the freezer compartment of a frost-free refrigerator gradually disappear. Why does this occur? Is leaving ice cubes in such a freezer an efficient use of energy? Explain.

SUMMARY Using Energy Wisely

- Governments, industries, and individuals can all work toward using energy wisely.

Section 5.4 Questions

Understanding Concepts

1. Describe what you can do to improve the efficiency of the ways in which you use energy.

Making Connections

2. Which technique would be more effective at conserving automobile fuel — fuel rationing or higher taxes? Justify your choice.
3. Should consumers pay more or less when they increase the rate of electrical energy they consume? What is the current pricing policy in your area?

CAREER

Careers in Energy, Work, and Power

There are many different types of careers that involve the study of energy, work, and power in one form or another. Have a look at the careers described on this page and find out more about one of them or another career in energy, work, and power that interests you.

Building Systems and Engineering Technologist

For admission to this three-year community college diploma course, a high school diploma with mathematics and English is required; physics and chemistry are an asset. These technologists manage the electrical, water, and mechanical systems of a building, and access land for potential development, particularly with regard to soil quality. They work for school boards, municipalities, and major property development or property management companies to provide the best energy-use options and to stabilize contracts for energy suppliers.

Meteorologist

An entry-level position in meteorology or atmospheric sciences requires a bachelor's degree in meteorology with a strong background in mathematics, physics, and computer science. Meteorologists study the atmosphere and must be familiar with the physical characteristics and the motion of Earth. They assess temperature, wind velocity, and humidity in order to forecast the weather. They also study air pollution and climate trends, such as global warming and ozone depletion. Meteorologists work with sophisticated computer models, weather balloons, satellites, and radar.

Heating, Refrigeration, and Air-Conditioning Technician

To gain entry to this two-year community college program, you would need a high school diploma with an advanced mathematics credit, or you could be a mature student. Many of the on-the-job technicians are sponsored by their employers to study for this program. These technicians usually work for developers, work in factories and office buildings, or work for government agencies. They work with a variety of hand tools and must be able to read diagrams.

Practice

Making Connections

1. Identify several careers that require knowledge about energy, work, and power. Select a career you are interested in from the list you made or from the careers described above. Imagine that you have been employed in your chosen career for five years and that you are applying to work on a new project of interest.
   (a) Describe the project. It should be related to some of the new things you learned in this unit. Explain how the concepts from this unit are applied in the project.
   (b) Create a résumé listing your credentials and explaining why you are qualified to work on the project. Include in your résumé
   - your educational background—what university degree or diploma program you graduated with, which educational institute you attended, post-graduate training (if any);
   - your duties and skills in previous jobs; and
   - your salary expectations.

   Follow the links for Nelson Physics 11, 5.4.

www.science.nelson.com

Chapter 5 Summary

Key Expectations

Throughout this chapter, you have had opportunities to

- analyze, in qualitative and quantitative terms, simple situations involving work, gravitational potential energy, kinetic energy, thermal energy, and heat, using the law of conservation of energy; (5.2, 5.3)
- apply quantitatively the relationships among power, energy, and time in a variety of contexts; (5.2)
- analyze, in quantitative terms, the relationships among percent efficiency, input energy, and useful output energy for several energy transformations; (5.1, 5.2, 5.3, 5.4)
- communicate the procedures, data, and conclusions of investigations involving work, mechanical energy, power, thermal energy, and heat, and the law of conservation of energy, using appropriate means (e.g., oral and written descriptions, numerical and/or graphical analyses, tables, diagrams); (5.2)
- analyze, using your own or given criteria, the economic, social, and environmental impact of various energy sources (e.g., wind, tidal flow, falling water, the Sun, thermal energy, and heat) and energy transformation technologies (e.g., hydroelectric power plants and energy transformations produced by other renewable sources, fossil fuel, and nuclear power plants) used around the world; (5.3, 5.4)
- identify and describe science and technology-based careers related to energy, work, and power; (career feature)

Key Terms

doubling time
energy transformation technology
energy converter
energy resource
renewable energy resource
non-renewable energy resource
hydrocarbons
bitumen
nuclear fission
solar energy
passive solar heating
active solar heating
hydraulic energy
wind energy
tidal energy
biomass energy
geothermal energy
nuclear fusion
fuel cell
atmosphere
heat pump
cogeneration
local consumption

Make a Summary

Draw a map of your area of the province and show the approximate location of your home and school as well as several other features, such as electrical generating stations, garbage dumps, bodies of water, rivers, and recycling depots. On your map, show how electricity is generated and distributed to your home and school. Add alternative ways that electrical energy could be generated in the future. Use as many of the concepts and key words from this chapter as possible.

Reflect on your Learning

Revisit your answers to the Reflect on Your Learning questions at the beginning of this chapter.

- How has your thinking changed?
- What new questions do you have?

Chapter 5 Review

Understanding Concepts

1. Discuss reasons why Canada's rate of energy consumption is much higher than the world average.
2. If the cost of natural gas increases at 3.6%/a, by what factor will the price have increased after
   (a) 20 a? (b) 40 a? (c) 100 a?
3. Assume that our federal government wants to be sure that 50 years will elapse before the number of cars in Canada doubles. What growth rate per annum should the government advocate?
4. Explain why scientists and politicians should understand growth rate when dealing with energy use.
5. What is the final form of energy that renders all energy transformation technologies less than 100% efficient?
6. Why are automobiles so energy inefficient?
7. Metal cooking pots can be made totally of aluminum or with aluminum sides and copper bottoms. (Copper is a better heat conductor than aluminum.) A chef puts 1.5 kg of 15°C water into an all-aluminum pot, which is then placed on a stove burner rated at 1.8 kW. The water takes 6 min 25 s to reach the boiling point, at which time the chef adds the pasta and some salt.
   (a) Determine the efficiency of the stove burner in heating the water to the boiling point.
   (b) Where does the wasted energy go?
   (c) How would using a copper-bottomed pot affect the efficiency?
8. What is a fossil fuel? What are its main components?
9. Contrast and compare the use of falling water with the use of fossil fuels for generating electricity.
10. State the main advantage and main disadvantage of using each of the following non-renewable energy resources:
    (a) oil
    (b) natural gas
    (c) coal
    (d) tar sands
    (e) uranium
11. State the main advantage and main disadvantage of each of the following renewable energy resources:
    (a) solar energy
    (b) hydraulic energy
    (c) wind
    (d) tides
    (e) biomass
    (f) geothermal energy
    (g) nuclear fusion
    (h) the atmosphere
12. You are given a 20.0-L bucket, several other smaller containers, an eye-dropper that dispenses 1.0 mL of water in 50 drops, and access to water.
    (a) If you were to add one drop of water from the eye-dropper to the empty bucket at time 0.0 s and then cause the number of drops inside the bucket to double every 10.0 s, estimate the time you would need to fill the bucket.
    (b) Calculate the number of drops of water needed to fill the bucket.
    (c) How full will the bucket be after 2.0×10^2 s? Show your calculations.
13. Describe ways you could minimize wasting energy when heating water in the kitchen. In your answer, apply what you learned in Lab Exercise 5.2.1 and related questions.
14. Suggest how the use of cars could become more efficient.
15. Assume that the world's annual energy consumption is 3.5×10^{20} J and that Canadians consume an average of 1.0×10^3 MJ per person per day. To answer the questions below, you will need to know the approximate population of Canada and the world.
    (a) How much energy does each Canadian consume *per annum*?
    (b) How much energy does all of Canada consume *per annum*?
    (c) What percent of the world's population lives in Canada?
    (d) What percent of the annual world energy consumption does Canada consume?
    (e) Calculate the ratio of the answer in (d) to the answer in (c). What do you conclude?
16. Engineers have long dreamed of harnessing the tides in the Bay of Fundy. Although in some places the difference between high tide and low tide can be over 15 m, the average change in height for the entire bay is about 4.0 m. The bay has the same area as a rectangle that is about 3.0×10^2 km long and 65 km wide. Water has a density of 1.0×10^3 kg/m^3.
    (a) Calculate the volume of water and the mass of water that flows out of the bay between high tide and low tide.
    (b) Determine the loss in gravitational potential energy when the water flows out of the bay. Assume that the decrease in gravitational potential energy is equal to that of the mass calculated in (a) being lowered a distance of 2.0 m.
    (c) If half the gravitational potential energy lost when the tide flows out could be converted to electricity over a 6-h period, determine the amount of electrical power that would be generated.

Applying Inquiry Skills

17. **Figure 1** shows a graph of the world's energy use from 1800 to 2000. From the graph, estimate the doubling time between the years 1900 and 2000.

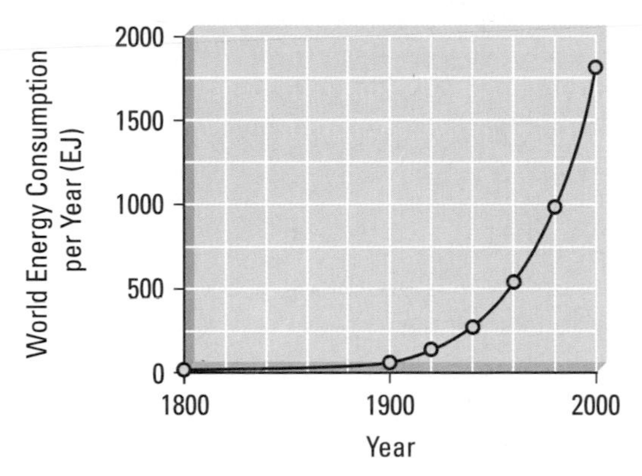

Figure 1

Making Connections

18. State immediate objectives that all of us can pursue to help alleviate the problems of energy use in our country.
19. Suppose you are a planner for a Canadian electrical utility company. You wish to build a hydraulic generating station. Describe briefly the main factors you would have to consider in selecting a site for such a development.
20. Describe long-term objectives that governments in Canada should pursue to ensure that we have a plentiful supply of low-pollution energy in the future.

Exploring

21. Research and report on current statistics on the annual growth rate of energy consumption in Canada and around the world. Follow the links for Nelson Physics 11, Chapter 5 Review.

GO TO www.science.nelson.com

22. **Figure 2** shows one example of a use of geothermal energy besides producing electrical energy. Use the Internet or other resources to research other such uses of geothermal energy. Two countries where these applications are found are Iceland and New Zealand.

Figure 2
For centuries, Maori natives in New Zealand have used geothermal energy for cooking food. Tourists in Rotorua can enjoy corn-on-the-cob cooked in the heat from the steam vent in the geothermal field called Whakarewarewa.

23. The Toyota Prius (**Figure 3**) was the world's first production car to combine a gasoline engine with an electric motor that never needs to be plugged in. The electric motor gains energy and becomes charged whenever brakes are applied in an energy transformation which converts kinetic energy back into stored electrical energy. Find out more about the Prius, and explain why it is more efficient and less polluting than most cars.

Figure 3

24. According to **Table 2** in section 5.3, Canada has large reserves of uranium (enough to meet a significant portion of foreseeable energy requirements). Research why Canada has not made more of a commitment to developing this energy potential. Follow the links for Nelson Physics 11, Chapter 5 Review.

GO TO www.science.nelson.com

Energy Cost-Benefit Analysis

Energy transformation technologies that produce electrical energy apply the concepts presented in Unit 2 in a variety of ways. All of the energy transformations involved obey the law of conservation of energy. The efficiencies of the transformations and the power of the generation systems can be determined and compared.

Studying energy resources and energy transformation technologies becomes increasingly important in our society as our population grows, our dependence on electrical energy increases, and our fossil-fuel supplies diminish. Thus, making a cost-benefit analysis of our energy resources and energy transformation technologies is important to our lives. Numerous sources of information are available, including books, encyclopedias, magazines, and CD-ROMs. Follow the links for Nelson Physics 11, Unit 2 Performance Task.

GO TO www.science.nelson.com

A Unique Example

One unique example of producing electrical energy is found in the southwest region of New Zealand in an area called Fiordland National Park, a region known for huge rainfalls—more than 8.5 m per year in some locations! An underground hydroelectric generating station, the Manapouri Power Station, was designed to allow water from Lake Manapouri to flow through seven shafts to the generating station 170 m below the lake, as illustrated in **Figure 1**. The water discharged from the station flows out through a tunnel to an inlet from the Tasman Sea, which is at sea level. To build this 10-km tunnel, workers needed almost 5 years to blast through some of the hardest rock in the world.

The history of this energy transformation technology is even more interesting than the design. If you choose to research this technology, you will discover how the New Zealand citizens banded together to prevent the designers from raising the water level in Lake Manapouri, which would have destroyed a large area of precious natural beauty. You will also find that changes to the design had to be made more recently to improve the efficiency of the generating system.

Studying this example or any of the many other technologies will help you appreciate the process of technological development of a major project. A need is identified, a plan is drawn up, a cost-benefit analysis is carried out, the technology is built and set into operation, problems are observed and corrected, and the entire project is studied by others.

The Task

For this task, you will be expected to analyze the costs and benefits of an energy resource or energy transformation technology used anywhere in the world. You should consider the economic, social, and environmental impact of the resource or technology you choose to analyze. In your process and final product, you should demonstrate an understanding, in qualitative and quantitative terms, of the concepts of work, energy, energy transformations, efficiency, and power.

Some of the energy resources to choose from are wind, tidal flow, falling water, the Sun, and biomass. Some of the energy transformation technologies to choose from are fossil-fuel power plants, nuclear power plants, or electric plants that use a renewable energy resource, such as hydroelectric power plants. Because so much information is available, you should focus on only one choice.

(a)

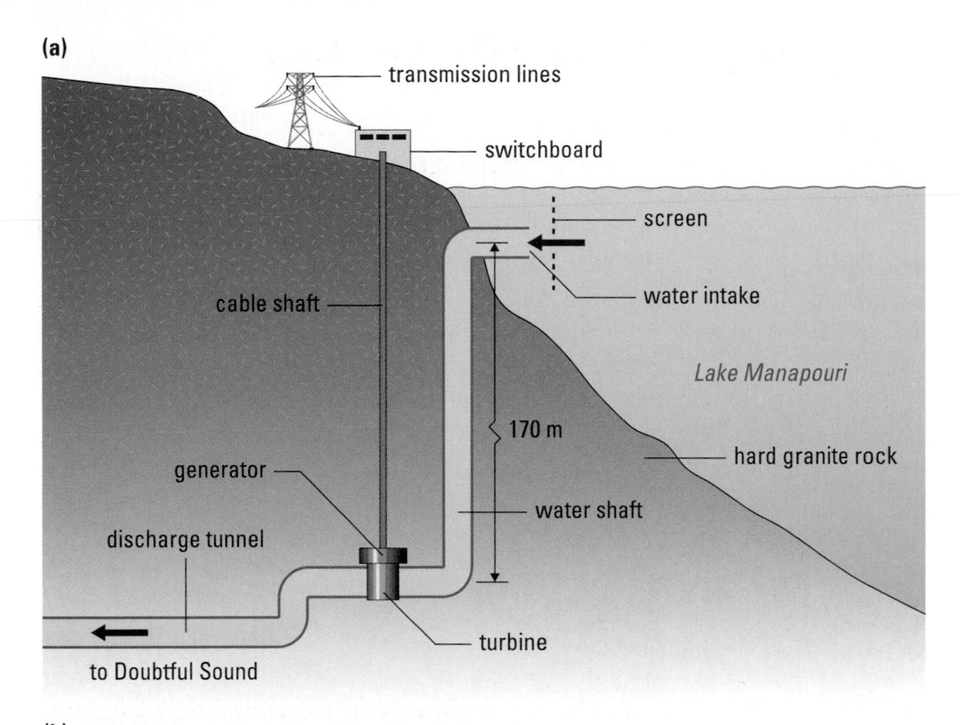

(b)

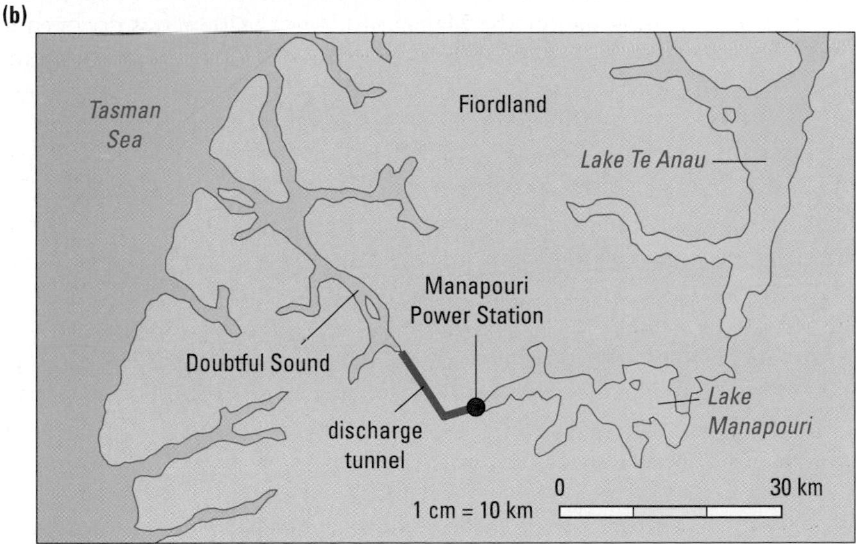

Analysis

You should consider the criteria listed below in your analysis.

- What are the physics principles applied in the design and/or use of the resource or technology?
- What are the economic, social, and environmental costs of the resource or technology?
- What are the benefits of the resource or technology?

Once you have chosen the resource or technology to research, you can design more detailed criteria based on the three main criteria given above. For instance, with the New Zealand example, you could analyze the environmental cost that would have ensued if the original design had been used, and the economic and environmental costs if an earthquake were to occur in the future at or near the generating station.

Assessment

Your completed task will be assessed according to the following criteria:

Process

- Develop detailed criteria for your cost-benefit analysis.
- Choose appropriate research tools, such as books, magazines, and the Internet.
- Carry out the research and summarize the information found with sufficient detail and appropriate organization.
- Analyze the physical principles and the costs and benefits of the energy resource or energy transformation technology.
- Evaluate the research and reporting process.

Product

- Demonstrate an understanding of the related physics concepts, principles, laws, and theories.
- Prepare a suitable research portfolio.
- Use terms, symbols, equations, and SI metric units correctly.
- Produce a final communication, such as an audiovisual presentation, to summarize the analysis.

Figure 1
The Manapouri Power Station in New Zealand is unique because it is built almost 200 m beneath the surface. Most hydroelectric plants are built at waterfalls or on dammed rivers.
(a) The diagram shows the basic design of the underground station that allows the gravitational potential energy of the lake water to be transformed into the kinetic energy of spinning turbines, which is then transformed into electrical energy in the generators. Only one of the seven vertical shafts is shown. Each shaft is 3.7 m in diameter and 170 m deep. Each generator produces electrical energy at a rate of 100 MW. Sadly, 16 people died during the construction of this project in the 1960s.
(b) The station is built on the west end of Lake Manapouri, located in the southwest corner of New Zealand's South Island. Water discharging from the generating station passes through a discharge tunnel, which is 9.5 m in diameter, and dumps into Doubtful Sound.

Understanding Concepts

1. Starting with radiant energy from the Sun, write the energy transformation equation for each situation.
   (a) Wind is used to produce electrical energy.
   (b) Biomass is burned to produce electrical energy.
   (c) A home uses passive solar heating.
   (d) Waterfalls are used to produce electrical energy.
   (e) Which of the situations above involve renewable energy?
2. Give an example of a situation in which
   (a) a force is acting, but there is no motion and therefore no work done
   (b) a force is acting, but the displacement is perpendicular to the force and therefore no work is done
   (c) there is motion, but since no force is acting to cause the motion, no work is done
3. Give an example in which positive work is done and an example in which negative work is done.
4. An electric motor does 1.7 MJ of work on a roller coaster to raise it and its riders to the top of the first hill. If the magnitude of the force needed to raise the coaster at a constant velocity along the ramp is 25 kN, how long is the ramp?
5. Which, if any, of the graphs shown in **Figure 1** applies to the equation $W = F\Delta d$? Explain why the others do not apply.

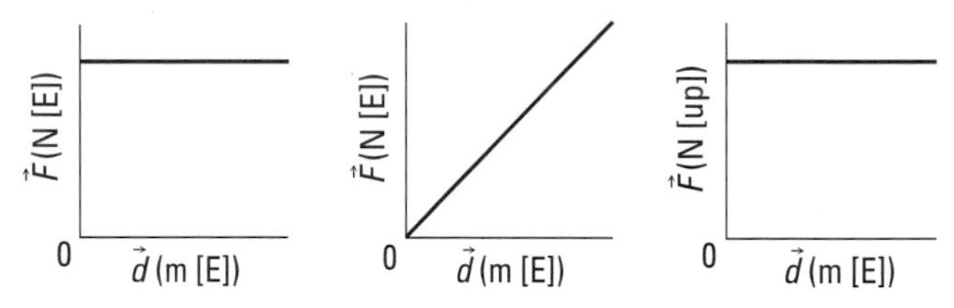

Figure 1

6. At the bottom of a hill, a snowboarder with a velocity of 16 m/s [forward] skids to a stop after moving a distance of 9.4 m. The mass of the board and rider is 71 kg, and the magnitude of the force of kinetic friction during the skid is 9.5×10^2 N.
   (a) Draw an FBD of the snowboarder during skidding.
   (b) Determine the work done by the force of friction on the snowboarder during skidding.
   (c) Determine the coefficient of kinetic friction during skidding.
   (d) Why is the value of the coefficient of kinetic friction so much higher than the typical value of a snowboard on snow?
7. Estimate how high a typical highrise elevator, loaded to its safe capacity, can raise its passengers for each megajoule of energy output. Assume that this output work goes entirely to working against force of gravity to raise the passengers only. Show your reasoning.
8. A black-backed gull cannot crack open an mollusk, so it carries the mollusk up to a height of 18 m and drops it to the pebbly shore below. If the gravitational potential energy of the mollusk is 7.9 J relative to the shore, what is its mass?
9. Before jumping off the platform, the bungee jumper in **Figure 2** has 2.5×10^4 J of gravitational potential energy relative to the river below. Assume the jumper's mass is 59 kg. How high is the platform above the river?

Figure 2
Bungee jumping in New Zealand

10. When the speed of running water increases by a factor of 3.1, by what factor does its kinetic energy change?
11. A 5.5×10^4 kg airplane, travelling at an altitude of 9.9 km, has a speed of 260 m/s relative to the ground. Nearing the end of the flight, the plane slows to a speed of 140 m/s while descending to an altitude of 2.1 km. Determine the plane's total loss of mechanical energy during this change of speed and altitude.
12. A player spikes a 270 g volleyball, giving it 24 J of kinetic energy. How fast is the ball travelling after it is spiked?
13. When the magnetic levitation train built in Germany is travelling at its maximum speed of 125 m/s, its kinetic energy is 875 MJ. What is its mass?

14. A 65-kg student and a 45-kg friend sit on identical swings. They are each given a push so that the two swings move through the same angle from the vertical. How will their speeds compare as they swing through the bottom of the cycle? Explain your answer.

15. A pendulum is drawn aside so that the centre of the bob is at position A as shown in **Figure 3**. A horizontal rod is positioned at B so that when the pendulum is released, the string catches at B, forcing the bob to swing in an arc which has a smaller radius. How will the height to which the bob swings on the right side of B compare to h? Explain your answer.

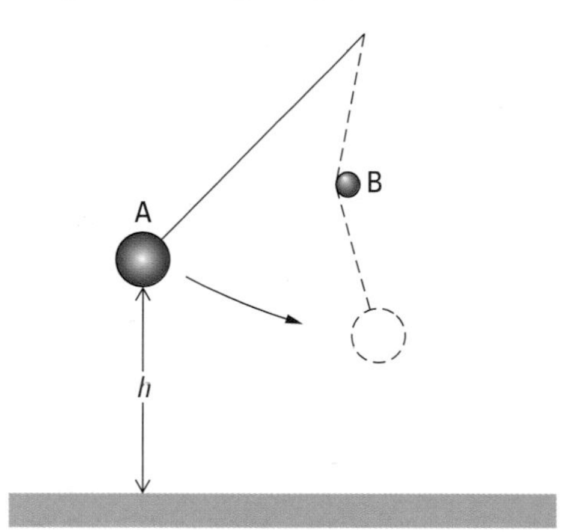

Figure 3

16. An engineer uses a single car to test the roller coaster track shown in **Figure 4**. In answering the following questions, assume that friction can be ignored and the speed at A is zero. In each case, give a reason for your answer.
    (a) Where is the gravitational potential energy the greatest?
    (b) Where is the kinetic energy the greatest?
    (c) Where is the speed the greatest?
    (d) Given a written description of what happens to the speed of the car as it rolls from A to B and so on to E.

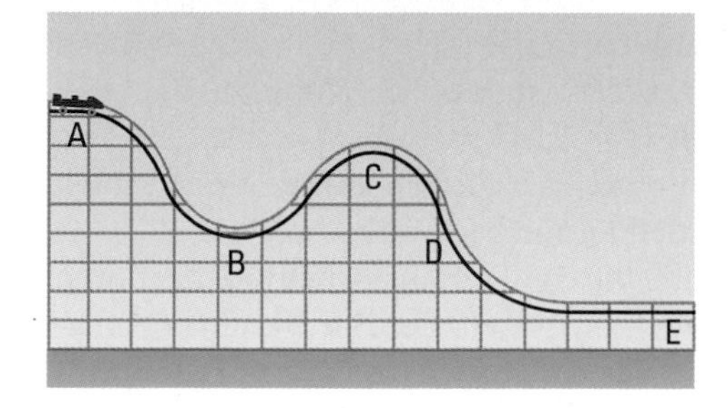

Figure 4

17. Most satellites circle Earth in elliptical orbits so that they are not always the same distance from Earth. (As shown in **Figure 5**, an ellipse is an oval shape.) A satellite in a stable elliptical orbit has a total mechanical energy which remains constant. At what point in the orbit would the speed of the satellite be the greatest? At what point would it be the least? Explain your reasoning.

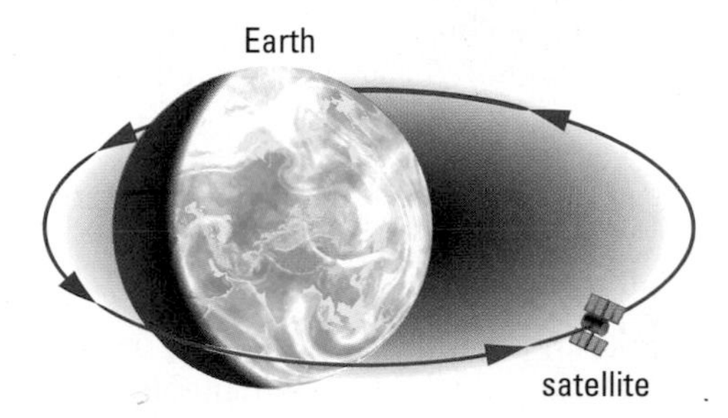

Figure 5

18. Some children go tobogganing on an icy hill. They start from rest at the top of the hill as shown in **Figure 6**. The toboggan and children have a combined mass of 94 kg. If friction is small enough to be ignored, apply the law of conservation of energy to determine
    (a) the total mechanical energy of the toboggan at A relative to B
    (b) the speed of the toboggan at B
    (c) the speed of the toboggan at C

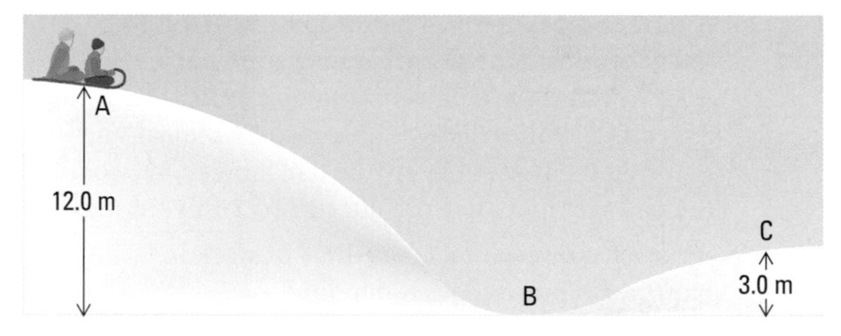

Figure 6

19. A high jumper of mass 55 kg wishes to jump over a bar 1.8 m above the ground. The centre of mass of the jumper is located 1.0 m above the ground. (We can imagine that all of the mass is located at this point for calculation purposes.) Use the law of conservation of energy to solve these questions.
    (a) If the jumper wishes to clear the bar while travelling at a speed of 0.40 m/s, how fast must the jumper be travelling the instant the jumper's feet leave the ground?
    (b) The jumper lands on the back on a foam pad that is 0.40 m thick. At what speed will the jumper be travelling upon first contact with the pad?

(c) Shortly after landing, the jumper will be at rest. What has happened to the mechanical energy the jumper had moments earlier?

20. It is estimated that 1.0 kg of body fat will provide 3.8×10^7 J of energy. A 67-kg mountain climber decides to climb a mountain 3500 m high.
    (a) How much work does the climber do against gravity in climbing to the top of the mountain?
    (b) If the body's efficiency in converting energy stored as fat to mechanical energy is 25%, determine the amount of fat the climber will use up in providing the energy required to work against the force of gravity.

21. Explain why it is impossible to have a motor that is 100% efficient.

22. What features distinguish radiant energy from other forms of energy?

23. Compare the thermal energies of each pair of substances described:
    (a) 500 g of water at 11°C and 500 g of water at 22°C
    (b) 400 g of water at 33°C and 800 g of water at 33°C

24. Describe an example of each of the three methods of heat transfer found in a typical modern kitchen.

25. Determine the heat gained or heat lost in each case:
    (a) 85 g of water is heated from 15°C to 79°C in 2.0 min
    (b) 2.5 kg of aluminum cools from 185°C to 12°C

26. A farmer drives a 0.10-kg iron spike with a 2.0-kg sledgehammer. The sledgehammer moves at a speed of 3.0 m/s and comes to rest on the spike after each swing. Assuming all the energy is absorbed by the nail and ignoring the work done by the nail, how much would the nail's temperature rise after 10 successive swings?

27. A power mower does 9.00×10^5 J of work in 0.500 h. How much power does it develop?

28. How long would it take a 0.500 kW-electric motor to do 1.50×10^5 J of work?

29. How much work can a 22-kW car engine do in 6.5 min?

30. The motor for an elevator can produce 14 kW of power. The elevator has a mass of 1100 kg, including its contents. At what constant speed will the elevator rise?

31. Explain why generating electrical energy by taking advantage of gravitational potential energy is a more efficient energy transformation technology than by using fossil fuels.

32. The equation for efficiency is defined using the energy input and energy output, but it could also be defined using the power input and power output. Show why, stating any assumptions you need to make.

33. The output of a certain electric generating station is 2500 MW.
    (a) Convert this quantity to watts, writing your answer in scientific notation.
    (b) Determine the energy output of the station each second, hour, day, and year.
    (c) If the station is 41% efficient, how much energy input is required during a single day's operation?
    (d) Assume that 1.5×10^{10} kg of water from a nearby lake circulates each day to cool the operating parts of the station. How much warmer is the return water than the intake water, assuming that all the waste energy goes to heating the returned water?

Applying Inquiry Skills

34. Draw a graph to show in each case how the first variable named depends on the second one.
    (a) work and applied force
    (b) the kinetic energy of an object and the object's speed
    (c) the gain in temperature of a liquid sample and the specific heat capacity of the liquid
    (d) power and time interval
    (e) doubling time and growth rate

35. Sketch a graph to show the general relationship between the efficiency of a moving energy transformation device and the friction the device experiences.

36. A student performs an activity to determine the specific heat capacity of liquid L.
    (a) List the materials needed to perform the activity.
    (b) List the steps you would use to perform the activity.
    (c) Describe sources of error in the activity.
    (d) The student finds that, to two significant digits, the specific heat capacity of liquid L is 4300 J/(kg•°C). What is likely the identity of liquid L?

37. At Paramount Canada's Wonderland, The Bat is a roller coaster that takes its riders both forward and backward (**Figure** 7). A motor is used to pull the coaster from the loading platform to the top of the starting side. After the coaster is released, it travels along the tracks, through the loops, and partway up the second side. There, the coaster is pulled by a motor to the top before being released for the backward trip.
    (a) Describe the energy transformations that occur from the time the coaster starts at the loading platform to when it stops at the same platform.
    (b) Why is the roller coaster unable to get to the top of the second side without the aid of a motor?

(c) Describe how you would experimentally determine the amount of energy lost to friction on this ride. Assume that measurements would have to be taken from outside the area of the ride.
(d) How would you estimate the efficiency of the ride?

Figure 7
In order to travel through the loops, this roller coaster must be released from the highest position of the track.

Making Connections

38. List advantages and disadvantages of using fossil fuels as an energy resource to
    (a) operate automobiles
    (b) generate electrical energy
39. (a) What are the advantages of the cogeneration of electricity?
    (b) Does using cogeneration have any effect on the efficiency of the energy transformation technology involved? Explain your answer, showing that you understand the law of conservation of energy.
40. Create a concept map related to your own energy consumption. Start with the word "Energy" in the middle of the page, and then branch off to show "Sources" and "Uses" of energy. Continue the map by indicating direct and indirect sources and uses, and adding as many details about your energy consumption as you can.
41. Describe how night-vision goggles and infrared detectors allow us to "see in the dark."
42. One day while using an electric blow drier, a hair designer notices that the air from the drier is hotter than usual. Suddenly the drier quits working. Several minutes later, the hair designer tries the drier again and it works, although it is hotter than usual.
    (a) What is the cause of the overheating of the drier? What must be done to eliminate the overheating?
    (b) How does the efficiency of the drier described compare to its efficiency when it was new? Where does the wasted energy go?
    (c) Based on this example, make a suggestion for the proper maintenance of electric appliances that have air moving across or through them.
43. In your opinion, are the following groups of people doing enough to aid in conserving energy? Justify each response.
    (a) officials at all levels of government
    (b) local school officials
    (c) automobile manufacturers
    (d) individual Canadians, on average
44. When a head-on traffic mishap between two vehicles occurs, the initial kinetic energy of each vehicle transforms into other forms of energy. To determine the total kinetic energy before any collision, assume that a compact car with a total mass (including passengers) of 1300 kg collides head-on with a sports utility vehicle (SUV) with a total mass (including the same number of passengers) of 3500 kg. Both vehicles are travelling at the same speed before each collision.
    (a) Set up a table of data to show the kinetic energy of each vehicle and the total kinetic energy for the following speeds of both vehicles: 0.0 m/s, 5.0 m/s, 10.0 m/s, 15 m/s, . . ., 35 m/s. Calculate the values needed to complete the table.
    (b) Plot a graph of kinetic energy as a function of speed. Plot all three sets of values on the same graph.
    (c) Use your graph to explain why high-speed crashes are far more dangerous than low-speed ones.
45. One suggestion that has been made for efficient travel between two cities is to link them with a straight tunnel bored through a portion of Earth's core, as shown in **Figure 8**. Passengers would then travel through the tunnel, rather than travel along the circumference. Using energy relationships, explain why this approach should greatly reduce the fuel required to get from one city to the other.

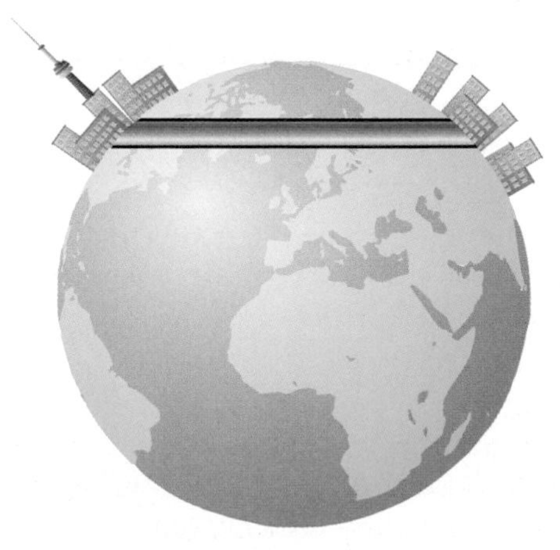

Figure 8

Unit 3

Waves and Sound

Whistles, squeaks, squawks, and screams are sounds used by orca whales for echolocation and social communication. Dr. John Ford, senior marine mammal scientist at the Vancouver Aquarium, studies orca whale calls using underwater microphones (hydrophones) and a sound analyzer that converts sounds into computer images. A hydrophone in Johnstone Strait off Vancouver Island is connected to a radio transmitter called ORCA FM. People within 15 km of the transmitter can listen to whales in the area on 88.5 FM.

"A live feed to the lab at the Vancouver Aquarium Marine Science Centre allows us to record the sounds and identify which pods are in the area without having to actually be there," Dr. Ford explains. "We hope to eventually have a system of hydrophones along the coast."

Dr. John Ford, marine scientist

Dr. Ford studies how noise from whale-watching boats affects the orca's ability to communicate, navigate, or hunt prey. These studies may lead to more stringent regulations for whale watching to ensure the orca whale's survival.

Overall Expectations

In this unit, you will be able to

- demonstrate an understanding of the properties of mechanical waves and sound and the principles underlying the production, transmission, interaction, and reception of mechanical waves and sound
- investigate the properties of mechanical waves and sound through experiments or simulations, and compare predicted results with actual results
- describe and explain ways in which mechanical waves and sound are produced in nature, and evaluate the contributions of technologies that make use of mechanical waves and sound to entertainment, health, and safety

Unit

3

Waves and Sound

Are You Ready?

Knowledge and Understanding

1. If you are positioned outdoors about 20 m from a friend, think of all the ways in which you can attract your friend's attention. Make two lists to classify the ways according to whether or not they involve the sense of hearing or any of your friend's other senses.
2. **Figure 1** shows waves being produced on a rope that has one end tied tightly to a post.
   (a) Where does the energy that produces the waves come from?
   (b) Which wave was produced first, the one in segment A or the one in segment B?
   (c) Did the waves in the two segments take the same amount of time to be produced? How do you know?
   (d) In your notebook, draw a sketch to show how the hand moved to create the waves shown. How many cycles were used?
   (e) What name would you give to the top part of each wave? the bottom part of each wave?
   (f) Use a ruler to estimate in centimetres the "wavelength" of the waves.

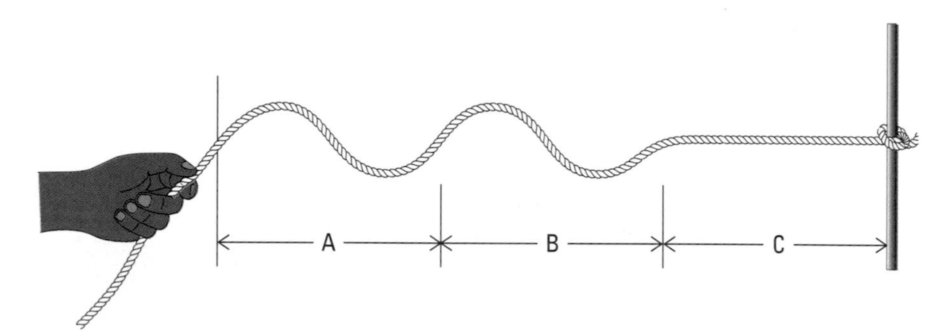

Figure 1
The waves on the rope are moving to the right.

3. Can sound travel in a solid? a liquid? a gas? a vacuum? Give an example of each "yes" answer, and explain any "no" answers that you give.
4. Why are you able to see lightning before you hear the sound of the thunder caused by the lightning strike?
5. You are standing 85 m from a batter who is hitting a baseball. You see the bat touch the ball, but you do not hear the sound made by the hit until 0.24 s after. What is the speed of the sound in air?

Inquiry and Communication Skills

6. (a) Describe in your own words the differences between noise and music.
   (b) Do you consider music to be a subjective or an objective topic? Why?
7. Some of the terms you will encounter in this unit are: infrasonic sounds, ultrasonic sounds, sound barrier, sonic boom, subsonic speed, and supersonic speed. What do you think each of these terms mean?

8. **Figure 2** shows a child on a playground swing.
   (a) What factor(s) do you think affects the amount of time for each back-and-forth cycle of the swing?
   (b) Describe how you would set up a controlled experiment to determine how the factor(s) you named in (a) affects the time for each cycle.

Figure 2

9. Which line on the graph in **Figure 3** best represents what happens to the speed of sound in air as the temperature increases? Give a scientific reason for your choice.

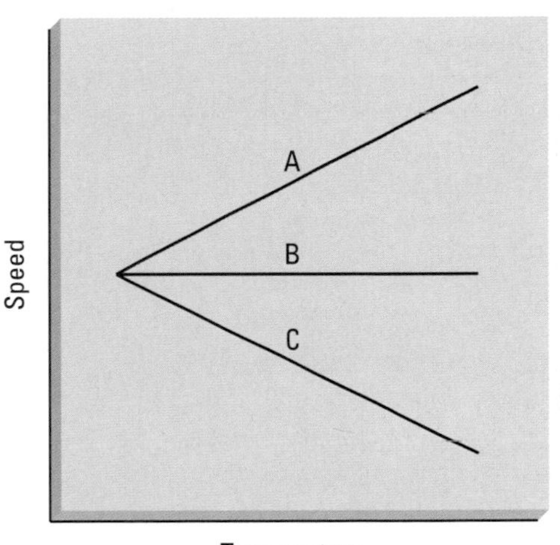

Figure 3
For question 9

Making Connections

10. List some musical instruments and sound equipment that you enjoy listening to. What are some of the features that these instruments have which help to improve the sounds they produce?
11. Some auditoriums and other "sound halls" have good acoustics, while others are not as high quality. What are some of the features of good acoustics that you like?
12. If you were designing a device to help a deaf person experience music, what features would the device have? (**Figure 4** shows one way this is done — in a sensory perception facility called a "Snoezelen room.")

Figure 4
A Snoezelen room is designed to help people experience sensations related to all five senses. In this photograph, a child is experiencing sound by touching a device that vibrates.

Chapter

6

Vibrations and Waves

In this chapter, you will be able to

- use a variety of activities to study the behaviour of vibrating objects and measure their frequency and period
- distinguish between objects vibrating in-phase and out-of-phase
- describe the various parts of and the behaviour of transverse and longitudinal waves
- use equations relating the frequency, period, wavelength, and speed of a wave
- determine the result of two waves interfering, using the principle of superposition
- determine the wavelength of interfering waves, using the standing wave pattern
- draw a diagram of the interference pattern between two point sources in phase

A moving object can transmit energy from one place to another. For example, when a pitcher throws a baseball to a catcher, the kinetic energy given to the ball by the pitcher is transferred to the catcher. However, energy is also transferred from the source to the receiver by means of a wave. A water-ski boat creates a large wake that travels across the surface of the water and crashes into a dock. Such a travelling disturbance is called a wave.

We live in a world surrounded by waves. Some waves are visible; others are not. We can see water waves (**Figure 1**) and the waves in a rope or spring. We cannot see sound waves and radio waves. A wave is a transfer of energy over a distance, in the form of a disturbance. For example, when you shake a spiral "Slinky," the energy you impart is transferred from coil to coil down the spring. When you throw a rock into a pool of water, creating a splash, circular waves radiate out from the point of contact. Energy is transferred as a wave from one water molecule to the next by the forces that hold the water molecules together. By observing the visible waves in ropes, springs, and water, you can discover characteristics that all waves have in common.

Reflect on your Learning

1. List similarities and differences between sound waves and water waves.
2. What is the difference between a transverse wave and a longitudinal wave?
3. List two medical and two non-medical technologies involving wave phenomena.
4. (a) State the law of reflection.
   (b) Draw a diagram illustrating the law of reflection.
   (c) Distinguish between reflection and refraction.

Throughout this chapter, note any changes in your ideas as you learn new concepts and develop your skills.

Wave Action

In this activity, you will observe the transfer of energy through a medium. You will also observe the differences in how the medium moves as the wave passes through it.

- Line up at least six students side by side, with their arms linked (**Figure 2(a)**). Gently push the first student back and forth at right angles to the line. Watch as energy passes up the line as a transverse wave.
- Line up some students single file, each with their hands on the shoulders of the student in front (**Figure 2(b)**). Gently push and pull the last student forward and back. Watch as the motion is passed up the line from student to student as a longitudinal wave.
- Tie a long length of light rope to a door knob or other rigid point. After stretching the rope taut, shake the free end back and forth sending a series of waves down the rope.
  - (a) In each case, how does the medium move with respect to the direction in which the wave is moving?

(a)

(b)

Figure 2
A human wave

6.1 Vibrations

wave: a transfer of energy over a distance, in the form of a disturbance

periodic motion: motion that occurs when the vibration, or oscillation, of an object is repeated in equal time intervals

transverse vibration: occurs when an object vibrates perpendicular to its axis

longitudinal vibration: occurs when an object vibrates parallel to its axis

torsional vibration: occurs when an object twists around its axis

Waves are disturbances that transfer energy over a distance. Water waves, sound waves, waves in a rope, and earthquake waves all originate from objects that are vibrating. For example, a water wave can result from the vibration caused by a boat rocking on the water, while sound waves could originate from a vibrating tuning fork or a vibrating guitar or piano string. In each case, the vibrating source supplies the energy that is transferred through the medium as a wave. Often the objects are vibrating so rapidly that they are difficult to observe with our eyes. For the purpose of studying the properties of vibrating objects, a slowly moving device, such as a mass bouncing up and down on a spring, or a swinging pendulum is ideal.

When an object repeats a pattern of motion—as a bouncing spring does—we say the object exhibits **periodic motion**. The vibration, or oscillation, of the object is repeated over and over with the same time interval each time.

There are three basic types of vibration. A **transverse vibration** occurs when an object vibrates perpendicular to its axis at the normal rest position. An example of a transverse vibration is a child swinging on a swing. A **longitudinal vibration** occurs when an object vibrates parallel to its axis at the rest position. An example is a coil spring supporting a vehicle. A **torsional vibration** occurs when an object twists around its axis at the rest position. An example occurs when a string supporting an object is twisted, causing the object to turn or vibrate around and back. The three kinds of vibration are shown in **Figure 1**. Throughout this chapter, the term oscillation could also be used instead of vibration.

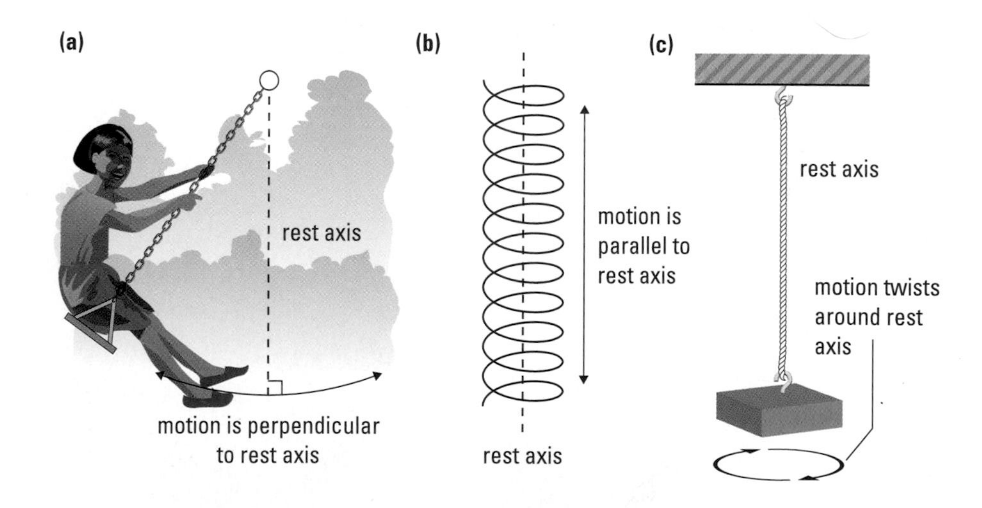

Figure 1
The three basic types of vibration
(a) Transverse vibration
(b) Longitudinal vibration
(c) Torsional vibration

cycle: one complete vibration or oscillation

frequency: (f) the number of cycles per second;

$$f = \frac{\text{number of cycles}}{\text{total time}}$$

hertz: 1 Hz = 1 cycle/s or 1 Hz = 1 s^{-1}, since cycle is a counted quantity, not a measured unit

period: (T) the time required for one cycle;

$$T = \frac{\text{total time}}{\text{number of cycles}}$$

When we describe the motion of a vibrating object, we call one complete oscillation a **cycle** (**Figure 2**). The number of cycles per second is called the **frequency** (f). The SI unit used to measure frequency is the **hertz** (Hz), named after Heinrich Hertz (1857–1894), the German scientist who first produced electromagnetic waves in the laboratory.

Another term used in describing vibrations is the **period** (T). The period is the time required for one cycle. Usually the second (s) is used for measuring the period, but for a longer period, like that of the rotation of the Moon, the day (d) or the year (yr) is used.

Since frequency is measured in cycles per second and period is measured in seconds per cycle, frequency and period are reciprocals of each other. Thus,

$$f = \frac{1}{T} \text{ and } T = \frac{1}{f}$$

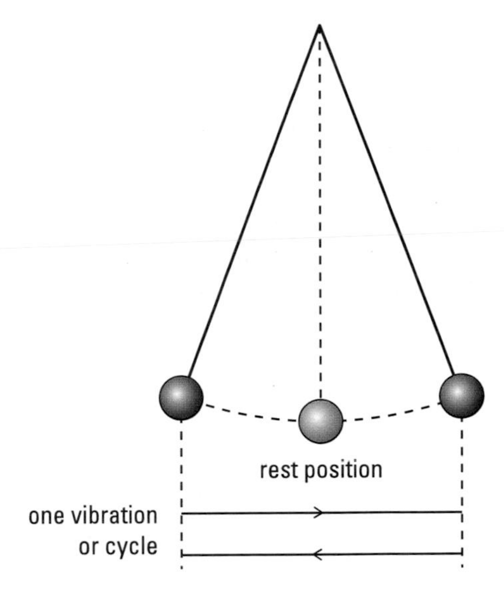

Figure 2
One cycle is equal to one complete vibration.

As a pendulum swings, it repeats the same motion in equal time intervals. The distance in either direction from the equilibrium, or rest, position to maximum displacement is called the **amplitude** (A) (**Figure 3**).

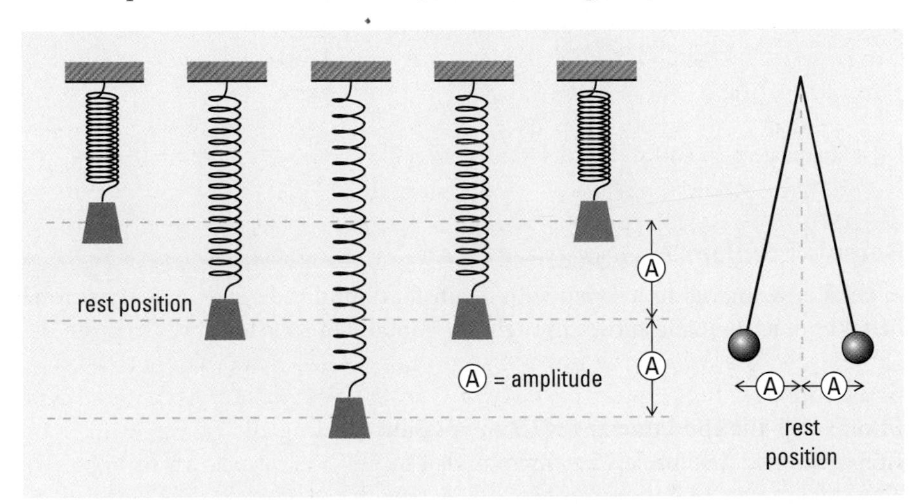

Figure 3
The rest position is where the object will remain at rest. An object can move through its rest position.

amplitude: distance from the equilibrium position to maximum displacement

Two identical pendulums are said to be vibrating **in phase** if they have the same period and pass through the rest position at the same time (**Figure 4**). Two identical pendulums are vibrating **out of phase** if they do not have the same period or if they have the same period but they do not pass through the rest position at the same time.

in phase: objects are vibrating in phase if they have the same period and pass through the rest position at the same time

out of phase: objects are vibrating out of phase if they do not have the same period or if they have the same period but they do not pass through the rest position at the same time

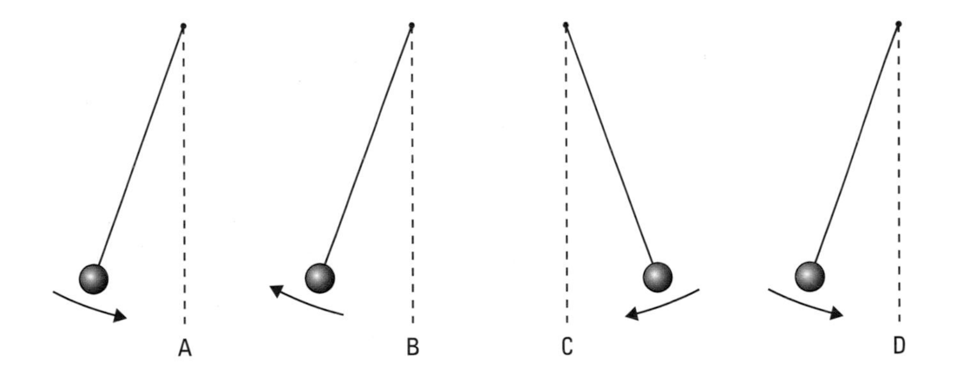

Figure 4
Pendulums A and D are in phase. Pendulums B and C are not. Why?

Sample Problem 1

A mass hung from a spring vibrates 15 times in 12 s. Calculate (a) the frequency and (b) the period of the vibration.

Solution

(a) number of cycles = 15 cycles
total time = 12 s
$f = ?$
$T = ?$

$$f = \frac{\text{number of cycles}}{\text{total time}}$$

$$= \frac{15 \text{ cycles}}{12 \text{ s}}$$

$$f = 1.2 \text{ cycles/s}$$

The frequency is 1.2 Hz.

(b)

$$T = \frac{\text{total time}}{\text{number of cycles}}$$

$$= \frac{12 \text{ s}}{15 \text{ cycles}}$$

$$T = 0.80 \text{ s/cycle, or } 0.80 \text{ s}$$

or

$$T = \frac{1}{f}$$

$$= \frac{1}{1.2 \text{ Hz}}$$

$$T = 0.80 \text{ s}$$

The period is 0.80 s.

Sample Problem 2

A child is swinging on a swing with a constant amplitude of 1.2 m. What total distance does the child move through horizontally in 3 cycles?

Solution

In one cycle the child moves $4 \times 1.2 \text{ m} = 4.8 \text{ m}$.
In 3 cycles the child moves $3 \times 4.8 \text{ m} = 14.4 \text{ m}$.
The child moves 14.4 m in 3 cycles.

Sample Problem 3

The frequency of a wave is 6.0×10^1 Hz. Calculate the period.

Solution

$f = 6.0 \times 10^1$ Hz
$T = ?$

$$T = \frac{1}{f}$$

$$= \frac{1}{6.0 \times 10^1 \text{ Hz}}$$

$$T = 0.017 \text{ s}$$

The period is 0.017 s.

Practice

Understanding Concepts

1. State the type of vibration in each of the following:
   (a) a tree sways in the wind
   (b) a sewing-machine needle moves up and down
2. Calculate the period in seconds of each of these motions:
   (a) a pulse beats 25 times in 15 s
   (b) a woman shovels snow at a rate of 15 shovelsful per minute
   (c) a car motor turns at 2450 rpm (revolutions per minute)

Answers

2. (a) 0.60 s
   (b) 4.0 s
   (c) 2.4×10^{-2} s

3. A stroboscope is flashing so that the time interval between flashes is $\frac{1}{80}$ s. Calculate the frequency of the strobe light's flashes.
4. A child on a swing completes 20 cycles in 25 s. Calculate the frequency and the period of the swing.
5. Calculate the frequency and period of a tuning fork that vibrates 2.40×10^4 times in 1.00 min.
6. Calculate the frequency of the following:
   (a) a violin string vibrates 88 times in 0.20 s
   (b) a physics ticker-tape timer produces 3600 dots in 1.0 min
   (c) a CD player rotates 4.5×10^3 times in 1.0 minute
7. If the Moon orbits Earth six times in 163.8 d, what is its period of revolution?
8. As you walk, describe the movement of your arms and legs as in-phase or out-of-phase oscillations.

Answers

3. 80 Hz
4. 0.80 Hz; 1.2 s
5. 4.00×10^2 Hz; 2.5×10^{-3} s
6. (a) 4.4×10^2 Hz
   (b) 60 Hz
   (c) 75 Hz
7. 27.30 d

Investigation 6.1.1

The Pendulum

INQUIRY SKILLS

- ○ Questioning
- ○ Hypothesizing
- ● Predicting
- ○ Planning
- ● Conducting
- ● Recording
- ● Analyzing
- ● Evaluating
- ● Communicating

A pendulum swings with a regular period, so it is a useful device for measuring time. In fact, early drawings of a pendulum clock were developed by Galileo Galilei in 1641. His ideas for a pendulum clock were based on his observations of the regular period of vibration of a lamp hanging in a church in Pisa, Italy. He then postulated laboratory experiments similar to this one to determine the factors that affected the period and frequency of a swinging pendulum (**Figure 5**).

Question

What are the relationships between the frequency of a simple pendulum and its mass, amplitude, and length?

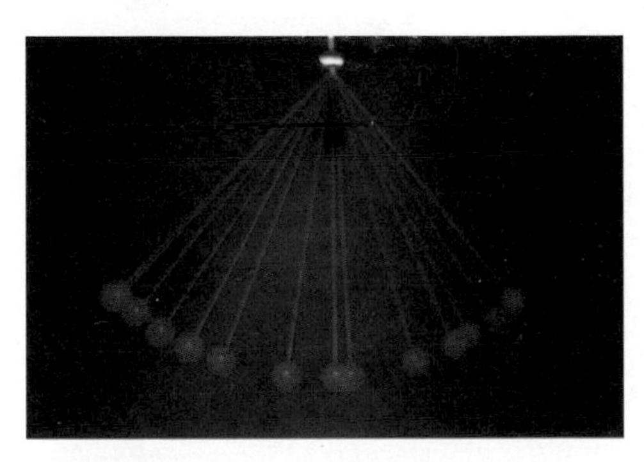

Figure 5
A strobe photograph of an oscillating pendulum

Materials

utility stand
clamp
test-tube clamp
split rubber stopper
string
stopwatch
metre stick
metal masses (50 g, 100 g, and 200 g)

Prediction

(a) Predict what will happen to the frequency of the pendulum in the following situations:
   (i) the mass increases, but the length and amplitude remain constant
   (ii) the amplitude increases, but the mass and length remain constant
   (iii) the length increases, but the mass and amplitude remain constant

Procedure

1. Set up a chart as shown in **Table 1**.

Table 1

Length (cm)	Mass (g)	Amplitude (cm)	Time for 20 cycles (s)	Frequency (Hz)
100	200	10		

2. Set up the utility stand as illustrated, ensuring that it is clamped securely and there is clearance on either side of at least 30 cm.
3. Obtain a string about 110 cm long and securely attach a 200-g mass to one end. Place the other end of the string into the split rubber stopper (**Figure 6**), adjust the pendulum length to 100 cm, and clamp the rubber stopper firmly so the string does not slip. Remember that the length is measured to the centre of the mass.
4. Keeping the string fully extended, pull the mass to one side for an amplitude of 10 cm. Release the mass, making sure not to push it as you let it go.
5. Measure the time taken for 20 complete cycles. Repeat once or twice for accuracy, then calculate the frequency. Enter the data in your observation table.
6. Repeat step 5 using amplitudes of 20 cm and 30 cm. Tabulate your data.
7. Determine the time taken for 20 complete cycles of the pendulum using a pendulum length of 100 cm, an amplitude of 10 cm, and 50 g, 100 g, and 200 g masses. Make sure you measure the length to the middle of each mass. Tabulate your data.
8. Determine the time taken for 20 complete cycles of the pendulum using an amplitude of 10 cm and a constant mass. Use pendulum lengths of 100 cm, 80 cm, 60 cm, 40 cm, and 20 cm. Tabulate your data.

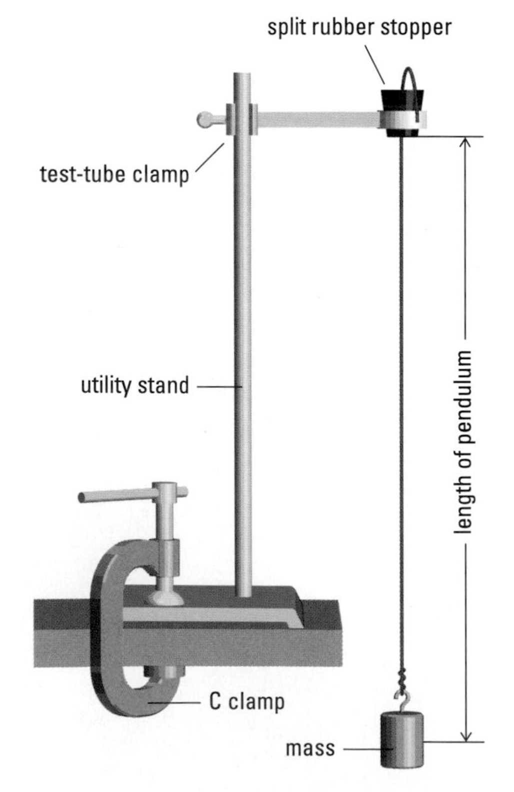

Figure 6
Setup for Investigation 6.1.1

Analysis

(b) With frequency as the dependent variable, plot graphs of frequency versus
   (i) amplitude, for a length of 100 cm and a constant mass
   (ii) mass, for a length of 100 cm and a constant amplitude
   (iii) length, for a constant amplitude and a constant mass

(c) Answer the Question by describing the relationship between frequency and amplitude, frequency and mass, and frequency and length, both in words and mathematically.

(d) For each of the different lengths, calculate the period of vibration. Plot a graph of period as a function of the length of the pendulum. Replot the graph to try to obtain a straight line. (*Hint:* You can either square the values of the period or find the square roots of the values of the length.)

Evaluation

(e) Evaluate the predictions you made regarding the frequency, mass, amplitude, and length of a pendulum.

(f) Describe the sources of error in the investigation and evaluate their effect on the results. Suggest one or two improvements to the experimental design.

Synthesis

(g) Calculate the maximum speed of the 100 g pendulum mass when it has a length of 100 cm and an amplitude of 50 cm. (*Hint:* You can apply the law of conservation of mechanical energy to this problem. The maximum vertical displacement of the mass above its rest position can be found by means of a scale diagram, by actual measurement using the pendulum, by applying the Pythagorean theorem, or by using trigonometry.)

As you will have discovered in the investigation, the frequency and period of a pendulum are affected only by its length. The frequency is inversely proportional to the square root of the length ($f \alpha \sqrt{\frac{1}{l}}$), or we could say the period is directly proportional to the square root of the length ($T \alpha \sqrt{l}$). The equation for the period of a pendulum involves only one other variable, Earth's gravitational field. Since Earth's gravitational field does not vary significantly, the pendulum's period remains constant for a specific length. Changes in the mass or amplitude have no effect, as long as the length remains constant. For these reasons the first accurate clocks usually involved a pendulum of some kind. Today the grandfather clock uses a long pendulum to keep accurate time, and the force of gravity on suspended weights provides the energy to maintain the amplitude of the vibration.

DID YOU KNOW ?

Calculating Acceleration Due to Gravity

The equation $T = 2\pi\sqrt{\frac{l}{g}}$ was used historically to determine the acceleration due to gravity. How? (In this equation, l is the length of the pendulum, and g is the magnitude of the gravitational field.)

SUMMARY Characteristics of Vibrations and Waves

- Most waves originate from a vibrating source.
- A wave is a transfer of energy over a distance in the form of a disturbance.
- In a transverse vibration, the object vibrates perpendicular to its length, while in a longitudinal vibration it vibrates parallel to its length.
- The frequency (f) is the number of cycles per second.
- The period (T) is the time required for one cycle.
- The frequency and the period are inversely related $\left(f = \frac{1}{T}\right)$.
- Objects are vibrating in phase if they have the same period and pass through the rest position at the same time.
- The period of a pendulum is directly proportional to the square root of its length.

Section 6.1 Questions

Understanding Concepts

1. State the type of vibration in each of the following:
   (a) a diving board vibrates momentarily after a diver jumps off
   (b) a woodpecker's beak pecks a tree trunk
   (c) the shock absorbers on a mountain bike vibrate as it travels over a rough trail
2. For the pendulum shown in **Figure 7**, state
   (a) the type of vibration
   (b) the amplitude
   (c) the total distance the mass moves through horizontally in five cycles

(continued)

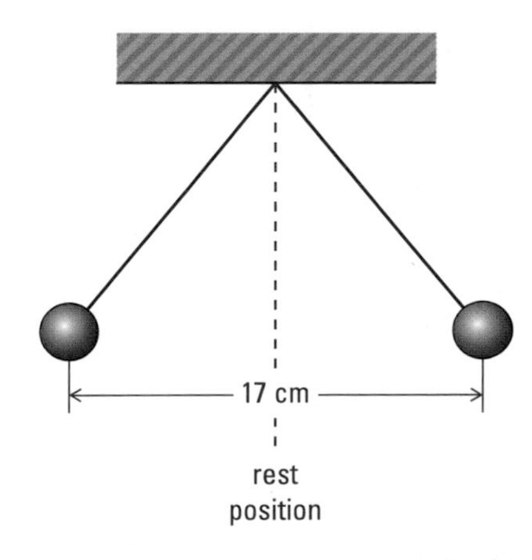

Figure 7

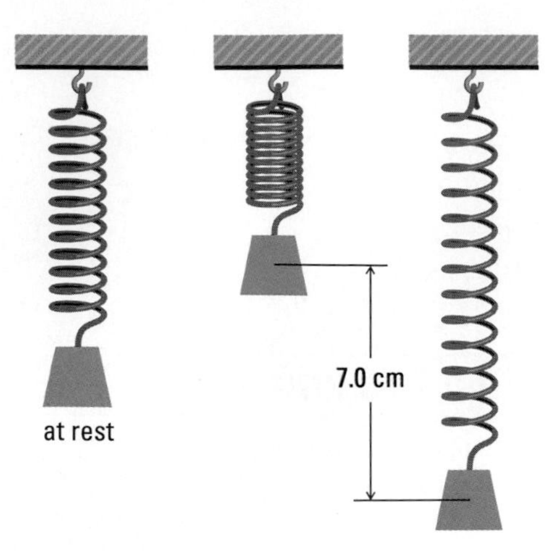

Figure 8

3. Calculate the frequency in hertz and the period in seconds for each of the following:
   (a) a television displays 1800 pictures each minute
   (b) a South African bird, the horned sungem, has the fastest wing-beat of any bird, at 1800 beats in 20.0 s
   (c) a butterfly beats its wings between 460 and 640 times per minute (In this case, find a range for the answer.)
   (d) the second hand of a watch
   (e) the minute hand of a watch
4. **Figure 8** shows a mass at rest on a spring and a mass vibrating. State
   (a) the type of vibration
   (b) the amplitude
   (c) the total distance the mass moves through in 3.5 cycles
5. State the relationship (if any) between these pairs of variables:
   (a) period and frequency of a vibration
   (b) period and length of a pendulum
   (c) mass and frequency of a pendulum
6. Musicians use a metronome, an inverted pendulum that produces ticks at various frequencies. The frequency can be altered by moving the mass on the metronome closer to or farther from the pivot point. What should a music student do to increase the metronome's frequency of vibration?
7. The accuracy of pendulum clocks is adjusted by moving the weight up and down on the pendulum, changing its effective length. If you move to another city where the force of gravity is slightly lower, will you have to move the weight up or down to keep accurate time? (*Hint:* See the equation for the period of a pendulum.)
8. In some clocks a spring mechanism is used with a balance wheel to keep accurate time. Why could these clocks also be called pendulum clocks?
9. If you knew the period of a pendulum measured at Earth's surface, how would its period change if you measured it on
   (a) the Moon? (b) Jupiter?
10. A pendulum was traditionally used to measure the acceleration due to gravity (g) at various points on Earth's surface. Using the equation for the pendulum $\left(T = 2\pi\sqrt{\frac{l}{g}}\right)$, determine a value for g if the pendulum length is 1.00 m and its period is 2.00 s.
11. In the individual medley, swimmers cover the four swimming styles in the following order: butterfly, backstroke, breaststroke, and freestyle. Describe each stroke as having an in-phase or out-of-phase arm motion.

Applying Inquiry Skills

12. Describe a method for determining the length of a swing without physically measuring it.

Making Connections

13. Many watches today are digital with a small battery that provides the energy. Research to determine what is vibrating to keep accurate time. Follow the links for Nelson Physics 11, 6.1.

GO TO www.science.nelson.com

6.2 Wave Motion

A high-wire artist kicks one end of the wire before starting to cross. She sees a small transverse movement dart along the wire and reflect back from the far end. The time taken for this round trip will tell her if the tension is correct. A football coach blows a whistle, creating fluctuations in the positions of air molecules and air pressure within it that make a shrill sound. Children drop pebbles into a pond; the surface of the water oscillates up and down, and concentric ripples spread out in ever-expanding circles. Electrons shift energy levels at the surface of the Sun, sending fluctuating electric and magnetic fields through the vacuum of space. These are all examples of wave motion, or transmission.

We should be quite clear about what is being transmitted. It is a disturbance from some normal value of the medium that is transmitted, not the medium itself. For the wire, it was a small sideways displacement from the normal equilibrium position. For the sound, it was a slight forward and backward motion of air molecules about their normal average position. In the water, the disturbance was a raising and lowering of the water level from equilibrium. The activity within the sunshine is a little harder to imagine, but here the disturbance is a fluctuating electromagnetic field.

Investigation 6.2.1

Wave Transmission: Pulses on a Coiled Spring

INQUIRY SKILLS

- ● Questioning
- ○ Hypothesizing
- ● Predicting
- ○ Planning
- ● Conducting
- ● Recording
- ● Analyzing
- ● Evaluating
- ● Communicating

If you hold a piece of rope in your hand and you move your hand up and down, a wave will travel along the rope away from you. Your hand is the vibrating source of energy and the rope is the material medium through which the energy is transferred. By moving your hand through one-half of a cycle, as shown in **Figure 1**, you can create a pulse.

During an investigation, it is somctimes easier to observe a single pulse in a spring than to try to study a wave consisting of a series of pulses. The knowledge gained by studying pulses on a spring can be applied to all types of waves. Thus, the concepts learned in this investigation are important to the study of waves.

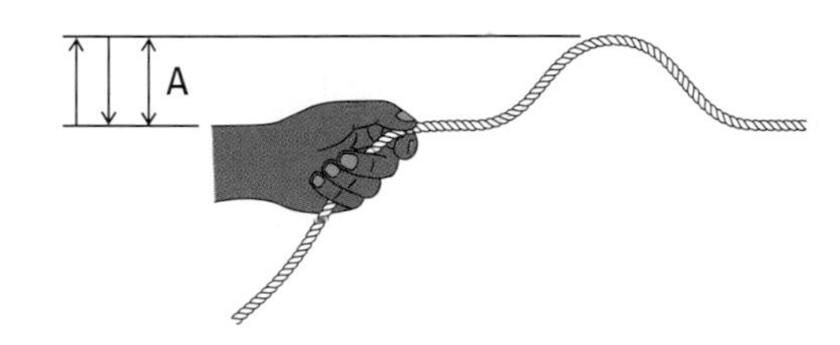

Figure 1
Producing a pulse

Questions

(i) How do pulses move along a coiled spring?
(ii) How are pulses reflected from a fixed end and a free end?

Materials

coiled spring (such as a Slinky toy)
masking tape
metre stick
piece of paper
stopwatch
string at least 4 m long

Procedure

Hold the spring firmly and do not overstretch it. Observe from the side, in the case of an accidental release.

1. Attach the masking tape to a coil near the middle of the spring. Stretch the spring along a smooth surface (such as the floor or a long table) to a length of 2.0 m. With one end of the spring held rigidly, use a rapid sideways jerk at the other end to produce a transverse pulse (**Figure 2(a)**). Describe the motion of the coils of the spring. (*Hint:* Watch the tape attached to the spring.)

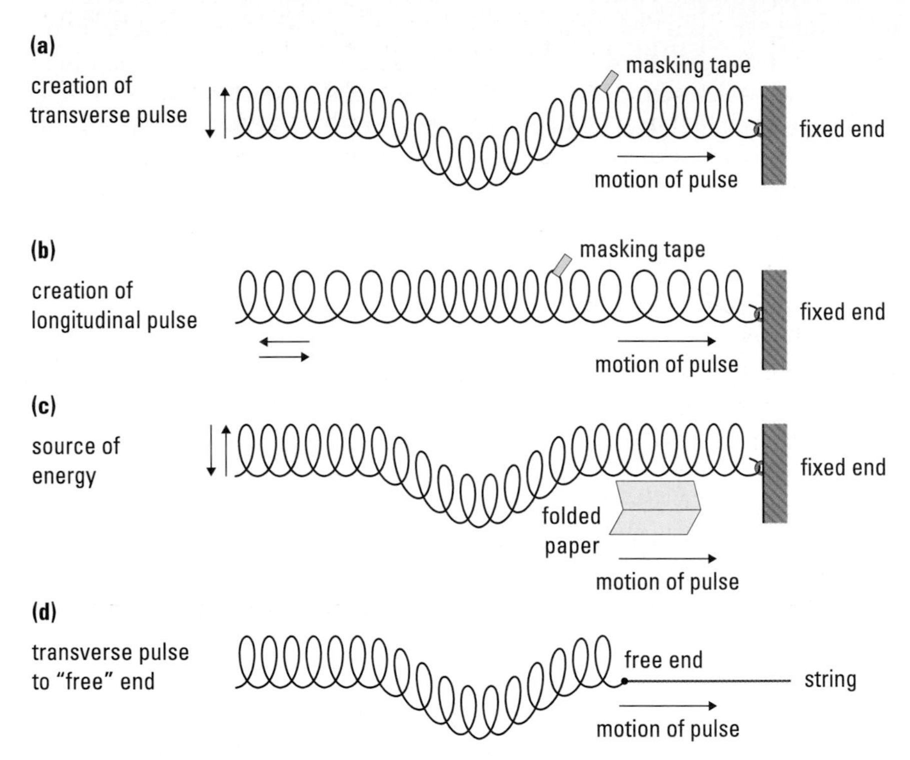

Figure 2
Procedure for Investigation 6.2.1

2. Use a rapid forward push to produce a longitudinal pulse along the spring (**Figure 2(b)**). Describe the motion of the coils of the spring.
3. Stand a folded piece of paper on the floor close to the end of the spring as shown in **Figure 2(c)**. Use the energy transferred by a transverse pulse to knock the paper over. Describe where the energy came from and how it was transmitted to the paper.
4. Holding one end of the spring rigid and stretching the spring as in step 1, send a transverse pulse toward the other end. Determine whether or not the pulse that reflects off this fixed end returns on the same side of the rest axis as the original or incident pulse.
5. Tie a piece of string at least 4 m in length to one end of the spring. Using the string to stretch the spring to approximately 2.0 m, send a transverse pulse toward the string as shown in **Figure 2(d)**. Determine whether or not the pulse that reflects off this free end of the spring returns on the same side as the incident pulse. (Note that while the free end is not truly "free" because a string is attached, it is a good approximation.)
6. Remove the string and stretch the spring to an appropriate length (e.g., 2.0 m). Measure the time taken for a transverse pulse to travel from one end of the spring to the other. Repeat the measurement several times for

accuracy while trying to keep the amplitude constant. Then find the time taken for the same type of pulse to travel from one end of the spring to the other and back again. Determine whether or not the reflected pulse takes the same time to travel the spring's length as the incident pulse.

7. Determine whether the time taken for a transverse pulse to travel from one end to the other and back again depends on the amplitude of the pulse, by sending pulses of different amplitudes and estimating the times. Repeated tests will be necessary, since the pulse moves so quickly.
8. Predict the relationship between the speed of the pulse and the stretch of the spring. Test your prediction experimentally by stretching the spring to various lengths and noting the time for a transverse pulse to travel down and back again.
9. If different types of springs are available, compare the speed of a transverse pulse along each spring.

Analysis

(a) Based on the observations in this experiment, discuss whether the following statements are true or false:
  (i) Energy may move from one end of a spring to the other.
  (ii) When energy is transferred from one end of a spring to the other, the particles of the spring are also transferred.

(b) State what happens to the speed of a pulse in a material under the following circumstances:
  (i) The condition of the material changes. For instance, stretching a spring changes its condition.
  (ii) The amplitude of the pulse increases.
  (iii) The pulse is reflected off one end of the material.

(c) A reflected pulse that is on the same side as the incident pulse is said to be in phase with the incident pulse. A reflected pulse on the opposite side of the incident pulse is out of phase with it. Is the reflected pulse in phase or out of phase for
  (i) fixed-end reflection?
  (ii) free-end reflection?

Evaluation

(d) Evaluate the prediction you made in step 8 of the procedure.

Transverse Waves

When a water wave moves across an ocean or a lake, it moves at a uniform speed. But the water itself remains in essentially the same position, merely moving up and down as the wave goes by. Similarly, when a rope is being vibrated at one end, the rope itself does not move in the direction of the wave motion; sections of the rope move back and forth or up and down as the wave travels along it.

Water waves and waves in a rope are examples of transverse waves (**Figure 3**). In a **transverse wave** the particles in the medium vibrate at right angles to the

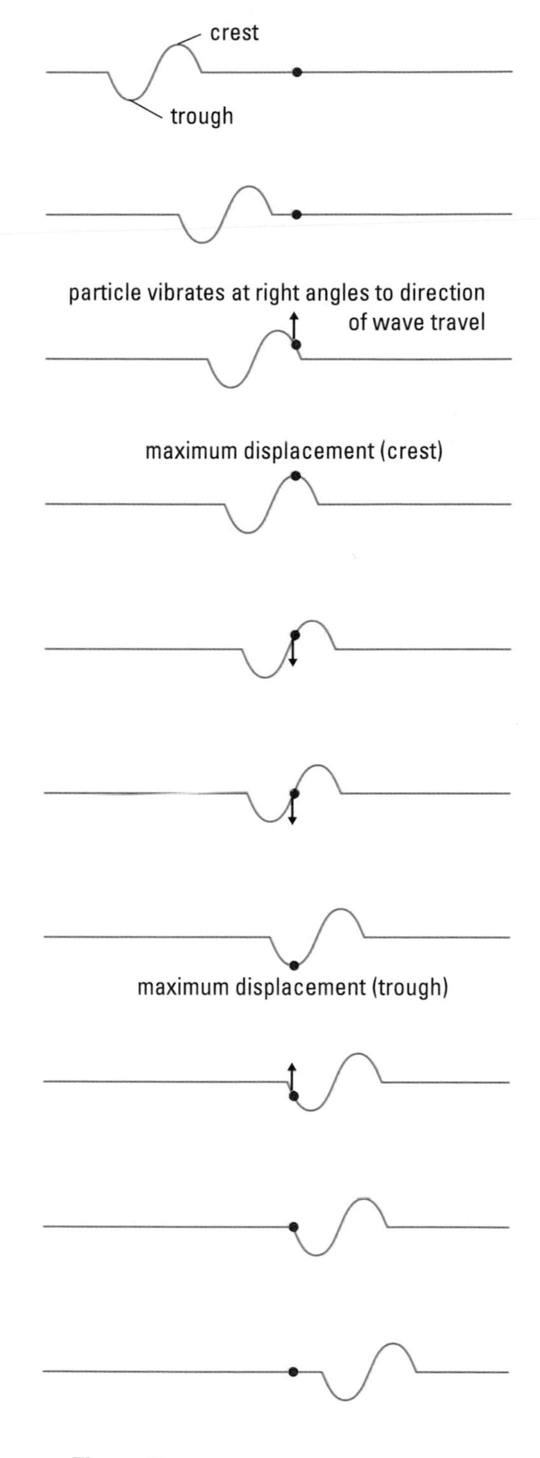

Figure 3
A crest moves through a rope from right to left.

transverse wave: particles in the medium move at right angles to the direction in which the wave travels

crest or **positive pulse:** high section of a wave

trough or **negative pulse:** low section of a wave

periodic waves: originate from periodic vibrations where the motions are continuous and are repeated in the same time intervals

pulse: wave that consists of a single disturbance

wavelength: (λ) distance between successive wave particles that are in phase

direction in which the wave travels. The high section of the wave is called a **crest** and the low section is called a **trough**. Since the crest lies above and the trough below the rest position (equilibrium), a crest can be referred to as a **positive pulse** and a trough as a **negative pulse**.

Periodic waves are waves where the motions are repeated at regular time intervals. However, a wave can also consist of a single disturbance called a **pulse**, or shock wave.

In periodic waves, the lengths of successive crests and troughs are equal. The distance from the midpoint of one crest to the midpoint of the next crest (or from the midpoint of one trough to the midpoint of the next) is called the **wavelength** and is represented by the Greek letter λ (lambda).

We have said that the amplitude of a wave is the distance from the rest position to maximum displacement. For a simple periodic wave, the amplitude is the same on either side of the rest position. As a wave travels through a medium, its amplitude usually decreases because some of its energy is being lost to friction. If no energy were required to overcome friction, there would be no decrease in amplitude and the wave would be what is called an ideal wave (**Figure 4**).

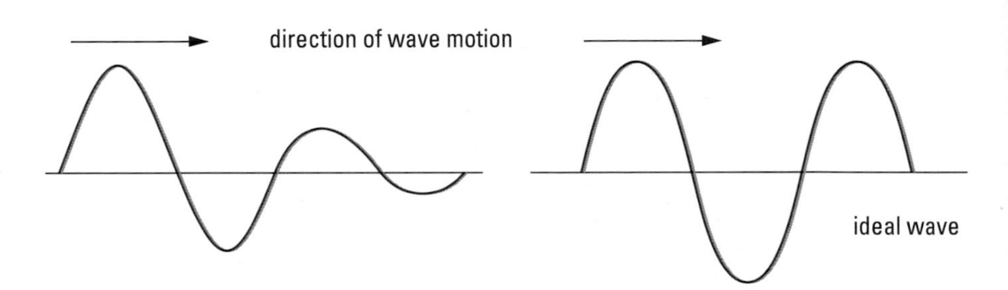

Figure 4
To make analysis easier, we will assume that the waves we are examining are ideal waves.

Longitudinal Waves

longitudinal wave: particles vibrate parallel to the direction of motion of the wave

compression: region in a longitudinal wave where the particles are closer together than normal

rarefaction: region in a longitudinal wave where the particles are farther apart than normal

In some types of waves the particles vibrate parallel to the direction of motion of the wave, and not at right angles to it. Such waves are called **longitudinal waves.** Longitudinal waves can be produced in "slinky" springs by moving one end of the spring back and forth in the direction of its length (**Figure 5**).

The most common longitudinal waves are sound waves, where the molecules, usually air, are displaced back and forth in the direction of the wave motion. In a longitudinal wave, the regions where the particles are closer together than normal are called **compressions**; the regions where they are farther apart are called **rarefactions.**

In longitudinal waves, one wavelength is the distance between the midpoints of successive compressions or rarefactions (refer to **Figure 5**). The maximum displacement of the particles from the rest position is the amplitude of the longitudinal wave. This will be discussed in more detail in Chapter 7.

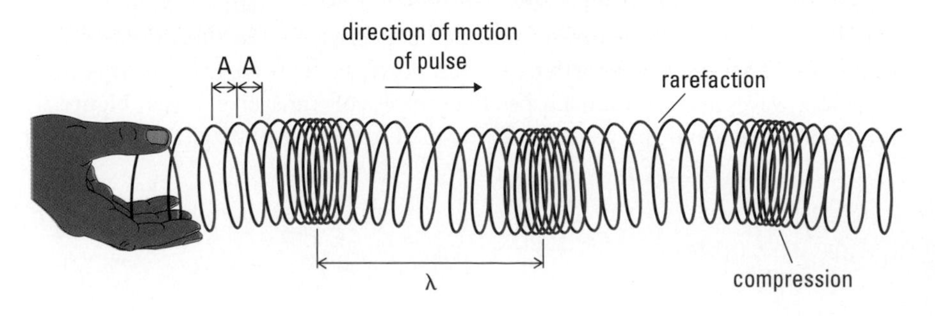

Figure 5
Longitudinal waves in a coiled spring

Sample Problem

Draw a periodic transverse wave consisting of two wavelengths with $A = 1.0$ cm and $\lambda = 2.0$ cm.

Solution

Draw the rest axis (PQ), then draw two lines 1.0 cm above and below PQ as shown in **Figure 6.** Label a starting point (B) and mark the points where the wave will cross the rest axis (at D, F, H, and J). Between B and D, mark the top of the crest (C) and mark all other crests and troughs in a similar fashion. Finally, draw a smooth curve joining the outlined points.

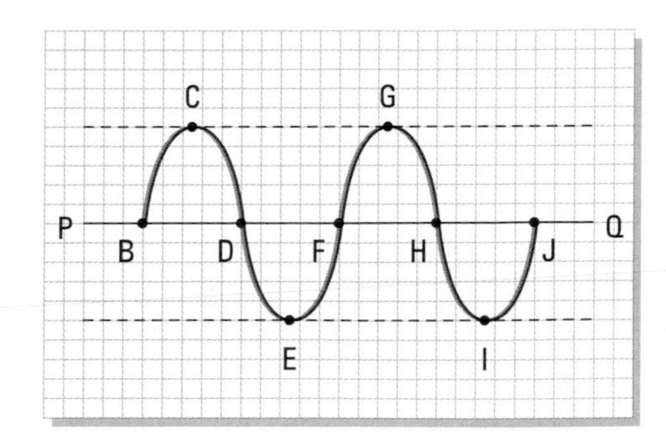

Figure 6
For Sample Problem

Practice

Understanding Concepts

1. A cross-section of a wave is shown in **Figure 7**. Name the parts of the wave indicated by the letters on the diagram.
2. Measure the amplitude and wavelength of the periodic transverse wave in **Figure 7**.
3. Measure the wavelength of the periodic longitudinal wave in **Figure 8**.
4. Draw a periodic wave consisting of two complete wavelengths, each with $\lambda = 4.0$ cm, for
   (a) a transverse wave (use $A = 0.5$ cm)
   (b) a longitudinal wave
5. **Figure 9** shows the profile of waves in a ripple tank.
   (a) Find the wavelength and the amplitude of the waves.
   (b) If crest A takes 2.0 s to move to where crest C is now, what is the speed of the waves?

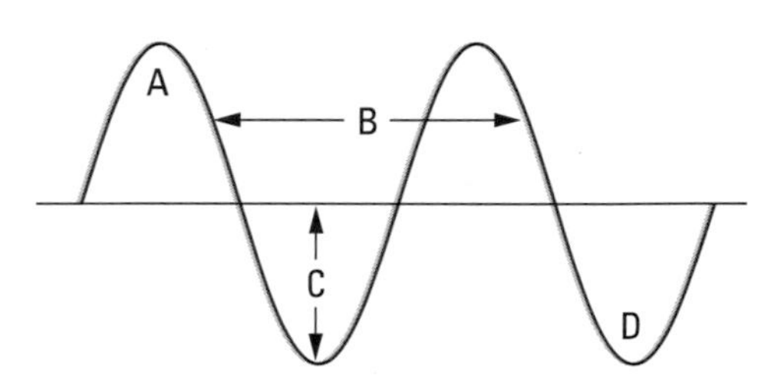

Figure 7
For questions 1 and 2

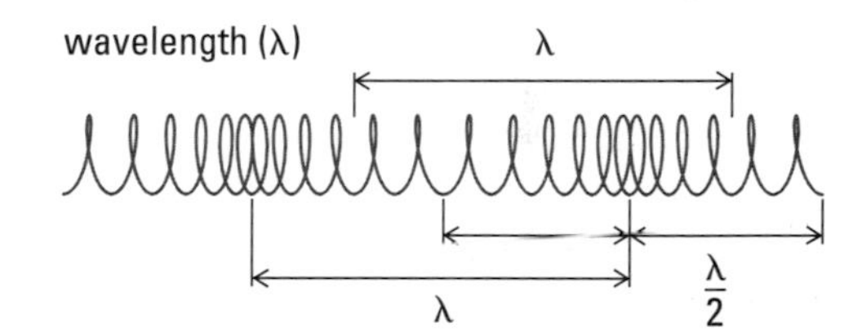

Figure 8
For question 3

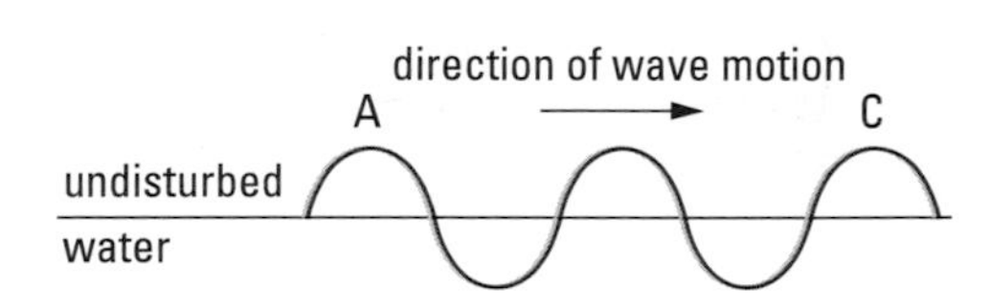

Figure 9
For question 5

SUMMARY Wave Motion

- Periodic waves originate from periodic vibrations where the motions are continuous and are repeated in the same time intervals.
- In a transverse wave, the particles of the medium move at right angles to the direction of the wave motion.
- In a longitudinal wave, the particles in the medium vibrate parallel to the direction in which the wave is moving.
- A transverse wave consists of alternate crests and troughs.
- A longitudinal wave consists of alternate compressions and rarefactions.
- One wavelength is the distance between equivalent points on successive crests or troughs in a transverse wave.
- In a longitudinal wave, one wavelength is the distance between the mid-points of successive compressions or rarefactions.
- The speed of a wave is unaffected by changes in the frequency or amplitude of the vibrating source.

Section 6.2 Questions

Understanding Concepts

1. Explain the difference between transverse and longitudinal waves.
2. Define the terms amplitude and wavelength.
3. Examine **Figure 10**.
   (a) List all pairs of points that are in phase.
   (b) Determine the wavelength, in centimetres, by measurement.
   (c) Determine the speed of the waves, if they take 0.50 s to travel from X to Y.

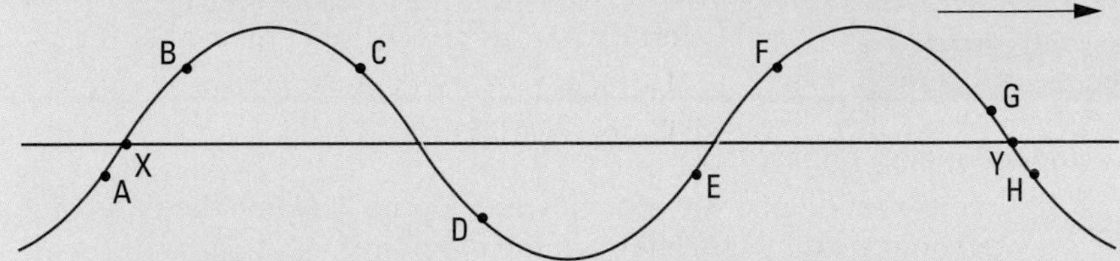

Figure 10

4. If you send a pulse down a long taut rope, its amplitude diminishes the farther it travels until eventually the pulse disappears. Explain why, using the law of conservation of energy.
5. A popular activity at a stadium sporting event is the "audience wave." The people in one section stand up quickly, then sit down; the people in the adjacent sections follow suit in succession, resulting in a wave pattern moving around the stadium. What type of wave is this? Why?
6. Describe one or more situations in your everyday life that involve waves of some kind.

Applying Inquiry Skills

7. How could you demonstrate a pulse given six billiard balls and a flat billiard table?

Making Connections

8. In 1883, a tsunami wave hit the coast of Java and Sumatra in the Pacific Ocean. Do some research on tsunamis to answer the following questions.
   (a) What type of wave is a tsunami?
   (b) What was the amplitude or range of amplitudes of the 1883 wave when it hit the shore?
9. Lithotripsy is a common treatment for kidney stones. Do some research on lithotripsy to answer the following questions. Follow the links for Nelson Physics 11, 6.2.
   (a) What are kidney stones?
   (b) How does lithotripsy help remove kidney stones?
   (c) What is the traditional method for removing kidney stones?
   (d) Why is lithotripsy the preferred method today?

GO TO www.science.nelson.com

6.3 The Universal Wave Equation

If you hold a piece of rope in your hand, you can create a crest along the rope by moving your hand through one-half of a cycle (**Figure 1**). When you move your hand in the opposite direction, a trough is produced that also travels along the rope behind the crest. If the motion is continued, a series of crests and troughs moves along the rope at a uniform speed. One cycle of the source produces one crest and one trough. The frequency of the wave is defined as the number of crests and troughs, or complete cycles, that pass a given point in the medium per unit of time (usually 1 s). The frequency of the wave is exactly the same as that of the source. It is the source alone that determines the frequency of the wave. Once the wave is produced, its frequency never changes, even if its speed and wavelength do change. This behaviour is characteristic of all waves.

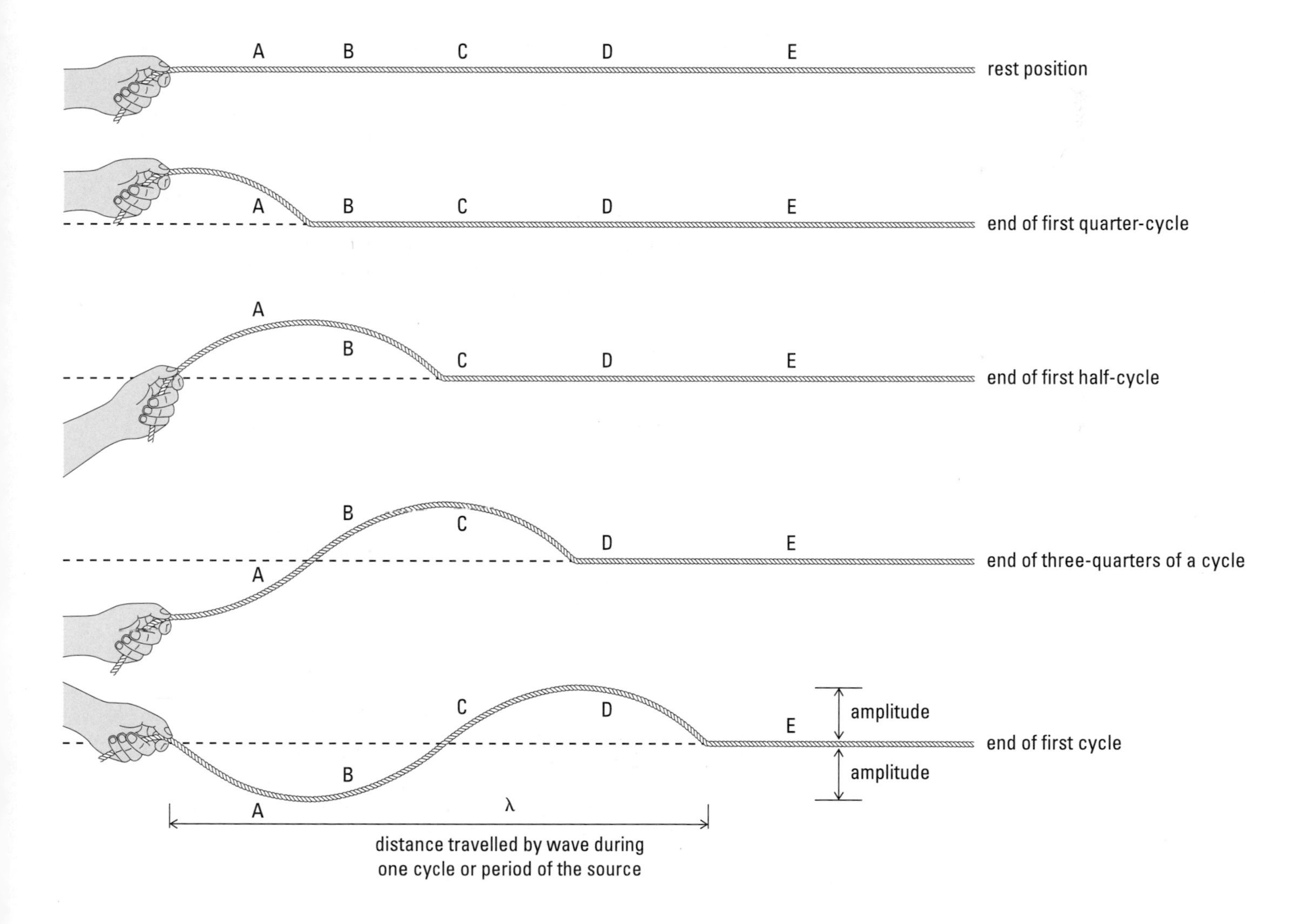

Figure 1
Creating a crest and a trough along a string

universal wave equation: $v = f\lambda$

When a wave is generated in a spring or a rope, the wave travels one wavelength (λ) along the rope in the time required for one complete vibration of the source. Recall that this time is defined as the period (T) of the source. Since speed (v) $= \frac{\text{distance}}{\text{time}}$, we can say

$$v = \frac{\lambda}{T} \quad \text{or} \quad v = f\lambda$$

This equation is known as the **universal wave equation.** It applies to all waves, visible and invisible.

Universal Wave Equation

$$v = f\lambda$$

Sample Problem 1

The wavelength of a water wave in a ripple tank is 0.080 m. If the frequency of the wave is 2.5 Hz, what is its speed?

Solution

$\lambda = 0.080$ m
$f = 2.5$ Hz
$v = ?$

$$\begin{aligned} v &= f\lambda \\ &= (2.5\text{Hz})(0.080\text{ m}) \\ v &= 0.20\text{ m/s} \end{aligned}$$

The speed of the wave is 0.20 m/s.

Sample Problem 2

The distance between successive crests in a series of water waves is 4.0 m, and the crests travel 9.0 m in 4.5 s. What is the frequency of the waves?

Solution

$\Delta d = 9.0$ m
$\Delta t = 4.5$ s
$\lambda = 4.0$ m
$f = ?$

$$\begin{aligned} v &= \frac{\Delta d}{\Delta t} \\ &= \frac{9.0\text{ m}}{4.5\text{ s}} \\ v &= 2.0\text{ m/s} \end{aligned}$$

$$\begin{aligned} v &= f\lambda \\ f &= \frac{v}{\lambda} \\ &= \frac{2.0\text{ m/s}}{4.0\text{ m}} \\ f &= 0.50\text{ Hz} \end{aligned}$$

The frequency of the waves is 0.50 Hz.

Sample Problem 3

The period of a sound wave from a piano is 1.18×10^{-3} s. If the speed of the wave in the air is 3.4×10^2 m/s, what is its wavelength?

Solution

$T = 1.18 \times 10^{-3}$ s
$v = 3.4 \times 10^2$ m/s
$\lambda = ?$

$$v = \frac{\lambda}{T}$$
$$\lambda = vT$$
$$= (3.4 \times 10^2 \text{ m/s})(1.18 \times 10^{-3} \text{ s})$$
$$\lambda = 0.40 \text{ m}$$

The wavelength is 0.40 m.

Practice

Understanding the Concepts

1. Calculate the speed (in metres per second) of the waves for each of the following:
   (a) $f = 18$ Hz, $\lambda = 2.7$ m
   (b) $f = 2.1 \times 10^4$ Hz, $\lambda = 2.0 \times 10^5$ cm
   (c) $T = 4.5 \times 10^{-4}$ s , $\lambda = 9.0 \times 10^4$ m
   (d) $T = 2.0$ ms, $\lambda = 3.4$ km
2. Write an equation for each of the following:
   (a) f in terms of v and λ
   (b) T in terms of v and λ
   (c) λ in terms of v and f
   (d) λ in terms of v and T

Answers

1. (a) 49 m/s
   (b) 4.2×10^7 m/s
   (c) 2.0×10^8 m/s
   (d) 1.7×10^6 m/s

SUMMARY The Universal Wave Equation

- One vibration of the source produces one complete wavelength.
- The frequency and the period of a wave are the same as those of the source, and they are not affected by changes in the speed of the wave.
- The universal wave equation, $v = f\lambda$, applies to all waves.

Section 6.3 Questions

Understanding Concepts

1. A vibrator in a ripple tank with a frequency of 20.0 Hz produces waves with a wavelength of 3.0 cm. What is the speed of the waves?
2. A wave in a skipping rope travels at a speed of 2.5 m/s. If the wavelength is 1.3 m, what is the period of the wave?
3. Waves travel along a wire at a speed of 10.0 m/s. Find the frequency and the period of the source if the wavelength is 0.10 m.
4. A crest of a water wave requires 5.2 s to travel between two points on a fishing pier located 19 m apart. It is noted in a series of waves that 20 crests pass the first point in 17 s. What is the wavelength of the waves?

(continued)

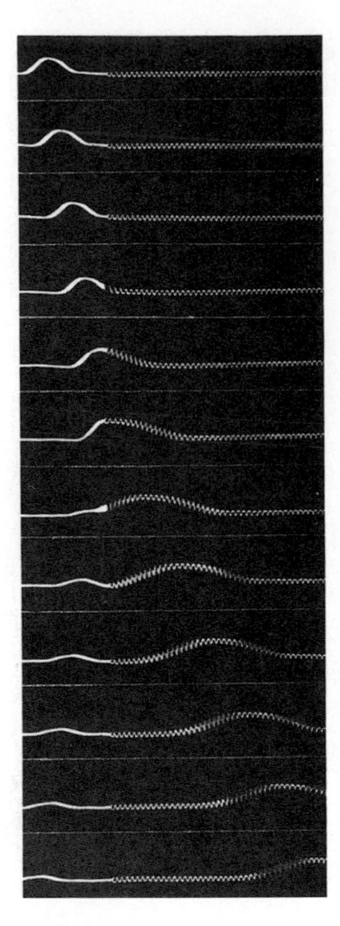

Figure 1
A pulse from a heavy spring (left) to a light spring (right)

5. The period of a sound wave emitted by a vibrating guitar string is 3.0×10^{-3} s. If the speed of the sound wave is 343 m/s, what is its wavelength?
6. Bats emit ultrasonic sound to help them locate obstacles. The waves have a frequency of 5.5×10^4 Hz. If they travel at 350 m/s, what is their wavelength?
7. What is the speed of a sound wave with a wavelength of 3.4 m and a frequency of 1.0×10^2 Hz?
8. An FM station broadcasts radio signals with a frequency of 102 MHz. These radio waves travel at a speed of 3.00×10^8 m/s. What is their wavelength?

6.4 Transmission and Reflection

Water waves and waves in long springs travel at a uniform speed as long as the medium they are in does not change. But if two long springs that differ in stiffness are joined together, the speed of a wave changes abruptly at the junction between the two springs (**Figure 1**). The speed change corresponds to a wavelength change. This wavelength change is predicted by the universal wave equation. Since the frequency of a wave remains constant, the wavelength is directly proportional to the speed; that is, $\lambda \propto v$.

When a wave travels from a light rope into a heavy rope having the same tension, the wave slows down and the wavelength decreases. On the other hand, if the wave travels from a heavy rope to a light rope, both the speed and the wavelength increase. These properties are true of all waves. A change in medium results in changes both in the speed of the wave and in its wavelength. In a medium where the speed is constant, the relationship $v = f\lambda$ predicts that $\lambda \propto \frac{1}{f}$. In other words, if the frequency of a wave increases, its wavelength decreases, a fact easily demonstrated when waves of different frequencies are generated in a rope or spring.

As you saw in Investigation 6.2.1, one-dimensional waves, such as those in a spring or rope, behave in a special way when they are reflected. In the case of reflection from a rigid obstacle, usually referred to as **fixed-end reflection**, the pulse is inverted. A crest is reflected as a trough and a trough is reflected as a crest (**Figures** 2 and 3). If the reflection occurs from a **free end**, where the medium is free to move, there is no inversion—crests are reflected as crests and troughs as troughs (**Figure** 2). In both fixed-end and free-end reflection there is no change in the frequency or wavelength. Nor is there any change in the speed of the pulse, since the medium is the same.

fixed-end reflection: reflection from a rigid obstacle when a pulse is inverted

free-end reflection: reflection where the new medium is free to move and there is no inversion

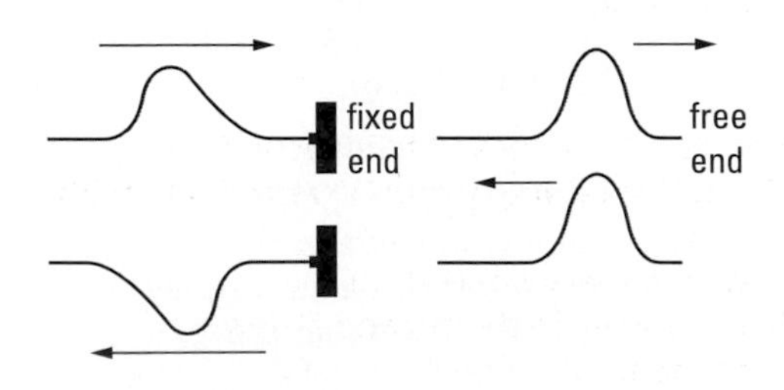

Figure 2
Fixed-end and free-end reflection

What happens when a wave travels into a different medium? At the boundary between the two media, the speed and wavelength change, and some reflection occurs. This is called **partial reflection** because some of the energy is transmitted into the new medium and some is reflected back into the original medium. This phenomenon is shown in **Figure 4** for a wave passing from a fast medium to a slow medium. Since the particles of the slower medium have greater inertia, this medium acts like a rigid obstacle, and the reflected wave is inverted. However, the transmitted wave is not inverted.

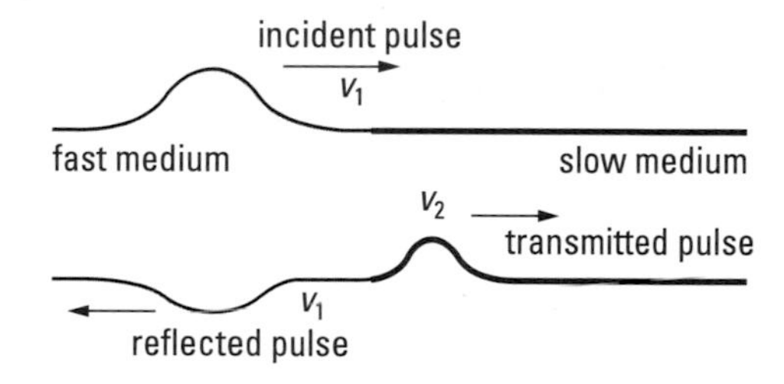

Figure 4
Wave travelling from fast to slow medium

When a wave travels from a slow medium to a fast medium, the fast medium acts like a "free-end" reflection. No inversion occurs in either the reflected or transmitted wave, but there are changes in the wavelength and in the speed of the transmitted wave, as shown in **Figure 5**.

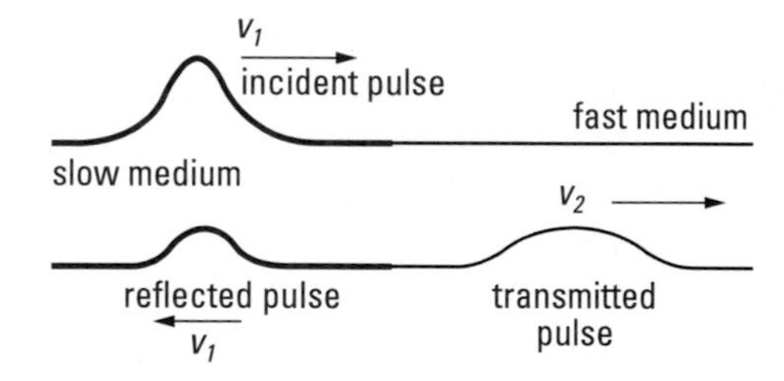

Figure 5
Wave travelling from slow to fast medium

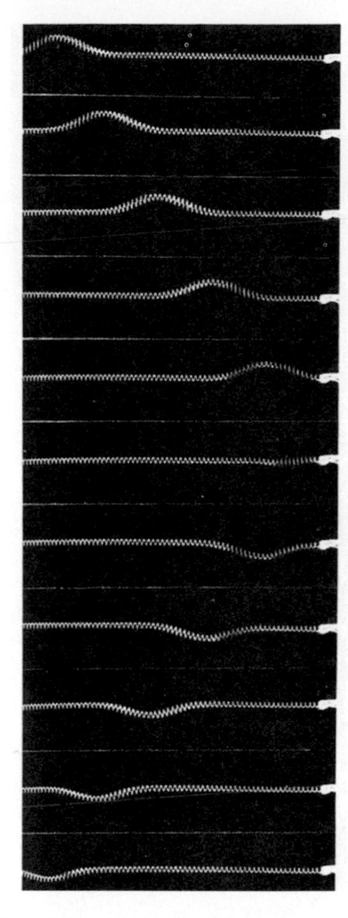

Figure 3
A crest, moving left to right, is reflected from a rigid obstacle as a trough.

partial reflection: some of the energy is transmitted into the new medium and some is reflected back into the original medium

Practice

Understanding Concepts

1. How is a negative pulse reflected from
   (a) a fixed end?
   (b) a free end?
2. You send a wave down a string that is attached to a second string with unknown properties. The pulse returns to you inverted and with a smaller amplitude.
   (a) Is the speed faster or slower in the second string?
   (b) Is the wavelength smaller or larger in the second string?
   Explain your reasoning in each case.

SUMMARY Transmission and Reflection

- Pulses reflected from a fixed end are inverted.
- Pulses reflected from a free end are not inverted.
- When a pulse enters a new medium, no inversion occurs.
- When a wave enters a slower medium, its wavelength decreases; in a faster medium the wavelength increases.
- When waves strike the boundary between two different media, partial reflection occurs.
- The phase of transmitted waves is unaffected in all partial reflections, but inversion of the reflected wave occurs when the wave passes from a fast medium to a slow medium.

Section 6.4 Questions

Understanding Concepts

1. Copy **Table 1** into your notebook and complete. (Note: + means a crest, – means a trough.)

Table 1

Incident pulse	Reflected pulse	Transmitted pulse	Medium change
+	?	?	fast to slow
+	–	?	?
–	?	?	slow to fast

Applying Inquiry Skills

2. Rope A, rope B, and rope C are made of different materials and are attached to one another in a linear fashion forming one continuous strand. Classify ropes A, B, and C as fast, intermediate, and slow media, based on the following observations.
   - A positive pulse in rope A is reflected as a positive pulse and is transmitted as a positive pulse into rope B.
   - The positive pulse transmitted into rope B is reflected as a positive pulse and is transmitted as a positive pulse into rope C.

6.5 Waves in Two Dimensions

We have been looking at waves in a stretched spring or in a rope to understand some of the basic concepts of wave motion in one dimension. How do waves move in two dimensions?

Using a ripple tank, we can study the behaviour of waves in two dimensions. A ripple tank is a shallow glass-bottomed tank. Light from a source above the tank passes through the water and illuminates a screen on the table below. The light is converged by wave crests and diverged by wave troughs, as shown in **Figure 1**, creating bright and dark areas on the screen. The distance between successive bright areas caused by crests is one wavelength (λ).

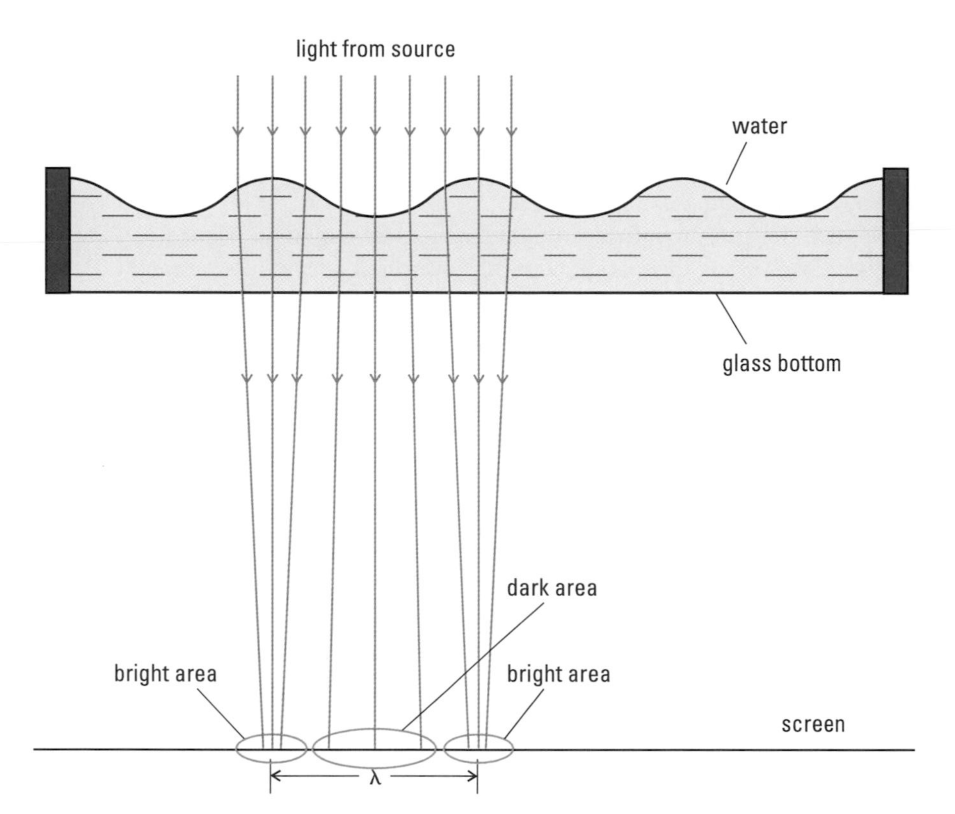

Figure 1
Bright lines occur on the screen where light rays converge.

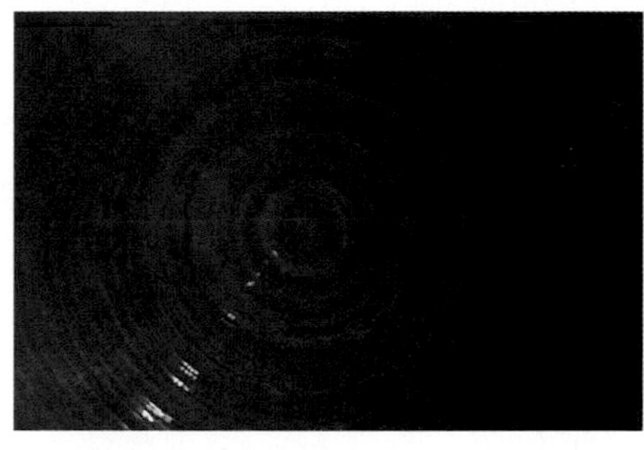

Figure 2
Periodic circular waves

Transmission and Reflection of Water Waves

Water waves coming from a point source in a ripple tank are circular (**Figure 2**), whereas waves from a linear source are straight (**Figure 3**). As a series of periodic waves moves away from its source, the spacing between successive crests and troughs remains the same as long as the speed does not change. That is, the wavelength does not change as long as the speed remains constant. This is predicted by the wave equation ($v = f\lambda$), because when f is constant, $\lambda \propto v$. When the speed decreases, as it does in shallow water, the wavelength also decreases (**Figure 4**).

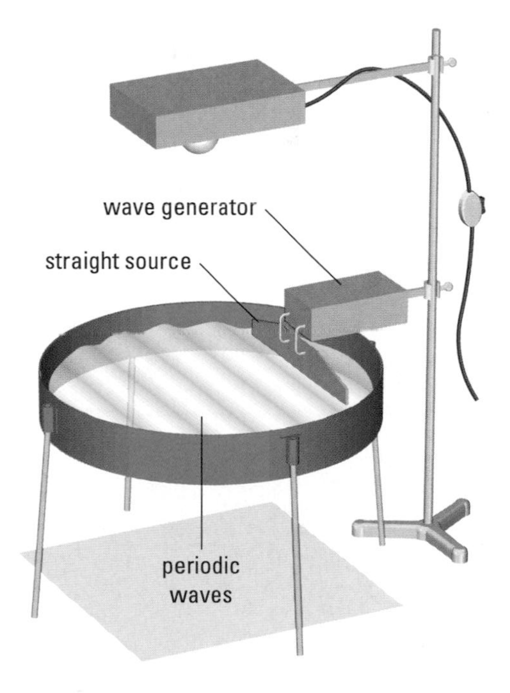

Figure 3
Straight waves

Figure 4
Waves travelling into shallow water have a slower speed and a shorter wavelength.

wavefront: the leading edge of a wave

wave ray: a straight line drawn perpendicular to a wavefront that indicates the direction of transmission

normal: a straight line drawn perpendicular to a barrier struck by a wave

When the frequency of a source is increased, the distance between successive crests becomes smaller. In other words, waves with a higher frequency have a shorter wavelength if their speed remains constant. Although the wavelength and the speed of a wave may change as the wave moves through a medium, the frequency will not change. The frequency can be changed only at the source, and not by the medium.

When waves run into a straight barrier, as shown in **Figure 5**, they are reflected back along their original path. However, if a wave hits a straight barrier obliquely (**Figure 6**), the **wavefront** is also reflected at an angle to the barrier.

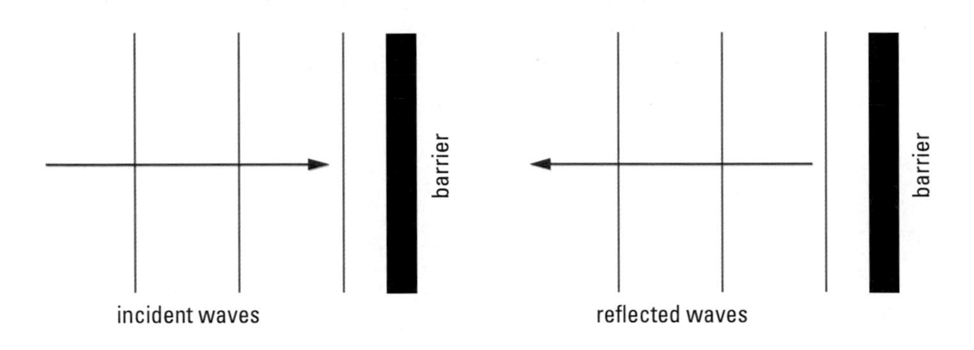

Figure 5
The frequency does not change as a wave moves through a medium.

focal point: a specific place where straight waves are reflected to

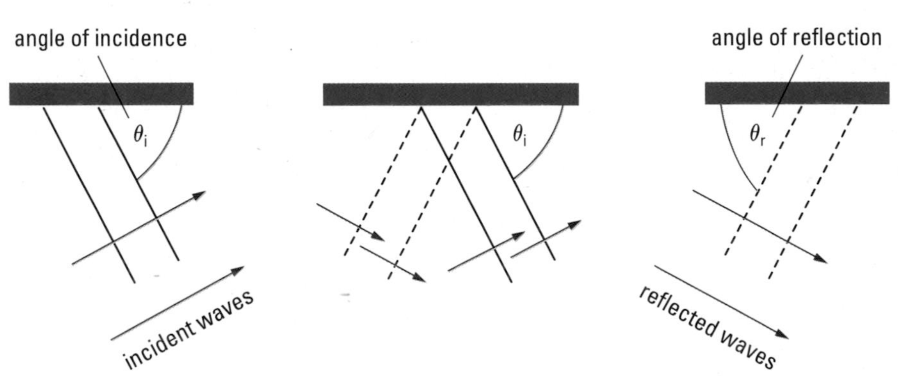

Figure 6
The angle of incidence equals the angle of reflection for a wave that hits a straight barrier obliquely.

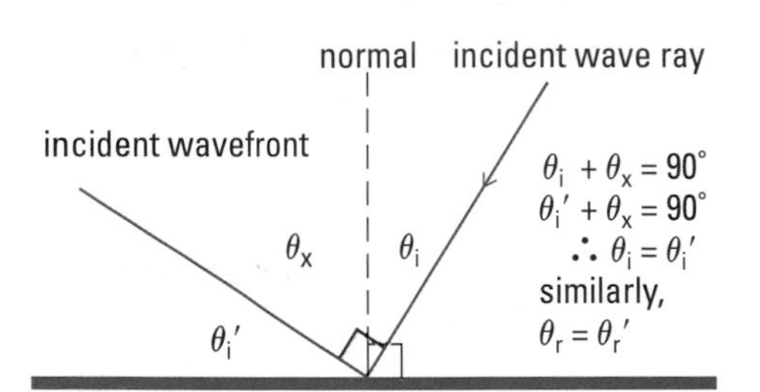

Figure 7

The angles formed by the incident wavefront and the barrier and by the reflected wavefront and the barrier are equal. These angles are called the angle of incidence (θ_i) and the angle of reflection (θ_r), respectively. In neither case does the reflection produce any change in the wavelength or in the speed of the wave.

When describing the reflection of waves, we use the term **wave rays.** Wave rays are straight lines perpendicular to wavefronts indicating the direction of travel. The angles of incidence and reflection for wave rays are measured relative to a straight line perpendicular to the barrier, called the **normal.** This line is constructed at the point where the incident wave ray strikes the reflecting surface. As may be seen from the geometrical analysis shown in **Figure 7**, the angle of incidence has the same value whether wavefronts or wave rays are used to measure it. In both cases the angle of incidence equals the angle of reflection. This is one of the laws of reflection from optics, and is covered in more detail in Unit 4.

Straight waves can be reflected by a parabolic reflector to one point, called the **focal point.** This could have been predicted by means of the laws of reflection using wave rays as shown in **Figure 8**.

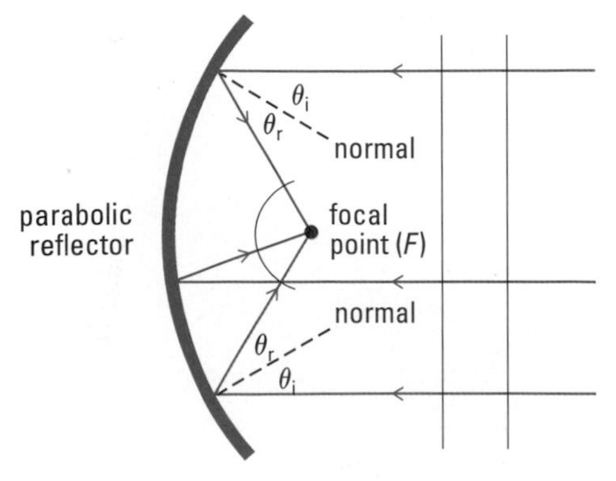

Figure 8

When a water wave enters a medium in which it moves more slowly, such as shallow water, its wavelength decreases as well (**Figure 9**). This could have been predicted using the universal wave equation, since we can conclude that $\lambda \alpha v$.

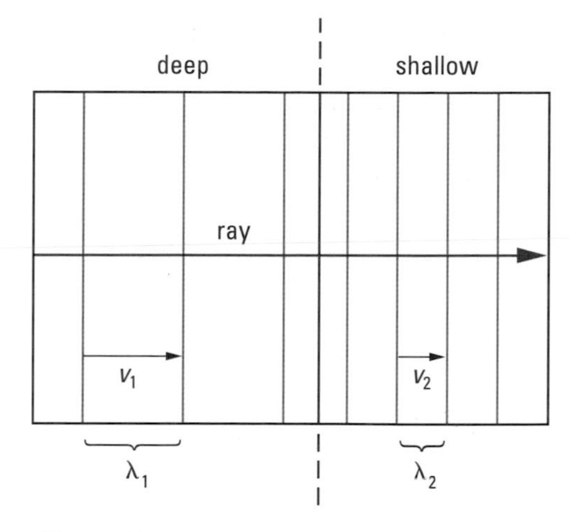

Figure 9

Refraction and Diffraction

When the wave travels from deep to shallow water, in such a way that it crosses the boundary between the two depths straight on, no change in direction occurs. On the other hand, if a wave crosses the boundary at an angle, the direction of travel does change (**Figure 10**). This phenomenon is called **refraction.**

Periodic straight waves travel in a straight line across a ripple tank as long as the depth doesn't change. But if the waves pass by the sharp edge of an obstacle or through a small opening in the obstacle, the waves change direction, as shown in **Figure 11**. This bending is called **diffraction.**

Figure 11
Diffraction of water waves

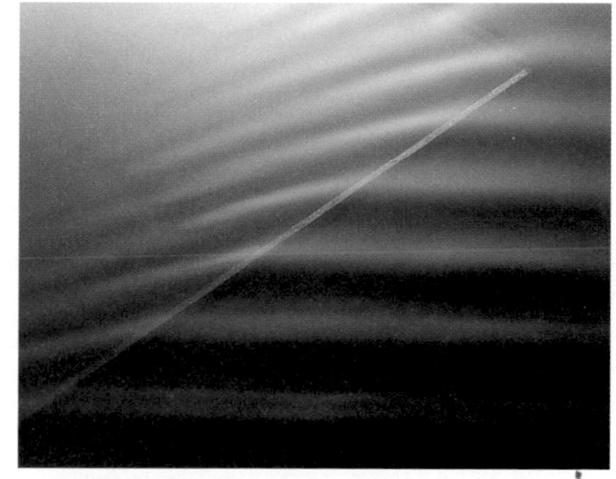

Figure 10
Waves are refracted as they cross a boundary between deep and shallow water if the boundary is at an angle.

How much the wave bends depends on both the wavelength and the size of the opening in the barrier. For the same opening, long-wavelength waves are diffracted more than short-wavelength waves. Decreasing the size of the opening will increase the amount of diffraction, as shown in **Figure 12**.

refraction: the bending effect on a wave's direction that occurs when the wave enters a different medium at an angle

diffraction: the bending effect on a wave's direction as it passes through an opening or by an obstacle

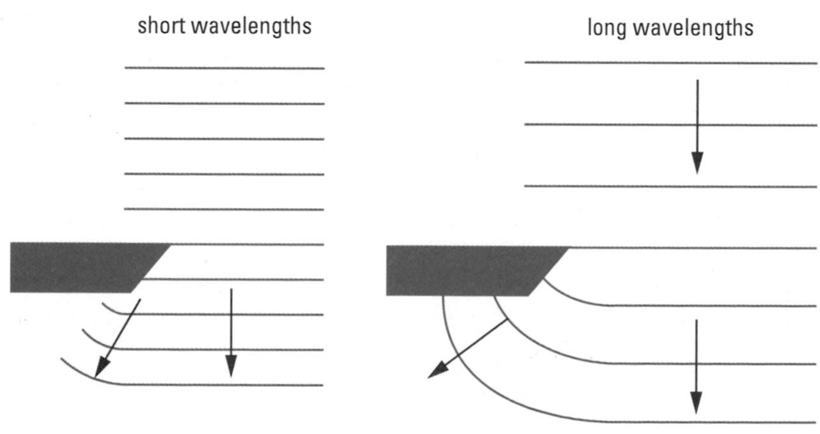

Figure 12

Practice

Understanding Concepts

1. (a) Do the bright areas on the screen of a ripple tank represent the crests or troughs of the water waves?
   (b) Are the bright areas formed by converging or diverging rays of light?
2. What happens to the wavelength and the speed of water waves when they reflect off a barrier?
3. When water waves move from deep to shallow water, they slow down. What effect does this have on the wavelength and on the frequency?
4. State the law of reflection as it applies to waves.
5. Distinguish between refraction and diffraction.

Applying Inquiry Skills

6. Predict the shape of the wave fronts that would result if straight periodic water waves went through a very small opening in a barrier. Draw lines to indicate the shape on the right-hand side of **Figure 13**.

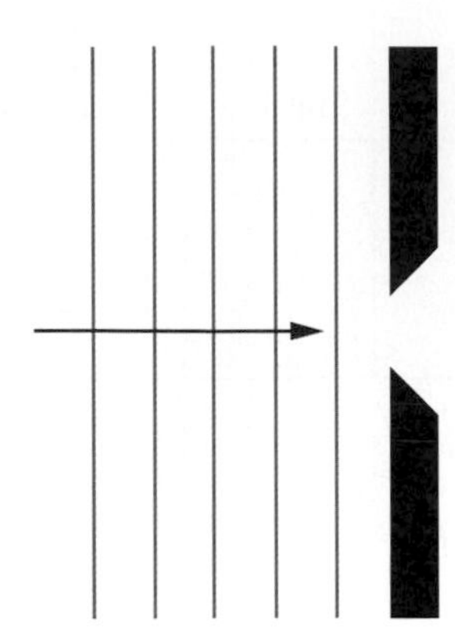

Figure 13
For question 6

SUMMARY Properties of Waves

- A wave ray is a straight line drawn at right angles to the wavefront, indicating the direction of the wave motion.
- When waves are reflected from a solid obstacle, the angle of incidence is always equal to the angle of reflection.
- When a wave enters a medium in which it moves more slowly, its wavelength decreases.
- When a water wave enters a slower medium at an angle, its direction of transmission changes; the wave has undergone refraction.
- When a wave passes by a barrier or through a small opening, it tends to diffract or change direction.

Section 6.5 Questions

Understanding Concepts

1. Draw a sketch illustrating how straight waves behave when
   (a) the frequency of the wave decreases
   (b) the wave encounters a straight barrier straight on
   (c) the wave encounters a straight barrier at an angle
   (d) the wave's speed is reduced
   (e) the wave enters shallow water at an angle
   (f) the wave passes through a small opening
2. What type of sound waves, high frequency (treble) or low frequency (bass), do you predict will diffract better around corners? Explain your answer.

Applying Inquiry Skills

3. Light is transmitted as a high-frequency wave. Predict the size of slit that would be necessary to observe diffraction. Test your prediction by observing a point source of light through a slit plate.

Making Connections

4. Actual water waves do not behave exactly like the ideal transverse waves described in this section. For example, ocean waves increase in amplitude and decrease in wavelength as they approach a beach, but when they near the shore, they "break." Research this issue and, on one page, describe with a diagram how the water particles in a water wave actually move and why this affects their behaviour when waves crash on the beach. Follow the links for Nelson Physics 11, 6.5.

GO TO www.science.nelson.com

6.6 Interference of Waves

Up to this point, we have been dealing with one wave at a time. What happens when two waves meet? Do they bounce off each other? Do they cancel each other out? When pulses travel in opposite directions in a rope or spiral spring, the pulses interfere with each other for an instant and then continue travelling unaffected. This behaviour is common to all types of waves.

Types of Interference

Wave interference occurs when two waves act simultaneously on the same particles of a medium. There are two types of interference: constructive and destructive. For transverse pulses, **destructive interference** occurs when a crest meets a trough. If the crest and trough have equal amplitude and shape, their amplitudes cancel each other for an instant. Then the crest and trough continue in their original directions, as shown in **Figure 1**. For longitudinal pulses, destructive interference occurs when a compression meets a rarefaction.

Constructive interference occurs when pulses build each other up, resulting in a larger amplitude (**Figure 2**). This occurs for transverse pulses when a crest

wave interference: occurs when two or more waves act simultaneously on the same particles of a medium

destructive interference: occurs when waves diminish one another and the amplitude of the medium is less than it would have been for either of the interfering waves acting alone

constructive interference: occurs when waves build each other up, resulting in the medium having a larger amplitude

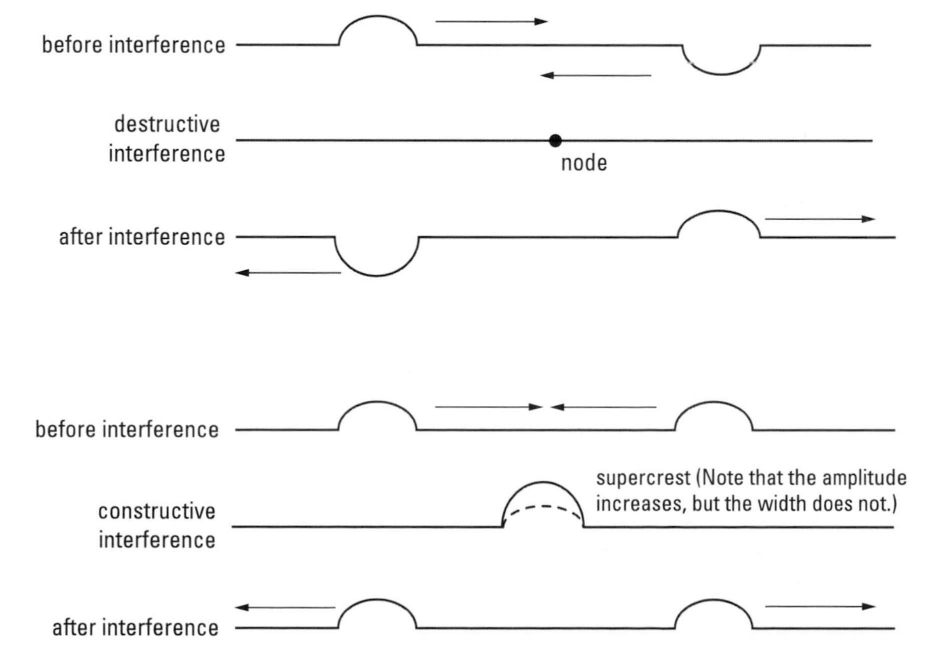

Figure 1
Destructive interference of pulses

Figure 2
Constructive interference of pulses

supercrest: occurs when a crest meets a crest

supertrough: occurs when a trough meets a trough

principle of superposition: At any point the resulting amplitude of two interfering waves is the algebraic sum of the displacements of the individual waves.

meets a crest, causing a **supercrest**, or a trough meets a trough, causing a **supertrough**. Constructive interference also occurs for longitudinal pulses.

In **Figure 1**, the interference in each case is shown at the instant that the pulses overlap or become superimposed on one another. If we call the amplitudes on one side of the rest axis positive and the amplitudes on the opposite side negative, then the superimposed amplitude is the addition of the individual amplitudes. For example, amplitudes of +1.0 cm and –1.0 cm are added to produce a zero amplitude. The concept of amplitude addition is summarized in the **principle of superposition**, which states that at any point the resulting amplitude of two interfering waves is the algebraic sum of the displacements of the individual waves.

This principle is especially useful for finding the resulting pattern when pulses that are unequal in size or shape interfere with one another. To learn how to apply the principle of superposition in one dimension, refer to **Figure 3**. In **Figure 3(a)** two straight-line pulses are added; in **Figure 3(b)** two curved-line pulses are added.

The principle of superposition may be used to find the resultant displacement of any medium when two or more waves of different wavelengths interfere. In every case, the resultant displacement is determined by an algebraic summing of all the individual wave displacements. These displacements may be added together electronically and the resultant displacement may be displayed on an oscilloscope (**Figure 4**). The resultant wave is the only one seen, not the individual interfering waves.

DID YOU KNOW ?

Principle of Superposition

The principle of superposition applies only to displacements that are reasonable in size because large waves that interfere may cause distortion of the medium through which they are travelling.

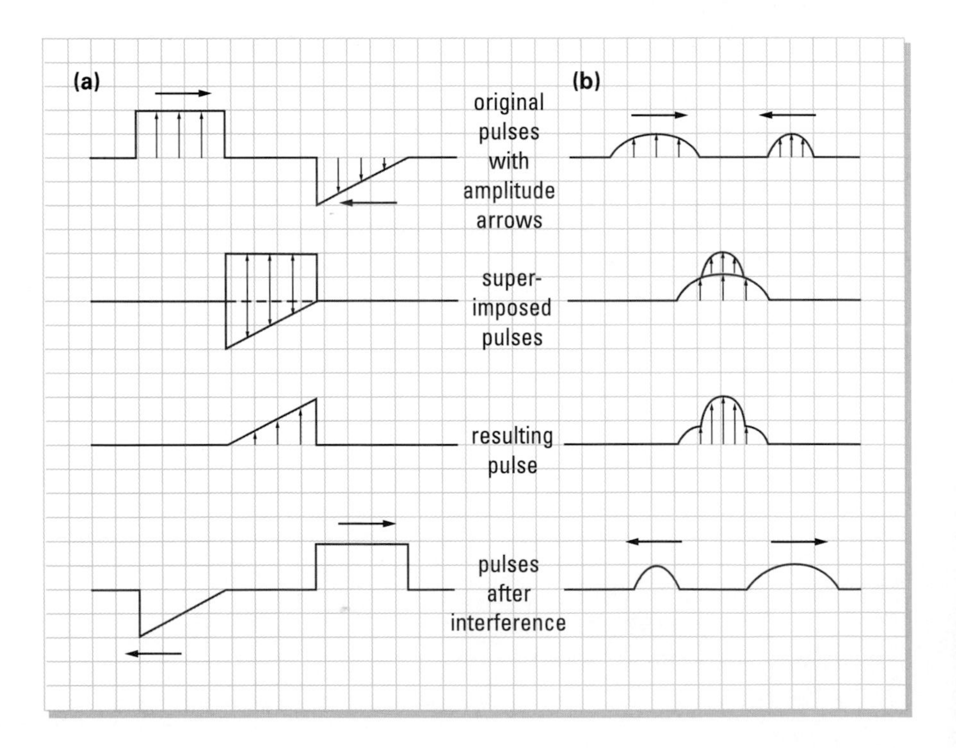

Figure 3
Applying the principle of superposition
(a) Straight-line pulses
(b) Curved-line pulses

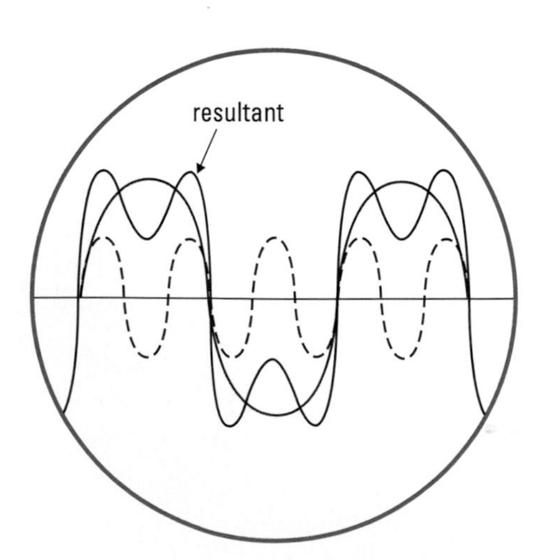

Figure 4
Resultant displacement as displayed on an oscilloscope. Note that only the solid black line would be observed.

Demonstrating Interference with Springs

Both constructive and destructive interference can be demonstrated using the Slinky spring.

- Tape a small tab to a coil at the middle of the spring.
- With a student at each end, stretch the spring to an appropriate length (e.g., 2.0 m).
- With one end held firm, send a positive pulse down the spring, noting the displacement of the tab.
- Simultaneously, send a positive pulse from each end of the spring, each with the same amplitude. Note the displacement of the tab.
- Repeat the previous step, except send a positive pulse from one end of the spring and a negative pulse from the other.

Hold the spring firmly and do not overstretch it.
Observe from the side, in case of an accidental release.

Practice

Understanding Concepts

1. State whether the interference is constructive or destructive when
   (a) a large crest meets a small trough
   (b) a supertrough is formed
   (c) a small compression meets a large compression
2. Use the principle of superposition to determine the resulting pulse when the pulses shown in **Figure 5** are superimposed on each other. (The point of overlap should be at the horizontal midpoints of the pulses.)

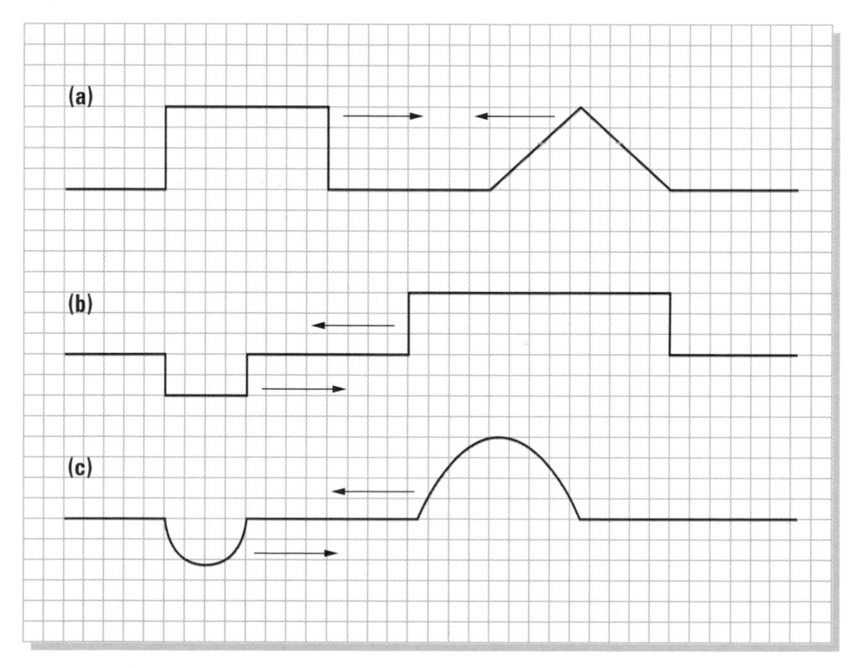

Figure 5

SUMMARY Interference of Waves

- Waves can pass through each other in a medium without affecting each other; only the medium particles are momentarily affected.
- The resultant displacement of a particle is the algebraic sum of the individual displacements contributed by each wave.
- If the resultant displacement is greater than that caused by either wave alone, constructive interference is occurring; if it is smaller, destructive interference is occurring.

Section 6.6 Questions

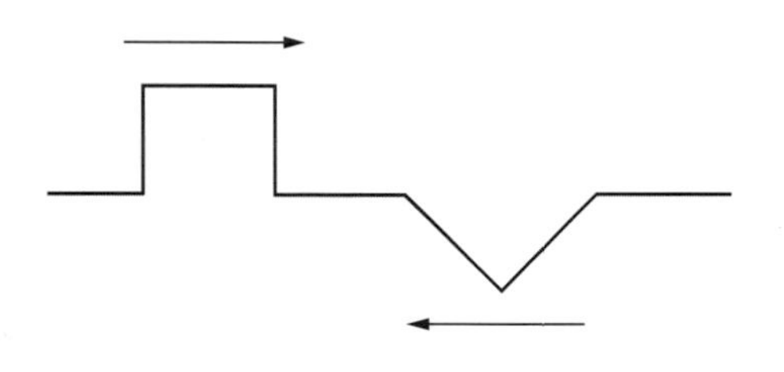

Figure 6
For question 1

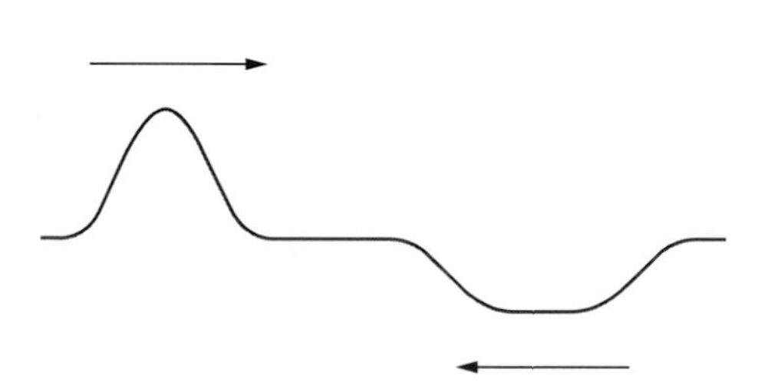

Figure 7
For question 2

Understanding Concepts

1. **Figure 6** shows two pulses approaching one another. Sketch the appearance of the medium when the two pulses overlap, centres coinciding.
2. Two pulses move toward each other as shown in **Figure 7**. Sketch the resultant shape of the medium when the two pulses overlap, centres coinciding.
3. Trace the pulses illustrated in **Figure 8** into your notebook and determine the resultant displacement of the particles of the medium at each instant, using the principle of superposition.

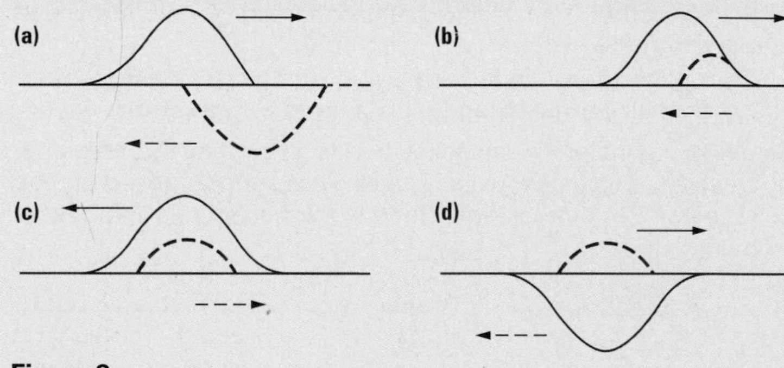

Figure 8

4. When waves crash into a seawall, the incoming waves interfere with the reflected waves, causing both constructive and destructive interference. How would you identify each type of interference?
5. What happens when two billiard balls, rolling toward one another, collide head on? How does this differ from two waves or pulses that collide head on?

Making Connections

6. (a) What is the difference between AM and FM radio signals?
   (b) Describe one advantage of FM broadcasting over AM broadcasting. Follow the links for Nelson Physics 11, 6.3.

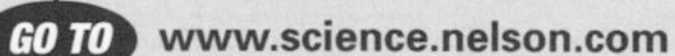

GO TO www.science.nelson.com

6.7 Mechanical Resonance

Every object has a natural frequency at which it will vibrate. To keep a child moving on a swing, we must push the child with the same frequency as the natural frequency of the swing. We use a similar technique to "rock" a car that is stuck in the snow. When a large truck passes your house, you may have noticed that the windows rattle.

These are all examples of a phenomenon called **resonance**, which is the response of an object that is free to vibrate to a periodic force with the same frequency as the natural frequency of the object. We call this phenomenon **mechanical resonance** because there is physical contact between the periodic force and the vibrating object.

resonance or **mechanical resonance:** the transfer of energy from one object to another having the same natural frequency

Resonance can be demonstrated with a series of pendulums suspended from a stretched string (**Figure 1**). When A is set in vibration, E begins to vibrate in time with it. Although B, C, and D may begin to vibrate, they do not continue to vibrate nor do they vibrate as much. When B is set in vibration, D begins to vibrate in sympathy, but A, C, and E vibrate spasmodically and only a little. The pairs A and E, and B and D each have the same lengths and, thus, have the same natural frequencies. They are connected to the same support, so the energy of A, for example, is transferred along the supporting string to E, causing it to vibrate. This occurs only if E is free to vibrate. The periodic vibratory force exerted by one pendulum moves through the supporting string to the other pendulums, but only the pendulum with the same natural frequency begins to vibrate in resonance. When an object vibrates in resonance with another, it is called a **sympathetic vibration.**

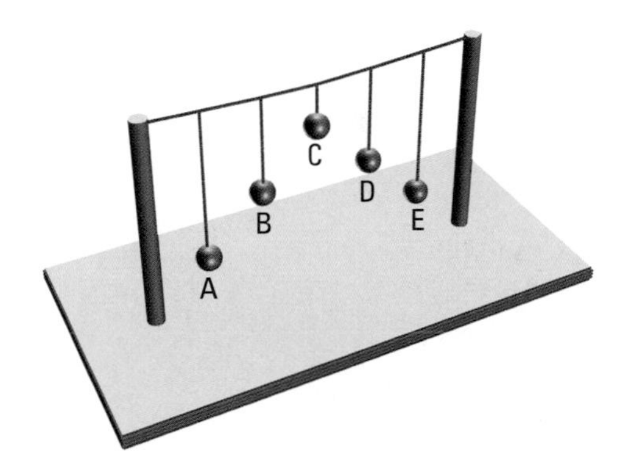

Figure 1
A series of pendulums suspended from a string can be used to show resonance.

sympathetic vibration: the response to a vibration with the same natural frequency

Mechanical resonance must be taken into account when designing bridges, airplane propellers, helicopter rotor blades, turbines for steam generators and jet engines, plumbing systems, and many other types of equipment. A dangerous resonant condition may result if this is not done. For example, in 1940 the Tacoma Narrows suspension bridge in Washington State collapsed when wind caused the bridge to vibrate (**Figure 2**). In 1841, a troop of British soldiers marched in step across a bridge, which created a periodic force that set the bridge in resonant vibration and caused the bridge to collapse. If an opera singer sings a note with the same natural frequency as that of a wineglass, the glass will begin to vibrate in resonance. If the sound has a high enough intensity, the wineglass could vibrate with an amplitude large enough that it shatters.

The human body also has resonant frequencies. Experiments have shown that the entire body has a mechanical resonant frequency of about 6 Hz, the head of between 13 Hz and 20 Hz, and the eyes of between 35 Hz and 75 Hz. Large amplitude vibrations at any of these frequencies could irritate or even damage parts of the body. In transportation and road construction occupations, efforts are made to reduce the effects of mechanical vibrations on the human body.

Even very large structures, such as the CN Tower and skyscrapers, can resonate, whether the external source of energy is as small as a gust of wind or as large as an earthquake. Engineers must consider specific structural features to minimize damage by earthquakes.

Radio and television provide another example of resonance. When you tune a radio or television, you are actually adjusting the frequency of vibration of particles in the receiver so that they resonate with the frequency of a particular signal from a radio or television station.

Figure 2
(a) The Tacoma Narrows Bridge begins to vibrate.
(b) The centre span of the bridge vibrates torsionally before collapsing.
(c) Vibrations eventually cause the bridge to collapse.
(d) The Tacoma Narrows Bridge today; notice the structural changes that were made.

(a)

(b)

(c)

(d)

Practice

Understanding Concepts

1. (a) Every playground swing has its own resonant or natural frequency. What does that frequency depend on?
   (b) To obtain a large amplitude of vibration, how does the frequency of the input force compare with the resonant frequency? Explain.
2. When crossing a footbridge, some students notice that the bridge moves up and down. One student suggests that they all jump up and down on the bridge at the same time. Explain why this is a dangerous idea.

Applying Inquiry Skills

3. (a) How would you design an experiment to determine how the resonant frequency of a long-stemmed glass depends on the amount of water in the glass?
   (b) Predict the relationship in (a).
   (c) With your teacher's approval, try your experimental design and describe what you discover.

SUMMARY Mechanical Resonance

- Resonance occurs when an object that is free to vibrate is acted on by a periodic force that has the same frequency as the object's natural frequency.
- Mechanical resonance must be considered when designing any object that has some freedom of movement and has some physical contact with a source of vibration.

Section 6.7 Questions

Understanding Concepts

1. Provide some examples of mechanical resonance.
2. Your car is stuck in ruts in the snow. How could you use the principle of resonance to free it?

Making Connections

3. Research to find out what structural features are used in modern construction to minimize earthquake damage in large buildings located near fault lines in areas such as Vancouver, San Francisco, and Japan.

6.8 Standing Waves — A Special Case of Interference in a One-Dimensional Medium

standing wave: created when waves travelling in opposite directions have the same amplitude and wavelength

The amplitude and the wavelength of interfering waves are often different. However, if conditions are controlled so that the waves have the same amplitude and wavelength, yet travel in opposite directions, the resultant interference pattern is particularly interesting. It is referred to as a standing wave interference pattern, or a **standing wave**. In most cases of interference, the resultant displacement remains for only an instant, making it difficult to analyze the interference. Standing wave interference remains relatively stationary, which makes it much easier to analyze.

Activity 6.8.1

Standing Waves in a One-Dimensional Medium

At one time or another, you have probably repeatedly shaken a long rope tied to a rigid object, such as a tree. You may have found that if you got the frequency just right, you could produce a fixed pattern of loops in the rope. What you produced was a standing wave interference pattern. A similar technique will be used in this investigation to study this phenomenon and to see how it can be used to measure wavelength.

Questions

(i) What are the conditions necessary to produce standing waves in a one-dimensional medium?
(ii) How is the pattern of interference affected by the frequency of the interfering waves?
(iii) What constitutes a wavelength in a standing wave interference pattern?
(iv) How can one calculate the speed of a wave, using the standing wave interference pattern?

Materials

long spiral spring
long string
electric vibrator

Procedure

Hold the spring firmly and do not over stretch it. Observe from the side in case of an accidental release.

1. Stretch a long spiral spring across a smooth floor, with a student at each end of the spring.
2. Simultaneously generate a series of waves of equal frequency and amplitude from each end of the spring. Draw a sketch of the resultant displacement of the whole spring.
3. Change the frequency of the waves, achieving a fixed interference pattern at a lower and a higher frequency. Sketch the pattern for each case.

4. At the lower frequency, place markers on the floor between two fixed points in the pattern. Compare this with the wavelength of the waves you are generating from each end.
5. Fix the spring rigidly at one end and from the free end generate waves with a constant frequency. Sketch the pattern of interference, labelling one wavelength on the pattern.
6. Attach a long string to a rigid point at one end and an electric vibrator at the other. Start the vibrator and adjust the tension and frequency until a standing wave is produced.
7. Measure the distance between series of at least four fixed points in the pattern. Calculate the wavelength of the wave.
8. Record the frequency of the vibrator and calculate the speed of the waves in the string.
9. Repeat steps 7 and 8 for two other frequencies.
10. Find an average for the speed of the waves in the string.

Analysis

(a) On your sketches, label the areas of constructive and destructive interference.
(b) For complete destructive interference to occur, what must be true of the wavelengths and amplitudes of the two waves?
(c) In a standing wave, what distance constitutes a wavelength? What constitutes half a wavelength?
(d) What type of interference always occurs at the fixed end, in a medium where standing waves are created using reflection?
(e) Why is it easier to produce standing waves using reflection than by generating identical waves from each end of the medium?
(f) Why is it easier to measure the wavelength of a wave using the standing wave interference pattern than it is to measure it directly?
(g) What was the average speed of the waves in the string?
(h) If the speed of a wave in a medium were known, how would you use standing waves to determine the frequency of the wave?
(i) Answer the Questions.

Nodes and Antinodes

When waves of equal amplitude and shape travelling in opposite directions interfere, there is a point, or points, that remains at rest throughout the interference of the pulses. This point is called a **node**, or **nodal point** (N). In **Figure 1**, two identical waves, A and B, are interfering. The resultant displacement caused by their interference consists of areas of constructive and destructive interference. Note that the nodes are equidistant and that their spacing is equal to one-half of the wavelength of the interfering waves. Midway between the nodes are points where double crests and double troughs occur. These points are called **antinodes**.

node or **nodal point:** (N) point that remains at rest

antinode: point midway between the nodes where maximum constructive interference occurs

Figure 1
The sequence of events that occurs when two identical waves travelling in opposite directions interfere. Diagrams **(a)**, **(c)**, and **(e)** show destructive interference occurring at every point in the medium. Diagrams **(b)** and **(d)** show constructive interference. Diagram **(f)** shows the resulting "standing wave interference pattern" created as the waves continually pass through one another.

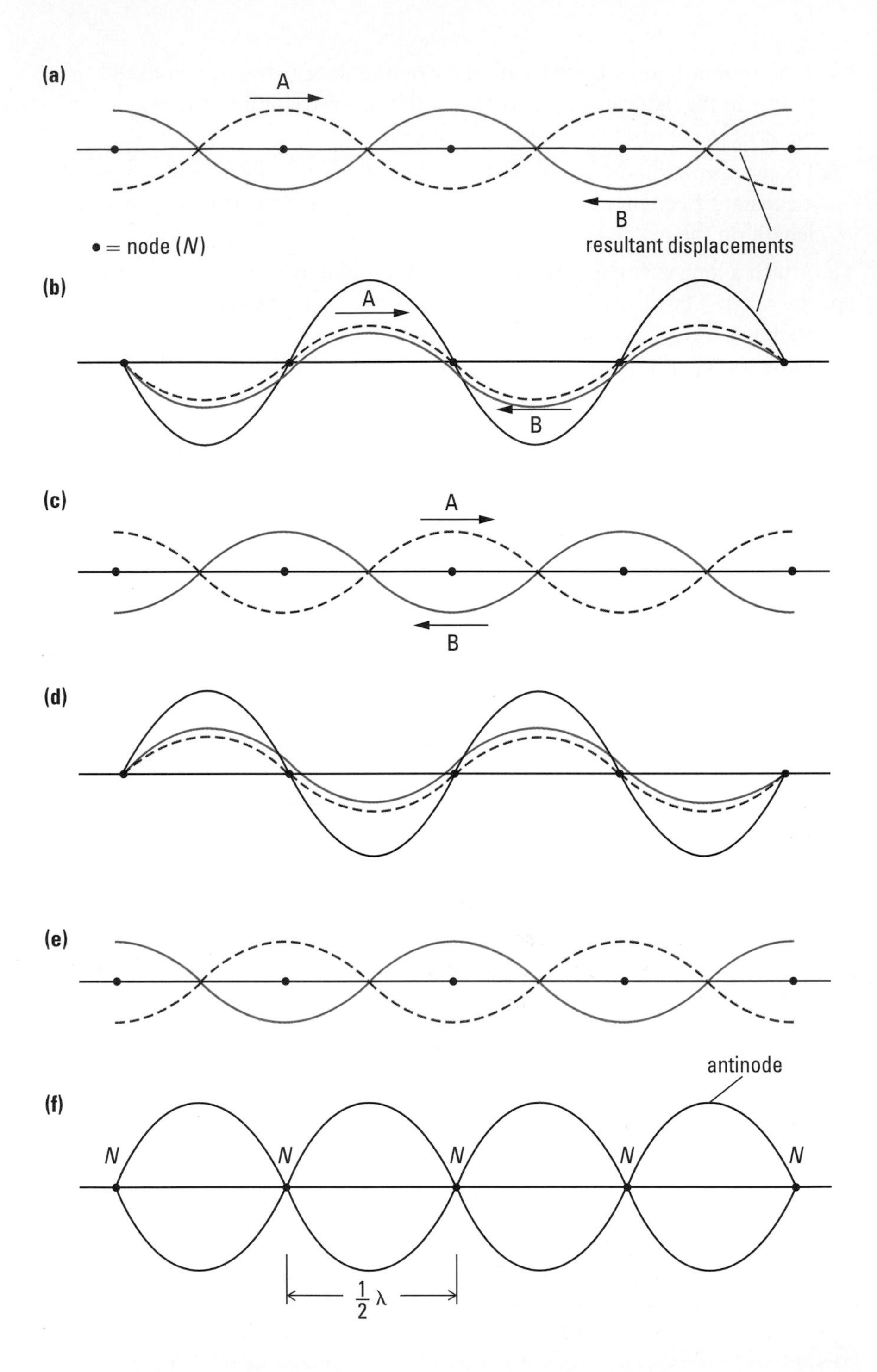

Standing waves may be produced by means of a single source (**Figure 2** and **Figure 3**). Reflected waves, for instance, will interfere with incident waves, producing standing waves. Since the incident waves and the reflected waves have the same source and cross the same medium with little loss in energy, they have the same frequency, wavelength, and amplitude. The distance between nodal points may be altered by changing the frequency of the source. However, for a given length of rope or of any other medium, only certain wavelengths are capable of maintaining the standing wave interference pattern because the reflecting ends must be nodes.

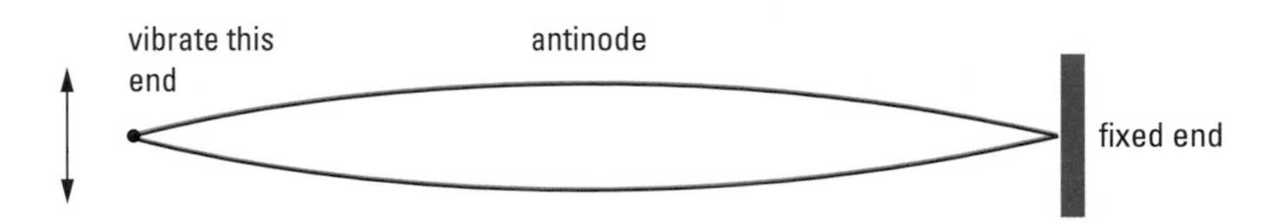

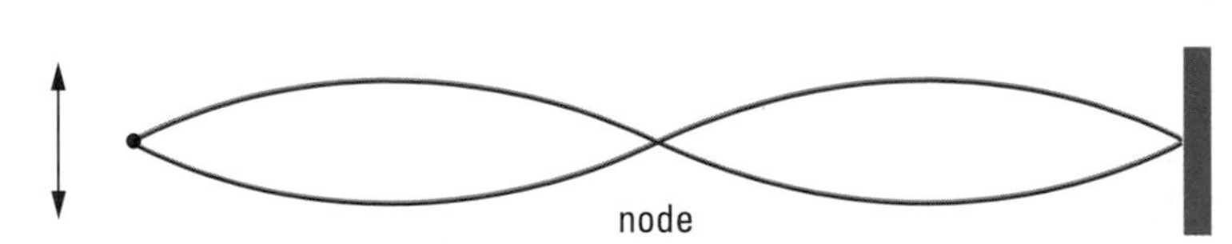

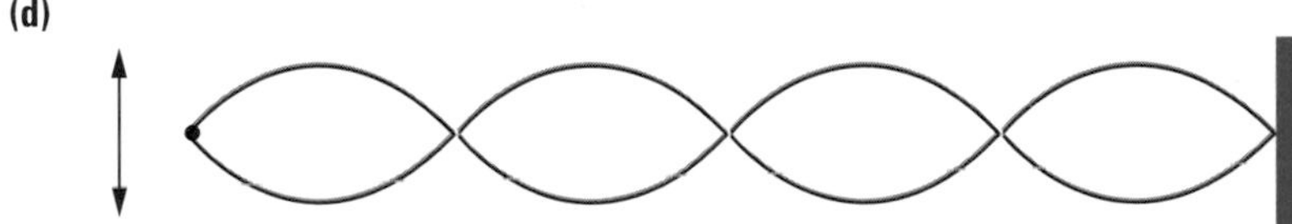

Figure 2
Producing standing waves
(a) Low frequency, long wavelength, zero nodes between ends
(b) Higher frequency, shorter wavelength, one node between ends
(c) Two nodes between ends
(d) Three nodes between ends

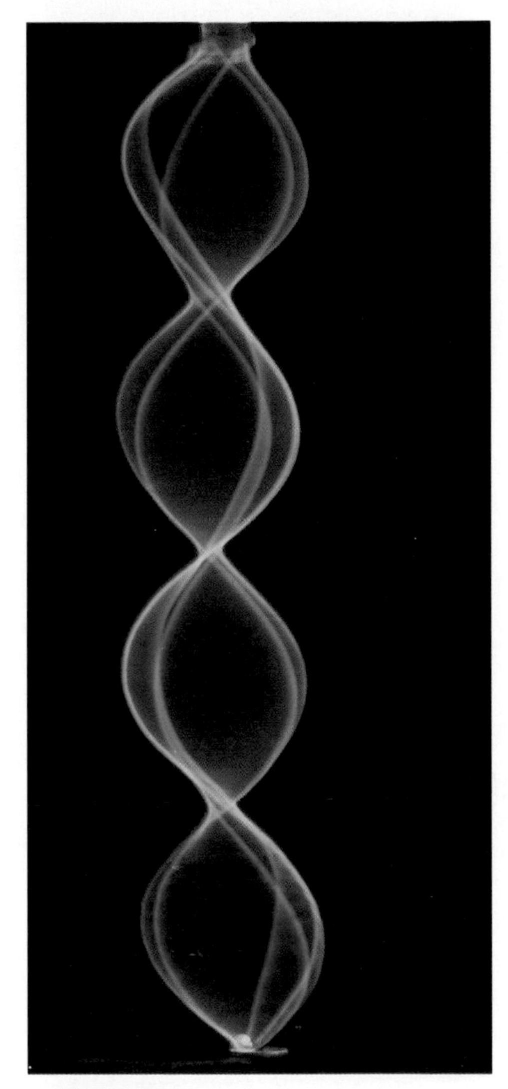

Figure 3
Standing waves in a vibrating string.

Sample Problem

The distance between two successive nodes in a vibrating string is 10 cm. The frequency of the source is 30 Hz. What is the wavelength? What is the speed of the waves? (Assume two significant digits.)

Solution

$f = 30$ Hz
$\lambda = ?$
$v = ?$

The distance between successive nodes is $\frac{1}{2}\lambda$.

$$\frac{1}{2}\lambda = 10 \text{ cm}$$
$$\lambda = 20 \text{ cm}$$

$$v = f\lambda$$
$$= (30 \text{ Hz})(20 \text{ cm})$$
$$v = 6.0 \times 10^2 \text{ cm/s}$$

Practice

Understanding Concepts

1. A standing wave interference pattern is produced in a rope by a vibrator with a frequency of 28 Hz. If the wavelength of the waves is 9.5 cm, what is the distance between successive nodes?
2. The distance between the second and fifth nodes in a standing wave is 59 cm. What is the wavelength of the waves? What is the speed of the waves if the source has a frequency of 25 Hz?

Answers

1. 19 cm
2. 39 cm; 9.8×10^2 cm/s

Answers

3. (a) 50.0 cm
   (b) 1.0×10^4 cm/s

3. The distance between adjacent nodes in the standing wave pattern in a piece of string is 25.0 cm.
   (a) What is the wavelength of the wave in the string?
   (b) If the frequency of the vibration is 2.0×10^2 Hz, calculate the speed of the wave.

Applying Inquiry Skills

4. You have one long piece of rope and a stopwatch. Devise a procedure to calculate the period and frequency of a standing wave with one, two, and three antinodes.

SUMMARY Standing Waves

- For total destructive interference to occur, the waves interfering must have identical wavelengths and amplitudes.
- Nodes are points in a medium that are continuously at rest; that is, the resultant displacement of the particles at these points is always zero.
- Points in a medium at which constructive interference occurs are called antinodes.
- A stationary interference pattern of successive nodes and antinodes in a medium is called a standing wave interference pattern.
- The distance between successive nodes or antinodes in a standing wave interference pattern is one-half the wavelength of the interfering waves.

Section 6.8 Questions

Understanding the Concepts

1. Draw a scale diagram of a standing wave pattern on an 8.0-m rope with four antinodes between the ends. What is the wavelength of the waves that produced the pattern?
2. The speed of a wave in a 4.0 m rope is 3.2 m/s. What is the frequency of the vibration required to produce a standing wave pattern with
   (a) 1 antinode?
   (b) 2 antinodes?
   (c) 4 antinodes?
3. Standing waves are produced in a string by two waves travelling in opposite directions at 6.0 m/s. The distance between the second node and the sixth node is 82 cm. Determine the wavelength and the frequency of the original waves.
4. Predict the frequencies of a 2.0-m rope that will produce a standing wave with one antinode, two antinodes, and three antinodes. (The speed of the wave is 2.8 m/s.) Test your prediction by performing this experiment and recording your results.

6.9 Interference of Waves in Two Dimensions

In a one-dimensional medium, such as a spring, successive regions of constructive or destructive interference may occur, sometimes producing fixed patterns of interference called a standing wave interference pattern. What patterns of interference occur between two waves interfering in a two-dimensional medium, such as a ripple tank?

In a ripple tank, the nodal lines of destructive interference appear grey on the screen since the water is nearly level and the light is not refracted. Between the nodal lines are the regions of constructive interference. These appear as alternating bright (double crests) and dark (double troughs) areas on the screen (see **Figure 1**).

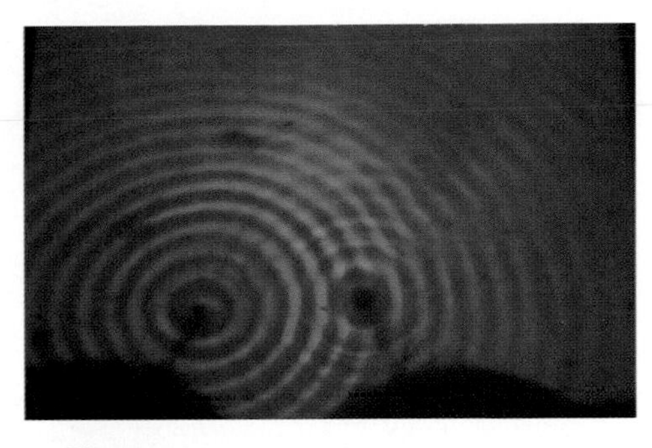

Figure 1

Two-Point Interference Pattern

When two vibrating point sources are attached to the same generator, they have identical wavelengths and amplitudes and they are also in phase. As successive crests and troughs travel out from each source, they interfere, sometimes crest-on-crest, sometimes trough-on-trough, and sometimes crest-on-trough. Thus, areas of constructive and destructive interference are produced. These areas move out from the sources in symmetrical patterns, producing nodal lines and areas of constructive interference as shown in **Figure 2**.

Although the nodal lines appear to be almost straight when they move away from the sources, they are actually curved lines in a mathematical shape called a hyperbola. When the frequency of the sources is increased, the wavelength decreases, bringing the nodal lines closer together and increasing their number. This two-point-source interference pattern is of importance in the study of the interference of sound and light waves.

Having discovered many of the properties of waves, we can apply this knowledge in the next chapter where we investigate the properties of sound, another wave phenomenon.

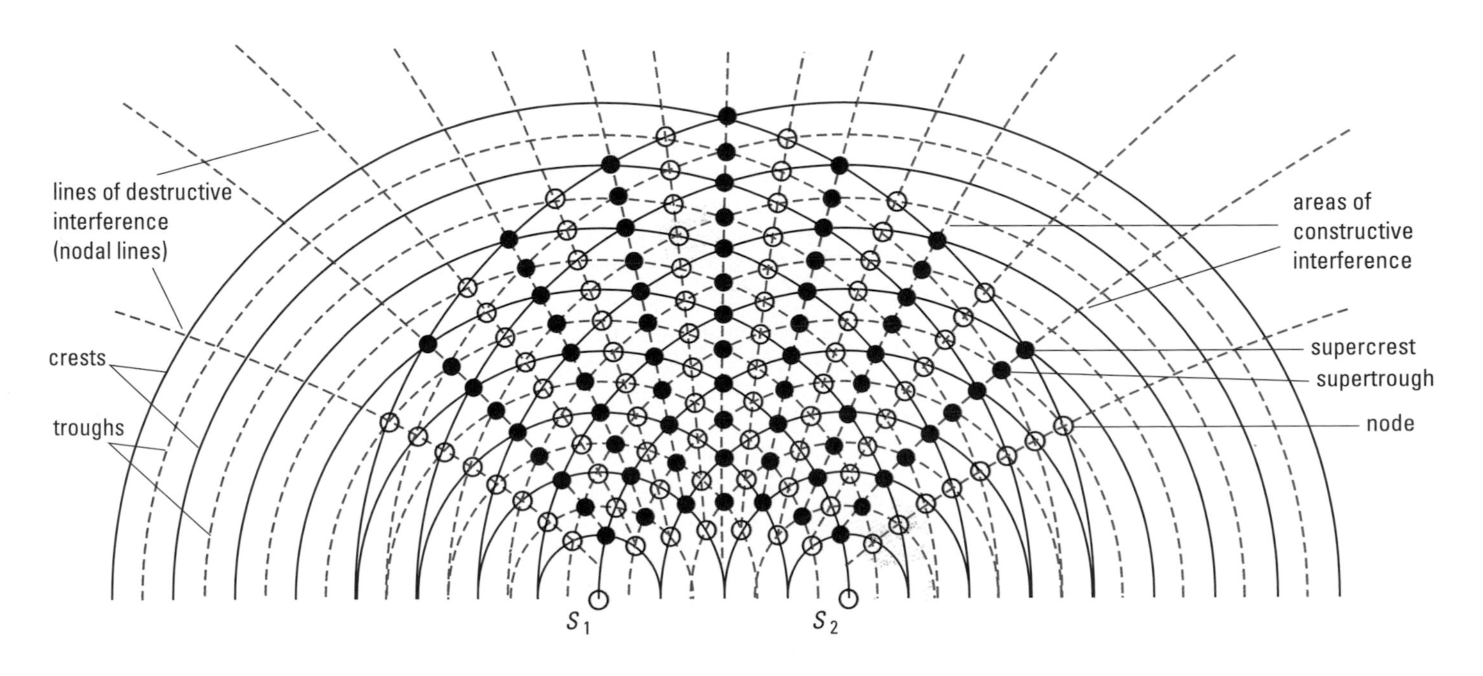

Figure 2
The interference pattern between two identical point sources (S_1 and S_2) is a symmetrical pattern of nodal lines and areas of constructive interference in the shape of hyperbolas.

Practice

Understanding Concepts

1. In a two-point interference pattern, what single line of constructive interference is not curved, but straight? Where is it located relative to the two sources?
2. If the frequency of the two sources was increased, why would it be more difficult to see the nodal line?
3. Assuming that frequency remains constant, predict what would happen to the number of nodal lines if the distance between the sources is increased. Explain.

Applying Inquiry Skills

4. In the middle of a page in your notebook, mark two points 4.0 cm apart. Using a compass, draw in circular wavefronts originating at the points with 2.0-cm wavelengths. Use solid lines for crests and dotted lines for troughs. Mark all the nodes and points of maximum constructive interference. Draw in the nodal lines. Describe the shape of the nodal lines.

SUMMARY Two-Point Interference Pattern

- The interference pattern between two identical point sources, vibrating in phase, is a symmetrical pattern of alternating areas of destructive and areas of constructive interference radiating out from the point sources.

Section 6.9 Questions

Understanding Concepts

1. Assume that the sources in **Figure 2** are loudspeakers producing exactly the same (single-frequency) sound. What would you hear if you were standing on any one of the nodal lines? Explain.
2. In a two-point interference pattern, are all lines of constructive interference curved? Explain.

Applying Inquiry Skills

3. Describe an experimental setup that would yield a two point interference pattern, using one source and a reflecting barrier, as in **Figure 2**.

Chapter 6 Summary

Key Expectations

Throughout this chapter, you have had opportunities to

- define and describe the concepts and units related to mechanical waves (e.g., longitudinal wave, transverse wave, cycle, period, frequency, amplitude, phase, wavelength, speed, superposition, constructive and destructive interference, standing waves, resonance); (6.1, 6.2, 6.3, 6.4, 6.5, 6.6, 6.7, 6.8, 6.9)
- describe and illustrate the properties of transverse and longitudinal waves in different media, and analyze the speed of waves travelling in those media in quantitative terms; (6.2, 6.3)
- explain and graphically illustrate the principle of superposition, and identify examples of constructive and destructive interference; (6.6, 6.9)
- analyze the components of resonance and identify the conditions required for resonance to occur in vibrating objects; (6.7)
- identify the properties of standing waves and explain the conditions required for standing waves to occur; (6.8)
- draw, measure, analyze, and interpret the properties of waves (e.g., reflection, diffraction, and interference, including interference that results in standing waves) during their transmission in a medium and from one medium to another, and during their interaction with matter; (6.4, 6.5, 6.6, 6.8, 6.9)

Key Terms

wave
periodic motion
transverse vibration
longitudinal vibration
torsional vibration
cycle
frequency
hertz
period
amplitude
in phase
out of phase
transverse wave
crest
positive pulse
trough
negative pulse
periodic waves
pulse
wavelength
longitudinal wave
compression
rarefaction
universal wave equation
fixed-end reflection
free-end reflection
partial reflection
wavefront
wave ray
normal
focal point
refraction
diffraction
wave interference
destructive interference
constructive interference
supercrest
supertrough
principle of superposition
resonance
mechanical resonance
sympathetic vibration
standing wave
node
nodal point
antinode

Make a Summary

The main concepts in this chapter relate to the properties of vibrations and waves. Summarize as many ideas from this chapter as possible by drawing labelled diagrams of pendulums, springs, ropes, and water waves. On the diagrams, include Key Terms and equations wherever you can.

Reflect on your Learning

Revisit your answers to the Reflect on Your Learning questions at the beginning of this chapter.

- How has your thinking changed?
- What new questions do you have?

Chapter 6 Review

Understanding Concepts

1. Classify the following examples as transverse, longitudinal, or torsional vibrations.
   (a) A ball attached to a spring is bouncing up and down.
   (b) An agitator moves in an upright washing machine.
   (c) A child swings on a swing.
   (d) A child sits on a swing with the ropes twisted, so the child is spinning as the ropes are unwinding.
2. The mass of a pendulum moves a total horizontal distance of 14 cm in one cycle. What is the amplitude of the vibration?
3. A Canada goose flaps its wings 16 times in 21 s. What is the frequency and period of the wing beat?
4. The world record for pogo jumping is 122 000 times in 15 h 26 min.
   (a) What type of vibration is occurring in the spring of the pogo stick?
   (b) Calculate the average period of vibration.
   (c) Calculate the average frequency of vibration.
5. Points A, B, C, D, and E are marked on a stretched rope (**Figure 1**). Pulses X and Y have the same shape and size, and travel through this rope at the same speed.
   (a) Which point(s) would have an instantaneous upward motion if both pulses moved to the right?
   (b) Which point(s) would have an instantaneous downward motion if pulse X moved to the right as pulse Y moved to the left?
   (c) Which point(s) would have no motion, provided the pulses moved in opposite directions?

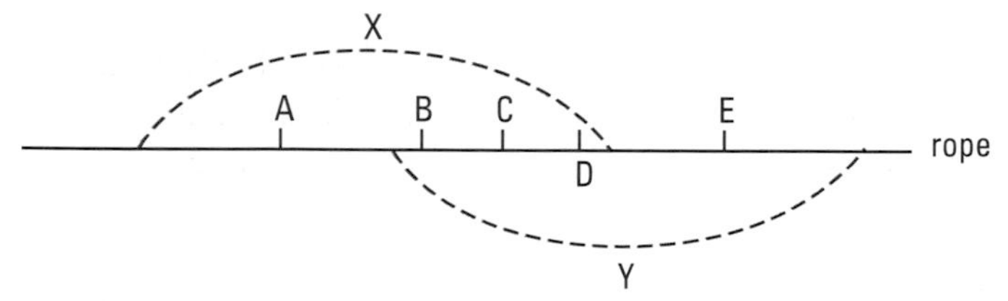

Figure 1

6. A sonar signal (sound wave) of 5.0×10^2 Hz travels through water with a wavelength of 3.0 m. What is the speed of the signal in water?
7. Water waves with a wavelength of 6.0 m approach a lighthouse at 5.6 m/s.
   (a) What is the frequency of the waves?
   (b) What is their period?
8. The distance between successive crests in a series of water waves is 5.0 m; the crests travel 8.6 m in 5.0 s. Calculate the frequency of a block of wood bobbing up and down in the water.
9. Two people are fishing from small boats located 28 m apart. Waves pass through the water, and each person's boat bobs up and down 15 times in 1.0 min. At a time when one boat is on a crest, the other one is in a trough, and there is one crest between the two boats. What is the speed of the waves?
10. The wavelength of a water wave is 3.7 m and its period is 1.5 s.
    (a) What is the speed of the wave?
    (b) What is the time required for the wave to travel 1.0×10^2 m?
    (c) What is the distance travelled by the wave in 1.00 min?
11. A boat at anchor is rocked by waves whose crests are 32 m apart and whose speed is 8.0 m/s. What is the interval of time between crests striking the boat?
12. The wavelength of a water wave is 8.0 m and its speed is 2.0 m/s. How many waves will pass a fixed point in the water in 1.0 min?
13. When a car moves at a certain speed, there is an annoying rattle. If the driver speeds up or slows down slightly, the rattle disappears. Explain.
14. A 6.0-m rope is used to produce standing waves. Draw a scale diagram of the standing wave pattern produced by waves having a wavelength of
    (a) 12 m (b) 6.0 m (c) 3.0 m
15. Standing waves are produced in a string by sources at each end with a frequency of 10.0 Hz. The distance between the third node and the sixth node is 54 cm.
    (a) What is the wavelength of the interfering waves?
    (b) What is their speed?
16. You send a pulse down a spring that is attached to a second spring with unknown properties. The positive pulse returns to you as a positive pulse, but with a smaller amplitude. Is the speed of the waves faster or slower in the second string? Explain your reasoning.
17. When a stone is dropped into water, the resulting ripples spread farther and farther out, getting smaller and smaller in amplitude until they disappear. Why does the amplitude eventually decrease to zero?
18. When standing waves are produced in a string, total destructive interference occurs at the nodes. What has happened to the wave energy?

19. Each diagram in **Figure 2** shows an incident pulse travelling toward one end of a rope. Draw a diagram showing the reflected pulse in each case.

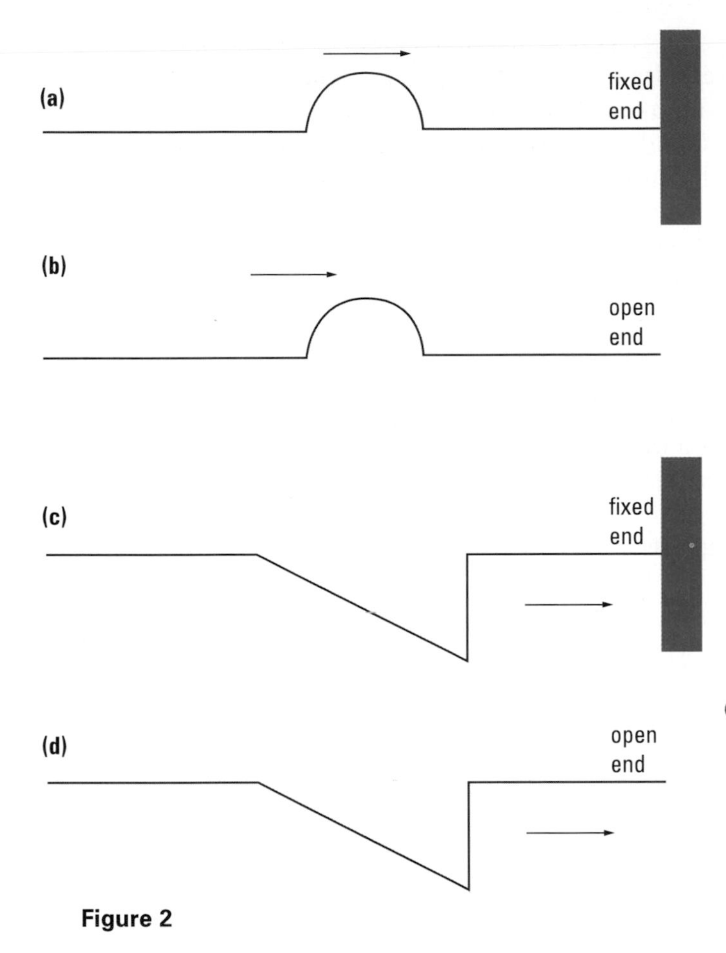

Figure 2

20. Apply the principle of superposition to draw the resultant shape when each of the sets of pulses shown in **Figure 3** interferes. (Draw the diagrams so that the horizontal midpoints of the pulses coincide.)

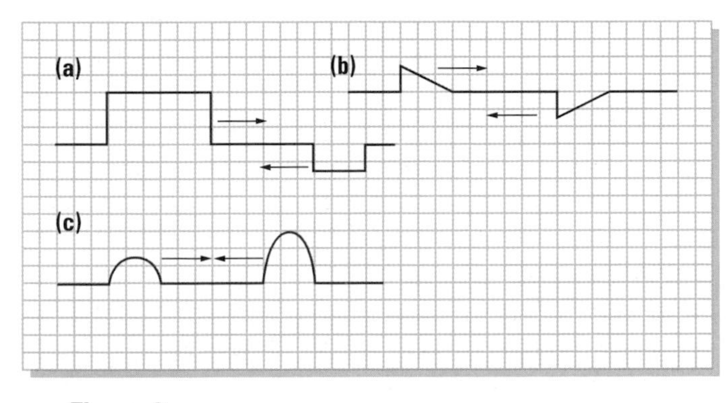

Figure 3

Applying Inquiry Skills

21. Draw a diagram illustrating structural features an offshore oil-drilling platform might possess to withstand very high waves.

Making Connections

22. A tsunami is a fast-moving surface wave that travels on an ocean after an underwater earthquake or volcanic eruption. In the deep ocean the wavelength might be over 250 km, the amplitude only about 5 m, and the speed up to 800 km/h. The wave might pass under a ship and not even be noticed, but it can strike a shore with an amplitude of perhaps 30 m and do severe damage. How is it possible that a wave that is seemingly harmless at sea can do such damage onshore?

23. The energy of earthquakes is transmitted by waves that are both longitudinal and transverse. Only one of these types of waves travels in the interior of Earth. Do some research to answer the following questions. Follow the links for Nelson Physics 11, Chapter 6 Review.
    (a) Which type of wave is it?
    (b) Why is this type of wave the only one that travels in the interior of Earth?

GO TO www.science.nelson.com

24. Find out the frequency at which your favourite local television station broadcasts its signal. If the speed of the electromagnetic waves emitted by the station tower is 3.0×10^8 m/s, what is the wavelength of the waves?

Exploring

25. Prepare a written report explaining why the development of the pendulum was a scientific landmark. Follow the links for Nelson Physics 11, Chapter 6 Review.

GO TO www.science.nelson.com

26. At the 2000 Summer Olympics in Sydney, Australia, the construction of the swimming pool and the lane ropes led to less water turbulence for the swimmers, improving their times. Find out what changes were made and why they reduced turbulence. Follow the links for Nelson Physics 11, Chapter 6 Review. Summarize your findings in a report with a maximum length of 200 words.

GO TO www.science.nelson.com

Chapter

7

Properties of Sound Waves

In this chapter, you will be able to

- describe and understand sound, its transmission, and its production
- understand why the speed of sound varies
- give examples of sound intensity on the decibel scale
- explain how sounds are produced and transmitted in nature
- understand sound interference
- explain the Doppler effect

The characteristics and nature of waves were the topics studied in Chapter 6. Since sound waves form one of our major sensory links to the world, it is important that we understand the properties of sound waves and hearing. Without sound we would be denied the ability to communicate by speech with one another, the pleasures of musical sound, and the ability to know when someone has approached us from behind.

Imagine you are asked to describe sound and hearing to someone who has been unable to hear since birth. Perhaps you would begin by describing the sounds of voices, music, radio, television, animals, and machines. You might then describe how sound is produced and transmitted. You could explain the relationship between vibrations and waves as discussed in the previous chapter. You could describe how a wineglass could be shattered by the amplified sound of a human voice (**Figure 1**). Finally, you could describe the physical functions of the human ear. In this chapter, we will study the characteristics of sound waves, and how the human ear receives and interprets sound energy.

Reflect on your Learning

1. What is sound?
2. Explain how the human ear receives and interprets sound.
3. Why does the speed of sound differ in different media?
4. How can echoes be used to locate objects?
5. Why do planes make a large noise when they break the sound barrier?

Throughout this chapter, note any changes in your ideas as you learn new concepts and develop your skills.

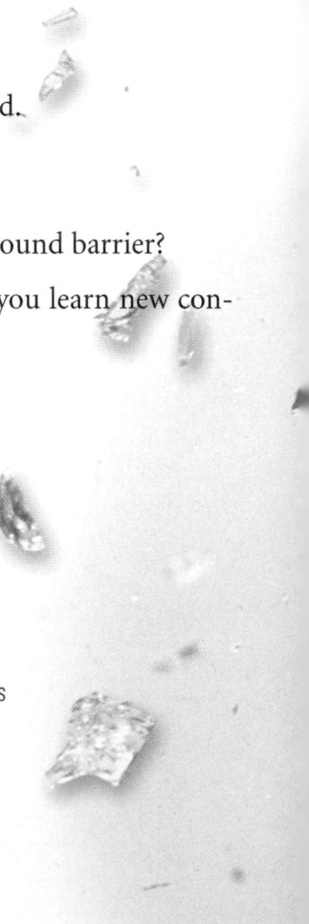

Figure 1
The energy of amplified sound waves causes the wine glass to shatter.

Vibrating Tuning Fork

Tuning forks are often used as a source of sound energy in scientific activities. You will use a tuning fork, a Styrofoam cup, a rubber hammer, and a pith ball to look at some properties of sound waves.

(a) Strike a low-frequency tuning fork with a rubber hammer, and touch the prongs to the surface of water in a cup. Describe what happens and why.

(b) Touch a vibrating tuning fork to a suspended pith ball (**Figure 2**). Describe what happens and why.

Figure 2
What happens when the vibrating tuning fork is brought close to a suspended pith ball?

Figure 1

sound: a form of energy, produced by rapidly vibrating objects, that can be heard by the human ear.

infrasonic: any sound with a frequency lower than the threshold of human hearing (approximately 20 Hz)

ultrasonic: any sound with a frequency above the range of human hearing (approximately 20 000 Hz)

7.1 What Is Sound?

From our earliest years, we become accustomed to a great variety of sounds: our mother's voice, a telephone ringing, a kitten purring, a piano being played, a siren, a jet engine roaring, a rifle shot, the blaring of a rock band (**Figure 1**). Some of these sounds are pleasant to the ear and some are not. **Sounds** are a form of energy produced by rapidly vibrating objects. We hear sounds because this energy stimulates the auditory nerve in the human ear.

In the 18th century, philosophers and scientists debated the question, "If a tree falls in the forest and no one is there to hear it, will there be sound?"

"Of course there will," said the scientists, "because the crash of the tree is a vibrating source that sends out sound waves through the ground and the air." To them, sound was the motion of the particles in a medium caused by a vibrating object.

"Of course not," said the philosophers, "because no observer is present." To them, sound was a personal sensation that existed only in the mind of the observer.

This debate could never be resolved because one group was defining sound objectively in terms of its cause, and the other was defining it subjectively in terms of its effects on the human ear and brain. In physics, we study the transmission of sound objectively, leaving the subjective interpretation of the effects of sound waves on the human ear and brain to the philosophers.

The ears of most young people respond to sound frequencies of between 20 Hz and 20 000 Hz (**Table 1** and **Figure 2**). Frequencies of less than 20 Hz are referred to as **infrasonic** and those higher than 20 000 Hz are called **ultrasonic**.

Table 1 Frequencies of Commonly Heard Sounds

Source	Frequency (Hz)
lowest piano note	27.50
male speaking voice (average)	120
female speaking voice (average)	250
middle C of piano	261.63
A above middle C	440.00
highest piano note	4186.01

Figure 2
Range of hearing

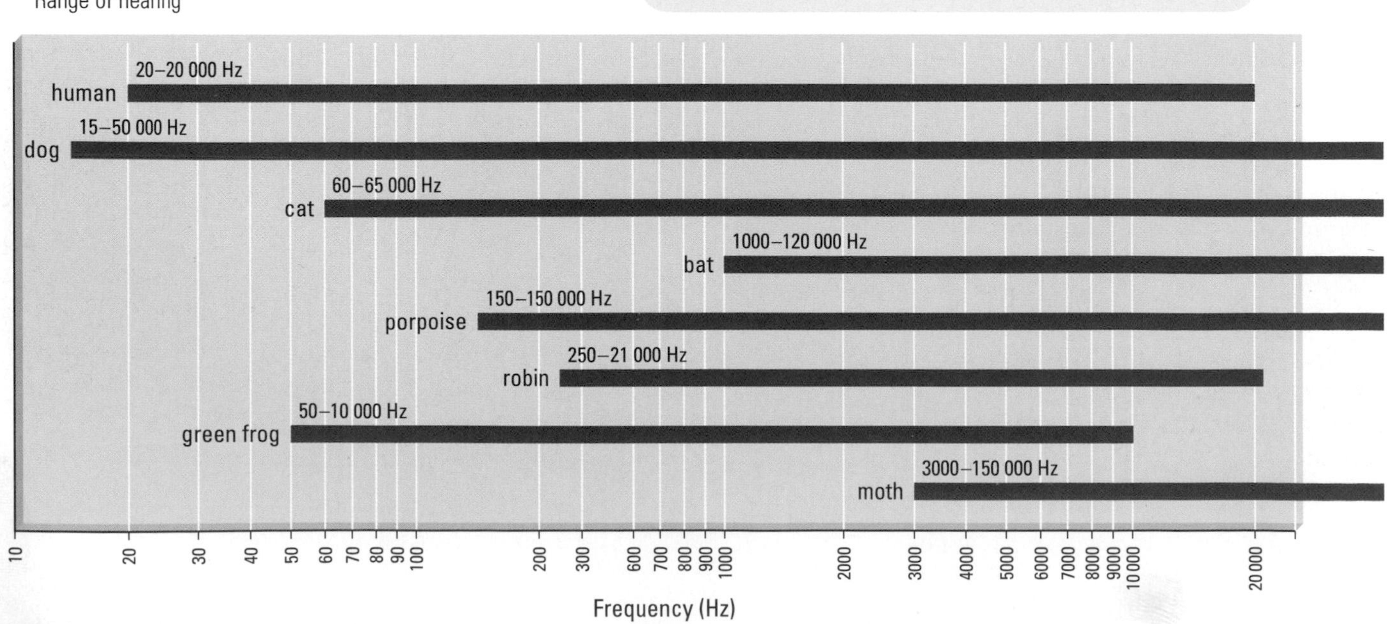

SUMMARY What Is Sound?

- All sound waves are caused by vibrations.
- Sound stimulates the auditory nerve.
- Humans can hear a range between 20 and 20 000 Hz.
- Infrasonic sounds have a frequency below 20 Hz.
- Ultrasonic sounds have a frequency above 20 000 Hz.

Section 7.1 Questions

Understanding Concepts

1. You have been hired to invent a whistle to call a trained pet dog back to its owner. This whistle must be heard by the dog, but not by humans. Predict the range of frequencies you could use for your whistle.

Making Connections

2. Artillery shells are often used to start avalanches before they become a danger to skiers. Sometimes the sound of the gun is sufficient to start an avalanche. What does this suggest to you about sound? If an artillery gun is not available, what warning signs would you post in an avalanche-prone area?
3. When some natural disasters such as earthquakes take place, some animals display unusual behaviours. Scientists are trying to find out if a particular frequency of sound precedes the earthquake.
   (a) What type of frequency do you think these scientists are trying to pinpoint for the sound preceding an earthquake?
   (b) If this preceding frequency sound wave is discovered, why do you think it was not discovered earlier?

7.2 Production and Transmission of Sound Energy

We can describe the sounds we hear in many ways. For example, leaves rustle, lions roar, babies cry, birds chirp, corks pop, and orchestras crescendo. However, all these types of sounds originate from vibrating objects. The definitions and concepts related to mechanical vibrations studied in Chapter 6 also apply to vibrations that cause sound.

Some vibrations are visible. If you pluck a guitar string (**Figure 1**) or strike a low-frequency tuning fork, you can see the actual vibrations of the object. Similarly, if you watch the low-frequency woofer of a loudspeaker system you can see it vibrating.

There are vibrations that are not visible, however. When you speak, for instance, parts of your throat vibrate. When you make a whistling sound by blowing over an empty pop bottle, the air molecules in the bottle vibrate to produce sound.

Figure 1
Visible vibrations of guitar strings

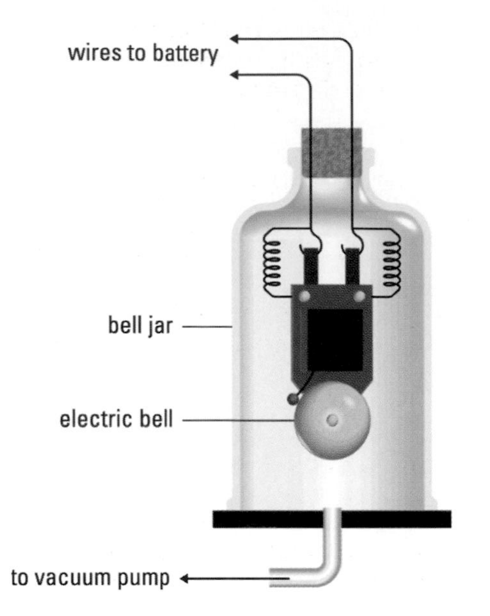

Figure 2
An electric bell in a vacuum

How Does Sound Travel?

In 1654, Otto von Guericke (1602–86), inventor of the air pump, discovered that the intensity of the sound from a mechanical bell inside a jar decreased steadily as the air was removed from the jar. We can demonstrate this effect by means of an electric bell in a bell jar from which air is drawn out by a vacuum pump (**Figure 2**). When most of the air is removed, no sound can be heard.

Von Guericke also showed that sound could travel in materials other than air. For example, in another experiment, Von Guericke learned that the sound of a ringing bell could be transmitted clearly through water. In his experiment, fish were attracted by the sound of a bell in the water. An underwater swimmer sometimes hears an approaching motorboat before a swimmer on the surface does. Most of us have experienced vibrations in the ground from a passing truck or train. If you put your ear against a steel fence, you can hear the sound of a stone being tapped against it a considerable distance away.

These examples all point to the conclusion that sound needs a material medium for its transmission. Sound cannot travel through a vacuum.

When a tuning fork is struck, it vibrates. As each tine moves, it pushes air molecules out until they bump against their neighbours. This creates a steadily moving area of collision (a compression). When the tine moves back it creates a region of emptiness (a rarefaction) into which the displaced air molecules rebound. This follows the compression outwards. In other words, sound travels out from its source as a longitudinal wave. As the tines move back and forth, rarefactions and compressions follow one another as the sound waves travel through the air away from the tuning fork (**Figure 3**).

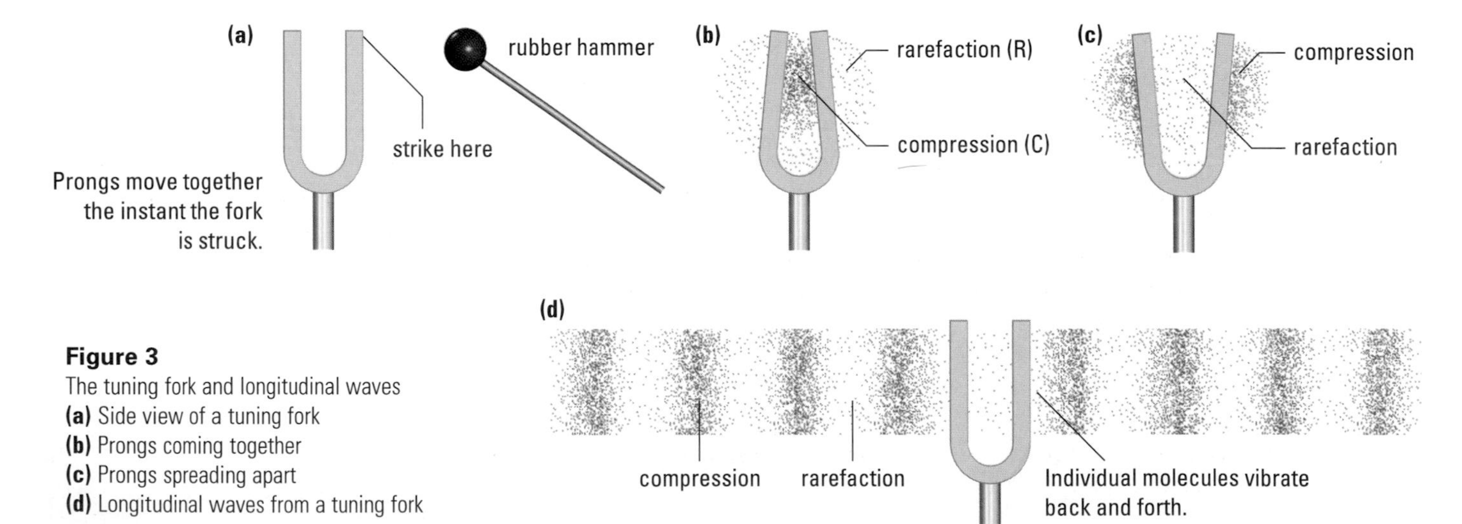

Figure 3
The tuning fork and longitudinal waves
(a) Side view of a tuning fork
(b) Prongs coming together
(c) Prongs spreading apart
(d) Longitudinal waves from a tuning fork

C R C R C
high
air pressure
normal
distance
low

Figure 4
Schematic representation of the density of air molecules

Note that the particles of air only vibrate locally; they do not move from the source to the receiver. In sound waves, a compression is an area of higher than normal air pressure, and a rarefaction is an area of lower than normal air pressure. **Figure 4** illustrates these variations in pressure as a sound wave moves away from its source. Longitudinal waves are difficult to represent visually and it is, at times, more helpful to use a graph of pressure variations.

All the terms used to describe waves in Chapter 6 can be used to describe sound waves. For example, one wavelength is the distance between the midpoints of successive compressions or successive rarefactions (**Figure 5**). The amplitude of the sound wave describes the displacement of the air molecules from the rest position. The larger the amplitude, the louder the sound. The term **pitch** is related to the frequency of sound waves. As the pitch of a sound increases, the frequency increases; the lower the frequency, the lower the pitch. We can also use the universal wave equation to describe sound waves, as shown in the following sample problem.

pitch: the perception of the highness or lowness of a sound; depends primarily on the frequency of the sound

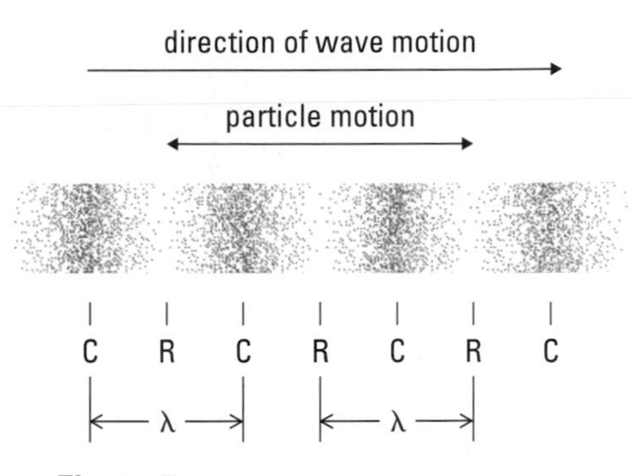

Figure 5
One wavelength of sound is the distance between the midpoints of successive compressions or rarefactions.

Sample Problem

The sound from a trumpet travels at a speed of 3.5×10^2 m/s in air. If the frequency of the note played is 3.0×10^2 Hz, what is the wavelength of the sound wave?

Solution

$v = 3.5 \times 10^2$ m/s

$f = 3.0 \times 10^2$ Hz

$\lambda = ?$

$$v = f\lambda$$

$$\lambda = \frac{v}{f} = \frac{3.5 \times 10^2 \text{ m/s}}{3.0 \times 10^2 \text{ Hz}}$$

$$\lambda = 1.2 \text{ m}$$

The wavelength is 1.2 m.

Practice

Understanding Concepts

1. An A-string on a violin is played at 8.8×10^2 Hz, and the wavelength of the sound wave is 4.1×10^{-1} m. At what speed does the sound travel in air?
2. Calculate the frequency of a sound wave from a clarinet if the speed of sound in air is 3.40×10^2 m/s, and the wavelength of the sound wave is 1.7×10^{-1} m.
3. Do compressions and rarefactions of a longitudinal sound wave travel in the same or opposite directions? Explain.
4. As mentioned in this section, pitch is different from frequency. Pitch is subjective, unlike frequency, which is objective. What is one subjective property of pitch?

Applying Inquiry Skills

5. Design a procedure to measure the speed at which a thunder cloud is travelling from point A to point B. Assume the speed of sound in air is 3.5×10^2 m/s, the cloud is travelling at a constant speed in a straight line from A to B, and the time for the lightning to travel from one point to the observer is instantaneous. You cannot see the cloud because it is nighttime, but you can see the lightning and hear the thunder at point A and point B. In your procedure, point A, point B, and the observer must be collinear (in a straight line).

Answers

1. 3.6×10^2 m/s
2. 2.0×10^3 Hz

SUMMARY Production and Transmission of Sound

- Sound originates from a vibrating source.
- Sound requires a physical medium for its transmission.
- Sound travels as a series of compressions and rarefactions.
- Pitch is related to frequency; the higher the frequency, the higher the pitch.

Section 7.2 Questions

Understanding Concepts

1. What vibrates to produce sound from the following?
   (a) banjo
   (b) drum
   (c) coach's whistle
   (d) crackling fire
2. Are the following statements true? Explain.
   (a) All sound is produced by vibrating objects.
   (b) All vibrating objects produce sound.
3. Calculate the frequency of a sound wave if its speed and wavelength are
   (a) 340 m/s and 1.13 m
   (b) 340 m/s and 69.5 cm
4. An organ pipe emits a note of 50.0 Hz. If the speed of sound in air is 350 m/s, what is the wavelength of the sound wave?
5. If a 260-Hz sound from a tuning fork has a wavelength of 1.30 m, at what speed does the sound travel?
6. You are standing on a straight road and see lightning strike the ground ahead of you. Three seconds later, you hear the thunderclap. If the speed of sound is 330 m/s, how far will you have to walk to reach the point where the lightning struck? (Assume that light travels almost instantly.)
7. An explosion occurs at an oil refinery, and several seconds later windows in an adjacent neighbourhood shatter. Why do the windows shatter?

Making Connections

8. Some helicopters have been designed to be quieter than others. Research the differences in the blades of these helicopters as compared with other "noisier" helicopters. Follow the links for Nelson Physics 11, 7.2.

   GO TO www.science.nelson.com
9. Barn owls have some feathers that are unlike those of any other bird. Research the first primary feather of a barn owl and explain how these feathers allow this owl to catch its prey with ease. Follow the links for Nelson Physics 11, 7.2.

   GO TO www.science.nelson.com

Reflecting

10. Even though we cannot see sound waves, we believe they exist. From your own personal experience, state four physical pieces of evidence observed in different situations that convince you that sound waves exist.

7.3 The Speed of Sound

Sound seems to move very quickly. However, during a thunderstorm, you see the lightning before you hear the thunder it causes. A similar phenomenon occurs at a high school track meet where the timer of a 100-m sprint stands at the finish line and watches for a puff of smoke from the starter's pistol. Shortly after the smoke is seen, the sound of the gun is heard. Light travels extremely fast (3.0×10^8 m/s in air); sound travels much more slowly.

Accurate measurements of the speed of sound in air have been made at various temperatures and air pressures. At normal atmospheric pressure and 0°C, the speed of sound in air is 332 m/s. If the air pressure remains constant, the speed of sound increases as the temperature increases. For every rise in temperature of 1°C, the speed of sound in air increases by 0.59 m/s. The speed of sound in air at normal atmospheric pressure can be calculated using the equation

$$v = 332 \text{ m/s} + \left(0.59 \frac{\text{m/s}}{\text{°C}}\right)t$$

where t is the air temperature in degrees Celsius. If the temperature drops below 0°C, the speed of sound in air is less than 332 m/s.

DID YOU KNOW ?

Calculating the Speed of Sound in Air

The speed of sound in air was first accurately measured in 1738 by members of the French Academy. Cannons were set up on two hills approximately 29 km apart. By measuring the time interval between the flash of a cannon and the "boom," the speed of sound was calculated. Two cannons were fired alternately to minimize errors due to the wind and to delayed reactions in the observers. From their observations, they deduced that sound travels at about 336 m/s at 0°C.

Sample Problem

Calculate the speed of sound in air when the temperature is 16°C.

Solution

$t = 16°\text{C}$

$v = ?$

$$\begin{aligned} v &= 332 \text{ m/s} + \left(0.59 \frac{\text{m/s}}{\text{°C}}\right)t \\ &= 332 \text{ m/s} + \left(0.59 \frac{\text{m/s}}{\text{°C}}\right)(16\text{°C}) \\ &= 332 \text{ m/s} + 9.44 \text{ m/s} \\ v &= 341 \text{ m/s} \end{aligned}$$

The speed of sound in air at 16°C is 341 m/s.

Practice

Understanding Concepts

1. What is the speed of sound in air when the temperature is
   (a) 21°C? (b) 24°C? (c) −35°C?
2. Rearrange the equation $v = 332 \text{ m/s} + \left(0.59 \frac{\text{m/s}}{\text{°C}}\right)t$ to express t by itself.
3. A violin string is vibrating at a frequency of 440 Hz. How many vibrations does it make when the sound produced travels 664 m through air at temperature 0°C?
4. A 200-m dash along a straight track was timed at 21.1 s by a timer located at the finish line who used the flash from the starter's pistol to start the stopwatch. If the air temperature was 30.0°C, what would the time have been if the timer had started the watch upon hearing the sound of the gun?

Answers

1. (a) 344 m/s
   (b) 346 m/s
   (c) 311 m/s
3. 880
4. 20.5 s

Activity 7.3.1

Measuring the Speed of Sound Outside

To perform an experiment outside to determine the speed of sound in air, you will need a loud source of sound such as two pieces of hardwood to be struck together, a stopwatch, and a thermometer.

Procedure

1. Locate a high wall with a clear space at least 150 m deep in front of it.
2. Clap two boards together and listen for the echo. Repeat until you have determined the approximate time interval between the original sound and the echo.
3. Now clap the boards after the echo so that the time between the previous clap and the echo is the same as the time between the echo and the next clap. In other words, clap the boards so that a regular rhythm is set up: clap ... echo ... clap ... echo.
4. When you have achieved the correct timing, have your partner record, with the stopwatch, the number of seconds required for 20 or more clap intervals. (Remember to start counting with zero!)
5. Determine the average interval between claps. This interval is equal to the time taken for the sound to travel four times the distance between you and the wall.
6. Measure the distance to the wall in metres and the temperature of the air in degrees Celsius.
7. Calculate the speed of sound in air using two different methods.

Analysis

(a) How does your value for the speed of sound in air compare with the value you would expect for air temperature when you collected your data? What was the percentage difference between the two values?

Investigation 7.3.1

Measuring the Speed of Sound in the Classroom

INQUIRY SKILLS

- ● Questioning
- ○ Hypothesizing
- ● Predicting
- ○ Planning
- ● Conducting
- ● Recording
- ● Analyzing
- ● Evaluating
- ● Communicating

Now that you have measured the speed of sound outside, we can apply the same concepts to the measurement of the speed of sound in the classroom. But, in the classroom the space is much smaller and the apparatus used to measure the reflected sound needs to be much more accurate. This is achieved by using an oscilloscope and microphone or a microphone connected to the appropriate computer interface. Your instructor will give you specific instructions on the use of the equipment.

Question

What is the speed of sound in the air in the classroom?

Prediction

(a) Predict the speed of sound in the classroom. You may use the equation for the speed of sound given at the beginning of this section.

Materials

cardboard mailing tube (closed at one end)
thermometer
button microphone
tape measure
oscilloscope and amplifier, or computer interface

Procedure

1. The tube can be mounted securely either horizontally or vertically.
2. Place the button microphone securely at the open end of the tube.
3. Prepare the equipment so the microphone can measure both the source of the sound and the reflected sound.
4. Snap your fingers or create a sharp sound adjacent to the microphone.
5. Record the time between the incident and reflected sounds.
6. Repeat steps 3 and 4 for at least three more trials.
7. Eliminate any aberrant readings and, if necessary, do more trials.
8. Measure the distance from the microphone to the bottom of the tube.
9. Measure the temperature of the air in the tube.

Analysis

(b) Find the average value for the time it takes the sound to travel up and down the tube.
(c) Calculate the speed of sound in the tube.

Evaluation

(d) Determine the percentage difference between your predicted value and the experimental value.
(e) Account for the sources of error in this investigation and explain how they could affect the results.
(f) Suggest changes to the procedure that would help reduce error.

Synthesis

(g) How would you use the same apparatus to measure the speed of sound in carbon dioxide? Try this investigation using the same equipment and dry ice.
(h) Design and carry out an experiment to measure the speed of sound in wood, using the same equipment, the wooden top of a lab desk, and a small hammer.

The Speed of Sound in Various Materials

Sound can be transmitted through any medium—gas, liquid, or solid. Children at play may discover that sound travels very easily along a metal fence. Swimmers notice that they can hear a distant motorboat better with their ears under the water than in the air. In both examples, sound is travelling in a material other than air.

Table 1 The Speed of Sound in Common Materials

State	Material	Speed at 0°C (m/s)
solid	aluminum	5104
	glass	5050
	steel	5050
	maple wood	4110
	bone (human)	4040
	pine wood	3320
solid/liquid	brain	1530
liquid	fresh water	1493 (at 25°C)
	sea water	1470 (depends on salt content)
	alcohol	1241
gas (at atmospheric pressure)	hydrogen	1270
	helium	970
	nitrogen	350 (at 20°C)
	air	332
	oxygen	317
	carbon dioxide	258

Sound travels most rapidly in certain solids, less rapidly in many liquids, and quite slowly in most gases. Thus, the speed of the sound depends not only on the temperature of the material, but also on the characteristic properties of the material. **Table 1** lists the speed of sound in different materials.

Scientists use knowledge of the speed of sound in various materials to search for oil and minerals, study the structure of the earth, and locate objects beneath the surface of the sea, among many other applications.

SUMMARY The Speed of Sound

- The speed of sound in air at 0°C is 332 m/s.
- The speed of sound increases when the temperature increases.
- The speed of sound can be measured using echoes.
- The speed of sound changes if the medium changes.

Section 7.3 Questions

Understanding Concepts

1. How much time is required for sound to travel 1.4 km through air if the temperature is 30.0°C?
2. A 380-Hz tuning fork is struck in a room where the air temperature is 22°C.
   (a) What is the speed of sound in the room?
   (b) What is the wavelength of the sound from the tuning fork?
3. How long does it take sound to travel 2.0 km in
   (a) aluminum? (b) hydrogen, at 0°C ?
4. Compare the time it takes sound to travel 6.0×10^2 m in steel with the time to travel the same distance in air, at 0°C.
5. A fan at a baseball game is 100.0 m from home plate. If the speed of sound is 350 m/s, how long after the batter actually hits the ball does the fan hear the crack? Assume that there is no wind.
6. A lightning flash is seen 10.0 s before the rumble of the thunder is heard. Find the distance to the lightning flash if the temperature is 20°C.
7. A pistol is used at the starting line to begin a 5.0×10^2-m race along a straight track. At the finish line, a puff of smoke is seen and 1.5 s later the sound is heard. What is the speed of the sound?
8. A vibrating 400.0-Hz tuning fork is placed in fresh water. What is the frequency in hertz and the wavelength in metres
   (a) within the water at 25°C?
   (b) when the sound waves move into the air at 25°C?

Making Connections

9. Marchers in a parade sometimes find it difficult to keep in step with the band if it is some distance from them. Explain why. How can this be avoided?

Reflecting

10. Have there been any situations where you have been in the same place outdoors and the speed of sound on one occasion seems faster or slower than on another occasion? Describe the conditions of both occasions and explain why there was a difference in the speed of sound.

7.4 The Intensity of Sound

There is a difference in loudness between a soft whisper and the roar of nearby thunder. However, the loudness of a sound you hear is a subjective evaluation that depends on several factors, including the objective quantity known as intensity.

sound intensity: power of a sound per unit area in watts per metre squared

Frequency, wavelength, and speed are all properties of sound that can be measured accurately. **Sound intensity** is more difficult to measure because the amount of energy involved is small in comparison with other forms of energy and because the potential range of sound intensity is great. For example, the thermal energy equivalent of the sound energy emitted over a 90-minute period by a crowd of 50 000 at a football game is only enough to heat one cup of coffee! A stereo amplifier with a maximum power output of 50 W per channel has an actual sound output from the speakers of less than 2.5 W—more than enough to fill an auditorium with sound. Note that sound energy depends on the vibrational energy of the source.

Sounds audible to humans can vary in intensity from the quietest whisper (10^{-12} W/m^2) to a level that is painful to the ear (10 W/m^2)—a difference of a factor of 10^{13}. One unit used to measure the intensity level of sound is the bel (B), named after Alexander Graham Bell (**Figure 1**). The **decibel** (dB) is more common than the bel (1 dB = 10^{-1} B). On the decibel scale, 0 dB is the threshold of hearing (10^{-12} W/m^2). The scale is not linear, but is a logarithmic scale. Every change of 10 units on the decibel scale represents a tenfold effect on the intensity level. For example, a sound 10 times more intense than 0 dB is 10 dB, a sound 100 times more intense than 0 dB is 20 dB, and a sound 1000 times more intense than 0 dB is 30 dB. The level of sound that is painful to the human ear (130 dB) is 10^{13} times more intense than the level at the threshold of hearing. The sound intensity levels for common sources are listed in **Table 1**.

Figure 1
Alexander Graham Bell was born in Scotland in 1847 and moved to Canada in 1870. One year later, he moved to Boston, where he developed a visible speech system for the deaf. Working with Tom Watson, he invented the telephone in 1876. Bell died in Baddeck, Nova Scotia, in 1922.

decibel: (dB) unit used to measure sound intensity level (0 dB = 10^{-12} W/m^2)

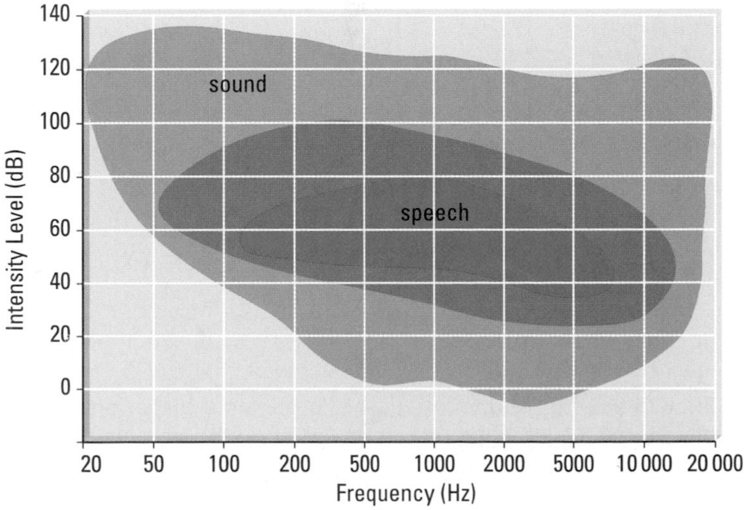

The loudness of the sounds humans perceive relates to the intensity of the sound. However, the two measures are not the same because the human ear does not respond to all frequencies equally. **Figure 2** shows a graph of sound level as a function of frequency. From the graph, it is evident that the average human ear is most sensitive to sound frequencies between about 1000 Hz and 5000 Hz. Lower frequencies must have a higher sound level or intensity to be heard.

Figure 2
Sensitivity of the average human ear to different frequencies

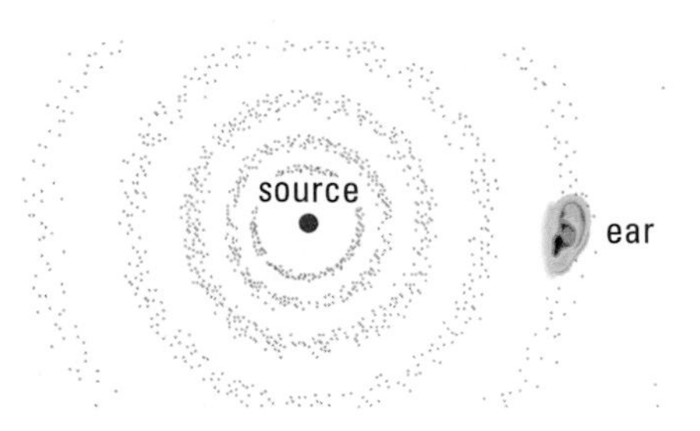

Figure 3
As a sound wave moves out from a source, its energy is spread more and more thinly.

Table 1 Sound Intensity Levels for Various Sources

Source	Intensity level (dB)	Intensity (W/m²)
threshold of hearing	0	10^{-12}
normal breathing	10	10^{-11}
average whisper at 2 m	20	10^{-10}
empty theatre	30	10^{-9}
residential area at night	40	10^{-8}
quiet restaurant	50	10^{-7}
two-person conversation	60	10^{-6}
busy street traffic	70	10^{-5}
vacuum cleaner	80	10^{-4}
at the foot of Niagara Falls	90	10^{-3}
loud stereo in average room	90	10^{-3}
passing subway train	90	10^{-3}
maximum level in concert hall (13th row)	100	10^{-2}
pneumatic chisel	110	10^{-1}
maximum level at some rock concerts	120	1
propeller plane taking off	120	1
threshold of pain	130	10^{1}
military jet taking off	140	10^{2}
wind tunnel	150	10^{3}
space rocket	160	10^{4}
instant perforation of the eardrum	160	10^{4}

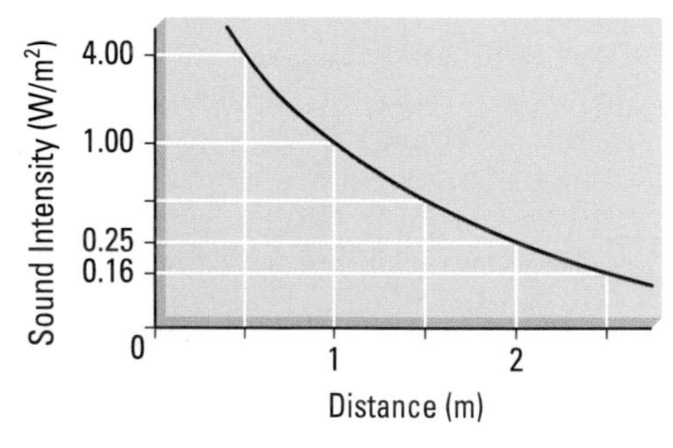

Figure 4
Graph of sound intensity versus distance

The intensity of a sound received by the human ear depends on the power of the source and the distance from the source. Like water waves, sound waves moving out from a point source spread their wave energy over an increasing area (**Figure 3**). Thus, the intensity of a sound decreases as the wave moves from the source to the receiver (**Figure 4**). The reading on the decibel scale also decreases as the distance increases. For example, a reading of 100 dB at 1 m becomes 60 dB at 100 m.

Practice

Understanding Concepts

1. Why are pitch and frequency similar to loudness and intensity?
2. Normal breathing has an intensity level of 10 dB. What would the intensity level be if a sound were
   (a) 10 times more intense?
   (b) 100 times more intense?
3. A bell is rung at a sound intensity of 70 dB. A trumpet is blown at an intensity level that is greater by a factor of 10^3. What is the intensity level of the trumpet?
4. Explain why the sound intensity of a person whispering in your ear is the same as a jet flying at an altitude of approximately 6000 m.

Answers

2. (a) 20 dB
   (b) 30 db
3. 100 dB

SUMMARY Sound Intensity

- Sound intensity is measured in watts/metre squared, and intensity level is measured in decibels.
- For humans the threshold of hearing is 0 dB and the threshold of pain is 130 dB.
- The intensity of sound decreases as the distance from the source increases.

Section 7.4 Questions

Understanding Concepts

1. On what factors does the loudness of a sound depend?
2. For each of the following pairs, state the number of times the intensity level of the first sound exceeds the second:
   (a) 4 B, 3 B (b) 8 B, 6 B
   (c) 100 dB, 70 dB (d) 120 dB, 80 dB
3. A 5.0×10^1 dB sound is increased in intensity level by a factor of 10^4. What is its new intensity level?
4. Refer to **Figure 2** and state the approximate threshold of hearing in decibels at the following frequencies:
   (a) 50.0 Hz (b) 1.0×10^3 Hz
5. Give an example of a sound that is intense, but not loud.
6. Investigate the possibility that hearing loss occurs as a result of loud music at concerts or nightclubs. What intensity levels are experienced that cause this hearing loss, and over what length of time does it take to permanently lose some hearing?

7.5 The Human Ear

The human ear consists of three sections: the outer ear, the middle ear, and the inner ear. The outer ear consists of the external ear (pinna) and the auditory canal. The external ear is shaped to collect sounds, which then travel down the auditory canal to the eardrum (**Figure 1**). The audible hearing range of a healthy young adult is approximately 20 to 20 000 Hz; the structure of the auditory canal magnifies frequencies between 2000 and 5500 Hz by a factor of nearly 10, emphasizing these frequencies.

The middle ear is separated from the outer ear by the eardrum, a very tough, tightly stretched membrane less than 0.1 mm thick. The eardrum is forced to vibrate by successive compressions and rarefactions coming down the ear canal. Compressions force the eardrum in and rarefactions cause it to move out, and the resultant vibration has the same frequency as the source of the sound waves (**Figure 2**).

Attached to the inside of the eardrum are three small interlocking bones: the hammer (malleus), the anvil (incus), and the stirrup (stapes). These bones transmit the vibrations of the eardrum to the inner ear, mechanically magnifying

Figure 1
The human ear

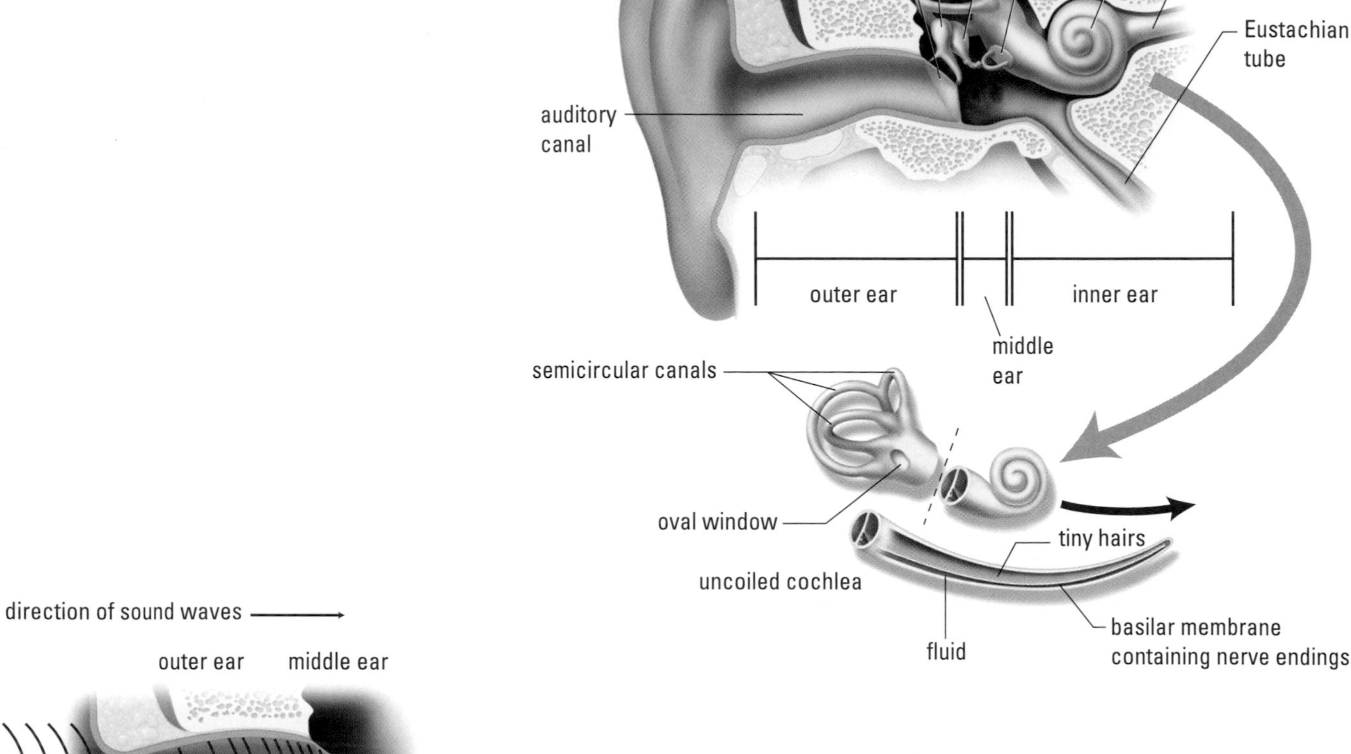

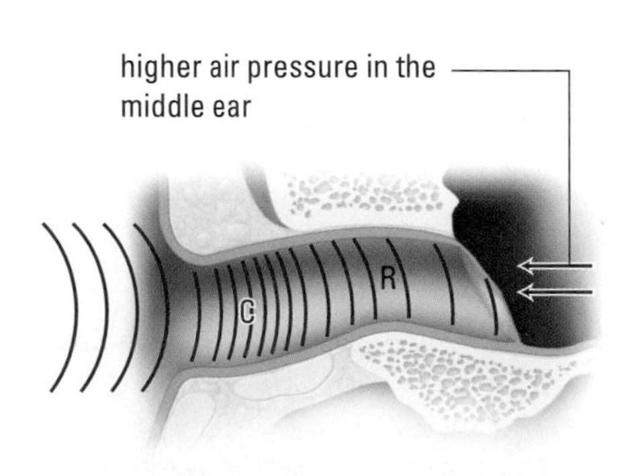

Figure 2
A rarefaction is actually an area of reduced atmospheric pressure. Thus, when a rarefaction approaches the external side of the eardrum, the higher atmospheric pressure in the middle ear causes the eardrum to move out.

the pressure variations by a factor of 18. The cavity containing the middle ear is filled with air and is connected to the mouth by the Eustachian tube. This tube is normally closed, but it opens during swallowing or yawning, equalizing the air pressure in the middle ear. If the Eustachian tube becomes blocked, because of a cold for example, pressure equalization cannot take place and the result is pressure in the middle ear, which can be quite painful and can affect hearing.

The stirrup transmits the eardrum vibrations to the threshold of the inner ear at the oval window. The vibrations set up pressure waves in the fluid that fills the inner ear's cochlea. The cochlea is a snail-shaped organ approximately 3.0 cm long, divided into two equal sections by a partition for most of its length. Waves are transmitted down one side of the cochlea, around the end of the partition, and back almost to the point of origin. As these waves move, they cause approximately 23 000 microscopic hairs to vibrate. Each hair is connected to a cell that converts the mechanical motion of the hair into an electrical signal, which in turn is transmitted to the brain by the auditory nerve. How the brain interprets these codified electrical signals is a wondrous thing we know very little about.

The inner ear also contains three hard, fluid-filled loops, called the semicircular canals, which are situated more or less at right angles to one another. These canals act as miniature accelerometers, transmitting, to the brain, electrical signals necessary for balance.

The human ear is not equally sensitive to all frequencies. **Figure 3** shows a graph of intensity level as a function of frequency for the human ear. The top line represents the threshold of pain. Sounds with intensities above this curve can actually be felt and cause pain. Note that the threshold of pain does not vary much with frequency. The middle line represents the threshold of hearing for the majority of the population. Sounds below this line would be inaudible. Note that the ear is most sensitive to sounds with a frequency between 2000 Hz and 3400 Hz. A 1000-Hz sound can be heard at a level of 20 dB, whereas a 100-Hz sound must have an intensity of at least 50 dB to be audible. The dashed line represents the hearing of a very good ear, usually found in less than 1% of the population. In general, only the young have such a low threshold of hearing.

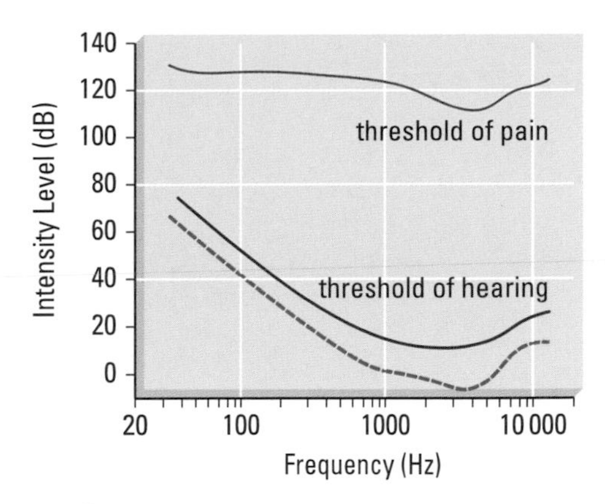

Figure 3
Sensitivity of the human ear as a function of frequency

Frequency Range of Hearing

Your ears are very sensitive organs that react to a wide range of frequencies. With an audio frequency generator connected to an appropriate loudspeaker or earphone, check your frequency range of hearing and compare it with the ranges of other students. Most students have an audible range of about 20 Hz to 20 kHz.

Exposure to loud sounds for long periods of time can contribute to premature loss of hearing in humans.

Figure 4 shows that all people suffer some loss of hearing as they grow older. The loss is significantly higher for those living in an industrial setting—whether or not they are employed in noisy occupations. Loud sounds do not, as a rule, harm the eardrum, although an exploding firecracker may cause it to burst. Such damage can be repaired, but permanent damage may be inflicted by intense sounds on the microscopic, hairlike cells in the inner ear. A particularly loud sound may rip away these delicate cells. A single blast of 150 dB or more can cause permanent damage to the ear, and levels of more than 90 dB over a prolonged period can produce the same effect. Persons employed in noisy places usually wear ear protectors as a precaution against this danger. For example, ground personnel who work near airplanes wear ear protection of the headphone design. In other environments, specially designed foam earplugs are easier to use and are inexpensive. In both cases, sound energy is reduced by absorbing the sound and blocking the ear canal. Many people with premature deafness have had prolonged exposure to loud sounds. Examples include rock musicians, music teachers, skeet shooters, factory workers, miners, construction workers, and people who listen to music at high levels.

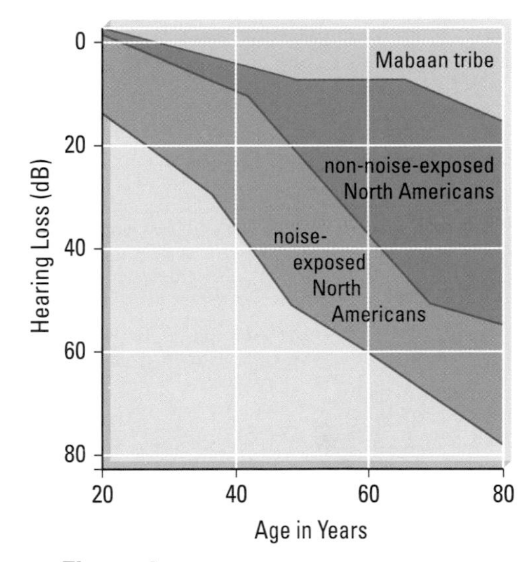

Figure 4
Hearing loss as a function of age. Tests carried out on people in remote Sudanese villages showed that the normal deterioration of their hearing was less than the average North American's, and much less than that of North American industrial workers.

Approximately one person in twenty in North America is either deaf or hard of hearing. Deafness may occur if signals cannot travel through the auditory nerve to the brain. There is no cure for this type of deafness. Deafness may also be caused by damage to the eardrum or the middle ear. This problem may be addressed by surgery or by the use of a hearing aid.

A hearing aid is an electronic device that amplifies sounds for people with hearing impairments. Hearing aids have the same basic components as any home entertainment system, except all the components are miniaturized and the amplified sound is delivered directly to the ear. The microphone, amplifier (consisting of transistors and integrated electronic circuits), miniature receiver (speaker), and battery are enclosed in a shell, which is worn behind or within the ear. A small tube directs the amplified sound from the receiver into the ear canal. Most hearing aids have adjustable controls; some have directional microphones. If the hearing loss has been caused by malformation of the ear canal or impaired

DID YOU KNOW?

Stereo Systems and the Human Ear

The variation in the sensitivity of the ear to both high and low frequencies is compensated for in many stereo and high-fidelity systems. A "loudness" control circuit boosts the intensities of the low and high frequencies relative to the middle frequencies. The compensation is greatest at low volumes. As the volume level increases, the compensation decreases until there is little change at moderate volume levels.

function of the middle ear, a small vibrator may be clamped against the mastoid bone behind the ear with a headband. In this case, the sound is conducted by the vibrator through the bones of the head to the inner ear. In some applications, the vibrator is connected directly to the bones in the inner ear. **Figure 5** shows three basic hearing aid designs.

A more recently developed cochlear implant is now available for deaf people whose auditory nerves remain functional. The device consists of electrodes, which are embedded in the cochlea of the inner ear to stimulate the auditory

DECISION MAKING SKILLS

- ○ Define the Issue
- ● Identify Alternatives
- ● Research
- ● Analyze the Issue
- ● Defend the Proposition
- ○ Evaluate

Explore an ***Issue***

Noise in an Urban Setting—The Use of Gasoline-Powered Blowers

Gasoline-powered blowers (**Figure 6**) commonly employ two-cycle engines, which are noisy and inefficient. Some municipalities in California have banned their use within the city limits, much to the distress of the landscaping industry and others, but with the support of many residents.

Take a Stand

Should gasoline-powered blowers be banned in urban settings?

Proposition

Gasoline-powered blowers are too noisy and pollute the environment.

There are arguments for banning them:

- The intensity of the sound could damage the hearing of both the residents and the operators.
- The sound is annoying, is a nuisance, and causes increased stress.
- The two-cycle engine creates excessive pollution.

There are arguments against banning them:

- It is difficult to determine the appropriate loudness that is annoying and/or could cause hearing loss.
- It is impractical to use electric-powered blowers.
- It will increase the workload of the landscapers and, thus, increase the cost of lawn maintenance for house owners.
- The municipality should fine noise and air polluters under existing bylaws.

Forming an Opinion

- Read the arguments described above and add your own ideas.
- Research the issues so you understand them better. Resources include newspaper and media accounts, literature from manufacturers of blowers and two-cycle engines, local noise bylaws, and indexes for magazines, such as *Popular Science*.
- In a small group, share your research and discuss the ideas.
- Create a position paper in which you state your opinion, supporting your arguments from your research and discussion. The "paper" can be a Web page, a video, a scientific report, or some other creative way of communicating. Make your position paper as persuasive as possible.

Figure 6
Gasoline-powered blower in operation

nerve, connected to a receiver surgically placed beneath the skin. A microphone near the ear relays sound signals to a microprocessor that converts them into electric signals; these signals are sent to a transmitter behind the ear, and then to the receiver and cochlear electrodes. The cochlear implant provides a crude approximation of normal sound, but cannot yet fully reproduce the sounds experienced by someone with full hearing. With further research and development, it may someday better reproduce a fuller range of frequencies.

(a)

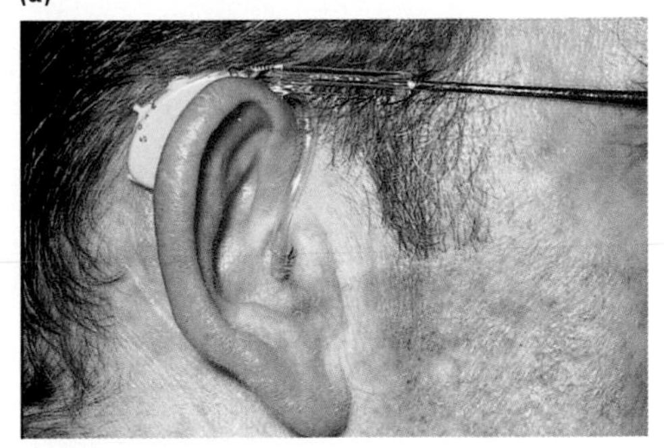

(b)

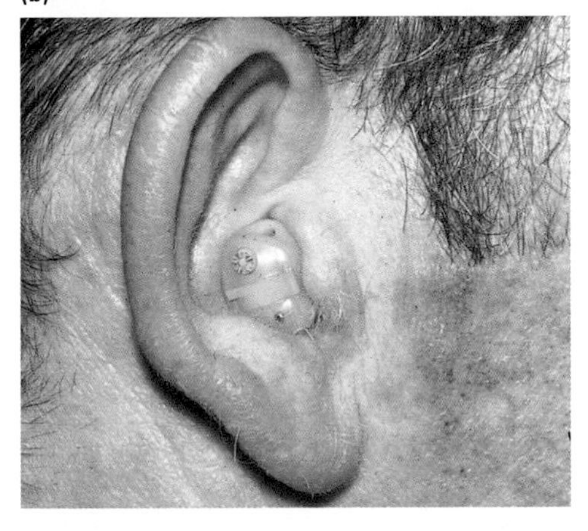

(c)

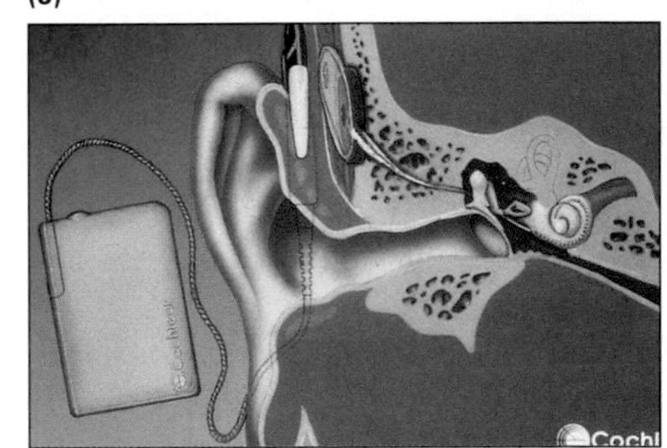

Figure 5
Hearing aids come in three basic designs
(a) Behind the ear
(b) In the ear canal
(c) Cochlear implant

Practice

Understanding Concepts

1. Hearing loss as a function of age is described earlier in **Figure 4** of this section. Explain the difference between the average hearing loss for an average member of each group from age 20 to age 60 in terms of intensity level.

SUMMARY The Human Ear and Hearing

- The human ear magnifies and transforms sound into electrical nerve pulses.
- The range of hearing for most young people is 20 Hz to 20 000 Hz.
- The range of hearing diminishes with age and/or with the exposure to loud sounds.
- People in noisy locations should protect their hearing.
- Hearing loss can be improved by electronic devices, such as hearing aids.

Section 7.5 Questions

Understanding Concepts

1. Assume that a person's eardrum vibrates with an amplitude of 2.0×10^{-10} m when listening to a 3.0 kHz sound. Through what total distance will the eardrum vibrate in one minute?
2. Why do some hunters have a loss of hearing in one of their ears?
3. One person has a threshold of hearing of 10 dB; another has a threshold of 30 dB at the same frequency. Which person has better hearing at this frequency?

Making Connections

4. Research and report on any two of the following:
   (a) the sound bylaws in your municipality
   (b) recent lawsuits against employers by employees who have had hearing loss because of their working environment
   (c) the types of ear protectors available for working in noisy environments
   (d) the noise abatement procedures in force at your local airport
   (e) the labour laws and regulations regarding sound levels in the workplace

(continued)

5. Hearing aids will not restore normal hearing, but they can help deafness. Do some research to answer the following questions. Follow the links for Nelson Physics 11, 7.5.
   (a) What are some limitations of hearing aids?
   (b) What factors affect the choice of an appropriate hearing aid?
   (c) What are the differences between conventional aids, programmable aids, and aids with digital circuitry?
   (d) How much do hearing aids cost?
   (e) A hearing health care professional is usually an audiologist or hearing instrument specialist. What education and training is required for licensing and certification?

GO TO www.science.nelson.com

7.6 The Reflection of Sound Waves

Just as a mirror reflects light, when sound waves radiating out from a source strike a rigid obstacle, the angle of reflection of the sound waves equals the angle of incidence.

Simple Reflection

In this activity, you will look at how sound is reflected.

- Line up two cardboard tubes so that they are at angles to each other and directed toward a flat surface, such as a blackboard or wall (**Figure 1**).
- Place a ticking source of sound near the end of one tube.
- Adjust the position of the second tube so that the ticking sound can be heard clearly by an observer's ear at the other end.
- Measure and compare the angles the tubes make with the flat surface when the sound is the loudest.

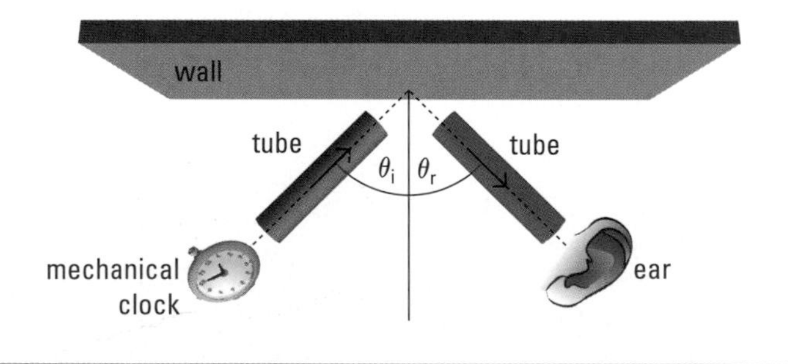

Figure 1

Sound waves conform to the laws of reflection you have studied previously (which are also covered in detail in Chapter 9). As was the case for waves in a ripple tank, sound waves emitted at the focal point of a curved reflector are reflected in a specific direction. If the sound waves encounter a curved surface, they are reflected to a specific area and, thus, are concentrated. For example, at the Ontario Science Centre in Toronto, two large metal concave reflectors are located at opposite ends of a large room (**Figure 2**). A person standing at the focus of one reflector can talk in a normal voice and be heard clearly by a person standing at

Figure 2
Concave reflectors at the Ontario Science Centre in Toronto

the focus of the other reflector 50 m away. Band shells are concave to reflect the sound of a band or orchestra to the audience; the performers are located in the focal plane of a hard concave reflector. Some concert halls use large acrylic discs, or sound panels, to reflect sound from the ceiling of a concert hall (**Figure 3**).

Figure 3
The Centre in the Square in Kitchener, Ontario, uses large sound panels to reflect sound.

echo: reflected sound waves

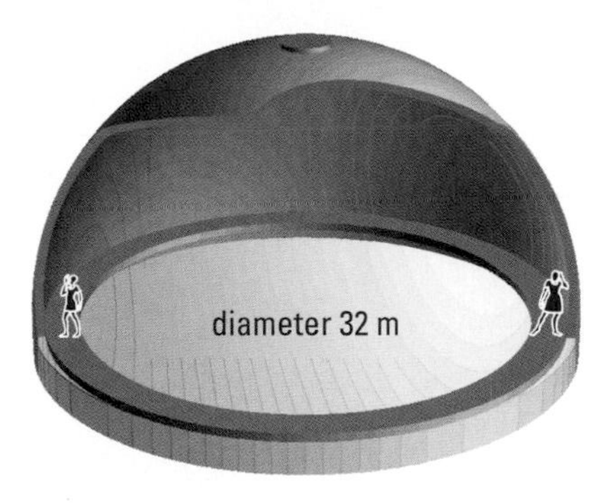

Figure 4
In the dome of St. Paul's Cathedral, London, whispers can be heard clearly 32 m away because the dome focuses the echoes together.

Echoes are produced when sound is reflected by a hard surface, such as a wall or cliff. An echo can be heard distinctly only if the time interval between the original sound and the reflected sound is greater than 0.1 s. The distance between the observer and the reflecting surface must be greater than 17 m for an echo to be heard (**Figure 4**).

The echo-sounder is a device that uses sound reflection to measure the depth of the sea (**Figure 5**). Similar equipment is used in the fishing industry to locate schools of fish. More sophisticated equipment of the same type is used by the armed forces to locate submarines. All such devices are called **sonar** (sound navigation and ranging) devices (**Figure 6**).

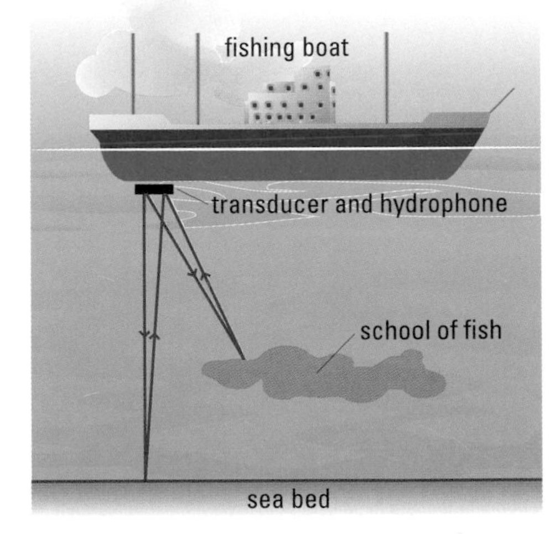

Figure 5
An echo-sounder. A transducer at the bottom of a ship converts electrical energy into sound energy and sends out a series of equally spaced sound pulses with a frequency of approximately 30 kHz. The pulses are reflected from the sea bed back to the ship, where they are received by an underwater microphone, called a hydrophone. The time interval between the emission of the signal and the reception of the echo is measured by a computer, which calculates the depth of the water.

sonar: system that uses transmitted and reflected underwater sound waves to locate objects or measure the distance to the bottom of the water body

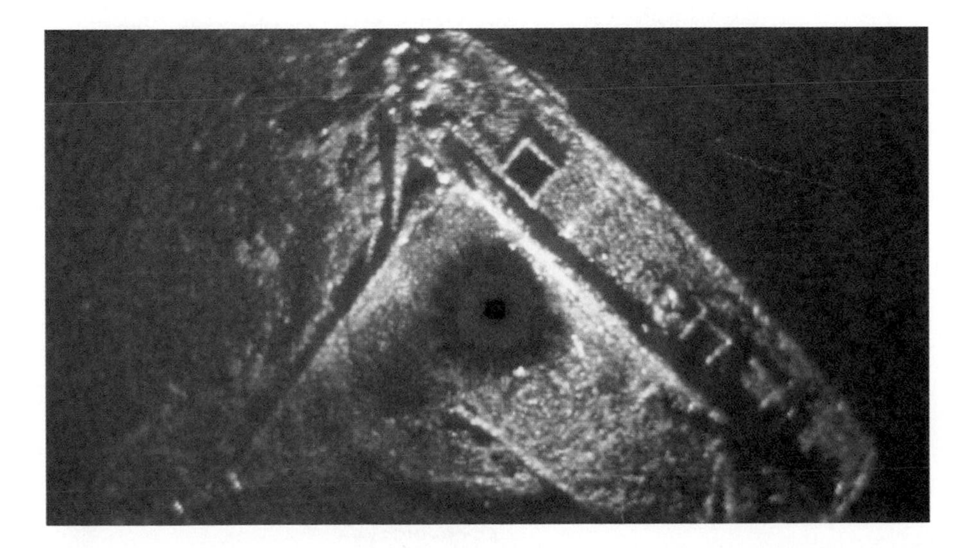

Figure 6
Sonar image of a sunken ship

Echolocation—Dolphins, Orca Whales, and Bats

Dolphins and orca whales rely on the production and reflection of sound to navigate, communicate, and hunt in dark and murky waters (**Figure 7**). The location of an object using reflected sound is called **echolocation**. Both animals produce clicks, whistles, and other sounds that vary in intensity, frequency, and pattern. Lower frequency sounds (0.5–50 kHz) probably function mainly for social communication, while higher frequencies (40–150 kHz) are probably used for echolocation.

Figure 7
Dolphins use echolocation to find their way in murky water

echolocation: location of objects through the analysis of reflected sound.

In the dolphin, the clicks pass through the melon (see **Figure 8**), which acts as an acoustical lens focusing the sound waves in a beam in front of the dolphin. These sound waves bounce off objects in the water and return to the dolphin as an echo. The major areas of sound reception are the fat-filled cavities in the lower jawbones that conduct the vibrations to the ear. The brain receives this information as nerve impulses, enabling the dolphin to interpret the meaning of the echoes. Echolocation is most effective in the range of 5 to 200 m. Although there are some differences in how the sound is produced by the orca whale and in the frequencies of the emitted sounds, the process of echolocation is similar.

Most bats use echolocation for navigation in the dark and for finding food. The bat can identify an object by the echo and can even tell the size, shape, and texture of a small insect. If the bat detects a prey, it will generally fly toward the source of the echo, continually emitting high frequency pulses until it reaches its target and scoops the insect up into its wing membranes and into its mouth.

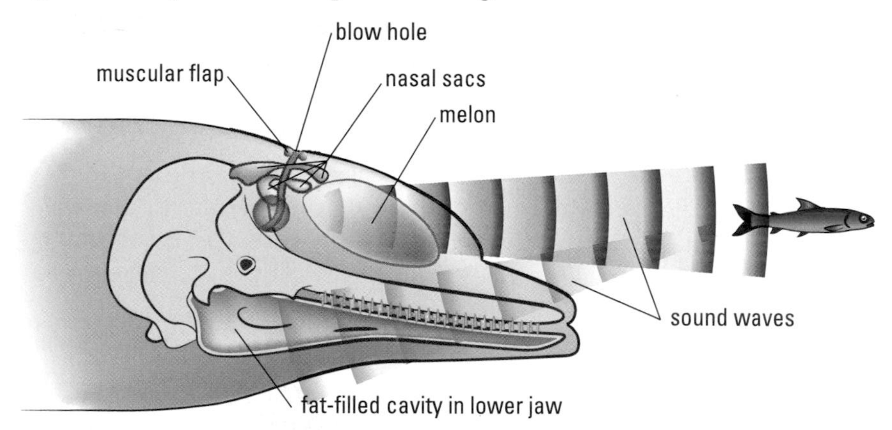

Figure 8
Dolphins produce high-frequency sound clicks that are directed by the melon. These clicks echo off objects in the water.

Sample Problem

A ship is anchored where the depth of water is 120 m. An ultrasonic signal sent to the bottom of the lake returns in 0.16 s. What is the speed of sound in water?

Solution

$\Delta t = 0.16$ s
$d = 120$ m
$\Delta d = 2d$, or 240 m
$v = ?$

$$v = \frac{\Delta d}{\Delta t}$$

$$= \frac{240 \text{ m}}{0.16 \text{ s}}$$

$$v = 1500 \text{ m/s}$$

The speed of the sound in water is 1.5×10^3 m/s.

Practice

Understanding Concepts

1. A student stands 86 m from the foot of a cliff, claps her hands, and hears the echo 0.50 s later. Calculate the speed of sound in air.
2. A pulse is sent from a ship to the floor of the ocean 420 m below the ship; 0.60 s later the reflected pulse is received at the ship. What is the speed of the sound in the water?
3. A sonar device is used in a lake, and the interval between the production of a sound and the reception of the echo is found to be 0.40 s. The speed of sound in water is 1500 m/s. What is the water depth?
4. Ultrasonic sound is used to locate a school of fish. The speed of sound in the ocean is 1.48 km/s, and the reflection of the sound reaches the ship 0.12 s after it is sent. How far is the school of fish from the ship?

Making Connections

5. Why is a parabolic microphone (**Figure 9**) sometimes used to pick up remote sounds at a sports event or to record birdcalls? How can you reproduce this effect with body parts only?

Answers

1. 3.4×10^2 m/s
2. 1.4×10^3 m/s
3. 3.0×10^2 m
4. 89 m

Figure 9
A parabolic microphone

Ultrasound Medical Applications

Ultrasound, like any wave, carries energy that can be absorbed or reflected by the medium in which it is travelling. Unlike X rays, ultrasound is used in medical diagnosis and therapy with no side effects.

In ultrasound imaging, the probe consists of a transducer, which emits the ultrasonic sound wave, and a microphone that receives the reflected waves. Ultrasound echoes are produced when the waves are partially reflected by tissues with differing densities. Typically, the probe sweeps across the area, the echoes are timed, and the distances are calculated. The results are computerized and viewed on a screen or digitally stored for future use. Such distance information reveals the thickness of tissue, the presence of a foreign body or tumour, or the image of a fetus (**Figure 10**).

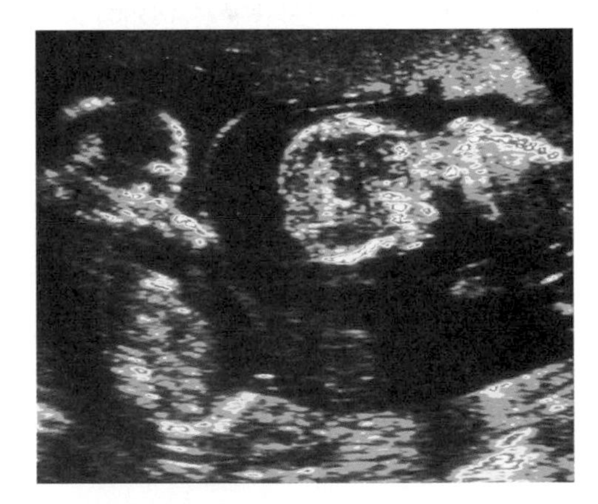

Figure 10
Ultrasound photo of twins

When high-energy ultrasound is directed at human tissue, the energy is converted to thermal energy, which can promote healing. To relieve pain, promote flexibility, or break down scar tissue, the therapist can apply ultrasound to injured or overextended muscles. This deep-heat treatment is known as ultrasound diathermy. Skill is necessary in the use of this therapy since overheating can result in damage to tissue or bones.

Focused, high-energy ultrasound can be used to destroy unwanted tissue, where the tip of a probe oscillates at 22 kHz with such a large amplitude it pulverizes the tissue. Gall and kidney stones can be shattered when a short pulse of high-energy ultrasound is focused on the stone.

SUMMARY The Reflection of Sound Waves

- Sound waves obey the laws of reflection.
- Echolocation is used in sonar applications, including navigation and food collection by some animals.
- Ultrasound has many medical applications, both for diagnosis and treatment.

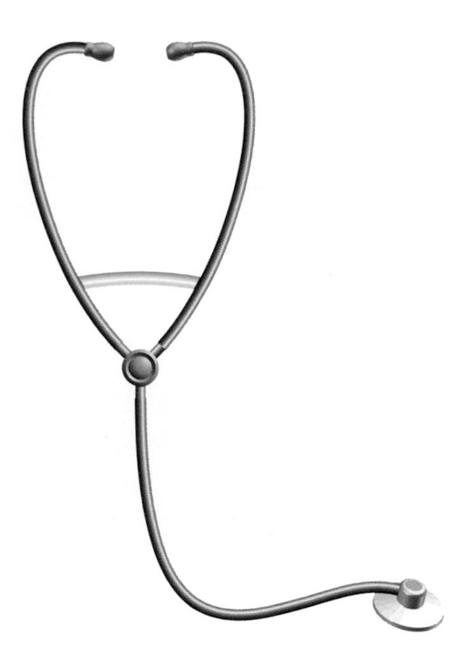

Figure 11
A stethoscope

Section 7.6 Questions

Understanding Concepts

1. A woman makes a sound and, 3.5 s later, the echo returns from a nearby wall. How far is the woman from the wall, assuming that the speed of sound is 350 m/s?
2. A man drops a stone into a mineshaft 180 m deep. If the temperature is 20.0°C, how much time will elapse between the moment when the stone is dropped and the moment when the sound of the stone hitting the bottom of the mineshaft is heard?
3. A ship is travelling in a fog parallel to a dangerous, cliff-lined shore. The boat whistle is sounded and its echo is heard clearly 11.0 s later. If the air temperature is 10.0°C, how far is the ship from the cliff?

Making Connections

4. Since the human ear is not very sensitive to low frequencies, such as heartbeats, the doctor usually uses a stethoscope (**Figure 11**). Research and explain how the construction of the stethoscope intensifies the sounds coming out of the chest.
5. A submarine's sonar receives strong echoes from ships and other submarines, but weak echoes from thermal currents in the ocean. Why?
6. Describe how a bat uses sound energy to locate its next meal.
7. Ultrasound has many applications in the medical field, but there are other uses. Research and report on industrial and other applications of ultrasound.

Reflecting

8. Ask family members or friends if they have had any ultrasound medical examinations performed on them. Describe the definition of ultrasound to them and write down the types of procedures they have had.

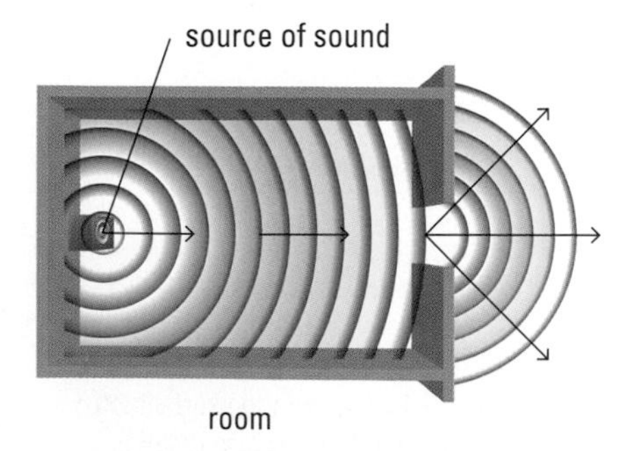

Figure 1
Sound waves are diffracted as they pass through a doorway from one room into the next.

7.7 Diffraction and Refraction of Sound Waves

Have you noticed that you can always hear the sounds of a classroom through an open door, even though the other students may be out of sight and separated by a wall? You can also easily hear sounds travelling through an open window and even around the corner of a building. The "sound around the corner" effect is so familiar to us that we do not give it a second thought. Sound waves can travel around corners because of diffraction. Diffraction describes the ability of waves to move around an obstacle or to spread out after going through a small opening (**Figure 1**).

As was discussed in Chapter 6, waves with relatively long wavelengths diffract more than those with short wavelengths. The diffraction of a wave depends on both the wavelength and the size of the opening. Lower frequency sound waves have relatively long wavelengths compared with the size of the openings they commonly encounter and, thus, are more likely to diffract. Higher frequency sounds have shorter wavelengths and diffract less. For example, when sound travels out from a loudspeaker, the bass notes (lower frequency) are diffracted more and travel easily into the next room, while the treble notes (higher frequency) do not. Test this out on your home stereo!

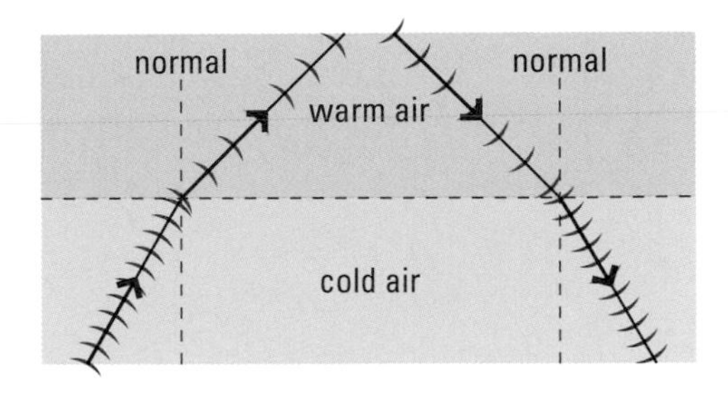

Figure 2
Sound waves are refracted in the same manner as light rays.

You know that the speed at which a sound travels through air can be affected by the temperature of the air; sound waves travel faster in warm air than they do in cold air. What happens when sound waves travel from air at one temperature to air at a different temperature? When sound waves move at an angle from air at one temperature to air at a different temperature, they are refracted—that is, they change their direction (**Figure 2**).

Suppose there is a layer of cold air over a layer of warm air. As shown in **Figure 3**, sound waves tend to be refracted upward from an observer, which decreases the intensity of the sound heard by the observer. However, at night, the cooler air near the surface of the Earth tends to refract sound waves toward the surface, so they travel a greater horizontal distance. This is particularly true over flat ground or on water since there is also less absorption of the sound. Watch what you say outdoors at night—you may be surprised at how far your voice carries!

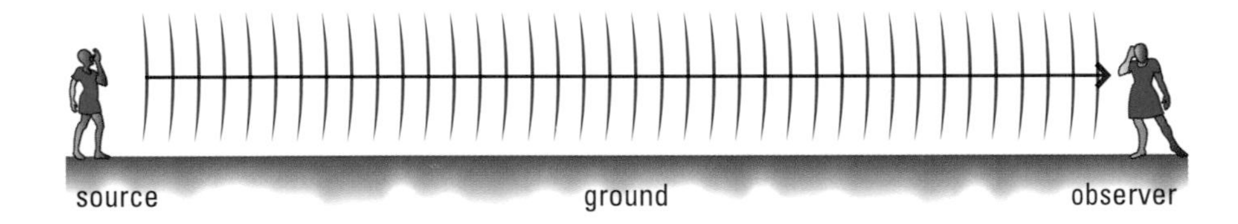

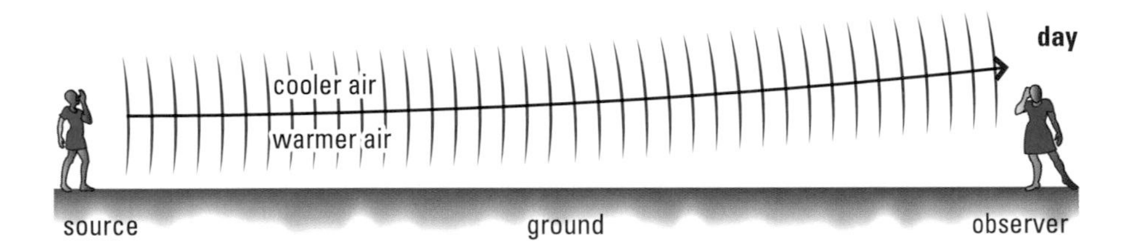

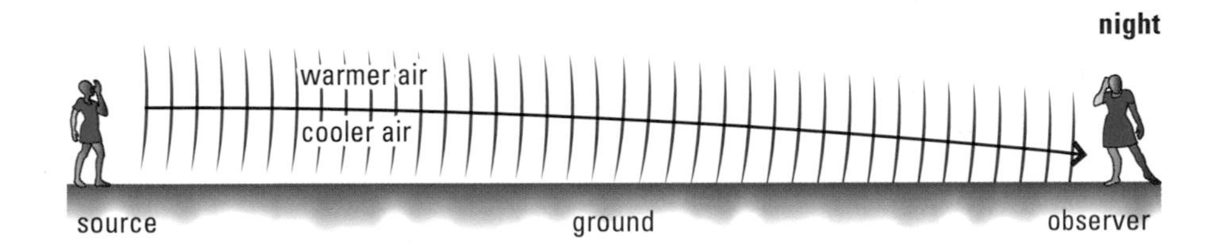

Figure 3
The sound wave produced by a person's voice travels faster in warm air and "gains" on the part of the wave in the cooler layer. The wave is refracted away from the surface on a warm day and toward the surface when a cooler layer of air is near the surface.

Practice

Understanding Concepts

1. Define diffraction of sound waves.
2. Define refraction of sound waves.
3. Why are lower frequency sound waves more likely to diffract than higher frequency sound waves?

SUMMARY Diffraction and Refraction of Sound Waves

- Sound waves can be diffracted and refracted.
- Diffraction is greater when the sound wavelengths are larger.

Section 7.7 Questions

Understanding Concepts

1. Which sound waves from a home entertainment centre will be easier to hear in the next room through an open doorway, those from a woofer (low frequency) or a tweeter (high frequency)? Why?
2. For a person with equal sensitivity in both ears, would the direction from which a sound is coming be more easily identified if the sound has a high pitch or a low pitch? Explain your reasoning.

Applying Inquiry Skills

3. **Figure 4** shows noise barriers that are present on many highways. Propose a design that would improve on these barriers to make them more effective. Draw an example of your barrier and explain why it is more effective, using the concepts of diffraction and reflection.

Making Connections

4. (a) Explain how the noise barriers in **Figure 4** affect highway noises of differing frequencies.
   (b) Why is it not feasible to install the barriers shown in **Figure 4** on all urban roads?
   (c) What else can be done to limit the spread of road noise?

Figure 4
Noise barriers

Reflecting

5. "People who choose to purchase or rent a home near a noisy highway should not expect the government to erect sound barriers." Take a position for or against this statement and prepare arguments in support of your position.

7.8 The Interference of Sound Waves

It is quite common for two or more sound waves to travel through a medium at the same time. When two or more sound waves act on the same air molecules at the same time, interference occurs. In Chapter 6, you examined the interference of transverse water waves and found that water waves can interfere constructively or destructively. Do sound waves exhibit many of the properties of wave interference?

Investigation 7.8.1

Interference of Sound Waves from a Tuning Fork and Two Loudspeakers

INQUIRY SKILLS

- ○ Questioning
- ○ Hypothesizing
- ● Predicting
- ○ Planning
- ● Conducting
- ● Recording
- ● Analyzing
- ● Evaluating
- ● Communicating

In section 6.9, you examined the interference of two water waves originating from two point sources in a ripple tank. Since the water waves were identical in frequency and amplitude, a stationary interference pattern was produced consisting of a symmetrical pattern of alternating nodal lines and regions of constructive interference.

In this investigation, you will examine the interference pattern created by sound waves originating from two identical sources: a rotating tuning fork (**Figure 1**), and two loudspeakers producing identical sound waves (**Figure 2**). Review the operation of a tuning fork in section 7.2.

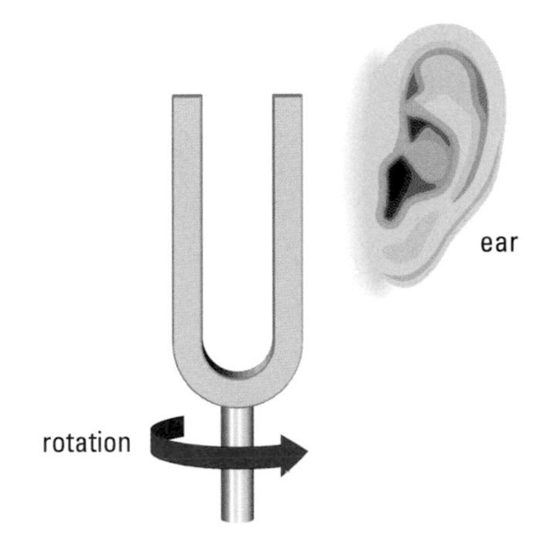

Figure 1
Position of tuning fork

Questions

Where are the areas of destructive interference located in the area surrounding the prongs of a tuning fork? Where are they located in the areas in front of two loudspeakers producing identical sound waves?

Materials

tuning fork
rubber hammer or rubber stopper
amplifier
audio generator
two identical loudspeakers

Prediction

(a) Use diagrams to predict where areas of destructive interference are located in the area surrounding a vibrating tuning fork (**Figure 1**), and in the areas in front of two loudspeakers producing identical sound waves (**Figure 2**).

Procedure

1. Strike a tuning fork with a rubber hammer or on a rubber stopper. Hold the fork vertically near your ear and slowly rotate it (**Figure 1**). Listen carefully for loud and soft sounds, and have your partner help you locate their exact positions. Repeat until you are certain of the results, and then draw a diagram of the top view of the tuning fork showing the positions of the loud and soft sounds. (To minimize reflection from walls, this should be performed in as large a room as possible.)
2. Set up the amplifier, generator, and speakers as shown in **Figure 2**, with the speakers approximately 2.0 m apart and raised about 1.0 m from the floor. Ensure that the speakers are in phase. Adjust the frequency of the generator to approximately 500 Hz, with the sound at a moderate intensity level.
3. Slowly walk along a path parallel to the plane of the speakers. Sketch the positions of the lowest sound intensity using your ears as detectors or a decibel meter, if available.
4. Slightly increase the frequency of the sound emitted from the speakers and repeat step 3.

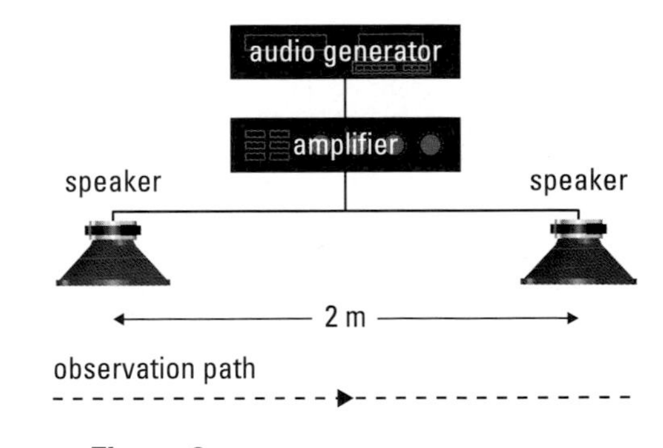

Figure 2
Setup for Investigation 7.8.1

Very high or very low frequencies can cause discomfort or harm.

Analysis

(b) Use a diagram to illustrate how interference occurs near a single tuning fork. (*Hint:* What type of interference occurs where a compression interacts with a rarefaction?)

(c) What changes in intensity occurred when you walked from one speaker to the other? Relate your observation to the interference between two identical sources in the ripple tank.

(d) What changes in the intensity pattern occurred when the speakers were adjusted to emit a sound of higher frequency? Why?

Evaluation

(e) How did your results compare with your predictions?

(f) List possible sources of error in this investigation.

(g) Suggest changes to the procedure that would help reduce experimental errors.

Interference Between Identical Sound Waves

When the tines of a tuning fork vibrate, a series of compressions and rarefactions is emitted from the outer sides of the tines and from the space between them. Since the tines are out of phase, the compressions and rarefactions interfere destructively, producing nodal lines that radiate out from the corners of the tines. In the area between the tines, constructive interference occurs and a normal sound wave emanates from the tines. When the tuning fork is rotated near the ear, the relative sound intensity alternates between loud (normal sound intensity) and soft (destructive interference).

The interference pattern between the two loudspeakers in phase (**Figure 3**) is similar to the pattern that is observed in water waves between two point sources. Areas of constructive and destructive interference are located symmetrically about the midpoint of the pattern, midway between the speakers. If the

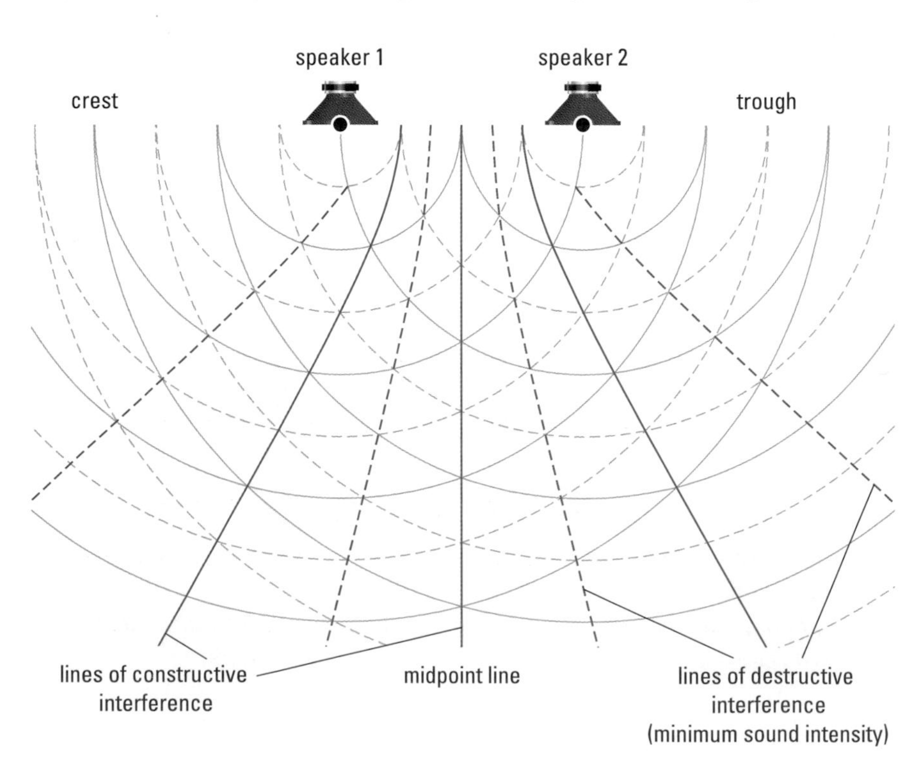

Figure 3
The interference of sound waves between two loudspeakers vibrating in phase

loudspeakers are in phase, there is an area of constructive interference (maximum sound intensity) at the midpoint. When the frequency is increased, the wavelength decreases. This produces more areas of destructive and constructive interference as shown, but the symmetry of the interference pattern does not change. It is difficult to produce areas of total destructive interference because sound waves are reflected from the walls and other surfaces in the room.

Interference in sound waves from a single source may be demonstrated with an apparatus called a Herschel tube (**Figure 4**). Sound waves from a source such as a tuning fork enter the tube and split, travelling along two separate paths. If the paths are of the same length, the waves will meet on the other side in phase; that is, compression will meet compression, rarefaction will meet rarefaction, and the intensity will be at a maximum. If the tube is longer on one side, the waves on that side will have to travel farther. At some point, compressions will emerge with rarefactions and interfere destructively to produce a minimum sound intensity. Further extension of the tube on one side will reveal other positions in which constructive and destructive interference will occur.

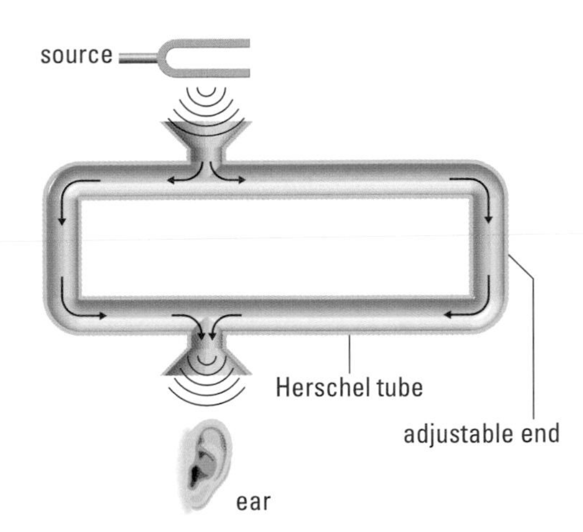

Figure 4
A Herschel tube

Noise cancellation headphones use a computer to cancel noise using destructive interference (**Figure 5**). The computer receives sound from a microphone, isolates the noise, and creates a sound wave exactly out of phase with the incoming sound. This can be more effective than normal noise suppression earphones used in noisy environments. When used with a CD player, a person can hear the music clearly even while cutting the lawn or riding in a crowded bus since the headphones electronically cancel up to 70% of external noise. Pilots use similar devices to suppress the external noise of a plane's engines, allowing for better communication and protection of the pilot's hearing. Other applications include voice recognition software on computers, communication systems for fast food drive-throughs, and mobile telephones.

Figure 5
Noise cancellation headphones

Practice

Understanding Concepts

1. Explain in your own words why there are loud and soft sound intensities in the area around a tuning fork.
2. When you extend one side of a Herschel tube, you reveal other positions in which constructive and destructive interference occur. Draw two diagrams to illustrate this situation.
3. Why is there a line of constructive interference (maximum sound intensity) at the midpoint line between the two speakers shown in **Figure 3** of this section?

Applying Inquiry Skills

4. Obtain a tuning fork from your teacher, build your own Herschel tube, and demonstrate it to someone in your class.

SUMMARY The Interference of Sound Waves

- Sound waves interfere, producing areas of constructive and destructive interference.
- The interference pattern between two identical sources of sound is similar to that produced by identical point sources in a ripple tank.

Section 7.8 Questions

Understanding Concepts

1. Stereo speakers have colour-coded terminals (usually black and red) in the back that are hooked up to the wire from the amplifier in the stereo system. If the connections are reversed, the second speaker moves out when the first speaker moves in. Why is it important to wire both speakers so they move in and out together?
2. Distinguish between destructive and constructive interference.

Making Connections

3. Theatres and concert halls are designed to eliminate "dead spots." Research the answers to the following questions.
   (a) What are "dead spots?"
   (b) How do engineers eliminate the "dead spots" from these facilities?

7.9 Beat Frequency

We have been examining the interference of sound waves with identical frequencies and wavelengths. Now we will consider the interference of sound waves with slightly different frequencies and wavelengths. Consider a tuning fork that has one tine "loaded" with Plasticine or an elastic band wrapped around it. If this fork is struck at the same time as an "unloaded," but otherwise identical, tuning fork, the observed sound will alternate between loud and soft, indicating alternative constructive and destructive interference. Such periodic changes in sound intensity are called **beats**.

beats: periodic changes in sound intensity caused by interference between two nearly identical sound waves

Activity 7.9.1

Beats from Nearly Identical Tuning Forks

This activity will allow you to observe sound beats produced by two tuning forks of nearly identical frequencies.

Procedure

1. Place two mounted tuning forks close to and facing each other. Wrap an elastic band tightly around a prong on one of the tuning forks (**Figure 1**).
2. Sound the two forks together and describe the resulting sound.
3. Repeat the procedure using two elastic bands on the same prong.
4. Finally, remove the bands and repeat the process a third time.
5. Demonstrate beats with an oscilloscope and sound generator, if they are available, or use a computer program to demonstrate beats.

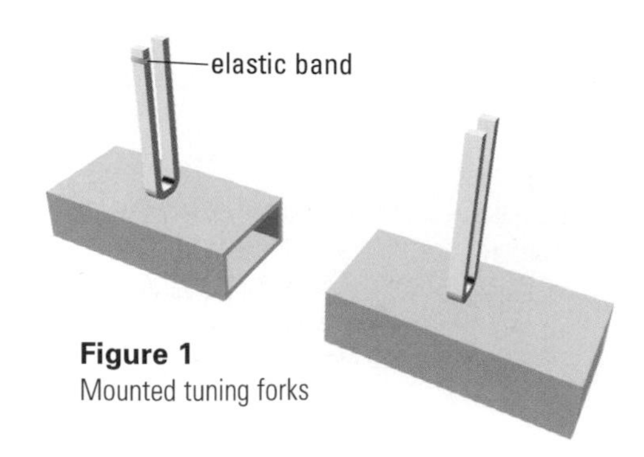

Figure 1
Mounted tuning forks

Analysis

(a) Did the frequency of a tuning fork increase or decrease when elastic bands were added to a prong? Explain your answer.
(b) Predict a relationship between the frequency of beats produced and the frequencies of the sources producing the beats.
(c) Explain how this activity demonstrates that sound energy travels by means of waves.

Interference Between Nearly Identical Sound Waves

The interference that occurs between two sources with slightly different frequencies is shown in **Figure 2**. The wavelengths are not equal and hence the distances between successive compressions and rarefactions are not the same. At certain points, a compression from one source coincides with a rarefaction from the other, producing destructive interference and minimum sound intensity. When compression and compression coincide and rarefaction and rarefaction coincide, constructive interference results and maximum sound intensity occurs. The number of maximum intensity points that occur per second is called the **beat frequency**.

beat frequency: the number of beats heard per second in hertz

To determine the beat frequency, the lower frequency is subtracted from the higher frequency. For example, if a tuning fork of 436 Hz is sounded with a 440-Hz tuning fork, the beat frequency will be 4 Hz.

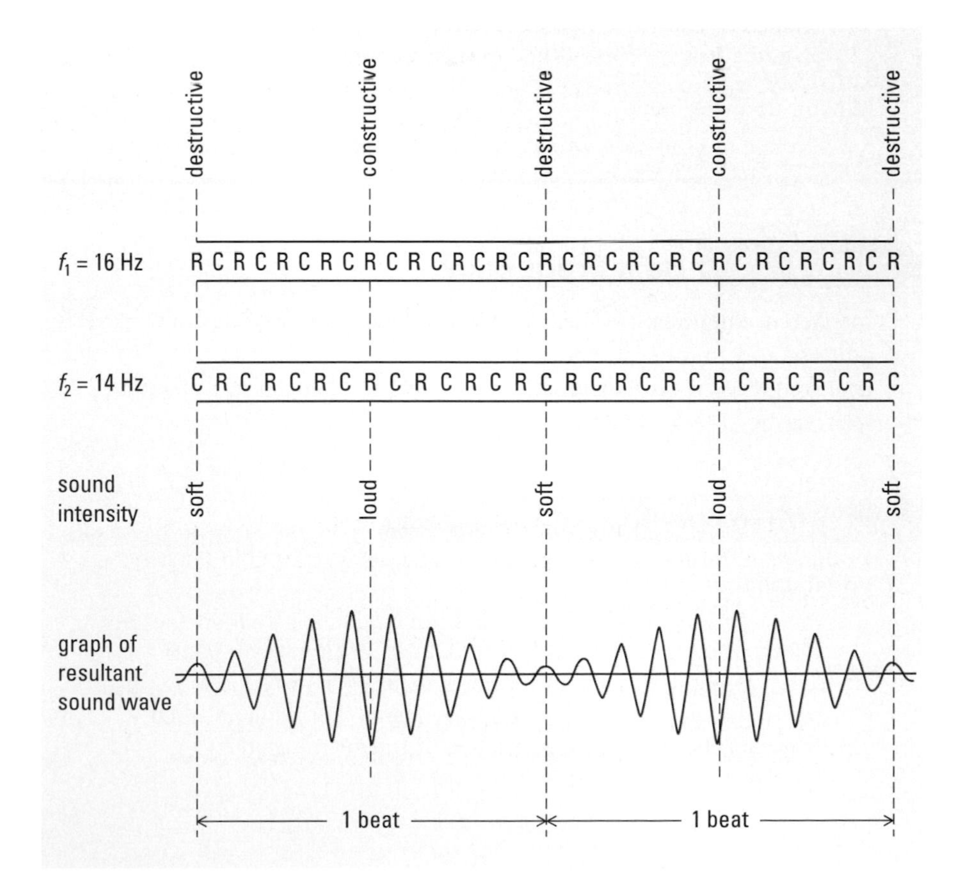

Figure 2
A source of sound with a frequency of 16 Hz interferes with a source with a frequency of 14 Hz to produce two beats in one second. The difference in frequency between the two sources is equal to the beat frequency.

DID YOU KNOW?

Human Detection of Beat Frequencies

In practice, the human ear can only detect beat frequencies less than 7 Hz.

Sample Problem

A tuning fork with a frequency of 256 Hz is sounded together with a note played on a piano. Nine beats are heard in 3 s. What is the frequency of the piano note?

Solution

$$\text{beat frequency} = \frac{\text{number of beats}}{\text{total time}}$$

$$= \frac{9 \text{ beats}}{3 \text{ s}}$$

$$\text{beat frequency} = 3 \text{ Hz}$$

$$\text{beat frequency} = |f_1 - f_2|$$

$$3 \text{ Hz} = |256 \text{ Hz} - f_2|$$

$$f_2 = 253 \text{ Hz} \quad \text{or} \quad 259 \text{ Hz}$$

Note that there are two possible answers. Without more information, there is no way of knowing which is correct.

Practice

Answers

1. 2 Hz
2. 193 Hz, 207 Hz

Understanding Concepts

1. What is the beat frequency when a 512 Hz and a 514 Hz tuning fork are sounded together?
2. Two tuning forks are sounded together. One tuning fork has a frequency of 200 Hz. An observer hears 21 beats in 3.0 s. What are the possible frequencies of the other tuning fork?

Making Connections

3. Explain in your own words how beats can be used to tune a guitar.

SUMMARY Beat Frequency

- Interference between two nearly identical sources of sound results in periodic changes in intensity, called beats.
- Beat frequency is the difference between the frequencies of two sound sources.

Section 7.9 Questions

Understanding Concepts

1. You sound two tuning forks together. One has a frequency of 300 Hz, the other a frequency of 302 Hz. What do you hear?
2. When a tuning fork with a frequency of 256 Hz is sounded at the same time as a second tuning fork, 20 beats are heard in 4.0 s. What are the possible frequencies of the second fork?

3. State the beat frequency when the following pairs of frequencies are heard together:
   (a) 202 Hz, 200 Hz (b) 341 Hz, 347 Hz (c) 1003 Hz, 998 Hz
4. Use the principle of superposition to "add" these two sound waves together on a piece of graph paper turned sideways:

   Wave A: λ = 4.0 cm; A = 1.0 cm; 5 wavelengths
   Wave B: λ = 5.0 cm; A = 1.0 cm; 4 wavelengths

   Describe how the resulting pattern relates to the production of beats.
5. A tuning fork with a frequency of 4.0×10^2 Hz is struck with a second fork, and 20 beats are counted in 5.0 s. What are the possible frequencies of the second fork?
6. A third fork with a frequency of 410 Hz is struck with the second fork in question 5, and 18 beats are counted in 3.0 s. What is the frequency of the second fork?
7. A 440-Hz tuning fork is sounded together with a guitar string, and a beat frequency of 3 Hz is heard. When an elastic band is wrapped tightly around one prong of the tuning fork, a new beat frequency of 2 Hz is heard. Determine the frequency of the guitar string.

Making Connections

8. How would a piano tuner use a tuning fork or pitch pipe to tune a piano by adjusting the tension of the strings?

7.10 The Doppler Effect and Supersonic Travel

If you've ever been to an automobile race, you probably noticed that when a racing car streaks past you, you can detect a change in frequency of the sound from the car. As the car approaches, the sound becomes higher in frequency. At the instant the car passes you, the frequency drops noticeably. The apparent changing frequency of sound in relation to an object's motion is called the **Doppler effect**, named after Christian Doppler (1803–53), an Austrian physicist and mathematician who first analyzed the phenomenon. You've probably heard this effect from train whistles, car horns, or sirens on fire trucks, ambulances, or police cruisers.

Doppler effect: when a source of sound approaches an observer, the observed frequency increases; when the source moves away from an observer, the observed frequency decreases

To understand why the Doppler effect occurs, look at **Figure 1**. **Figure 1(a)** shows sound waves travelling outward from a stationary source. **Figure 1(b)** shows the source of sound waves travelling to the left. As the waves approach observer A, they are closer together than they would be if the source were not moving. Thus, observer A hears a sound of higher frequency. Observer B, however, hears a sound of lower frequency because the source is travelling away, producing sounds of longer wavelength. A similar effect occurs when the source of sound is stationary and the observer is moving toward or away from it.

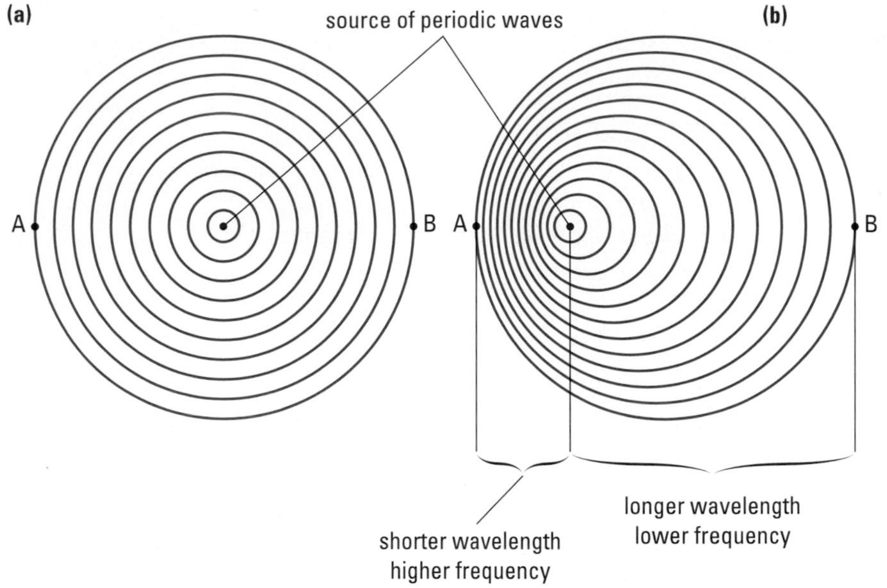

Figure 1
The Doppler effect
(a) The source is stationary. Both observers A and B hear the same frequency of sound.
(b) The source is moving to the left. Observer A hears a higher frequency and Observer B hears a lower frequency.

It can be shown that the following relationship describes the effect on frequency. When a source either moves toward or away from a stationary observer,

$$f_2 = f_1\left(\frac{v}{v \pm v_s}\right)$$

where v is the speed of sound in the medium and v_s is the speed of the source through the medium. The denominator $v + v_s$ is used if the source is moving away from the observer, the denominator $v - v_s$ is used if the source is moving toward the observer.

Sample Problem 1

A car travelling at 100.0 km/h sounds its horn as it approaches a hiker standing on the highway. If the car's horn has a frequency of 440 Hz and the temperature of the air is 0°C, what is the frequency of the sound waves reaching the hiker

(a) as the car approaches?
(b) after it has passed the hiker?

Solution

$t = 0°C$
$v = 332$ m/s
$v_s = 100.0$ km/h or 27.8 m/s
$f_2 = ?$

(a)
$$f_2 = f_1\left(\frac{v}{v \pm v_s}\right)$$
$$= 440\text{ Hz}\left(\frac{332\text{ m/s}}{332\text{ m/s} - 27.8\text{ m/s}}\right)$$
$$f_2 = 4.8 \times 10^2\text{ Hz}$$

The frequency as the car approaches is 4.8×10^2 Hz.

(b) $$f_2 = f_1\left(\frac{v}{v \pm v_s}\right)$$

$$= 440 \text{ Hz}\left(\frac{332 \text{ m/s}}{332 \text{ m/s} + 27.8 \text{ m/s}}\right)$$

$$f_2 = 4.0 \times 10^2 \text{ Hz}$$

The frequency after the car passes the hiker is 4.0×10^2 Hz.

Practice

Understanding Concepts

1. You are standing at a railway crossing. A train approaching at 125 km/h sounds its whistle. If the frequency of the whistle is 442 Hz and the air temperature is 20.0°C, what frequency do you hear when the train approaches you? when the train has passed by you?
2. A car sounds its horn (502 MHz) as it approaches a pedestrian by the side of the road. The pedestrian has perfect pitch and determines that the sound from the horn has a frequency of 520 Hz. If the speed of sound that day was 340 m/s, how fast was the car travelling?

Answers

1. 492 Hz; 401 Hz
2. 13 m/s

Doppler Shift

Although the Doppler effect was first explained in relation to sound waves, it may be observed in any moving object that emits waves. The change in frequency and resulting change in wavelength is called the Doppler shift. Astronomers use the Doppler effect of light waves to estimate the speed of distant stars and galaxies relative to that of our solar system.

Short-range radar devices, such as those used by the police, work on the Doppler shift principle to determine the speed of a car (**Figure 2**). Radar waves from a transmitter in the police car are reflected by an approaching car and arrive back at a radar receiver in the police car with a slightly higher frequency. The original waves and the reflected waves are very close together in frequency, and beats are produced when the two are combined. The number of beats per second is directly related to the speed of the approaching car. This beat frequency is electronically translated into kilometres per hour and displayed on a meter or paper chart in the police car. Radar devices automatically correct for the movement of the police car. Similar techniques are used to track the path of satellites circling the Earth, to track weather systems, in ultrasonic and infrared detectors used in home security systems, and to measure the speed of a pitched baseball.

(a)

(b)

Figure 2
The Doppler effect is applied in the use of radar to determine the speed of vehicles on a highway. The radar system in a moving police cruiser can determine the speed of a car ahead or behind, travelling in **(a)** the same direction or **(b)** in opposite directions.

Supersonic Travel

subsonic speed: speed less than the speed of sound in air

Mach number: the ratio of the speed of an object to the speed of sound in air; Mach number = $\frac{\text{speed of object}}{\text{speed of sound}}$

supersonic speed: speed greater than the speed of sound in air

Objects travelling at speeds less than the speed of sound in air have **subsonic speeds.** When the speed of an object equals the speed of sound in air at that location, the speed is called Mach 1. The **Mach number** of a source of sound is the ratio of the speed of the source to the speed of sound in air at that location. Thus, at 0°C near the surface of Earth, Mach 2 is 2×332 m/s = 664 m/s. Speeds greater than Mach 1 are **supersonic.** Speeds for supersonic aircraft, such as the Concorde (**Figure 3**) and fighter aircraft, are given in terms of Mach number rather than kilometres per hour.

Figure 3
A Concorde passenger airplane

Sample Problem 2

The speed of sound at sea level and 0°C is 332 m/s, or approximately 1200 km/h. At an altitude of 10 km, it is approximately 1060 km/h. What is the Mach number of an aircraft flying at an altitude of 10 km with a speed of 1800 km/h?

Solution

$$\text{Mach number} = \frac{\text{speed of object}}{\text{speed of sound}}$$
$$= \frac{1800 \text{ km/h}}{1060 \text{ km/h}}$$
$$\text{Mach number} = 1.7$$

The Mach number of an aircraft flying at 1060 km/h is 1.7.

Practice

Understanding Concepts

3. What is the Mach number of an aircraft travelling at sea level at 0°C with a speed of
   (a) 1440 km/h? (b) 920 km/h?
4. A military interceptor airplane can fly at Mach 2.0. What is its speed in kilometres per hour at sea level and at 0°C?
5. What is the Mach number of a plane travelling at each of the following speeds at sea level in air with a temperature of 12°C?
   (a) 1020 m/s (b) 170 m/s (c) 1836 km/h

Answers

3. (a) 1.2
   (b) 0.77
4. 2.4×10^3 km/h
5. (a) 3.0
   (b) 0.50
   (c) 1.5

DID YOU KNOW ?

Mach Number

This ratio is named after Ernst Mach (1838–1916), an Austrian physicist and philosopher.

Breaking the Sound Barrier

sound barrier: a high-pressure region produced as an airplane approaches a speed of Mach 1

A static, or stationary, source radiates sound waves in concentric spheres (**Figure 4(a)**). An airplane radiates spheres of sound waves from successive positions. Sphere 1 in **Figure 4(b)** was produced by a subsonic airplane at position 1, sphere 2 at position 2, and so on. Note that because the aircraft was moving, the wavefronts were farther apart behind it than they were in front of it.

When an airplane is flying at the speed of sound, the wavefronts in front of the airplane pile up, producing an area of very dense air, or intense compression, called the **sound barrier.** To exceed the speed of sound, extra thrust is needed until the aircraft "breaks through" the sound barrier. Unless the aircraft has been designed to cut through this giant compression, it will be buffeted disastrously. Only specially constructed aircraft can withstand the vibrations caused in

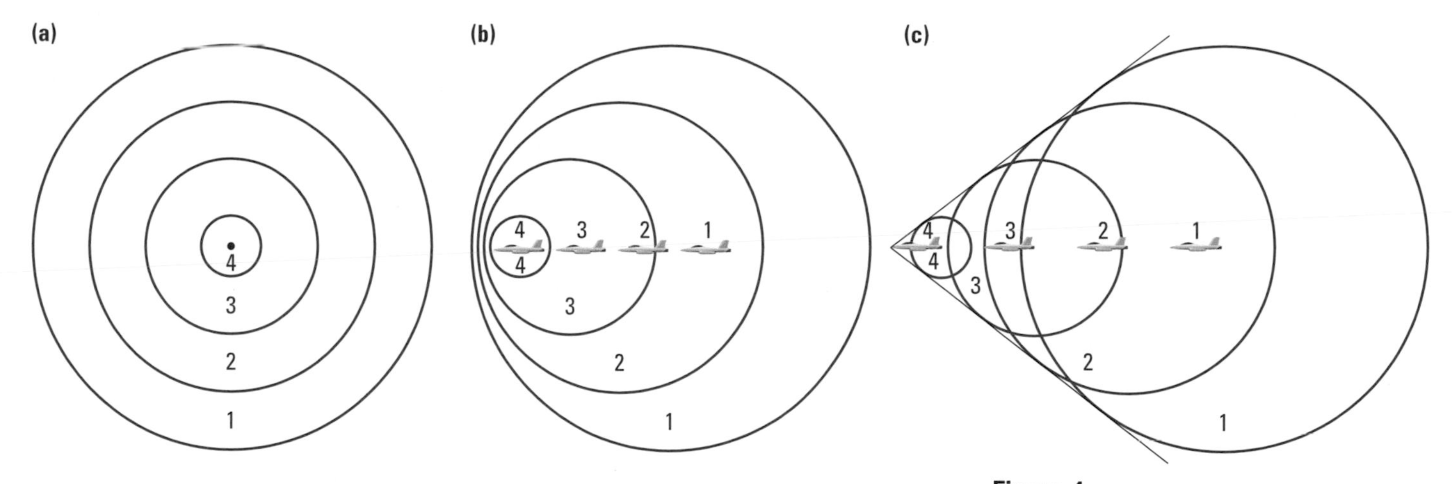

Figure 4
(a) Stationary source
(b) Subsonic source
(c) Supersonic source

breaking through the sound barrier to reach supersonic speeds. In present-day supersonic aircraft, such as the Concorde, only slight vibration is noticed when the sound barrier is crossed.

At supersonic speeds, the spheres of sound waves are left behind the aircraft (**Figure 4**(c)). These sound waves interfere with one another constructively, producing large compressions and rarefactions along the sides of an invisible double cone extending behind the airplane, from the front, and from the rear. This intense acoustic pressure wave sweeps along the ground (**Figure 5**) in a swath having a width of approximately five times the altitude of the aircraft. This is

DID YOU **KNOW ?**

Record-Breaking Speed

The record high speed of a fixed-wing aircraft is approximately Mach 25. It is held by the space shuttle *Columbia* and was achieved where the speed of sound was about 300 m/s.

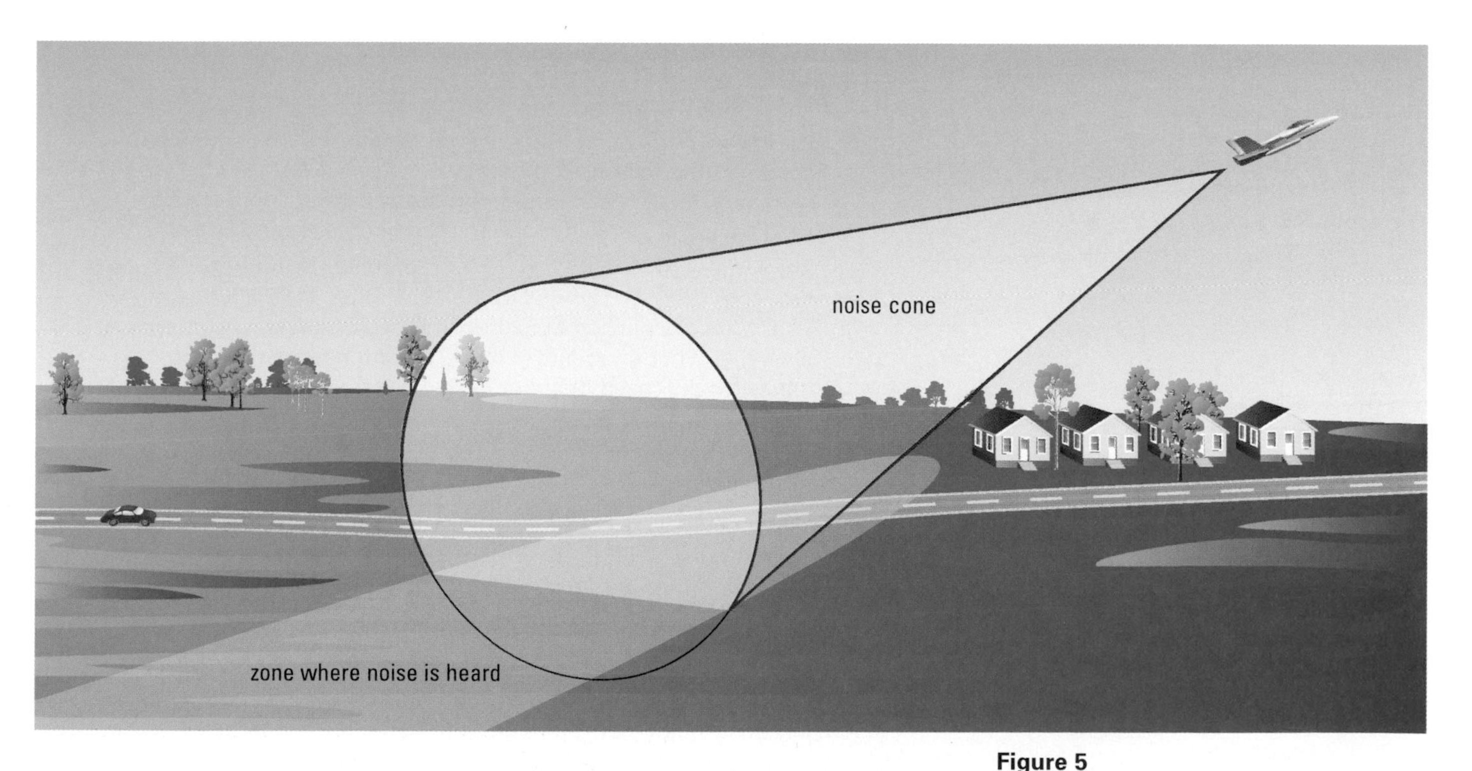

Figure 5
The acoustic pressure waves trail behind a supersonic aircraft.

sonic boom: an explosive sound that radiates from an aircraft travelling at supersonic speeds

DID YOU KNOW ?

Cracking the Whip

When a circus or rodeo performer snaps a whip quickly enough, a loud cracking sound is produced. This sound is actually a small sonic boom, created when the tip of the whip moves through the air faster than the speed of sound.

usually referred to as a **sonic boom**. The sonic boom is heard as two sharp cracks, like thunder or a muffled explosion.

For an airplane flying faster than the speed of sound at a height of 12 km, the sonic boom is produced for 30 km on either side of the flight path. Unless it comes from a supersonic aircraft at a low altitude, the sonic pressure wave is not strong enough to cause any damage on the ground, although the sudden noise may startle or frighten human beings and animals.

It is believed that most ecosystems can tolerate random sonic booms. Recurring booms over a long period, on the other hand, might upset them. Supersonic commercial aircraft, as a result, are restricted by many countries to subsonic speeds except over water. For example, the Concorde flying from London to New York does not go faster than the speed of sound until it is well over the ocean and it reduces speed below Mach 1 when it approaches land. Most of the training of fighter pilots occurs in areas where there are few inhabitants or over water. Aside from the annoyance and distraction factor, the breaking of the sound barrier can also break windows!

SUMMARY Doppler Effect and Supersonic Travel

- Doppler shift is the perceived frequency shift when a source of sound moves relative to an observer.
- Doppler shift is used to measure the speed of moving objects.
- The Mach number is used to measure supersonic speeds.
- Objects exceeding the speed of sound break the sound barrier, creating sonic booms.
- Sonic booms create disturbances for animals and humans.

Section 7.10 Questions

Understanding Concepts

1. State what happens to the apparent frequency of a sound source in each of the following situations:
   (a) The listener is stationary and the source is approaching.
   (b) The listener is stationary and the source is receding.
   (c) The source is stationary and the listener is approaching.
   (d) The source is stationary and the listener is receding.
2. A car's horn is pitched at 520 Hz. If the car travels by someone at 26 m/s, at what frequency will the person hear it as the car moves away (speed of sound is 344 m/s)?
3. Assuming that the speed of sound at a certain altitude is 330 m/s, calculate the speed of an airplane that is travelling at
   (a) Mach 0.70 (b) Mach 4.2
4. A plane travelling at 260 m/s has a speed of Mach 0.80. What is the speed of sound in the air?
5. Why is it difficult for an aircraft to break through the sound barrier?

Applying Inquiry Skills

6. You are travelling in a car near a speeding train. The train whistle blows, but you fail to hear the Doppler effect. What conditions might prevent you from hearing it?

Chapter 7 Summary

Key Expectations

Throughout this chapter, you have had opportunities to

- define and describe the concepts and units related to mechanical waves (e.g., longitudinal wave, cycle, amplitude, wavelength); (7.1, 7.2, 7.3, 7.4, 7.5, 7.6, 7.7, 7.9, 7.10)
- describe and illustrate the properties of longitudinal waves in different media, and analyze the velocity of waves travelling in those media in quantitative terms; (7.2, 7.3, 7.6)
- compare the speed of sound in different media, and describe the effect of temperature on the speed of sound; (7.3)
- explain and graphically illustrate the principle of superposition, and identify examples of constructive and destructive interference; (7.8, 7.9)
- explain the conditions required for standing sound waves to occur; (7.9)
- explain the Doppler effect, and predict in qualitative terms the frequency change that will occur in a variety of conditions; (7.10)
- draw, measure, analyze, and interpret the properties of waves (reflection, diffraction, and interference) during their transmission in a medium and from one medium to another, and during their interaction with matter; (7.6, 7.7, 7.8, 7.9)
- explain how sounds are produced and transmitted in nature, and how they interact with matter in nature; (7.1, 7.5, 7.6, 7.9)
- describe how knowledge of the properties of waves is applied to the design of buildings (e.g., with respect to acoustics) and to various technological devices, as well as in explanations of how sounds are produced and transmitted in nature, and how they interact with matter in nature (e.g., how organisms produce or receive infrasonic, audible, and ultrasonic sounds); (7.6)
- evaluate the effectiveness of a technological device related to human perception of sound; (7.5)
- identify sources of noise and explain how such noise can be reduced by the erection of highway noise barriers or the use of protective headphones; (7.4, 7.5, 7.7, 7.8, 7.6)

Key Terms

sound
infrasonic
ultrasonic
pitch
sound intensity
decibel
echo
sonar
echolocation
beats
beat frequency
Doppler effect
subsonic speed
Mach number
supersonic speed
sound barrier
sonic boom

Make a Summary

In this chapter, you learned how sound energy is produced, transmitted, and received, and how sound can undergo reflection, refraction, and interference. You have also learned about various applications, such as ultrasonic sounds and supersonic speeds. To summarize the ideas found in this chapter, draw two quarter-sized circles far apart on a page. In one circle put *"SOURCES"* and in the other put *"RECEIVERS."* Add as many concepts as possible to the page, and show how the concepts link to one another. For example, you can start by showing the production of sound from a tuning fork (the source), the transmission of the sound through the air, and the receiving of the sound in a human ear (the receiver).

Reflect on your Learning

Revisit your answers to the Reflect on Your Learning questions at the beginning of this chapter.

- How has your thinking changed?
- What new questions do you have?

Chapter 7 Review

Understanding Concepts

1. What vibrates to produce the sound originating from each of the following?
   (a) an acoustic guitar
   (b) an electric doorbell
   (c) a stereo system
2. The tine of a vibrating tuning fork passes through its central (rest) position 600 times in 1.0 s. How many of each of the following has the prong made?
   (a) complete cycles
   (b) complete waves
   (c) compressions
   (d) rarefactions
3. Describe how sound energy is transferred through air from a source to a listener.
4. Calculate the speed of a sound wave in air at temperature 0°C if its frequency and wavelengths are
   (a) 384 Hz and 90.0 cm
   (b) 256 Hz and 1.32 m
   (c) 1.50 kHz and 23.3 cm
5. Find the wavelength that corresponds to each of the following frequencies. (Assume that the speed of sound is 342 m/s.)
   (a) 20.0 Hz
   (b) 2.0×10^4 Hz
6. What is the speed of sound in air if the temperature is
   (a) 15°C?
   (b) −40.0°C?
7. Thunder is heard 3.0 s after the lightning that caused it is seen. If the air temperature is 14°C, how far away is the lightning?
8. Find the period of a sound wave with a wavelength of 0.86 m, if the temperature is 20.0°C.
9. A man sets his watch at noon by the sound of a factory whistle 4.8 km away. If the temperature of the air is 20.0°C, how many seconds slow will his watch be?
10. At a baseball game, a physics student with a stopwatch sits behind the centre-field fence marked 136 m. He starts the watch when he sees the bat hit the ball and stops the watch when he hears the resulting sound. The time observed is 0.40 s.
    (a) How fast is the sound energy travelling?
    (b) What is the air temperature?
11. A marching drummer strikes the drum every 1.0 s. At what distance from the drummer will a listener hear the drum at the instant when the drumstick is farthest from the drum? The air temperature is 20.0°C.
12. Orca whales make sounds so loud that hydrophones can pick them up from a great distance in seawater. Assume that an orca travelling at 55 km/h is approaching an observer using a hydrophone. How long after the sound is heard will the whale take to reach the observer from the initial distance of 8.0 km?
13. The oboe sounds the note A (440 Hz) when the orchestra tunes up before a performance. How many vibrations does the oboe make before the sound reaches a person seated in the audience 45 m away? Assume the temperature in the concert hall is 20.0°C.
14. A ship sends a sound signal simultaneously through the air and through the salt water to another ship 1.0 km away. Using 336 m/s as the speed of sound in air and 1450 m/s as the speed of sound in salt water, calculate the time interval between the arrival of the two sounds at the second ship.
15. A 21-gun salute is about to be given by a Navy ship anchored 3.0 km from a person who is swimming near the shore. The swimmer sees a puff of smoke from the gun, quickly pops her head under water, and listens. In 2.0 s she hears the sound of the gun; she then lifts her head above water. In another 8.6 s, she hears the sound from the same shot coming through the air. Find the speed of the sound in the water and in the air.
16. By what factor is a sound intensity level of 70 dB greater than one of 30 dB?
17. A 120-dB sound is reduced to 10^{-7} times the intensity level after 0.5 s. What is the new intensity level?
18. An armed forces ship patrolling the ocean receives its own sound signals back by underwater reflection 4.5 s after emitting them. How far away is the reflecting surface in metres?
19. A ship is 2030 m from an above-water reflecting surface. The temperature of the air and water is 25°C.
    (a) What is the time interval between the production of a sound wave and the reception of its echo in air?
    (b) What would the interval be if the reflecting surface was under fresh water?

20. Two sources with frequencies of 300.0 Hz and 306 Hz are sounded together. How many beats are heard in 4.0 s?
21. Two tuning forks are sounded together, producing three beats per second. If the first fork has a frequency of 300.0 Hz, what are the possible frequencies of the other fork?
22. Four tuning forks with frequencies of 512 Hz, 518 Hz, 505 Hz, and 503 Hz are available. What are all the possible beat frequencies when any two tuning forks are sounded together?
23. Plasticine (which lowers the pitch) is added to one tine of the tuning fork of unknown frequency referred to in question 21. The number of beats decreases to one. What was the frequency of the unknown fork before Plasticine was added to one of its tines? What are the possible new frequencies of the fork with Plasticine?
24. A 400.0-Hz string produces 10 beats in 4.0 s when sounded at the same time as a tuning fork. When Plasticine is placed on one tine of the tuning fork, the number of beats increases. What was the frequency of the fork before the Plasticine was added to it?
25. Two nearly identical tuning forks, one of which has a frequency of 384 Hz, produce seven beats per second when they vibrate at the same time. When a small clamp is placed on the 384 Hz fork, only five beats per second are heard. What is the frequency of the other fork?
26. At a certain altitude the speed of sound is 1066 km/h. Jet A is travelling at Mach 1.2 [W] and jet B is travelling at Mach 2.0 [E] at a safe distance from A. Both velocities are relative to the ground.
    (a) What is the velocity of each jet relative to the ground in kilometres per hour?
    (b) What is the speed of one jet relative to the other?
27. Estimate how long it would take the space shuttle *Endeavour*, travelling at Mach 25, where the speed of sound is 296 m/s, to travel once around the Earth at an altitude of 3.5×10^2 km. Assume that the Earth's radius is 6.4×10^3 km.
28. An oncoming ambulance moving at 100.0 km/h with siren blaring at a steady 850 Hz approaches a person standing on a sidewalk. If the speed of sound is 340 m/s, what frequency will she hear as the ambulance approaches and as the ambulance moves away from her?
29. Why would you not be surprised to learn that the speed of sound in seawater is the same as in human tissue? Would the speed of sound in bone be higher or lower than it is in human tissue? Why?
30. The power of the sound emitted from an average conversation between two people is 1.0×10^{-6} W. How many two-people conversations would it take to keep a 60-W bulb glowing, assuming that all the sound energy could be converted into electrical energy?

Applying Inquiry Skills

31. Rasa and Jamie are rehearsing for a piano recital where they will play a piece requiring two pianos. Rasa strikes the middle C key on her piano and, a little while later, Jamie strikes the same key on his piano. The pitch of the two notes sounds the same.
    (a) Without using electronic devices, how can Rasa and Jamie determine whether the pitch is exactly the same?
    (b) If the pitch turns out to be slightly different, what steps can be taken to correct the problem?

Making Connections

32. Geologists, in their search for oil and gas, use the speed of sound in various materials in their search. Write a short report (one page) based on your research into this topic. Follow the links for Nelson Physics 11, Chapter 7 Review.

GO TO www.science.nelson.com

33. Discuss the advantages and disadvantages of supersonic air travel and make a prediction about the future for planes like the Concorde.

Exploring

34. There is no air on the Moon. How do you think astronauts on the Moon hear each other speak? You can check the accuracy of your prediction by searching the NASA Web site. Follow the links for Nelson Physics 11, Chapter 7 Review.

GO TO www.science.nelson.com

35. The areas surrounding Goose Bay, Labrador are used by many NATO nations for pilot training in high-speed aircraft. Sonic booms are a common daily occurrence. After researching the issues, be prepared to take the side of the government or that of the people who live in the area.

Chapter

8

Music, Musical Instruments, and Acoustics

In this chapter, you will be able to

- apply the terms pitch, consonance, dissonance, and octave
- distinguish between the scientific musical scale and the musicians' scale
- identify and explain certain properties of a vibrating string
- use an air column to investigate certain properties of sound
- describe how vibration and resonance are applied in stringed, wind, and percussion instruments
- describe the function of the parts of the human body that affect the quality of the voice
- describe how the components of an audio system affect the quality of the amplified sound
- describe the function of the main parts of an electronic instrument
- explain how the acoustics of a room affect how well a sound is heard

You have learned about the properties of vibrations, waves, and sound in this unit. You will now use your knowledge to analyze the scientific differences between music and noise. You will also apply what you know about the behaviour of waves to the study of music and musical instruments. Among the important questions you will learn to answer are these: What are the characteristics of musical sounds? How do strings vibrate? How does resonance enhance sound? How do musical instruments produce pleasant sounds? How is music changed or altered electronically? How does the study of acoustics help improve the sound in auditoriums (**Figure 1**)?

Reflect on your Learning

1. What is the difference between noise and music?
2. What major components are involved with producing sound in the following instruments: guitar, violin, piano, trombone, flute, and drum?
3. As you fill a bottle with water, the pitch of the sound heard from the bottle gradually increases. Why?
4. How do you account for the noise that you hear when you place a large seashell to your ear?
5. The piano sounds produced by a synthesizer are usually not exactly the same as the sounds from a traditional piano. Why?
6. When the crowd is cheering for a basketball team in a gymnasium, the sound can become garbled and deafening. This is not usually the case in an auditorium with a similar-sized crowd. Why?

Throughout this chapter, note any changes in your ideas as you learn new concepts and develop your skills.

Figure 1
Roy Thomson Hall, home of the Toronto Symphony Orchestra, has modern design features to provide a balance of reflection and absorption of sound for good acoustics.

Try This **Activity**

Seeing Sound

The study of music, whether or not you have a "musical ear," can be made objective by means of a setup that allows you to "see" sound as you hear it. One setup is a combination of a microphone, amplifier, and oscilloscope (**Figure 2**); another setup is to use a personal computer where the software, activated by a microphone, can replicate the oscilloscope. In either case, the sound received by the microphone can be displayed as a transverse wave, which can be analyzed.

Materials: oscilloscope, microphone, amplifier, tuning fork

- Strike a tuning fork and place it near to, but not touching, the microphone.
- Make adjustments to the oscilloscope so that when the tracing stops, at least one complete wavelength is displayed.
- Sketch the tracing you see on the screen.
- Repeat the previous steps for a variety of single sounds and noises.
  (a) Periodic sounds repeat the same pattern over and over and have a definite frequency and wavelength. Random sounds do not repeat and do not have definite frequencies and wavelengths. Classify each of the sounds you see.

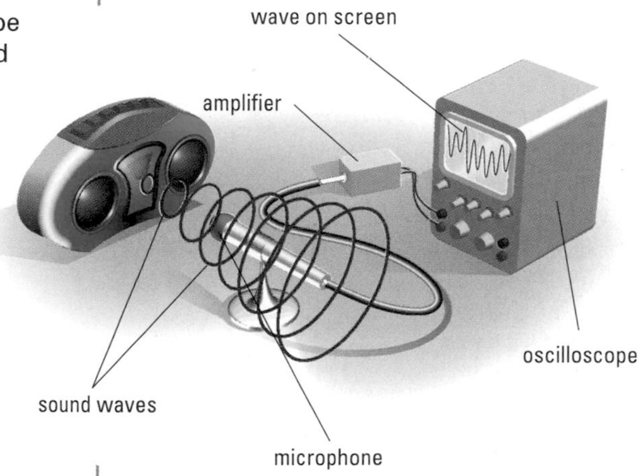

Figure 2
Using an oscilloscope to "see" sound

8.1 Music and Musical Scales

How would you describe the difference between noise and music? You could say that noise is sound that is unpleasant and annoying, while music is sound that is pleasant and harmonious. However, such a description of the difference between the two depends largely on individual judgement. There is a scientific difference between music and noise.

music: sound that originates from a source with one or more constant frequencies

noise: sound that originates from a source where the frequencies are not constant

A musical note originates from a source vibrating in a uniform manner with one or more constant frequencies. **Music** is the combination of musical notes. With an oscilloscope, music is displayed as a constant waveform, as illustrated with the sound from a tuning fork (**Figure 1**(a)). In contrast, **noise** originates from a source where the frequencies are constantly changing in a random manner. Displayed on an oscilloscope, noise does not have a constant waveform (**Figure 1**(b)).

(a)

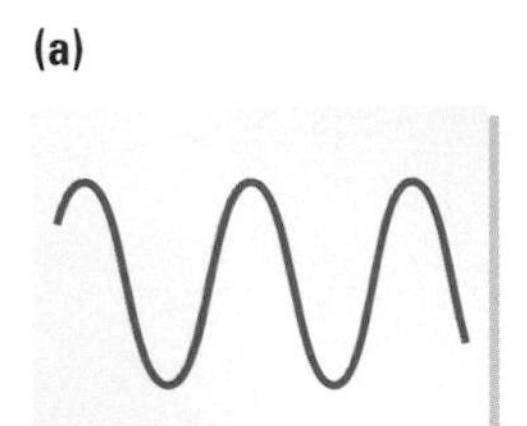

(b)

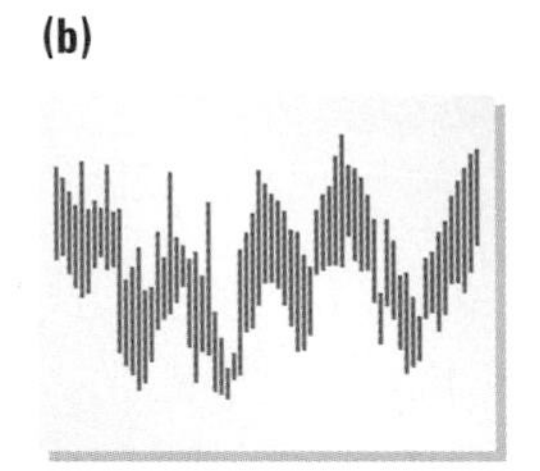

(c)

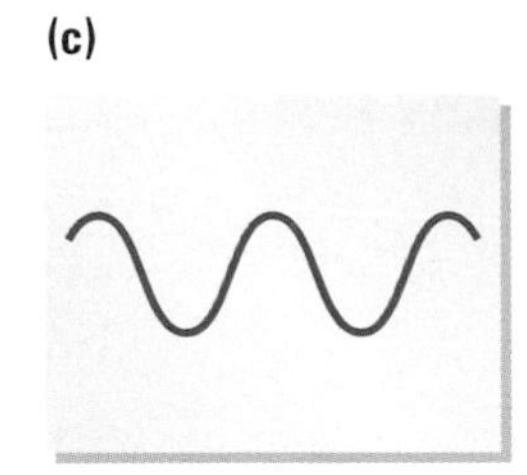

(d)

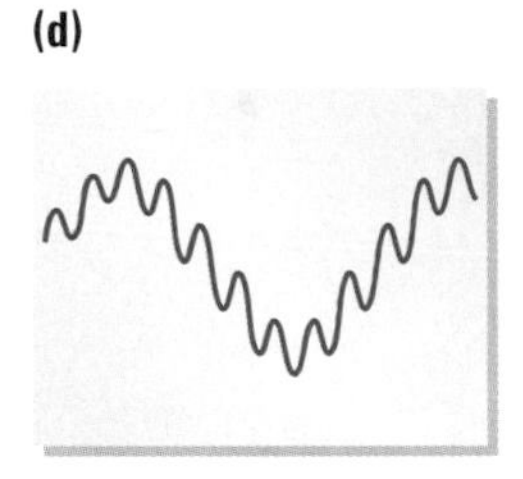

(e)

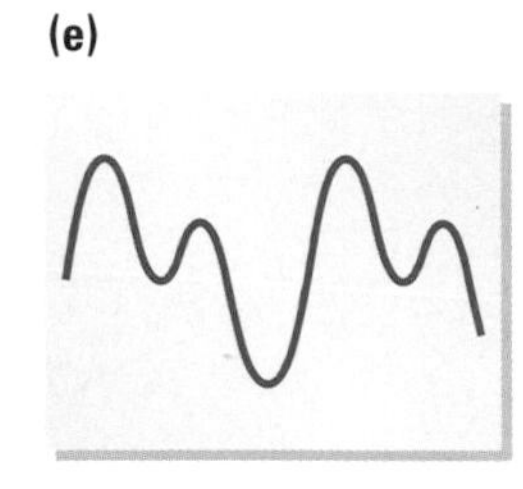

Figure 1
Sound traces on an oscilloscope. What do you think produces the "ooooo" sound?
(a) Pure sound from a tuning fork
(b) Random noise
(c) "oooooo" sound
(d) "eeeee" sound
(e) "awwww" sound

There are three main characteristics of musical sounds: pitch, loudness (discussed in Chapter 7), and quality. Each of these characteristics depends not only on the source of the musical sound, but also on the listener. Thus, they are called subjective characteristics. **Figure 2** shows some oscilloscope displays of sound waves that can be used to help us analyze the characteristics of musical sounds in a scientific, objective way.

(a)

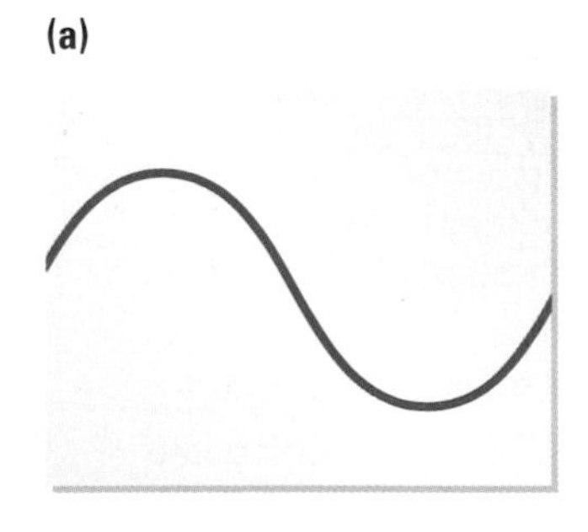

(b)

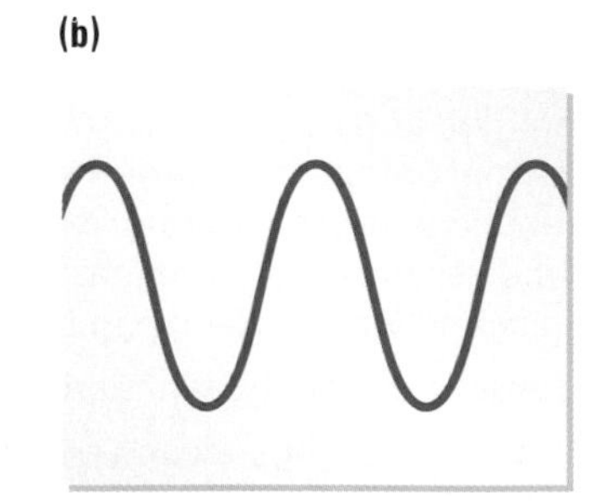

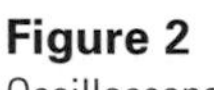

Figure 2
Oscilloscope traces of pitch and wavelength
(a) Low frequency, low pitch, long wavelength
(b) High frequency, high pitch, short wavelength

If you have ever been near a pond on a summer evening, you might have heard the crickets chirping and the bullfrogs croaking. You could easily distinguish between the sounds because cricket sounds have a high pitch and bullfrog sounds have a low pitch. These sounds are different because their frequencies are different.

A pitch wheel, shown in **Figure 3**, is a device that demonstrates the relationship between pitch and frequency of vibration. It consists of a set of three or four wheels of equal diameter, each with a different number of teeth. As the device is spun by an electric motor, a piece of paper is held up to each wheel in turn.

The paper vibrates with the lowest frequency when it touches the wheel with the fewest teeth. This action produces the sound of lowest pitch. The more teeth a wheel has, the higher the pitch of the sound produced. Thus, we find that pitch increases as frequency increases.

As the pitch of a sound changes, the waveform also changes. As pitch increases, wavelength decreases. Although there is an objective relationship between pitch and frequency, the actual pitch a person hears depends on other factors, including the observer, the complexity of the sound, and, to a certain extent, the loudness of the sound. Thus, pitch is a subjective characteristic.

In music, a **pure tone** is a sound where only one frequency is heard. Musical sounds are not normally pure tones—they usually consist of more than a single sound. In general, two or more sounds are harmonious if their frequencies are in a simple ratio. Harmonious pairs of sounds have high **consonance**; unpleasant pairs of sounds have high **dissonance**, or low consonance. Unison is a set of sounds of the same frequency. An **octave** has sounds with double the frequency of the sounds in another frequency. For example, a 200-Hz sound is one octave above a 100-Hz sound.

Figure 3
A pitch wheel

pure tone: a sound that consists of one frequency

consonance: combinations of sounds of specific frequencies that are pleasing to the ear; the frequencies are often in a simple ratio

dissonance: combinations of sounds of specific frequencies that have a harsh effect; the frequencies are not in a simple ratio

octave: sounds that differ in frequency by a factor of two

Practice

Understanding Concepts

1. Describe the difference between objective and subjective properties or variables. Give examples of each relative to sound and music.
2. (a) Which of the sound traces shown in **Figure 1** are pure tones?
   (b) Which of the traces has the highest dissonance? Why?

Musical Scales

A musical scale is a set of pure tones with increasing or decreasing frequency. There are two common musical scales: the scientific musical scale and the musicians' scale. On the scientific musical scale, the standard frequency is 256 Hz and is based on the number 2^8. The standard frequency can be multiplied by simple ratios to give the entire scale. Tuning forks in classrooms are often labelled with the frequencies of this scale. The notes G and C sound pleasant together because their frequencies are in the ratio 3 : 2. Another pleasant-sounding pair is B and E because the ratio of their frequencies ($\frac{480}{320}$) is 3 : 2. Because they are simple ratios, we say the notes have high consonance.

The musicians', or equitempered, scale has a standard frequency of 440 Hz, which is the frequency of the note A above middle C on the piano. An octave below has a frequency of 220 Hz and an octave above has a frequency of 880 Hz.

Figure 4 shows parts of the piano scale, including the notes, their frequencies, and their staff notations. This is the scale used in tuning most musical instruments.

The calculations for the frequencies of the musicians' scale are based on two facts: (1) a note one octave above another is double the frequency of the other and (2) there are exactly 12 equal intervals per octave. On a piano keyboard, for example, there are five black keys and eight white keys per octave—13 keys, 12 frequency intervals. (Note that the eighth white key is the last note of the first octave and the first note of the next octave.)

DID YOU KNOW ?

A Standard Frequency

The standard frequency A = 440 Hz is used in many countries throughout the world. To allow the public to check for that frequency, government agencies broadcast it by telephone, on the radio, and on the Internet.

Figure 4
The musicians' scale illustrated on a piano keyboard

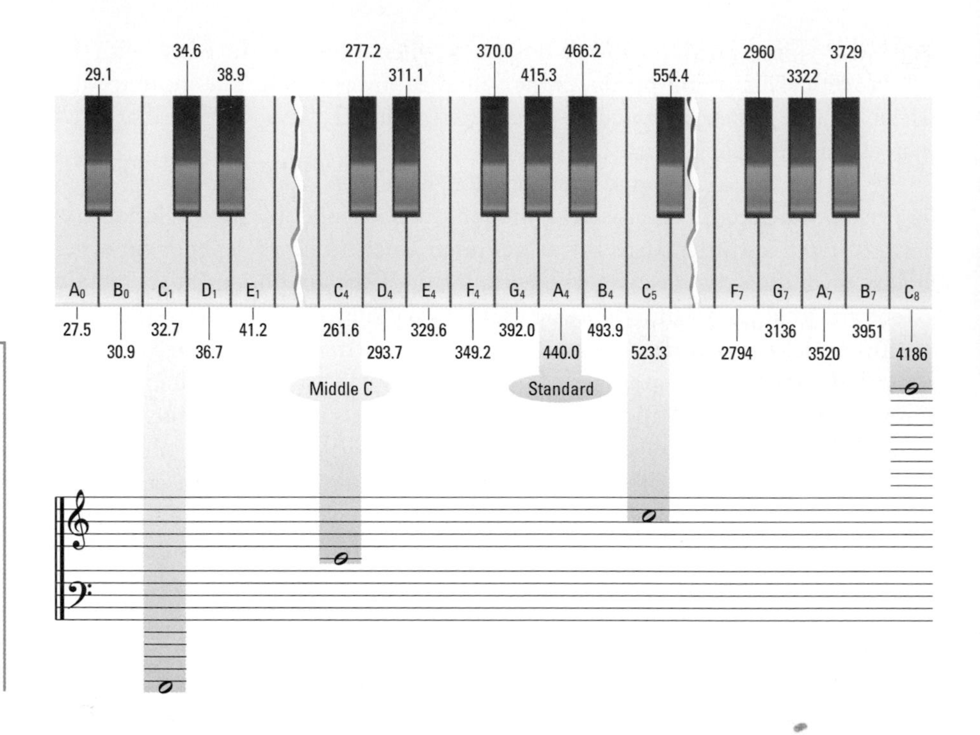

DID YOU KNOW ?

The Musicians' Scale

The musicians' scale was standardized in 1953. The advantage of the scale is that it allows musicians the freedom to change from one key to another within one piece of music or to play a variable-frequency instrument, such as a trumpet, along with a fixed-frequency instrument, such as a piano. This flexibility is possible because all whole-tone intervals are equal in size and two semitone intervals are exactly equal to one whole tone.

Practice

Understanding Concepts

3. For a sound with a frequency of 220 Hz, determine the frequency of a sound that is
   (a) one octave higher
   (b) one octave lower
4. For each pair of notes listed, determine the ratio of their frequencies as a simple fraction, then determine which pair has the higher consonance.
   (a) 600 Hz and 400 Hz
   (b) 800 Hz and 700 Hz

Answers

3. (a) 440 Hz
   (b) 110 Hz
4. (a) 3 : 2
   (b) 8 : 7

SUMMARY Sound Waves and Music

- Longitudinal sound waves can be displayed on a computer monitor or oscilloscope as transverse waves for easier analysis.
- Music consists of notes with constant frequencies; noise is sound with constantly changing frequency.
- Pitch, loudness, and quality are all subjective characteristics of sound since they depend on the perception of the listener.
- If the frequency of a sound increases, pitch of the note increases.
- Sounds are harmonious and have consonance if their frequencies are simple ratios.
- A musical interval of an octave has a sound that is double the frequency of the original sound.
- The two common musical scales are the scientific musical scale, based on 256 Hz, and the musicians' scale, based on 440 Hz.

Section 8.1 Questions

Understanding Concepts

1. What does the pitch of a musical sound depend on?
2. A portable electric saw is used to cut through a wooden log. What happens to the pitch of the sound from the saw as the blade cuts farther into the log? Why?
3. State the frequency of a note one octave below and one octave above the note with the following frequency:
   (a) 310 Hz
   (b) 684 Hz
   (c) 1.2×10^3 Hz
4. For each pair of notes listed, determine the ratio of their frequencies as a fraction and state whether their consonance when sounded together is high or low.
   (a) 750 Hz and 500 Hz
   (b) 2000 Hz and 1000 Hz
   (c) 820 Hz and 800 Hz

Applying Inquiry Skills

5. Predict and draw the shape of an oscilloscope trace for
   (a) a person whistling
   (b) an impact sound such as a clap
   (c) a plucked guitar string

   Check your predictions by displaying the sounds on an oscilloscope or computer monitor.

8.2 Vibrating Strings

As you have seen, vibrating objects produce sound. Each string on a guitar, for example, vibrates to produce sounds of different frequencies. The tuning pegs alter the tension of the strings, which changes the frequency. It is possible to have two strings of the same length and the same tension, but with different frequencies. This can occur because the frequency is also affected by the diameter of the string and the material of which it is made. Guitar strings can be made of nylon, steel, or aluminum. The pitch of a guitar string may, of course, also be altered by changing the effective length of the string. When you press a string against the neck of a guitar, the effective part of the string is shortened and the frequency is increased. The frequency of a vibrating string is determined by four factors: its length, its tension, its diameter, and the density of the material.

Since the frequency of a string is affected by its length, tension, diameter, and density (**Table 1**), all of these factors are taken into consideration when designing stringed musical instruments, such as the piano, guitar, and violin. When a piano is made, for example, it is theoretically possible to obtain the range of frequencies necessary (27 to 4096 Hz) by altering the length of the strings. However, this would be absurd because the shortest string would be 1.0 m in length and the longest string would be 150 m! This is avoided by winding wire around the bass strings, increasing both their diameter and density. To avoid having the treble strings too short, they are placed under greater tension. In a typical piano, the total tension on the steel frame by the more than 200 strings is over 2.7×10^4 N, which is equivalent to the gravitational force on a mass of 2700 kg.

Table 1 Relationship between Frequency and Length, Tension, Diameter, and Density of a String

Variable	Relationship to frequency (f)	Alternative expression
length (l)	$f \propto \frac{1}{l}$	$\frac{f_1}{f_2} = \frac{l_2}{l_1}$
tension (F)	$f \propto \sqrt{F}$	$\frac{f_1}{f_2} = \frac{\sqrt{F_1}}{\sqrt{F_2}}$
diameter (d)	$f \propto \frac{1}{d}$	$\frac{f_1}{f_2} = \frac{d_2}{d_1}$
density (D)	$f \propto \frac{1}{\sqrt{D}}$	$\frac{f_1}{f_2} = \frac{\sqrt{D_2}}{\sqrt{D_1}}$

Sample Problem 1

The D string of a violin is 30.0 cm long and has a natural frequency of 288 Hz. Where must a violinist place his or her finger on the string to produce a B note (384 Hz)?

Solution

$l_1 = 30.0$ cm
$f_1 = 288$ Hz
$f_2 = 384$ Hz
$l_2 = ?$

$$\frac{f_2}{f_1} = \frac{l_1}{l_2}$$

$$l_2 = \frac{f_1 l_1}{f_2}$$

$$= \frac{(288 \text{ Hz})(30.0 \text{ cm})}{384 \text{ Hz}}$$

$$l_2 = 22.5 \text{ cm}$$

The violinist's finger should be placed 7.5 cm from the end of the string.

Sample Problem 2

A piano string with a pitch of A (440.0 Hz) is under a tension of 1.4×10^2 N. What tension would be required to produce a high C (523.0 Hz)?

Solution

$f_1 = 440.0$ Hz
$f_2 = 523.0$ Hz
$F_1 = 1.4 \times 10^2$ N
$F_2 = ?$

$$\frac{f_2}{f_1} = \frac{\sqrt{F_2}}{\sqrt{F_1}}$$

$$\sqrt{F_2} = \frac{f_2}{f_1}\sqrt{F_1}$$

$$F_2 = \left(\frac{f_2}{f_1}\right)^2 F_1$$

$$= \frac{(523.0 \text{ Hz})^2(1.4 \times 10^2 \text{ N})}{(440.0 \text{ Hz})^2}$$

$$F_2 = 2.0 \times 10^2 \text{ N}$$

A tension with a force of 2.0×10^2 N would be required.

Practice

Understanding Concepts

1. Sketch a graph to show how the frequency of a vibrating string depends on the string's (a) length and (b) tension.
2. State what happens to the frequency of a vibrating string if there is a four-fold increase in the string's (a) diameter and (b) density.

Answers

2. (a) 4 times lower
   (b) 2 times lower

3. A 38-cm string on a violin under a tension of 24 N vibrates at 4.4×10^2 Hz. Determine the new frequency of the string in each case.
   (a) The length remains the same but the tension increases to 28 N.
   (b) The tension remains the same but the length changes to 32 cm.

Answers

3. (a) 4.8×10^2 Hz
   (b) 5.2×10^2 Hz

SUMMARY Vibrating Strings

- The frequency of a vibrating string is determined by its length, tension, diameter, and density.
- The frequency of a string is inversely proportional to its length $\left(f \alpha \frac{1}{l}\right)$.
- The frequency of a string is directly proportional to the square root of its tension $(f \alpha \sqrt{F})$.
- The frequency of a string is inversely proportional to its diameter $\left(f \alpha \frac{1}{d}\right)$.
- The frequency of a string is inversely proportional to the square root of its density $\left(f \alpha \frac{1}{\sqrt{D}}\right)$.

Section 8.2 Questions

Understanding Concepts

1. A note on a guitar string sounds "flat." What must be done to obtain the proper frequency?
2. A 1.0-m string has a frequency of 220 Hz. If the string is shortened to 0.80 m, what will its frequency become?
3. A string under a tension of 150 N has a frequency of 256 Hz. What will its frequency become if the tension is increased to 300 N? (Assume three significant digits.)
4. Two strings have the same diameter, length, and tension. One is made of brass (density = 8.70×10^3 kg/m^3) and the other is made of steel (density = 7.83×10^3 kg/m^3). If the frequency of the brass string is 440 Hz, what is the frequency of the steel string?
5. A steel string of diameter 1.0 mm produces a frequency of 880 Hz. What is the frequency of a steel string with the same length and tension if its diameter is
   (a) 2.0 mm?
   (b) 0.40 mm?

Applying Inquiry Skills

6. A *sonometer* is a sounding board with at least one string whose tension and length can be varied. (A diagram of one type of sonometer is shown in **Figure 3** of section 8.3.)
   (a) Describe how you would use a sonometer to demonstrate the relationship between the frequency of a vibrating string and the length of the string.
   (b) If your school has a sonometer, look at its features and accessories, and describe how you would control variables to illustrate how the frequency of the string depends on other factors.

8.3 Modes of Vibration—Quality of Sound

In section 6.8 you studied the interference of a standing wave pattern in a rope fixed at one end. A series of equally spaced antinodes (areas of maximum constructive interference) and nodes (points of destructive interference) were formed as the waves interfered after being reflected from the fixed end. Recall that at the fixed end, a node always occurs (**Figure 1**).

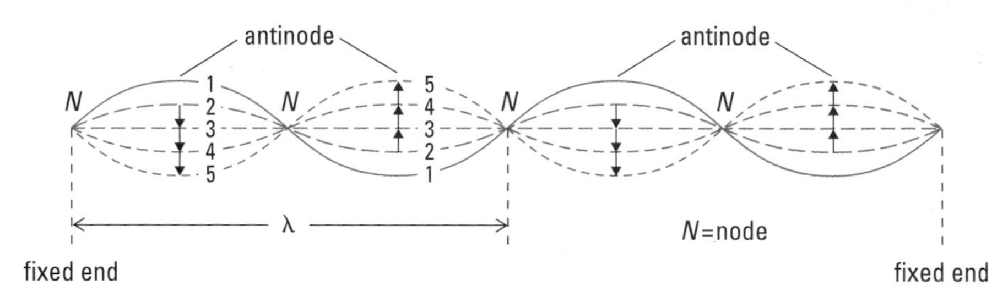

Figure 1
Motion of a string in which there is a standing wave

fundamental mode: simplest mode of vibration producing the lowest frequency

fundamental frequency: (f_0) the lowest natural frequency

overtones: the resulting modes of vibration when a string vibrates in more than one segment emitting more than one frequency

harmonics: whole-number multiples of the fundamental frequency

In a vibrating string stretched between two fixed points, nodes occur at both ends. Different frequencies may result depending on how many nodes and antinodes are produced. In its simplest, or **fundamental mode** of vibration, the string vibrates in one segment (**Figure 2**). This produces its lowest frequency, called the **fundamental frequency** (f_0). If the string vibrates in more than one segment, the resulting modes of vibration are called **overtones.** Since the string can only vibrate in certain patterns and always with nodes at each end, the frequencies of the overtones are simple, whole-numbered multiples of the fundamental frequency called **harmonics**, such as $2f_0$, $3f_0$, $4f_0$, and so on. Thus, the fundamental frequency (f_0) is the first harmonic, $2f_0$ (the first overtone) is the second harmonic, $3f_0$ (the second overtone) is the third harmonic, etc.

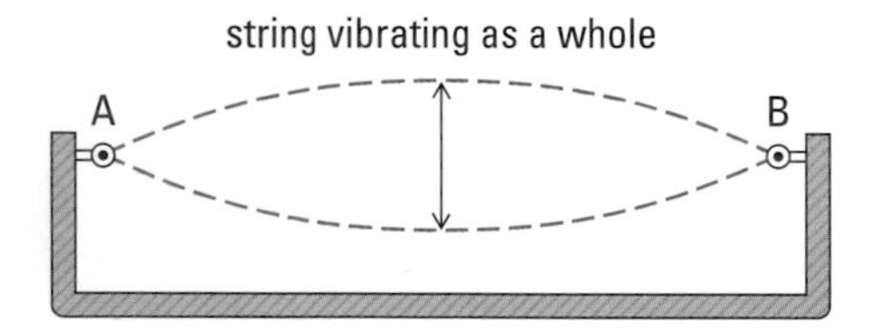

Figure 2
A string vibrating in its fundamental mode

The fundamental frequency and overtones can be demonstrated with a sonometer. In this device, the string is touched at its exact centre with a feather or lightly with a finger, and simultaneously stroked with a bow between the centre and the bridge. The string is able to vibrate in only two segments, producing the first overtone, which has twice the frequency of the fundamental. By adjusting the position of the feather or the bow, the string can be made to vibrate in three or more segments, which produces frequencies that are simple multiples of the fundamental frequency (**Figure 3**).

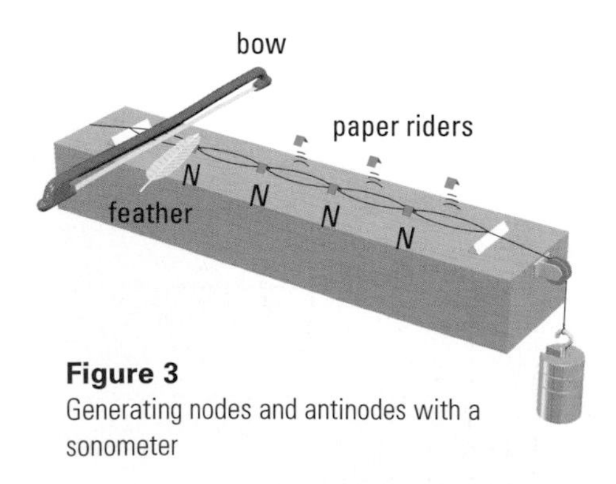

Figure 3
Generating nodes and antinodes with a sonometer

The strings of violins and other stringed instruments vibrate in a complex mixture of overtones superimposed on the fundamental (**Figure 4**). Very few vibrating sources can produce a note free of overtones. An exception is the tuning fork, but even it has overtones when first struck. However, because the overtones disappear quickly, the tuning fork is valuable in studying sound and tuning musical instruments.

The **quality of a musical note** depends on the number and relative intensity of the overtones it produces along with the fundamental. It is the element of quality that enables us to distinguish between notes of the same frequency and intensity coming from different sources; for example, we can easily distinguish between middle C on the piano, on the violin, and in the human voice.

quality of a musical note: a property that depends on the number and relative intensity of harmonics that make up the sound

(a) (b) (c)

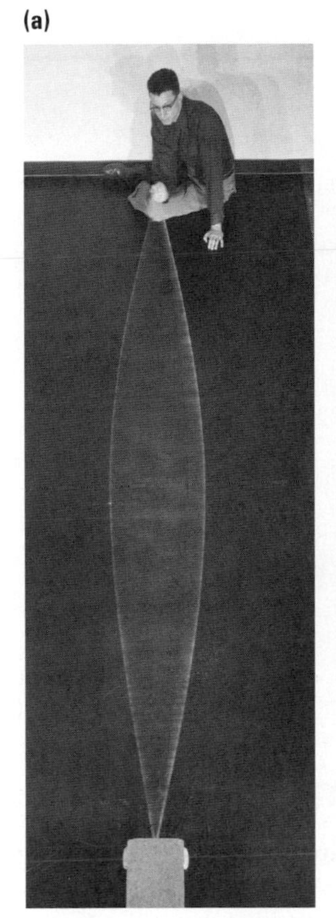

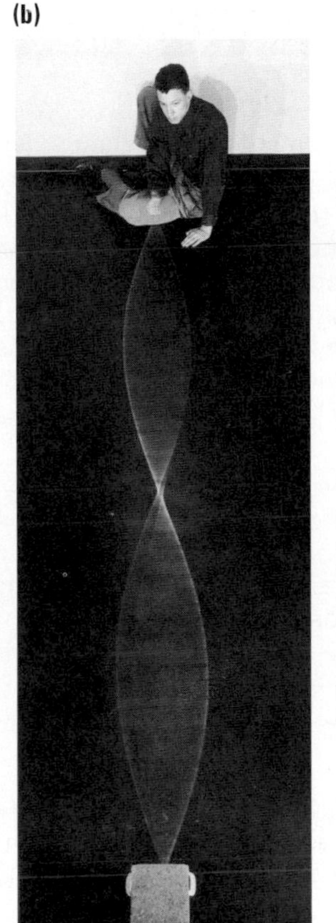

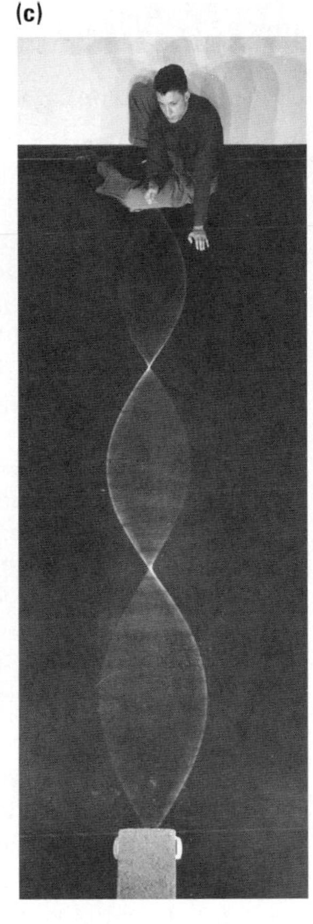

Figure 4
A string vibrating in a mixture of overtones
(a) Fundamental (first harmonic)
(b) First overtone (second harmonic)
(c) Second overtone (third harmonic)

Using an oscilloscope, you can see that the wave pattern produced by a tuning fork struck lightly is symmetrical because the fundamental is the only frequency present. If the tuning fork is struck sharply, it will also produce overtones, which will interfere with the fundamental. Oscilloscope patterns show that the resultant wave for a given frequency is unique for different instruments (**Figure 5**). The fundamental frequency sets the pitch of a musical note, but in some cases the overtones may be more intense than the fundamental. The overtone frequency structures for various instruments are shown in the figure.

(a)

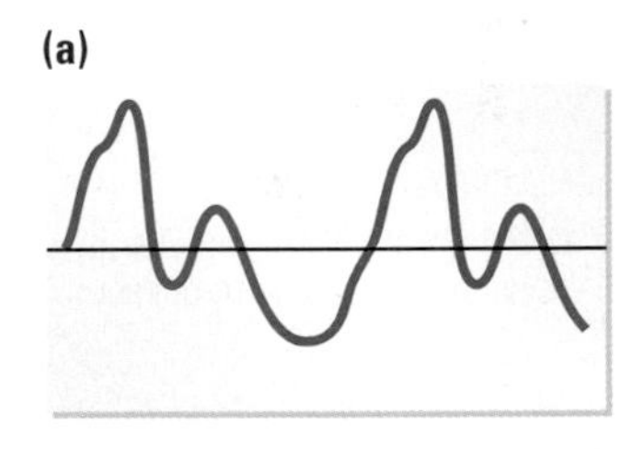

(b)

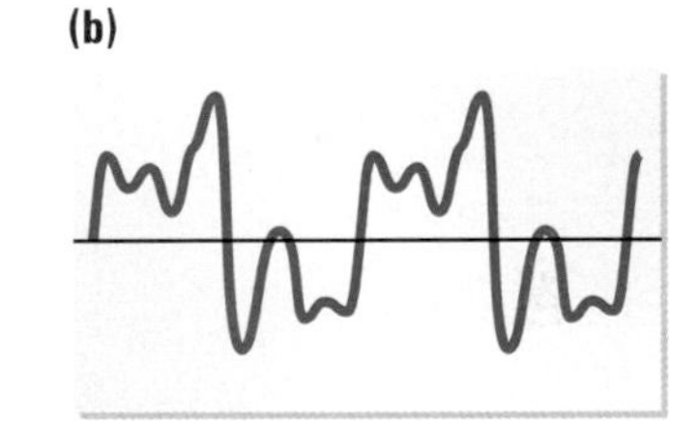

(c)

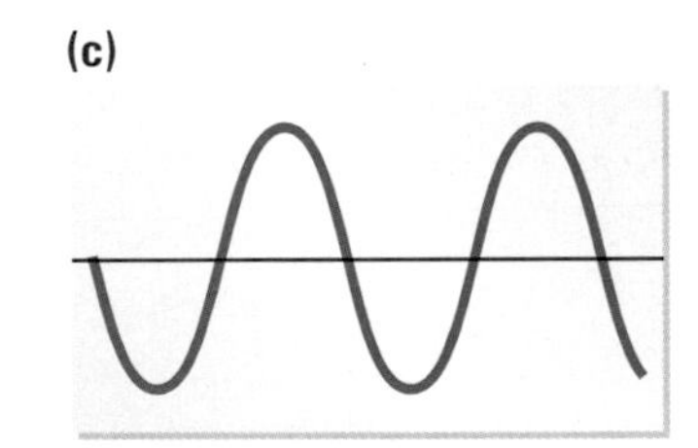

(d)

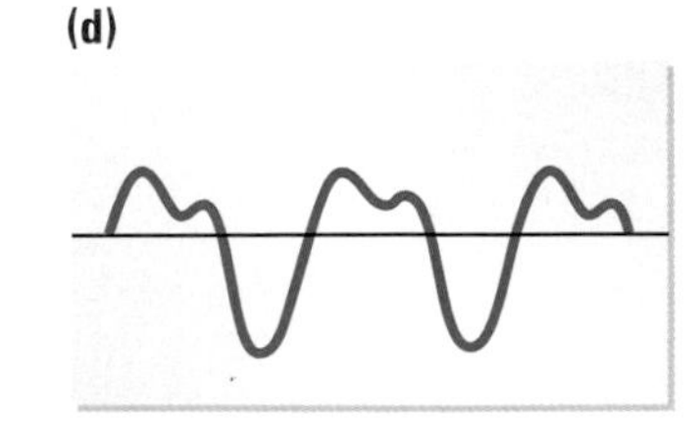

Figure 5
Oscilloscope tracings of
(a) A violin
(b) A clarinet
(c) A tuning fork
(d) An organ pipe

Try This **Activity**

Fundamentals and Overtones

The resultant waveforms created by various sources of sound can be viewed on an oscilloscope or computer interface connected to a microphone.

- View the waveforms created by various sources of musical sound pitched to approximately the same frequency, such as a tuning fork, a plucked string, and other musical instruments.
  (a) For each source of sound, sketch and label the waveform.
  (b) Can you label a waveform where the fundamental is not the frequency with the largest intensity?
  (c) Refer to the first step above and determine the harmonic that has the largest intensity. Explain your analysis.

Answers

3. (b) 6.2×10^2 Hz; 9.3×10^2 Hz
   (c) 12 cm; 24 cm

Practice

Understanding Concepts

1. What does the quality of a musical note depend on?
2. For each pair of sounds named below, choose the one that has the higher quality of sound.
   (a) a high consonance sound; a high dissonance sound
   (b) a pair of sounds with a ratio of frequencies of 9 : 8; a pair of sounds with a ratio of frequencies of 3 : 2
   (c) a pair of sounds consisting of f_0 and $5f_0$; a pair of sounds consisting of f_0 and $2f_0$
3. A certain string of length 36 cm has a fundamental frequency of 3.1×10^2 Hz.
   (a) Draw sketches of this string when it is vibrating in the fundamental mode, the first overtone, and the second overtone. Label the sketches.
   (b) Determine the frequency of the string in the first and second overtone modes.
   (c) Where are the nodes located on the string when the string is vibrating in the second overtone mode?

SUMMARY Vibrations and Sound Quality

- A string may vibrate along its whole length in a single segment or in segments that are simple (whole-number) fractions of its length.
- When an object vibrates in its fundamental mode of vibration, it produces its lowest possible frequency, called the fundamental frequency.
- Frequencies of overtone modes are simple (whole-number) multiples of the fundamental frequency called harmonics.
- The quality of a musical note depends on the number and relative intensity of the overtones it produces along with the fundamental.

Section 8.3 Questions

Understanding Concepts

1. The fundamental frequency of a string is 220 Hz. Determine the frequency of the
   (a) second harmonic
   (b) fourth harmonic
2. The note A (f = 440 Hz) is struck on a piano. Determine the frequency of the following:
   (a) the second harmonic
   (b) the third harmonic
   (c) the fifth harmonic
3. Use the principle of superposition to add the following waves. Describe how the resulting waveform relates to the quality of musical sound.

   Wave X: λ = 6.0 cm, A = 2.0 cm (Draw one wavelength.)

   Wave Y: λ = 2.0 cm, A = 1.0 cm (Draw three wavelengths to coincide with Wave X.)
4. Assume that a 900-Hz note is the third harmonic of a sound. What are the first, second, and fourth harmonics?

8.4 Resonance in Air Columns

Sound waves from one source can cause an identical source to vibrate in resonance. Just as with mechanical resonance (section 6.7), a small force produces a large vibration. For example, look at **Figure 1**. When the first tuning fork is struck and then silenced, a sound of the same frequency comes from the second fork even though it was not struck. Resonance has occurred because the forks have the same natural frequency. Energy has been transferred from one fork to the other by sound waves.

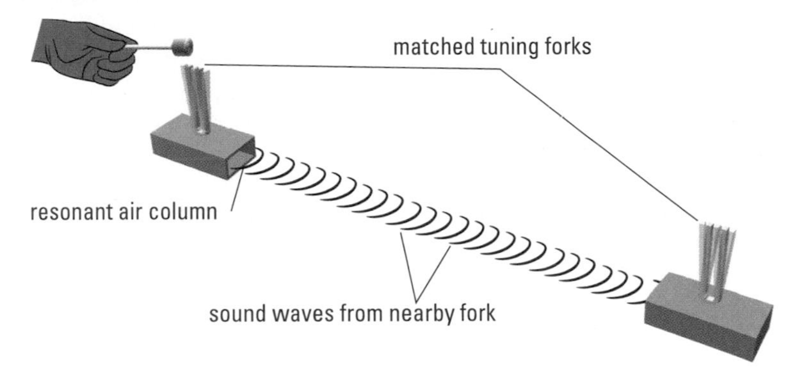

Figure 1
Two tuning forks with identical frequency are mounted on identical wooden boxes open at one end and placed about 1 m apart.

But why is the tuning fork mounted on a wooden box? Why is the box open at one end and closed at the other, and why is the box designed to be of a specific length? You will look at these variables in the next investigation.

Investigation 8.4.1

Resonance in Closed Air Columns

INQUIRY SKILLS

- ○ Questioning
- ○ Hypothesizing
- ● Predicting
- ● Planning
- ● Conducting
- ● Recording
- ● Analyzing
- ● Evaluating
- ● Communicating

An air column that is closed at one end and open at the other is called a **closed air column**. When a vibrating tuning fork is held over the open end of such a column and the length of the column is increased, the loudness increases sharply at very specific lengths. If a different tuning fork is used, the same phenomenon is observed except the maxima occur at different lengths. In this investigation, you will examine the relationship between the frequency of the tuning fork and the resonant length of a closed air column.

closed air column: column closed at one end and open at the other

Question

What lengths of a closed air column will resonate in response to a tuning fork of a known frequency?

Materials

80 cm of plastic pipe
large graduated cylinder
at least two tuning forks (e.g., 512 Hz and 1024 Hz)
metre stick
thermometer

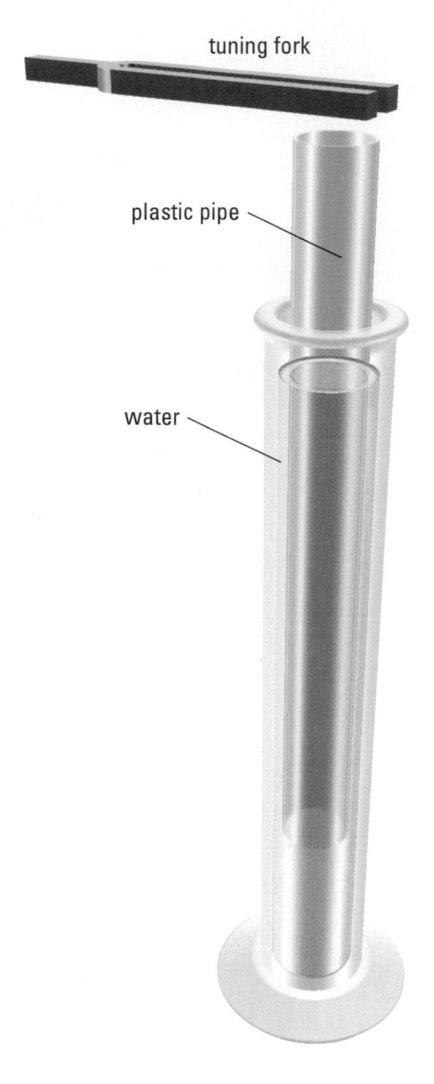

Figure 2
Setup for Investigation 8.4.1

Prediction

(a) Use the diagrams in **Figure 3** to help you predict two lengths of a closed air column that will resonate in response to a 512-Hz tuning fork and a 1024-Hz tuning fork.

Procedure

1. Place the plastic pipe in the graduated cylinder, as shown in **Figure 2**. Fill the graduated cylinder with water as close to the top as possible.
2. Sound the tuning fork and hold it over the mouth of the plastic pipe. Have your partner move the pipe slowly out of the water and listen for the first resonant point. At points of resonance, the intensity of the sound originating from the tuning fork will increase dramatically. Ignore points of slightly increased intensity that are not of the same frequency as the tuning fork.
3. Use the metre stick to measure the length of the air column for the first resonant point. Record your measurements in a chart similar to **Table 1**.

Table 1

	Tuning fork 1 (f = 512 Hz)		Tuning fork 2 (f = 1024 Hz)	
Resonant point	**Length (cm)**	**Length (wavelengths)**	**Length (cm)**	**Length (wavelengths)**
first				
second				
third				

4. Continue to raise the pipe, finding and measuring other resonant points.
5. Repeat steps 2 through 4 with a tuning fork of a higher frequency.
6. Record the air temperature in the room.

Analysis

(b) What is the speed of sound at the air temperature you recorded?
(c) What is the wavelength of the sound wave emitted by each tuning fork used in the investigation?
(d) For each tuning fork, what is the relationship between the length of the closed air column for the first resonant point you encountered and the wavelength of the tuning fork?
(e) For each tuning fork, what is the relationship between the length of the closed air column for the second resonant point you encountered and the wavelength of the tuning fork?
(f) As a general rule, what are the resonant lengths, expressed in wavelengths, for a closed air column?

Evaluation

(g) How did your measured lengths compare with your predicted lengths? Determine a percent difference in each case.
(h) Describe the sources of error in this investigation and suggest improvements.

Resonance in Closed Air Columns

As you saw in section 6.8, when a series of transverse waves was sent down a rope to a fixed end, the wave was reflected back and interfered with the incident waves. A node always formed at the fixed end where the reflection occurred.

In a similar way, when longitudinal sound waves are emitted by a tuning fork, some of them travel down the closed air column. The end of the tube reflects the sound waves back in the same way that the waves in a rope are reflected from the fixed end. A node is formed at the bottom of the column (**Figure 3**).

Resonance first occurs when the column is $\frac{1}{4}\lambda$ in length, since a single node is formed. The next possible lengths with a node at one end are $\frac{3}{4}\lambda$, $\frac{5}{4}\lambda$, etc. Thus, the resonant lengths in a closed air column occur at $\frac{1}{4}\lambda$, $\frac{3}{4}\lambda$, $\frac{5}{4}\lambda$, $\frac{7}{4}\lambda$, and so on. The resonant length of a wooden box that is open at one end and attached to a tuning fork is $\frac{1}{4}\lambda$. For a 256-Hz tuning fork, $\frac{1}{4}\lambda$ would be approximately 34 cm at room temperature (20°C).

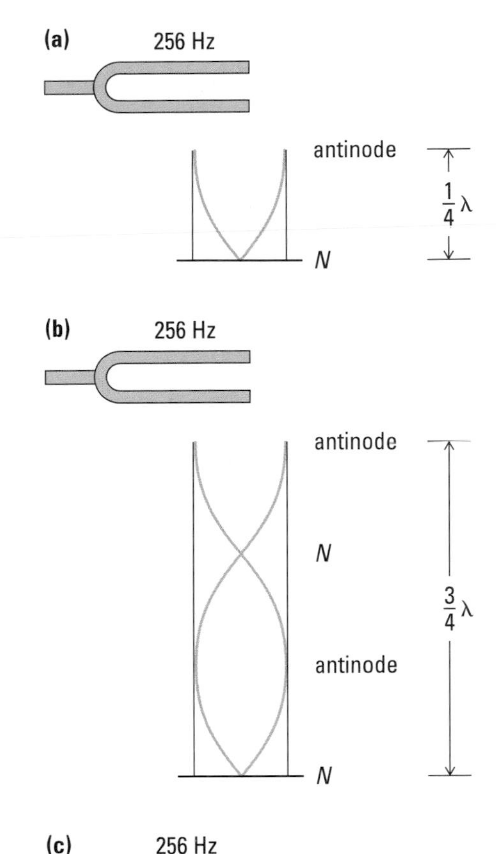

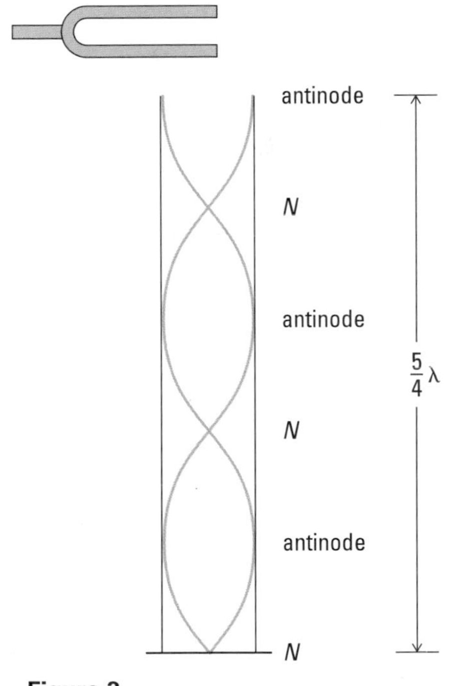

Figure 3
Resonant lengths of a closed air column with a sound of given frequency. The diagrams show the nodes and antinodes for the longitudinal displacement sound wave.
(a) First resonant length
(b) Second resonant length
(c) Third resonant length

Sample Problem 1

A vibrating tuning fork is held near the mouth of a column filled with water. The water level is lowered, and the first loud sound is heard when the air column is 9.0 cm long. Calculate the following:

(a) the wavelength of the sound from the tuning fork
(b) the length of the air column for the second resonance

Solution

(a) $\frac{1}{4}\lambda = 9.0\text{ cm}$

$\lambda = 36\text{ cm}$

(b) The air column length is increased by $\frac{1}{2}\lambda$, or 18 cm, to obtain the second resonance. Thus, the total length is 9.0 cm + 18 cm = 27 cm, or $\frac{3}{4}\lambda$.

Sample Problem 2

The first resonant length of a closed air column occurs when the length is 16 cm.

(a) What is the wavelength of the sound?
(b) If the frequency of the source is 512 Hz, what is the speed of sound?

Solution

(a)
$$\text{first resonant length} = \frac{1}{4}\lambda$$
$$\frac{1}{4}\lambda = 16\text{ cm}$$
$$\lambda = 64\text{ cm or }0.64\text{ m}$$

The wavelength is 0.64 m.

(b)
$$v = f\lambda$$
$$= (512\text{ Hz})(0.64\text{ m})$$
$$v = 328\text{ m/s}$$

The speed of sound is 3.3×10^2 m/s.

Answers

1. 30 cm; 90 cm; 150 cm
2. 15 cm; 45 cm
3. 20.0 cm
4. 1.7×10^3 Hz
5. 17.5 cm; 52.4 cm; 87.5 cm
6. 1.08 m; 346 m/s
7. (a) 92.0 cm; 1.20×10^2 cm; 152 cm
   (b) 371 Hz; 284 Hz; 224 Hz

Practice

Understanding Concepts

1. The first resonant length of a closed air column occurs when the length is 30.0 cm. What will the second and third resonant lengths be?
2. The third resonant length of a closed air column is 75 cm. Determine the first and second resonant lengths.
3. What is the shortest air column, closed at one end, that will resonate at a frequency of 440.0 Hz when the speed of sound is 352 m/s?
4. A signalling whistle measures 5.0 cm from its opening to its closed end. Find the wavelength of the sound emitted and the frequency of the whistle if the speed of sound is 344 m/s.
5. The note B_4 (f = 494 Hz) is played at the open end of an air column that is closed at the opposite end. The air temperature is 22°C. Calculate the length of the air column for the first three resonant sounds.
6. A tuning fork causes resonance in a closed pipe. The difference between the length of the closed tube for the first resonance and the length for the second resonance is 54.0 cm. If the frequency of the fork is 320 Hz, find the wavelength and speed of the sound waves.
7. An organ pipe resonates best when its length is $\frac{1}{4}\lambda$. Three pipes have lengths of 23.0 cm, 30.0 cm, and 38.0 cm.
   (a) Find the wavelength of the sound emitted by each pipe.
   (b) Find the frequency of each pipe, if the speed of sound is 341 m/s.

INQUIRY SKILLS

- ○ Questioning
- ○ Hypothesizing
- ● Predicting
- ● Planning
- ● Conducting
- ● Recording
- ● Analyzing
- ● Evaluating
- ● Communicating

Investigation 8.4.2

Speed of Sound in a Closed Air Column

You now have knowledge of the resonant lengths and resonant frequencies for closed air columns. You can use a similar approach to that used in Investigation 8.4.1 to design and carry out a method to determine the speed of sound in a closed air column.

Question

What is the speed of sound in a closed air column?

Prediction

(a) Using information from section 7.3, predict the speed of sound in a closed air column.

Design

Discuss with your group and teacher the method you will follow to perform your experiment. Decide on the materials and how you will use them to obtain data.

Write out your experimental design, including your procedure, and prepare a table to record your observations. Draw a labelled diagram showing the proposed experimental setup. Submit your procedure to your teacher for approval before commencing with the investigation.

Never touch a glass cylinder with a vibrating tuning fork—the cylinder might shatter.

Materials

The materials will depend on your design. Materials that you may consider include plastic tube open at both ends, tall glass cylinder, metre stick, thermometer, tuning fork, and striking pad for the tuning fork. Prepare a list of materials and get your teacher's approval.

Analysis

(b) Show the relevant equation(s) and calculations used to determine the numerical solution to the problem.

Evaluation

(c) Answer the original question and describe how you determined the speed of sound in a closed air column.
(d) Discuss the accuracy of your answer, including percentage error, and identify possible sources of error.
(e) Calculate the percentage difference between your predicted value and the experimental value for the speed of sound in a closed air column.
(f) Evaluate your design for this experiment. If you were to perform it again, what changes would you make to improve it?
(g) Compare the value you obtained for the speed of sound with the value found in a standard source (such as a textbook) and determine the percent difference.
(h) Why is your procedure an easier and more accurate way to determine the speed of sound in air than to measure it directly?

Synthesis

(i) How would you use the same approach to measure the speed of sound in a gas, such as carbon dioxide or helium?

Resonance in Open Air Columns

Resonance may also be produced in an **open air column**, that is, a column that is open at both ends. If a standing wave interference pattern is created by reflection at a free end, an antinode occurs at the free end. Since a pipe is open at both ends, antinodes occur at both ends. The first length at which resonance occurs is $\frac{1}{2}\lambda$. Succeeding resonant lengths will occur at λ, $\frac{3}{2}\lambda$, 2λ, and so on (see **Figure 4**).

open air column: column open at both ends

Sample Problem 3

An organ pipe, 3.6 m long and open at both ends, produces a musical note at its fundamental frequency.

(a) What is the wavelength of the note produced?
(b) What is the frequency of the pipe if the speed of sound in air is 346 m/s?

Solution

(a) The fundamental frequency corresponds to the simplest resonance pattern, which has a resonant length of $\frac{1}{2}\lambda$.

$$\frac{1}{2}\lambda = 3.6 \text{ m}$$
$$\lambda = 7.2 \text{ m}$$

The wavelength of the note is 7.2 m.

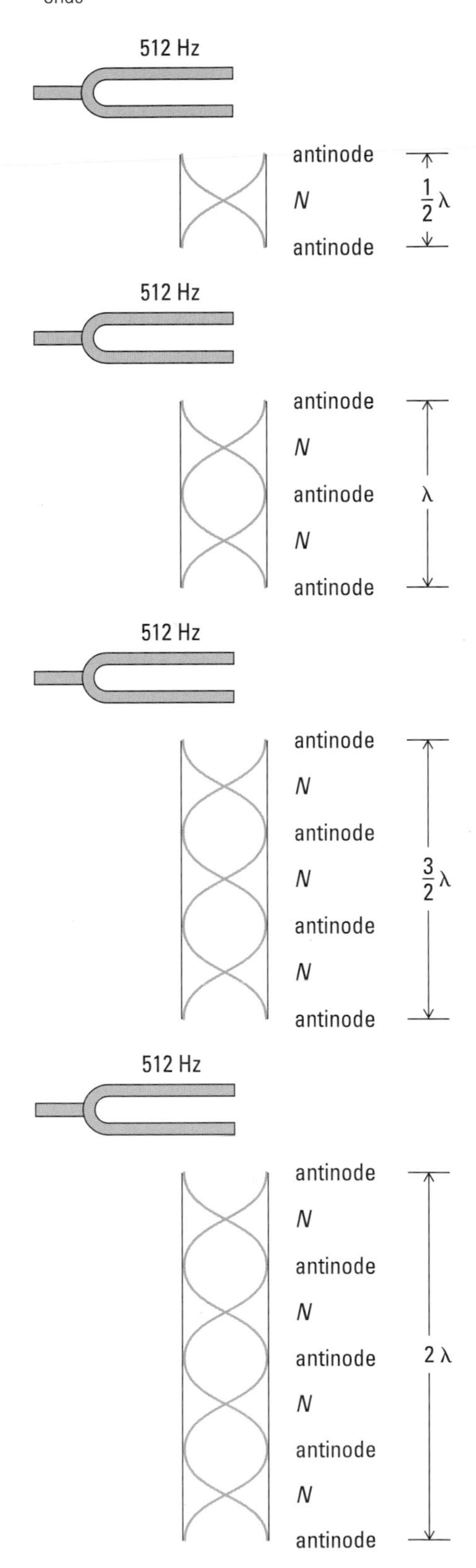

Figure 4
Resonant lengths of an open air column with a sound of a given frequency

(b) $v = f\lambda$

$$f = \frac{v}{\lambda}$$

$$= \frac{346 \text{ m/s}}{7.2 \text{ m}}$$

$$f = 48 \text{ Hz}$$

The frequency of the pipe is 48 Hz.

SUMMARY Resonance in Air Columns

- Resonance occurs in closed air columns at lengths of $\frac{1}{4}\lambda$, $\frac{3}{4}\lambda$, $\frac{5}{4}\lambda$, and so on of the original sound wave.
- The resonant lengths of open air columns are $\frac{1}{2}\lambda$, λ, $\frac{3}{2}\lambda$, 2λ, and so on of the original sound wave.

Section 8.4 Questions

Understanding Concepts

1. What is the length of an open air column that resonates at its first resonant length with a frequency of 560 Hz? (The speed of sound is 350 m/s.)
2. The second resonant length of an open air column is 48 cm. Determine the first and third resonant lengths.
3. An organ pipe, open at both ends, resonates at its first resonant length with a frequency of 128 Hz. What is the length of the pipe if the speed of sound is 346 m/s?
4. A 1.0×10^3-Hz tuning fork is sounded and held near the mouth of an adjustable column of air open at both ends. If the air temperature is 20.0°C, calculate the following:
   (a) the speed of the sound in air
   (b) the wavelength of the sound
   (c) the minimum length of the air column that produces resonance
5. A closed air column is 60.0 cm long. Calculate the frequency of forks that will cause resonance at
   (a) the first resonant length
   (b) the third resonant length (The speed of sound is 344 m/s.)
6. In an air column, the distance from one resonance length to the next is 21.6 cm. What is the wavelength of the sound producing resonance if the column is
   (a) closed at one end?
   (b) open at both ends?
7. What vibrates to create sound in a column of air?
8. How would a higher air temperature affect the lengths of the resonating air columns in Investigation 8.4.1? Why?
9. When water is added to a bottle, what happens to the pitch of the sound as the water is added? Explain why.

Applying Inquiry Skills

10. Explain how the external ear and ear canal magnify sounds entering the ear.

8.5 Musical Instruments

Musical sounds, like all sounds, originate from a vibrating object. All musical instruments can be grouped depending on how the vibrations are produced. Music may be produced by plucking, bowing, or striking stringed instruments; by blowing air across a hole or a reed, or through a special mouthpiece in wind instruments; or by striking various surfaces in percussion instruments.

Stringed Instruments

Stringed instruments, which include guitars, pianos, and violins, consist of two main parts—the vibrator (string) and the resonator. The **resonator** is the case, box, or sounding board that the string is mounted on. A string by itself does not produce a loud or even a pleasant sound. It must be attached to a resonator, through which vibrations are forced, improving the loudness and quality of the sound. Even a tuning fork has a louder and better sound if its handle touches a desk, wall, or resonance box.

Stringed instruments can be played by plucking, striking, or bowing them. The quality of sound is different in each case and depends on what part of the string is plucked, struck, or bowed. Stringed instruments that are usually plucked include the banjo, guitar, mandolin, ukulele, harp (**Figure 1**), and sometimes the members of the violin family. A string plucked gently in the middle produces a strong fundamental frequency, or first harmonic (**Figure 2(a)**). A string plucked at a position one-quarter of its length from one end vibrates in several modes, two of which are shown in **Figure 2(b)**. In this case, the second harmonic is added to the first harmonic, changing the sound quality. **Figure 2(c)** shows the third harmonic superimposed on the first harmonic, again resulting in a different quality of sound.

	Vibration of string	Sound wave
(a)	pluck here	
(b)	pluck here	
(c)	pluck here	

Figure 2
Changing the quality of sound of a vibrating string
(a) First harmonic
(b) Second harmonic superimposed on the first harmonic
(c) Third harmonic superimposed on the first harmonic

DID YOU KNOW?

Power of Strings

Compared with other types of instruments, stringed instruments do not give out a great amount of power. The maximum amount of acoustic power from a piano is 0.44 W, while a base drum can produce 25 W. This explains why an orchestra needs many more violins than drums or trumpets.

resonator: an object, usually a hollow chamber, that vibrates in resonance with a source of sound

Figure 1
A harp

Figure 3
The strings and hammers inside a piano

The best-known stringed instrument that is struck is the piano. A piano key is connected by a system of levers to a hammer that strikes the string (or strings) to produce a certain note (**Figure 3**). A modern piano has 88 notes with a frequency range from 27.5 Hz to 4186 Hz. The short, high-tension strings produce high-pitched notes and the long, thick wires produce low-pitched notes. The sounds from the strings are increased in loudness and quality by the wooden sounding board of the piano.

Stringed instruments that are usually bowed belong to the violin family. This family consists of the viola, cello, bass, and violin (**Figure 4**). One side of each bow consists of dozens of fine fibres that have been rubbed with rosin to increase the friction when stroked across a string. Each instrument has four strings and wooden sounding boards at the front and back of the case. The members of the violin family have no frets and, thus, the frequency can be changed gradually, not necessarily in steps as in the guitar family. The only other modern Western instrument capable of changing frequency gradually is the trombone, a wind instrument.

Practice

Understanding Concepts

1. You can make a string or a fishing line under tension vibrate, producing a sound. But that sound has no musical quality. Why not?
2. Classify the stringed instruments into categories that
   (a) have frets
   (b) have no frets
   (c) are struck
3. The note A_4 (440 Hz) is played on a violin string. Describe ways in which the quality of the sound emitted can be altered.

Figure 4
The violin, the smallest instrument in the violin family, produces high-frequency sounds; the largest instrument, the bass, produces low-frequency sounds.

Activity 8.5.1

Waveforms of Stringed Instruments

We can study the modes of vibration of a string using an oscilloscope or a computer.

Procedure

1. Using a sonometer or a stringed instrument, pluck a string at various positions; for example, pluck the string in the middle, etc.
2. Compare the resulting frequencies and qualities of the sounds.
3. Place a microphone near to, but not touching the string. View the waveforms on an oscilloscope or computer.

Analysis

(a) Sketch a diagram of each screen tracing.
(b) Analyze the tracing of one complex vibration and determine the number of harmonics present.

Wind Instruments

All wind instruments contain columns of vibrating air molecules. The frequency of vibration of the air molecules, and thus the fundamental frequency of the sound produced, depends on whether the column is open or closed at the ends. As is the case with all vibrating objects, large instruments create low-frequency sounds, and small instruments create high-frequency sounds.

In some wind instruments such as the pipe organ in **Figure 5**, the length of each air column is fixed. However, in most wind instruments, such as the trombone, the length of the air column can be changed.

To cause the air molecules to vibrate, something else must vibrate first. There are four general mechanisms for forcing air molecules to vibrate in wind instruments.

First, in air reed instruments, air is blown across or through an opening. The moving air sets up a turbulence inside the column of the instrument. Examples of such instruments are the pipe organ in **Figure 5**, the flute, the piccolo, the recorder, and the fife. The flute and piccolo have keys that are pressed to change the length of the air column. The recorder and fife have side holes that must be covered with fingers to change the length of the air column and thus control the pitch.

Figure 5
The pipe organ of Notre Dame Basilica, Montreal

In single-membrane reed instruments, moving air sets a single reed vibrating, which in turn sets the air in the instrument vibrating. Examples of these instruments include the saxophone, the clarinet, and the bagpipe. In the bagpipe, the reeds are located in the four drone pipes attached to the bag, not in the mouth pipe. Again, the length of the air column is changed by holding down keys or covering side holes.

Next, in double-mechanical reed instruments, moving air forces a set of two reeds to vibrate against each other. This causes air in the instrument to vibrate. Examples include the oboe, the English horn, and the bassoon. Keys are pressed to alter the length of the air column.

The final mechanism is found in lip reed instruments, also called brass instruments. In this type of instrument, the player's lips function as a double reed. The lips vibrate, causing the air in the instrument to vibrate. None of the air escapes through side holes, as with other wind instruments; rather, the sound waves must travel all the way through the brass instrument. Examples include the bugle, the trombone, the trumpet, the French horn, and the tuba. The length of the air column is changed either by pressing valves or keys that add extra tubing to the instrument, or, in the case of the trombone, by sliding the U-tube.

The quality of sound from wind instruments is determined by the construction of the instrument and the experience of the player. However, as with stringed instruments, the quality of sound also depends on the harmonics produced by the instrument. In section 8.4 you learned that a standing wave pattern is set up when resonance occurs in air columns. The standing waves not only create the fundamental frequencies or first harmonics, but also higher harmonics.

Figure 6(a) shows an open air column with a length one-half the wavelength of the resonating sound. This situation produces the first harmonic (f_0); the corresponding waveform is shown adjacent to the air column. When the length of the column is constant, as is the case at any given instant with a wind instrument, a second harmonic can be created by various means, such as blowing harder. **Figure 6(b)** shows that the wave must be half the length of the first wave to produce a standing wave pattern. Thus, the frequency is double ($2f_0$), and we have the second harmonic. **Figure 6(c)** shows the first and second harmonics added together to give sound of higher quality. In open air columns, other harmonics may also be present, further enhancing the quality.

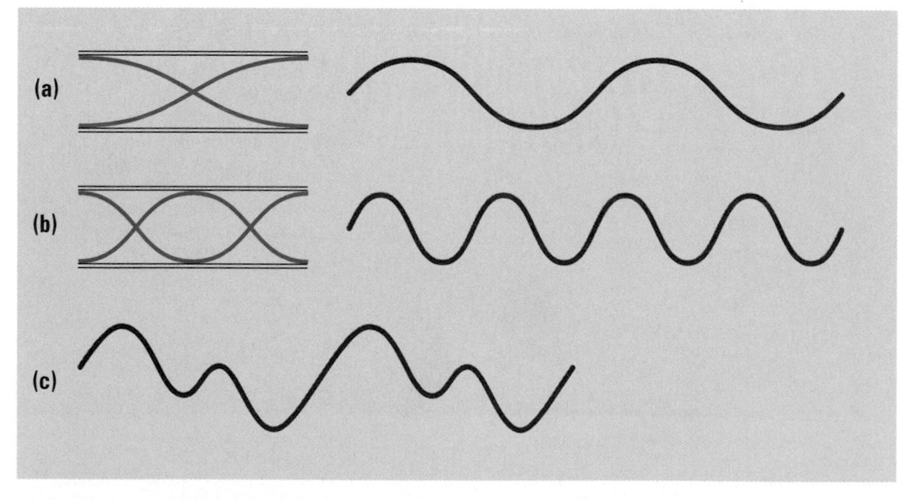

Figure 6
Improving the quality of sound in wind instruments
(a) First harmonic (f_0)
(b) Second harmonic ($2f_0$)
(c) Resulting waveform (f_0 and $2f_0$)

An air column, closed at one end, has only odd numbers of harmonics (f_0, $3f_0$, $5f_0$, and so on). Thus, the quality of this sound differs from the sound from an air column open at both ends. Every instrument, whether it is a wind instrument or a stringed instrument, has its own harmonic structure and its own unique quality. This fact is used to create "artificial" music on music synthesizers.

Practice

4. Describe the different possible ways in which air can be made to vibrate in a wind instrument, in each case naming two instruments as examples.

Percussion Instruments

Percussion is the striking of one object against another. Percussion instruments are usually struck by a firm object such as a hammer, bar, or stick. These musical instruments were likely the first invented because they are relatively easy to make. Percussion instruments can be divided into three categories.

Single indefinite pitch instruments are used for special effects or for keeping the beat of the music. Examples include the triangle (**Figure 7**), the bass drum, and castanets.

Multiple definite pitch instruments have bars or bells of different sizes that, when struck, produce their own resonant frequencies and harmonics. Examples are the tuning fork, orchestra bells, the marimba, the xylophone, and the carillon.

Variable pitch instruments have a device used to rapidly change the pitch to a limited choice of frequencies. An important example is the timpani, or kettledrums (**Figure 8**), which have a foot pedal for quick tuning.

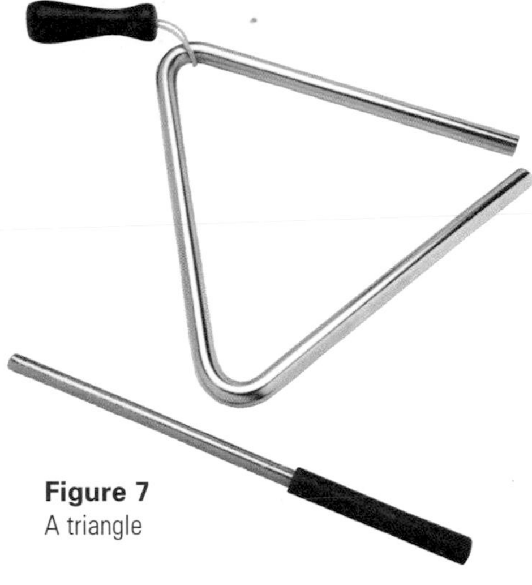

Figure 7
A triangle

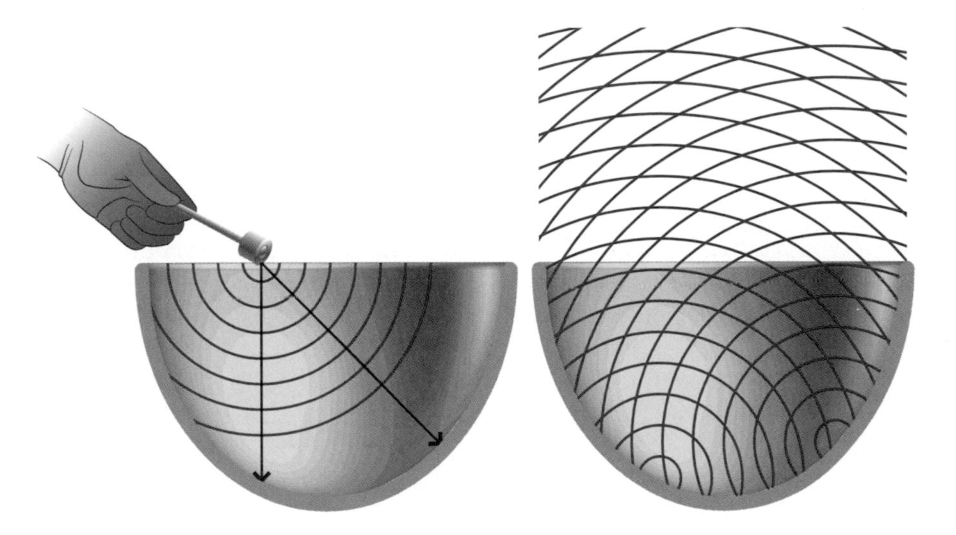

Figure 8
A sound produced at the head of a kettle-drum is reflected from a multitude of points on the inner surface of the metal kettle. The reflection from only two points is shown. Note that all the sound is emitted from the top. Interference of the reflected waves creates the drum's rich boom.

Analyzing the Sound from Various Musical Instruments

We can study the sounds emitted from musical instruments by ear and oscilloscope or computer software, including simulation software.

- Strike a tuning fork at various positions along one prong, and listen carefully. Describe the sounds produced.
  (a) Is the quality of the fundamental frequency enhanced or lessened by changing the striking position?
  (b) Where approximately is the best position to strike the fork to eliminate the high-frequency sound?
- Obtain a single reed from a clarinet, an oboe, or a saxophone.
- Demonstrate that sound can be produced by blowing across each reed.
- Place a microphone near to, but not touching, each reed. View the waveforms produced on an oscilloscope or computer.
- Repeat the above steps for a brass instrument, such as a trumpet or trombone.
- Repeat for a percussion instrument, such as a triangle.

DID YOU KNOW ?

Percussion at the Doctor's Office

Doctors use percussion when they tap a patient's chest or back and listen for sounds that indicate either clear or congested lungs.

SUMMARY Musical Instruments

- Most musical instruments consist of a vibrating source and a structure to enhance the sound through mechanical and acoustical resonance.
- A stringed instrument consists of a vibrating string and resonating sound board, or hollow box.
- Wind instruments contain either open or closed air columns of vibrating air; the initial vibration is created by a reed or by the player's lips.
- Percussion instruments involve striking one object against the other.

Section 8.5 Questions

Understanding Concepts

1. When the resonator of a stringed instrument vibrates, does it do so as a result of sympathetic vibrations? Explain your answer.
2. The fundamental frequency of a string is 220 Hz. What is the frequency of the
   (a) second harmonic?
   (b) fourth harmonic?
3. A 440-Hz tuning fork is sounded along with the A-string on a guitar and a beat frequency of 3 Hz is heard. When the tension in the string is decreased, a new beat frequency of 4 Hz is heard.
   (a) What is the frequency of the guitar string when the beat frequency is 4 Hz?
   (b) What should be done to tune the string to 440 Hz?
4. From the pairs of instruments listed, choose the one with the higher range of frequencies. (*Hint:* Consider the size of the instruments.)
   (a) piccolo, flute
   (b) bassoon, English horn
   (c) oboe, English horn
   (d) tuba, trumpet
5. Use diagrams to show that an air column closed at one end has only odd numbers of harmonics.
6. Describe factors that you think affect the quality of sound from percussion instruments.
7. You tune your woodwind instrument in a house at room temperature. When you go outside to play in the cold the pitch has changed. Will the pitch increase or decrease? Why?
8. How many fundamental notes can be produced by a three-valve trumpet? Why?
9. A piano is classified as a stringed instrument. Could it also be described as a percussion instrument? Why or why not?

8.6 The Human Voice as a Musical Instrument

The human voice is a fascinating instrument; it is also the oldest and most versatile of all musical instruments. The main parts of the body that help produce sound are shown in **Figure 1**. The human voice consists of three main parts: the source of air (the lungs); the vibrators (the vocal folds or vocal cords); and the resonators (the lower throat or pharynx, mouth, and nasal cavity).

To create most sounds, air from the lungs passes by the vocal cords, causing them to vibrate. The vocal cords are two bands of skin that act like a double reed. Loudness is controlled by the amount of air forced over the vocal cords. The pitch is controlled by muscular tension as well as by the size of the vibrating parts. Since, as you have seen, larger instruments have lower resonant frequencies, in general, male voices have a lower frequency range than the female voice (see **Table 1**).

DID YOU KNOW ?

Infant Sounds

Human infants cannot produce most sounds that we understand as articulate speech until after they are one year old when the larynx drops down in the throat.

Table 1 Approximate Frequency Ranges of Singers

Type of singer	Frequency range (Hz)
bass	82–294
baritone	110–392
tenor	147–523
alto	196–698
soprano	262–1047

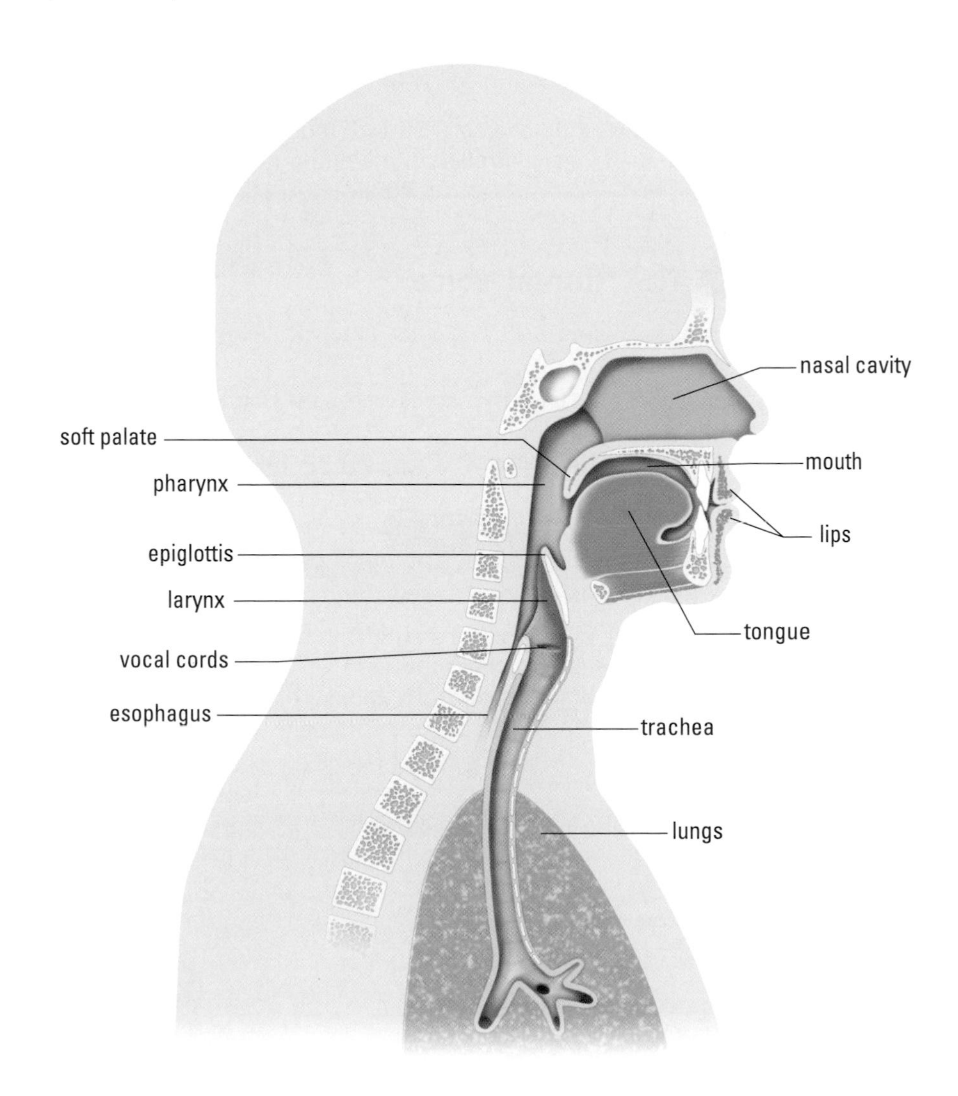

Figure 1
The human voice

The quality of sound from the voice is controlled by the resonating cavities (pharynx, mouth, and nasal cavity) as well as the tongue and lips. The quality of sound may be improved by proper training. Good singers can control such effects as vibrato and tremolo. Vibrato is a slight, periodic changing of frequency (frequency modulation, or FM), and tremolo is a slight, periodic changing of amplitude (amplitude modulation, or AM).

Waveforms of certain sounds from the human voice were shown in the first section of this chapter. If you try to re-create those waveforms, you will likely be unsuccessful. Human voice waveforms are too complex to analyze or to reproduce accurately. However, the use of computers now makes it possible to recognize and reproduce many sounds. Computers can be programmed to respond to voice commands. Similar software is used in voice recognition programming for word processing and controlling other computer programs. Police use "voice recognition" software to identify recorded voices. It is expected that the use of voice-commanded computers will be common in the near future.

Practice

Understanding Concepts

1. Describe how each of the characteristics of musical sounds (pitch, loudness, and quality) is controlled in the human voice.

SUMMARY The Human Voice

- In the human voice, air from the lungs causes the vocal chords to vibrate, initiating a sound.
- The throat, mouth, and nasal cavity create a resonant chamber that affects the quality of the resulting sound.

Section 8.6 Questions

Understanding Concepts

1. Discuss whether the human voice should be classified as a stringed, wind, or percussion instrument.
2. When you whistle through your lips, how do you control the pitch of the whistle?

Making Connections

3. Research and report on how "voice prints" are created and used for identification purposes. Follow the links for Nelson Physics 11, 8.6.

GO TO www.science.nelson.com

8.7 Electrical Instruments and Audio Reproduction

Electrical instruments are made of three main parts—a source of sound, a microphone, and a loudspeaker. At hockey and football games, for instance, the announcer's voice directs sound energy into a microphone. The microphone changes the sound energy into electrical energy that, after amplification, causes vibrations in a loudspeaker. These vibrations reproduce the original sound with an amplified loudness.

Many musical instruments can be made into electrical instruments through the addition of a microphone and a loudspeaker. A microphone can be attached directly to the body of a stringed instrument, which normally gives out low amounts of power. In some cases, the design of the instrument is altered; an electric guitar, for example, may have a solid body rather than the hollow body of acoustic guitars (**Figure 1**).

Figure 1
An electric guitar

Loudspeakers are important in determining the quality of sound from an electrical instrument. A single loudspeaker does not have the same frequency range as our ears, so a set of two or three must be used to give both quality and frequency range. **Table 1** lists the details of three common sizes of loudspeakers used in electrical sound systems. **Figure 2** shows a typical set of loudspeakers.

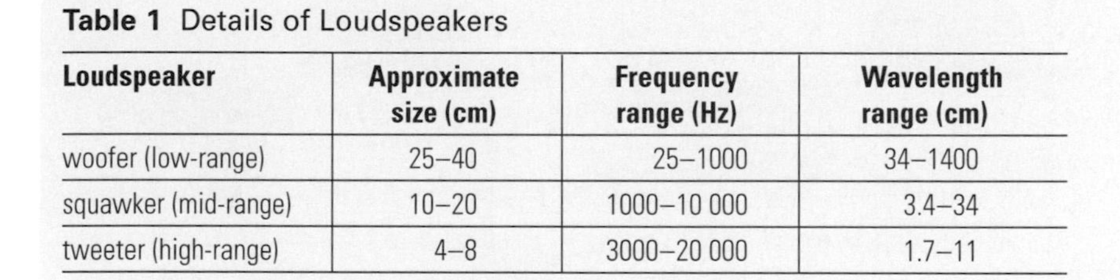

Table 1 Details of Loudspeakers

Loudspeaker	Approximate size (cm)	Frequency range (Hz)	Wavelength range (cm)
woofer (low-range)	25–40	25–1000	34–1400
squawker (mid-range)	10–20	1000–10 000	3.4–34
tweeter (high-range)	4–8	3000–20 000	1.7–11

Figure 2
Loudspeakers improve the sound quality from an electrical instrument.

As you can see from the table, the sound waves from the tweeter have much shorter wavelengths than those from the woofer. Long wavelengths are diffracted easily through doorways and around furniture and people. However, the short waves from a tweeter are not diffracted around large objects; their sound tends to be directional. As a result, a listener must be in front of the tweeter to get the full sensation of its sound, especially in the very high frequency range.

For speakers manufactured to reproduce low frequencies, the enclosure design is important. When the speaker cone moves out it produces a compression. At the same time, a rarefaction is created at the back of the cone. If this back wave interferes with the front wave, destructive interference will occur, diminishing the intensity of the sound. Two methods are used to correct this problem. In the bass reflect speaker enclosure, the design delays the back wave sufficiently that it reinforces the front wave. In the acoustic resistance speaker enclosure, the back wave is absorbed so there is no interference at all. You will study how the speaker cone moves in Chapter 13.

Practice

Understanding Concepts

1. When an electric guitar is played in the "unplugged" mode, it can be considered an acoustic guitar. Why?
2. Why must woofers be larger in size than tweeters?

Digital Sound

With the introduction of the compact disc in the 1980s, digital sound has become the standard for most sound reproduction. In a process known as analog-to-digital conversion (**Figure 3**), the amplitude of a sound wave is measured 44 100 times per second and the data are stored on a magnetic tape. In a sense, the sound wave is "chopped" into tiny segments and then stored as binary numbers using 0s and 1s, in the same manner that data are stored in a computer. During playback, the numerical values are used to reconstruct the original sound wave and the wave is "smoothed out" electronically in a process called digital-to-analog conversion.

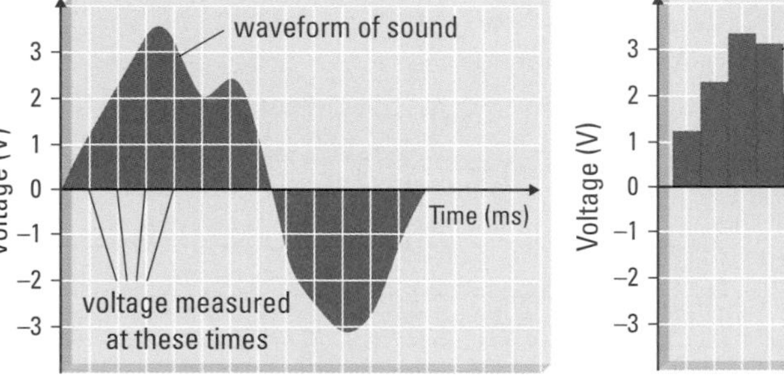

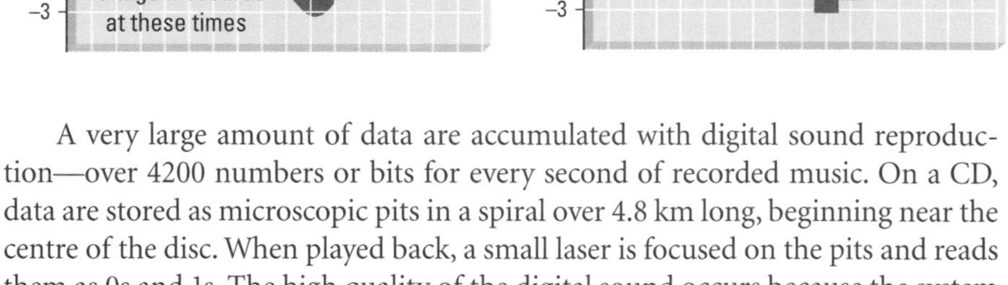

Figure 3
Digital sound reproduction

A very large amount of data are accumulated with digital sound reproduction—over 4200 numbers or bits for every second of recorded music. On a CD, data are stored as microscopic pits in a spiral over 4.8 km long, beginning near the centre of the disc. When played back, a small laser is focused on the pits and reads them as 0s and 1s. The high quality of the digital sound occurs because the system only looks for numbers and can ignore extraneous sounds, such as the "hiss" you can hear on cassette tapes. An electronic error-correction system can even compensate for garbled or missing numbers. Because there is no physical contact between the disc and the laser, there is little wear and tear and the CD can be used over and over with little deterioration.

Analog-to-digital and digital-to-analog conversions are now common and are used extensively in telephone networks; wireless communication, including satellite; and digital videos (DVD). Digital television (HDTV) and digital recordings are totally digital, eliminating the conversion process and, thus, ensuring even higher quality.

DID YOU KNOW?

Harmonics and Headphones

An interesting phenomenon occurs with headphones used on portable tape and CD players. Some of these headphones do not create the low frequencies usually associated with the woofer, although the listener actually "hears" low-frequency sounds. The higher harmonics produced by the headphones cause a sensation that makes us believe we are hearing the absent first and second harmonics.

Practice

Understanding Concepts

3. You hear the same song recorded by the same artist, one recording made 50 years ago and the other re-mastered very recently using digital sound reproduction. Describe both the process used to re-master the music and the quality of the new sound compared with the original sound.

Activity 8.7.1

Evaluating Headphones

Headphones are quite commonly used in many sound devices, such as Walkmans, Discmans, MP3 players, portable radios, and home entertainment systems. Headphones have many different designs and functions, depending on the application. As a result, many questions must be answered when making a selection as an intelligent consumer.

One of the criteria used to evaluate quality in sound systems is frequency response. Frequency response is the range of frequencies that the headphones can reproduce effectively. Sound systems and headphones may be advertised as having "a smooth, flat response curve." To find out what this means we will measure the frequency response of a single headphone.

Procedure

1. Your teacher will provide you with an audio oscillator for a sound source. This will be connected to an amplifier and the headphone will be plugged into the appropriate jack or apparatus provided by the teacher. You will use a decibel sound level meter to measure the sound level at different frequencies of the audio spectrum.
2. Obtain a headphone from home or use one supplied.
3. Set up the sound intensity meter and headphone so the meter measures only the sound from one headphone. This may involve using a Styrofoam block and/or sound-absorbing material. Your teacher will provide assistance and direction.
4. Make sure the oscillator and amplifier are set for a "flat" response. In the case of the amplifier, any "loudness" circuit should be off and there should be no equalization.
5. Select at least fifteen frequencies to measure in the range 20 Hz–20 kHz.
6. Create an observation table to record the frequency (Hz) and the sound intensity level (dB).
7. When you have completed your measurements, share them with at least two other teams.
8. To graph the data, use semi-logarithmic graph paper. The vertical axis for the sound intensity level is linear and the horizontal axis is logarithmic. (Refer to Appendix A5 for help with logarithmic graphs.)
9. Graph the results, including those obtained from other groups.

Analysis

(a) Describe the frequency response for your headphone, noting where the frequency response is not flat.

(b) Compare the frequency response with that obtained by the other groups. Determine which headphone had the "best" frequency response and explain why.

(c) When selecting a headphone, there are a number of features to be considered, beyond frequency response. Listed below are most of the important areas. Write a short note for each section so as a consumer you understand what you are looking for when making a purchase. After researching the Nelson Science Web site, you should have all of the information you require. Follow the links for Nelson Physics 11, 8.7.

GO TO www.science.nelson.com

Criteria for Evaluation

- sound quality (frequency response, distortion, sensitivity, field equalization)
- power requirements and maximums
- feel and fit
- durability
- operation (dynamic, electrostatic, etc.)
- earpiece construction (open air, closed air, in the ear)
- wireless or corded
- 3-D and surround sound
- noise cancellation/reduction
- cost (basic, moderate, expensive)

SUMMARY Audio Reproduction

- Electrical instruments amplify and alter the vibrations for reproduction by a loudspeaker.
- Loudspeakers are designed to reproduce the full human audio range (20 Hz–20 kHz).
- Digital sound recording is a method of storing sound as a series of binary numbers, not as waves.

Section 8.7 Questions

Understanding Concepts

1. Which sounds are more directional, those from a tweeter or those from a woofer? Why?
2. In an analog-to-digital conversion, scientists perform experiments to test the quality of the sounds produced. Would you expect a higher quality from samples at 40 000 times per second or samples at 20 000 times per second? Explain your answer.

8.8 Electronic Musical Instruments

You are probably familiar with music synthesizers or electronic synthesizers (**Figure 1**). These instruments, in contrast with stringed, wind, and percussion instruments, produce vibrations using electronic components, such as resistors and transistors.

Figure 1
A portable electronic synthesizer

An electronic instrument consists of four main parts. An oscillator creates the vibrations. The filter circuit selects the frequencies that are sent to the mixing circuit. The mixing circuit adds various frequencies together to produce the final signal. The amplifier and speaker system make the sound loud enough to be heard.

The shape of the sound waves produced by a synthesizer can be controlled; as a result, synthesizers can emit sounds that resemble the sound of almost any musical instrument. The basic shapes of the waves used to create more complex waves are shown in **Figure 2**.

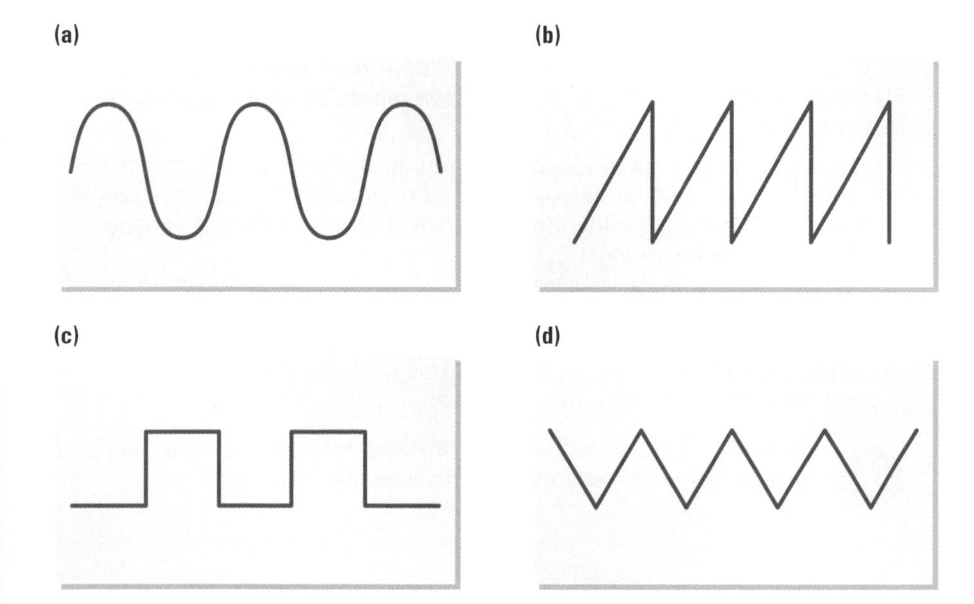

Figure 2
The waveforms shown can be used to generate various other waves.
(a) Sinusoidal wave
(b) Sawtooth wave
(c) Square wave
(d) Triangular wave

Electronic instruments can also control the attack and decay patterns of a sound. The attack occurs when the sound is first heard. It may be sudden, delayed, or overshot, as illustrated in **Figure 3(a)**. The decay occurs when the sound comes to an end; it may be slow, fast, or irregular (**Figure 3(b)**).

(a)

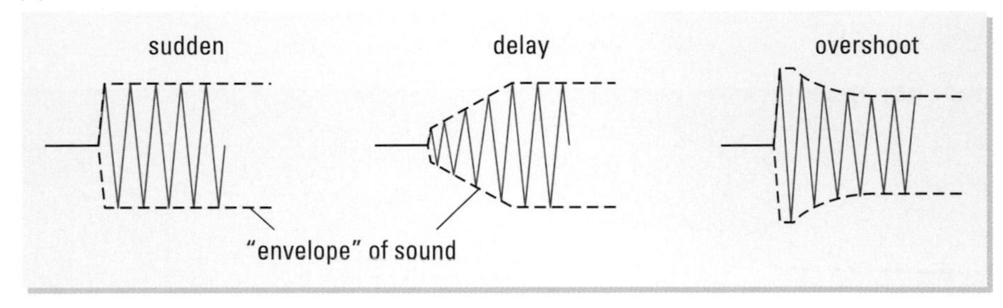

(b)

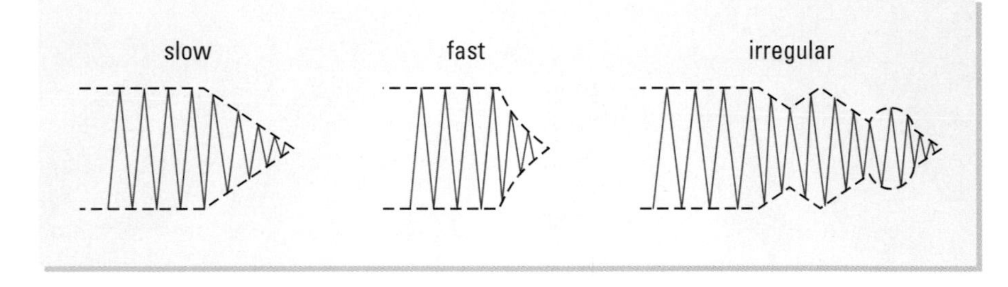

Figure 3
Control patterns
(a) Growth patterns
(b) Decay patterns

Practice

Understanding Concepts

1. Electronic music is considered to be very versatile. Explain why.

Electronic Sound

Artificial sound, generated electronically, is different from that produced by a vibrating object. Although we can sometimes detect by ear, the differences between the actual sound of an instrument and that generated electronically, it is easier to analyze the differences using an oscilloscope or a computer.

- Listen to the sounds produced by an audio frequency generator that creates sine waves, square waves, and triangular waves all at a constant frequency. Describe the differences in the sounds and explain why the differences exist.
- If a musical synthesizer is available, have someone demonstrate its capabilities.
- Listen to a note played on an instrument with the same note played on a synthesizer recreating the instrument. Compare the differences you hear by ear and the tracings on an oscilloscope.

Keep sound waves within a comfortable zone for listeners. Very high or very low frequencies can cause discomfort or harm.

Section 8.8 Questions

Understanding Concepts

1. A synthesizer adds a positive square wave, S (λ = 4.0 cm, A = 2.0 cm), and a positive triangular wave, T (λ = 4.0 cm, A = 1.0 cm), to obtain a new sound. Draw two wavelengths of S and T, and use the principle of superposition to show their addition.

Making Connections

2. Describe both the advantages and the disadvantages of electronic music. Consider your own ideas as well as the opinions of others. (Show that you realize that music is both objective and subjective.)

8.9 Acoustics

Some people claim that their singing voice is better in the shower than anywhere else. This may be true as a result of the many sound reflections that occur in a small room. The qualities of a room or auditorium that determine how well sound is heard are called **acoustics.**

acoustics: the total effect of sound produced in an enclosed space

The acoustics of a room depend on the shape of the room, the contents of the room, and the composition of the walls, ceiling, and floor. You have probably noticed how your voice echoes in a large, empty room, but when rugs and furniture are put in the room, the acoustics improve.

Without echoes, music would sound flat. The acoustics of auditoriums and theatres are designed to use echoes to ensure that everyone in the audience can hear and to create the best possible sound. If you have ever been to a concert, you have probably noticed that when the musicians stop playing, the music can still be heard for a fraction of a second. This is called **reverberation time** and is defined as the time required for the intensity of the sound to drop to 10^{-6} (one millionth) of its original value or until the sound is inaudible.

reverberation time: the time required for the intensity of the sound to drop to 10^{-6} (one millionth) of its original value or until the sound is inaudible

The most important acoustic property of a concert hall is its reverberation time. The reverberation time depends on the materials used on the walls, the height of the ceiling, the length of the hall, and the presence or absence of an audience. The comparative absorption values of various construction materials are given in **Table 1**. Notice that the absorption of sound by a given material

Table 1 Sound Absorption Coefficients for Various Substances

Substance	Frequency (512 Hz)	Frequency (2048 Hz)
concrete	*0.025	0.035
brick	0.03	0.049
wood (pine)	0.06	0.10
carpet	0.02	0.27
fibreglass	0.99	0.86
acoustic tile	0.97	0.68
theatre seats	1.6–3.0	—
seated audience	3.0–4.3	3.5–6.0

* To simplify the chart, units have not been supplied. Substances with larger coefficients have better sound-absorption qualities, and hence shorter reverberation times.

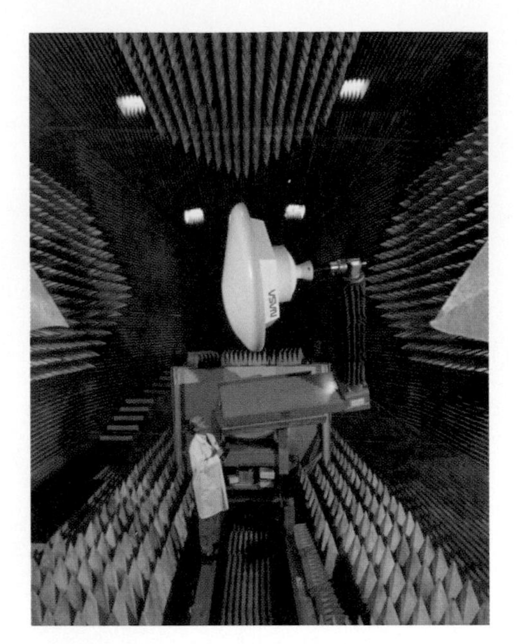
Figure 1
NASA scientist testing an antenna in an anechoic chamber

varies with the frequency. Devices hung from the ceiling can help control both the amount and direction of the sound reflected to the audience. Even the seats may be designed so that if unoccupied, the seat will absorb the same amount of sound as a seated person. Well-designed concert halls tend to have a reverberation time of between 1 s and 2 s; halls that are best for choral music have a reverberation time of 2 s to 5 s. These times vary for different frequencies of sounds, so you can likely appreciate the difficulty of designing an auditorium with excellent acoustics. By the careful choice of materials, a room can be made acoustically "dead," with a reverberation time near 0 s. Such rooms are called anechoic, and they are used for studying the performance of sound devices such as telephones, microphones, and loudspeakers (**Figure 1**).

A band shell in a park directs sound to an outdoor audience. In some cases, spheres of different sizes are used to reflect the short-wavelength, high frequency sounds in several directions. **Figure 2** shows an open-air theatre built over 2000 years ago. A person standing at the centre of the stage who speaks with an ordinary voice can be heard everywhere in the theatre. Although the Romans could not explain acoustics scientifically, they certainly could design theatres with excellent sound characteristics.

Practice

Understanding Concepts

1. Two auditoriums, A and B, are identical except that A has wooden-backed seats and B has cloth-covered seats. Compare the acoustic properties of the auditoriums when they are (a) empty, and (b) filled with people. Explain your reasoning.
2. A drum creates a sound of intensity level 88 dB in an auditorium. After the reverberation time, what is the intensity level of the sound?

Making Connections

3. An auditorium with adjustable features is set up to accommodate choral music. What adjustments could be made so this auditorium is better suited for a concert with an orchestra?

Answers

2. 28 dB

Figure 2
This 25 000-seat outdoor theatre was built in Ephesus, Turkey, by the Romans about 2000 years ago. The theatre has excellent acoustical properties.

Activity 8.9.1
Reverberation Time

A typical graph illustrating the experimental determination of the reverberation time in a concert hall is shown in **Figure 3**. As you can see in the graph, the sound intensity level builds up and reaches a maximum value of 70 dB. This level is sustained briefly, then the sound is suddenly turned off (at $t = 6.4$ s). Within 1.8 s (at $t = 8.2$ s), the intensity level has dropped by 60 dB to 10 dB (i.e., to 10^{-6} of the maximum value). Thus, the reverberation time is 1.8 s.

Question

What is the reverberation time of a large room?

Procedure

1. Attach a microphone to an oscilloscope or to a computer in a large room, such as the school auditorium or gym.
2. Make a loud noise by clapping two boards together. (It may be easier to record the sound and then analyze it in the classroom.)
3. Determine the reverberation time of the room.

Analysis

(a) Look at the physical structure of the room. What changes might you make to decrease and to increase the reverberation time?

(b) What would you do to the reverberation time to improve the acoustics? Why?

(c) Does the placement of the microphone or the type of noise affect your results? How?

(d) To validate your prediction in (c), try a different placement of the sound and/or the microphone, and measure the reverberation time.

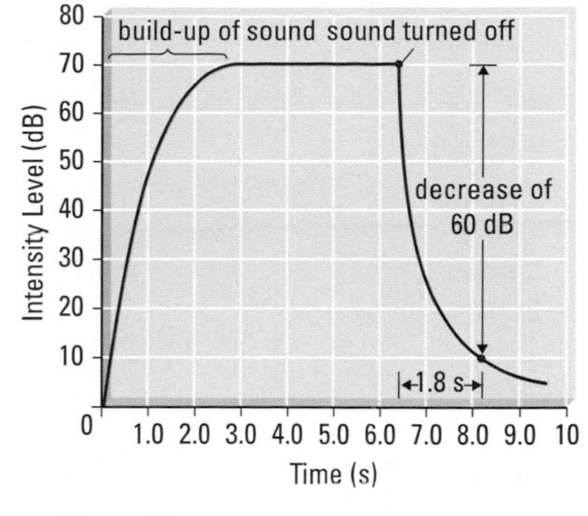

Figure 3
Experimental determination of reverberation time

SUMMARY Acoustics

- Acoustics are the total effect of sound produced in an enclosed space.
- Reverberation time is the time for the intensity of a sound in a room to diminish to the point that it is inaudible.
- The types and locations of sound reflectors in a building determine its acoustical properties, including reverberation time.

Section 8.9 Questions

Understanding Concepts

1. Discuss the acoustics of both your physics classroom and the school auditorium.
   (a) What has been done to provide good acoustics?
   (b) What could be done to improve the acoustics?
2. By how many decibels does the intensity level of a sound decrease after the reverberation time in any room has elapsed?
3. Determine the intensity level of a sound after reverberation time if the original loudness is
   (a) 76 dB (b) 67 dB (c) 60 dB
4. The human ear can distinguish an echo from the sound that caused the echo if the two sounds are separated by an interval of about 0.10 s or more. At an air temperature of 22°C, what is the shortest length for an auditorium that could produce a distinguishable echo of a sound emanating from one end?

CAREER

Careers in Waves and Sound

There are many different types of careers that involve the study of waves and sound in one form or another. Have a look at the careers described on this page and find out more about one of them or another career in waves and sound that interests you.

Geophysicist
This profession requires a B.Sc. degree, with a major in geophysics, plus an MA and Ph.D. in geophysics. Research must also be done in the person's area of specialty. A geophysicist studies Earth's magnetic, electric, and gravitational fields. Using a wide variety of tools including computers and remote sensing equipment such as radar, geophysicists work in universities, research centres, and private companies.

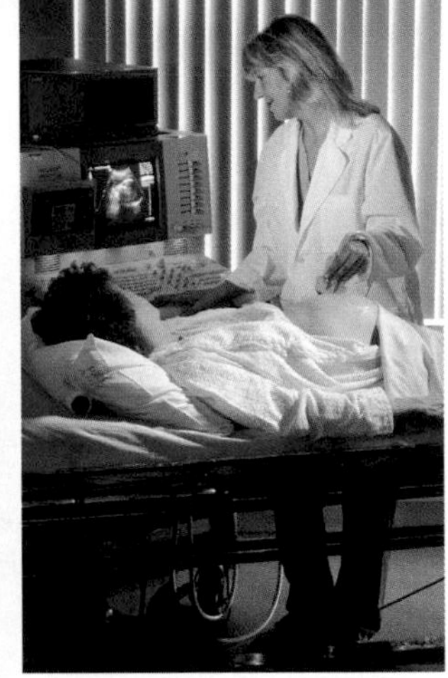

Audiologist
Working in clinics, universities, and speech and hearing centres, as well as for school boards, audiologists are trained in the identification of hearing loss and its treatment and prevention. Audiologists also prescribe and fit hearing aids. They measure hearing with computer-linked calibrated audiometers and use otoscopes to examine the inner ear. To be an audiologist, you need to complete a three-year master's program after obtaining a university degree in either arts or science.

Ultrasound Technician
An ultrasound technician works in a clinic, hospital obstetrician's office, or a research facility. Using ultrasound waves and computer imaging, the ultrasound technician scans patients' bodies to detect tumours or to gauge the progress of a developing baby. In order to train for this job, you need to complete two years of study as an undergraduate in a science degree course, and then study in a program from a university or college for three more years. Nurses or doctors who have one full-year of practical experience may also be admitted to this program.

Practice

Making Connections

1. Identify several careers that require knowledge about waves and sound. Select a career you are interested in from the list you made or from the careers described above. Imagine that you have been employed in your chosen career for five years and that you are applying to work on a new project of interest.
   (a) Describe the project. It should be related to some of the new things you learned in this unit. Explain how the concepts from this unit are applied in the project.
   (b) Create a résumé listing your credentials and explaining why you are qualified to work on the project. Include in your résumé
   - your educational background—what university degree or diploma program you graduated with, which educational institute you attended, post-graduate training (if any);
   - your duties and skills in previous jobs; and
   - your salary expectations.

   Follow the links for Nelson Physics 11, 8.9.

GO TO **www.science.nelson.com**

Chapter 8 Summary

Key Expectations

Throughout this chapter, you have had opportunities to

- identify the conditions required for resonance to occur in various media; (8.4)
- describe and illustrate the properties of transverse and longitudinal waves in different media, and analyze the velocity of waves travelling in those media in quantitative terms; (8.4)
- explain and graphically illustrate the principle of superposition, and identify examples of constructive and destructive interference; (8.3, 8.8)
- identify the properties of standing waves and explain the conditions required for standing waves to occur in strings and air columns; (8.3)
- analyze in quantitative terms the conditions needed for resonance in air columns, and explain how resonance is used in a variety of situations (e.g., analyze resonance conditions in air column in quantitative terms, identify musical instruments using such air columns, and explain how different notes are produced); (8.4, 8.5)
- design and conduct an experiment to determine the speed of waves in a medium, compare theoretical and empirical values, and account for discrepancies; (8.4)
- analyze through experimentation the conditions required to produce resonance in vibrating objects and/or in air columns (e.g., in stringed instruments, tuning forks, wind instruments), predict the conditions required to produce resonance in specific cases, and determine whether the prediction is correct through experimentation; (8.4, 8.5)
- describe how knowledge of the properties of waves is applied in the design of buildings (e.g., with respect to acoustics) and of various technological devices (e.g., musical instruments, audiovisual and home entertainment equipment); (8.1, 8.2, 8.3, 8.5, 8.6, 8.7, 8.8, 8.9)
- evaluate the effectiveness of a technological device related to human perception of sound (e.g., hearing aid, earphones, cell phone), using given criteria; (8.7)
- identify and describe science and technology-based careers related to waves and sound; (career feature)

Key Terms

music
noise
pure tone
consonance
dissonance
octave
fundamental mode
fundamental frequency
overtones
harmonics
quality of a musical note
closed air column
open air column
resonator
acoustics
reverberation time

Make a Summary

Design an entertainment venue for your school that includes the ideas presented in this chapter. Begin with an appropriate setting (such as an auditorium) with a stage. Add an orchestra that has all of the types of musical instruments (stringed, wind, percussion, electrical, and electronic). Include a choir or other ensemble to show off singing voices, with a loudspeaker system to amplify the voices. Add some "sound engineers" who are using an oscilloscope to analyze the sounds produced in the venue. Show the features of the venue that help to create good acoustics.

You can summarize your design with labelled diagrams on a single piece of paper, or you can create a poster or use computer software to generate the summary. In the design, include as many of the concepts and key words from this chapter as possible.

Reflect on your Learning

Revisit your answers to the Reflect on Your Learning questions at the beginning of this chapter.

- How has your thinking changed?
- What new questions do you have?

Chapter 8 Review

Understanding Concepts

1. The diagrams in **Figure 1** show sound waveforms displayed on an oscilloscope. In each case, state a probable source of the sound and describe the sound heard.

(a)

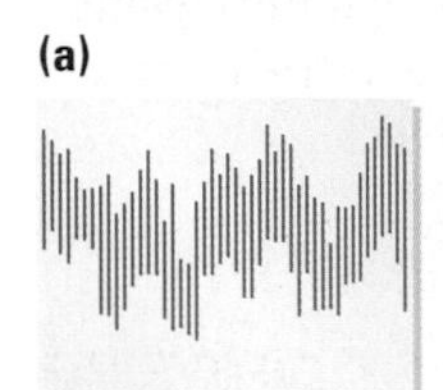

(b)

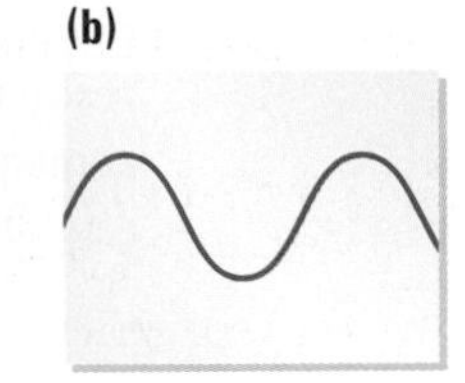

(c)

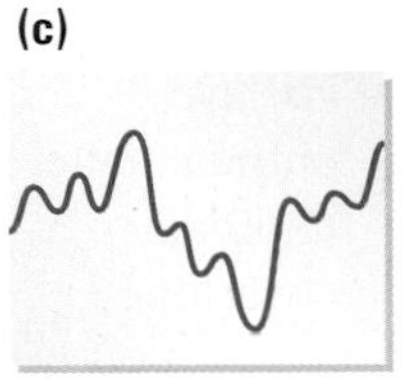

(d)

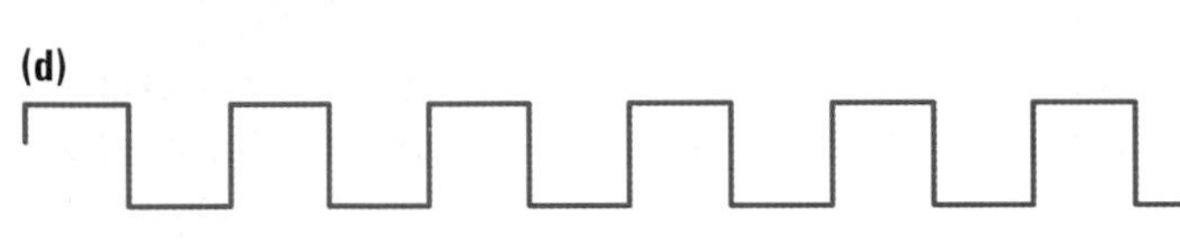

(e)

Figure 1

2. Name the three characteristics of musical sounds and state what each depends on.
3. For each pair of notes, determine the ratio of their frequencies as a fraction and state whether their consonance is high or low.
   (a) 1000 Hz, 800 Hz
   (b) 820 Hz, 800 Hz
4. For a frequency of 800 Hz, state the frequency of the following:
   (a) a note two octaves higher
   (b) a note one octave lower
5. Why are the frets on the neck of a guitar spaced closer together as you move toward the bridge?
6. Sketch a graph to show how the frequency of a vibrating string depends on the string's
   (a) length
   (b) tension
7. A 50.0-cm string under a tension of 25 N emits a fundamental note having a frequency of 4.0×10^2 Hz. What must the new frequency be if the tension is changed to 36 N?
8. A vibrating string has a frequency of 2.0×10^2 Hz. What will its frequency be if
   (a) its length is decreased to $\frac{1}{4}$ of its original length?
   (b) its tension is quadrupled?
9. A string of given length, diameter, and density is under a tension of 1.0×10^2 N and produces a sound of 2.0×10^2 Hz. How many vibrations per second will be produced if the tension is changed to
   (a) 9.0×10^2 N?
   (b) 25 N?
10. Describe how the quality of a musical sound is affected by the harmonics structure of the sound.
11. If the fundamental frequency produced by a guitar string is 4.0×10^2 Hz, what is the frequency of the second overtone?
12. What is the beat frequency produced by the first overtones of two strings whose fundamental frequencies are 280 Hz and 282 Hz?
13. A closed air column is 60.0 cm long. If the speed of sound is 344 m/s, calculate the frequency of a tuning fork that will cause resonance at
    (a) the first resonant length
    (b) the second resonant length
    (c) the third resonant length
14. A pipe that is closed at one end can be made to resonate by a tuning fork at a length of 0.25 m. The next resonant length is 0.75 m (speed of sound is 338 m/s).
    (a) Calculate the wavelength of the sound emitted by the tuning fork.
    (b) Calculate the frequency of the tuning fork.
15. An open organ pipe has a fundamental frequency of 262 Hz at room temperature (20°C). What is the length of the pipe?
16. The first resonant length of an open air column in resonance with a 512-Hz fork is 33.0 cm. What is the speed of sound?
17. An air column has a fundamental frequency of 330 Hz. What are the next two harmonics if the air column is
    (a) open at both ends?
    (b) closed at one end?
18. You are constructing a wooden resonance box open at both ends on which you want to mount a 320-Hz tuning fork. What is the minimum length of the box? (Assume a room temperature of 20°C.)
19. In section 8.5, a piano was described as a stringed instrument. Do you think it could also be described as a percussion instrument? Why or why not?
20. Contrast and compare whistling with playing a trombone.

21. For a person with equal sensitivity in both ears, would the direction from which a sound is coming be more easily identified if the sound came from a woofer or a tweeter? Explain your reasoning. (*Hint:* Consider the diffraction of sound waves.)
22. Electrical and electronic instruments differ in the way in which they create sound vibrations. Explain the difference.
23. **Figure 2** shows the top view of a speaker system having a woofer and a tweeter facing to the right. Compare the sounds heard by an observer at positions A, B, and C.

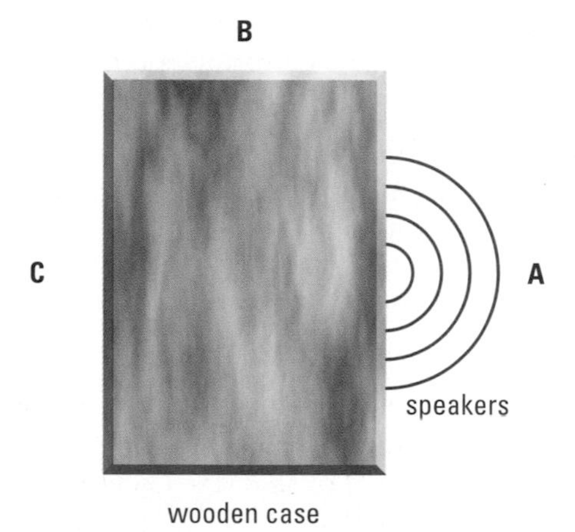

Figure 2

24. The impact sound from a drum registers 72 dB on a sound intensity level meter in an auditorium. After the auditorium's reverberation time has elapsed, what is the reading on the meter?
25. Why are some of the catgut strings on the bass violin wrapped with very fine wire?
26. Singers are reputed to be able to shatter delicate wine glasses by singing high, loud notes. Explain.
27. If you press the loudness pedal on a piano and sing a note, a sound usually comes from the piano. Explain this phenomenon. Why might more than one piano string respond?
28. For a column of air that is closed at one end, compare the amplitude of vibration of the air molecules at the two ends of the column.

Applying Inquiry Skills

29. Sound tests are performed in a concert hall where the echoes seem to be too powerful. The reverberation time for sounds in the frequency range of 2000 Hz to 4000 Hz is found to be 2.9 s, longer than the optimum value. What techniques could be used to decrease the reverberation time?
30. A synthesizer produces a constant-frequency sound with a delayed attack, then a sustained amplitude, and finally a slow decay. Draw a diagram of the waveform of the sound.
31. Listen to the sound produced by a small loudspeaker connected to an audio-frequency generator. Now cut a hole the size of the speaker in a large board or piece of cardboard. Hold the speaker at the hole and listen again. Use the concepts studied in this chapter to explain what you hear.
32. A car muffler uses the principle of resonance to reduce engine exhaust noise. Its internal construction consists of a series of open tubes of different lengths. How might a tube be used to reduce the intensity of sounds with specific frequencies?

Making Connections

33. When 80 students in the concert band practise in the music room, the sound level could exceed 100 dB at times. One of the occupational hazards for music teachers is premature hearing loss. The proper design and construction of the music room can reduce this. List as many design features as you can for a "good" music room. Evaluate the music room in your school and make some suggestions for improvement, if necessary.
34. Write a brief (200 to 300 words) report on the Moog Synthesizer. As well as outlining the history of its development, briefly explain the physical principles of its operation. Follow the links for Nelson Physics 11, Chapter 8 Review.

GO TO www.science.nelson.com

35. Some music synthesizers operate with a personal computer. Write a brief report describing how this device can be effectively used by musical composers and arrangers. Follow the links for Nelson Physics 11, Chapter 8 Review.

GO TO www.science.nelson.com

Figure 1
Many properties of waves had to be considered in the design of the water-pool theatre used for the amazing Cirque du Soleil show. Sound is delivered to the audience in an acoustically designed theatre, but it is also delivered in the pool to the swimmers and the divers who help the swimmers under the water. The designers even had to find a way of absorbing the energy of the waves created when the components of the segmented stage rise and fall.

Sound Quality Testing

Architects, engineers, scientists, and entertainers apply the properties of waves in order to create sound and help people hear sound. Much of this knowledge can be acquired by analyzing the application of the properties of sound waves to the design and construction of musical instruments, audio-visual equipment, home-entertainment systems, hearing aids, cell phones, earphones, auditoriums, and theatres.

Consider, for example, the theatre in the Bellagio in Las Vegas, Nevada, which is home to a unique Cirque du Soleil® show titled "O™" (from the French word for water, eau), shown in **Figure 1**. The 1800-seat theatre has a pool that holds 5.7 million litres of water. In the pool there are seven underwater lifts that are used to change the floor elevation as the acrobats perform various acts. The lifts are lowered for high dives into the water. Some or all of the lifts can be moved to create a segmented stage where the performers can be at any required depth in the pool or they can be on the stage with no water at all. As the lifts move the various segments of the stage, the energy of the created waves is absorbed by perforations drilled through the deck as the water falls on numerous different-sized pebbles placed in a gutter. To help the performers hear the music and timing cues while under the water, a revolutionary underwater sound system has been designed and constructed. In the auditorium, more than 30 surround-sound loudspeakers are built into the balcony and side walls to deliver the music from the live orchestra to the audience. The timing of the music, and thus each act, is computer controlled, while sound adjustment is monitored by the sound mixing engineer during the show.

However, the many properties of waves can also be learned by designing and building much less sophisticated devices than those described above. A steel band (**Figure 2**) uses metal components to create a variety of pleasant, rhythmic sounds. Some musicians use wine glasses filled to varying levels with water to create sounds. Some "physics bands" have entertained their audience by blowing into pop bottles containing water at different levels to produce music.

The Task

You may choose one of the three options that follow.

Option 1: Research the Creation and Control of Sound

Your task is to research the design and function of a technological device (such as a musical instrument, an audio-visual instrument, or a home-entertainment system), or a theatre (such as an IMAX or OMNIMAX theatre or another entertainment structure). You will be expected to analyze the design and function to show that you understand how the device or structure applies the properties of waves.

Option 2: Research Devices That Create Sound

Your task is to research a technological device related to the human perception of sound (such as a hearing aid, an earphone, a noise-cancellation device, or an electronic insect repellent) and evaluate the effectiveness of the device using the criteria described in Analysis. To help with your research, follow the links for Nelson Physics 11, Unit 3 Performance Task.

GO TO www.science.nelson.com

Figure 2
A typical steel band

Option 3: Create a "Physics Band"

Your task is to join others in a group to create a physics band. You will need to design, construct, test, and modify a set of musical instruments made of simple household materials such as tubes, pipes, spoons, cans, bottles, jars, sticks, wires, wooden boards, etc. The set should consist of samples of stringed, wind, and percussion instruments. After constructing the instruments, your band should learn to play at least one musical tune to entertain the class. You will be expected to analyze the physics principles related to the instruments in the band.

Analysis

Your analysis will depend on the option you chose.

Option 1 and Option 2

(a) How does your device apply physics principles, laws, and theories?
(b) How has the basic design of this type of device evolved from earlier times to the present?
(c) What are the main uses of the device, and who are the primary users?
(d) Identify one or two companies that produce this type of device. Select two models that are currently on the market and compare their features, prices, and availability.
(e) What impact has this device had on the individual user and society in general? Discuss societal concerns related to the use of the technology.
(f) What careers are related to the production or use of the device?
(g) What research is being carried out to try to change or improve the device?
(h) Suggest some innovations related to your choice that you would like to have implemented in the future.
(i) List problems that you encountered during your research and explain how you solved them. (You will get ideas as you research your chosen topic.)

Option 3

(a) How do the instruments apply physics principles, laws, and theories?
(b) How can the tuneable instruments be tuned to play various musical notes?
(c) How did your group test your musical instruments to try to improve how they sounded together?
(d) How could your knowledge gained in designing the instruments be applied to the design of commercial instruments?
(e) What advice would you give to other physics students when they design their own band?
(f) List problems that you encountered during your creation of the physics band and explain how you solved them. (You will get ideas as you test and modify the instruments.)

Assessment

Your completed task will be assessed according to the following criteria:

Process

- Choose appropriate research tools, such as books, magazines, the Internet, and experts in the field OR choose appropriate apparatus and materials for the instruments.
- Carry out the research and summarize the information with appropriate organization and sufficient detail OR apply physics principles to the design, construction, and testing of the musical instruments.
- Analyze the physical principles of the device or structure, including those related to the properties of waves OR evaluate the design, construction, testing, and modification process.
- Evaluate the research and reporting process (for Option 1 and Option 2).

Product

- Demonstrate an understanding of the related physics principles, laws, and theories.
- Use terms, symbols, equations, and SI metric units correctly.
- For Option 1 and Option 2, prepare a suitable research portfolio in print and/or audio-visual form.
- For Option 3, prepare a formal report and a live or videotaped audio-visual presentation.

Unit 3 Review

Understanding Concepts

1. Two students are on two adjacent swings. One student completes 24 complete swings in 1.00 min while the other completes 20 swings in the same minute.
   (a) What is the period of each person's swing?
   (b) Who is on the longer swing? Why?
   (c) If the students start out together, how many times will they be swinging together in that minute?
2. A stroboscope is flashing at a frequency of 5 Hz. A white pendulum starts to swing in a dark room and the pendulum appears to be stopped every half cycle. How long does it take for the pendulum to make a full cycle?
3. A low frequency woofer speaker has a diameter of 38.0 cm. What is the frequency of a sound that has a wavelength equal to this diameter if the speed of sound is 344 m/s?
4. The frequencies on the FM radio band range from 88 MHz to 108 MHz and these radio waves travel at 3.0×10^8 m/s. What is the range of their wavelengths?
5. Each of the two diagrams in **Figure 1** shows two waves interfering. For each case, trace the diagram on a piece of paper and determine the resultant wave.

(a)

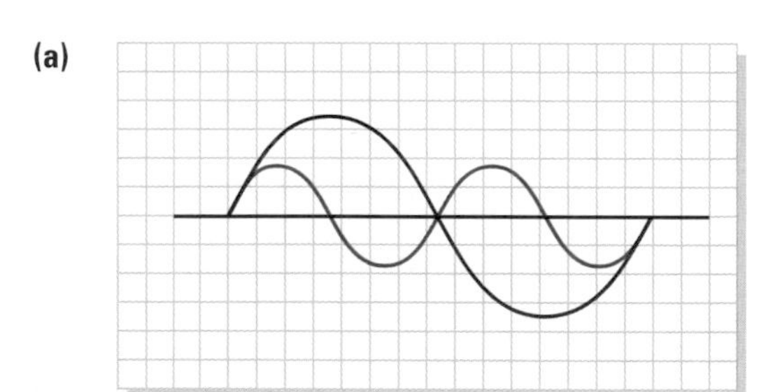

(b)

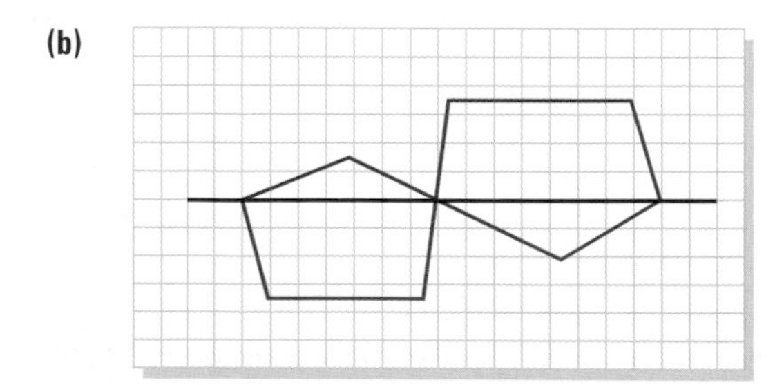

Figure 1

6. A thick rope and a thin rope joined together have the same tension when a wave comes along, as shown in **Figure 2**. Draw a sketch of the rope shortly after the wave has passed the junction. Be careful to draw the following characteristics carefully: height and length of pulse, and distance the pulse is away from where the two ropes meet.

Figure 2

7. Use the principle of superposition of waves to draw the resultant waveform for the following superimposed waves:
   (a) two waves of the same frequency, in phase, with amplitude ratio 2 : 1
   (b) two waves of the same frequency, in opposite phase, with amplitude ratio 2 : 1
   (c) two waves with frequency ratio 2 : 1, in phase, with equal amplitudes
8. The speed of waves travelling in opposite directions in a string is 4.0×10^2 m/s. If the frequency of a standing wave produced is 286 Hz, how far apart are the nodes?
9. Two water waves are generated by two point sources S_1 and S_2, each with a wavelength of 2 cm. The point sources are 6 cm apart and are in phase with one another.
   (a) Using a compass, draw on a scale diagram the resulting interference pattern.
   (b) Indicate on the diagram four nodes, four supercrests, four supertroughs, areas of constructive interference, and lines of destructive interference (nodal lines).
10. **Figure 3** shows the sound wave produced by a tuning fork.

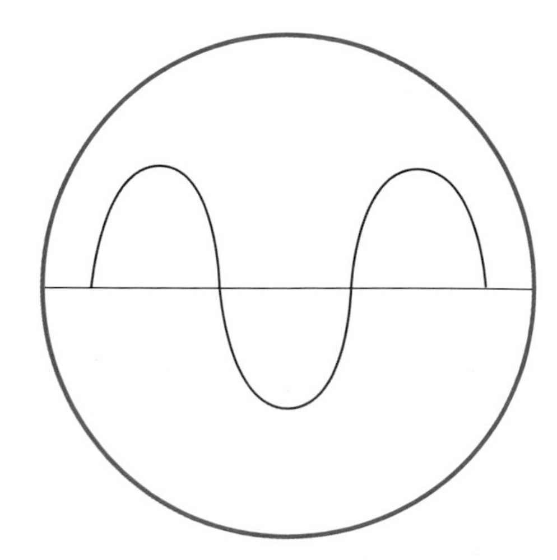

Figure 3

Copy the diagram on a sheet of paper and then draw a graph to show each of the following:

(a) a sound wave with a higher pitch, but the same loudness
(b) a sound wave of the same pitch, but louder
(c) a sound wave of the same pitch, but emitted by a different instrument

11. Give two reasons why circular water waves decrease in amplitude as they move away from a source.
12. Find the speed of sound in air at 22°C and at −22°C.
13. If a brass instrument is tuned to 440 Hz at 20°C, what will its frequency be at 0°C ?
14. A sound wave in a steel railway track has a frequency of 10 Hz and a wavelength of 0.51 km. How long will it take the wave to travel 2.0 km down the track?
15. One end of an iron pipe is struck with a hammer and an observer hears two sounds, one heard at 0.50 s before the other. What is the length of the pipe if the speeds of sound in air and in iron are 3.4×10^2 m/s and 5.0×10^3 m/s, respectively?
16. A Formula One car is moving at Mach 0.16 when the temperature of the air is 20°C. What is the speed of the car in kilometers per hour?
17. A student stands 180 m from a flat wall and starts to clap hands to hear clap-echo-clap-echo at a steady rate of 1 clap per second. What result does the student get for the speed of sound?
18. A student ignites a firecracker. The sound is reflected from a wall located at a distance of 275 m away and the echo is heard 1.6 s after. What is the speed of sound?
19. A member of a school band blows a note into a flute. The band member then blows harder. What characteristics of the sound wave will change as a result?
20. A sound wave travels through a metal rod into the surrounding air. Compare the wavelength, λ, and frequency, f, of the sound wave before it left the metal and after it entered the air.
21. Tuning fork A has a frequency of 256 Hz and is sounded at the same time as another 256-Hz tuning fork, B, that is loaded to reduce its frequency. If the beats are heard at a frequency of 10.0 Hz, what is the frequency of fork B?
22. An oboe plays A (440 Hz) at 20°C. At the same time, an identical oboe plays the same note at 0°C, and beats are heard. What is the best frequency?
23. Two sound waves of wavelengths 2.29 m and 2.33 m interfere to produce 13 beats in 5.0 s.
    (a) To what wavelength does the lower frequency respond?
    (b) Find an expression for the frequency in (a), and determine the speed of sound.
24. A violin string vibrates at 196 Hz. What frequency will it vibrate with if it is fingered at one quarter of its length?
25. If the frequency of a violin vibrates at 256 Hz in its fundamental mode, what are the first and fourth harmonics?
26. With some musical instruments, such as a violin, a skilled player will produce an attractive sound, whereas an unskilled player will produce an unattractive sound while trying to play the same note. Using the terms dissonance, consonance, and harmonics, explain why the two sounds are so different.
27. An instrument emits harmonics of 2.80 m and 3.10 m in air at 15°C.
    (a) What will the beat frequency be?
    (b) How far apart in space are the regions of loudest sound?
28. At certain definite engine speeds, parts of a car such as the door panel may vibrate strongly. Briefly explain the physical phenomenon this illustrates and give three other examples.
29. A buret is filled with water and the water is allowed to slowly drain out until the column resonates to a tuning fork of 512 Hz.
    (a) How far is the water from the top of the buret, if the temperature is 21°C?
    (b) If the buret is 1.0 m long, at what points will other resonances occur?
30. With a tuning fork of 384 Hz, resonance tube lengths are achieved at 0.647 m and 1.09 m. What is the speed of sound?
31. If the speed of sound in air is 344 m/s, what will the first two resonant lengths in an adjustable closed air column be, if a tuning fork with a frequency of 440 Hz is struck at the open end?
32. An open organ pipe is 0.50 m long.
    (a) Find the frequency of its fundamental and first overtone.
    (b) What would the answers for (a) be if the organ pipe were closed (speed of sound = 340 m/s)?
33. A trumpeter plays a note that is the first overtone.
    (a) What is the bore length of the instrument if the wavelength is 0.50 m?
    (b) If the temperature is 20.0°C, what would the frequency of the fundamental be for the same bore length?

34. A cork in a ripple tank oscillates up and down as the ripples pass beneath it. If the ripples travel at 20.0 cm/s, have a wavelength of 15 mm, and have an amplitude of 5.0 mm, what is the maximum speed of the cork?
35. If a person inhales a lungful of helium, his or her voice will sound very high-pitched. Why?
36. The flash of a gun is seen due north 4.0 s before the sound is heard. If the temperature is 20.0°C, find
    (a) the distance to the observer if there is no wind
    (b) how much longer the sound would take if the wind is blowing from the south at 48 km/h
37. The lower frequency sounds (0.50–50 kHz) emitted by dolphins probably function mainly for social communication, while higher frequencies (40–150 Hz) are probably used for echolocation. (Assume two significant digits.)
    (a) What are the periods for the range of sounds used for echolocation?
    (b) If the speed of sound in salt water is given as 1.5×10^3 m/s, what is the range of wavelengths for echolocation?
38. Bats send out ultrasonic sounds of approximately 3.0×10^4 Hz from their mouth and nose. When a sound wave from a bat hits an object, it echoes back.
    (a) What is the period of the emitted sound?
    (b) If the temperature is 30.0°C, what is the wavelength of the emitted sound?
    (c) If the time for the echo to return from a moth is 4.6×10^{-2} s, how far away is the moth from the bat?
39. For a vibrating string, which harmonics have a node exactly in the middle?
40. Why will organ pipes have a different pitch when the auditorium is at a lower temperature?
41. In a church, the reverberation time for an organ note is 6.0 s. If the organ is located at the front of the church, and the back wall is 160.0 m away, how many reflections occur before the sound becomes inaudible? Assume the temperature of the air in the church is 21°C.
42. The earthquake that struck Eastern India on January 26, 2001 measured 7.9 on the Richter scale. The Richter scale is logarithmic, just like the intensity scale used in sound. On the Richter scale, each whole number increase in magnitude represents a tenfold increase in the measured amplitude. How much larger was the measured amplitude for the Indian earthquake than an earthquake of 5.9 and of 6.9 on the Richter scale?
43. Using Newton's laws of motion, explain why placing an elastic band on the prong of a tuning fork lowers its pitch.
44. A wine glass can be made to vibrate by gently rubbing the rim with a moistened finger. Assume that a standing wave produces a resonant vibration in the rim.
    (a) If four nodes are produced on the rim, draw a sketch of how the rim would vibrate.
    (b) What is the wavelength of the vibration if the diameter of the rim is 6.0 cm?
45. When filling a gas tank, you notice a pitch change as the tank reaches its full capacity. What is the pitch change and what causes the change?
46. The Doppler effect shows that a change in frequency is precisely related to the relative speed of the source and the observer. The spectra of stars outside of our galaxy appear to be shifted to the red end of the spectrum (longer wavelength).
    (a) Does this mean a perceived increase or decrease in frequency? Explain your reasoning.
    (b) Are the stars outside our galaxy, moving away or toward us? Why?
47. A megaphone horn makes sounds louder in one direction than it would without the horn. Why?
48. During the latter part of the Second World War, V2 rockets were used extensively to bombard England. During a bomb attack, the sound of the blast was heard before the noise of the rocket. Why?
49. By rocking a tub of water at just the right frequency, the water rises and falls alternately at each end. If the frequency producing the standing wave in a 60.0-cm tub is 0.71 Hz, what is the speed of the water waves?

Applying Inquiry Skills

50. What evidence can you cite to show that the speed of sound in air does not change significantly for different frequencies?
51. Television and FM radio signals cannot be received behind a hill, whereas AM radio signals can. What does this tell you about their respective wavelengths? Explain.
52. Why is fine wire wrapped around the thin tough cord made from animal intestines for some guitar strings?

Making Connections

53. Examine the auditorium designs in **Figure 4** in terms of focusing of sound, reflections, and echoes. Which design would have better acoustics? Explain your reasoning.

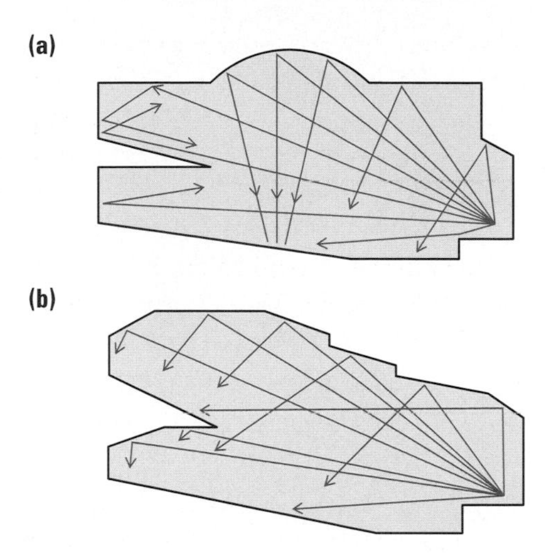

Figure 4
Two auditorium designs

54. Why is it usually preferable to have a longer reverberation time in an auditorium for music than one for speeches?
55. There are three types of seismic waves. The first and fastest is called the compression, primary, or P wave. It moves through the Earth's interior, compressing and expanding the rock in the direction of the wave's movement. The second type is known as the shear, secondary, or S wave. It causes the Earth to vibrate at right angles to the direction the wave is travelling. The slowest type, which moves along the Earth's surface, is called the surface, or L wave, and is further divided into two types. Rayleigh waves displace the ground vertically, whereas Love waves move the ground from side to side. It is the surface waves that cause the havoc, death, and destruction inflicted by a major earthquake.
    (a) Classify each of the waves described as longitudinal or transverse.
    (b) Which type of wave is most likely to cause
        (i) tall buildings to shake from side to side?
        (ii) folding and cracking of the earth's surface?
        Explain your reasoning in each case.
    (c) P and S waves travel at different speeds, and this fact is used to determine the origin or epicentre of the earthquake. Assuming that the P and S waves have speeds of 9.0 km/s and 5.0 km/s, respectively, how far away would an earthquake occur if the seismic station measured the two waves to be 1.20×10^2 s apart?
    (d) Explain why more than one seismic station is needed to precisely locate the epicentre.
56. Research, using the Internet or another resource, how the sound wave pitch that hits the eardrum is transmitted to the brain. Follow the links for Nelson Physics 11, Unit 3 Review.

GO TO www.science.nelson.com

57. Research the purpose of a crossover in a speaker. Write a short report (about half a page) on why crossovers are used in some speakers and not others. Follow the links for Nelson Physics 11, Unit 3 Review.

GO TO www.science.nelson.com

58. Find the fundamental frequency of a wind or stringed instrument. You may ask your music teacher, ask a friend, or research the information. If possible, bring the instrument to class and demonstrate the first, second, and third harmonics. Be prepared to explain why these are the first, second, and third harmonics.
59. For the human voice,
    (a) What determines the pitch?
    (b) Why are women's voices higher than men's?
    (c) Why do the voices of boys change when they become teenagers?
60. When you hear your own voice on a tape recorder, it sounds thin to you, although your friends say your recorded voice sounds normal. Why?
61. In areas where deer are numerous, many drivers put deer whistles on their vehicle to frighten the deer away. Research how deer whistles work and report your findings in one or two paragraphs.

Exploring

62. When you whistle through your lips, how do you change the pitch?
63. If you twirl a flexible, corrugated plastic tube or a vacuum cleaner hose around your head, a musical note is produced.
    (a) What produces the musical note?
    (b) Why does the pitch increase if the tube is whirled at higher frequencies?
64. Investigate the standing wave patterns created on the surface of a drum by sprinkling fine sand on the surface and causing the sand to vibrate by tapping the drum gently in different places.

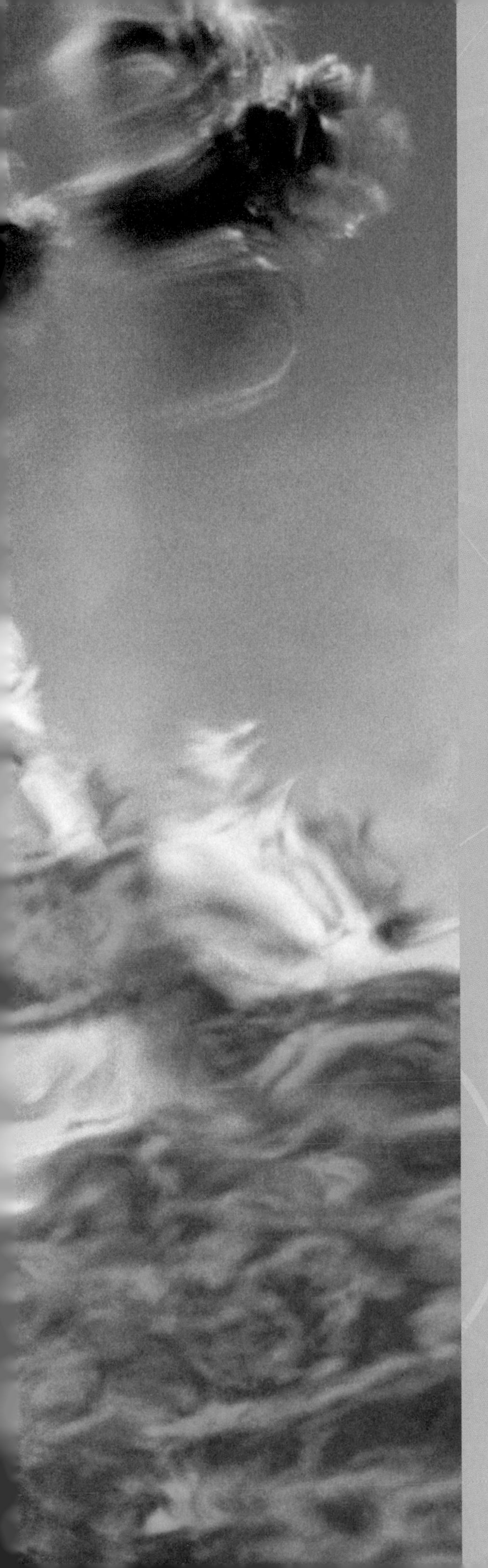

Unit 4 Light and Geometric Optics

Age-related blindness and laser eye surgery are hot topics relating to optics. Dr. Melanie Campbell explores the optical elements of the eye, the image produced on the retina, and how this image changes with age and refractive surgery.

"We can measure the blur of the optical image on the retina by imaging the light coming from the eye with an array of tiny lenslets, called a Hartmann-Shack array. We image the back of the eye using a scanning laser beam (confocal scanning laser ophthalmoscope), which allows us to take optical slices of the rear of the eye and create three-dimensional images."

Dr. Melanie Campbell, Professor at the University of Waterloo

Dr. Campbell examines ways to treat macular degeneration, a leading cause of blindness in older people. She also researches refractive surgeries in which the cornea is shaped with a laser to correct near- and far-sightedness.

Overall Expectations

In this unit, you will be able to

- demonstrate an understanding of the properties of light and the principles underlying the transmission of light through a medium and from one medium to another
- investigate the properties of light through experimentation, and illustrate and predict the behaviour of light through the use of ray diagrams and algebraic equations
- evaluate the contributions made by the development of optical devices and other technologies that make use of light to areas such as entertainment, communications, and health

Unit

4

Light and Geometric Optics

Are You Ready?

Knowledge and Understanding

1. **Figure 1** shows a device that produces light energy.
   (a) How is light energy produced by this device?
   (b) List some sources of light energy in your daily life.
   (c) Which light source is most important to life on Earth? Why is it so important?

Figure 1

2. The three mirrors in **Figure 2** have different shapes.
   (a) What are the names of these types of mirrors?
   (b) Why do images of the same object appear to be so different in different mirrors?
   (c) State some specific uses of the three types of mirrors shown.

Figure 2

3. Hold your hand a few centimetres above a flat sheet of paper on your desk and observe the shadows cast onto the paper.
   (a) Describe the different parts of the shadow you see.
   (b) What property of light is responsible for causing shadows?
4. **Figure 3** shows a typical "ray box" used by students in performing experiments in geometric optics. Draw three diagrams to show light rays coming from such a box that are
   (a) parallel
   (b) converging
   (c) diverging
5. The human eye is a unique and important optical device. Describe what you know about its structure and function.
6. Visible light forms a small part of the electromagnetic spectrum.
   (a) List the colours of the visible spectrum.
   (b) Describe the properties of the waves that form the electromagnetic spectrum.
7. Name examples of materials or devices that
   (a) absorb most of the light that strikes them
   (b) reflect most of the light that strikes them
   (c) both absorb and reflect the light that strikes them
   (d) allow most of the light that strikes them to transmit through
8. How does the speed of light in air compare with some other common speeds, such as the speed of sound in air?

Figure 3
For question 4

Inquiry and Communication Skills

9. Lenses and curved mirrors have focal points. If you were given a magnifying glass, what experimental procedure would you use to locate the focal point?
10. Communicating mathematically is important in geometric optics. Applying trigonometric ratios is an example of such communication. To review the basic trigonometric ratios (sine, cosine, and tangent), refer to the right-angled triangle ABC in **Figure 4**.
    (a) Using symbols only, write each of the following trigonometric ratios: sin A, sin C, cos A, cos C, tan A, and tan C.
    (b) Use your calculator keys for trigonometric ratios to determine the measures of angles A and C to two significant digits. In each case, use more than one ratio, and compare the results.

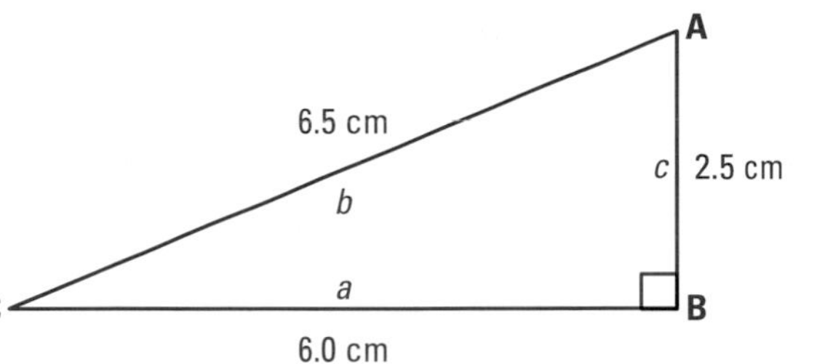

Figure 4
For question 10

Making Connections

11. Think of as many optical devices as you can, including those that have a lens, a mirror, a prism, a laser or other specialized light source, colour filters, or a video screen. Classify the optical devices according to which of the following they are best suited for:
    (a) entertainment
    (b) communication
    (c) health applications
    (d) scientific discovery

Chapter

9

In this chapter, you will be able to

- define and describe concepts and units of wavelength, frequency, and speed of light
- describe the ray model of light and use it to explain natural phenomena such as apparent depth, shimmering, mirages, and rainbows
- demonstrate and illustrate, using light-ray diagrams, reflection, refraction, and partial reflection and refraction
- describe how virtual images are produced
- predict the refraction of light as it passes from one medium to another, using Snell's law
- analyze, describe, and explain applications of total internal reflection such as periscopes, retroreflectors, and optical fibres

Light Rays, Reflection, and Refraction

If you have ever been outside on a cool, clear evening, away from town and city lights, you were no doubt impressed by the night sky. Stars too dim to see in the city due to light pollution become visible. But how much light does Earth actually receive from stars millions of light years away? Since our nights are dark we can say, very little. Yet, using telescopes like the one in **Figure 1**, we are able to capture and produce beautiful images from such small amounts of light (**Figure 2**).

If we study the basic properties of light, we will understand what light is and how humans use it.

Figure 1
The Very Large Telescope in Paranal, Chile (built in 1998), uses an 8.2-m diameter concave mirror that collects light from stars 15 million times dimmer than the faintest star visible to the unaided eye. Optical, mechanical, and software systems compensate for distortions in image quality. The resulting images reveal incredible details.

Reflect on your Learning

1. There are some similarities and some major differences between sound and light. Reflect on your understanding of the properties of sound. Compare these properties to what you currently know about the properties and behaviour of light.
2. Light can be reflected and refracted.
   (a) Describe examples in your home and in nature in which light reflects off objects. Include more than just light reflecting off mirrors.
   (b) Describe examples in which light is refracted, or bent, as it goes from one material into another.

Throughout this chapter, note any changes in your ideas as you learn new concepts and develop your skills.

Figure 2
Many images like this are captured by the Very Large Telescope every day.

Capturing Light

Materials: 40-W or 60-W light bulb, assorted optical components, light meter, and a metre stick.

- Ensure that your light meter works by turning on the light bulb and measuring the intensity of the light 10 cm from the centre of the bulb.
- Position the light meter 30 cm from the centre of the light bulb (**Figure 3**).
- Take a reading and record your results.
- Use some or all of the optical components provided by your teacher to direct as much of the light from the light bulb as possible to the sensor of the light meter.
  - (a) When you have obtained your best result, record the positions of the components you have used relative to the light bulb.
  - (b) How do the optical components direct the light to the light sensor? Draw a ray diagram to support your answer. Make sure that the maximum intensity is clearly written on your drawing.
  - (c) Circulate around your class to see other drawings and intensity values. Identify the arrangement of components that produced the maximum intensity and try to reproduce the results at your station. If you have difficulty, ask the students who obtained the maximum value to demonstrate their technique.

Allow all light bulbs to cool for 5 min before handling them.

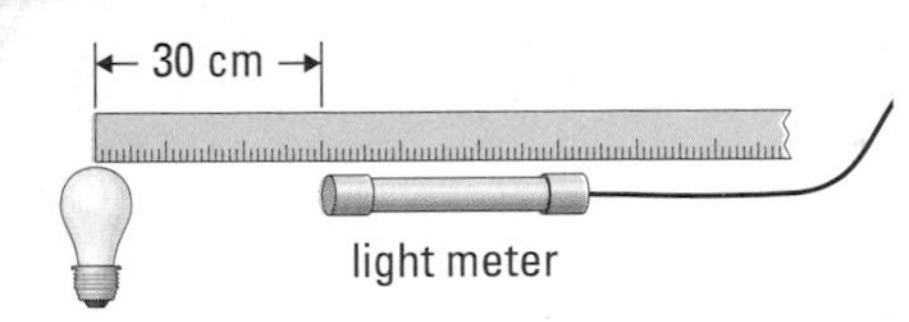

Figure 3

9.1 What Is Light?

One of the earliest written accounts of light and optics comes from the Greek philosopher and mathematician Pythagoras (ca. 582–500 B.C.), who theorized that light was made up of a stream of particles emitted from an object. Later the Greek philosopher Aristotle (384–322 B.C.) proposed that light moved as a wave, like ripples on water.

In 1801, the English physician and physicist Thomas Young (1773–1829) performed an experiment that showed conclusively that light has wave characteristics. By the mid-19th century, English physicist James Clerk Maxwell (1831–1879) showed that light is a type of electromagnetic wave, just like radio waves, microwaves, and X rays (**Figure 1**).

light: a form of energy that is visible to the human eye

We now know that **light** is a form of energy that is visible to the human eye. It *sometimes* displays properties of waves and *sometimes* displays properties of particles.

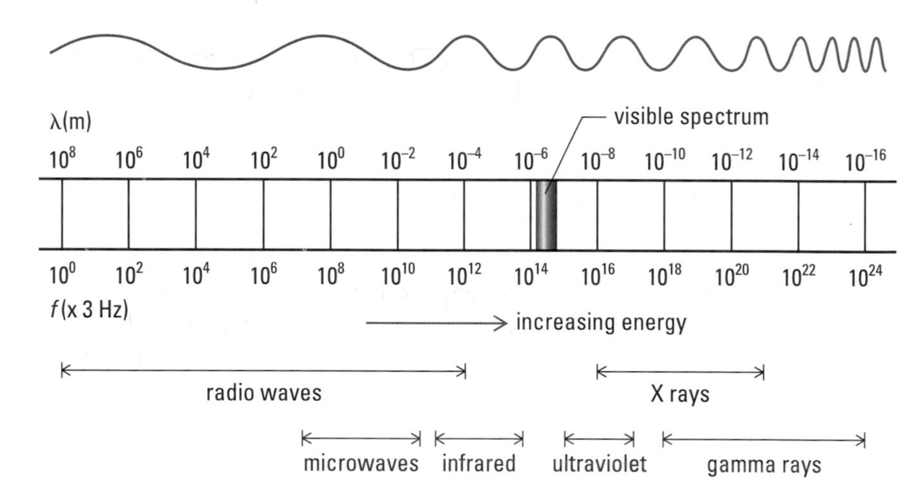

Figure 1
Radio waves, visible light, microwaves, and X rays are all forms of electromagnetic radiation.

Practice

Understanding Concepts

1. Based on the information in **Figure 1**, describe how visible light differs from other components of the electromagnetic spectrum.

The Transmission of Light

When sunlight falls on a solid obstacle, a shadow appears on the ground. The sharp edges of the shadow remind us that light travels in straight lines. This property of light is called **linear propagation.**

linear propagation: one of the properties of light; describes light as travelling in straight lines

Sometimes, if there is dust in the air, we see "rays" of sunlight streaming into the room. In everyday language, "ray" means a narrow stream of light energy, but in physics a **ray** is defined as the path taken by light energy. A beam of light is a group of light rays, which can be *converging, diverging* (**Figure 2**), or *parallel.*

ray: the path taken by light energy

A converging beam is a beam of light whose rays move closer together as the beam propagates through a medium. A diverging beam is a beam of light whose rays move farther apart as the beam propagates through a medium. A parallel beam is a beam of light whose rays remain parallel to each other as the beam propagates through a medium.

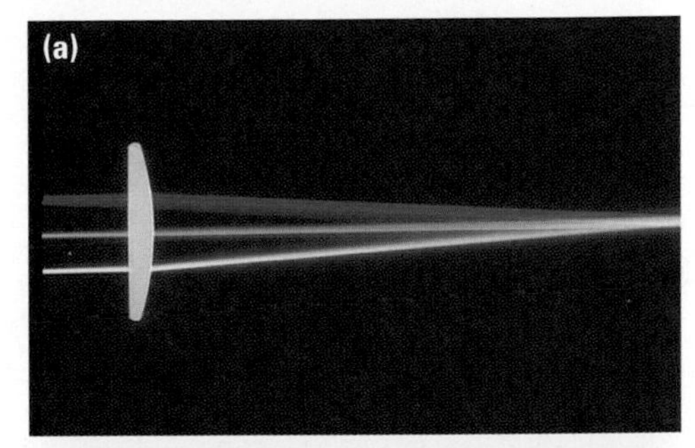

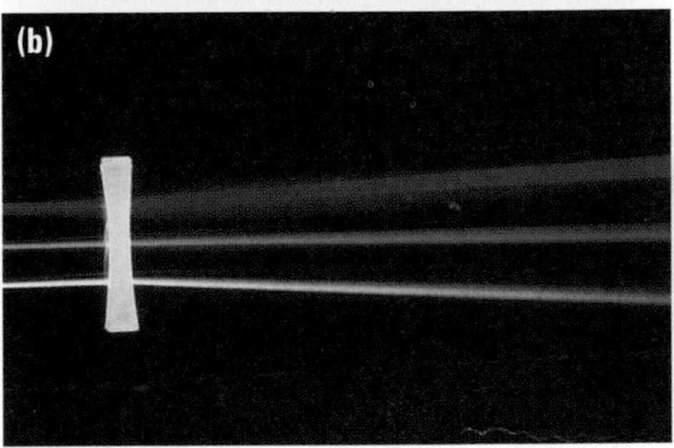

Figure 2
(a) A converging beam
(b) A diverging beam

The Speed of Light

The first successful measurement of the speed of light was reported by Danish astronomer Olaus Roemer (1644–1710). Roemer observed eclipses of Io (one of Jupiter's moons) and calculated the speed of light correct within 25% by timing the appearance of light at different times in Earth's orbit.

In the late 19th century, the German-born American physicist Albert Michelson (1852–1931) obtained more accurate results using an ingenious arrangement of mirrors. Light from a very bright source was reflected from a surface, on an eight-sided, rotatable mirror, to a mirror located about 35 km away. The distant mirror reflected the light back to the rotating mirror where it was reflected and observed through a telescope.

The **speed of light** is such an important quantity in physics that it is given its own symbol, *c*. The speed of light in a vacuum is defined as

$$c = 299\,792\,458 \text{ m/s}$$

or $c = 3.00 \times 10^8$ m/s (to three significant digits)

From your study of waves, it will come as no surprise that light waves also obey the universal wave equation. In the equation $v = f\lambda$, v is replaced with the symbol for light, so we use $c = f\lambda$ for all electromagnetic waves. The sample problem below demonstrates how to use the universal wave equation to calculate the frequency of light.

speed of light: (*c*) the speed of light in a vacuum with a value of 299 792 458 m/s, or 3.00×10^8 m/s

Sample Problem

A laser based on yttrium aluminum garnet (called a YAG laser) emits a wavelength of 1064 nm. Calculate the frequency of the laser.

Solution

$\lambda = 1064 \text{ nm} = 1.06 \times 10^{-6}$ m

$c = 3.00 \times 10^8$ m/s

$f = ?$

$$c = f\lambda$$

$$f = \frac{c}{\lambda}$$

$$= \frac{3.00 \times 10^8 \text{ m/s}}{1.06 \times 10^{-6} \text{ m}}$$

$$f = 2.83 \times 10^{14} \text{ Hz}$$

The frequency of the laser is 2.83×10^{14} Hz.

DID YOU KNOW ?

Measuring the Speed of Light

In 1676, Roemer was observing Io, Jupiter's innermost moon, as it was being eclipsed by Jupiter. When Earth and Jupiter were farthest apart, Roemer noticed that his prediction of eclipse start times was 22 min behind start times when the two planets were closest. From this, Roemer deduced that light took longer to travel the greater distance—the difference in distances being the diameter of Earth's orbit—and calculated the speed of light to be 2.14×10^8 m/s. Roemer's calculation was off because he didn't have an accurate measurement of the diameter of Earth's orbit. Prior to this time, many believed that the speed of light was infinite.

Practice

Understanding Concepts

2. Determine the frequency of laser light with a wavelength of 622 nm.
3. A certain yellow light has a frequency of 5.44×10^{14} Hz. Determine its wavelength in metres and nanometres.

Answers

2. 4.82×10^{14} Hz
3. 5.51×10^{-7} m, 551 nm

SUMMARY What Is Light?

- Light is a form of energy that is visible to the human eye.
- Light sometimes behaves as a wave and sometimes as a particle.
- Rays of light travel in straight lines under most circumstances.
- Rays of light converge, diverge, or remain parallel as they propagate through a uniform medium.
- The speed of light is represented as $c = 3.00 \times 10^8$ m/s.

Section 9.1 Questions

Understanding Concepts

1. In communicating with a space probe, radio signals travelling at the speed of light must travel a distance of 8.7×10^9 m each way. How long does it take for a radio signal to travel to the probe and back?
2. If light takes an average of 5.0×10^2 s to reach us from the Sun, what is the average radius of Earth's orbit?
3. Calculate the wavelength of an X ray with a frequency of 2.0×10^{18} Hz.
4. Determine the frequency of a microwave that has a wavelength equal to your height. (Assume three significant digits.)

Applying Inquiry Skills

5. Choose six local radio stations, three FM and three AM stations. Determine the frequency of the radio waves used by each station (for instance, 101.4 MHz and 961 kHz). Set up and complete a table to record the stations' call letters, frequencies, and wavelengths. How do the wavelengths of AM radio stations compare with the wavelengths of FM radio stations?

Making Connections

6. A laser made from indium, gallium, arsenic, and phosphorus produces light with a wavelength of about 808.5 nm. Study the electromagnetic spectrum in **Figure 1** of this section to explain why this laser is dangerous to use.
7. Determine, to two significant digits, the distance from Earth to the Andromeda Galaxy, a distance of 2.0×10^6 light years. (A light year is the distance light travels in a vacuum in one Earth year.)

9.2 Reflection and the Formation of Images

Mirrors and highly polished opaque surfaces reflect light in predictable ways. The terms used by physicists to describe the reflection of light are illustrated in **Figure 1**.

incident ray: the approaching ray of light

reflected ray: the ray of light reflected from the reflecting surface

point of incidence: the point at which the incident ray strikes the reflecting surface

normal: the line drawn at right angles to the reflecting surface at the point of incidence

angle of incidence: (θ_i) the angle between the incident ray and the normal

angle of reflection: (θ_r) the angle between the reflected ray and the normal

laws of reflection: The angle of incidence is equal to the angle of reflection. The incident ray, the reflected ray, and the normal all lie in the same plane.

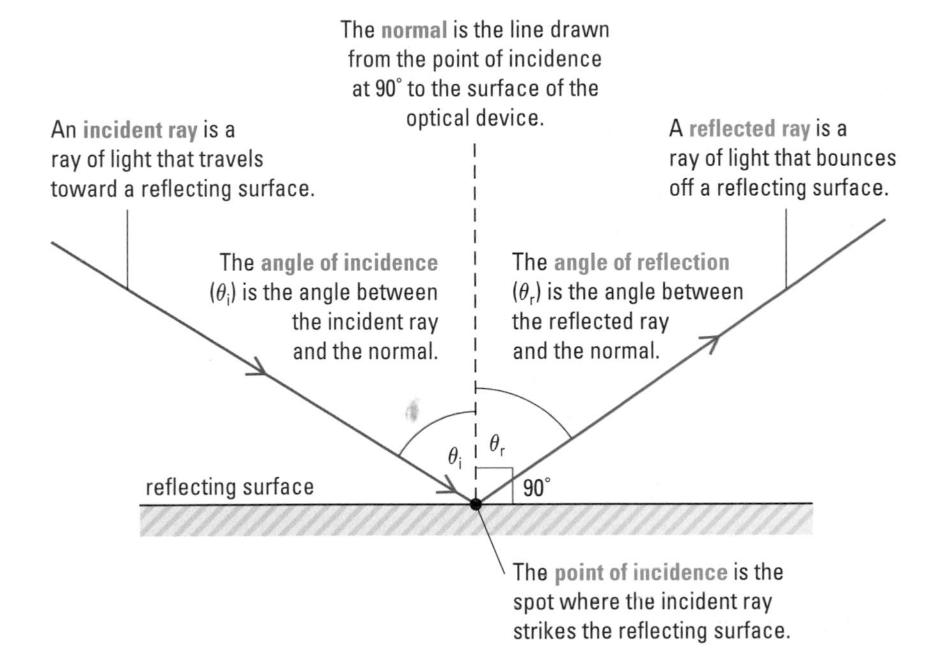

Figure 1
Reflection of light from a plane surface

For any ray directed toward a plane mirror, the angle of incidence equals the angle of reflection; this is true even when a ray strikes a plane mirror straight on, since the value of both angles is zero. Also, the incident ray, the normal, and the reflected ray always lie in the same plane. This gives us two **laws of reflection:**

Laws of Reflection

- The angle of incidence is equal to the angle of reflection ($\theta_i = \theta_r$).
- The incident ray, the reflected ray, and the normal all lie in the same plane.

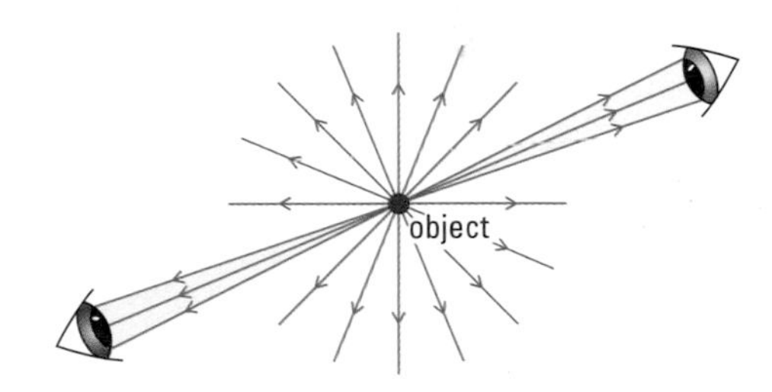

Figure 2
Although a lighted object gives off light in all directions, your eye sees only the particular diverging cone of rays that is coming toward it. If you go to the other side of the object, a different cone of rays will enter your eye.

Images in a Plane Mirror

When you look into a plane mirror, your image appears to be located somewhere behind the mirror. To find its position, you must consider how your eye sees light rays coming from an object (**Figure 2**). When you see an **object point** in a plane mirror, a cone of rays is reflected by the mirror (**Figure 3**). Since light does not originate behind the mirror (but only appears to), the image in a plane mirror is called a **virtual image point.**

object point: a point representing a source of rays, whether by production or reflection

virtual image point: point from which rays from an object point appear to diverge

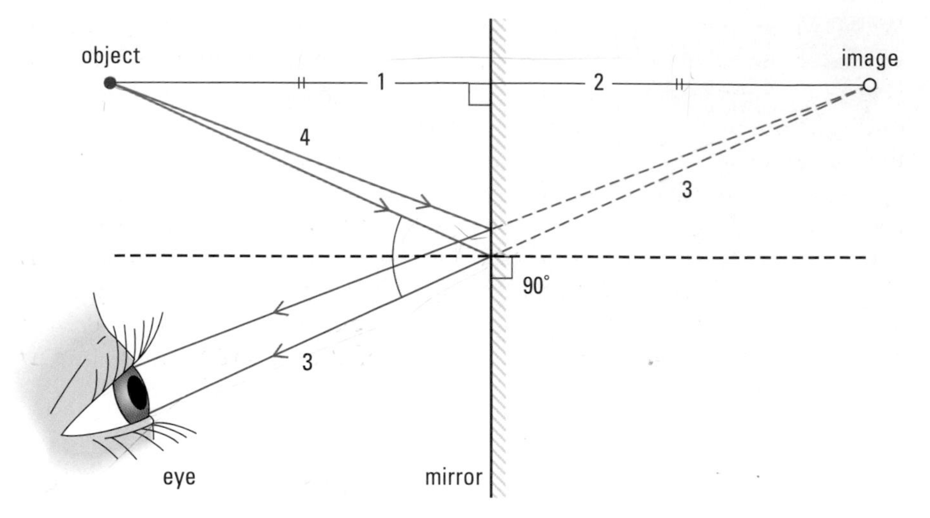

Figure 3
Notice that the image appears to be the same distance behind the mirror as the object is in front of the mirror.

The following sample problem uses ray diagrams to show how your eye perceives the virtual image.

Sample Problem

Given an object located in front of a plane mirror, locate the image and show how the eye "sees" it.

Solution

(Each step is numbered in **Figure 3**.)

1. Draw a line from the object to the mirror such that the line is perpendicular to the mirror.
2. Extend this line an equal distance behind the mirror to locate the image.
3. The eye perceives that the light originates from a point source image behind the mirror. The light rays from a point source travel out in all directions, and they include a cone of rays travelling toward the eye.
4. Rays actually originate from the object, as illustrated. Note that the angle of incidence equals the angle of reflection for each ray.

Practice

Understanding Concepts

1. Trace **Figure 4** into your notebook and locate the image in the mirror. You only need to consider the rays of two object points to construct the image. Which two points should you choose?
2. In **Figure 5**, which eye(s) would see the image in the mirror?

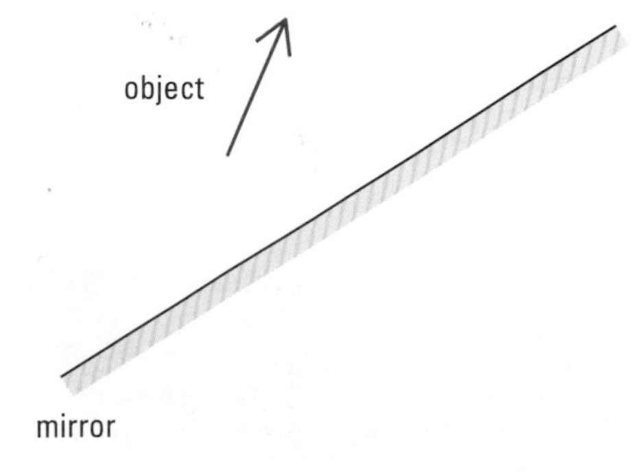

Figure 4
For question 1

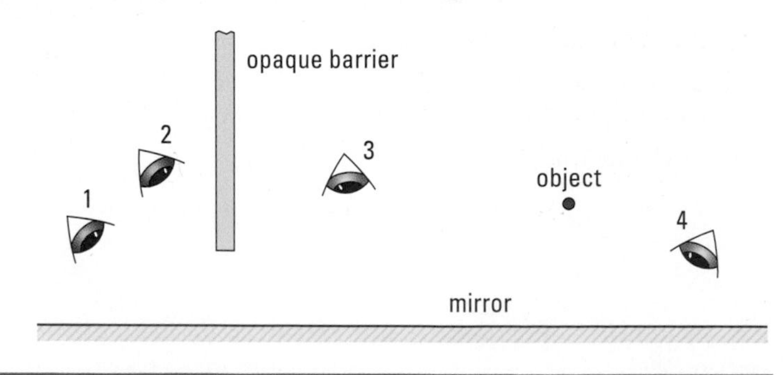

Figure 5

When two mirrors are mounted at right angles, not only are the two expected virtual images I_1 and I_2 formed, but an extra image is produced as well (**Figure 6**). Some of the light that enters the eye has been reflected twice, producing a third image. We can designate the third image $I_{2,1}$, since light reflects off the second mirror and then the first mirror (**Figure 7**).

If we add a third mirror so that we have three perpendicular mirrors, a ray coming from any direction will be reflected back outward at an angle equal to the entry angle. This arrangement of mirrors is called a retroreflector. If a laser is aimed at a retroreflector, its distance from the laser can be determined by measuring the time it takes for a pulse to return. The distance can then be calculated by substituting the time Δt into the equation of motion $d = v\Delta t = c\Delta t$, where Δt is half the total return time.

Figure 6

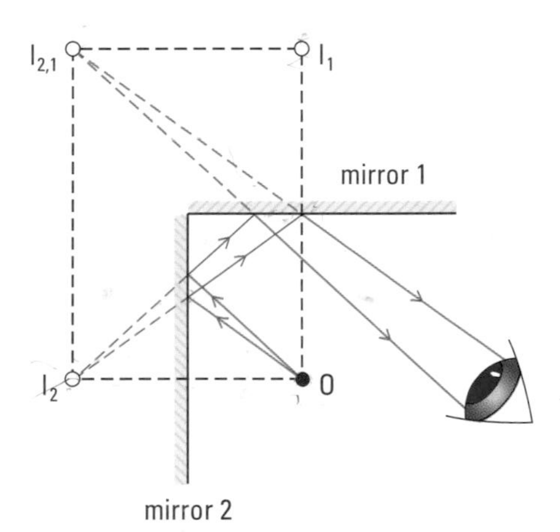

Figure 7
How your eye sees $I_{2,1}$. Geometrically, the object and the three images lie at the corners of a rectangle whose centre is at the intersection of the mirrors.

SUMMARY Reflection and the Formation of Images

- There are two laws of reflection:
  (i) The angle of incidence is equal to the angle of reflection ($\theta_i = \theta_r$).
  (ii) The incident ray, the reflected ray, and the normal all lie in the same plane.
- A virtual image point is the point from which rays from an object point appear to diverge.
- Multiple images can be produced from a single object point by the use of two or more mirrors.
- Ray diagrams are an effective way to locate image points.

Section 9.2 Questions

Understanding Concepts

1. Draw a ray diagram showing a plane mirror, an incident ray with an angle of incidence of 37°, and the reflected ray.
2. In your own words, explain how to geometrically locate the image of an object placed in front of a plane mirror.
3. A box is painted black on the inside, and a jar and a piece of glass are placed as in **Figure 8.**
   (a) Copy the figure into your notebook, and show where a candle should be placed to give an observer the illusion that it is in the jar.
   (b) Complete a ray diagram to explain your answer in (a).
4. A basketball bounce pass between teammates obeys the laws of reflection. However, there are exceptions. Can you think of them? Are they the same for a billiard ball striking the side walls of a billiard table? Explain.

Applying Inquiry Skills

5. When the Sun sets over a lake, instead of seeing one image of the Sun formed by the reflective water surface, you often see a long narrow band of light on a wavy surface of water. This band consists of multiple images of the Sun. Draw a diagram to explain these multiple images.

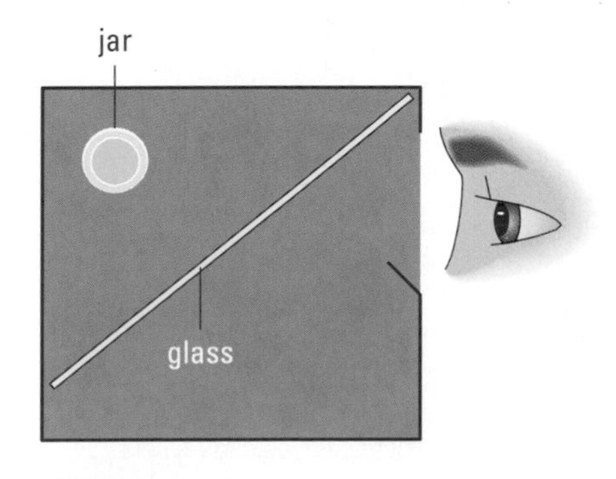

Figure 8
Creating an illusion

9.3 The Refraction of Light

refraction: the change in direction of light as it passes from one medium into another of differing density

Have you ever noticed a distorted view of your legs as you walk in clear, waist-deep water? This effect is caused by **refraction**, the change in direction of light as it passes from one medium into another of differing density (**Figure 1**).

Figure 1
Light travelling obliquely from air, a low density medium, into glass is bent toward the normal.

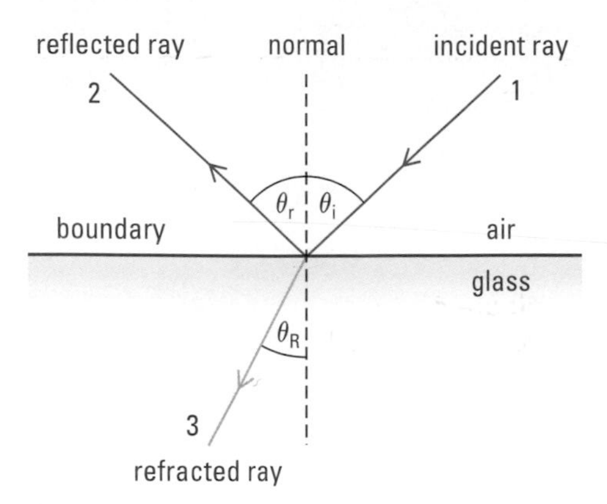

Figure 2
In this example of partial reflection and refraction, ray 2 is reflected back into the air, and ray 3 is refracted through the glass. If the incident ray were extended into the glass, we could see that the refracted ray is bent toward the normal. This means the angle of refraction is less than the angle of incidence.

angle of refraction: (θ_R) the angle between the refracted ray and the normal

partial reflection and refraction: a beam of light splitting into two rays; one is reflected; the other is refracted

The ray diagram in **Figure 2** illustrates this phenomenon and shows the angles used to describe refraction. The angle of incidence (θ_i) is the angle between the incident ray and the normal at the point of incidence. The angle of reflection (θ_r) is the angle between the reflected ray and the normal. The **angle of refraction** (θ_R) is the angle between the refracted ray and the normal.

In **Figure 2**, ray 1 (originating in air) strikes the plane glass surface. This ray divides into two, as shown by rays 2 and 3. When a beam of light splits into two rays like this, we refer to it as **partial reflection and refraction** of light.

In **Figure 3**, incident ray 1 originates in glass. Even in this case, light is reflected back into the original medium as illustrated by ray 2.

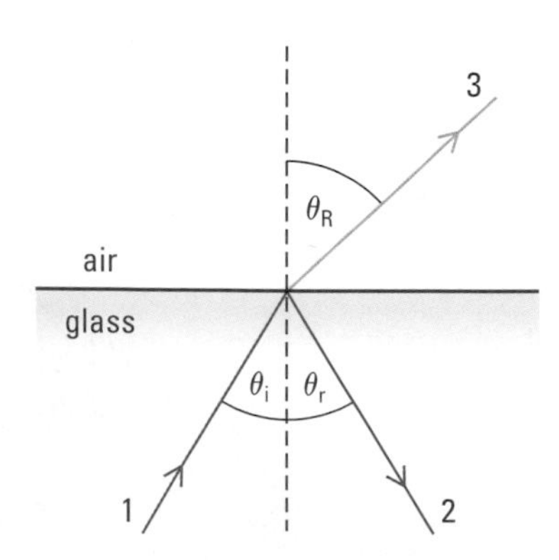

Figure 3
If the incident ray were extended into the air, we could see that the refracted ray is bent away from the normal. This means the angle of refraction is greater than the angle of incidence.

Whenever a ray is normal to a boundary, that is, where two media meet (in this case, air and glass), the transmitted ray continues in the same direction as the incident ray; however, partial reflection at the boundary still occurs. We can't see the partially reflected rays because the reflected ray lies exactly on the incident ray. When we want to show the reflected ray, it is drawn beside the incident ray, as in **Figure 4**.

Why does water produce a virtual image whose apparent depth is less than the actual depth (**Figure 5**)? Let the point R beneath the water surface represent a coin. Light that has reached the coin reflects and then refracts into your eye. Rays of light from point R pass from water into air, bending away from the normal because they are entering a less dense medium. When the rays are traced backward (see dashed lines), they seem to originate from point V and not point R. Since the light originated from point R, which is different from the perceived point V, the image is virtual.

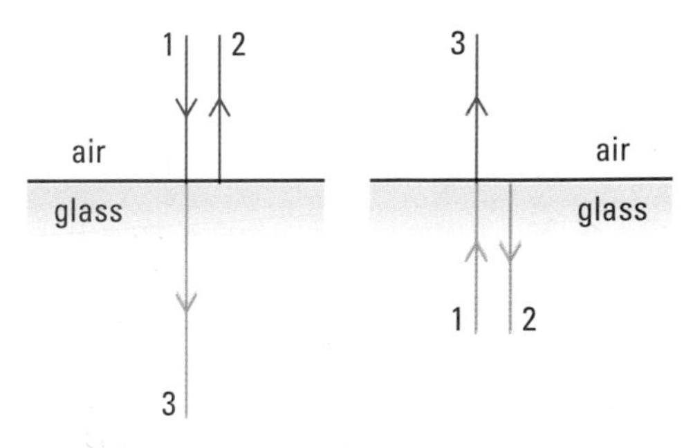

Figure 4
How can we prove that partial reflection is actually occurring?

Practice

Understanding Concepts

1. As light travels from one medium to another, state the condition in which
   (a) light refracts toward the normal
   (b) light refracts away from the normal
   (c) no refraction occurs

SUMMARY The Refraction of Light

- Refraction is the change of direction of light as it moves from one medium to another.
- A ray directed along a normal does not refract as it moves from one medium to another.
- A beam of light that reflects and refracts at the boundary between two materials is said to undergo partial reflection and refraction.

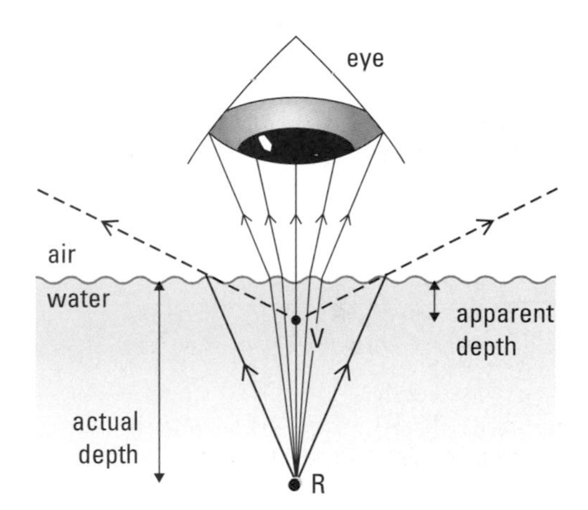

Figure 5
The image of an object beneath water is closer to the surface than the object.

Section 9.3 Questions

Understanding Concepts

1. Imagine you are in a darkened room looking out a window. Someone turns on the light in the room. Will this help you see out the window better? Explain using a ray diagram.
2. As light travels from medium A into medium B, the angle of incidence is 36° and the angle of refraction is 21°.
   (a) Does light bend toward or away from the normal as it travels into medium B? Use a diagram with a normal to justify your answer.
   (b) What is the angle of reflection of the partially reflected ray?
3. Your bicycle chain breaks off in a shallow pool of water and sinks to the bottom. Does the chain appear closer or farther than it actually is? Explain.
4. A stick in water appears bent when viewed from above. Copy **Figure 6** into your notebook. Find the image points of O_1 and O_2 and label them V_1 and V_2. Use V_1 and V_2 to draw the apparent position of the segment of the stick below the surface.
5. The principle of the reversibility of light (PRL) states that if the direction of a ray of light is reversed, it will trace back the path taken by the incident ray. Compare **Figures 2** and **3** of this section.
   (a) In **Figure 2,** ray 1 bends toward the normal to become ray 3. According to the PRL, how should ray 3 bend?
   (b) Study **Figure 3** to confirm your prediction in (a).
   (c) Compare **Figures 2** and **3.** Which rays does the PRL not take into account?

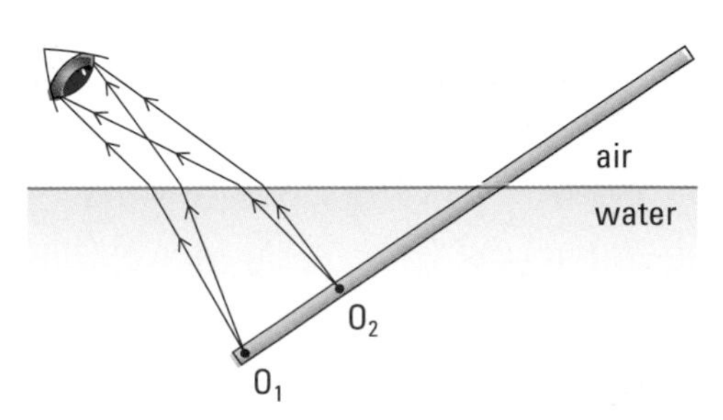

Figure 6
Why does a stick appear bent in water?

9.4 Index of Refraction

Light changes direction in passing from one medium to another because the speed of light changes as it travels from one medium to another. In 1862, the French physicist Jean Foucault (1819–1868) determined that light travels in water at approximately three-quarters the speed it travels in a vacuum. Other materials produce different values, but the speed is always less than the speed of light in a vacuum. The ratio of the speed of light in a vacuum (c) to the speed of light in a given material (v) is called the **index of refraction** (n) of that material:

index of refraction: (n) the ratio of the speed of light in a vacuum (c) to the speed of light in a given material (v);

$$n = \frac{c}{v}$$

$$n = \frac{c}{v}$$

Since c and v have the same units, n is dimensionless. The speed of light in a medium is often expressed in terms of c, the speed of light in a vacuum. For example, the speed of light in water can be written $0.75c$.

Table 1 shows the indexes of refraction for various materials.

The index n varies slightly with the colour of the light, except in a vacuum. The values in **Table 1** are for monochromatic (one-colour) light of a specific wavelength, in this case yellow light.

Table 1 Indexes of Refraction

Medium	Index of refraction (n)
vacuum	1.000000
gases (at 0°C, 101.3 kPa)	
air	1.000293
carbon dioxide	1.000450
hydrogen	1.000139
liquid (at 20°C)	
water	1.33
ethyl alcohol	1.36
glycerin	1.47
benzene	1.50
solid (at 20°C)	
ice (0°C)	1.31
glass (crown)	1.52
glass (flint)	1.65
sodium chloride	1.53
zircon	1.92
diamond	2.42

Note: Indexes are for yellow light (589 nm).

The term "optically dense" is used when referring to a medium in which the speed of light decreases. In ice, the speed of light is 2.29×10^8 m/s. In crown glass (a soda-lime glass that is exceptionally hard and clear with low refraction), the speed is 1.97×10^8 m/s; therefore, crown glass is optically denser than ice. Relatively speaking, ice is a "faster" medium and crown glass is a "slower" medium.

Note that the slower the medium, the greater the amount of refraction in that medium. Since the angle of refraction is measured from the normal, this actually means that the angle of refraction is smaller in "slower" media.

Sample Problem

Use **Table 1** to (a) calculate the speed of light in zircon and (b) express the speed in terms of c, the speed of light.

Solution

(a) $c = 3.00 \times 10^8$ m/s

$n = 1.92$

$v = ?$

$$n = \frac{c}{v}$$

$$v = \frac{c}{n}$$

$$= \frac{3.00 \times 10^8 \text{ m/s}}{1.92}$$

$$v = 1.56 \times 10^8 \text{ m/s}$$

The speed of light in zircon is 1.56×10^8 m/s.

(b)
$$v = \frac{c}{n}$$
$$= \left(\frac{1}{n}\right)(c)$$
$$= \left(\frac{1}{1.92}\right)(c)$$
$$v = 0.521c$$

The speed of light in zircon is $0.521c$.

Practice

Understanding Concepts

1. What is the index of refraction of a liquid in which light travels at 2.50×10^8 m/s?
2. The index of refraction of diamond is 2.42. What is the speed of light in diamond?
3. What is the speed of light in sodium chloride expressed in terms of the speed of light in a vacuum?

Answers

1. 1.20
2. 1.24×10^8 m/s
3. $0.654c$

SUMMARY Index of Refraction

- The ratio of the speed of light in a vacuum to the speed of light in a given material is the index of refraction, n.
- The index of refraction is given by $n = \frac{c}{v}$.
- An optically dense medium has a high index of refraction.
- The higher the index of refraction, the slower the speed of light and the smaller the angle of refraction.

Section 9.4 Questions

Understanding Concepts

1. In **Figure 1**, in which substance, A or B, is light travelling more slowly? Which substance has the lower optical density? Explain.
2. Which medium is more refractive, crown glass or diamond? Explain.
3. What is the speed of light in
   (a) a 30.0% sugar solution at 20.0°C whose index of refraction is 1.38?
   (b) ethyl alcohol expressed in terms of the speed of light in a vacuum?
4. Through which medium does light travel faster, ice or water? By how much in metres per second?
5. Light travels slightly slower in ethanol than in methanol.
   (a) Which medium has a higher index of refraction?
   (b) When light strikes the boundary between the two media, it bends away from the normal. Which is the incident medium and which is the refractive medium? Answer with a labelled sketch.

(continued)

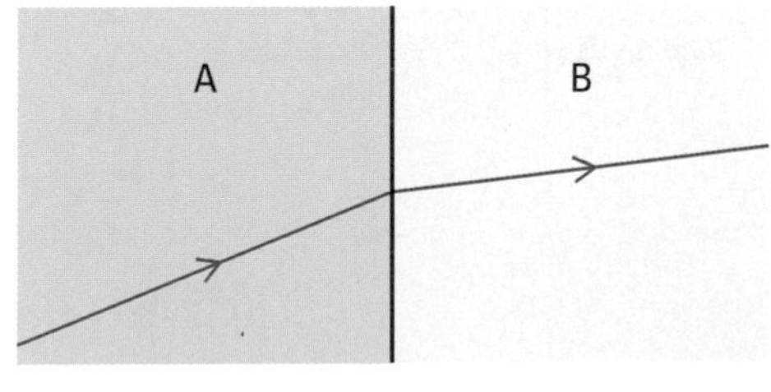

Figure 1

Making Connections

6. Use the Internet or other sources to find out the index of refraction of the lens of the human eye. Use this to explain why your vision is blurry when you open your eyes under water. Include drawings with your explanation. Follow the links for Nelson Physics 11, 9.4.

GO TO www.science.nelson.com

9.5 Laws of Refraction

In 1621, the Dutch mathematician Willebrod Snell (1580–1626) determined the exact relationship between the angle of incidence and the angle of refraction. Snell determined that there was a direct proportion between the sine of the angle of incidence and the sine of the angle of refraction:

$$\sin \theta_i \propto \sin \theta_R$$

Snell's law: The ratio of the sine of the angle of incidence to the sine of the angle of refraction is a constant.

The expression above, called **Snell's law**, can be converted into an equation by introducing a constant of proportionality. It has been found that the constant of proportionality and the index of refraction (n) are one and the same. Consequently, the Snell's law relationship may be rewritten as

$$\sin \theta_i = n \sin \theta_R \quad \text{or}$$

$$\frac{\sin \theta_i}{\sin \theta_R} = n$$

This relationship assumes that light originates in air or a vacuum and is incident on a medium of higher index of refraction. As was the case with reflection, the incident ray, the refracted ray, and the normal all lie in the same plane.

laws of refraction: The ratio of $\sin \theta_i$ to $\sin \theta_R$ is a constant (Snell's law). The incident ray and the refracted ray are on opposite sides of the normal at the point of incidence, and all three are in the same plane.

The **laws of refraction** can now be summarized:

Laws of Refraction

- The ratio of the sine of the angle of incidence to the sine of the angle of refraction is a constant (also known as Snell's law).
- The incident ray and the refracted ray are on opposite sides of the normal at the point of incidence, and all three are in the same plane.

Suppose light is incident in one material (n_1) and refracts in another (n_2), as shown in **Figure 1.** Here is one way to generalize Snell's law for this situation. Consider the materials separately. Assume that a ray with the angle of incidence θ_1 in material 1 bends away from the normal at θ_{air} (**Figure 1(a)**) and bends at θ_2 in material 2 (**Figure 1(b)**).

For material 1 and material 2, respectively, Snell's law gives

$$\frac{\sin \theta_{air}}{\sin \theta_1} = n_1 \quad \text{and} \quad \frac{\sin \theta_{air}}{\sin \theta_2} = n_2$$

Rearranging both equations to isolate $\sin \theta_{air}$ yields

$$\sin \theta_{air} = n_1 \sin \theta_1 \quad \text{and} \quad \sin \theta_{air} = n_2 \sin \theta_2$$

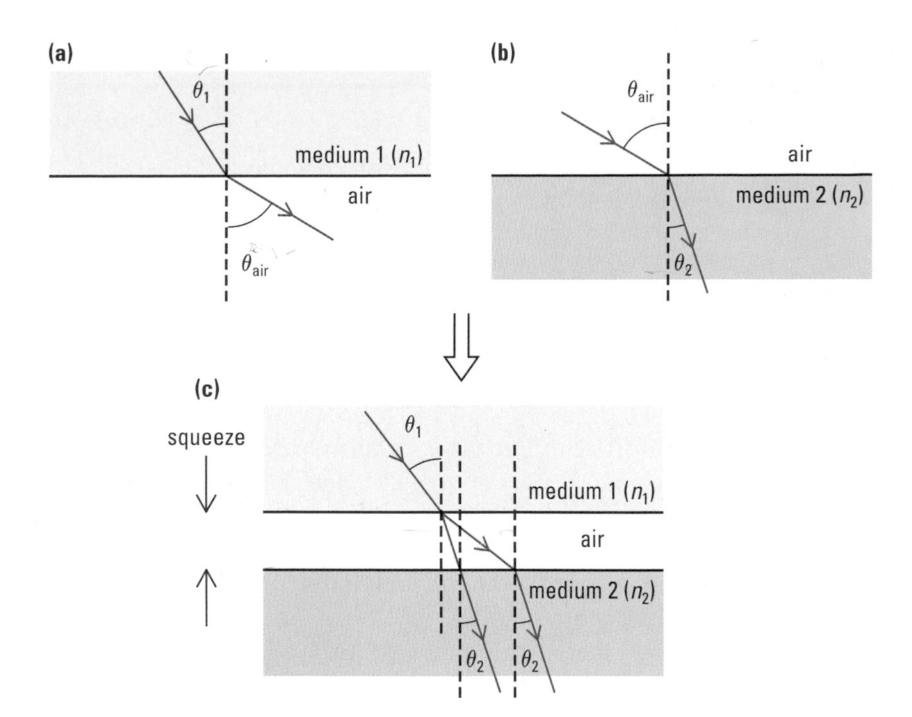

Figure 1
Generalizing Snell's law

Now, squeeze the two media together to push out the air (**Figure 1(c)**). As the air is pushed out, the incident ray in material 1 refracts as it enters material 2. The result is the **general equation of Snell's law** for any two media:

general equation of Snell's law: $n_1 \sin \theta_1 = n_2 \sin \theta_2$

General Equation of Snell's Law

$$n_1 \sin \theta_1 = n_2 \sin \theta_2$$

The following sample problem illustrates the use of this general equation.

Sample Problem

Light travels from crown glass (g) into water (w). The angle of incidence in crown glass is 40.0°. What is the angle of refraction in water?

Solution

$n_g = 1.52$

$n_w = 1.33$

$\theta_g = 40.0°$

$\theta_w = ?$

$$n_g \sin \theta_g = n_w \sin \theta_w$$

$$\sin \theta_w = \frac{n_g \sin \theta_g}{n_w}$$

$$= \frac{(1.52)(\sin 40.0°)}{1.33}$$

$$\sin \theta_w = 0.735$$

$$\theta_w = \sin^{-1}(0.735)$$

$$\theta_w = 47.3°$$

The angle of refraction in water is 47.3°.

Practice

Understanding Concepts

1. Light passes from air into diamond with an angle of incidence of 60.0°. What is the angle of refraction?
2. What is the index of refraction of a material if the angle of incidence in air is 50.0° and the angle of refraction in the material is 40.0°?
3. A diamond ring is dropped into water. The index of refraction for diamond is 2.42. If light strikes the diamond at an angle of incidence of 60.0°, what will be the angle of refraction in diamond?

Answers

1. 21.0°
2. 1.19
3. 28.4°

When a light ray passes from air into glass and then back into air, it is refracted twice. If the two refracting surfaces are parallel, Snell's law can prove that the emergent ray is parallel to the incident ray but it is no longer moving in the same path. Such sideways shifting of the path of a ray is called **lateral displacement** (**Figure 2**). This means that whenever you look through a window at an angle, the entire world is shifted a tiny amount.

lateral displacement: sideways shifting of the path of a refracted ray

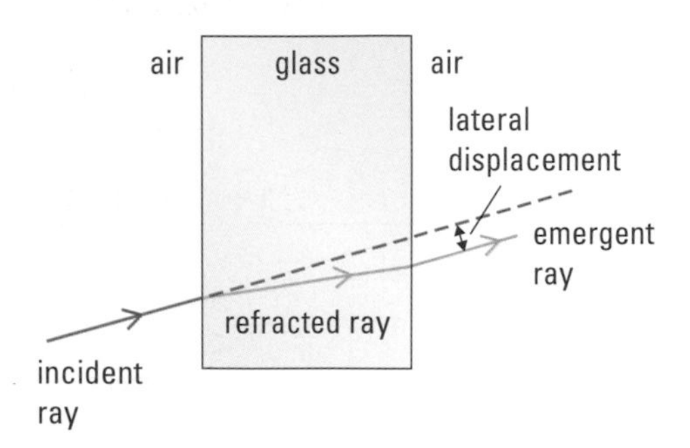

Figure 2

In circle geometry, a tangent at a point on the circumference of a circle is perpendicular to the radius of the circle at that point (**Figure 3**). Consider the point P on the circle in **Figure 3**. The radius from the centre C to the point P is the normal at point P. The tangent drawn at P is considered to be the boundary at that point between the two media. Observe the effect that this circular **prism** has on the ray incident at point P.

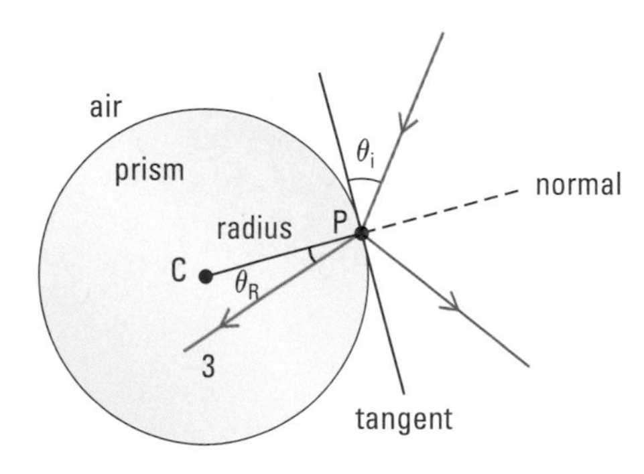

Figure 3
Light entering a circular prism

prism: a transparent solid that has at least two plane refracting surfaces

Activity 9.5.1

Reflection and Refraction at Curved Surfaces

The laws of reflection and refraction are not limited to plane mirrors or rectangular prisms. Applying these laws to curved surfaces is important in the study of curved mirrors and lenses.

Materials

coloured pencils
circular prism
ray box
protractor
ruler

Procedure

1. Predict the directions of partial reflection and refraction that would result from shining light along the rays illustrated in **Figure 4**. Draw a ray diagram that shows your predictions.
2. Shine the ray box along the incident rays on your ray diagram.

The bulb of a ray box can become very hot. Turn off your ray box when it is not in use. Allow your ray box to cool before putting it away.

3. Trace the actual directions of the rays using a colour different from the one used for your predictions.
4. After drawing normals for all three incident rays, measure the angles of incidence, angles of reflection, and angles of refraction, recording your observations in an appropriate table.
5. Compare the angles of incidence with the angles of reflection in your table.
6. Use your observations with Snell's law to determine the index of refraction of the circular prism.

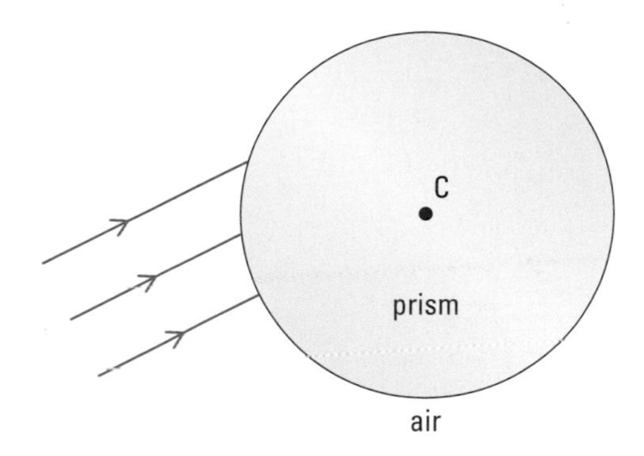

Figure 4
The direction of reflected and refracted rays can be predicted from the laws of reflection and refraction.

Investigation 9.5.1

Refraction of Light—Air into Glass

The purpose of this investigation is to test Snell's law of refraction.

INQUIRY SKILLS

- ○ Questioning
- ○ Hypothesizing
- ● Predicting
- ○ Planning
- ● Conducting
- ● Recording
- ● Analyzing
- ● Evaluating
- ● Communicating

Question

How can Snell's law of refraction be verified?

Materials

ray box (single-slit)
semicircular prism of known index of refraction
polar coordinate paper (optional)

The bulb of a ray box can become very hot. Turn off your ray box when it is not in use. Allow your ray box to cool before putting it away.

Prediction

(a). Use Snell's law to predict the angle of refraction for angles of incidence of 0.0° to 60.0° in intervals of

(i) 5.0°, if using a spreadsheet or programmable calculator to speed up this prediction process;

(ii) 10.0°, if calculating with a regular calculator.

Record your predictions in a table like **Table 1.**

Table 1

Angle of incidence (θ_i)	Predicted angle of refraction (θ_R)	Measured angle of refraction (θ_R)	Difference	$\frac{\sin\theta_i}{\sin\theta_R}$
0.0°				—
5.0°				
10.0°				

Procedure

1. Place the prism on a blank sheet or polar coordinate paper, as illustrated in **Figure 5.** (If polar graph paper is available for this procedure, your teacher will explain its use.)

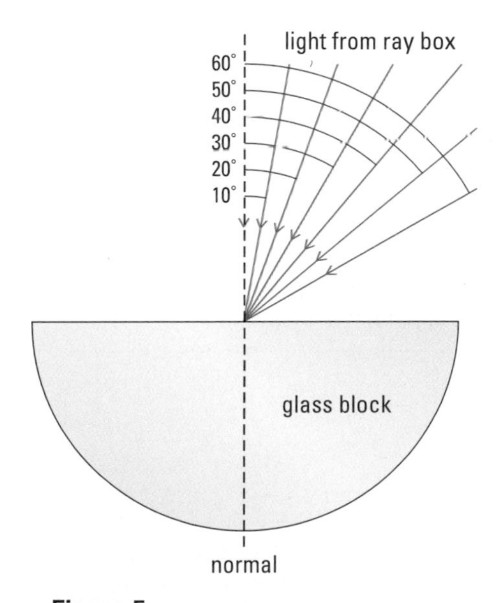

Figure 5
Testing Snell's law

2. Test your first ray by directing a single ray of light at the centre of the flat surface of the glass along the normal. This will ensure that the ray does not refract at the second curved surface. Your angle of refraction in this case should be 0.0°.
3. Measure the angle of refraction for each of your predictions and record it in your chart. Always aim your ray at the centre of the flat surface.
4. Calculate the differences between your predictions and your measurements. Identify angles that don't agree with Snell's law. Check your calculations and measurements for these angles.

Analysis

(b) When light travels from air into glass with an angle of incidence of 0°, that is, along the normal, what happens to it?

(c) Why must the rays be shone at the centre of the flat surface? Illustrate the difficulty that results if this is not done.

(d) How does the angle of refraction compare with the angle of incidence in each case?

(e) What do you notice about the ratio $\left(\frac{\sin \theta_i}{\sin \theta_R}\right)$ for all angles of incidence greater than 0°?

(f) Plot a $\sin \theta_i$ versus $\sin \theta_r$ graph using the data you've collected. Calculate the slope of the line. What does it represent? Explain.

Evaluation

(g) What modifications would you have to make to this experiment if, instead of a semicircular prism, only a rectangular prism were available?

SUMMARY Laws of Refraction

- The ratio of the sine of the angle of incidence to the sine of the angle of refraction is a constant.
- The incident ray and the refracted ray are on opposite sides of the normal at the point of incidence, and all three are in the same plane.
- Snell's law is represented, in general, by $n_1 \sin \theta_1 = n_2 \sin \theta_2$.

Section 9.5 Questions

Understanding Concepts

1. A ray of light leaves a piece of glass (n = 1.52) at an angle, entering water (n = 1.33). Does the ray bend toward or away from the normal? Include a sketch with your answer.
2. A ray of light passes from air into water (n_w = 1.33) at an angle of incidence of 50.0°. What is the angle of refraction?
3. Light travels from air into water. If the angle of refraction is 30.0°, what is the angle of incidence?
4. A diver shines her flashlight upward from beneath the water at an angle of 30.0° to the vertical. At what angle relative to the surface of the water does the beam of light emerge?
5. Study the rays that enter the semicircular prism in **Figure 6.** Ignoring partial reflection, we know that each ray will pass through the flat surface and then pass through the curved surface.

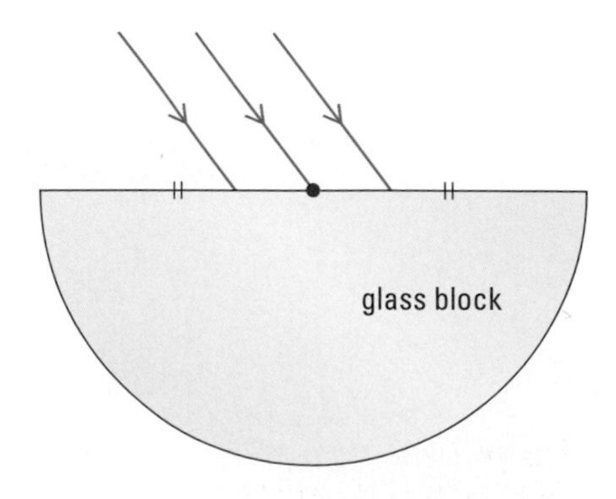

Figure 6
Only one ray undergoes a single refraction.

Which of the rays refracts only once? Explain. Use a ray diagram in your answer.

Applying Inquiry Skills

6. Design an experiment to determine a relationship between the index of refraction of salt water and the concentration of a salt solution. Express the relationship in the form of a graph.

9.6 Total Internal Reflection

When light travels from one medium to another in which its speed changes, some of the light is reflected and some is refracted. When the light originates in the optically denser medium, for example from glass to air, the amount of reflection can be greater than the amount of refraction.

As the angle of incidence increases, the intensity of a reflected ray becomes progressively stronger and the intensity of a refracted ray progressively weaker. As the angle of incidence increases, the angle of refraction increases, eventually reaching a maximum of 90°. Beyond this point, there is no refracted ray at all, and all the incident light is reflected at the boundary, back into the optically denser medium (**Figure 1**). This phenomenon is called **total internal reflection.** It can occur only when light rays travel in the optically denser medium toward a medium in which the speed of the light increases, that is, when the angle of refraction is greater than the angle of incidence.

total internal reflection: the reflection of light in an optically denser medium; it occurs when the angle of incidence in the more dense medium is equal to or greater than a certain critical angle

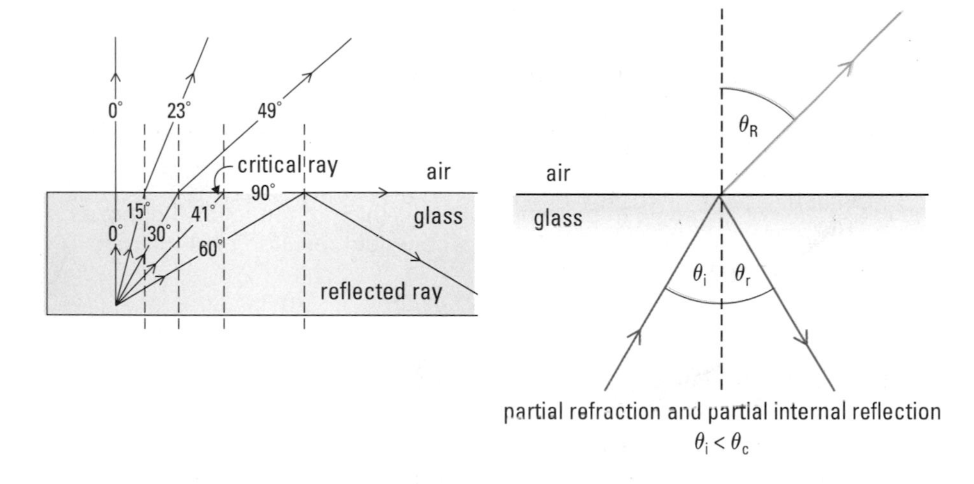

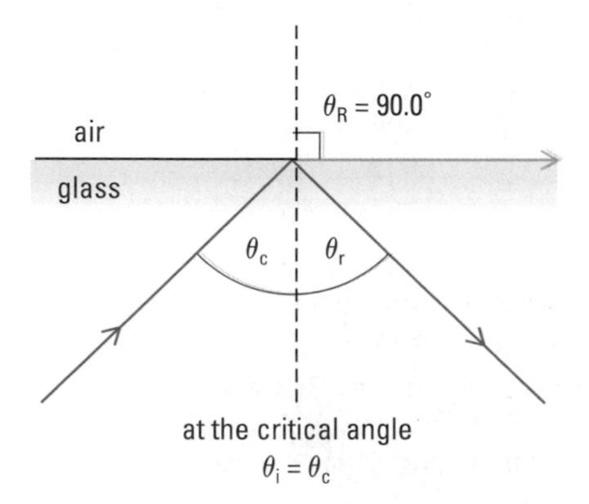

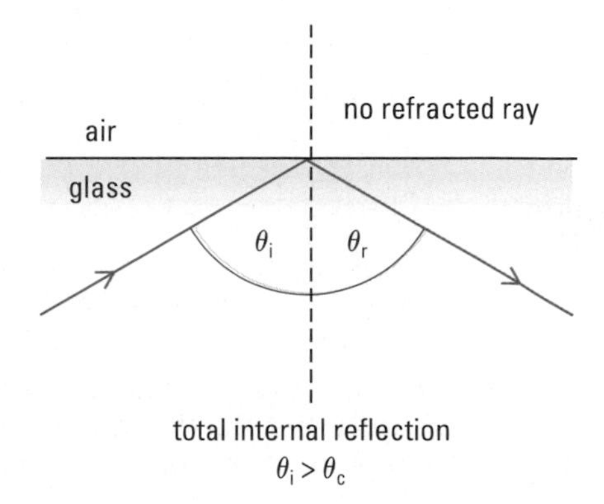

Figure 1
What happens when light rays pass from glass into air?

critical angle: (θ_c) the angle in an optically denser medium at which total internal reflection occurs; at this angle the angle of refraction in the less dense medium is 90°

When the angle of refraction is 90°, the incident ray forms a unique angle of incidence called the "critical angle of incidence," or simply the **critical angle** (θ_c). Unless a second material is mentioned, always assume that the material under study is in air. The following sample problem illustrates how to determine the critical angle.

Sample Problem 1

What is the critical angle in water? Include a diagram in your solution.

Solution

The diagram for this solution is shown in **Figure 2**.

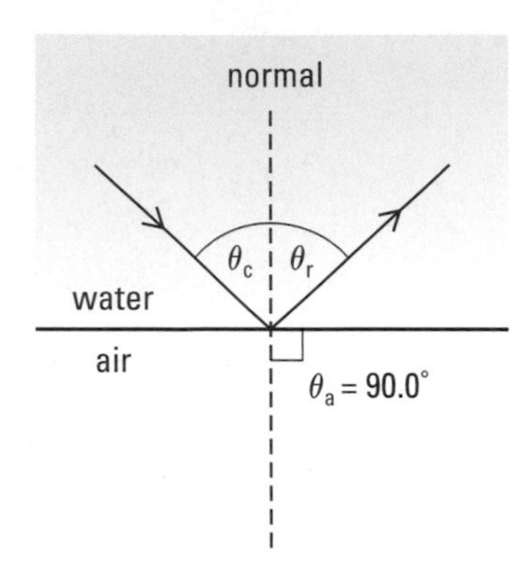

Figure 2
For Sample Problem 1

$n_w = 1.33$

$\theta_w = \theta_c$

$n_{air} = 1.00$ (Assume that the second medium is air.)

$\theta_{air} = 90.0°$

$$n_w \sin \theta_c = n_{air} \sin \theta_{air}$$

$$\sin \theta_c = \frac{n_{air} \sin \theta_{air}}{n_w}$$

$$\sin \theta_c = \frac{(1.00)(\sin 90.0°)}{1.33}$$

$$\theta_c = \sin^{-1}(0.752)$$

$$\theta_c = 48.8°$$

The critical angle in water is 48.8°.

Practice

Understanding Concepts

1. The critical angle in benzene is 41.8°. Which of the following angles of incidence of light rays in benzene would result in total internal reflection?
   (a) 35.1° (b) 50.5° (c) 42.0° (d) 3.0°
2. What is the critical angle for zircon?
3. The critical angle for glass is 41°. What is the index of refraction of the glass? What type of glass is it?
4. Light incident in water (n_w = 1.33) strikes a layer of ice (n_i = 1.31) that has formed on top of the water. What is the critical angle in the water?

Applying Inquiry Skills

5. You are given a sample of each of the four liquids mentioned in **Table 1**, section 9.4. Describe how you would apply the concept of internal reflection to experimentally determine the identity of each liquid.

Answers

2. 31.4°
3. 1.5 (likely crown glass)
4. 80.1°

Activity 9.6.1

Critical Angle

In this activity you will

(a) determine the critical angle in various materials surrounded by air
(b) study an application of total internal reflection

Materials

ray box with single-slit and double-slit windows
semicircular solid prism (made of glass or plastic)
semicircular plastic dish for water
2 triangular solid prisms having angles of 45° and 90°
polar graph paper (optional)

Procedure

1. Place the semicircular solid prism on a piece of paper and draw its outline. (If polar graph paper is available for this procedure, your teacher will explain its use.) Aim a single light ray ($\theta_i = 10°$) through the curved side of the prism directly toward the middle of the flat edge, as in **Figure 3(a)**. This will ensure that no refraction takes place at this curved surface. Trace the direction of the reflected and refracted rays.
2. Slowly increase the angle of incidence from 10° until the refracted ray in the air just disappears, as in **Figure 3(b)**. Mark the rays, remove the prism, and measure the critical angle, θ_c, for the solid prism.
3. Predict the critical angle θ_c for water ($n_w = 1.33$). Repeat steps 1 and 2 to verify your prediction, using water in a semicircular plastic dish.
4. Make a periscope from two 45° right-angle glass prisms as in **Figure 4**. To observe how light travels in a periscope, aim two rays along X and Y as shown. Draw a ray diagram of your results.

Analysis

(a) Account for any differences between the predicted critical angle for water and the measured critical angle.
(b) Assume that the prisms in **Figure 4** have an index of refraction of $n_g = 1.5$. Explain why light entering and leaving each triangular prism is not refracted and why light rays striking the hypotenuse side of the prism at an angle of 45° are totally internally reflected.
(c) Explain why the arrangement of triangular prisms is considered to be a periscope.

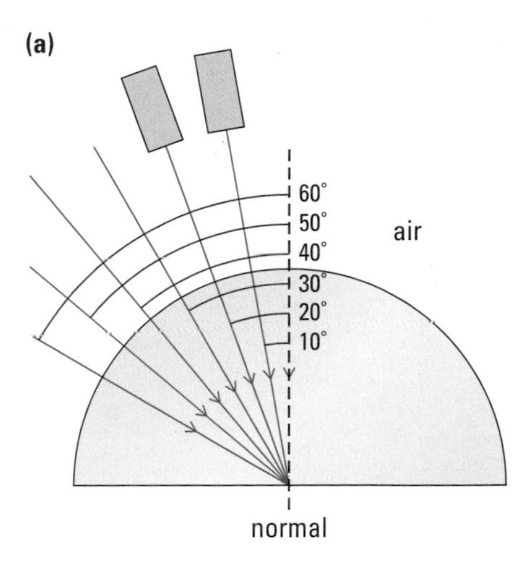

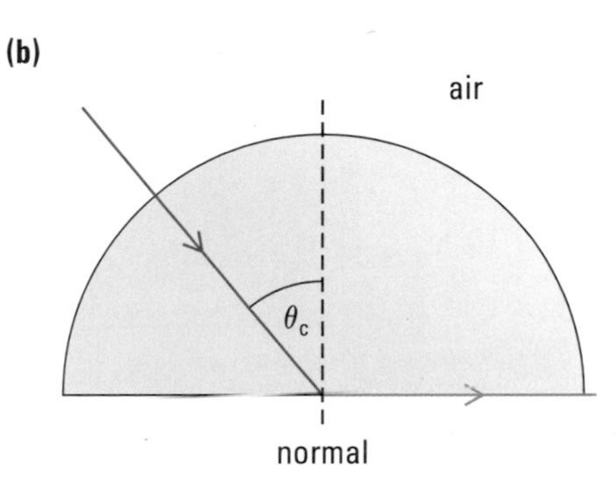

Figure 3
Determining the critical angle in a prism or water

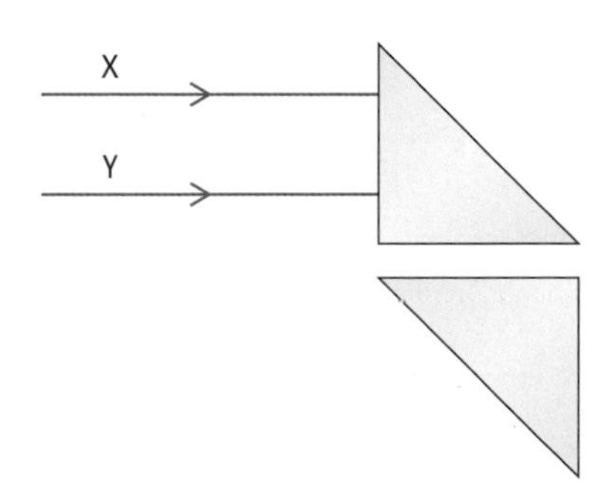

Figure 4
Why is this arrangement of prisms a periscope?

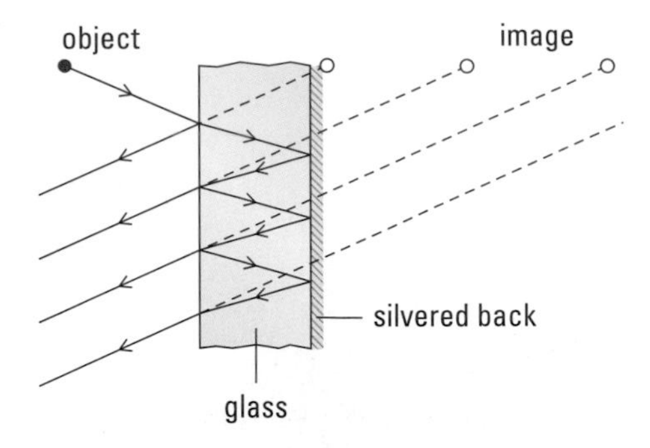

Figure 5
Rear surface mirrors cause multiple images and dispersion. Extra images are usually much weaker than the image formed by the silvered surface at the back; therefore, they go unnoticed.

Applications of Total Internal Reflection

Mirrors and Prisms

Only 90% of the light energy is reflected by most metallic reflectors, the other 10% being absorbed by the reflective material. Plane mirrors, made by silvering the back of the glass, can produce extra images due to reflection from the front surface of the glass and internal reflections in the glass (**Figure 5**). For this reason, the curved mirrors in telescopes are silvered on the front surface. Front surface mirrors, however, are easily damaged and require special handling, cleaning, and care.

To remedy some of the difficulties with front surface mirrors, total internal reflection prisms are used. They reflect nearly all the light energy and have non-tarnishing reflective surfaces. For example, total internal reflection prisms are used in combination with a series of lenses in binoculars (**Figure 6**).

In some cases, different types of prisms are used to change or rotate the attitude of images (**Figure 7**). A single 45° right-angle prism can be used to deflect an image 90° or 180°. Right-angle prisms are used in many laser applications.

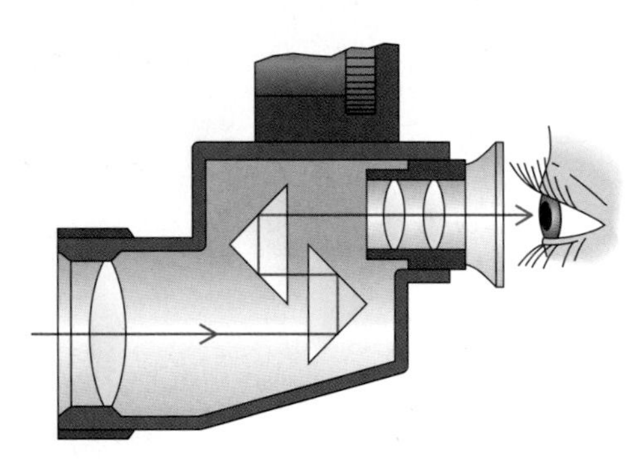

Figure 6
Using total internal reflection in binoculars

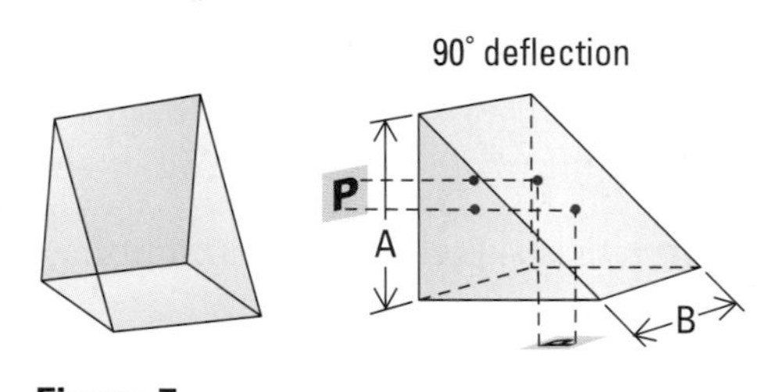

Figure 7

Sample Problem 2

An incident ray of light enters a 45°–90°–45° flint prism as shown in **Figure 8**. Trace the ray through the prism, indicating the angles of incidence, reflection, and refraction at each boundary until the light leaves the prism. Record your answers in an appropriate chart (see **Table 1**).

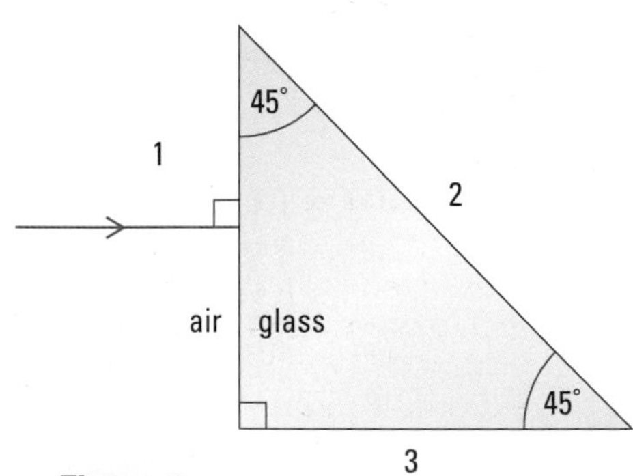

Figure 8
For Sample Problem 2

Solution

Table 1

Boundary	Angle of incidence	Angle of reflection	Angle of refraction
1	0°	0°	0°
2	45°	45°	no refraction
3	0°	0°	0°

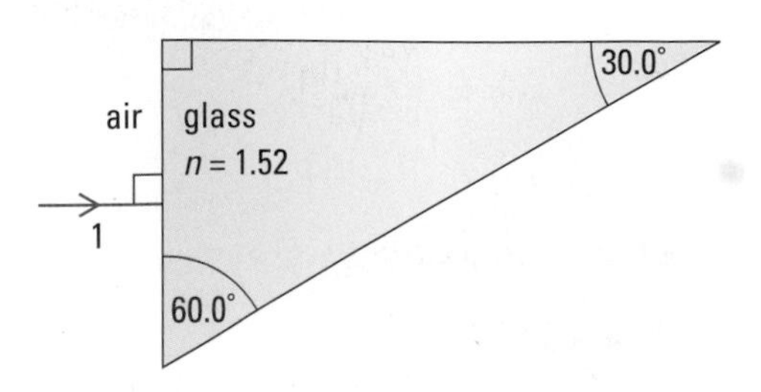

Figure 9
For question 6

Practice

Applying Inquiry Skills

6. An incident ray of light enters a crown glass prism as shown in **Figure 9**. Trace the ray through the prism, indicating the angles of incidence, reflection, and refraction at each boundary until the light leaves the prism. Take into account the possibility of partial reflection and refraction. Record your answers in an appropriate chart.

Fibre Optics and Communication

Another important application of total internal reflection is communication by the transmission of light along fibre optic cables. The most basic optic fibre is a transparent glass or plastic rod.

In **Figure 10** you can see how ray 1 refracts out of the rod, but rays 2, 3, and 4 don't. They strike the internal surface of the rod at angles greater than the critical angle and so undergo total internal reflection.

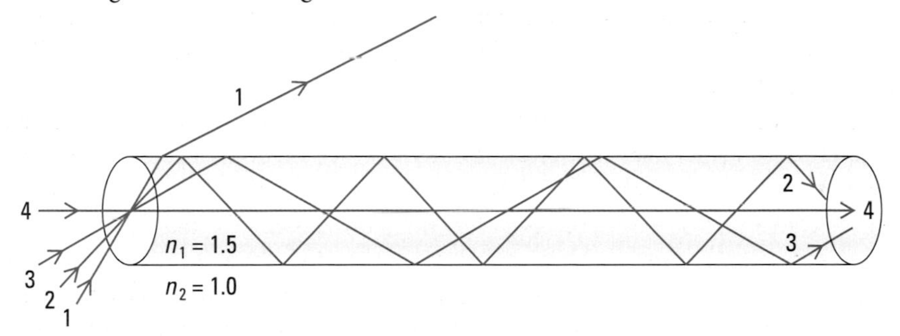

Figure 10
As long as light enters the rod close to the rod's axis, the light remains trapped in the rod by total internal reflection.

Even if the fibres are gently bent or twisted, the critical angle is not usually exceeded, and the light energy is "trapped" by reflecting every time it encounters an internal surface (**Figure 11**).

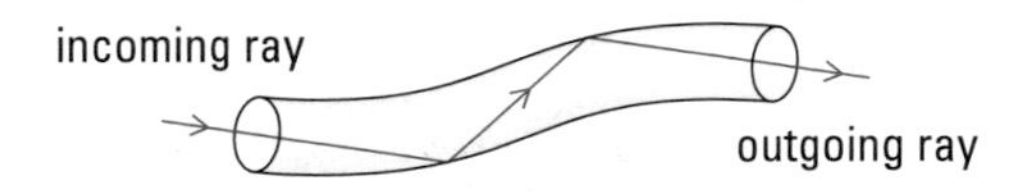

Figure 11
Light will follow a fibre even around bends and twists.

Fibre optic technology is gradually replacing the thousands of copper wires that connect telephone substations (**Figure 12**). Eventually optical fibres will be the backbone of the information superhighway, transmitting voice, video, and data to businesses, schools, hospitals, and homes.

Figure 12
Laser light transmitted through thin glass fibres carries many more telephone calls at a time, with much less energy loss, than is possible with thick copper wire. Also, laser light is not affected by solar flares, lightning flashes, or other forms of electronic interference.

SUMMARY Total Internal Reflection

- When the angle of incidence is equal to the critical angle, the angle of refraction is 90.0°.
- Use θ_c = 90.0° in the general mathematical relationship of Snell's law to determine the critical angle of the denser medium in the less optically dense medium.
- Light remains confined to fibre optic cables due to total internal reflection.

Section 9.6 Questions

Understanding Concepts

1. State two conditions necessary for total internal reflection.
2. The critical angle for a medium is 60.0°. What is the index of refraction of the medium?
3. A layer of olive oil (n = 1.47) floats on water. What is the critical angle for a ray directed from the oil to the water?
4. The glass cores used in fibre optics are surrounded by a thin, transparent film called a *cladding* (**Figure 13**).
   (a) Should the cladding have a lower or higher optical density than the fibres in the core? Explain your answer.
   (b) What do you think is the function of the cladding? (*Hint:* Hundreds of fibres are bundled together to form a fibre tube.)

(continued)

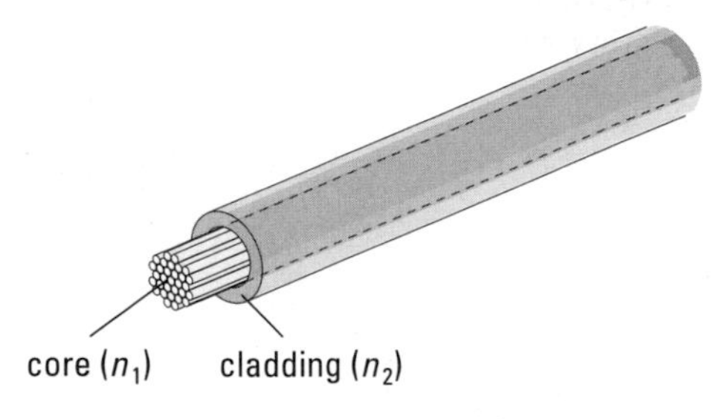

Figure 13
Individual glass fibres are surrounded by a transparent protective cladding.

5. In **Figure 10** of this section, ray 1 is incident on the inside surface of the glass fibre at an angle less than the critical angle.
   (a) What is the critical angle for this fibre if the index of refraction for the glass is 1.50?
   (b) With what maximum angle relative to the normal of the fibre's end can it strike the end of the fibre without later escaping?

Applying Inquiry Skills

6. Design an experiment to determine whether light will undergo total internal reflection within gases, for example, when travelling in carbon dioxide ($n = 1.000450$) toward air ($n = 1.000293$). What is the critical angle of light in carbon dioxide, according to Snell's law? How will this affect the design of your experiment?

Making Connections

7. Research which fields, other than communications, have been greatly affected by the use of fibre optics.

DID YOU KNOW ?

Medical Application of Total Internal Reflection

An endoscope is a device that transmits images using bundles of fibres. Endoscopes probe parts of the body that would otherwise require exploratory surgery.

9.7 Applications of Refraction

Geometric optics is a good model of light for explaining many natural optical phenomena. Recall that a point is perceived by the eye when a cone of diverging light enters the eye. The diverging light is traced backward and where the rays meet, an image of the object is perceived.

Atmospheric Refraction

For many purposes, we assume the index of refraction of air to be 1.000. Although variation of the index of refraction of air from this value is just a small fraction of a percent, it is this small fraction that causes many optical effects. The atmosphere's index of refraction varies, especially along vertical lines. Refractive effects such as mirages and shimmering are due to variations in the index of refraction caused by temperature variations.

Twinkling and Shimmering

Earth's atmosphere consists of flowing masses of air of varying density and temperature; therefore, the refractive index varies slightly from one region of the atmosphere to another. When light from a star enters the atmosphere it is refracted as it moves from one mass of air to another, and, since the variable masses of air are in motion, the star seems to twinkle.

Under conditions of constant pressure, warm air has a slightly lower index of refraction than cold air. When light passes through a stream of warm air, as it does above a hot stove or barbecue, it is refracted away from the normal. This refraction is not uniform because the warm air rises irregularly, in gusts. Light from objects seen through the warm air is distorted by irregular refraction, and the objects appear to shimmer. You can see the same effect over hot pavement in the summer or at an airport as jets go by.

DID YOU KNOW ?

Observing the Sky

Most large telescopes are located high on mountains where there is less of Earth's atmosphere between them and the stars, and where the air is more uniform in temperature. This minimizes distortions caused by atmospheric refraction. Astronomers find that cold winter nights are usually best for celestial observation since vertical temperature variations are minimized. Any atmospheric distortions are corrected using specialized software.

Mirages

Two types of mirages can occur, depending on the temperature variation. An inferior mirage occurs when cooler layers of air lie above warmer layers. The viewer sees an image displaced downward. Inferior mirages are most often associated with deserts, but can occur over any hot, flat surface, such as a road on a summer day (**Figure 1**). As the light from the sky nears the ground it is progressively refracted by successive layers of warmer air, each with a lower index of refraction (**Figure 2**). The angle of incidence to the boundaries of these layers keeps increasing. Eventually, the light is totally internally reflected upward from one of the layers. What your eye sees is a virtual image of the sky below the road. It just appears that there is a layer of water on the road because you perceive that there is a reflecting surface.

Figure 1
Sometimes, what seems to be a sheet of water appears on the highway a short distance ahead of you, but you never reach it.

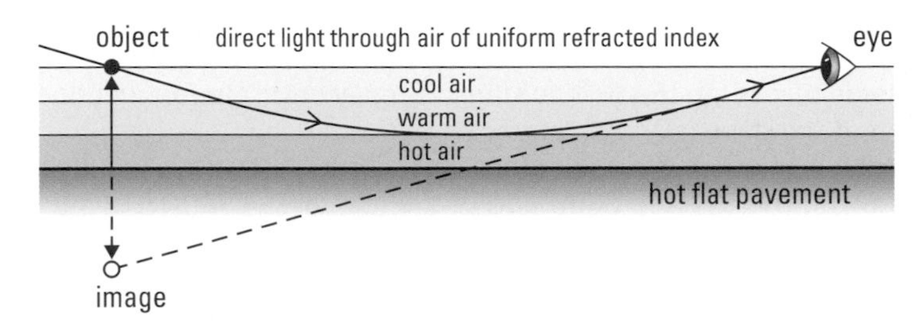

Figure 2
An inferior mirage

A superior mirage occurs when optically denser layers of air lie below less dense layers. The viewer sees an image displaced upward, as in **Figure 3**(a).

In this case, light refracts toward the denser air (**Figure 3(b)**).

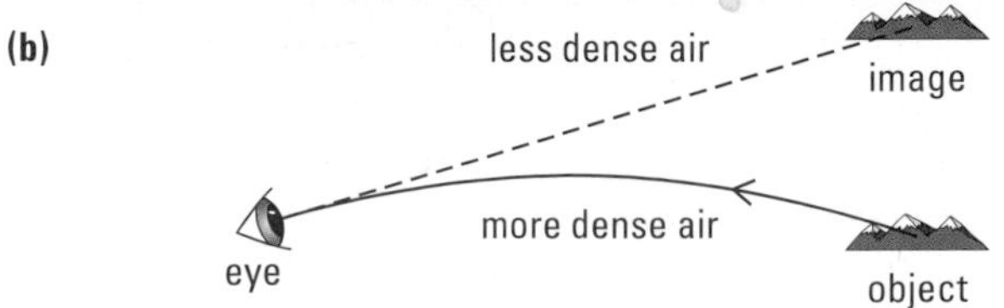

Figure 3
(a) A complex mirage of the Alaska Range with Mt. McKinley on the far right, 250 km away. In this photograph there are no clouds. This mirage is a result of a temperature inversion: layers of cold air lie beneath layers of warmer air.
(b) Light refracts toward denser air. A virtual image is perceived.

Figure 4
The shape of the Sun appears to change as it approaches the horizon.

When the Sun is very close to the horizon, the light from the lower part of the Sun is refracted more than the light from the upper part (**Figure 4**). This gives the impression that the Sun has a flattened bottom, making it appear oval rather than round.

Try This **Activity**

Mirages

A mirage can be set up in a laboratory setting using solutions of different densities and an empty fish tank.

Materials: a fish tank, solutions with different densities, and a simple line drawing.

- Pour the densest solution into the fish tank first.
- To prevent mixing, pour the second solution slowly onto a board that is placed on the first solution. The second solution will float on top of the denser first solution.
- At the end of the fish tank place a simple line drawing. Look at the drawing through the other end of the fish tank. Compare what you see through the fish tank with what you see at the same distance without the fish tank in place.
- Repeat the previous two steps by adding a third liquid, less dense than the other two.
  (a) If you see a mirage, is it an inferior or superior mirage? Explain.
- Demonstrate the device to your class.

dispersion: the spreading of white light into a spectrum of colours

Dispersion and Rainbows

The ancient Egyptians were probably the first to discover that fragments of clear, colourless glass and precious stones spread, or disperse, the colours of the rainbow when placed in the path of a beam of white light. We now know that the index of refraction of a medium changes according to the colour of light that is entering the medium. These changes in the index of refraction are the source of **dispersion**.

A rainbow is the Sun's spectrum produced by water droplets in the atmosphere (**Figure 5**). Light enters the spherical rain droplets, where it is refracted, dispersed, and reflected internally. The violet and red rays intersect internally, emerging with violet at the top, red at the bottom, and the other colours of the spectrum in between. If the rays that enter your eye are traced backward, you can see an image called the **primary rainbow** (**Figure 6**). Other orders of rainbows, which are much more difficult to observe than a primary rainbow, occur when the light is internally reflected more than once. For instance, a secondary rainbow, which is located higher in the sky than the primary (**Figure 5**), occurs after two internal reflections.

A rainbow arc appears at certain points in the sky because droplets of water along that arc reflect the spectrum at just the right angle into the eye of the observer. The angle is approximately 42° to the horizontal sunlight.

Figure 5
This photograph shows a rainbow over the onetime home of Sir Isaac Newton. When watching the dazzling colours of a rainbow, you likely make two basic observations. First, rainbows consist of the same colours of the spectrum as produced by a prism, with red on the outside of the bow and blue and violet on the inside. Second, whenever you see a rainbow, the Sun is at your back. This is why you see rainbows in the western sky in the morning and in the eastern sky before sunset. Both of these observations can be explained by the concepts of dispersion, refraction, and total internal reflection.

primary rainbow: the rainbow of visible spectral colours that results from a single internal reflection in rain drops

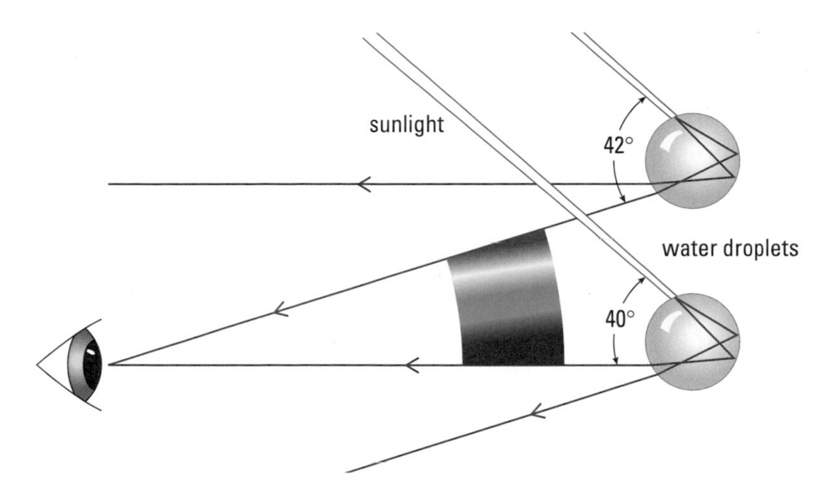

Figure 6
Formation of a primary rainbow. Looking at millions of drops, you see the spectrum in an arc of a semicircle with red on the outside and violet on the inside.

Activity 9.7.1

Modelling Atmospheric Refraction

There is a continuous decrease in the index of refraction of air as you move from the ground to the outermost layer of the atmosphere. Although the decrease in temperature increases the index of refraction, the decrease in pressure as you move up through the atmosphere has a greater effect on the index of refraction. This effect is simplified by representing the atmosphere by layers (**Figure 7**).

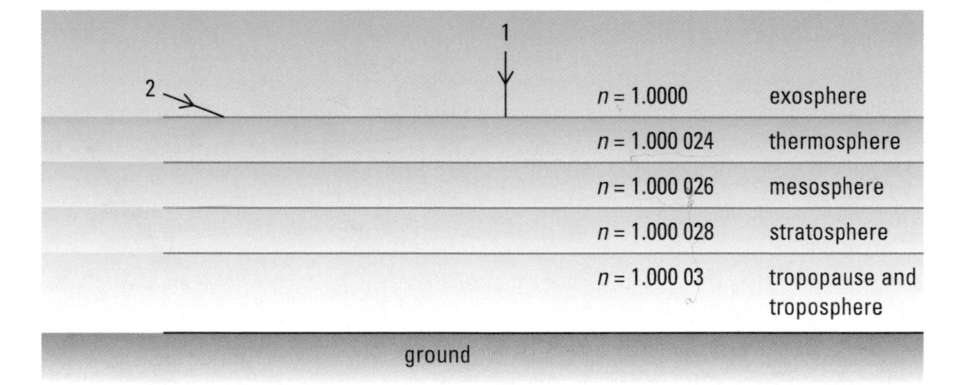

Figure 7
The indexes of refraction of the layers of the atmosphere. (Layers not to scale.)

Procedure

1. Copy **Figure 7** into your notebook.
2. Use your knowledge of refraction to predict the path that rays 1 and 2 take as they travel through the layers. (Calculations are not required for this model.)
3. Trace back refracted rays to support or refute the following statement: "To the observer, the Sun's light appears to be coming from a higher point in the sky than it actually is."
4. In **Figure 8**, a ray from the Sun enters the atmosphere. It refracts because of the changing index of refraction. Use **Figures** 7 and **8** to support or refute the following statement: "When the observer sees the Sun set, the light seen is coming from below the horizon. The Sun has already set!"

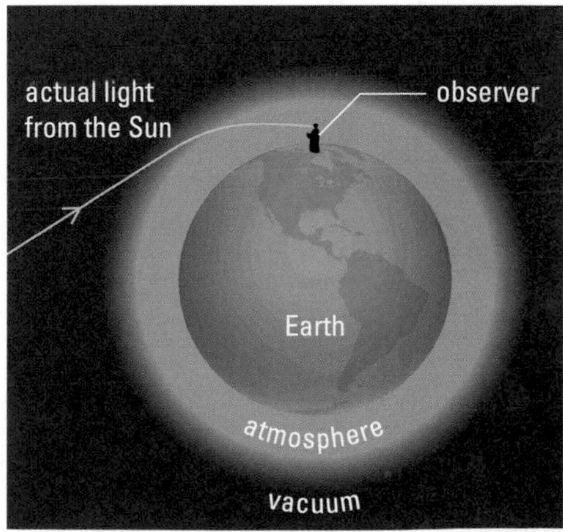

Figure 8
Refraction of sunlight in the atmosphere. (Diagram not to scale.)

SUMMARY Applications of Refraction

- Under constant pressure, the higher the temperature of the air, the lower the index of refraction; under constant temperature, the higher the pressure of air, the greater the index of refraction.
- Variations in the index of refraction of air cause many atmospheric refraction phenomena such as twinkling, shimmering, and mirages.
- Dispersion is due to small variations in the index of refraction for the different colours of light.

Section 9.7 Questions

Understanding Concepts

1. Describe the conditions needed for and examples of an inferior mirage and a superior mirage. In your answer, show that you understand why the mirages are called inferior and superior.
2. Draw a ray diagram to demonstrate the origin of the Sun's oval shape as the Sun approaches the horizon.

Applying Inquiry Skills

3. Design an activity to demonstrate the formation of a visible spectrum using a beam of white light aimed from a ray box or a projector toward one side of an equilateral glass or acrylic prism. Draw a diagram of the spectrum produced, and compare it with a primary rainbow.

Making Connections

4. Zircon (or zirconium) is used to make fake diamonds.
   (a) Determine the critical angle in both zircon and diamond. (Refer to Table 1 in section 9.4 to find the index of refraction of each medium.)
   (b) Use your answers in (a) to describe why a real diamond sparkles more than a zircon gem.
   (c) Find out how the cost of a zircon gem compares with the cost of a diamond gem of comparable size. Why is there such a difference? Follow the links for Nelson Physics 11, 9.7.

GO TO www.science.nelson.com

5. Canada is a major exporter of copper, an important metal in the manufacture of cables used for telephone lines. Canada also is in a world of rapidly increasing communications technology, where a laser pulse along a single glass fibre can carry about five million times as many phone messages as a single copper wire. As the use of glass fibres in fibre optics increases, the need for copper decreases. Should governments in Canada support the development of fibre optics or the maintenance of an economically important copper industry? Give reasons for your answer.

Reflecting

6. Dispersion effects are very common once you begin to look for them. List other colour effects that you have seen that you think can be explained by dispersion. Can you think of any colour effects that cannot be explained by dispersion? Describe them.

Chapter 9 Summary

Key Expectations

Throughout this chapter, you have had opportunities to

- define and describe concepts and units related to light; (9.1, 9.2, 9.3, 9.4, 9.5, 9.6, 9.7)
- demonstrate and illustrate, using light ray diagrams, the refraction, partial refraction, and reflection, critical angle, and total internal reflection of light at the interface of a variety of media; (9.2, 9.3, 9.4, 9.5, 9.6)
- carry out experiments involving the transmission of light, compare theoretical predictions and empirical evidence, and account for discrepancies; (9.5, 9.6, 9.7)
- describe the scientific model for light and use it to explain optical effects that occur as natural phenomena; (9.1, 9.2, 9.3, 9.4, 9.5, 9.6, 9.7)
- predict, in qualitative and quantitative terms, the refraction of light as it passes from one medium to another, using Snell's law; (9.5, 9.6, 9.7)
- carry out an experiment to verify Snell's law; (9.5)
- construct a prototype of an optical device; (9.6)
- explain the conditions required for total internal reflection, using light-ray diagrams, and analyze and describe situations in which these conditions occur; (9.6)

Key Terms

light
linear propagation
ray
speed of light
incident ray
reflected ray
point of incidence
normal
angle of incidence
angle of reflection
laws of reflection
object point
virtual image point
refraction
angle of refraction
partial reflection and refraction
index of refraction
Snell's law
laws of refraction
general equation of Snell's law
lateral displacement
prism
total internal reflection
critical angle
dispersion
primary rainbow

Make a Summary

Write the words "reflection" and "refraction" in the middle of a page. At different positions, write down all the phenomena that you can explain with these concepts. Underneath each phenomenon, list the additional terms that you would use to write a complete explanation.

Reflect on your Learning

Revisit your answers to the Reflect on Your Learning questions at the beginning of this chapter.

- How has your thinking changed?
- What new questions do you have?

Chapter 9 Review

Understanding Concepts

1. Light travels from medium A to medium B. The angle of refraction is greater than the angle of incidence.
   (a) Which medium has the higher index of refraction?
   (b) In which medium does the light travel at a lower speed?
2. Use your knowledge of refraction to describe how you would identify an unknown, clear substance.
3. The index of refraction for blue light in glass is slightly higher than that for red light in glass. What does this indicate about
   (a) the relative speeds of red light and blue light in glass?
   (b) the angles of refraction for each colour, for the same angle of incidence?
4. How would you use the index of refraction of salt water to find the concentration of a salt solution?
5. How does refraction lengthen the amount of sunlight we get at sunrise and sunset?
6. To successfully spear a fish, you must aim below the apparent position of the fish. Explain. Use a ray diagram in your explanation.
7. The speed of light in a vacuum is 3.00×10^8 m/s. The speed of light in a medium is 1.24×10^8 m/s. Determine the index of refraction. What might the medium be?
8. The index of refraction of crown glass for violet light is 1.53 and for red light, 1.51. What are the speeds of violet light and red light in crown glass?
9. One ray of light in air strikes a diamond ($n_d = 2.42$) and another strikes a piece of fused quartz ($n_q = 1.46$), in each case at an angle of incidence of 40.0°. What is the difference between the angles of refraction?
10. A ray of light strikes a block of polyethylene ($n_p = 1.50$) with angles of incidence of
    (a) 0.0°
    (b) 30.0°
    (c) 60.0°
    Determine the angle of refraction in each case.
11. An underwater swimmer looks up toward the surface of the water on a line of sight that makes an angle of 25.0° with a normal to the surface of the water. What is the angle of incidence in air for the light rays that enter the swimmer's eye?
12. Answer the following questions, assuming the second medium is air:
    (a) What is the critical angle if the index of refraction of a medium is 1.68?
    (b) What is the index of refraction of a medium if the critical angle is 40.0°?
13. In which medium does light travel faster: one with a critical angle of 27.0° or one with a critical angle of 32.0°? Explain. (For both cases, air is the second medium.)

Applying Inquiry Skills

Transfer the diagrams in **Figure 1** into your notebook. Draw in the direction of the partially reflected and refracted ray(s) in each case.

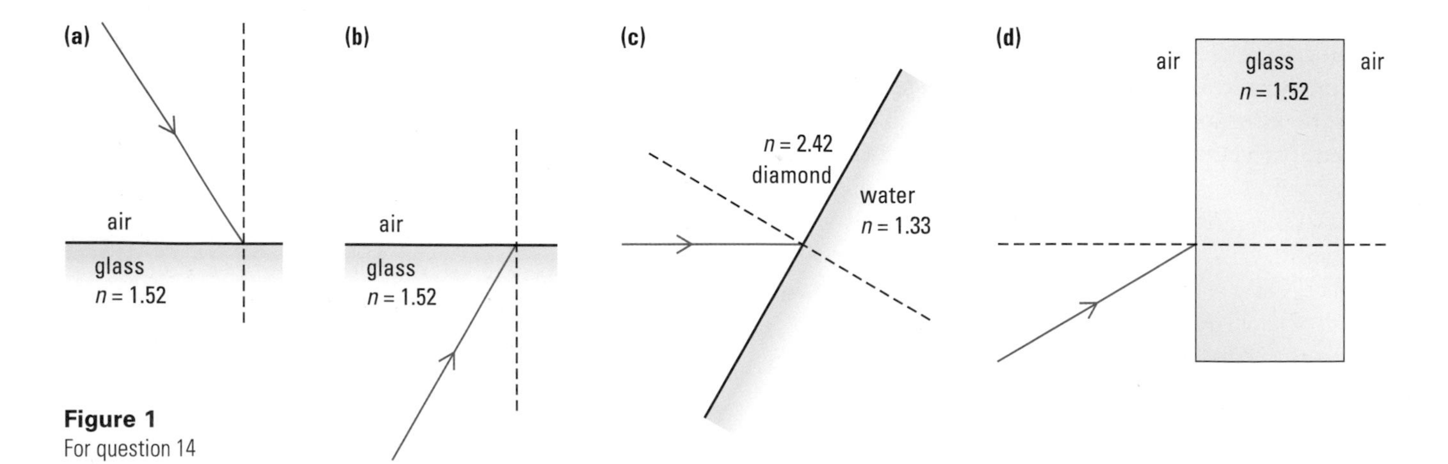

Figure 1
For question 14

15. Measure the appropriate angles in **Figure 2** to calculate the index of refraction of the second substance. Assume the incident ray is travelling in air.

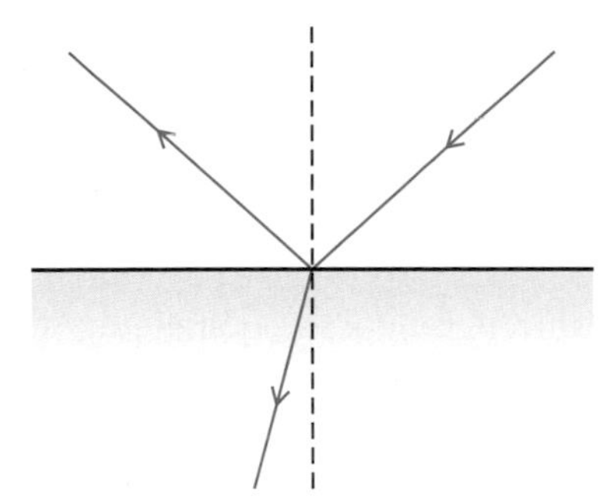

Figure 2

16. A right-angled periscope with a single prism can be used for seeing around corners. Draw a diagram showing how you would design such a periscope by applying the principle of internal reflection.

Making Connections

17. The index of refraction is an important physical characteristic of gemstones and concentrations of solutions. A device used to measure the index of refraction is called a *refractometer.* Investigate the following using the Internet or other resources. Follow the links for Nelson Physics 11, Chapter 9 Review.
    (a) How does a Pulfrich refractometer work?
    (b) How does a digital fibre optic refractometer work?

GO TO www.science.nelson.com

18. A pentaprism (**Figure 3**) is a five-sided prism containing two reflecting surfaces, in which the image is rotated 90°. Pentaprisms are used in surveying, alignment machinery, cinematography, and medical instruments. Each type of prism has a special geometry designed to achieve a particular type of reflection. Using the Internet and other sources, investigate one or more of the following types of prisms: wedge, Dove, rhomboid, Amici roof, reversion, corner cube, or direct vision prisms. In which optical devices are such prisms used? What effects do these prisms have on an image? Follow the links for Nelson Physics 11, Chapter 9 Review.

GO TO www.science.nelson.com

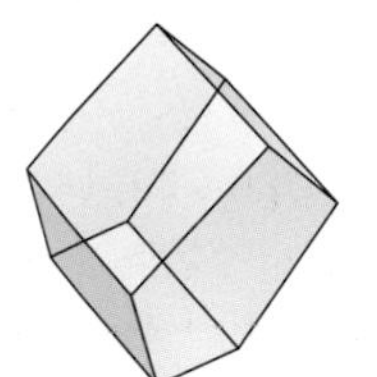
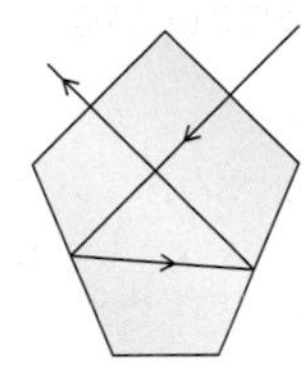

Figure 3

19. Large numbers such as the speed of light are difficult to grasp. Scientists enjoy translating such numbers into everyday situations that are easier to understand. For example, instead of stating that the speed of light is 3.00×10^8 m/s, we can say, "Light travels 300 000 km in one second," or we can say, "Light travels so fast that in one second it could travel around the circumference of Earth almost seven and a half times." Find ways to express the following numbers in everyday language:
    (a) The frequency of blue light is 6.7×10^{14} Hz.
    (b) The wavelength of yellow-green light is 570 nm.

Exploring

20. Research the following production processes and devices: fluorescence, tribo- luminescence, phosphorescence, bioluminescence, gas-discharge tubes, light-emitting diodes, and sono- luminescence. Give brief explanations of how light is produced in each case.

21. The automotive industry has developed vehicles that can use various fuels. These vehicles need a way to detect alcohol concentrations in the fuel system to make adjustments for efficient combustion. Optical sensors known as flexible-fuel vehicle sensors measure the differences in a fuel's index of refraction or critical angle. Use the Internet to research how these sensors work. Follow the links for Nelson Physics 11, Chapter 9 Review.

GO TO www.science.nelson.com

22. Research how glass fibres are made and write a brief report (200–300 words) of your findings.

23. The Sun and Moon appear larger when near the horizon. There is no way to trace rays to explain this effect. Research the origin of this effect.

Chapter

10

In this chapter, you will be able to

- define and describe concepts and units related to lenses such as focal point, real images, and virtual images
- describe and explain, with the aid of light-ray diagrams, the characteristics and positions of the images formed by lenses
- describe the effects of converging and diverging lenses on light, and explain why each type of lens is used in specific optical devices
- analyze the characteristics and positions of images formed by lenses
- evaluate the effectiveness of enhancements to human vision such as eyeglasses, 3-D glasses, contact lenses, and laser surgery
- describe lens aberrations, limitations, and their solutions

Lenses and the Eye

It's said that a picture is worth a thousand words, though sometimes a word is worth a thousand pictures. For example, the French word "fare" means lighthouse, and the Spanish word "farola" means headlights. Both words may find their origin from The Lighthouse of Alexandria, built around 300 B.C. on the Island of Pharos in Egypt (**Figure 1**). Probably the tallest building on Earth at the time it was built, it is now one of the Seven Wonders of the Ancient World.

By the 18th century, lighthouses had become important navigational tools for the growing shipping industry. At that time, lighthouses required huge lenses to focus light out to sea—some had a mass of up to 24 t. In 1822, the French physicist and engineer Augustin Fresnel (1788–1827) produced a lens that revolutionized the way lighthouses projected light. His largest design, now called a first-order Fresnel lens, contained over 1000 prisms, stood 4 m tall and 2 m wide, and had a mass of about 3 t.

A French-built Fresnel lens (**Figure 2**) was installed in the Sambro Island, Nova Scotia, lighthouse in 1906, making it one of Canada's major coastal beacons (**Figure 3**). It had a range of 28 km and flashed every 5 s. The Fresnel lens was removed from the lighthouse in 1967. The Sambro Island lighthouse now uses a modern 91-cm airport-type beacon, which has a range of about 40 km and flashes every 5 s.

The Fresnel lens is one example of the types and applications of lenses you will explore in this chapter. Another example is the lens found in the most important optical instrument of all, the human eye.

Reflect on your Learning

1. There are some situations where you must look directly at an optical device to see an image, for example, when you want to see yourself in a mirror.
   (a) Name other situations where you can see the image only by looking toward the optical device.
   (b) Describe situations where you do not have to look at the optical device to see the image it produces. Start by thinking of a situation in your classroom.
2. Do you think the image of this page at the back of your eye is upright or inverted? Explain your answer.
3. (a) When considering images in an optical device, what does "magnification" mean?
   (b) Based on your experience with microscopes or other optical devices, how would you calculate the magnification of the device?
4. (a) What are some features of the human eye that make it an extremely useful device?
   (b) Describe several problems with the human eye that affect vision.

Throughout this chapter, note any changes in your ideas as you learn new concepts and develop your skills.

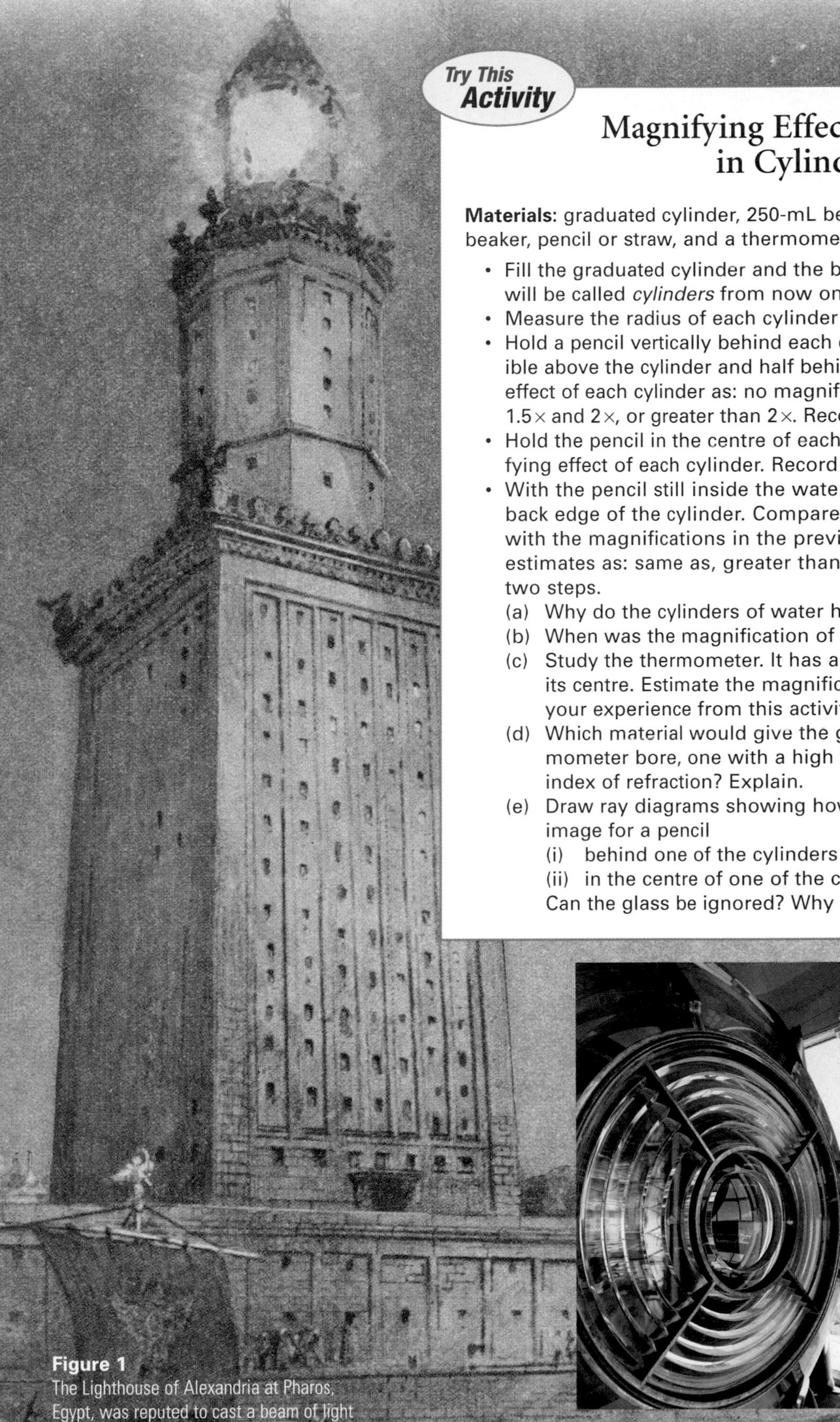

Figure 1
The Lighthouse of Alexandria at Pharos, Egypt, was reputed to cast a beam of light over 50 km away using a special arrangement of mirrors that reflected the light of a large fire at night and the Sun's rays during the day. In 1323, the lighthouse was destroyed by a series of earthquakes.

Try This **Activity**

Magnifying Effect of Liquids in Cylinders

Materials: graduated cylinder, 250-mL beaker, 500-mL beaker, 1000-mL beaker, pencil or straw, and a thermometer

- Fill the graduated cylinder and the beakers with water. (The beakers will be called *cylinders* from now on.)
- Measure the radius of each cylinder and record in a data table.
- Hold a pencil vertically behind each cylinder with half the pencil visible above the cylinder and half behind. Estimate the magnifying effect of each cylinder as: no magnification, less than 1.5×, between 1.5× and 2×, or greater than 2×. Record your estimates in your table.
- Hold the pencil in the centre of each cylinder. Estimate the magnifying effect of each cylinder. Record your estimates in your table.
- With the pencil still inside the water, move it so that it touches the back edge of the cylinder. Compare the magnification of the pencil with the magnifications in the previous two steps. Record your estimates as: same as, greater than, or less than in the previous two steps.

(a) Why do the cylinders of water have a magnifying effect?
(b) When was the magnification of the cylinder the greatest? Why?
(c) Study the thermometer. It has a thin bore filled with liquid down its centre. Estimate the magnification of the thin bore based on your experience from this activity.
(d) Which material would give the greater magnification of the thermometer bore, one with a high index of refraction or a low index of refraction? Explain.
(e) Draw ray diagrams showing how your eye sees a magnified image for a pencil
  (i) behind one of the cylinders
  (ii) in the centre of one of the cylinders
  Can the glass be ignored? Why or why not?

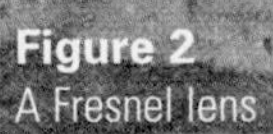

Figure 2
A Fresnel lens

Figure 3
Built in 1760, the lighthouse on Sambro Island near Halifax, Nova Scotia, is the oldest operational lighthouse in Canada

Harold Oakley

(a)

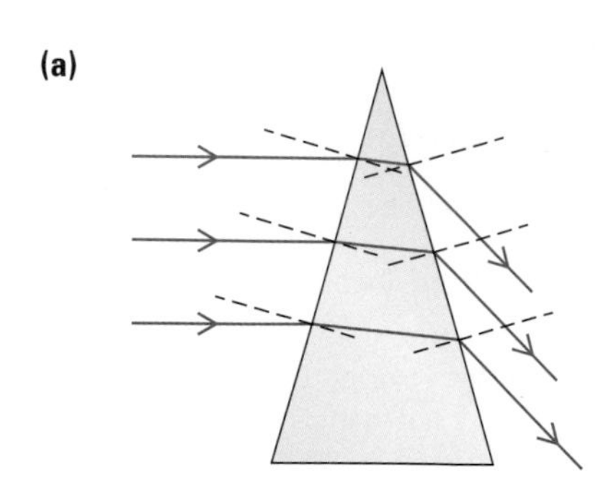

(b)

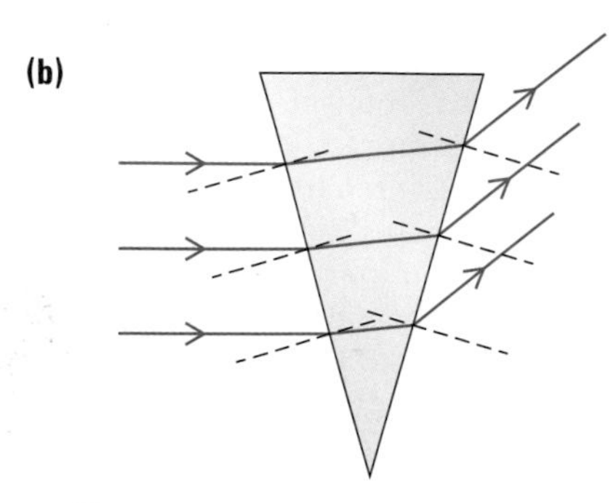

Figure 1
Light passing through a triangular prism refracts twice, obeying Snell's law of refraction: once as it enters the prism at the air-prism boundary and then again as it leaves at the prism-air boundary.
(a) The shape of the prism causes the light to be refracted downward.
(b) The light is refracted upward.

You can easily make very basic lenses yourself by combining two triangular prisms.

10.1 Refraction in Lenses

Lenses are not a recent invention. They were first used by the Chinese and the Greeks over 2000 years ago, and later, in medieval times, by the Arabs. Lenses of many different types are an essential part of optical devices such as cameras, microscopes, telescopes, and projectors. They also enable millions of people to see clearly and read comfortably.

In Chapter 9 you learned that white light changes direction and separates into colours when it goes through a triangular prism. This bending occurs because the two sides of the prism are not parallel (**Figure 1**).

A **lens** is a transparent device with at least one curved surface that changes the direction of light passing through it. A **converging lens** causes parallel light rays to come together so that they cross at a single **focal point**. A **diverging lens** spreads parallel light rays apart, as if they are emerging from a point called the virtual focal point. Some common shapes of converging and diverging lenses are shown in **Figure 2**.

(a)

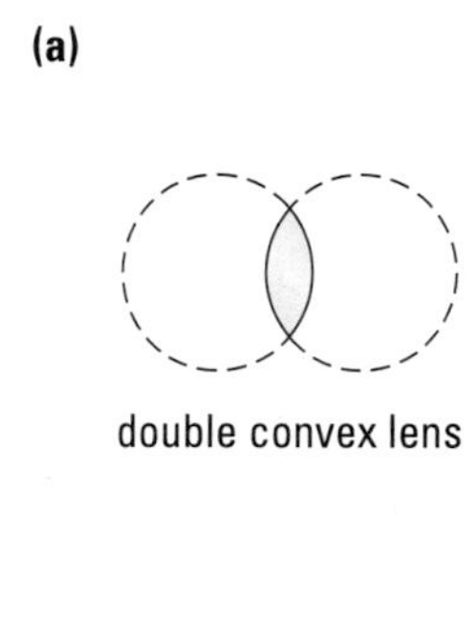

double convex lens

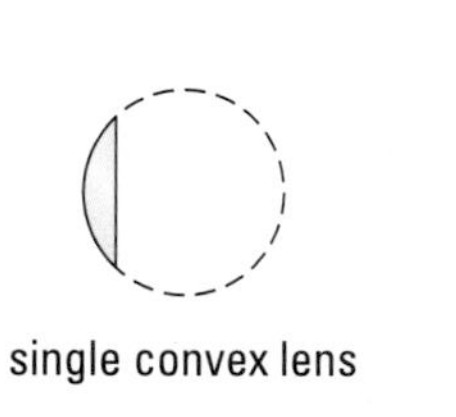

single convex lens

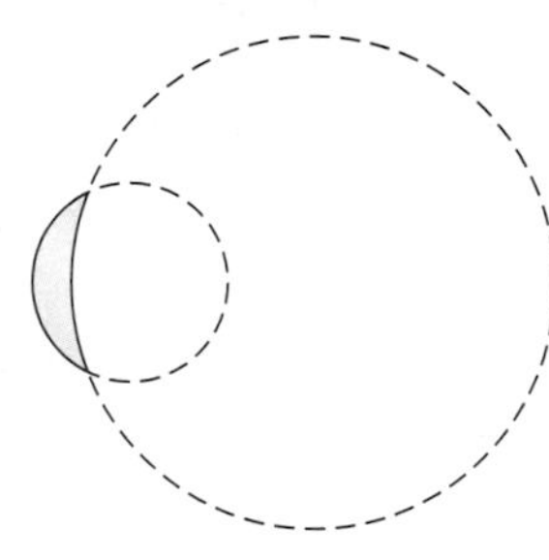

concavo-convex lens

(b)

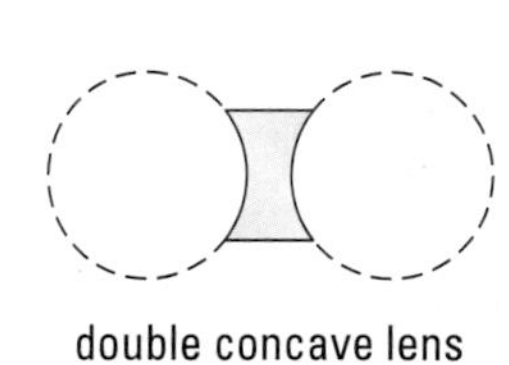

double concave lens

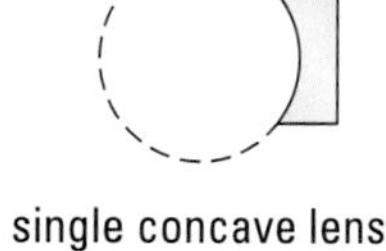

single concave lens

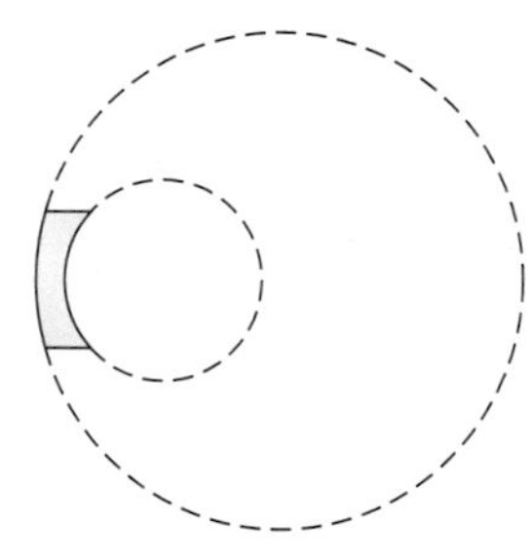

convexo-concave lens

Figure 2
The two types of lenses are easily distinguished by their shape.
(a) Converging lenses are thicker at the centre than they are at the edges.
(b) Diverging lenses are thinner at the centre than at the edges.

Practice

Understanding Concepts

1. (a) Considering the way light acts in each device in **Figure 2**, does a converging lens act more like a concave mirror or a convex mirror? Explain your answer.
   (b) Repeat (a) for a diverging lens.
2. Explain why light rays change direction when they pass through a lens. (Show that you understand what you learned in the previous chapter.)

lens: transparent device with at least one curved surface that changes the direction of light passing through it

converging lens: lens that causes parallel light rays to come together so that they cross at a single focal point

focal point: the position where parallel incident rays meet, or appear to come from, after they pass through a lens

diverging lens: lens that causes parallel light rays to spread apart so that they appear to emerge from the virtual focal point

SUMMARY Refraction in Lenses

- Converging lenses bring parallel rays together after they are refracted.
- Diverging lenses cause parallel rays to move apart after they are refracted.
- Rays are refracted at the surfaces of lenses according to Snell's law.

Section 10.1 Questions

Understanding Concepts

1. **Figure 3** shows light rays in air approaching three glass lenses. Draw diagrams of the same shape, but larger in size, in your notebook and draw the approximate direction of each light ray as it travels into the glass and back into the air. (*Hint:* Draw the appropriate normals wherever a ray strikes a surface.)
2. Draw a large double concave lens, then draw three parallel rays on one side of the lens. Apply the method you used in question 1 to show that light rays emerging from the lens diverge. Extend the rays straight back until they meet, or almost meet. Label this point appropriately.

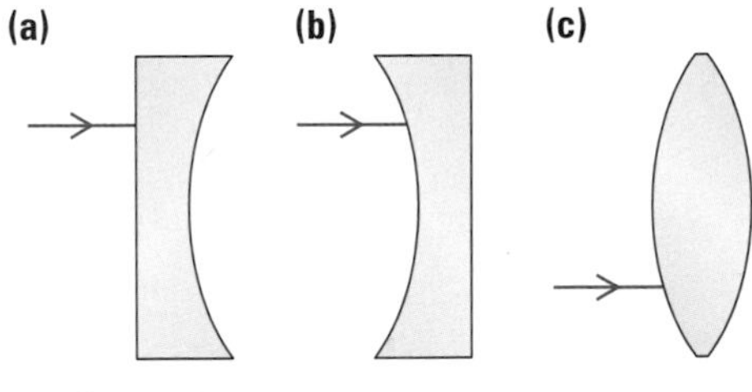

Figure 3
For question 1

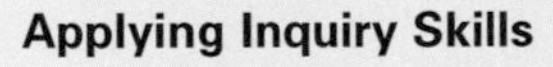

Applying Inquiry Skills

3. Pretend that the fingers of your wide-open hand, when spread out flat on a piece of paper, represent diverging light rays that originate at a single virtual focal point. Devise a way to measure the distance from the focal point to the tip of your middle finger.

Making Connections

4. The lenses in your science classroom are likely double convex and double concave lenses. These are not suited for eye glasses. Why? Which lenses in **Figure 2** are best suited for eye glasses? Why?

10.2 Images Formed by Lenses

In all lenses, the geometric centre is called the **optical centre** (O), as shown in Figure 1. A vertical line drawn through the optical centre is called the **optical axis** (OA) of the lens. A horizontal line drawn through the optical centre is

optical centre: (O) the geometric centre of all lenses

optical axis: (OA) a vertical line through the optical centre

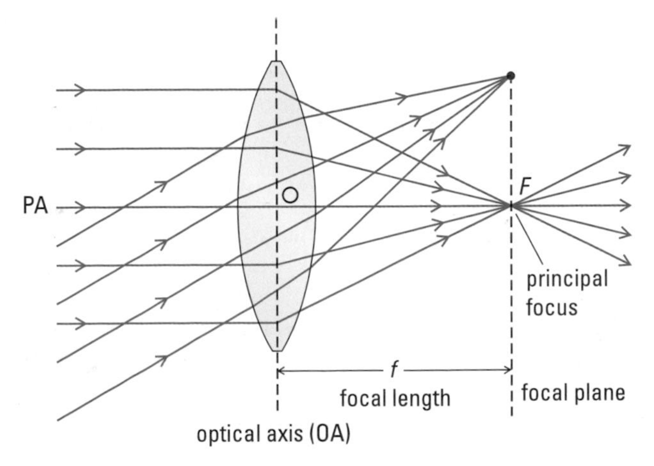

Figure 1

principal axis: (PA) a horizontal line drawn through the optical centre

principal focus: (F) the point on the principal axis through which a group of rays parallel to the principal axis is refracted

focal length: (f) the distance between the principal focus and the optical centre, measured along the principal axis

focal plane: the plane, perpendicular to the principal axis, on which all focal points lie

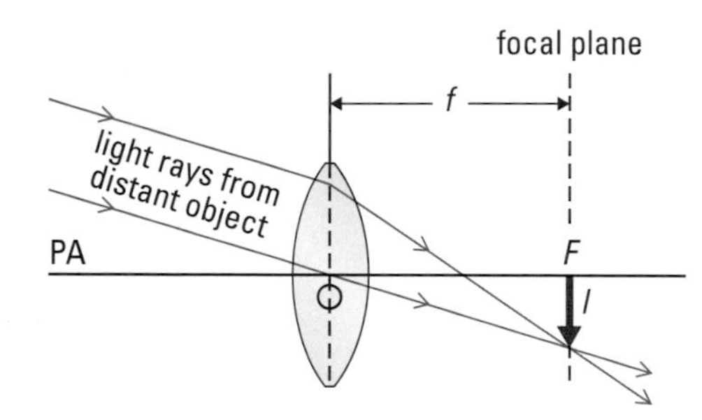

Figure 2

called the **principal axis** (PA). If a lens is thin, a group of rays parallel to the principal axis is refracted through a point on the principal axis called the **principal focus** (F). The **focal length** (f) is the distance between the principal focus and the optical centre, measured along the principal axis.

A beam of parallel rays that is not aligned with the principal axis converges at a focal point that is not on the principal axis (**Figure 1**). All focal points, including the principal focus, lie on the **focal plane**, perpendicular to the principal axis. When a converging lens refracts light from a distant object (**Figure 2**), the rays arriving at the lens are nearly parallel; thus, a real image is formed at a distance close to one focal length from the lens.

Since light can enter a lens from either side, there are two principal *foci*. The focal length is the same on both sides of the lens, even if the curvature on each side is different. The notation F is always given to the primary principal focus, the point at which the rays converge or from which they appear to diverge; the secondary principal focus is usually expressed as F'.

Characteristics of an Image

Once an image has been observed or located in a ray diagram, we can use four characteristics to describe it relative to the object: magnification, attitude, location, and type of the image.

Magnification of the Image

The height of an object is written as h_o; the height of the image is written as h_i. To compare their heights, the magnification (M) of the image is found by calculating the ratio of the image height to the object height:

$$M = \frac{h_i}{h_o}$$

Note that magnification has no units because it is a ratio of heights.

Attitude of the Image

The attitude of the image refers to its orientation relative to the object. For example, when an image forms on film in a camera, the image is inverted relative to the object that was photographed. An image is either upright or inverted, relative to the object.

Location of the Image

The distance between the subject of a photograph and the lens of a camera is the object distance, designated by d_o. An image of the object is formed on the film inside the camera. The distance between the image on the film and the lens is the image distance, d_i. The image is located either on the object side of the lens, or on the opposite side of the lens. When discussing object and image distances, we refer to the location as "between F and the lens," "between F and $2F$," or "beyond $2F$."

Type of the Image

An image can be either real or virtual. A real image can be placed onto a screen; a virtual image cannot. Recall that light diverges from a real object point (**Figure 3(a)**). An optical device can converge light from an object point to a point called the **image point.** A diverging beam from this image point must enter your eye in order for you to see the image point. Such an image point is called a real image point. By using a screen to scatter the light from the real image point, the image is visible from many angles.

image point: point at which light from an object point converges

An optical device can also change the direction of a diverging beam from an object point so that the rays appear to diverge from behind the object point

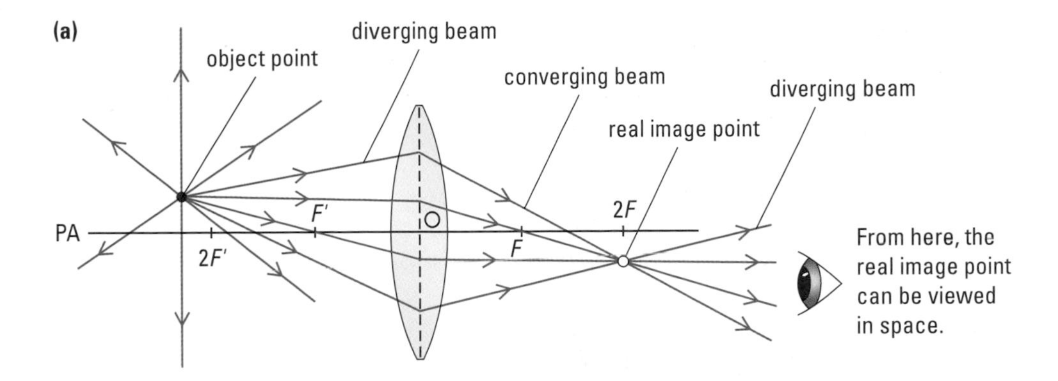

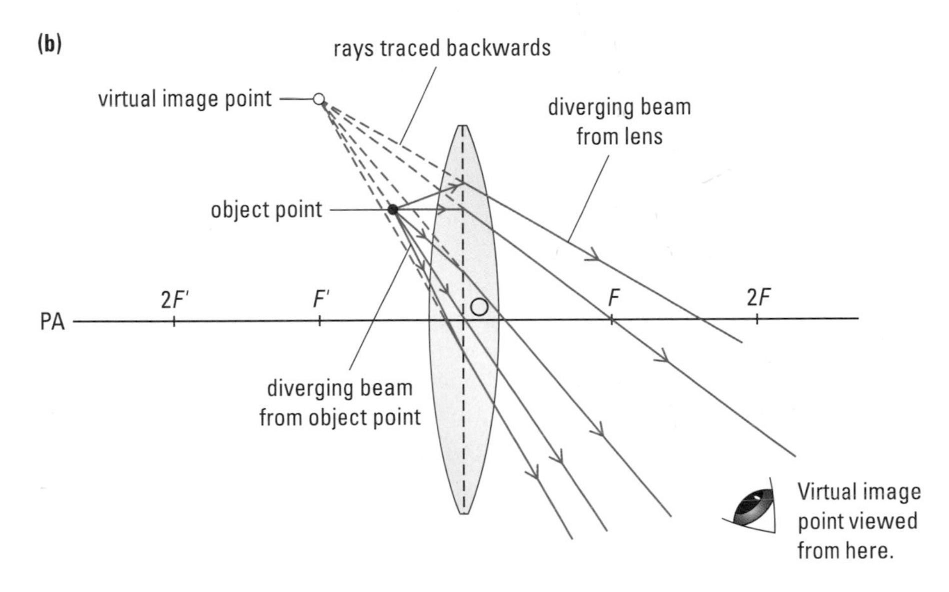

Figure 3
(a) Real image point
(b) Virtual image point

(**Figure 3(b)**). Such an image point is called a **virtual image point**. The virtual image point is located by extending the rays backward until they intersect. Since the light itself does not intersect at the virtual image point, a virtual image cannot be formed or captured on a screen. Whenever extending rays backward to locate virtual image points, use *dashed* lines on your ray diagrams.

virtual image point: point from which rays from an object point appear to diverge

Practice

Understanding Concepts

1. Create a chart, table, or detailed diagram to summarize the symbols and meanings of these properties or variables of lenses: optical centre, optical axis, principal axis, principal focus, focal length, object distance, object height, image distance, image height, magnification.
2. The word "MALT" can be used as a memory aid for the four characteristics of images. Print the word vertically. Add the complete word that corresponds to each letter. In each case, list the possible choices used to describe an image.

Images Formed by Converging Lenses

An object gives off light rays in all directions, but, for the purpose of locating its image, we are only interested in those rays that pass through the lens (**Figure 4**).

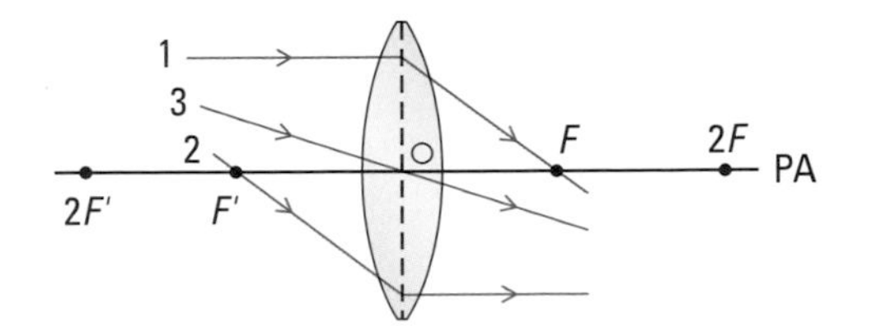

Figure 4
Three rays are particularly convenient for locating image points, since they either pass through the lens reference positions *F* and *F'* or are parallel to the principal axis. Once image points are located, we can predict the four characteristics of the image.

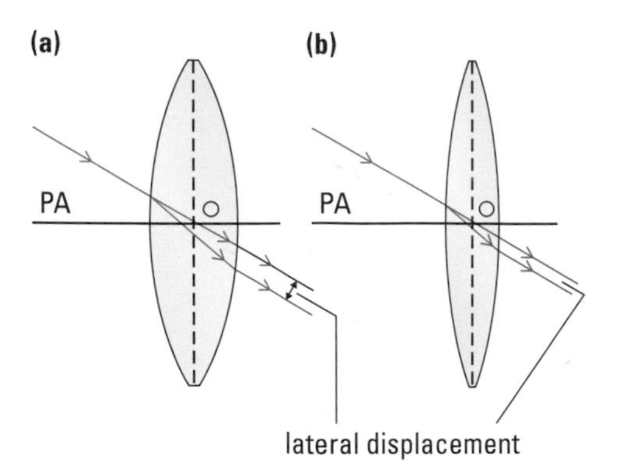

Figure 5

The following are the three rays and their rules that can be used:

Rules for Rays in a Converging Lens

1. A light ray travelling parallel to the principal axis refracts through the principal focus (F).
2. A light ray that passes through the secondary principal focus (F') refracts parallel to the principal axis.
3. A light ray that passes through the optical centre goes straight through, without refracting.

Note: Only two rays are needed to locate an image point. The third ray can be used as a check of accuracy.

It may seem strange that the ray that passes through the optical centre in **Figure 4** is not refracted, since most rays passing through the optical centre are laterally displaced. However, in thin lenses, the lateral displacement of the ray is so small that we can assume that the ray is not refracted (**Figure 5**).

In both diagrams in **Figure 6**, a construction line has been drawn through the optical centre perpendicular to the principal axis. The actual path of the light ray is indicated by a solid line.

An image can form either in front of a lens or behind it, and measurements are made either above or below the principal axis. Therefore, we need a sign convention to distinguish between real and virtual images and to interpret magnification calculations.

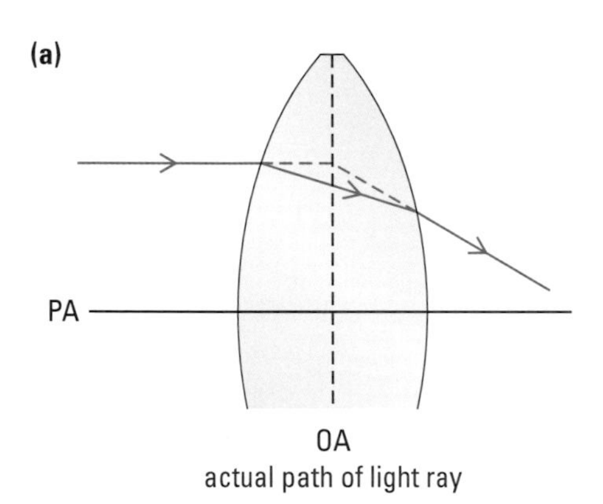

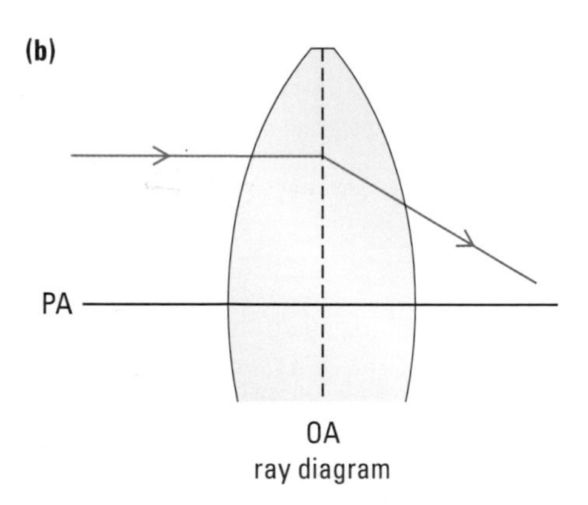

Figure 6
For simplicity, when drawing ray diagrams in lenses, we can represent all the refraction of light as occurring just once along the optical axis instead of twice at the curved surfaces of the lens. This results in the same image location.

Sign Convention

1. Object and image distances are measured from the optical centre of the lens.
2. Object distances are positive if they are on the side of the lens from which light is coming; otherwise they are negative.
3. Image distances are positive if they are on the opposite side of the lens from which light is coming; if on the same side, the image distance is negative. (Image distance is positive for real images; negative for virtual images.)
4. Object heights and image heights are positive when measured upward and negative when measured downward from the principal axis.

Using this convention, a converging lens has a real principal focus and a positive focal length. A diverging lens has a virtual principal focus and a negative focal length. The orientation of the image is predicted using the sign convention. Magnification is positive for an upright image and negative for an inverted image.

Sample Problem 1

A 1.5-cm-high object is 8.0 cm from a converging lens of focal length 2.5 cm.

(a) Draw a ray diagram to locate the image of the object and state its attitude, location, and type.

(b) Measure the image height from your diagram, and calculate the magnification.

Solution

(a) **Figure 7** shows an arrow resting on the principal axis. Three incident rays and three refracted rays are drawn, according to the rules for converging lenses. The resulting image, seen in the ray diagram, is real and inverted. The image is located on the opposite side of the lens between F and $2F$.

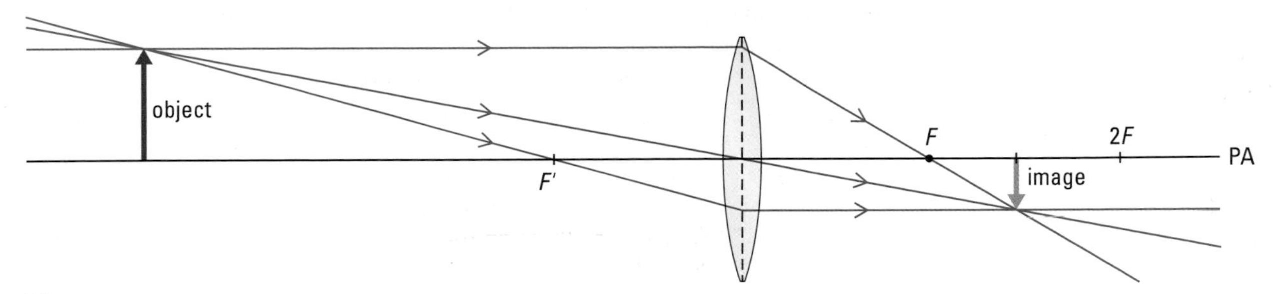

Figure 7

(b) $h_o = 1.5$ cm

$h_i = -0.70$ cm (negative because it was measured below the principal axis)

$$M = \frac{h_i}{h_o}$$

$$= \frac{-0.70 \text{ cm}}{1.5 \text{ cm}}$$

$$M = -0.47$$

The magnification of the image is -0.47. Note that a negative magnification is consistent with both the sign convention and the ray diagram.

The position of an object relative to a lens affects how the image is formed. There are five cases for a converging lens (**Figure 8**).

(a) object beyond $2F'$

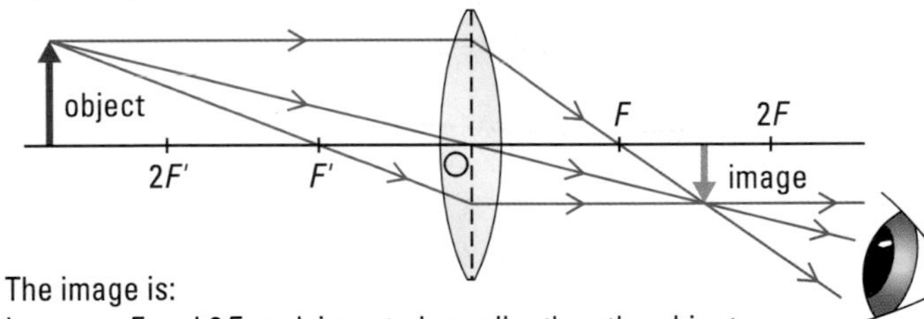

The image is:
between F and $2F$, real, inverted, smaller than the object.

(b) object at $2F'$

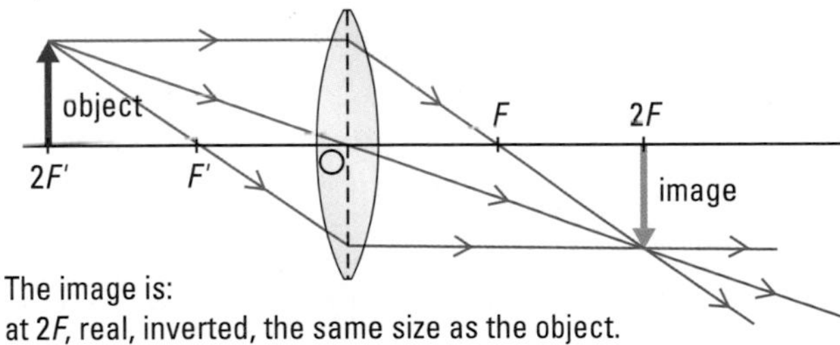

The image is:
at $2F$, real, inverted, the same size as the object.

(c) object between F' and $2F'$

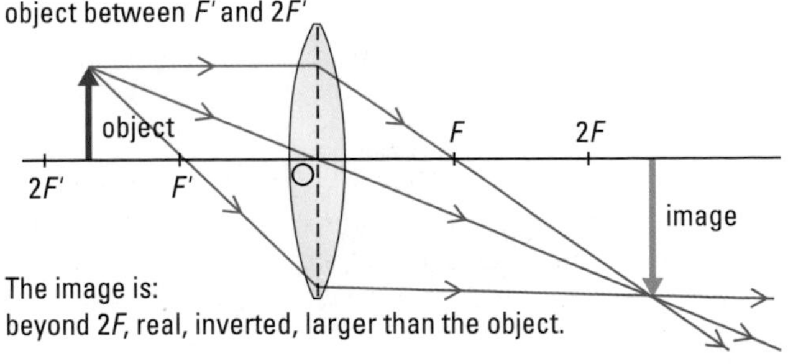

The image is:
beyond $2F$, real, inverted, larger than the object.

(d) object at F'

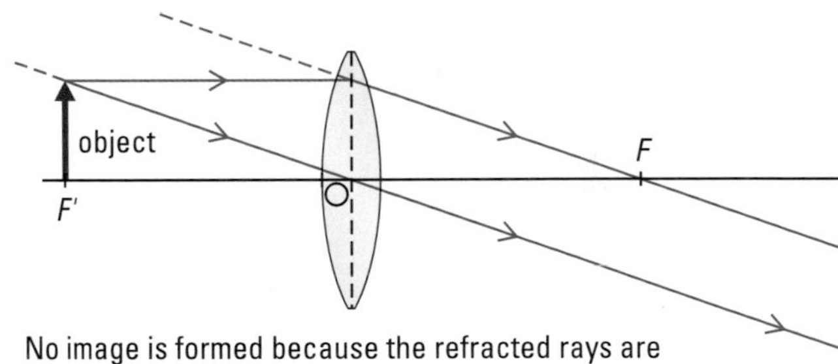

No image is formed because the refracted rays are parallel and never meet.

(e) object between lens and F'

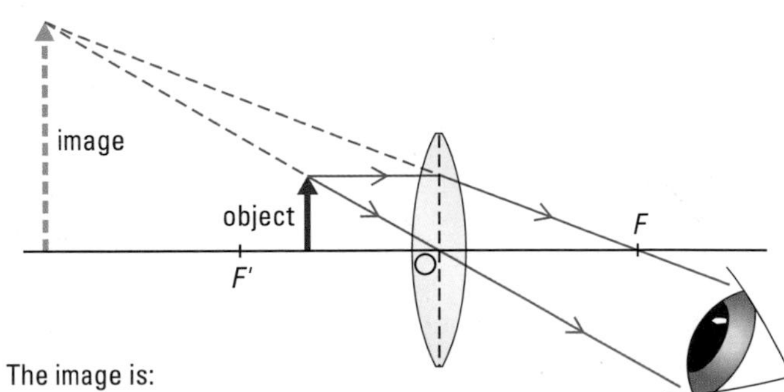

The image is:
behind the object, virtual, erect, larger than the object.

Figure 8

Answers

3. (a) $M = -1.0$
   (b) $M = -1.6$
   (c) $M = 2.0$

Practice

Understanding Concepts

3. A 15-mm-high object is viewed with a converging lens of focal length 32 mm. For each object distance listed below, draw a ray diagram using all the rules to locate the image of the object. State the characteristics of each image, including the magnification for
   (a) $d_o = 64$ mm
   (b) $d_o = 52$ mm
   (c) $d_o = 16$ mm

Making Connections

4. Below is a list of optical devices that match the arrangements of lenses and objects in **Figure 8**. For each of the following cases, explain where the lens must be placed relative to the object:
   (a) a copy camera produces an image that is real and the same size
   (b) a hand magnifier produces an image that is virtual and larger
   (c) a slide projector produces an image that is real and larger
   (d) a 35-mm camera produces an image that is real and smaller
   (e) a spotlight produces parallel light; there is no image
   (f) a photographic enlarger produces an image that is real and larger

Images Formed by Diverging Lenses

In a diverging lens, parallel rays are refracted so that they radiate outward from the principal focus (F) as shown in **Figure 9**.

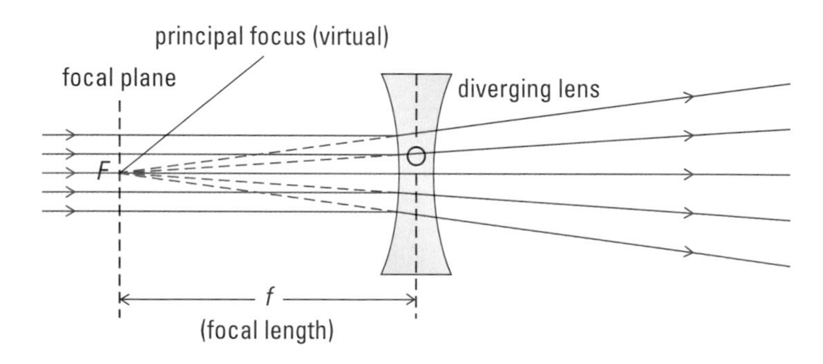

Figure 9
A diverging lens

The rays used to locate the position of the image in a diverging lens are similar to those used with converging lenses (**Figure 10**). As a result, one set of rules is used for all lenses. The important difference is that the principal focus in the converging lens is real, whereas in the diverging lens it is virtual.

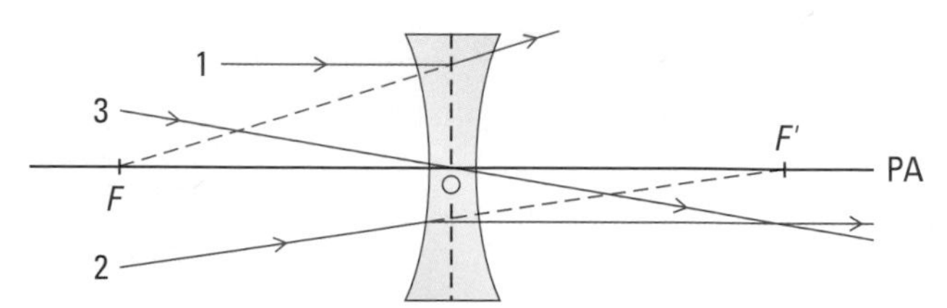

Figure 10
Rays refracting through a diverging lens

As with converging lenses, we assume with ray diagrams in diverging lenses that all refraction occurs at the optical axis of the lens. This makes the ray diagram easier to draw.

Rules for Rays in a Diverging Lens

1. A light ray travelling parallel to the principal axis refracts in line with the principal focus (F).
2. A light ray that is aimed toward the secondary principal focus (F') refracts parallel to the principal axis.
3. A light ray that passes through the optical centre goes straight through, without refracting.

Figure 11 illustrates the formation of an image by a diverging lens. For all positions of the object, the image is virtual, upright, and smaller. The image is always located between the principal focus and the optical centre.

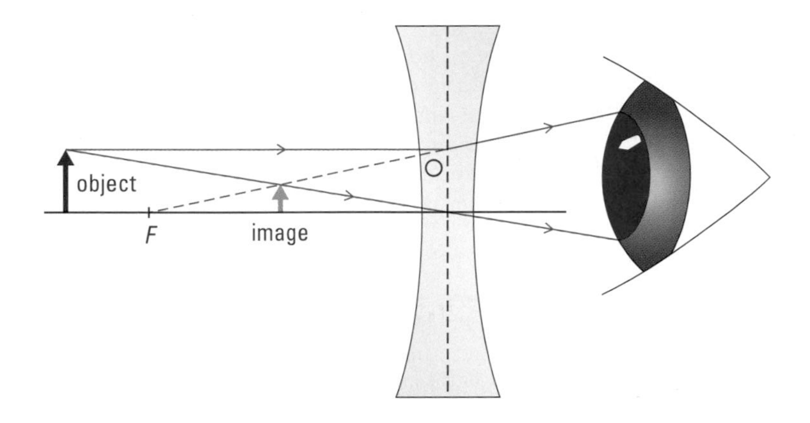

Figure 11
Image formation in a diverging lens

Sample Problem 2

A 1.5-cm-high object is located 8.0 cm from a diverging lens of focal length 2.5 cm.

(a) Draw a ray diagram to locate the image of the object and state its attitude, location, and type.

(b) Measure the image height from your ray diagram, and calculate the magnification.

Solution

(a) The refracted rays must be extended straight back to where they meet on the side of the lens where the object is located, as in **Figure 12**. The resulting image is virtual, upright, and located between F and the lens.

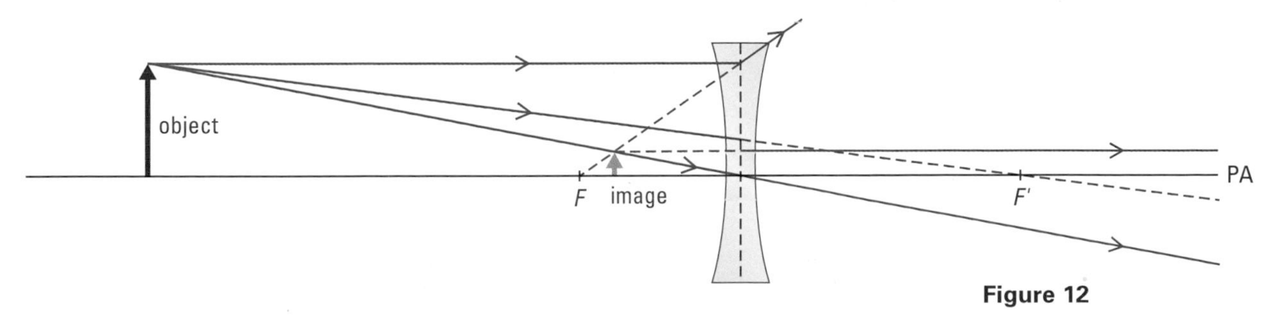

Figure 12

(b) $h_o = 1.5$ cm

$h_i = 0.40$ cm (positive because it was measured above the principal axis)

$$M = \frac{h_i}{h_o}$$

$$= \frac{0.40 \text{ cm}}{1.5 \text{ cm}}$$

$$M = 0.27$$

The magnification of the image is 0.27.

Practice

Understanding Concepts

5. A 12-mm-high object is viewed using a diverging lens of focal length −32 mm. Using all the rules, draw a ray diagram to locate the image of the object for each situation listed below. State the characteristics of each image, and write a conclusion describing what happens to the magnification of the image viewed in a diverging lens as the object distance decreases for
   (a) $d_o = 64$ mm
   (b) $d_o = 32$ mm

Answers

5. (a) $M = 0.33$
   (b) $M = 0.50$

INQUIRY SKILLS

- ○ Questioning
- ○ Hypothesizing
- ● Predicting
- ○ Planning
- ● Conducting
- ● Recording
- ● Analyzing
- ● Evaluating
- ● Communicating

Investigation 10.2.1

Images of a Pinhole Camera, Converging Lens, and Diverging Lens

The principle of the pinhole camera was described early in the 10th century by the Egyptian scholar Alhazen, who used it to indirectly view a solar eclipse. The purpose of this Investigation is to study the images formed by a pinhole camera and compare them with images produced by lenses.

Question

What are the differences among images formed by a pinhole camera, a converging lens, and a diverging lens?

Materials

opaque screen with a pinhole (pinhole camera)
small light source (miniature light bulb)
translucent screen (white paper)
converging lens ($f = 20$ cm)
diverging lens ($f = -20$ cm)
metre stick
optical bench

Prediction

(a) Predict the differences among the images formed by the pinhole camera, the converging lens, and the diverging lens.

Procedure

1. Position the light source at the end of the optical bench (**Figure 13**).

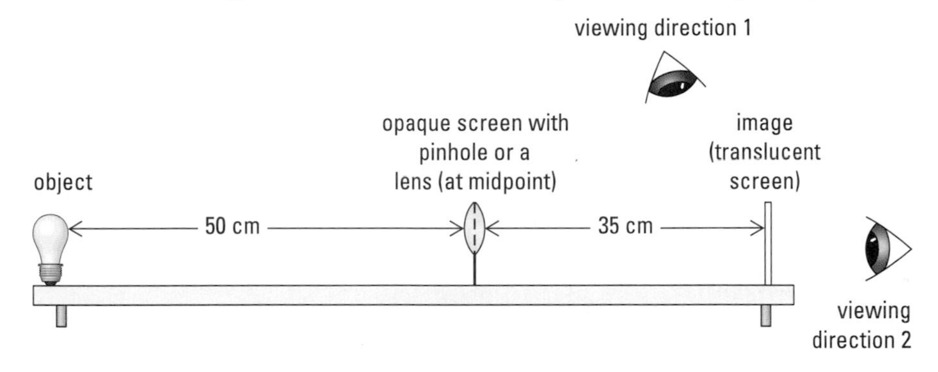

Figure 13
Setup for Investigation 10.2.1

2. Place the opaque screen 50 cm away from the light source.
3. In a dark part of the room, place the translucent screen about 35 cm from the opaque screen. Slowly move the translucent screen forward and backward. Make observations from both viewing directions at different positions along the principal axis. Record what happens to the image.
4. Remove the translucent screen and make observations from both viewing directions, as shown in **Figure 13**.
5. Replace the opaque screen with a converging lens.
6. (a) Replace the translucent screen and move it until you get a relatively sharp image on the screen. It may help to angle the converging lens up slightly. Slowly move the screen forward and backward along the principal axis. Record what happens to the image.
   (b) Cover the upper half of the lens with a piece of paper. Observe and record the effect on the image. Cover the left half of the lens. Observe and record the effect on the image.
7. Repeat step 4.
8. Replace the converging lens with a diverging lens.
9. Repeat step 4.
10. Place the converging lens between the opaque screen and the light source. Observe the resulting image with a screen. Move the screen back and forth along the principal axis. Change the relative positions of the lens, pinhole, and screen to obtain the "sharpest" image possible. Record the positions of the pinhole, lens, and screen.

Analysis

(b) Using the evidence you have collected, answer the Question. Your explanation should include neatly drawn ray diagrams and a discussion of linear propagation, formation of real and virtual images, and scattering of a real image by a screen.

(c) Which arrangement of pinhole camera, lens, and screen produced the sharpest image?

Evaluation

(d) Evaluate your predictions.

(e) Describe the sources of error in the investigation and evaluate their effect on the results. Suggest one or two improvements to the experimental design.

SUMMARY Images Formed by Lenses

- An image can be described by four characteristics: magnification, attitude, location, and type.
- The magnification of an image is the ratio of the image height to the object height.
- Images are real or virtual; real images can be scattered by a screen and viewed from many angles.

Section 10.2 Questions

Understanding Concepts

1. When a converging lens is used as an ordinary magnifying glass, is the image it produces real or virtual? Include a sketch with your explanation.
2. Copy **Figure 14** into your notebook and draw the path of each ray after it has been refracted.

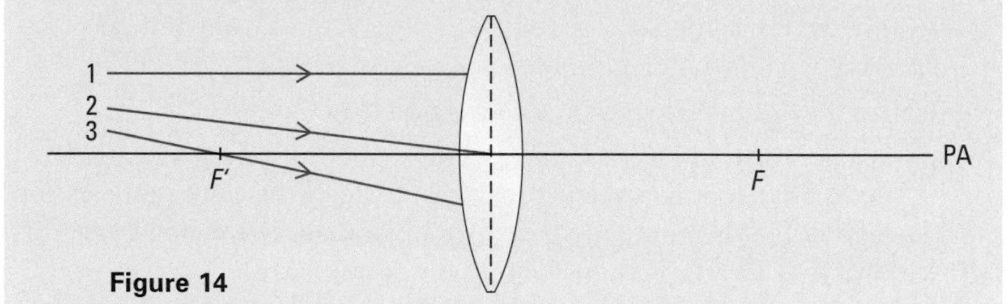

Figure 14

3. Copy **Figure 15** into your notebook and draw the path of each ray after it has been refracted.

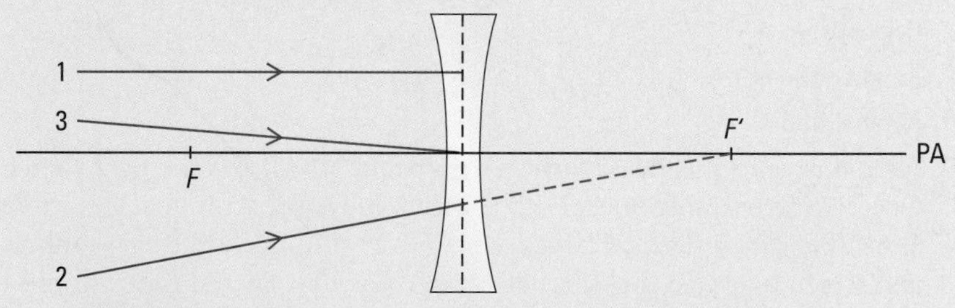

Figure 15

4. Images in a diverging lens are always virtual. When an object is placed between the focus and optical centre of a converging lens, the image is also virtual. What are the differences between these virtual images? Explain by studying the four characteristics of each image.
5. Use ray diagrams to explain why a half-covered lens still produces a complete image of an object.
6. For each situation described below, draw a ray diagram to locate the image of the object, then describe the four characteristics of each image. (In each case, try to use all three rules for drawing ray diagrams.)
   (a) $f = 3.0$ cm; $d_o = 7.5$ cm; $h_o = 2.0$ cm
   (b) $f = 4.4$ cm; $d_o = 2.2$ cm; $h_o = 1.2$ cm
   (c) $f = -2.8$ cm; $d_o = 5.0$ cm; $h_o = 2.2$ cm

Applying Inquiry Skills

7. (a) Draw a ray diagram of your own design of a variable-length pinhole camera made of common household materials, such as cardboard tubes or shoe boxes.
   (b) How would you use your design to determine the relationship between the magnification of the image and the distance between the pinhole and the image?

10.3 Mathematical Relationships for Thin Lenses

You have studied how a converging lens produces images for light sources placed at different positions along the principal axis. Now you will study the quantitative relationship between object and image positions.

Investigation 10.3.1

Predicting the Location of Images Produced by a Converging Lens

INQUIRY SKILLS

- ○ Questioning
- ○ Hypothesizing
- ● Predicting
- ○ Planning
- ● Conducting
- ● Recording
- ● Analyzing
- ● Evaluating
- ● Communicating

The purpose of this investigation is to determine the relationship among the focal length, the image distance, and the object distance of a converging lens.

Question

What is the relationship among the focal length, the image distance, and the object distance of a converging lens?

Materials

converging lens ($f = 10$ cm $\rightarrow$ 25 cm)
small light source (miniature light bulb)
translucent screen (white paper)
optical bench

Prediction

(a) Predict what will happen to the size and distance of an image as an object is moved closer to a converging lens.

Procedure

Part 1: Real Image

1. Create a table similar to **Table 1**.
2. In a dark part of the room, hold the lens so that light from a distant object passes through it and onto the screen, as in **Figure 1**. Move the screen back and forth until the image is clearly focused. Measure the focal length, *f*, of the lens, the distance between the lens and the screen.

Table 1

Observation	Object distance (d_o)	Image distance (d_i)	Characteristics			$\frac{1}{d_o}$	$\frac{1}{d_i}$	$\frac{1}{d_o}+\frac{1}{d_i}$	Magnification (*M*)
			size	attitude	type				
1	$2.5f=$								
2	$2.0f=$								
3	$1.5f=$								
4	$f=$								
5	$0.5f=$								

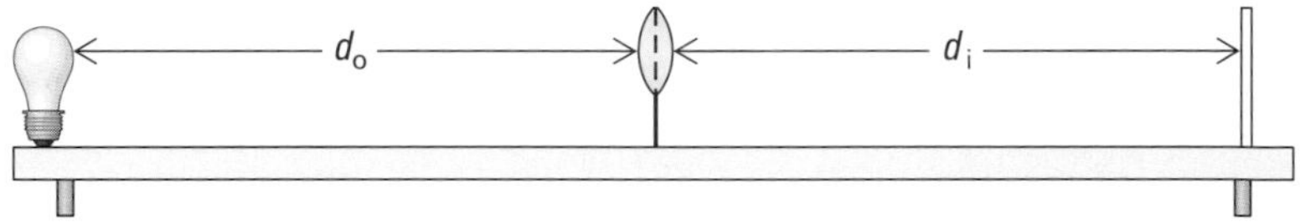

Figure 1
Setup for Investigation 10.3.1

3. Turn the lens around and repeat step 1. Calculate an average of the two measurements.
4. Using this average value, calculate the following object distances: $2.5f$, $2.0f$, $1.5f$, f, and $0.5f$. Record the information in your table.
5. Place the lens in the exact centre of the optical bench.
6. Using chalk or masking tape, mark on the optical bench the object distances calculated in step 4.
7. Place the object at $2.5f$. Move the screen back and forth until the image is focused clearly on the screen. Record the image distance and the characteristics of the image. Remember to look down along the principal axis to see if you can observe an image without using a screen. Be careful not to place your eye at a location where light is focused to a point.
8. Repeat step 7 for the other object distances that result in real images.
9. Complete columns $\frac{1}{d_o}$, $\frac{1}{d_i}$, and $\left(\frac{1}{d_o}+\frac{1}{d_i}\right)$ for the first three observations only.
10. Determine the value of the reciprocal of the focal length (f).
11. Calculate the magnification (M) for the first three observations and enter the values in the table.

Part 2: Virtual Image

12. Place the object at $0.5f$. Hold a pencil above the image that you see in the lens (you do not want to see the image of the pencil through the lens), then move the pencil until there is no relative motion between the pencil and the image when you move your head from side to side.
13. Measure the distance from the pencil to the lens and record the image distance d_i.

Analysis

(b) Why does the method in step 2 work for finding the focal length of a converging lens?
(c) How do the two focal lengths for both sides of the lens compare?
(d) As the object moves closer to the lens, what regular changes occur in the size of the image? the distance of the image? the attitude of the image?

(e) At what object distance was it difficult, if not impossible, to locate a clearly focused image?
(f) Where would you place an object in relation to the principal focus to form a real image? a virtual image?
(g) How does the value of $\frac{1}{f}$ relate to the value of $\left(\frac{1}{d_o} + \frac{1}{d_i}\right)$ for the cases involving real images?
(h) How does the value of $\frac{1}{f}$ relate to the value of $\left(\frac{1}{d_o} + \frac{1}{d_i}\right)$ for the case involving a virtual image?
(i) Answer the Question.

Evaluation

(j) Evaluate your predictions.
(k) Describe the sources of error in the investigation and evaluate their effect on the results. Suggest one or two improvements to the experimental design.

The Thin Lens Equation

To gain further insight into the mathematical relationships of lenses, we can derive a quantitative relationship for thin lenses that connects the image distance, the object distance, and the focal length. We do this by studying the geometry of two of the special rays that are used in constructing ray diagrams.

In **Figure 2**, the triangles AOF and EDF are similar.

Therefore, $\frac{\text{ED}}{\text{AO}} = \frac{\text{DF}}{\text{OF}}$

or $$\frac{h_i}{h_o} = \frac{d_i - f}{f}$$

$$\frac{h_i}{h_o} = \frac{d_i}{f} - 1$$

Also, triangles BCO and EDO are similar.

Therefore, $\frac{\text{DE}}{\text{BC}} = \frac{\text{DO}}{\text{CO}}$

or $$\frac{h_i}{h_o} = \frac{d_i}{d_o}$$

Therefore, $$\frac{d_i}{d_o} = \frac{d_i}{f} - 1$$

$$\frac{d_i}{d_o} + 1 = \frac{d_i}{f}$$

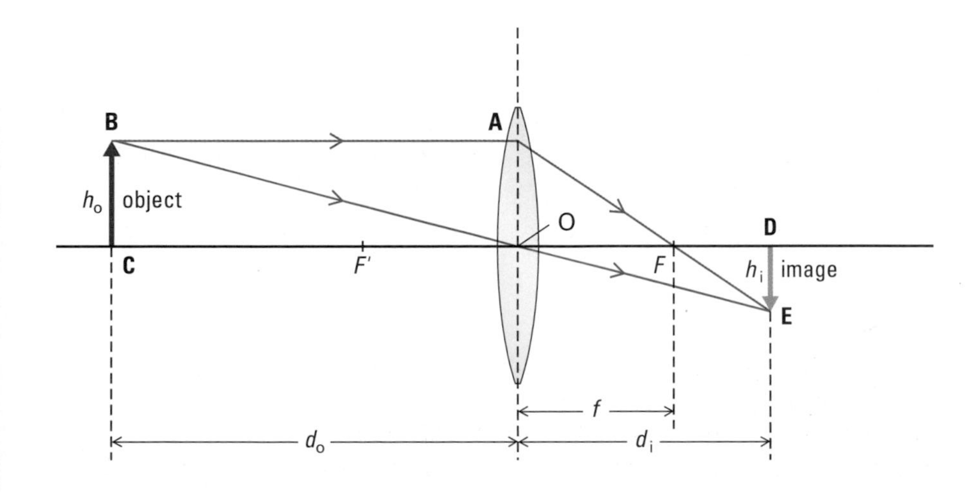

Figure 2
Using similar triangles to derive the thin lens equation

thin lens equation: the quantitative relationship for thin lenses; $\frac{1}{d_o} + \frac{1}{d_i} = \frac{1}{f}$

We now divide both sides by d_i to obtain the **thin lens equation:**

Thin Lens Equation

$$\frac{1}{d_o} + \frac{1}{d_i} = \frac{1}{f}$$

You saw from triangles BCO and EDO that

$$\frac{DE}{BC} = \frac{DO}{CO} \quad \text{or} \quad \frac{h_i}{h_o} = \frac{d_i}{d_o}$$

The first ratio was defined as the magnification. Using the sign convention for a real image formed by a converging lens, h_o is positive and h_i is negative, whereas d_o and d_i are both positive. For this reason, a negative sign must be added to the second ratio so that it agrees with the sign convention. This gives us two ways to calculate magnification:

$$M = \frac{h_i}{h_o} \quad \text{and} \quad M = -\frac{d_i}{d_o}$$

DID YOU KNOW ?

Calculator Exercises

Your calculator has a $\frac{1}{x}$ or x^{-1} function key that can be very useful for solving thin lens equation problems. To ensure that you know how to use this key, try keying in the following exercises:

1. To find the decimal equivalent of $\frac{1}{5}$, enter $5\frac{1}{x}$.
2. To find the sum $\frac{1}{2} + \frac{1}{4}$, enter $2\frac{1}{x} + 4\frac{1}{x} =$.
3. To solve for f, given $d_o = 20$ cm and $d_i = 5$ cm, enter $20\frac{1}{x} + 5\frac{1}{x} = \frac{1}{x}$.

Never rely on one method for computing. Try different methods to check your work. In how many different ways can you obtain the following answers to the three exercises above? Answers: 1. 0.2, 2. 0.75, 3. 4 cm

Sample Problem

(a) At what distance must a postage stamp be placed behind a magnifying glass with a focal length of 10.0 cm if a virtual image is to be formed 25.0 cm behind the lens?

(b) What are the magnification and attitude of the image?

Solution

(a) $f = 10.0$ cm (positive, since a magnifying glass is a converging lens)
$d_i = -25.0$ cm (negative, since it is a virtual image, on the object side of the lens)

The diagram for this problem would be similar to **Figure 8(e)** in section 10.2.

$$\frac{1}{d_o} + \frac{1}{d_i} = \frac{1}{f}$$

$$\frac{1}{d_o} = \frac{1}{f} - \frac{1}{d_i}$$

$$d_o = \left(\frac{1}{f} - \frac{1}{d_i}\right)^{-1}$$

$$= \left(\frac{1}{10.0 \text{ cm}} - \frac{1}{-25.0 \text{ cm}}\right)^{-1}$$

$$d_o = 7.1 \text{ cm}$$

The postage stamp must be placed 7.1 cm behind the magnifying glass.

(b)

$$M = -\frac{d_i}{d_o}$$

$$= -\frac{-25.0 \text{ cm}}{7.1 \text{ cm}}$$

$$M = 3.5$$

The magnification is 3.5. The image is upright, as indicated by a positive value for the magnification.

Practice

Understanding Concepts

1. Positive and negative signs are important when applying the thin lens and magnification equations. For each of the variables listed, state one situation in which the value is positive, and one in which it is negative. (For example, f is positive for a converging lens.) The variables are f, d_o, d_i, h_o, h_i, and M.
2. Go back to the ray diagrams you drew in Practice questions 3 and 5 in section 10.2. Use the appropriate equations to determine the image location and magnification in each diagram, then compare the values with the values you found using the diagrams. If the answers don't agree fairly closely, check your diagrams and your use of positive and negative signs in the equations.

SUMMARY The Thin Lens Equation

- The quantitative relationship for thin lenses is represented by the thin lens equation, $\frac{1}{d_o}+\frac{1}{d_i}=\frac{1}{f}$.

Section 10.3 Questions

Understanding Concepts

1. A ruler 30.0 cm high is placed 80.0 cm in front of a converging lens of focal length 25.0 cm.
   (a) Using a scale ray diagram, locate the ruler's image and determine its height.
   (b) Using the lens and magnification equations, determine the image position and its height. Compare your calculations with your ray diagram.
2. A lamp 20.0 cm high is placed 60.0 cm in front of a diverging lens of focal length −20.0 cm.
   (a) Using a scale ray diagram, locate the image and determine its height.
   (b) Using the appropriate equations, calculate the image position and the height of the image. Compare your calculations with your ray diagram.
3. The focal length of a slide projector's converging lens is 10.0 cm.
   (a) If a 35-mm slide is positioned 10.2 cm from the lens, how far away must the screen be placed to create a clear image?
   (b) If the height of a dog on the slide film is 12.5 mm, how tall will the dog's image on the screen be?
4. A lens with a focal length of 10.0 cm produces an inverted image half the size of a 4.0-cm object. How far apart are the object and image?

Applying Inquiry Skills

5. Do you think Investigation 10.3.1 could be done using a diverging lens instead of the converging lens suggested? If not, why not? If so, what method would you use to view the images as the object distances are changed?

10.4 The Human Eye and Vision

Anatomy of the Eye

Your eyes are amazing optical instruments (**Figure 1**). In fact, as you read this sentence, you are probably oblivious to the thousands of pieces of visual information that your eyes are gathering each second.

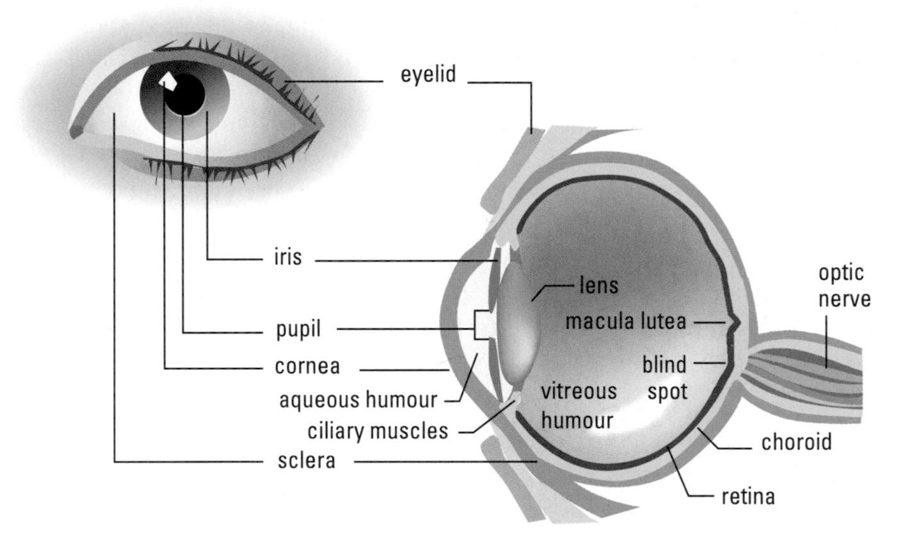

Figure 1
Along the back wall of the eye is a curved layer of light-sensitive cells called the retina. These cells react to light and send electric signals along the optic nerve to the brain, which, in turn, translates the signals into an image. The macula lutea (meaning "yellow spot") is a central part of the retina responsible for sensing fine detail and for looking straight ahead. At the location in each eye where the retinal nerves join the optic nerve, a blind spot occurs.

When you look at an object, three things must happen:

- The image must be reduced in size to fit onto the retina.
- The light from the object must focus at the surface of the retina.
- The image must be curved to match the curve of the retina.

This is done by the eye's lens between the retina and the pupil (the aperture, or opening, that allows light into the back of the eye). In front of the lens is a doughnut-shaped ring called the iris diaphragm, or the iris. The size of the pupil is controlled by the iris, which governs the amount of light entering the eye.

The lens of the eye is flexible so its shape, and hence its focal length, can be controlled by the ciliary muscles. This process of changing the focal length of the eye is called accommodation. In this way, the eye can keep objects in focus.

The eyeball itself has a tough white wall called the sclera. Its front portion is transparent and forms the cornea. The shape of the eye is maintained by the pressure of colourless, transparent fluids in the eye. The liquid between the cornea and the lens is a watery substance, the aqueous humour. The remainder of the eye is filled with a clear, jelly-like substance, the vitreous humour.

Figure 2
The additive method of colour production. When light beams of the primary colours (red, green, and blue) are projected onto a white screen, their mixtures produce the colours shown. By varying the intensities of the beams, most colours or hues can be produced. In the retina, red cone cells respond primarily to red light, green cone cells to green light, and blue cone cells to blue light.

Colour Vision

There are two types of light-sensitive cells in the retina: rod cells and cone cells. There are about 120 million rod cells that act as low-light sensors at night. Approximately 7 million cone cells work in greater light intensities and are able to detect colour. There are three types of cone cells, each sensitive to one of the three primary additive colours: red, green, and blue (**Figure 2**).

The cone cells contain light-sensitive pigment molecules called rhodopsin. When these pigments absorb light of sufficient energy, they break up and release electrical energy, sending messages to the neurons of the optic nerve. The pigment is then reassembled and can be used over and over again.

Stereoscopic Vision

A great deal of the information that you receive from your surroundings is due to the fact that you have two eyes. The distance between the pupils of your eyes gives you what is called binocular, or stereoscopic, vision. This short distance enables you to perceive the world in three dimensions (3-D); it gives you depth perception.

To reproduce a 3-D object as a 3-D image, one image is sent to the left eye and the other to the right eye. To do this using a camera, the object must be imaged by two separate cameras separated in the same way as your eyes. This was accomplished in the early 1800s using red- and blue-coloured filters that produced red and blue images called anaglyphs.

Colour filter glasses (**Figure 3**) were used to watch 3-D movies and some early computer games. The glasses are inexpensive and any standard colour monitor can display 3-D images in this way. However, the perception of depth comes at a price; practically all colour is lost, so most of the images seen this way appear grey.

A head-mounted display (HMD) is a helmet consisting of two separate display units (**Figure 4**) that send a different image to each eye. The 3-D images produced by HMDs are so realistic, they are known as virtual reality systems. Virtual reality technology is widely used in training programs, medical imaging, geographic mapping, and the entertainment industry.

Figure 3
Two images, one red and one blue, appear very close together without the 3-D glasses. When wearing the colour filter glasses, the red filter blocks out the blue image, and the blue filter blocks out the red image. In this way, your brain receives two different images and perceives depth even though the image is flat.

Practice

Understanding Concepts

1. Summarize the parts of the eye and their functions in tabular form using these headings:

Eye part	Function

2. Explain why the pupil of a human eye appears to be black.
3. Based on your own experience as you go from a bright room to a totally dark room or from a totally dark room to a very bright room, is the rate at which your iris reduces pupil diameter in bright light the same as it increases pupil diameter in dim light? Explain your answer.

Making Connections

4. Why would driving a car, riding a bicycle, or even walking up and down stairs be more difficult if you had one eye covered with a patch? In your answer, show that you understand depth perception.

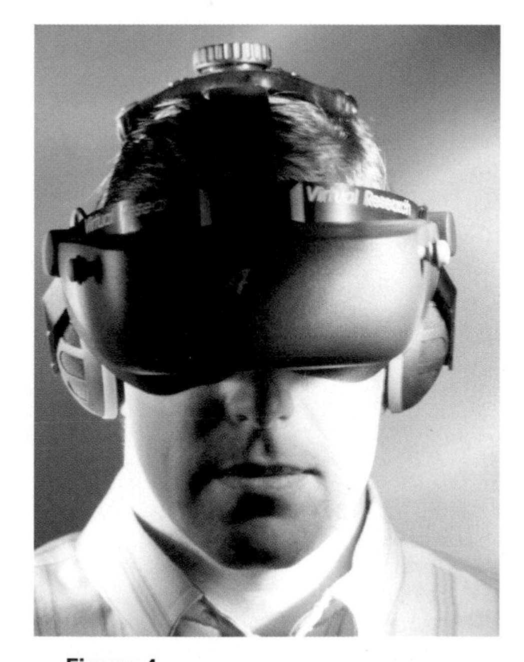

Figure 4
The optics of an HMD device makes it possible to project very realistic 3-D images to your brain. In addition, many HMD devices have a head motion feedback system. As you move your head, the image presented to your eyes changes, so that you feel you are inside a 3-D world.

SUMMARY The Human Eye and Vision

- The eye is a complex optical device that produces images on light-sensitive cells.
- Colour vision depends on red-, blue-, and green-sensitive pigments.
- Seeing with two eyes is called stereoscopic vision; this is what gives us depth perception.

Section 10.4 Questions

Understanding Concepts

1. Where does most of the refraction of light occur in the human eye?
2. Explain how you perceive depth.
3. (a) Use two ray diagrams to show that when the focal length of the lens in an eye remains constant, the image of a nearby object is farther from the lens than the image of a more distant object.
   (b) Knowing that the image in the eye must be located on the retina to be in focus, what must happen to the focal length of the eye's lens in order to view a nearby object? How is this accommodation accomplished?
   (c) When your eyes feel tired from looking at nearby objects, it will help to look at distant objects for a few minutes. Explain why, referring to your answers in (a) and (b) above.

Applying Inquiry Skills

4. The caption for **Figure 1** indicates that the eye has a "blind spot." To observe this phenomenon, draw two dark circles about 5 mm in diameter and 10 cm apart on a piece of paper. Label the left dot L and the right one R. Holding the paper at arm's length, cover your left eye and stare at dot L with your right eye. Move the paper back and forth slowly until dot R disappears. This shows the plane of the location of your blind spot. Describe how you would perform an experiment to determine the angle between this plane and your forward line of sight.

Making Connections

5. The random dots in **Figure 5** can actually be seen as 3-D images by focusing on a point behind the picture. These pictures are called *random dot stereograms*. The principles were invented in 1960 by Hungarian-born American physicist Dr. Bela Julesz at Bell Labs. Explain how these amazing pictures work.

Figure 5
A random dot stereogram can be seen as a 3-D image if viewed the right way.

10.5 Vision Problems and Their Treatments

Vision problems have many causes, including defective lenses, eye injury, disease, occupational stresses, heredity, and aging. Devices such as magnifiers, telescopes, large-print materials, special light fixtures, and talking calculators and computers can now help those with untreatable low-vision problems. Eyeglasses, or spectacles, were used as early as the late 13th century to correct vision problems. Nowadays we also have hard, soft, and disposable contact lenses.

You may also have heard about laser eye surgery to treat vision problems. The fundamental principle of laser surgery is to reshape the eye's cornea to enhance vision. Two procedures for nearsightedness are common today, both involving the use of an ultraviolet laser. Photorefractive keratotemy (PRK) reshapes the surface of the cornea. Laser in situ keratomileusis (LASIK) involves cutting away the surface of the cornea so that the layers underneath can be reshaped. Other procedures involve slipping plastic implants and lenses into or behind the cornea to change its shape.

By changing the shape of the lens, a healthy eye is able to keep close objects in focus. Light rays from distant objects enter the eye parallel to one another; those from close objects diverge. The lens becomes flatter to accomodate for distant objects and rounder for closer objects (**Figure 1**).

DID YOU KNOW?

Power of a Lens

The unit used by opticians to describe and prescribe lenses is the power (P) in diopters (d). The power of a lens is simply the inverse of the focal length of a lens in metres.

$$P = \frac{1}{f} \quad \text{or} \quad f = \frac{1}{P}$$

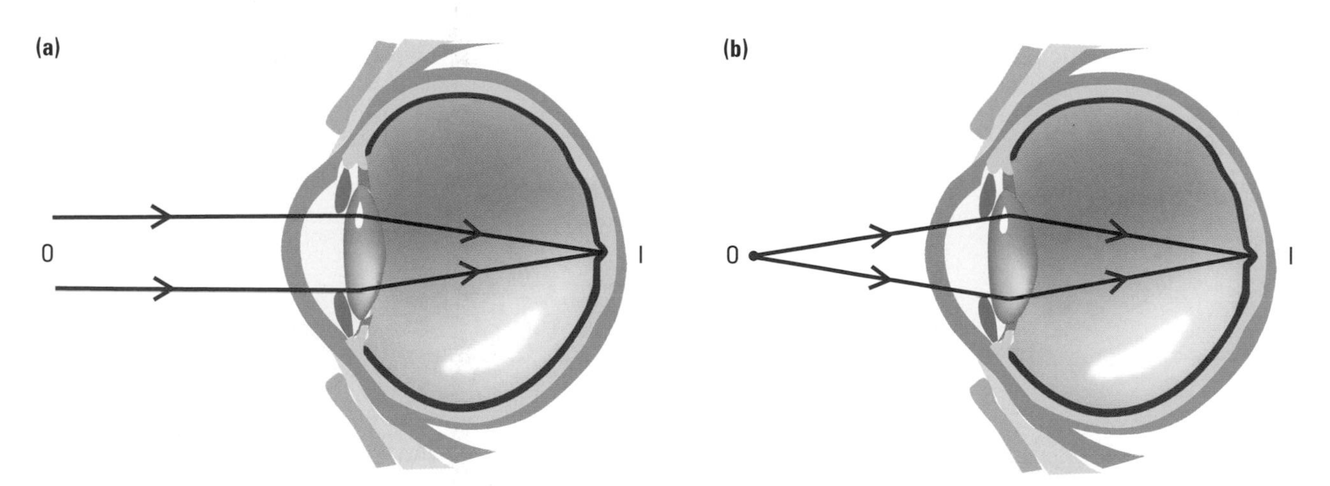

Figure 1
A healthy eye can focus light from objects onto the retina.
(a) Viewing a distant object
(b) Viewing a close object

The rest of this section will introduce you to some eye problems and the underlying physics that addresses them.

Nearsightedness (Myopia)

If the cornea-lens system converges light too strongly, or if the distance between the lens and retina is too great, the eye will not be able to see objects clearly (**Figure 2(a)**). This condition is called nearsightedness, or **myopia**. In a myopic eye, parallel light rays from distant objects are focused in front of the retina.

myopia: eye defect in which distant objects are not seen clearly; also called nearsightedness

One method of correcting myopia is eyeglasses with diverging lenses. These glasses diverge the light rays so that the eye lens can focus the image clearly on the retina (**Figure 2(b)**).

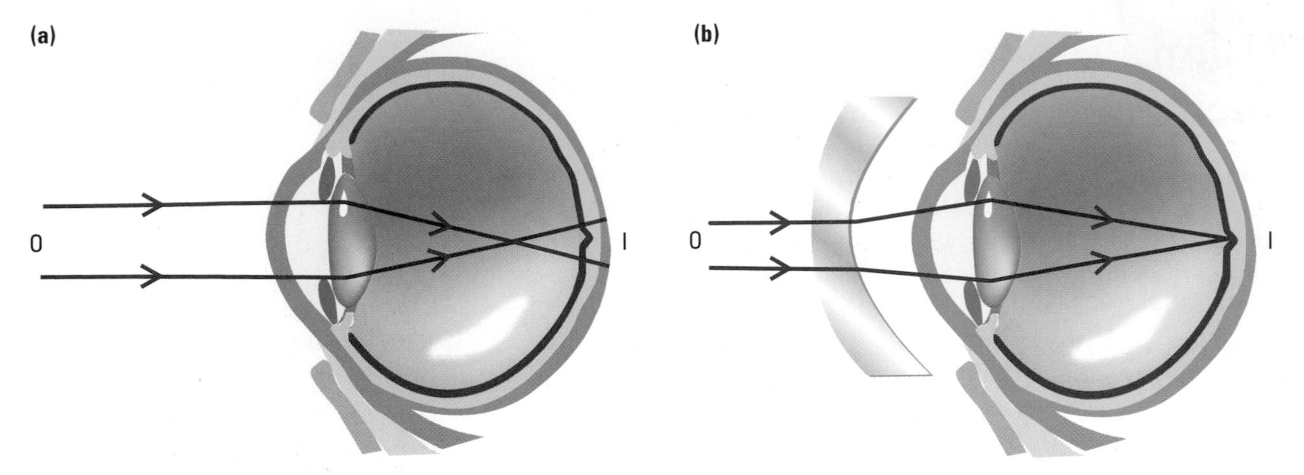

Figure 2
(a) A nearsighted (myopic) eye focuses the light rays from a distant object too strongly. The image is formed in front of the retina.
(b) Myopia can be corrected with diverging lenses.

hyperopia: eye defect in which nearby images are not seen clearly; also called farsightedness

Farsightedness (Hyperopia)

Farsightedness, or **hyperopia**, is an eye defect resulting in the inability to see nearby objects clearly. It usually occurs because the distance between the lens and the retina is too small, but it can occur if the cornea-lens combination is too weak to focus the image on the retina (**Figure 3(a)**).

This defect can be corrected by using a converging lens to bring the rays of light to a focus on the retina (**Figure 3(b)**).

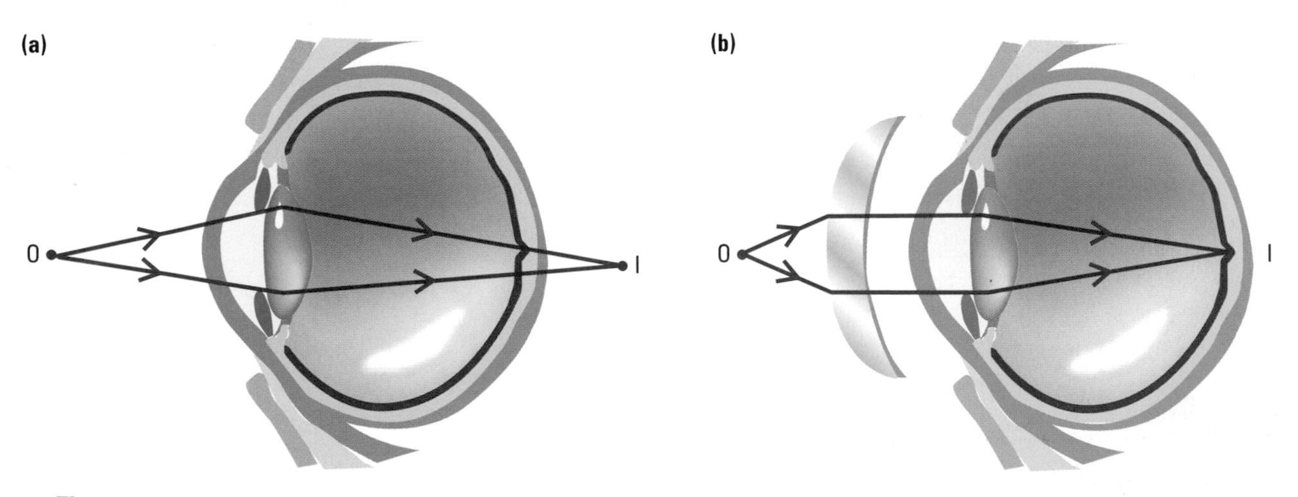

Figure 3
(a) A farsighted eye focuses the light rays of nearby objects behind the retina. The image is formed behind the retina.
(b) Hyperopia can be corrected with converging lenses.

presbyopia: eye defect in which the lens in the eye loses elasticity; a type of farsightedness

Loss in Accommodation (Presbyopia)

As you grow older, your eye lenses lose some of their elasticity, resulting in a loss in accommodation. This kind of farsightedness is known as **presbyopia**. This condition can be corrected by glasses with converging lenses. Distant vision is usually unaffected, so bifocals are used. Bifocals have converging lenses in the lower portion of each frame, convenient for reading and other close work for which the eyes are lowered.

Astigmatism

Astigmatism occurs when either the cornea or the lens of the eye is not perfectly spherical. As a result, the eye has different focal points in different planes. For example, the image may be clearly focused on the retina in the horizontal plane, but in front of the retina in the vertical plane. Once diagnosed (**Figure 4**), astigmatism is corrected by wearing glasses with lenses having different radii of curvature in different planes. They are commonly cylindrical lenses.

astigmatism: eye defect in which the cornea or the lens of the eye is not perfectly spherical

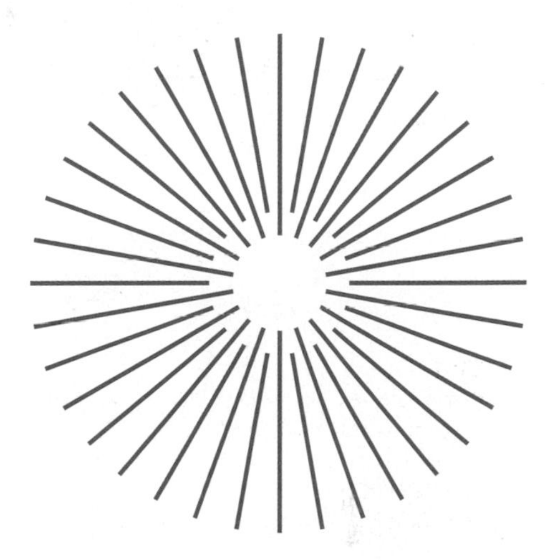

Figure 4
Astigmatism is tested for by looking with one eye at a pattern similar to this. Some of the lines will appear sharply focused, whereas others will appear blurred.

Glaucoma

Glaucoma involves damage to the optic nerve, leading to vision loss. It can be caused by a number of different eye diseases. In most cases, there is an elevated intraocular pressure. The increased pressure is not the disease itself, but it is the most important risk factor for developing glaucoma. The disease strikes without obvious symptoms, so the person with glaucoma is usually unaware of it until serious loss of vision has occurred.

In some cases, intraocular pressure can increase so suddenly that acute glaucoma develops. An attack like this can occur within a matter of hours. Symptoms include intense pain, red eyes, cloudy corneas, morning headaches, or pain around your eyes after watching TV. Treatment can involve one or all of eye drops, pills, and surgery to halt progress of the disease. There is currently no way to reverse damage caused by glaucoma.

glaucoma: eye defect in which the optic nerve is damaged

Cataracts

A **cataract** is an opaque, cloudy area that develops in the normally clear lens of the eye. If untreated, the cataract gradually blocks or distorts the light entering the eye, progressively reducing vision.

cataract: opaque, cloudy area that develops in a normally clear eye lens

Cataracts occur naturally in at least half of people over the age of 65. The lens of the eye needs a good supply of nutrients from the surrounding fluids. As the nutrient supply decreases—as commonly occurs in aging—the lens loses its transparency and flexibility, causing cataracts. Other causes include eye injuries or infections, diabetes, drugs, and excessive exposure to ultraviolet, X ray, and gamma radiation. Some types of cataracts seem to be hereditary.

In the initial stages little or no treatment is required, except eyeglasses. If cataracts progress to significantly impair sight, surgical removal of the lens is usually required. Cataract removal is the most common of all eye operations.

Practice

Understanding Concepts

1. State the type of lens used to correct each of the following:
   (a) myopia (b) astigmatism (c) hyperopia (d) presbyopia
2. A person with normal vision can see objects clearly from a "near point" of about 25 cm to a "far point" of an unlimited distance. For a person whose near point is 12 cm and far point is 72 cm, state
   (a) the type of defect
   (b) whether the person can see an object located at a distance from the eye of 10.0 cm, 50.0 cm, and 90.0 cm
   (c) how the defect can be corrected

Applying Inquiry Skills

3. Use **Figure 4** to determine whether you have astigmatism in either eye. If you have corrective lenses, check each eye with and without the use of the lenses. Draw and label a sketch of the astigmatism test to show which axes or planes appear to be sharply focused.

DECISION MAKING SKILLS

- Define the Issue
- Identify Alternatives
- Research
- Analyze the Issue
- Defend the Proposition
- Evaluate

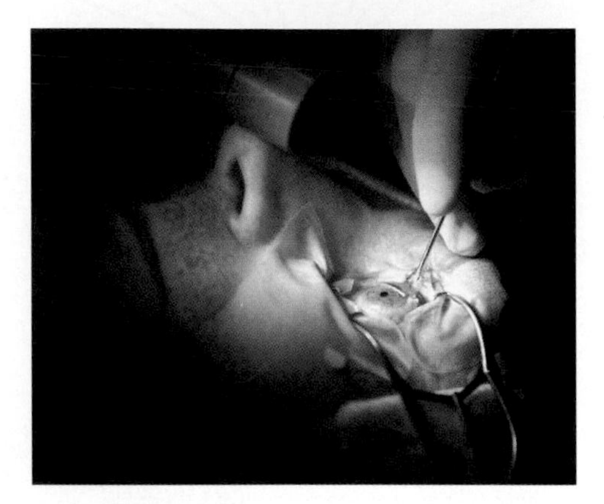

Figure 5

Explore an Issue

Treating Vision Problems

Lisa, Bruce, Ravi, and Karine are all experiencing vision problems. Laser eye surgery (**Figure 5**) is an option, but is it appropriate for everyone?

Understanding the Issue

Read their cases below and identify why each person may or may not want to have laser eye surgery.

- *Student:* Lisa is an eight-year-old girl who is suffering from headaches. An eye examination revealed that Lisa is nearsighted. She recently heard about laser eye surgery and wants to know if she can have an operation so that she doesn't have to wear glasses.
- *Journalist:* Bruce applied for a job as a writer at one of the major Canadian newspapers. He has an untreatable vision impairment that requires adaptations to the normal business routines at the newspaper. The editor is not sure if the newspaper can afford these adaptations in such a competitive market.
- *Virtual Reality Programmer:* Ravi specializes in developing software for the video arcade industry. His duties involve evaluating 3-D software on a regular basis, which involves wearing special virtual reality goggles and other 3-D enhancers. He currently uses disposable contact lenses for a worsening case of nearsightedness. He is afraid that glasses would really get in the way of his work.
- *Athlete:* Karine is an avid athlete. The occasional blur in her vision has become more distracting since she began training seriously. Now, she has been diagnosed with severe astigmatism. Wearing glasses during competition would greatly hinder her performance.

Take a Stand

If each of these four people were a close friend or relative, what advice would you give each of them concerning laser eye surgery?

Forming an Opinion

1. Formulate questions that will guide you in researching each case.
2. Using the Internet and other resources, research the latest developments in vision enhancement. Follow the links for Nelson Physics 11, 10.5.

GO TO www.science.nelson.com

3. List the options for each individual, outlining the advantages and disadvantages of each option.
4. Evaluate the different options, and state your opinion for the best course of action for each person. Discuss your opinions with your classmates.

SUMMARY Vision Problems and Their Treatments

- Most vision problems are due to refractive errors of the lens system.
- Nearsightedness is corrected by using a diverging lens. Farsightedness is corrected by using a converging lens.
- Loss in accommodation is corrected using a converging lens.
- Astigmatism is corrected by using cylindrical lenses.
- Glaucoma is irreversible; treatment can involve eye drops, pills, and surgery.
- Cataracts are treated by surgery.

Section 10.5 Questions

Understanding Concepts

1. Contrast nearsightedness and farsightedness and their corrections or treatments.
2. Use two ray diagrams as reference, one of myopia and the other of hyperopia, to help you explain why laser surgery is much more common for nearsightedness than for farsightedness.
3. Explain why cataracts are more easily treated than glaucoma.

Applying Inquiry Skills

4. (a) On a class list prepared by your teacher, indicate any vision problems that you have, including the power of lenses that you use, if any. Then using the class data, construct a pie chart of the vision categories according to groupings that are evident from the data.
   (b) Do you think your data are a good representation of the distribution of eye problems in the general population? Why or why not?

Making Connections

5. "Eat your carrots; they're good for your eyes!" This expression is often used to convince children to eat carrots. Research to find the connections between diet and vision.

10.6 Lens Aberrations and Limitations, and Their Solutions

chromatic aberration: coloured fringes around objects viewed through lenses

There is no such thing as a perfect lens. All lenses have imperfections called aberrations. There are many techniques to reduce aberrations, but they can never be eliminated. The fewer the aberrations in a lens, the more costly the lens is to produce.

Chromatic Aberration

As you have learned, when white light passes through a lens, the lens disperses it into its components, forming the colours of the spectrum. This creates coloured fringes around objects viewed through the lens, which can be annoying in optical instruments such as cameras and telescopes. The defect, called **chromatic aberration**, is usually corrected by joining converging and diverging lenses made of glass with differing optical densities (**Figure 1**).

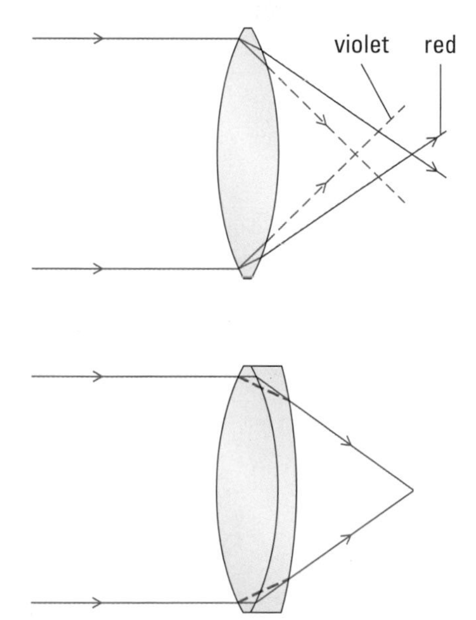

Figure 1
The dispersion of one lens is corrected by that of the other. Cameras of good quality usually use two or more components to correct for the effects of dispersion. These combinations are called achromatic lenses.

Spherical Aberration

A spherical lens does not focus parallel rays to a single point, but along a line on the principal axis. Light passing through the edge of the lens is focused closer to the lens than light entering the lens near the centre. Specially shaped lenses can reduce spherical aberration.

Coma

Off-axis objects produce images at the focal plane. However, off-axis points do not quite converge at the focal plane. Images of off-axis points may have blurry tails similar to those seen on comets, hence the name of this aberration.

Distortion

You may be familiar with computer monitor settings that allow characteristics like brightness and contrast to be changed. Two other common characteristics are called *pincushion* and *barrel*. These types of characteristics are categorized as distortion. The image of a square may have sides that curve in (pincushion) or curve out (barrel). This happens because magnification is not constant as one moves away from the principal axis.

Cost of Lenses

High-quality optical components account for a large part of the cost of most optical devices. Pressing hot glass, grinding, and polishing processes are time-consuming and expensive. In the last few decades, plastic moulding has made available inexpensive, high-quality plastic lenses and prisms that can be integrated into other plastic components far more easily than glass.

SUMMARY Lens Aberrations and Limitations, and Their Solutions

- All lenses have aberrations because there is no such thing as a perfect lens.
- Examples of aberration are chromatic and spherical aberration, coma, and distortion.

Section 10.6 Questions

Understanding Concepts

1. State the type of lens aberration that
   (a) causes a slightly blurred image along the principal axis
   (b) results in colour fringes
   (c) causes a blurry image relatively far from the principal axis
2. You can observe chromatic aberration in the classroom by looking at the edges of the light projected onto the screen by an overhead projector, especially when the projector is not properly focused.
   (a) What factors affect the chromatic aberration of the lens in an overhead projector?
   (b) Describe the chromatic aberration observed in theatres where bright white spotlights are aimed at light-coloured curtains or other objects.
   (c) Discuss other situations in which you have observed chromatic aberration.

Applying Inquiry Skills

3. Draw three squares with about 1 cm of space between each square. Use different colours for each square. Look at the squares through a converging lens and a diverging lens. Try various positions and angles of the lenses. Describe how many different types of aberration you observe.

Making Connections

4. Investigate how Fresnel lenses work. List other applications of Fresnel lenses and explain the advantage of the Fresnel lens in these applications.

Chapter 10 Summary

Key Expectations

Throughout this chapter, you have had opportunities to

- define and describe concepts and units related to light such as focal point, real images, and virtual images; (10.1, 10.2)
- describe and explain, with the aid of light-ray diagrams, the characteristics and positions of the images formed by lenses; (10.2)
- describe the effects of converging and diverging lenses on light, and explain why each type of lens is used in specific optical devices; (10.2)
- analyze, in quantitative terms, the characteristics and positions of images formed by lenses; (10.2)
- predict, using ray diagrams and algebraic equations, the image position and characteristics of a converging lens, and verify the predictions through experimentation; (10.2, 10.3)
- carry out experiments involving the transmission of light, compare theoretical predictions and empirical evidence, and account for discrepancies; (10.2, 10.3)
- evaluate the effectiveness of a technological device or procedure related to human perception of light; (10.4, 10.5)
- analyze, describe, and explain optical effects that are produced by technological devices; (10.6)

Key Terms

lens
converging lens
focal point
diverging lens
optical centre
optical axis
principal axis
principal focus
focal length
focal plane
image point
virtual image point
thin lens equation
myopia
hyperopia
presbyopia
astigmatism
glaucoma
cataract
chromatic aberration

Make a Summary

Draw a diagram of the eye. Label all the parts of the eye. Use this as the starting point of a concept map. Try to include small sketches all around the eye to highlight concepts developed in this chapter.

Reflect on your Learning

Revisit your answers to the Reflect on Your Learning questions at the beginning of this chapter.

- How has your thinking changed?
- What new questions do you have?

Chapter 10 Review

Understanding Concepts

1. A converging lens is sometimes called a "burning glass." Does the sunlight burn because the lens magnifies? Where, relative to the lens, must the object to be burned be placed?
2. A lens can be formed by a bubble of air in water. Is such a lens converging or diverging? Use a diagram in your answer.
3. Why do swimmers with normal eyesight see distant objects as blurry when swimming underwater? How do swimming goggles or a face mask correct this problem?
4. A converging lens of diamond and a lens of crown glass have the same shape. Which lens will have the greater focal length? Explain your answer.
5. To direct parallel light rays through a slide, the light source of a projector is located at the centre of curvature of the converging mirror and at the principal focus of the converging lens in the condenser, as shown in **Figure 1**. Why?

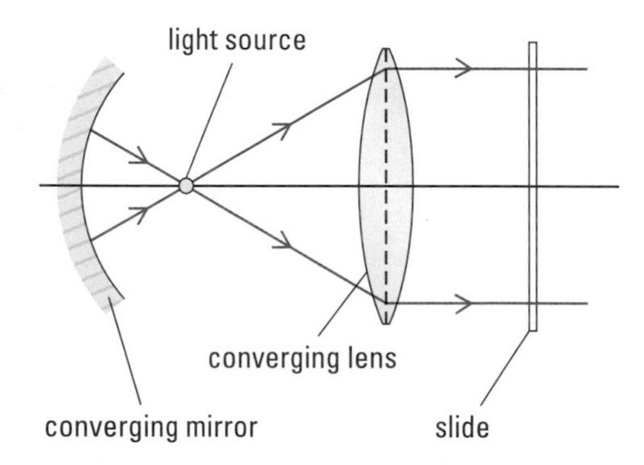

Figure 1

6. In your notebook, write the names of each of the numbered parts in **Figure 2**.

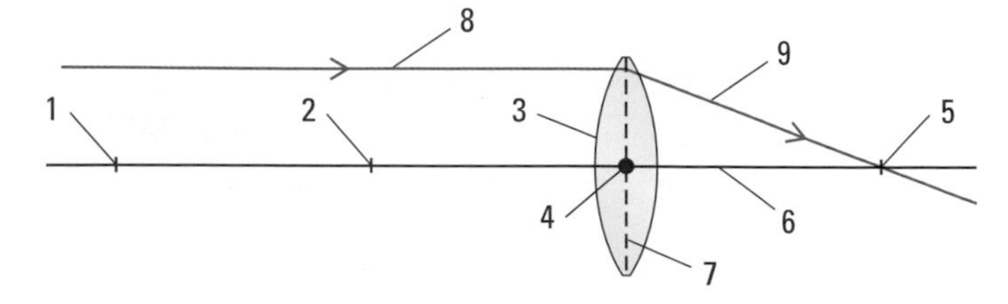

Figure 2

7. Which shape of lens in **Figure 3** has the greater focal length? Explain your answer. (The lenses are made of the same material.)

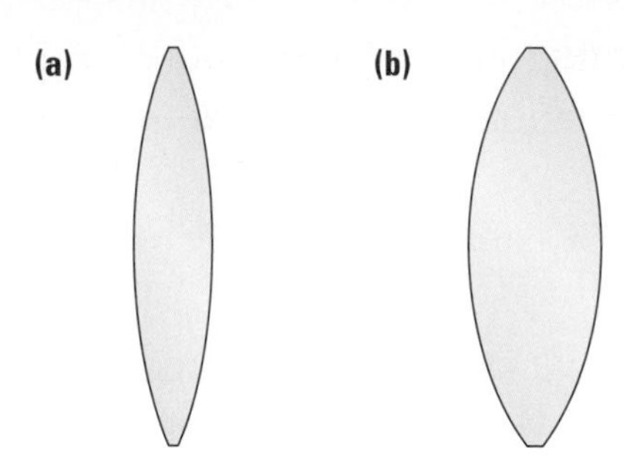

Figure 3

8. The lenses in **Figure 4** are made of the same material. Which lens has the greater focal length? Explain your answer.

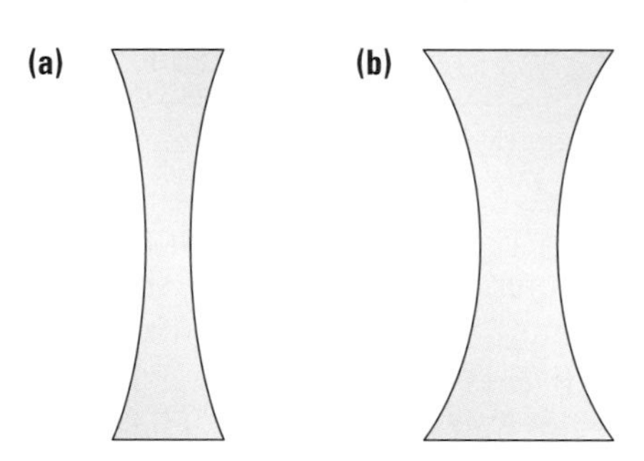

Figure 4

9. In **Figure 5** there are four lenses. Given four different focal lengths (−20 cm, −5 cm, 5 cm, and 20 cm), match each focal length with the correct lens. Explain your choices.

Lens 1	Lens 2	Lens 3	Lens 4

Figure 5

10. A converging lens has a focal length of 20.0 cm. A 5.0-cm object is placed
    (a) 10.0 cm from the lens
    (b) 20.0 cm from the lens
    (c) 30.0 cm from the lens

    Use scale drawings to locate the images in each situation, and state whether the image is real or virtual, upright or inverted, and smaller, larger, or the same size.

11. Each of the two diagrams in **Figure 6** shows an object and an image formed by a converging lens. Copy the diagrams into your notebook, and by means of ray diagrams locate the principal focus of each lens.

(a)

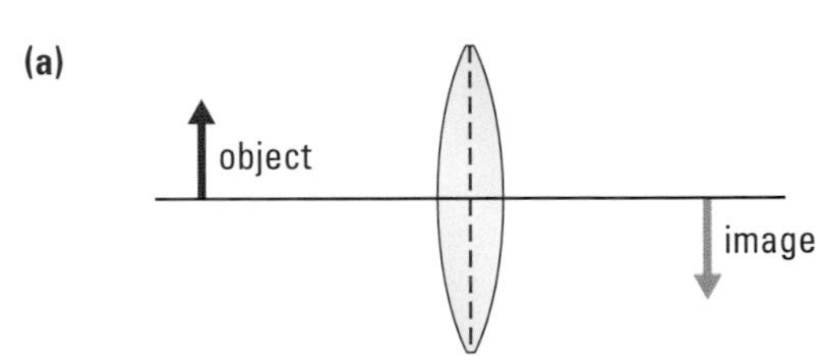

(b)

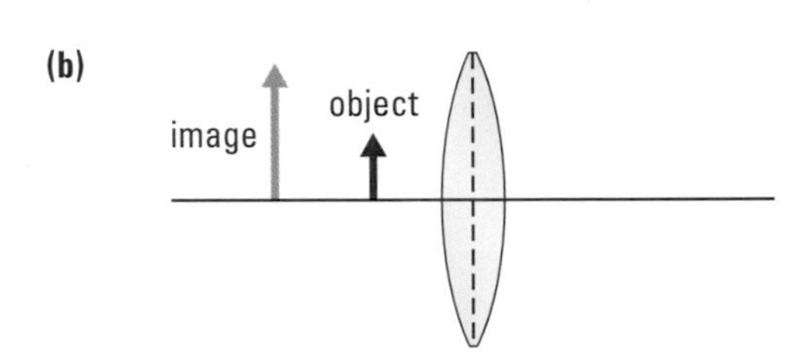

Figure 6

12. Using a scale ray diagram, locate the position of the image of a candle 10.0 cm high, placed 20.0 cm in front of a converging lens of focal length 25.0 cm. Check your answer using the lens equation.
13. Using a scale ray diagram, locate the image of an object 5.0 cm high that is 15 cm in front of a diverging lens of focal length –25 cm. Check your answer using the lens equation.
14. A candle is placed 36 cm from a screen. Where between the candle and the screen should a converging lens with a focal length of 8.0 cm be placed to produce a sharp image on the screen?
15. An object 5.0 cm high is placed at the 20.0-cm mark on a metre stick optical bench. A converging lens with a focal length of 20.0 cm is mounted at the 50.0-cm mark. What is the size and position of the image?
16. A lens has a focal length of 120 cm and a magnification of 4.0. How far apart are the object and the image?

Applying Inquiry Skills

17. Design an experiment to compare the intensity of a beam of light focused by a Fresnel lens with that of a regular glass lens, especially for beams not parallel to the principal axis.
18. (a) Describe how you would find the focal length of a solid glass or acrylic tube.
    (b) Describe a way to use solid plastic tubing to help gather energy in a solar heating project.
19. An interesting feature of the human eye is its ability to retain an image for about $\frac{1}{25}$ s after the object viewed is removed. If a new image at a slightly different location replaces the old one after a short time interval (for example, $\frac{1}{30}$ s), the image appears to be in constant motion.
    (a) How is this principle applied to make cartoon movies?
    (b) Design and construct a device to illustrate the retention of images by the human eye. Test your device on other students.

Making Connections

20. When you get your eyes examined, many devices are used to help determine how well you see. Find out how two of these devices work, and prepare a short presentation.
21. List ten occupations for which you would need a good understanding of optical instruments that use mirrors and lenses.

Exploring

22. Using the Internet and other resources, research virtual reality goggles that are made of polarizing glasses and those that are made of LCD (liquid crystal display) shutter glasses. In each case, how are two separate images produced and reproduced in each of your eyes? Follow the links for Nelson Physics 11, Chapter 10 Review.

GO TO www.science.nelson.com

23. If you hold a sunglass lens in front of your right eye and watch a movie, you will experience a mild 3-D effect. This is called the Pulfrich effect. Using the Internet and other sources, research how this effect works and write a one-page summary. Follow the links for Nelson Physics 11, Chapter 10 Review.

GO TO www.science.nelson.com

24. Research the latest techniques used to improve the vision for either a detached retina or glaucoma. Write a brief report (200–300 words) of your findings.

Chapter

11

Optical Instruments

In this chapter, you will be able to

- describe how images are produced in film cameras
- analyze quantitatively the characteristics and positions of images formed by lens systems
- describe how images are produced and reproduced digitally
- describe the role of optical systems in entertainment
- describe and explain, with the aid of light-ray diagrams, the characteristics and positions of the images formed by microscopes and telescopes
- construct, test, and refine one of the following optical instruments: periscope, binoculars, telescope, or microscope

Simple telescopes and microscopes, both originally two-lens optical devices, have allowed us to access the world beyond our normal vision (**Figure 1**). When Galileo Galilei documented his observations of Jupiter and its moons, he challenged the belief that Earth was the centre of the universe. In fact, his discoveries heavily influenced the change in view from the Sun revolving around Earth to that of Earth revolving around the Sun, as Copernicus had proposed many years earlier. Will further discoveries in optics continue to change our view of the world in such a dramatic way? To answer this question, we need to understand how optical devices work.

An optical instrument is a device that produces an image of an object. A film camera, for instance, produces permanent images that resemble the real world that we see with our eyes. A microscope helps us to view images of tiny cells and other objects invisible to the unaided eye. Binoculars produce images of distant objects, making them appear much closer. In this chapter, you will explore how these and other instruments apply the principles studied in the previous two chapters to produce images.

Reflect on your Learning

1. Why is there a mirror in a single-lens reflex (SLR) camera but not in a disposable camera?
2. (a) What is the main difference between a reflecting telescope and a refracting telescope?
   (b) What is the main difference between a terrestrial telescope and an astronomical telescope?
3. (a) What is a charge-coupled device (CCD)?
   (b) Name two optical instruments that contain a CCD.

Throughout this chapter, note any changes in your ideas as you learn new concepts and develop your skills.

Figure 1
The technology of a modern optical telescope has changed dramatically since Galileo made his early drawing shown above, yet the optical principles behind early optical telescopes and modern optical telescopes remains the same.

Try This **Activity**

Magnification

By how much do simple magnifying lenses actually magnify?

Materials: newspapers and magazines, a few converging lenses of different focal lengths.

- Devise a way to measure the maximum magnification of images observed with the lenses. Explain your method.
- Use your method to measure the maximum magnification of a particular letter for each lens.
- Record your observations in a table that includes the focal length and maximum magnification.
  (a) What is the best magnification that you can achieve?
  (b) How does focal length affect magnification? Explain.

11.1 Lens Cameras and Photography

By the middle of the 16th century, lenses were being used to help produce brighter and sharper images in pinhole cameras. In the early 18th century, scientists discovered that light caused silver salts to turn dark, making possible the invention of photographic film that was sensitive to light. These discoveries led to the evolution of a **lens camera**, a lightproof box inside which a converging lens directs light onto a film to capture a real image.

lens camera: lightproof box inside which a converging lens directs light onto a film to capture a real image

During much of the 19th century, lens cameras used a single photographic plate that was both bulky and fragile. In the late 1880s, the American inventor and manufacturer George Eastman (1854–1932) introduced flexible roll film and the Kodak camera in which to use it. Several years later, Eastman developed a process for colour photography; soon thereafter, daylight-loading roll-film cameras came into use.

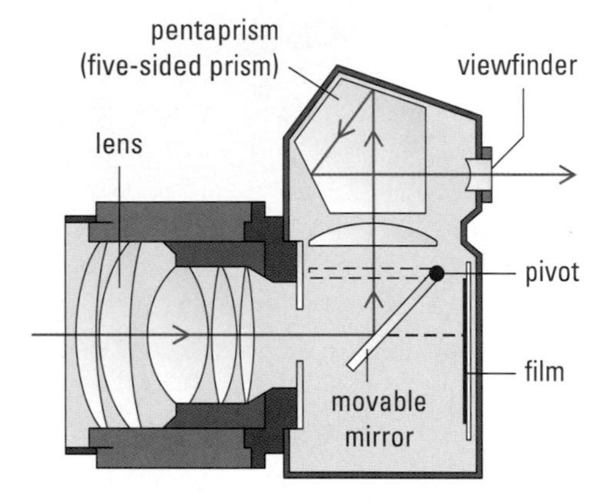

Figure 1
A typical SLR camera. Light enters the camera's lens system and strikes a plane mirror. The light is reflected up to the pentaprism (a five-sided prism), where total internal reflection guides the light through the viewfinder. When the photograph is snapped, the mirror lifts up, allowing light to strike the film.

Today, relatively inexpensive and easy-to-use photographic equipment is available in the form of disposable cameras, instant cameras, digital cameras, movie cameras, and, most commonly, roll-film cameras. Professional photographers use more sophisticated equipment in studios and for press and industrial photography. The movie and television industries use special equipment for high-speed, underwater, aerial, microscope, and infrared photography.

Modern lens cameras, with the exception of digital cameras, contain the following parts:

- a lightproof box, which supports the entire apparatus
- a converging lens, or system of lenses, which gathers light and focuses it onto a film
- a diaphragm, which controls the amount of light passing through the aperture when the picture is taken
- a shutter, which is the mechanical device that varies the aperture
- a film, which records the image on a light-sensitive mixture of gelatin and silver bromide that is later developed to give a permanent record of the scene photographed
- a viewfinder, which allows the photographer to see what he or she is photographing

The **single-lens reflex (SLR) camera** (Figure 1) is simple enough to understand yet complex enough to illustrate several principles of physics. The SLR camera is a 35-mm camera, so called because the film used is 35 mm wide, exposing a frame 36 mm × 24 mm. The advantage of this system is that the photographer is able to see an upright image that corresponds to the scene that will be exposed on the film, even though the image on the film is inverted.

single-lens reflex (SLR) camera: camera that uses a single lens and a movable mirror

Practice

Understanding Concepts

1. State the attitude and type of image on the film of a lens camera.

Photography

Using SLR cameras and certain other types of cameras, a photographer is able to control five important variables: film speed (by choosing a specific film), focus, focal length, exposure time, and aperture setting.

Film Speed

The film speed is a measure of the film's sensitivity to light. The film has an emulsion containing silver halide grains that react when struck by light. Thicker emulsions have larger silver halide grains so that more silver is affected by light at one time; that is, the film reacts "faster." An ISO number such as 100, 200, or 400 indicates the speed of the film. Film speeds are considered slow, medium, or fast (**Table 1**). Landscape photographers prefer slow film to produce pictures with rich colours. In contrast, to catch athletes in action, sports photographers use fast film.

Table 1 ISO Numbers

Slow	Medium	Fast
25	64	800
32	100	1600
50	200	3200
	400	

DID YOU KNOW ?

Halides

When Group VII elements such as chlorine, bromine, and iodine are found in compounds such as silver bromide, or more commonly sodium chloride, the Group VII elements gain an electron to become negative ions. A Group VII element in this state is called a halide.

Focus

The focus adjusts the distance from the lens system to the film. Adjusting the focus allows objects at various distances from the camera to be seen clearly ("in focus"). SLR cameras have a focus control that is operated by turning the lens, which is mounted on a spiral screw thread. Simpler cameras have a fixed focus in which everything beyond about 1.5 m is reasonably clear.

DID YOU KNOW ?

ISO

ISO is an acronym for the International Organization for Standardization. ISO develops worldwide production standards for industry. For example, credit cards have standard dimensions with a thickness of 0.76 mm so that machines that accept cards can be standardized around the world. Everything from nuts and bolts to film speed codes are universal because of the ISO.

Focal Length

The focal length of the lens system is controlled by changing lenses or by adjusting an attachable zoom lens. SLR camera lenses can easily be interchanged because they have bayonet (push and twist) fittings.

Exposure Time

Exposure time (also called the shutter speed) is the amount of time the shutter remains open when a photograph is taken. Exposure time is stated in seconds or fractions of a second, for instance, 1/60 s and 1/1000 s.

Aperture Setting

The aperture is the adjustable opening that limits the diameter (d) to which the shutter opens. It is controlled by the **f-stop number** on the camera, which is the ratio of the focal length (f) of the lens to the diameter of the aperture:

$$f\text{-stop} = \frac{f}{d}$$

For example, a 50.0-mm lens (i.e., focal length is 50.0 mm), where the aperture is set to open only at 25.0 mm in diameter, has an f-stop number written as $f/2$. The typical series of f-stop numbers is $f/1.4$, $f/2$, $f/2.8$, $f/4$, $f/5.6$, $f/8$, $f/11$, $f/16$, $f/22$, and $f/32$. Each change in f-stop indicates either a doubling or a halving of the amount of light. Thus, at $f/8$ there is approximately twice as much light as at $f/11$.

f-stop number: the ratio of the focal length of the lens to the diameter of the aperture;

$f\text{-stop} = \frac{f}{d}$

Figure 2 illustrates the effect of exposure time on a scene in which an object is moving. A short exposure time produces a sharp image, while a long exposure time produces a blurry image.

(a)

(b)

(c)

Figure 2
Controlling exposure time. As the exposure time increases, the sharpness of moving images decreases.
(a) 1/1000 s
(b) 1/60 s
(c) 1/15 s

Practice

Understanding Concepts

2. (a) Describe the difference between shutter speed and film speed.
   (b) In the same lighting conditions, how should a photographer change the shutter speed if a film of higher speed is used?
3. Calculate the aperture of a 50.0 mm lens when the *f*-stop number is set at (a) *f*/16 and (b) *f*/11.
4. (a) Show that when the aperture is doubled, the surface area allowing light to enter the lens is 4 times as great.
   (b) If an exposure time of 1/100 s works well at the first aperture setting, what exposure time is needed when the aperture is doubled?

Answers

3. (a) 3.1 mm
   (b) 4.5 mm
4. (b) 1/400 s

Activity 11.1.1

Dissection of a Disposable Camera

Even the simplest disposable camera consists of many systems that work together (**Figure 3**). In this activity, you will dissect a disposable camera and classify its parts as mechanical (gears, screws, levers, springs), optical (lenses, mirrors, prisms), electronic (batteries, resistors, capacitors, inductors, switches, transformers, diodes), or opto-electronic (electronic light sensors, LEDs), depending on the brand of camera used. You will likely find other parts made of metal or plastic. Classify them as best you can.

Figure 3
A typical disposable camera consists of over 40 parts.

Materials

disposable camera (preferably without a flash)
cup or bowl for small parts
safety goggles

Procedure

1. Check for the battery and remove it if it is still there.
2. While wearing goggles, remove as many pieces of the camera as you can without breaking anything. The plastic is very brittle, and some parts may break and present a danger to your eyes.
3. Remove all plastic lenses from the camera. Sketch the shape of each lens and briefly describe its location on the camera. Carefully place the lenses in a cup or bowl for later analysis.
4. Identify as many of the parts as you can from prior experience in mechanics, optics, and electronics.
5. Identify the converging lenses and the diverging lenses.
6. Determine the focal length of the main lens that focuses light onto the film.

Small parts may fly off the camera during the dissection. Wear safety goggles.

If your disposable camera has a flash, it contains a capacitor that stores a large electrical charge used for generating the flash. The large capacitor may still be charged. A shock from such a capacitor can be very dangerous. Ensure that your teacher discharges it by shorting out its leads before handling. After several minutes, measure the voltage across the leads of the capacitor. If the voltage is zero, the capacitor is discharged.

Analysis

(a) How is the camera able to focus on objects anywhere from 2 m to 100 m without an adjustable focal length?

(b) Which of the following variables can be controlled with a disposable camera? Explain each case.
 (i) exposure time
 (ii) focal length
 (iii) aperture
 (iv) speed of film

Synthesis

(c) Using the Internet, research flash circuitry. Formulate questions about the role or function of the electronic parts of your disposable camera and then find answers. Follow the links for Nelson Physics 11, 11.1.

GO TO www.science.nelson.com

Alternative Methods of Image Capture

CCD Cameras

In 1969, two American physicists working at Bell Labs, Willard Boyle and George Smith, used semiconductor technology to invent a light-sensitive device called a **charge coupled device** (CCD) as a way to store optical data. Since then, CCD technology has been incorporated into television cameras, telescopes, medical imaging systems, and hand-held digital cameras.

charge coupled device: (CCD) a light-sensitive device used to store optical data

While many optical devices produce images from lenses, mirrors, and light alone, others use light-sensitive materials, **semiconductors**, to convert light energy into electrical energy.

semiconductor: material used to transmit and amplify electronic signals

digital camera: camera that uses digital information instead of film to capture images

pixel: image-forming unit

A CCD array is at the heart of most **digital cameras.** Each CCD contains millions of cells of silicon (a semiconductor) arranged into rows and columns of squares—image-forming units—called **pixels.** Each pixel acts as a tiny electronic switch that either blocks the light or lets it through. The more pixels there are in the array, the greater the detail that can be captured.

Lenses in digital cameras focus light onto the CCD array (**Figure 4**). After exposure to light, each pixel records information about how much light has struck it, and the information is converted to digital form by a computer chip. This digital image information is then compressed and stored electronically to be retrieved later for printing, display on a monitor, or transfer over the Internet.

The cells on a CCD respond only to light—not to colour. In order to capture colour, the pixels are further divided into pixels sensitive to the three primary colours: red, green, and blue. Since each pixel can represent only one colour, the true colour is made by averaging the light intensity of the pixels around it.

This method of image capture has many advantages over photographic film. Unlike a traditional film-based camera, digital photography allows images to be captured in a format that is available instantly, with no need for a chemical development process. In a digital camera, images captured can be viewed before printing; in this way, only successful exposures are saved and printed. The CCD cells are fixed in place and can go on making photos for the lifetime of the camera. Another advantage is that the images are in digital form and can be mathematically manipulated by software for enhancement or special effects. They can also be stored on magnetic storage devices such as floppy disks or hard disks.

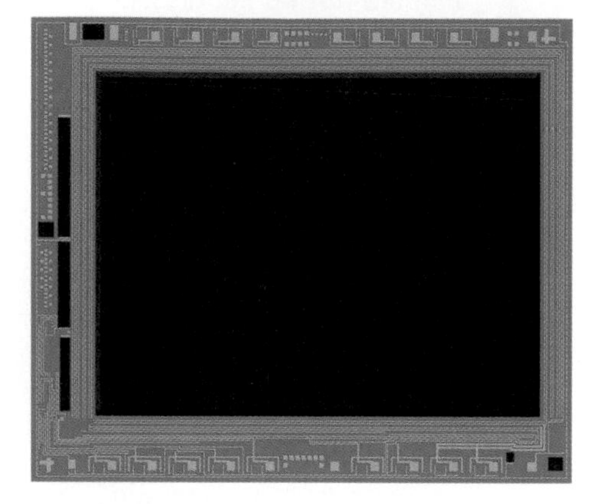

Figure 4
The location of each cell in the array has a unique address specified by the row and column of the pixel in which it is found. A typical rectangular CCD chip has an area less than a square centimetre and is capable of storing a megapixel of data. It is now possible to produce CCD arrays with over 4 megapixels of storage.

Night-Vision Goggles

Night-vision goggles, or infrared goggles, enhance vision in dim or dark conditions. Originally developed for military use, they are now also used for surveillance, search and rescue, and night-time hunting and birding.

Figure 5 shows a cross-section of a modern night-vision system. Incoming light is focused by an objective lens onto a fibre optic bundle that directs the light to a photocathode. The photocathodes are made of gallium arsenide, a semiconductor. Minute traces of visible light or infrared light will cause the gallium arsenide to emit electrons, which are then accelerated to a metal-coated glass disk containing millions of channels. Electrons enter the channels and knock out more electrons. These electrons then strike a screen similar to the one on your television or computer monitor. The green light produced by the screen is directed through another fibre optic bundle through an eyepiece.

DID YOU KNOW ?

Seeing Dark Matter

Physicists believe that an invisible form of matter, referred to as dark matter, makes up over 90% of the universe. In May 2000, astrophysicists at Bell Laboratories detected dark matter by analyzing what they believe is evidence of distortion of light from hundreds of thousands of galaxies. These distortions are so small that the researchers first had to design an incredibly sensitive CCD camera for imaging the galaxies with their telescope.

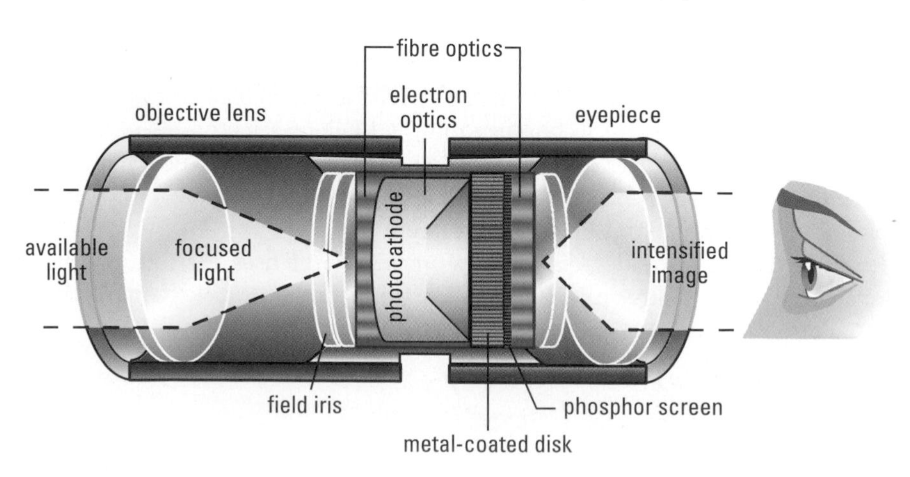

Figure 5
Cross-section of night-vision goggles. The amount of light leaving the eyepiece can be 30 000 times more intense than the light that entered.

Practice

Understanding Concepts

5. What is a semiconductor? Describe its use in digital cameras and night-vision goggles.
6. In night-vision goggles, living organisms, such as animals, are seen more easily than their non-living surroundings. Explain why.

Making Connections

7. What are the main advantages and disadvantages of using a CCD camera over a film camera?

SUMMARY Lens Cameras and Photography

- Five main factors affecting the recorded image are the film, focus, focal length of the lens system, exposure time, and aperture setting.
- The *f*-stop is directly proportional to the focal length and inversely proportional to the diameter of the aperture.
- Cameras may consist of an optical system and a mechanical system, and an electronic system or an opto-electronic system.

Section 11.1 Questions

Understanding Concepts

1. State the function of the following in an SLR camera:
   (a) shutter (b) viewfinder (c) silver halide (d) focus
2. You want to take two pictures of water spraying upward out of a fountain. In picture A you want the water to look like a fine, misty streak. In picture B you want to see details of the water droplets at the top of the spray. How will your choice of film speeds and/or shutter speeds compare for the two pictures? Explain.
3. Which setting allows more light to enter a lens, *f*/22 or *f*/2.8? Explain your answer.
4. Determine the *f*-stop number of a lens of focal length 85 mm when its aperture is set at 21 mm.
5. Assume that a distant object is in focus when viewed through a camera lens. You then aim the camera at a nearby object, which appears out of focus. Should the lens be moved closer to or farther from the film to obtain proper focus? Relate your answer to experimentation with lenses in the previous chapter. (*Hint:* Draw two light-ray diagrams using a lens of constant focal length. Use two distinctly different object distances, both of which must be greater than the focal length.)

Making Connections

6. Photographers can create a photograph with a blurry background by controlling the aperture of the camera lens. Such a photograph has a shallow *depth of field*, which is the range of object distance at which the image is clear.
   (a) Which *f*-stop number would create a more shallow depth of field, *f*/22 or *f*/2.8? Why? (*Hint:* The central part of a lens causes less distortion than the outer edges.)

(continued)

(b) Find examples of photographs with a shallow depth of field on a calendar, in a magazine, or in some other source. What are the advantages of this type of photography?

Reflecting

7. Think about a very disappointing photograph that you have seen or taken. Try to explain why the photograph turned out poorly and how you could have changed conditions (lighting, film, or focal length) to improve the photograph.
8. How would you rather receive a picture of a friend or family member, as a hard-copy photograph in the mail with a letter or as an attachment in an e-mail? Why?

Fi...
This op... ...on up to three times.

Figure 2
A student microscope

microscope: instrument that produces enlarged images of objects too small to be seen by the unaided eye

objective: in an optical instrument, the lens closest to the object

ocular: in an optical instrument, the lens the eye looks into

11.2 The Microscope

You have learned how lenses help record images and improve vision. Now you will learn how they produce enlarged images of objects too small to be seen by the unaided eye. A **microscope** uses lenses to perform this function.

In Chapter 10, you saw how a converging lens can be used as a magnifier. Typically, the image appears two or three times larger than the object. **Figure 1** shows a convenient device that can be worn by people who require magnified views, for example, jewellers viewing gemstones.

A magnification of only two or three times is insufficient for viewing small objects such as cells. To obtain larger images, two or more lenses are combined to make a simple compound microscope such as the one shown in **Figure 2**. The lens closest to the object is called the objective lens, or simply the **objective**, and the lens the eye looks into is called the eyepiece lens, or **ocular**.

To see how an image is produced in a two-lens instrument, refer to the three light-ray diagrams in **Figure 3**. **Figure 3**(a) shows an object located beyond the focal point of the objective lens. The real image formed by the objective lens then becomes the "object" for the eyepiece in **Figure 3**(**b**). This "object" is located

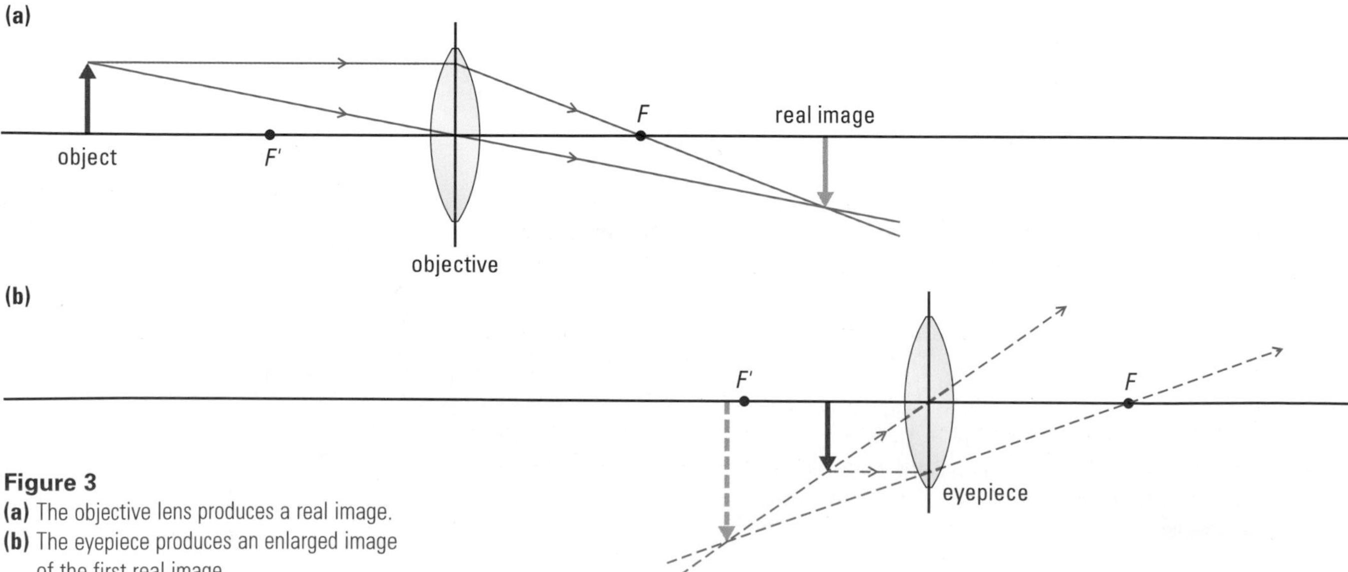

Figure 3
(a) The objective lens produces a real image.
(b) The eyepiece produces an enlarged image of the first real image.

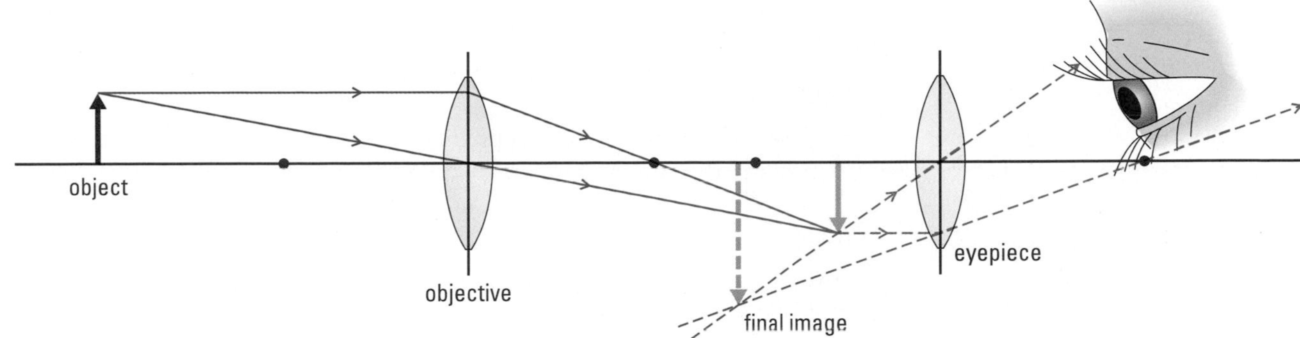

Figure 3
(c) The two light-ray diagrams are combined to show the effect of two lenses.

between the eyepiece and its secondary focal point. The final image, found by using the usual rules for ray diagrams, is virtual, inverted relative to the original object, larger than the original object, and located beyond the secondary focal point of the eyepiece. **Figures 3(a)** and **3(b)** are combined to give **Figure 3(c)**, a complete ray diagram of a two-lens system.

Practice

Understanding Concepts

1. The objective lens of a microscope has a focal length of 16.80 mm. A specimen on the stage is 18.48 mm from the lens. The objective lens produces a real image 22.50 mm away from an eyepiece with a focal length of 25.00 mm.
   (a) Sketch the positions of the specimen, the lenses, and the focal lengths. Verify with a calculation that the objective lens produces a real image.
   (b) What is the magnification of the objective lens?
   (c) What is the magnification of the eyepiece?
   (d) What is the overall magnification of the specimen?
   (e) What does the sign of the magnification of the image indicate about the image's attitude?

Answers

1. (b) −10.0×
   (c) 10.0×
   (d) −100.0×

Investigation 11.2.1

A Two-Lens Microscope

INQUIRY SKILLS

- ○ Questioning
- ○ Hypothesizing
- ● Predicting
- ○ Planning
- ● Conducting
- ● Recording
- ● Analyzing
- ● Evaluating
- ● Communicating

In any multi-lens optical instrument, the objective lens and all other lenses positioned before the eyepiece must produce a real image; then the eyepiece will enlarge that real image. For this to happen, the eyepiece should be fairly close to the image being magnified.

Converging lenses of different focal lengths are needed in this investigation. If the focal lengths of the lenses are not labelled, they can be found by aiming parallel light rays (from a ray box or a distant light source) toward the lens, finding the focal point on a screen, and measuring the focal length. The purpose of this investigation is to determine how the eyepiece and objective lenses are best arranged in a microscope.

Question

How are the eyepiece and objective lenses arranged to form a two-lens microscope?

Prediction

(a) Based on the various lenses, predict which arrangement of two lenses will result in the best microscope.

Materials

converging lenses of various focal lengths
optical bench apparatus with corresponding supports
screen holder
small letter or symbol

Procedure

1. Determine the focal lengths of two or three different lenses.
2. Determine how the magnification of a distant object viewed through a lens depends on the focal length of the lens. (Hold the lens about 30 cm from your eye.)
3. Determine how the magnification of an object near a lens depends on the focal length of the lens. (Again, hold the lens about 30 cm from your eye.)
4. Choose two lenses and place them in the supports on the optical bench so that they are separated by the sum of their focal lengths. Place the screen holder slightly beyond the focal point of the objective lens near one end of the optical bench. Insert a piece of paper with small print into the screen holder.
5. Look through the eyepiece and move the lenses and your eye back and forth until you find the clearest and largest image. The image should be inverted and larger than the object if the system is acting like a microscope. Estimate the magnification of your image.
6. Repeat steps 4 and 5 using various combinations of lenses; for example, objective $f = 5$ cm and eyepiece $f = 5$ cm; objective $f = 5$ cm and eyepiece $f = 10$ cm. Determine which combination gives the largest and clearest image.
7. When you have exhausted all possibilities with your lenses and have a successful arrangement of lenses, record the distance of the screen holder from the objective lens. Also record the focal lengths of the lenses and distance between the lenses.

DID YOU KNOW ?

Determining Focal Length

If an object is placed at the secondary focal point of a thin converging lens, no image will be formed because rays will form a beam of light parallel to the principal axis. If a mirror is held anywhere perpendicular to this beam of light, it should be reflected back on itself. This parallel beam will form an image at the same position as the object, but inverted. How can you use this fact to determine the focal length of your lens?

Analysis

(b) Calculate the image distance d_i produced by the objective lens. How do you know the image is real?
(c) Calculate the magnification of the specimen by the objective lens.
(d) Use the image distance d_i and the distance between the lenses to determine the distance of the real image from the eyepiece. This distance is the object distance d_o for the eyepiece.
(e) Use d_o for the eyepiece and the eyepiece's focal length to determine the position of the image produced by the eyepiece. How do you know from your calculation that the image is virtual?

(f) Determine the magnification of the eyepiece.
(g) What is the total magnification of your simple compound microscope?
(h) Answer the Question.

Evaluation

(i) How does your calculated value of the magnification compare with your estimation? Calculate a percent difference.
(j) Suggest improvements to the procedure.
(k) Look through your microscope again to compare what you see with what you calculated. How does your calculated value compare with what you see?

SUMMARY Lenses and the Microscope

- In a two-lens microscope, a high magnification of a nearby object is achieved.
- The objective lens produces a real image of a specimen.
- The eyepiece produces a virtual image of the objective lens's real image.

Section 11.2 Questions

Understanding Concepts

1. How is the magnification of a lens related to the distance of an object to that lens?

Applying Inquiry Skills

2. Two students drew a scale diagram of an optical bench microscope as follows:

   focal length of objective lens: 1.2 cm

   focal length of eyepiece: 5.7 cm

   distance between lenses: 7.8 cm

   position of specimen from objective lens: 1.7 cm

   height of specimen: 0.50 cm

   (a) Reproduce their scale diagram in your notebook.
   (b) Draw an accurate ray diagram to determine the location of both the real image produced by the objective lens and the virtual image produced by the eyepiece.
   (c) Measure the height of the virtual image.
   (d) Calculate the height of the virtual image using the thin lens equation and the values given above. Compare your ray diagram with your calculations and explain any discrepancies.

Making Connections

3. In groups of three, choose one question each from the list that follows for research in your library. Write brief answers to share with your group.

   (a) Some microscopes can download images directly to a computer. How do these computer microscopes work?

(continued)

(b) Most eyepieces are now made of two or more lenses. What kind of lenses are involved in Huygenian, Ramsden, Periplan, and Kellner eyepiece designs? Which of those eyepieces is the best one to use? Why?

(c) The Dutch naturalist Anton van Leeuwenhoek (1632–1723) is credited as one of the earliest significant contributors to the development of the microscope. However, his microscope consisted of only one lens. How was he able to achieve the high magnifications for which he was known?

Reflecting

4. Estimate the number of times microscope technology has been used to monitor your health in the last two or three years. Justify your estimate and then discuss it with a partner.

11.3 The Telescope

A single lens does not help the normal eye view distant objects. Two or more lenses must be used together to view distant objects such as stars, planets, and the Moon with detail. Images are smaller than the actual objects in space, but the images are much closer so they appear larger.

Opera glasses are conveniently small optical instruments that provide a low magnification, typically three to four times. They are manufactured in many forms, from simple toys to expensive status symbols. Their operation is based on the two-lens system invented by Galileo Galilei in 1609. That system, now called the "Galilean telescope" (**Figure 1**), consists of a converging lens as the objective and a diverging lens as the eyepiece. To produce an upright image, the diverging lens must intercept the light rays from the objective before a real image is formed.

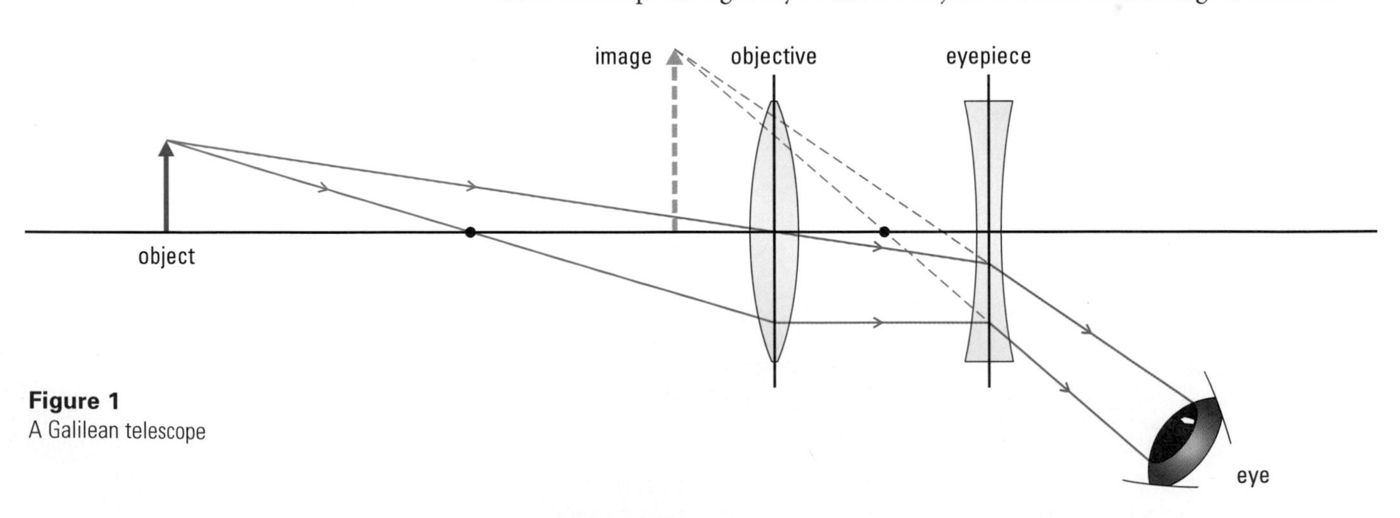

Figure 1
A Galilean telescope

Reflecting Telescopes

reflecting telescope: telescope that uses a parabolic mirror to focus light

Reflecting telescopes use parabolic mirrors to focus light onto the focal plane. The larger the diameter of the mirror, the stronger the concentration of light energy at the focus. This makes it possible for astronomers to see distant stars whose light energy is so low that they cannot be seen otherwise. An eyepiece is used to magnify and focus the image (**Figure 2**).

The first telescope of this type was made by Isaac Newton in 1668. To make it easier to see the image, Newton placed a plane mirror at 45° to the axis of the concave mirror, in front of the principal focus. This reflected the rays to one side, and the image could then be viewed through an eyepiece.

The mirror in the reflecting telescope at the David Dunlap Observatory in Richmond Hill, Ontario, once the largest in the world, has a diameter of 1.8 m.

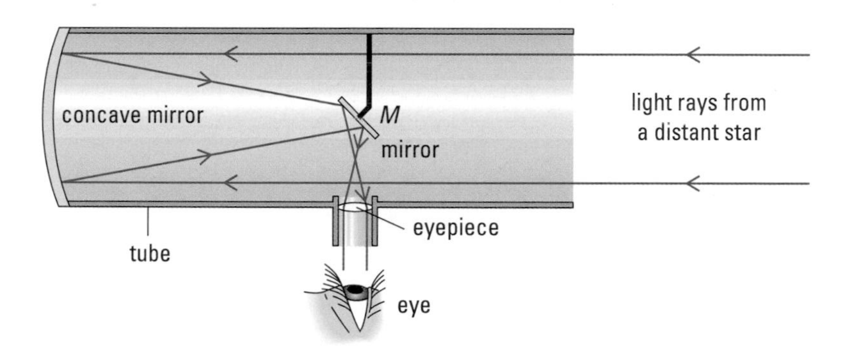

Figure 2
A reflecting telescope

Refracting Telescopes

refracting telescope: telescope that uses two converging lenses to focus light

Refracting telescopes, or refractors, use lenses. **Figure 3** shows how a refractor works: it is constructed from two converging lenses. The objective lens has a long focal length and the eyepiece has a short focal length. When the lens is used to view distant objects, the rays of light are nearly parallel when they enter the objective lens. The objective lens forms a real image (I) just inside the secondary principal focus (F') of the eyepiece. The eyepiece acts as a magnifying glass, producing a greatly magnified virtual image. The image is inverted, but when used for astronomical purposes, for example, this does not matter.

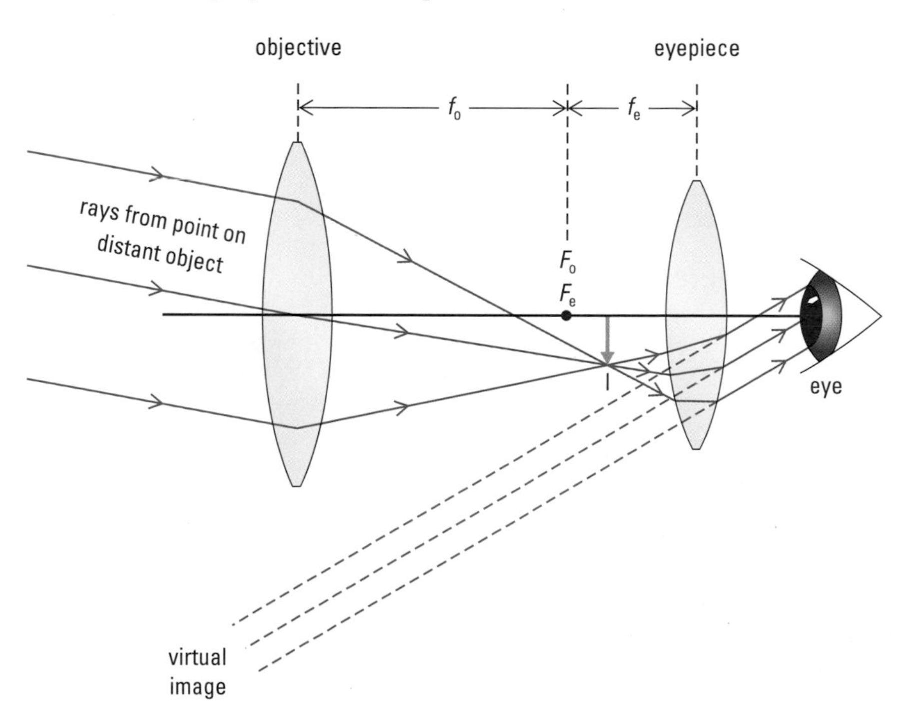

Figure 3
A refracting telescope

terrestrial telescope: telescope that is similar to a refractor, but with the addition of a third lens, the erector lens, to obtain upright images

Terrestrial Telescopes

The **terrestrial telescope** (Figure 4) is similar in construction to the refractor, except for an additional converging lens located between the objective lens and the eyepiece. The third lens is called the erector lens. Its purpose is to invert the image so that it has the same attitude as the object.

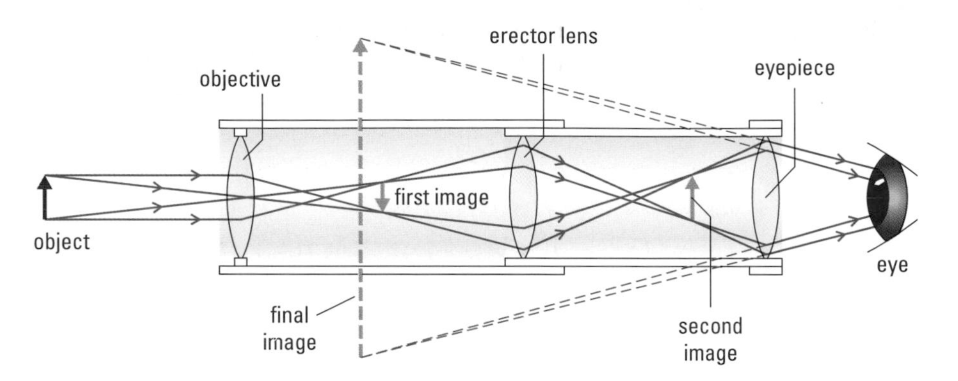

Figure 4
A terrestrial telescope

Practice

Understanding Concepts

1. Describe how a refracting telescope and a microscope are (a) similar and (b) different.
2. State the four characteristics of the final image relative to the original object seen in the terrestrial telescope shown in **Figure 4**. Show your calculation of the magnification, and describe how you can tell whether the final image is real or virtual.

Making Connections

3. Refracting telescopes make distant objects look much closer than opera glasses do. However, at a live theatre performance, opera glasses have advantages over a refracting telescope. What are these advantages?

Answers

2. 7.0 (approx.)

Activity 11.3.1

Model Telescopes

In this activity you will arrange lenses to simulate a refractor and a terrestrial telescope. An objective lens of a telescope should have a large diameter to collect as much light as possible, as well as a large focal length. The greater the focal length, the larger the image of an object located a great distance away.

Materials

various lenses
optical bench
distant target (greater than 5 m)

Procedure

Part 1: The Refractor Telescope

1. Obtain two lenses of different focal lengths and determine the focal length of each. Choose which lens will be the objective and which will be the eye-piece, based on what you've already learned in this chapter.
2. Place the eyepiece in a holder at one end of the optical bench. Place the objective at a distance equal to the sum of the focal lengths away from the eyepiece.
3. Determine the magnification of the telescope by drawing two sets of equally spaced lines on a piece of paper. View the lines from a distance through the telescope as well as with the unaided eye.
4. Try various combinations of lenses to discover which one gives the largest image of the same object. For instance, try objective $f = 5$ cm and eyepiece $f = 10$ cm; objective $f = 5$ cm and eyepiece $f = 5$ cm. Describe your observa-tions.
5. Look through the eyepiece at a distant object and move the lenses and your eye until you obtain the clearest and largest image of the object.

Part 2: The Terrestrial Telescope

6. Choose a third lens, keeping the two lenses that worked well in Part 1. Determine the focal length of the third lens.
7. On the optical bench, arrange the three lenses so that the erector lens is located between the objective and the eyepiece. Use your experience in the previous parts of this experiment to help you decide where to place the lenses.
8. Discover how to obtain a clear, upright, and enlarged image of a distant object. Describe what you discover.

Adaptations of Telescopes

Prism Binoculars

Prism binoculars are really just two refracting telescopes mounted side by side, one for each eye (**Figure 5**). Between each pair of lenses is a pair of prisms that invert the image and reduce the distance between the two lenses. Binoculars are much shorter than a telescope and easier to handle. Note that the distance the light travels between the two lenses in a telescope and between the two lenses on each side of a pair of binoculars is the same, although the telescope is longer.

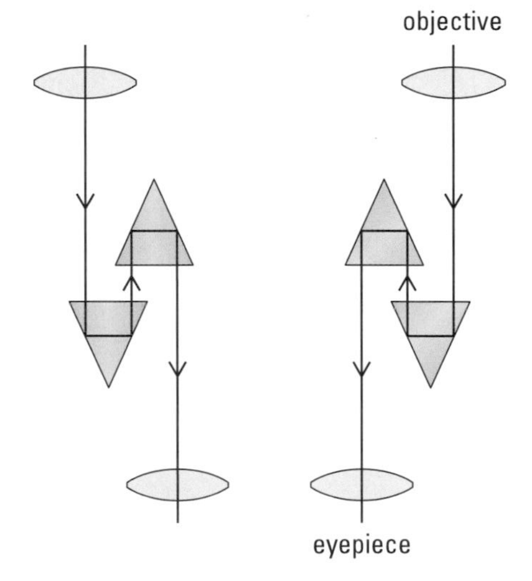

Figure 5
Prism binoculars

Zoom Lens

A zoom lens (**Figure 6**) allows a photographer to adjust the focal length. Lenses with a short focal length (about 25 mm to 35 mm) see a wide-angle view, while those with a long focal length (about 70 mm to 1000 mm) see a small angle but an enlarged image. Essentially, the zoom lens is a telescope mounted on a camera.

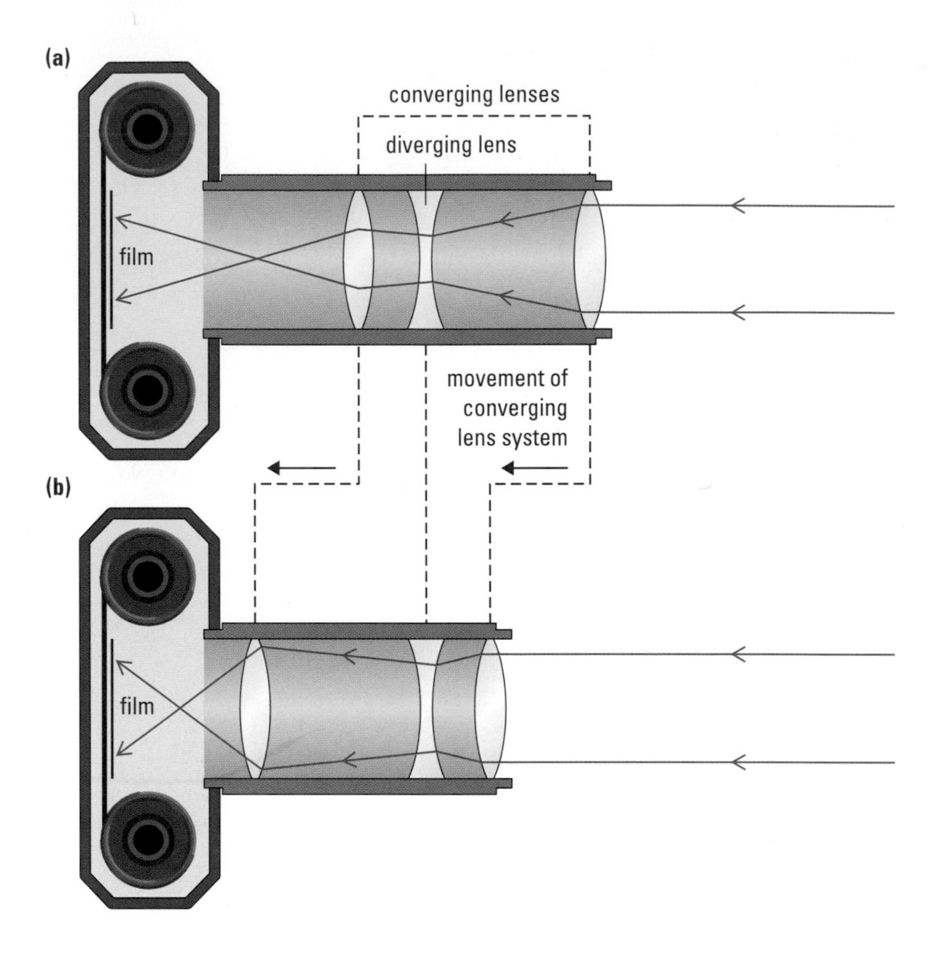

Figure 6
(a) Telephoto setting: longer focal length, larger image, smaller angle of view
(b) Wide-angle setting: shorter focal length, smaller image, larger angle of view

Practice

Understanding Concepts

4. Based on your understanding of zoom lenses, which focal length of binocular lenses provide greater magnification, 35 mm or 50 mm? Which provides a greater angle or field of view? Explain your answers.

SUMMARY Lenses and Telescopes

- A Galilean telescope, also called opera glasses, consists of a converging lens and a diverging lens.
- A reflecting telescope uses a curved mirror to focus light onto the focal plane.
- A refractor consists of two converging lenses.
- A terrestrial telescope is a refractor with an additional lens that ensures that images are upright.
- Binoculars and zoom lens systems are modified telescopes.

Section 11.3 Questions

Understanding Concepts

1. How do the images produced by a Galilean telescope differ from those of a refractor?
2. What is the purpose of the third lens in a terrestrial telescope?
3. (a) Why should a refractor's objective lens have a large diameter?
   (b) How does this limit the usefulness of lenses in telescopes? How have astronomers overcome this limitation?
4. Most single-lens reflex cameras come with a standard lens of focal length 50.0 mm. Suppose the lens is used to focus the image of your friend 10.0 m away.
   (a) Determine the image distance of your friend.
   (b) What is your friend's magnification?
   (c) The lens is replaced with a telephoto lens of focal length 20.0 cm. Determine the image distance and magnification now.
   (d) What is the ratio of the magnification of the 20.0-cm lens to that of the 50.0-mm lens?
   (e) If the 50.0-mm lens is replaced with a 28.0-mm lens, what is the ratio of the magnification of the 28.0-mm lens to that of the 50.0-mm lens? Lenses with focal lengths less than 28.0 mm are called *wide-angle* lenses. Why?

Making Connections

5. You have won a $1000 gift certificate toward the purchase of a telescope and telescope accessories for the essay you entered in the contest entitled "What will our telescopes allow us to see 100 years from now?" sponsored by the Canadian Space Agency. Investigate the current telescopes and accessories available on the market.
   (a) While collecting information, list and describe the criteria that become important in your decision.
   (b) Make your decision. Include a sample order form listing the items to be purchased, their cost, and applicable taxes. Discuss your decision by explaining how you used your list of criteria.
6. Using the Internet and other sources, research the Hubble Space Telescope. Follow the links for Nelson Physics 11, 11.3.
   (a) What is the Hubble Space Telescope?
   (b) Who was Hubble and why is the space telescope named after him?
   (c) Give a brief description of the Hubble's optics.
   (d) Describe a recent discovery made by the Hubble Space Telescope.

GO TO www.science.nelson.com

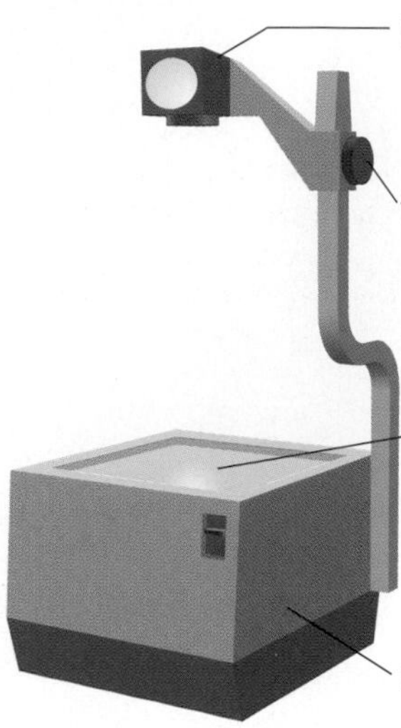

11.4 The Optics of Other Devices

Activity 11.4.1

Optics of an Overhead Projector

Figure 1
An overhead projector

Overhead projectors (**Figure 1**), like many optical systems, consist of three systems that work together: a mechanical system, an electronic system, and an optical system. Their function is to project an enlarged image from a transparent film onto a distant screen. In this activity, you will see how the different optical components of the projector work together.

Materials

overhead projector
appropriate screwdrivers

Procedure

1. Before turning on the overhead projector, open the optical stage to see inside the projector case. Sketch the arrangement of optical components by considering what a cross-section of the projector would look like. Note the arrangement of any bulbs, mirrors, or lenses that you find in the projector case. Add the optics of the projection head to your sketch.
2. Turn on the projector to project an image of a letter onto a screen nearby. Make adjustments to focus the image.
3. Use the focus knob to move the projection head upward. How does this affect the image? Refocus the image.
4. Use the focus knob to move the projection head downward. How does this affect the image? Refocus the image.
5. Move the projector farther from the screen. How does this affect the image?

Analysis

(a) Draw a ray diagram, with at least three different rays, showing how light travels from the bulb to the screen.
(b) In table form, describe the structure and function of each optical component of the overhead projector.
(c) In what ways do you think an overhead projector is similar to a microscope? In what ways is it different?

Slide or Movie Projector

Figure 2 shows the arrangement of lenses and mirrors in a typical slide or movie projector. The source of light is usually a tungsten filament lamp or a quartz iodide lamp that produces the bright light necessary to illuminate a slide or movie. Two converging lenses refract the light so that the object to be projected is uniformly illuminated. The object is placed a distance of one to two focal lengths in front of the projection lens, resulting in a large, real image.

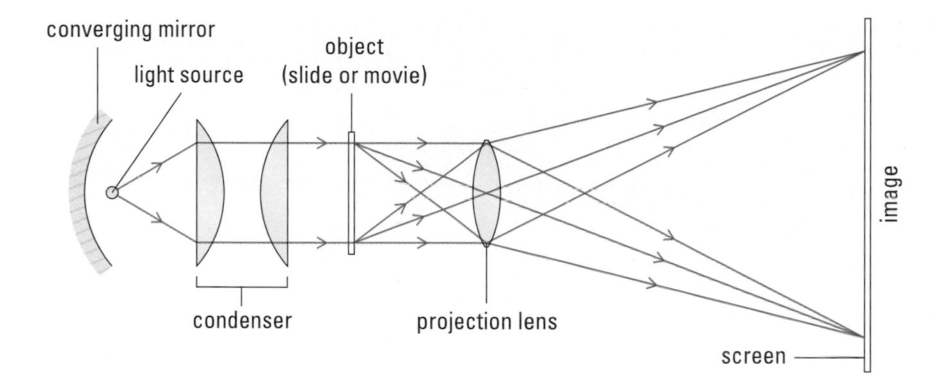

Figure 2
In a slide or movie projector the image is inverted, so the slide or film must be inverted vertically and horizontally when placed in the projector. The projection lens is mounted in a sliding tube or geared mount so that it can be moved to focus the image.

Large-Screen Projection Systems

While attending Expo 67 in Montreal, three Canadian filmmakers—Graeme Ferguson, Roman Kroitor, and Robert Kerr—were inspired by the effects created by the multi-projector systems being used. Their goal was to develop a cinematographic projection system that would give the audience a sensation of reality with the size and clarity of its images. That same year, they founded IMAX Corporation in Mississauga, Ontario. The first permanent giant-screen theatre opened at Ontario Place in Toronto in 1971.

To capture a large field of view, special wide-angle lenses had to be designed for a new camera system. The lens illustrated in **Figure 3** has a field of view of 150°. To obtain enough clarity in the projected image, each film frame (the storage medium) had to be much larger than the regular 35-mm and conventional 70-mm film formats that are still the standard for motion pictures.

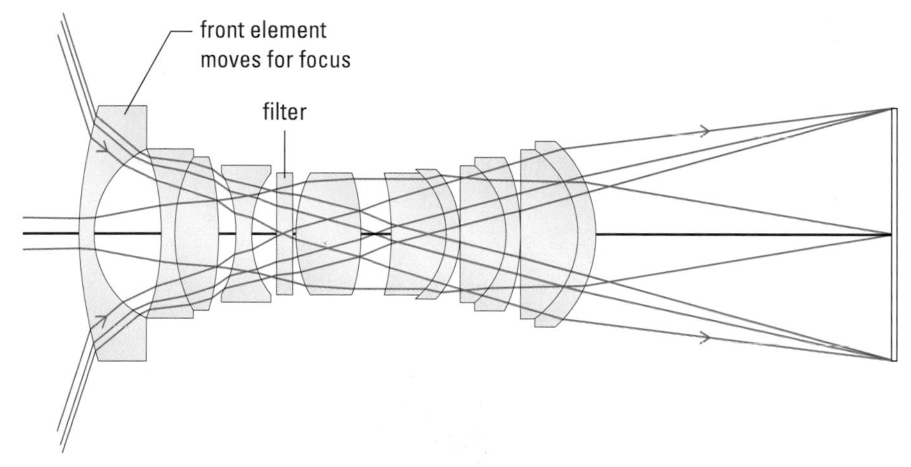

Figure 3

As shown in **Figure 4**, conventional 70-mm film runs vertically in conventional 70-mm cameras, and each frame is only five perforations long. (The perforations are the holes on each side of the film that keep the film on track as it runs through the camera.) In IMAX cameras, the film width is still 70 mm, but each frame is 15 perforations long, and the film runs horizontally through the camera. Since the film runs horizontally and the film frame is longer, each film frame used in IMAX photography is much larger than the film frames of conventional motion picture photography.

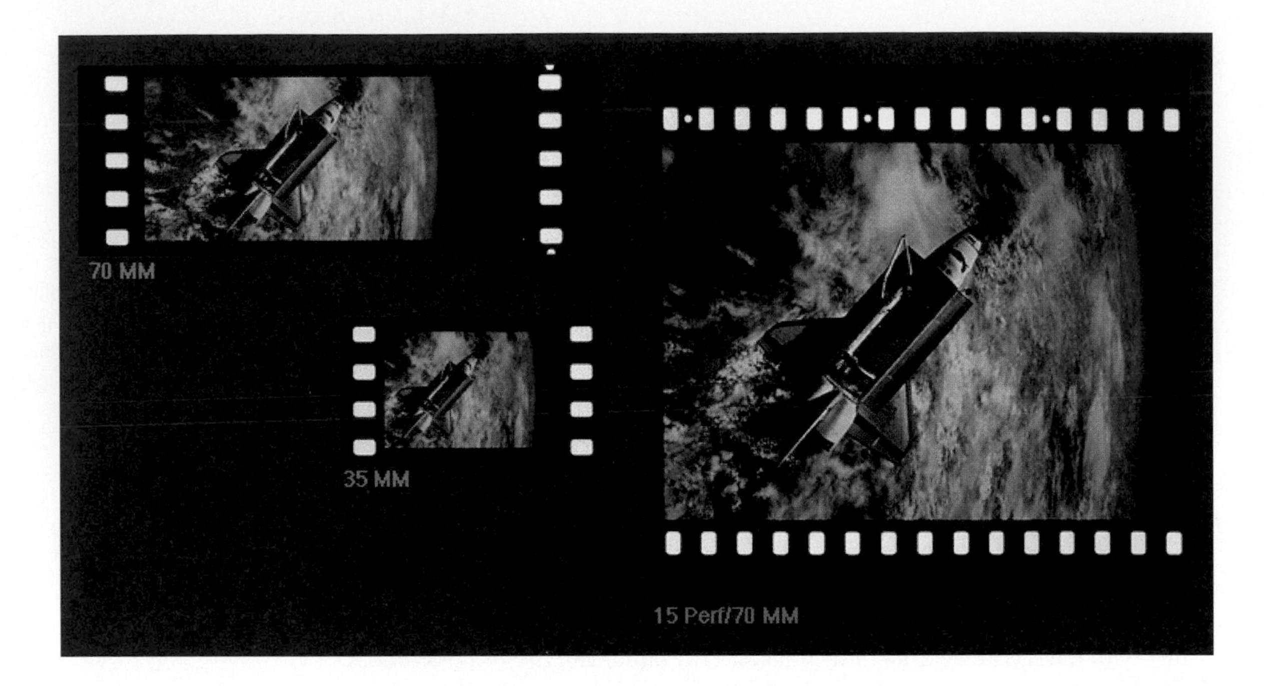

Figure 4
Standard film formats compared to IMAX film (far right). Such a large film format increased the weight of a film roll to the extent that a new method of moving the film across the projection lens was needed. During projection, the film is fed through the projector horizontally instead of vertically. To eliminate vibrations in the film as it moves, it is held firmly against the rear element of the lens by a vacuum.

To project images from this film onto a huge screen, a powerful light source is needed. Typical lamps for large-screen theatres are rated at an average of 15 000 W. A typical lens exposed to light of such high intensity would absorb so much heat that it could explode or crack; therefore, a special cooling system is used to remove some of the heat generated by these lamps. IMAX projection lenses are specially designed to withstand enormous amounts of heat.

Practice

Understanding Concepts

1. (a) What is the main function of all projectors?
   (b) Do all projectors produce the same type of image? Explain.

Applying Inquiry Skills

2. (a) Describe two different sets of measurements you could make to calculate the magnification of the overhead projector used in your classroom.
   (b) Estimate the magnification of the overhead projector when it is used at its typical location.
   (c) With your teacher's approval, apply your suggestions in (a) above to check your estimate.

SUMMARY Optics of Other Devices

- Different arrangements of optical components can be designed to manipulate light in many ways.
- Overhead projectors, slide and movie projectors, and large screen projection systems are examples of complicated optical devices.

Section 11.4 Questions

Understanding Concepts

1. State the four characteristics of the image produced by a movie projector. Does the image of a slide projector have the same characteristics?
2. Explain why the lens in an IMAX system must be a wide-angle lens.

Making Connections

3. OMNIMAX theatres, such as the one in the Ontario Science Centre in Toronto, use advanced optical technology.
   (a) What does the prefix "omni" mean? How does the meaning relate to an OMNIMAX movie presentation?
   (b) Use the Internet to find out about OMNIMAX theatres, film, and movie projectors. Describe how the features compare with IMAX features. Follow the links for Nelson Physics 11, 11.4.

GO TO www.science.nelson.com

11.5 Construction of Optical Instruments

By now you have seen ray diagrams for many optical instruments. You have also studied overhead projectors, cameras, and microscopes. Using two lenses, you observed how we can magnify objects that we can't see because they are either too small or too far away. In this section you will use your knowledge to construct and refine an optical instrument of your choice. But first we look at how instruments control the position of lenses and the amount of light.

Figure 1
Simple hand lenses use a threaded eyepiece to adjust the position of the lens.

Controlling the Position of Lenses

Recall that an object positioned between the focus and the optical centre of a converging lens produces a virtual image. If the lens were moved farther away, the object would be located at a distance longer than the focal length, producing a real, inverted, and smaller image. For this reason, there are many different methods of moving lenses to allow ease of focusing.

In inexpensive plastic microscopes and hand lenses (**Figure 1**), a lens is mounted into a hollow, threaded tube that is connected to a base. The object distance can be varied by twisting the lens clockwise or counterclockwise to move the lens up or down.

The simplest way to control the relative positions of two lenses is to mount the lenses in a sliding tube (**Figure 2**), like the slide of a trombone.

Figure 2
In an alignment telescope, object and image distances can be adjusted with a sliding tube.

When greater control is needed, or when the device contains heavy components, a rack-and-pinion gear system is used (**Figure 3**).

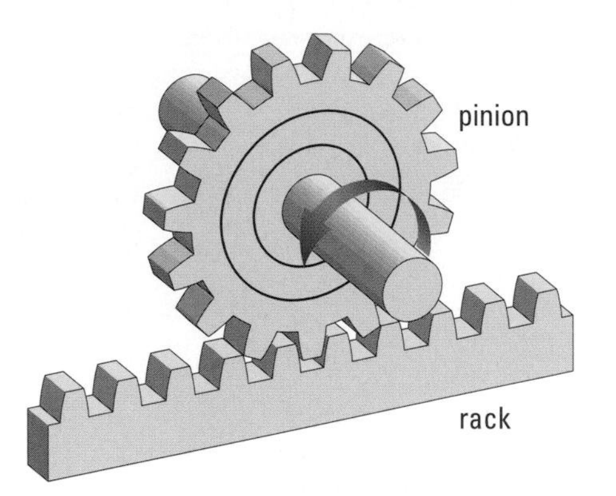

Figure 3
A rack-and-pinion system. In the compound microscopes that you have used in school, either the body tube or the stage is moved by a rack-and-pinion gear.

Controlling the Amount and Type of Light

Each type of optical instrument works best under different light conditions. Your classroom microscope may have a disk with different-sized holes positioned above the light source, allowing you to observe specimens under either bright or dim light. Other microscopes use an iris diaphragm, allowing smoother control of the intensity of light passing through the specimen.

In other applications, it is important to control the type of light. In many cases, coloured light filters are used. There are even filters that can control infrared and ultraviolet light.

Practice

Understanding Concepts

1. Which method of lens control
   (a) is easiest to move quickly?
   (b) provides the most sensitivity?
   (c) is acceptable for inexpensive optical instruments?
2. Explain why an iris diaphragm allows smoother control of the amount of light entering an optical instrument than a disk with different-size holes.

Activity 11.5.1

Constructing an Optical Instrument

Before designing an optical instrument, it is important to have a clear goal in mind. What is the purpose of the instrument? Who will use the instrument? Also, consider that every optical instrument has at least two basic systems: an optical system and a mechanical system. Keep thorough notes of your ideas, sketches, and drawings as you complete this activity.

Procedure

1. Decide which instrument you wish to build, then design your optics system using ray diagrams to model the effect of your system on incident light. If your school has optics bench software, you may be able to simulate your optical system.
2. Sketch your design, showing how you will hold all the optical components together.
3. In groups of three or four, present your designs to your peers. Give constructive comments to your peers, and record suggestions given to you.
4. Refine your plans based on suggestions or your own new ideas. Make new sketches showing your revisions.
5. Construct your optical instrument. Be sure to wear goggles when appropriate, and take proper safety precautions with glass.
6. Submit your plans and refinements along with your construction.
7. Demonstrate to your teacher how your optical instrument works.

DID YOU KNOW ?

Handling Glass

Cleanliness and smoothness of optical components are very important in order for optical instruments to function properly. If you wear glasses, you probably have first-hand knowledge of this fact. Recall from your work with microscopes that lenses should be cleaned with proper lens paper or ethanol. You should always clean across a lens, rather than using a circular motion. Why?

SUMMARY Optical Instruments

- There are many ways to change the position of lenses within an optical system.
- There are many ways to adjust light levels and colours used by optical devices.

Section 11.5 Questions

Understanding Concepts

1. List as many components of a student-made optical instrument as you can, and classify the components as optical, mechanical, or other (for example, electronic).

Making Connections

2. Photographers use filters to cover the objective lens of a camera for various purposes.
   (a) What type of filter is popular for protecting a camera's expensive lens? Does this filter have any other purpose?
   (b) What are the purposes of various other types of filters, such as infrared and ultraviolet filters? Follow the links for Nelson Physics 11, 11.5.

GO TO www.science.nelson.com

CAREER

Careers in Light and Geometric Optics

There are many different types of careers that involve the study of light and geometric optics in one form or another. Have a look at the careers described on this page and find out more about one of them or another career in light and geometric optics that interests you.

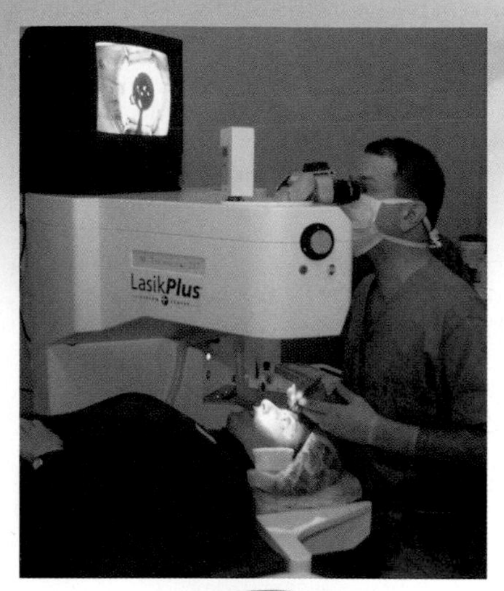

Ophthalmologist

Ophthalmology is a medical specialization. An ophthalmologist is a medical doctor who has undergone four years of training in a medical school, after first completing a bachelor's degree, often in science, and sometimes in arts. Ophthalmologists diagnose eye problems and diseases and are licensed to prescribe medications and conduct eye surgery. Tools of the trade include cameras, bioscopes, and lasers. These doctors work in private practice and in hospitals, or in the field of medical research.

Fibre Optics Installer

Working with fibre optic cables requires a solid technical understanding of construction techniques and practices. It takes years of on-the-job training to learn how to place cable. Fibre optics installers use a variety of construction tools such as aerial bucket trucks and machines that feed cable. To carry out their duties, fibre optics installers must be able to read architectural drawings. These installers generally work for communications companies.

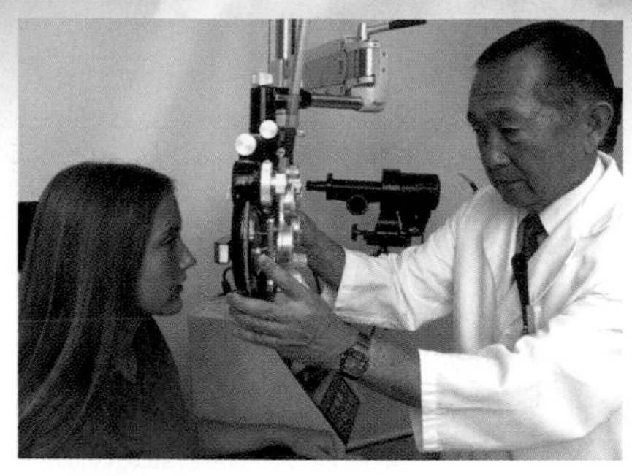

Optometrist

Optometry is a four-year degree course. Optometrists usually earn a bachelor's degree in science before they enrol in this course. Their work involves testing eyes for refractive disorders, such as near- or farsightedness, and for astigmatism. Most optometrists work in private practice. They must know how to handle different eye equipment, varying from microscopes to cameras. After testing a patient's eyes, optometrists prescribe corrective contact lenses or eyeglasses. Optometrists are always on the alert for eye problems, such as tumours, that are signs of a patient's ill health.

Practice

Making Connections

1. Identify several careers that require knowledge about light and geometric optics. Select a career you are interested in from the list you made or from the careers described above. Imagine that you have been employed in your chosen career for five years and that you are applying to work on a new project of interest.
   (a) Describe the project. It should be related to some of the new things you learned in this unit. Explain how the concepts from this unit are applied in the project.
   (b) Create a résumé listing your credentials and explaining why you are qualified to work on the project. Include in your résumé
   - your educational background — what university degree or diploma program you graduated with, which educational institute you attended, post-graduate training (if any);
   - your duties and skills in previous jobs; and
   - your salary expectations.

   Follow the links for Nelson Physics 11, 11.5.

   GO TO www.science.nelson.com

Chapter 11 Summary

Key Expectations

Throughout this chapter, you have had opportunities to

- analyze, in quantitative terms, the characteristics and position of images by lenses; (11.1, 11.2, 11.3, 11.4, 11.5)
- describe and explain, with the aid of light-ray diagrams, the characteristics and positions of the images formed by lenses; (11.1, 11.2, 11.3, 11.4, 11.5)
- predict, using ray diagrams and algebraic equations, the image position and characteristics of a converging lens, and verify the predictions through experimentation; (11.1, 11.2, 11.3, 11.4, 11.5)
- construct, test, and refine a prototype of an optical device; (11.5)
- describe how images are produced and reproduced for the purposes of entertainment and culture; (11.1, 11.4)
- evaluate, using given criteria, the effectiveness of a technological device or procedure related to human perception of light; (11.1, 11.2, 11.3, 11.4, 11.5)
- analyze, describe, and explain the optical effects that are produced by technological devices. (11.1, 11.2, 11.3, 11.4, 11.5)
- identify science and technology-based careers related to light and geometric optics; (career feature)

Key Terms

lens camera
single-lens reflex (SLR) camera
f-stop number
charge coupled device
semiconductor
digital camera
pixel
microscope
objective
ocular
reflecting telescope
refracting telescope
terrestrial telescope

Make a Summary

Choose three optical instruments: a camera, some type of projector, and either a microscope or telescope. Draw a sketch or a ray diagram of each instrument and label as many components and other features (object, images, light rays, etc.) as you can. On your sketches, show that you understand the principles of photography and image production.

Reflect on your Learning

Revisit your answers to the Reflect on Your Learning questions at the beginning of this chapter.

- How has your thinking changed?
- What new questions do you have?

Chapter 11 Review

Understanding Concepts

1. Place the advancements in photography listed below in chronological order.
   (a) introduction of colour film
   (b) use of lenses
   (c) invention of roll film
   (d) first use of lens cameras
   (e) discovery of the sensitivity of silver halides to light
2. State the main function of these parts of a lens camera:
   (a) diaphragm
   (b) zoom lens
   (c) aperture control
3. Describe the factors that can affect the quality of a photograph.
4. What is an *f*-stop number?
5. State whether the image viewed in each of the following instruments is real or virtual:
   (a) simple magnifier
   (b) microscope
   (c) television camera
   (d) opera glasses
   (e) computer data projector
   (f) refracting telescope
6. **Figure 1(a)** shows a lens of fixed focal length focused on an object about 4 m from the camera. **Figure 1(b)** shows the same lens focused on the object from a different distance. Is that distance greater or less than 4 m? Explain your answer.

(a)

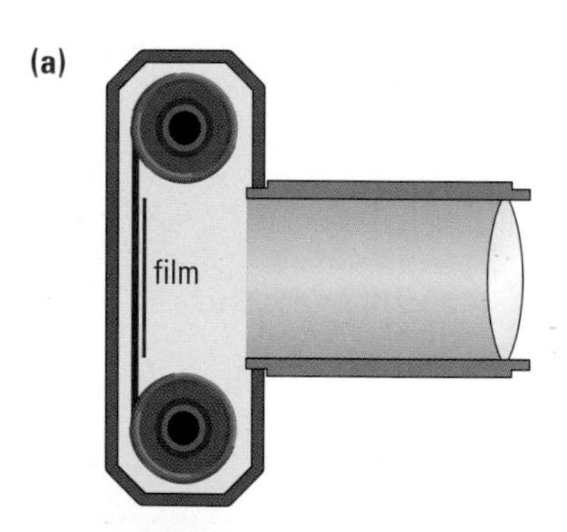

(b)

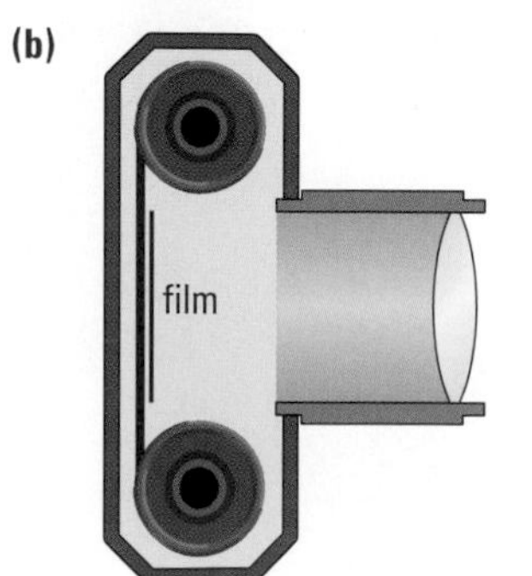

Figure 1

7. An amateur photographer takes a series of photographs of a lunar eclipse using a camera with a lens of focal length 6.0 cm. What is the size of the Moon's image on the film? (The diameter of the Moon is 3.5×10^5 m and the distance to the Moon is 3.7×10^8 m.)
8. In the case of a refractor, explain with the aid of ray diagrams what is gained by making the objective
   (a) of larger diameter
   (b) of shorter focal length
9. Describe the similarities and differences between a microscope and a refracting telescope.
10. What is the function of the erector lens in a terrestrial telescope? Does the erector lens produce a real or virtual image? Explain your answer.
11. A projector is required to make a real image, 0.55 m tall, of a 5.0-cm high object placed on a transparency. To produce a clear image, the transparency is placed 11 cm from the lens. What is the focal length of the lens?
12. Is the magnification of a telescope greater or smaller than 1.0? Explain your answer. Think about the equation(s) for magnification as well as looking at objects such as Mars or the Moon.
13. Which optical instrument applies the principle of total internal reflection? Why is this application an advantage?
14. A compound microscope has a 40× objective and a 10× ocular. What is the total magnification?
15. A microscope that provides a total magnification of 750× has a 10.0× ocular. What is the magnification of the objective?
16. In a slide or movie projector, what is the function of the converging mirror?
17. The high-power objective lens of a compound microscope has a focal length of 4.478 mm. A specimen on the stage is 4.590 mm from the lens. The objective lens produces a real image 22.5 mm away from an eyepiece with a focal length of 25.0 mm.
    (a) Sketch the positions of the specimen, lenses, and focal lengths. Verify with a calculation that the real image is produced 183.5 mm from the objective lens.
    (b) What is the magnification of the objective lens?
    (c) What is the magnification of the eyepiece?
    (d) What is the overall magnification of the specimen?

(e) Repeat the same question but this time round the numbers related to the objective lens to three significant digits. Which number in the question changes? By how much? What effect has this small change made on the overall calculated magnification of the specimen? Relate this to the difficulty students have when using the high-power objective lens of compound microscopes.

Applying Inquiry Skills

18. Draw a sketch to show how you would arrange a candle for an object, a metre stick for support, and your choice of lenses to demonstrate a
    (a) camera
    (b) compound microscope
    (c) refracting telescope
    (d) terrestrial telescope
19. **Table 1** shows the lenses available in a certain physics lab for experimenting with optical instruments.

Table 1 Lenses Available in a Physics Lab

Type of lens	Focal length (cm)	Number available
converging	50	2
converging	20	2
converging	5	2
diverging	–20	2
diverging	–5	2

State which of these lenses you would use to make each of the instruments listed below. Explain your choice.
    (a) a compound microscope
    (b) a refracting telescope
    (c) a Galilean telescope
    (d) a terrestrial telescope
    (e) a projector (Only the objective lens is required.)
20. Images with a "high resolution" are very clear.
    (a) In a lens camera, how does resolution depend on the aperture of the lens?
    (b) In a digital camera, how does resolution depend on the pixel arrangement?
    (c) Describe how you could use a computer printer to demonstrate the difference between high and low resolution.

Making Connections

21. Some cameras, lenses, and binoculars are fitted with an "image stabilizer" feature. Assume that a lens of focal length 75 cm is set for proper exposure at *f*/11 and a shutter speed of 1/15 s.
    (a) For a regular camera with no image stabilizer, what is the disadvantage of using an exposure time of 1/15 s?
    (b) If the exposure time is adjusted to 1/60 s, what aperture setting will provide the proper exposure?
    (c) Describe the conditions when using an image stabilizer would be advantageous.
22. Reflecting telescopes can be made much larger than refracting telescopes, so astronomers tend to use reflecting telescopes when gathering light from distant stars and galaxies. Why are refracting telescopes more limited in size than reflecting telescopes?

Exploring

23. Research and describe the steps used in developing black and white photographic paper. What advice would you give to a person attempting the developing process for the first time?
24. Some cameras offer autofocus as an option. Using the Internet and other sources, research how a camera focuses automatically. Write a three-paragraph summary; include diagrams if appropriate. Follow the links for Nelson Physics 11, Chapter 11 Review.

GO TO www.science.nelson.com

25. Research the use of radio telescopes in astronomy. Do these telescopes resemble refractors or reflectors? Follow the links for Nelson Physics, 11, Chapter 11 Review.

GO TO www.science.nelson.com

26. IMAX continues to develop new systems of projecting images, including 3-D projection systems. Using the Internet, research these systems and find out what they do and how they work. Prepare a short presentation. Follow the links for Nelson Physics 11, Chapter 11 Review.

GO TO www.science.nelson.com

Constructing an Optical Device

Visitors to science centres or science and technology museums who have interacted with the displays tend to enjoy the visit more and are able to remember a lot more of what they have learned (**Figure 1**). For the study of a human eye, a passive display, such as a drawing of the human eye, may be informative, colourful, and scientifically accurate. However, an active display is more fun and helps the viewer understand the effects of changing variables. For example, a working model of the human eye could be set up to show what happens when the muscles that control the lens make the lens thicker, or a light could be shone into the eye to allow the viewer to see what an eye doctor sees during an eye examination.

Figure 1
Science centres, such as Science North in Sudbury and the Ontario Science Centre in Toronto, have hands-on displays that involve visitors in "active learning."

In this Performance Task, you will design an interactive optical device or model. The device will have various optical components such as lenses, prisms, and light sources. To be interactive, these optical components need to be controlled by some built-in mechanical components. When you start brainstorming about how you are going to design the control features, you may think that optical devices are too complex to model. For instance, in a new auto-focus camera, the lens system is controlled by a battery-powered motor using distances that are judged by ultrasound waves. However, this degree of sophistication may not be necessary for the creation of an informative and interesting device. Looking at how optical devices were designed in the past will help you design an interesting but simple model. The first photographs were taken with a pinhole camera, a device much easier to model than today's automatic cameras.

The Task

Your task is to design, construct, test, and refine an optical device that operates on the principles presented in this unit. The device should be a working model with both optical and mechanical components. The mechanical components should control one of the major variables of the optical system, such as the position of a lens or the amount of light that enters a lens. (Simple electrical components, such as a light source, can be used with your teacher's approval.) Your device should be designed so it would be appropriate for a safe, hands-on display at a science centre or a science and technology museum. The design can be based on historical models, or it can incorporate some of the features of the most modern instruments.

There are many optical devices that you can choose for this task (see **Figure 2**). Telescopes, microscopes, and binoculars use lenses. Binoculars and periscopes use prisms that allow light to undergo internal reflection. Working pinhole cameras can take black-and-white photographs or produce negatives. Projectors place images onto screens. Other devices include those that can create a mirage, an optical illusion, or a shimmering effect.

Figure 2
What optical instruments, other than the ones shown, were featured in this unit?

You will be expected to design and construct the device, test the prototype, and then refine the device based on your tests. You will also be expected to analyze the physics principles involved in using the device and explain how the device could be used in a science centre or a science and technology museum. As you work on this task, keep thorough notes of your research, brainstorming ideas, diagrams, safety considerations, tests, and analysis, and develop a portfolio that you can submit with the final product.

Analysis

Your analysis should include answers to the questions below.

(a) How does your device apply physics principles, laws, and theories?
(b) How has the basic design of this type of device evolved from historical times to the present?
(c) What are the main uses of the device, and who are the primary users?
(d) What careers are related to the production or use of the device?
(e) List problems that you encountered during the construction of the device and describe how you solved them. (You will get ideas as you test and refine the device.)

Evaluation

(f) Evaluate the contributions of the device to such areas as entertainment, communication, and health.
(g) Evaluate the process you used in working on this task. If you were to begin this task again, what would you modify to ensure a better process and a better product?

Assessment

Your completed task will be assessed according to the following criteria:

Process
- Choose an appropriate optical device to suit your interest and desire to improve your skills.
- Draw up detailed plans of the technological design, tests, and refinements.
- Choose appropriate materials for the optical device.
- Carry out the construction, tests, and refining of the optical device.
- Analyze the task (as described in Analysis).
- Analyze the physical principles of the optical device.
- Evaluate the task (as described in Evaluation).

Product
- Demonstrate an understanding of the related physics principles, laws, and theories.
- Use terms, symbols, equations, and SI metric units correctly.
- Produce a working model of the optical device complete with instructions detailing how a viewer in a science centre or science and technology museum could interact with it.
- Prepare and submit a task portfolio with your research notes, ideas, diagrams, tests, and analyses.

Understanding Concepts

1. Determine the frequency of blue light with a wavelength of 400 nm (to two significant digits).
2. Light and sound travel in the form of waves. Light can travel through a vacuum, but sound cannot. Explain this apparent contradiction.
3. An actor stands in front of a plane mirror. The distance from the actor to the image in the mirror is 5.0 m. Sketch a diagram to show how far the actor is from the mirror.
4. Two students are standing at a perpendicular distance of 4.0 m in front of a plane mirror (**Figure 1**). Student A looks into the mirror and sees Student B. From Student B, the light rays travel for 10.0 m to reach the eyes of Student A. The angle of reflection of the light rays is 36.9°. How far apart are the two students?

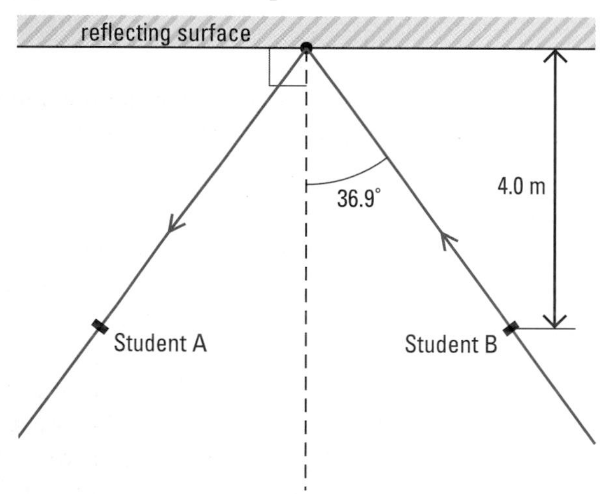

Figure 1

5. A flashlight is turned on and emits a light beam perpendicular to a flat piece of glass. The light beam passes through the flat piece of glass but is not dispersed into a spectrum. Explain why.
6. What is the index of refraction of a liquid in which light travels at a speed of 2.50×10^8 m/s?
7. Draw a labelled diagram of a ray that refracts. Include the terms: angle of incidence, normal, angle of refraction, incident ray, refracted ray, and reflected ray.
8. The index of refraction of a 30.0% sugar solution at 20.0° C is 1.38. What is the speed of light in this solution?
9. A parallel beam of light is directed onto the flat surface of a block of crown glass. Part of the beam is refracted with an angle of refraction of 30.0°. What is the angle of reflection?
10. Consider factors that affect the change in direction of a light ray as it travels through a sheet of glass.
    (a) Draw two diagrams to show that the lateral displacement is greater for thick refracting materials than for thin ones.
    (b) List two other factors that affect lateral displacement. Discuss how these factors affect lateral displacement.
    (c) Using Snell's law, prove that a ray of light entering a plate of glass always emerges in a direction parallel to the incident ray. (*Hint:* Draw a ray diagram and extend the incident ray to the far side of the glass.)
11. Which has the greater critical angle, glass with an index of refraction of 1.53 or glass with an index of refraction of 1.60? Would your answer be different if the second medium were water instead of air?
12. Diamond has a critical angle of 24°. Crown glass has a critical angle of 42°. Why does a piece of diamond sparkle more in bright light than a piece of crown glass, even if the two pieces are of the same shape?
13. If a piece of flint glass had facets like a typical diamond, much of the light would escape through its lower side and the flint glass would not sparkle nearly as much as a diamond would. Explain with a diagram and calculations.
14. When a canoe paddle is placed in the water, it looks bent when viewed from above. Sketch a diagram to explain this phenomenon.
15. When a light beam shines on a lens that is thicker in the middle, the light ray that passes through the middle comes out slightly behind the ones at the edges. Explain with the aid of a sketch.
16. Draw a fully labelled ray diagram to locate the image of the object in each situation described below. State the characteristics of each resulting image.
    (a) A detective is inspecting part of a broken toothpick with a magnifying glass (converging lens) of focal length 65.0 mm. The toothpick is 15.0 mm long and is held 30.0 mm from the middle of the magnifying glass.
    (b) A 30.0-cm tall ruler is located 90.0 cm from a diverging lens of focal length −40.0 cm.
17. A converging lens of focal length 4.0 cm is placed in front of another converging lens of focal length 40.0 cm. What is the distance between the lenses that will cause parallel rays of light entering the 4.0-cm lens to emerge from the second lens as parallel rays? Explain your answer.

18. A camera lens has a focal length of 6.0 cm and is located 7.0 cm from the film. How far from the lens is the object positioned if a clear image of the object has been produced on the film?

19. A lens of focal length 20.0 cm is held 12.0 cm from a grasshopper that is 7.0 mm tall. Find the size, position, and type of the image of the grasshopper.

20. A 2.0-cm-high object is placed 75.0 cm in front of a diverging lens of focal length –20.0 cm. Determine the following:
    (a) the distance the image is from the lens
    (b) the type of image formed
    (c) the attitude of the image
    (d) the height of the image

21. A 10.0-cm-high object is placed 15.0 cm in front of a converging lens with a focal length of 10.0 cm. Determine
    (a) the distance the image is from the lens
    (b) the type of image formed
    (c) the attitude of the image
    (d) the height of the image

22. A converging lens projects an image onto a screen 1.0 m away from the lens. The height of the image is ten times the height of the object. Calculate the focal length of the lens.

23. An optometrist prescribes a corrective lens for farsightedness that has a power of 2.5 d. What is the focal length of the prescribed lens?

24. On a bright sunny day arson is reported at an abandoned haystack. Two suspects wearing eyeglasses are detained at the scene. The two suspects are very close in build, hair colour, etc. Both of them profess their innocence. An eyewitness cannot be sure which suspect used the eyeglasses to set the fire. Upon further questioning, the police officer finds out that suspect A has myopia and suspect B has hyperopia. The police officer, having done well in high school physics, immediately knows the identity of the arsonist. Explain how the officer is able to identify the perpetrator.

25. A nearsighted tailor cannot focus on objects more than 10.0 cm away from his eyes. What must be the power of the lens to enable the tailor to see distant objects clearly?

26. A simple compound microscope consists of an objective lens and an eyepiece mounted 177.5 mm apart. A specimen is placed on the stage 38.8 mm from the objective lens that has a focal length of 31.0 mm. A real image of the specimen is produced by the objective lens in front of the eyepiece that has a focal length of 25.0 mm.
    (a) Sketch the positions of the specimen, the lenses, and the principal focus. Verify with a calculation that the objective lens produces a real image 154.2 mm away from it and hence 23.3 mm away from the eyepiece.
    (b) What is the magnification of the objective lens?
    (c) What is the magnification of the eyepiece?
    (d) What is the entire magnification of the specimen?
    (e) Why is the magnification in (b) negative? Why is the magnification in (c) positive?

27. Compare and contrast the design of a two-lens microscope and a refracting telescope.

28. You buy a map of the Moon to study its surface detail and you also study the Moon through your refracting telescope at night. That night, as you look at the map, you realize that it is printed upside down. You then go back to the store and explain the situation to the storeowner. The storeowner states that there is nothing wrong with the map. Explain why.

29. Contrast the specimen in a microscope and the film in a movie projector. Where are they located and what are the attitudes of the images?

30. Consider the following situation:

    A flat metal plate about 1 m long and 15 cm wide is supported by 2 or 3 hot plates close enough together so that the metal plate is uniformly heated. A 40-W light bulb is placed at about the same level as the metal plate and about 30 cm from one of its ends. A translucent piece of glass or tracing paper is positioned between the bulb and the plate to diffuse the light. A small drawing of trees or houses is placed very close to the end of the plate in front of the translucent glass. You then look at the drawing from the other end of the metal plate, at about the same level as the plate.

    (a) Sketch the arrangement described above.
    (b) You see a mirage. Is it an inferior or a superior mirage? Explain.
    (c) Air blows over the metal plate. Why do you see a shimmering effect?

31. A light ray is directed toward the centre of a semi-circular glass prism at different angles of incidence. The angles of incidence and refraction are recorded. Explain how the data can be used to determine the index of refraction of the glass prism. Why is this the best method?

32. A coin on the bottom of a swimming pool is 1.0 m from the edge of the pool and under 1.2 m of water (**Figure 2**). A flashlight beam is directed over the edge of the pool to illuminate the coin. At what angle relative to the ground must the flashlight be aimed?

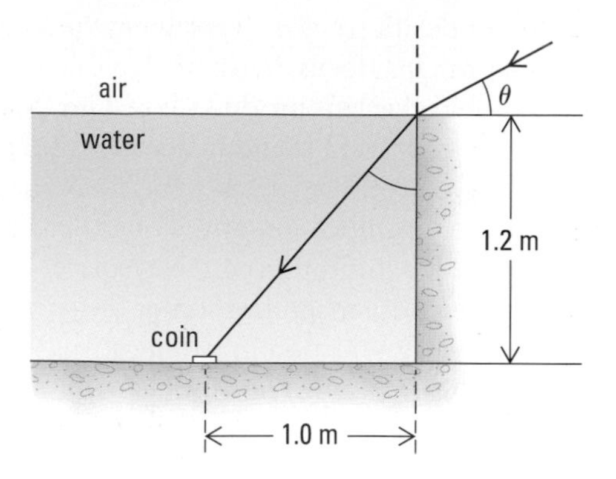

Figure 2
For question 32

33. The index of refraction for red light in glass or in water is slightly less than that for blue light or violet light, as illustrated in **Table 1**.

Table 1

Index of refraction	Glass	Water
red (660 nm)	1.512	1.331
orange (610 nm)	1.514	1.332
yellow (580 nm)	1.518	1.333
green (550 nm)	1.519	1.335
blue (470 nm)	1.524	1.338
violet (410 nm)	1.530	1.342

(a) What does this indicate about the relative speeds of red light and blue light in glass?
(b) Red light, with $n_r = 1.512$, and violet light, with $n_v = 1.530$, enter a piece of glass at an angle of incidence of 30.0°. Show that the difference between their angles of refraction is 0.24° by calculation.
(c) If the refracting surfaces of a material are not parallel, such as in the case of a prism, the emerging ray will take a completely different path. Such a change in the direction of a ray is measured in degrees. The angle between the incident ray and the emerging ray is called the angle of deviation. In **Figure 3**, an incident ray of white light enters a prism from the left. Explain why different colours of light will emerge at different angles of deviation.
(d) What observation shows that diamond has a slightly different index of refraction for each of the various colours of the spectrum?

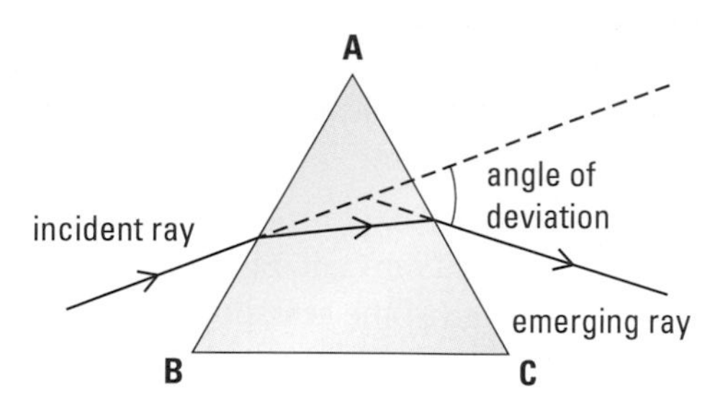

Figure 3
For question 33

Applying Inquiry Skills

34. You have discovered that when you place two mirrors at right angles to each other, three images appear in the mirrors. Try this activity. Are some of the images darker than others? Which ones are they? Why?

35. The rectangular face of a solid prism measures 3.00 cm by 5.00 cm. Draw a labelled diagram of a ray from air striking the surface along its long edge such that $\theta_i = 50.0°$ and $\theta_R = 30.0°$.
(a) At what angle does the ray leave the rectangular prism? Draw this ray in your diagram.
(b) Calculate the index of refraction of the material.
(c) What is the speed of light in the prism?
(d) Identify the substance from the table that lists various indexes of refraction (**Table 1** in section 9.4).
(e) What is the critical angle of light in the prism if the prism is surrounded by air?

36. The densities of ice, benzene (a component of motor fuel), ethyl alcohol, and water are listed along with their indexes of refraction in **Table 2**.

Table 2

	Density (g/cm^3)	Index of refraction (yellow light)
ice (0°C)	0.91	1.31
benzene	0.899	1.50
ethyl alcohol	0.791	1.36
water	1.00	1.33

(a) Reorganize the data from the lowest to the highest densities.
(b) Reorganize the data from the lowest to the highest indexes of refraction.

(c) Discuss any patterns that you discover. Formulate a hypothesis for the relationship between density and index of refraction. Find more data to test your hypothesis.

Making Connections

37. When does the Sun look oval? Research this and sketch a diagram to support your explanation.
38. The structure and function of the human eye and a simple camera have many similarities. Using the terms iris, retina, eyeball, cornea, and lens, identify corresponding camera structures and describe their function.
39. A cat's eye is similar to a human eye, but they are different in some ways. Research the similarities and differences between a cat's eye and the human eye, and prepare a short presentation for your class.
40. A baby or an infant has very little depth perception. Research the reason for this and write a short report, about two pages, on the subject. Include sketches that will help with your explanation.
41. Light that enters a camera lens perpendicularly undergoes less reflection than light that enters the lens or hits the lens at an angle. Therefore, the photograph taken would be darker near the corners and brighter at the centre. How is this problem resolved in cameras? Follow the links for Nelson Physics 11, Unit 4 Review.

GO TO www.science.nelson.com

42. When purchasing binoculars, the consumer has a great deal of choice. Prices of binoculars range from tens of dollars to thousands of dollars. Use the Internet to research different types of binoculars and answer the questions that follow. Follow the links for Nelson Physics 11, Unit 4 Review.

GO TO www.science.nelson.com

(a) What main factors govern the price of binoculars?
(b) Consider the following situations:
   - An ornithologist wishes to study the nesting habits of a rare bird from a distance of 20 m.
   - At a hockey game, a fan wishes to see the sweat on the brow of the most valuable player from the highest seats in the stadium.
   - You want to get a better look at the craters of the Moon on an upcoming camping trip.

For each situation, describe the binoculars that are best suited for the task, assuming you have a budget of $100 to $800. Explain your choice with respect to the main factors described in (a).

Exploring

43. Visit a jewellery store or do some research to find out why diamonds that look the same may have different values. Write a paragraph of your findings.
44. Laser surgery is becoming a popular solution for people with nearsightedness or myopia. Using the Internet and other sources, research the types of laser surgery used to treat myopia, how these types of surgery work, the advantages and disadvantages of the surgeries, and the percentage success in the industry. Prepare a three-page report on the results of your research, and include any diagrams or statistics that will support your findings. Follow the links for Nelson Physics 11, Unit 4 Review.

GO TO www.science.nelson.com

45. Light that passes through an eyeglass lens is partially reflected at the lens' surface and we can see a reflection of the surrounding room in eyeglasses. Research how this reflection can be greatly reduced. Prepare a diagram to illustrate the procedure and write a brief explanation of the diagram.

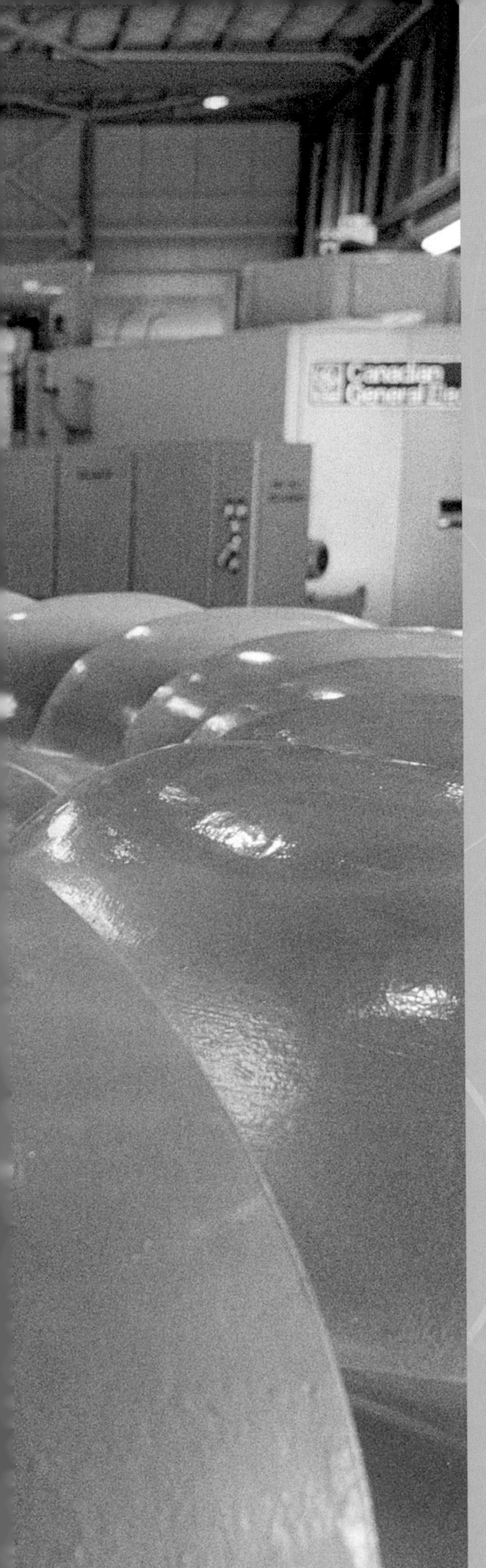

Unit 5

Electricity and Magnetism

In the future, will intelligent robots dispose of toxic waste or perform intricate operations? Dr. Mehrdad Moallem believes so. He works in the field of mechatronics. This area blends mechanics and electricity to design intelligent, autonomous machines to perform tasks that are dull, repetitive, hazardous, or that require skills beyond human capability.

Embedded computers are the "brains" of these machines. New sensor and actuator technologies, such as piezo-electric devices and shape-memory alloys, integrated with these embedded modules improve how the machines interface with the outside world. Dr. Moallem uses this technology in his research.

"In the study of mechatronic systems, we examine aspects of human function using mechanical components, sensors, actuators, and computers," explains Dr. Moallem. "It is a challenge to integrate these systems to perform tasks that will improve the quality of life."

Dr. Mehrdad Moallem, Assistant Professor at the University of Western Ontario

Overall Expectations

In this unit, you will be able to

- demonstrate an understanding of the properties, physical quantities, principles, and laws related to electricity, magnetic fields, and electromagnetic induction
- carry out experiments or simulations, and construct a prototype device, to demonstrate characteristic properties of magnetic fields and electromagnetic induction
- identify and describe examples of domestic and industrial technologies that were developed on the basis of the scientific understanding of magnetic fields

Unit

5

Electricity and Magnetism

Are You Ready?

Knowledge and Understanding

1. A simple electric circuit is shown in **Figure 1**. Copy the diagram into your notebook and answer the questions that follow.

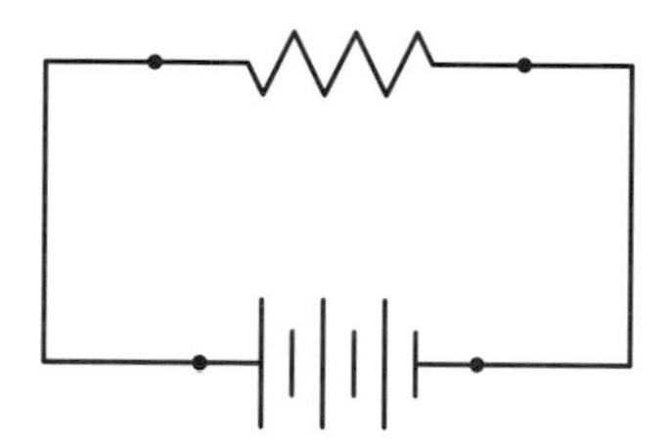

Figure 1

(a) Label all parts of the circuit (resistor, battery, wires).
(b) State the function of each part of the circuit.
(c) What are ammeters and voltmeters used for?
(d) Indicate on the diagram how an ammeter and a voltmeter should be connected.
(e) Distinguish between electric potential difference and electric current.

2. (a) **Figure 2** shows two types of electric circuits: series and parallel. Copy each circuit into your notebook, label each circuit as series or parallel, and explain how you can tell the difference between the two types of circuits.

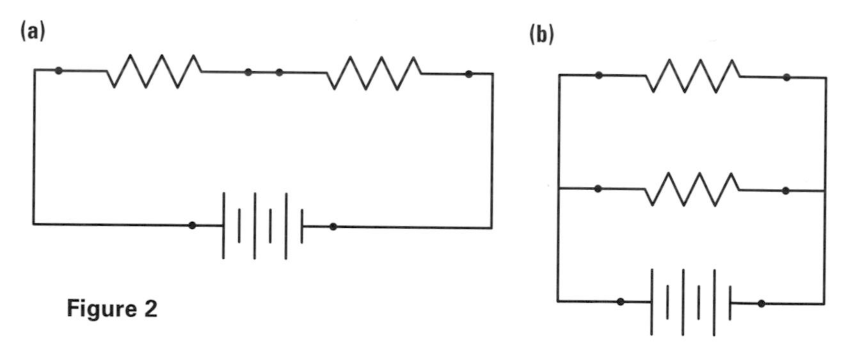

Figure 2

(b) Copy **Table 1** into your notebook and complete it by comparing the properties of series and parallel circuits.

Table 1

Property	Series circuit	Parallel circuit
electric current		
electric potential difference		

3. Each of the diagrams in **Figure 3** shows two bar magnets with poles close to each other. Copy the diagrams into your notebook and do the following:

(a) Indicate on each diagram whether the magnets will be attracted or repelled.
(b) State the law of magnetic poles.
(c) List some substances that are attracted to magnets.

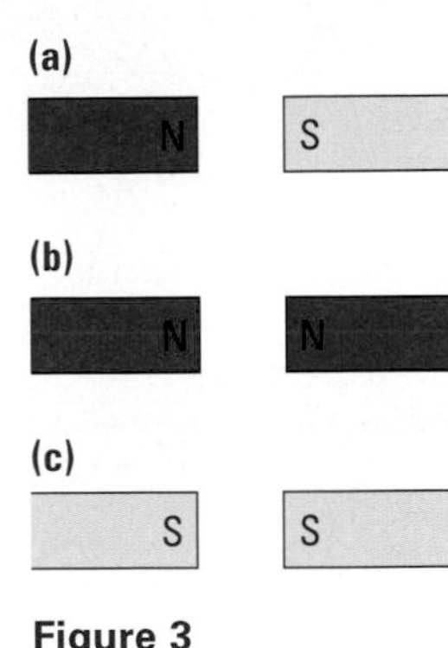

Figure 3
For question 3

Inquiry and Communication

4. Examine the graph of electric potential difference (V) versus electric current (I) in a circuit, shown in **Figure 4**.
   (a) According to the graph, how are electric potential difference and electric current related?
   (b) What does the slope of the graph represent?

Math Skills

5. Use the graph in **Figure 4** to do the following:
   (a) Find the slope of the graph.
   (b) Write an equation relating V and I.
   (c) Describe the significance of lines that have smaller and greater slopes.

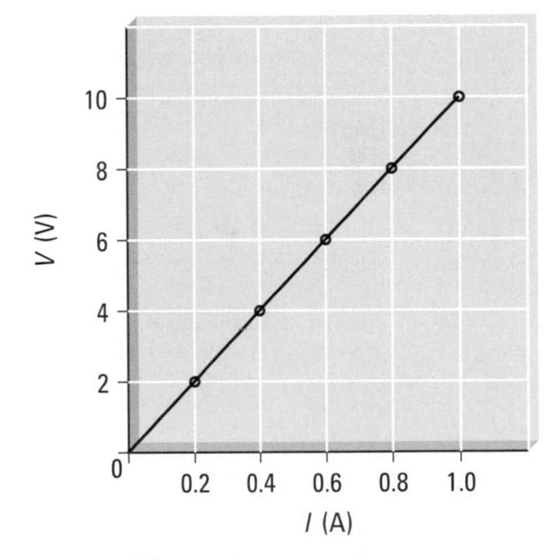

Figure 4
For question 5

Technical Skills and Safety

6. In **Figure 5**, there is something wrong with each circuit. Identify the problem and suggest a way to solve it.
7. A student is performing an experiment involving electricity. A conductor gets very hot and starts to smoke. List some possible steps that the student could take to safely resolve the situation.
8. Some permanent magnets are very strong, while others can pick up only a few metal paper clips.
   (a) How can magnets be safely stored?
   (b) What are some safety considerations when dealing with strong magnets?
   (c) List some equipment in the laboratory that can be damaged by strong magnets.

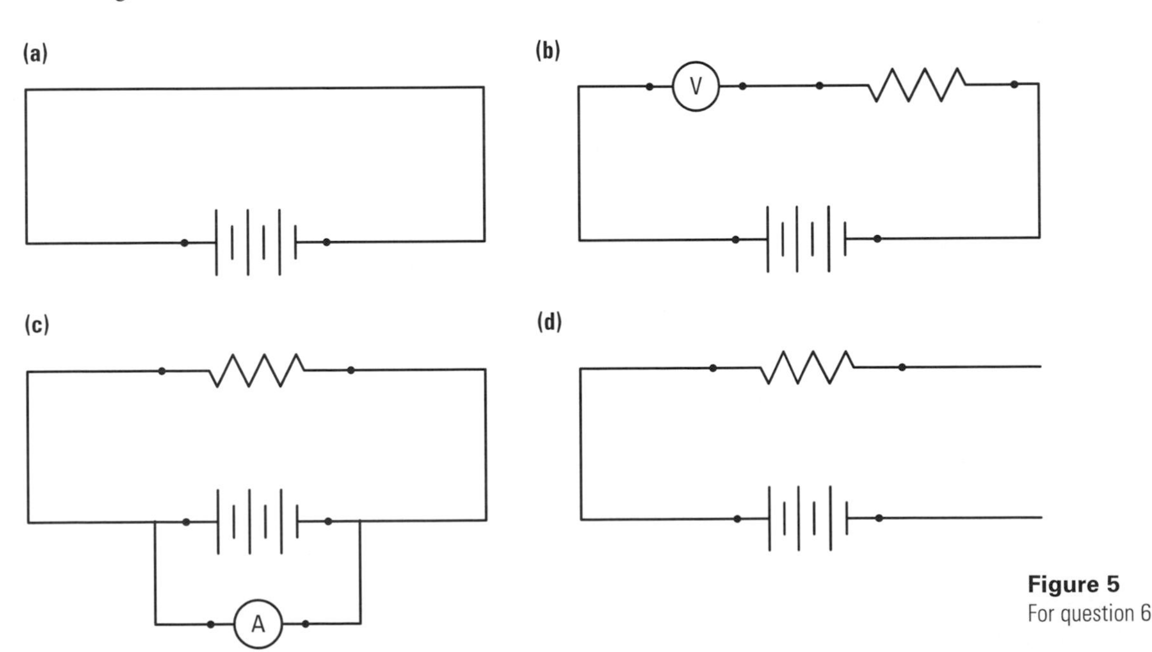

Figure 5
For question 6

Chapter

12

Electricity

In this chapter, you will be able to

- use a variety of technologies to investigate the electrical nature of matter
- derive, understand, test, and use equations related to electrical phenomena
- investigate technologies developed using the principles of electricity
- develop an understanding of how the different concepts in electricity are interrelated
- investigate how electrical principles are used at home and in industry
- develop an understanding of how physics concepts are created, tested, and used

How does the electrical nature of matter affect the way you live? When there's lightning outside (**Figure 1**) you probably tend to stay indoors. If the power goes out, many of your regular activities such as watching television, cooking, and working on a computer are no longer possible. You might not think about it all the time, but you depend on electricity for a lot. Just ask anyone in Eastern Canada who went without electricity for days after a major ice storm in January 1998 brought down hundreds of power lines (**Figure 2**).

To understand electricity, we must learn how it affects matter. Objects can develop an electric charge, and some objects allow a charge to flow through them. You've probably encountered static electricity in one form or another, like the shock you feel after touching a door knob on a cold, dry winter day. In **Figure 3** you can see a dramatic demonstration of some of the properties of static electricity, using a Van de Graaff generator.

Figure 1
Lightning is a phenomenon of nature to be respected. The lightning bolt is nature's most spectacular example of a static electric discharge.

Reflect on your Learning

1. What are some objects or devices that can do the following:
   (a) hold a static charge?
   (b) allow an electric charge to flow through them?
2. Identify five activities of everyday life that would change if the supply of electricity was turned off during the winter months. In each case, describe alternatives.
3. Distinguish between alternating current and direct current.
4. Describe the energy transformations that occur when
   (a) an electric fan is turned on
   (b) wood burns in a fireplace
5. Distinguish between potential difference (V) and current (I).

Throughout this chapter, note any changes in your ideas as you learn new concepts and develop your skills.

Safety and Electricity

From previous studies and personal experiences, list seven safety rules that should be followed when dealing with electricity. These rules should involve electricity in the home and in the lab. Beside each rule, write a brief explanation of why the rule is important.

(a) Use the following as a guide for your list of rules:
- electrical outlets
- electric cords and cables
- power supplies
- electric devices near water
- what to do when a device gets hot, begins to smoke, or catches fire
- storage of equipment
- electrical meters

(b) Be prepared to discuss the rules with the class and your teacher. All the rules should be clear to you before starting lab activities.

(c) Add any new rules or change any rules that need adjusting after the class discussion.

Figure 3

Figure 2
After the ice storm in Quebec and Eastern Ontario, in 1998, many hydro workers from all over Canada worked around the clock to repair power lines.

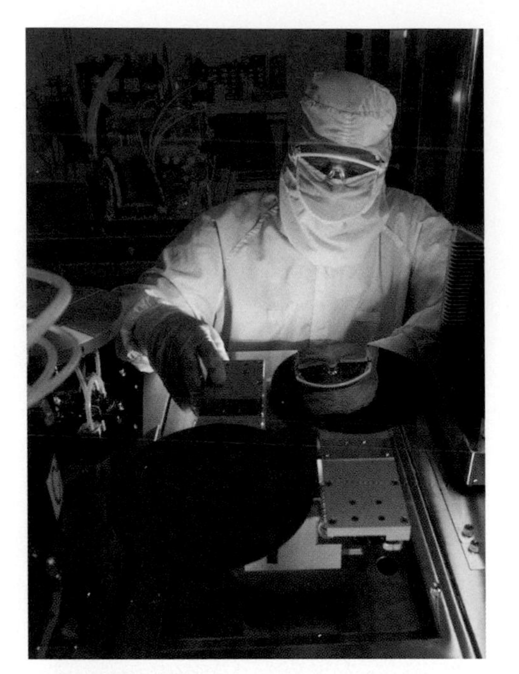

Figure 1
In a high-tech clean room, many precautions are taken to protect sensitive devices from electrostatic discharge.

12.1 Electrostatics

Electrostatics is the area of physics that deals with objects that have an **electric charge**. Electric charge is a property of matter that is responsible for all electric and magnetic forces and interactions. Electric charge is arbitrarily described as negative or positive. **Static electricity**, which is a buildup of stationary electric charge on a substance, is one of the most difficult problems encountered in a high-tech clean room, where electronic equipment is designed and assembled (**Figure 1**). Often the electronic equipment is so sensitive that even small electric discharges can cause permanent damage. Anyone working in such an environment needs a complete understanding of the properties of static electricity.

The Electrical Structure of Matter

The Bohr-Rutherford model of the atom (**Figure 2**) can help us understand electrical phenomena. The principal concepts in this atomic model are as follows:

1. Matter is composed of sub-microscopic particles called **atoms**.
2. Electric charges are carried by particles within the atom that are called **electrons** and **protons**.
3. Protons are found in a small central region of the atom called the **nucleus**. They are small, heavy particles, and each one carries a positive electric charge of a specific magnitude, called the **elementary charge** (*e*).
4. Electrons move in the space around the nucleus. They are small, very light particles (each with only slightly more than $\frac{1}{2000}$ the mass of a proton), yet each of them carries a negative electric charge equal in magnitude to that of the proton.
5. Atoms are normally electrically neutral, because the number of (positive) protons in the nucleus is equal to the number of (negative) electrons in the space around the nucleus.
6. **Neutrons** are small, heavy particles (each slightly heavier than a proton) found in the nucleus. They carry no electric charge.
7. If an atom gains an extra electron, it is no longer neutral but has an excess of electrons and a net negative charge. Such an atom is called a **negative ion**.
8. If an atom loses an electron, it will have a deficit of electrons and a net positive charge. Such an atom is called a **positive ion**.

The atoms of a solid are held tightly in place; their nuclei vibrate but are not free to move about within the solid. Since the nuclei contain all of the protons, the positive charges in a solid remain fixed. However, it is possible for negative charges within a solid to move because some of the outermost electrons can move from atom to atom.

The Basics of Electrostatics

When charged objects are brought close to each other they interact at a distance, causing them to attract or repel. Even some neutral objects are attracted to charged objects. Through experimentation, the American statesman, inventor, and scientist Benjamin Franklin (1706–1790) discovered that there are two types of charges; he gave them the names *positive* and *negative*. The **fundamental laws of electric charges** are as follows:

electric charge: a basic property of matter described as negative or positive

static electricity: a buildup of stationary electric charge on a substance

atom: sub-microscopic particle of which all matter is made

electron: negatively charged particle which moves around the nucleus of an atom

proton: positively charged particle found in the nucleus of an atom

nucleus: the central region of an atom, where protons and neutrons are found

elementary charge: (*e*) electric charge of magnitude equal to the charge on a proton and an electron

neutron: a neutral particle found in the nucleus

negative ion: an atom that has at least one extra electron and is negatively charged

positive ion: an atom that has lost at least one electron and is positively charged

fundamental laws of electric charges: Opposite charges attract each other. Similar electric charges repel each other. Charged objects attract some neutral objects.

Fundamental Laws of Electric Charges

- Opposite electric charges attract each other.
- Similar electric charges repel each other.
- Charged objects attract some neutral objects.

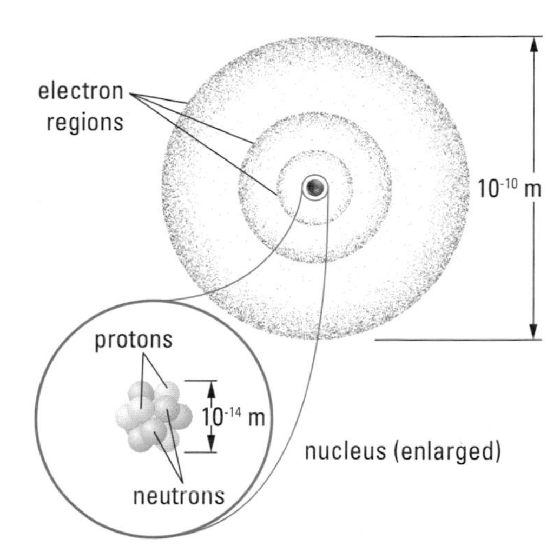

Figure 2
The Bohr-Rutherford model of the atom

Investigation 12.1.1

Electric Charges and Forces

In this investigation, you will explore the interactions between objects possessing different types of electric charges and the effects that they have on each other.

INQUIRY SKILLS

- ● Questioning
- ○ Hypothesizing
- ● Predicting
- ○ Planning
- ● Conducting
- ● Recording
- ● Analyzing
- ● Evaluating
- ● Communicating

Do not perform this investigation if you are allergic to animal fur.

Question

Can the Bohr-Rutherford model of the atom be used to predict the interaction of charged objects?

Prediction

(a) Using the Bohr-Rutherford model of the atom and the laws of electric charges, predict the type of electrical interaction that will occur between two ebonite (plastic) rods rubbed with fur and brought close to one another.

Materials

ebonite rods or polyethylene strips and fur
insulated rod hanger
glass rods or acetate strips and silk
pith ball suspended on thread
small bits of paper
sawdust
iron filings

Procedure

1. Place an ebonite rod on a hanger suspended by a thread. Bring another ebonite rod close to the suspended rod and note whether the suspended rod moves in response to the other rod's proximity.
2. Rub one of the ebonite rods vigorously with the fur and place it in the hanger. Then rub the other ebonite rod with fur and bring it near the suspended rod. Note the effect that the second rod has on the first.
3. Repeat steps 1 and 2 with the glass rods, rubbing them with the silk.
4. Suspend a glass rod rubbed with silk from the hanger and bring an ebonite rod rubbed with fur near it. Note the effect that the ebonite rod has on the glass rod.
5. Suspend a pith ball from a thread and touch it with your finger. Slowly bring a charged ebonite rod near to, but not touching, the pith ball, and note its effect. Note what happens after the pith ball and the ebonite rod have touched.
6. Repeat the procedure using a charged glass rod.
7. Using first a charged glass rod and then an ebonite rod, approach each of the following in turn and note the results: small bits of paper, sawdust, iron filings, and a thin, continuous stream of water from a tap (**Figure 3**).

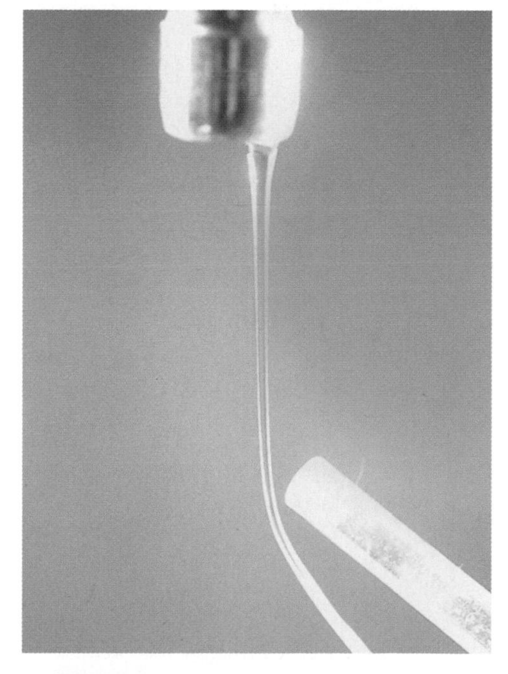

Figure 3
A charged rod is held near a water stream. Will other liquids such as cooking oil show the same behaviour?

Analysis

(b) How many different types of electric charge were you able to identify?
(c) What was the purpose of steps 1 and 3?
(d) Why did you touch the pith ball with your finger in step 5?
(e) Give simple descriptions of the interaction between similarly charged objects and between oppositely charged objects, and explain your observations using the Bohr-Rutherford model of the atom.
(f) What happened when a charged rod was brought near some neutral objects? Will the same thing occur with all neutral objects? Explain your observations using the Bohr-Rutherford model of the atom.
(g) What must be true about water molecules for them to behave as they do in the presence of a charged rod?

Evaluation

(h) Evaluate your predictions.
(i) Identify sources of error in the procedure and suggest a few changes that will help reduce error.

conductor: solid in which charge flows freely

insulator: solid that hinders the flow of charge

Charging an Object

Charging an object involves the addition or removal of electrons. If electrons are removed from an object, it will be charged positively; if electrons are added, it will be charged negatively. In both cases the number of protons in each object remains constant. The change in charge on the object is due to the loss or gain of electrons (**Figure 4**). The net charge in solids is due to an excess or deficit of electrons.

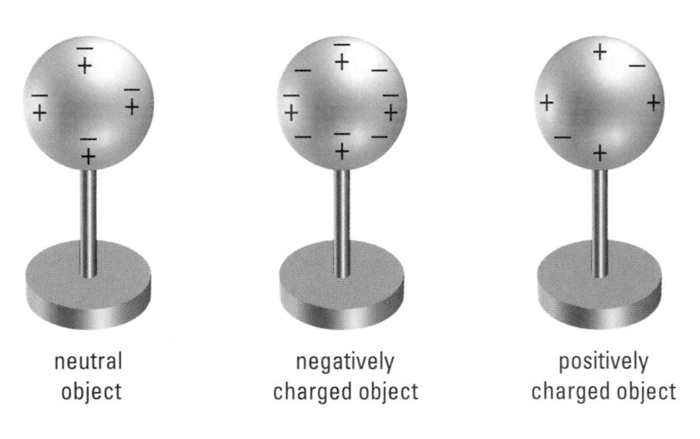

Figure 4
Neutral and charged objects can be represented by sketches with positive and negative signs marked on the objects.

(a)

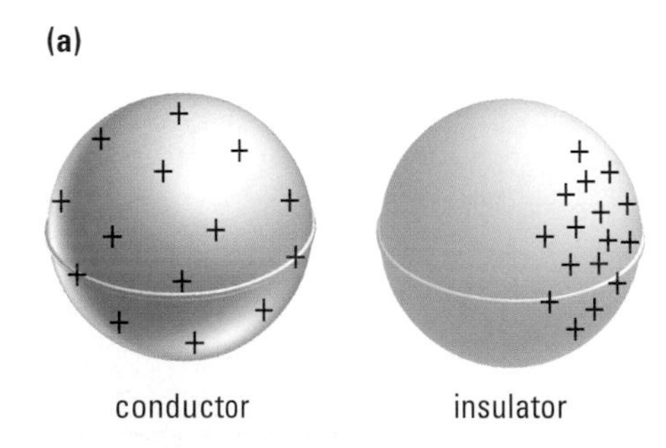

(b)

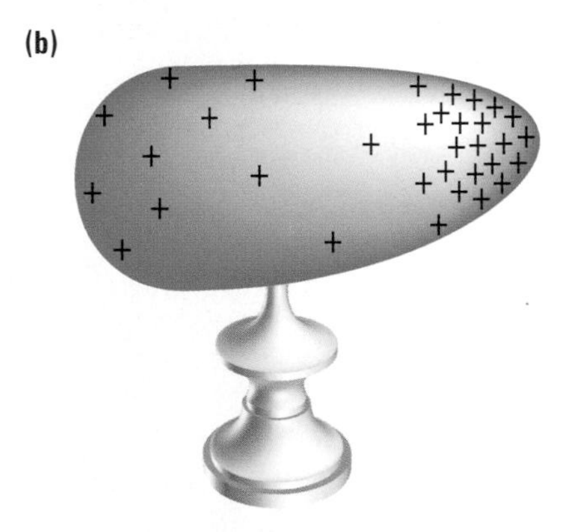

Figure 5
(a) On a spherical conductor, the charge spreads out evenly. On an insulator, the charge remains in the spot where it was introduced.
(b) On other shapes of conductors, the charges tend to repel one another toward the more pointed surfaces.

Most solids fall into one of two broad categories based on their electrical properties. They are either **conductors**, solids in which charge flows freely, or **insulators**, solids that hinder the flow of charge. Most metals are excellent conductors, the best being silver, gold, copper, and aluminum. Some of the electrons in these conductors are called "conduction electrons" because of their ability to move about. On a spherical conductor, charge spreads out evenly (**Figure 5(a)**). On other shapes, charges tend to repel one another toward the more pointed surfaces (**Figure 5(b)**). Plastic, cork, glass, wood, and rubber are excellent insulators. If you rub wool on one end of a plastic comb so it develops a charge, the other end will remain neutral due to the limited mobility of charges in insulators.

Charging by Friction

An atom holds onto its negative electrons by the force of electrical attraction of its positive nucleus. Some atoms exert stronger forces of attraction on their electrons than others (**Figure 6**).

When charging by friction, the type of charge that develops can be determined using a chart called the **electrostatic series** (Table 1). When rubbed together, substances at the top of the series will lose electrons to substances lower in the series, meaning that the substance higher in the table becomes positively charged while the substance from lower in the table becomes negatively charged.

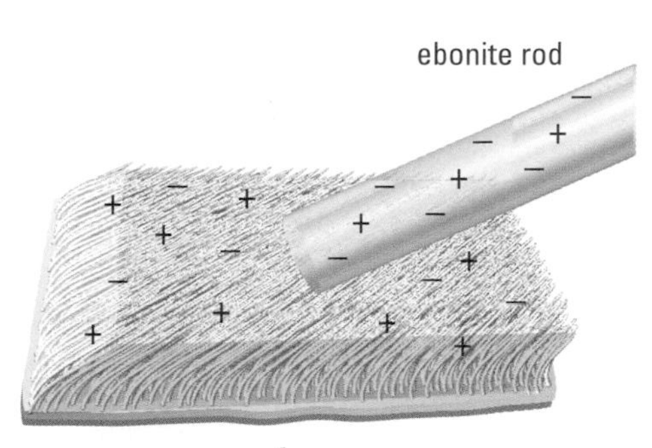

Figure 6
When ebonite and fur are rubbed together, some of the electrons from the atoms in the fur are "captured" by the atoms in the ebonite, which exert stronger forces of attraction on those electrons than do the atoms in the fur. After rubbing, the ebonite has an excess of electrons and the fur has a deficit.

electrostatic series: chart that shows a substance's tendency to gain or lose electrons

Table 1 The Electrostatic Series

Substance	
acetate	weak hold on electrons
glass	↓
wool	
cat fur, human hair	
calcium, magnesium, lead	
silk	
aluminum, zinc	increasing tendency
cotton	to gain electrons
paraffin wax	
ebonite	
polyethylene (plastic)	
carbon, copper, nickel	
rubber	
sulphur	↓
platinum, gold	strong hold on electrons

Electroscopes

An **electroscope** is a device that is used to detect the presence of an electric charge and to determine the charge's "sign" (that is, whether it is positive or negative).

electroscope: device that is used to detect the presence of an electric charge and to determine the charge's "sign" (that is, whether it is positive or negative)

Pith-Ball Electroscope

A pith-ball electroscope is a light, metal-coated ball suspended on an insulating thread (**Figure 7(a)**). If the ball is charged, it can be used to detect the presence of a charge on nearby objects. It will be repelled by a similarly charged object and attracted to an oppositely charged object or a neutral object.

(a) (b) negative charge positive charge

Figure 7
(a) Repulsion of a negative pith ball by a charged ebonite rod
(b) Similar net charges on a pair of metal leaves cause the leaves to move apart, indicating the presence of a charge on the electroscope.

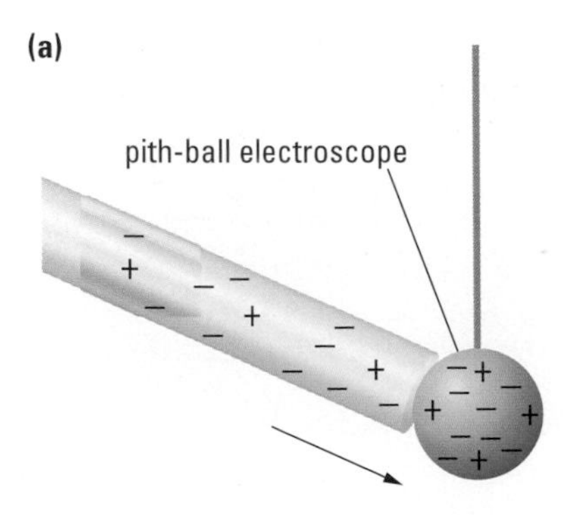

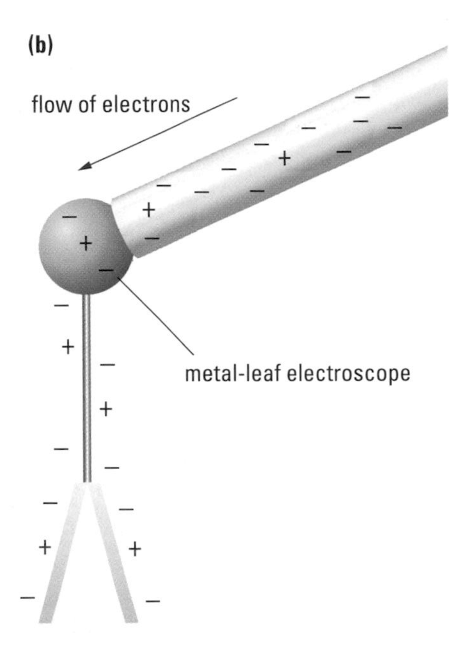

Figure 8
Charging by contact with a charged ebonite rod

Metal-Leaf Electroscope

A metal-leaf electroscope consists of two thin metal leaves suspended from a metal rod in a glass container. A metal knob or plate is attached to the top of the metal rod. Since the centre of the electroscope is made of a conducting material, any charge on it spreads out over the entire knob, rod, and leaves. Since the leaves are then charged similarly, they repel each other and move apart, indicating the presence of a charge (**Figure 7(b)**). The farther apart the leaves move, the greater the charges they are carrying.

Charging by Contact

When a charged ebonite rod is touched to a neutral pith-ball electroscope, some of the excess electrons on the ebonite rod are repelled by their neighbours and move onto the pith ball (**Figure 8(a)**). The pith ball and the ebonite rod share the excess of electrons that the rod previously had, and both end up with net negative charges. A similar sharing of electrons occurs when a charged ebonite rod touches the knob of a metal-leaf electroscope (**Figure 8(b)**).

When a positively charged glass rod is used (**Figure 9**), some of the electrons on the neutral pith-ball or metal-leaf electroscope are attracted to the glass rod until the electroscope shares the deficit of electrons that the rod previously had. Both end up with net positive charges. These examples show that an object that is charged by contact has the same charge as the charging object.

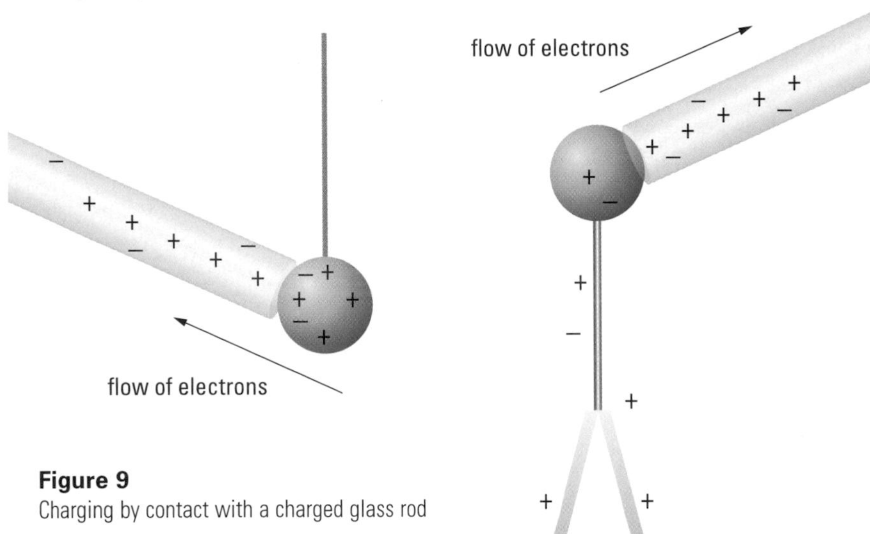

Figure 9
Charging by contact with a charged glass rod

Charging by Contact

How is an object charged by contact?

- Rub an ebonite rod with fur and touch it to the neutral pith ball or leaf electroscope. Bring the rod close to the pith ball again and observe what happens.
- Repeat the procedure for a glass rod charged with silk.
  (a) When the charged ebonite rod touched the pith ball, what type of charge did the pith ball acquire? Explain by discussing the way the electrons moved. Use sketches showing before, during, and after contact.
  (b) When the charged glass rod touched the pith ball, what type of charge did the pith ball acquire? Explain using diagrams.

Charging by Induction

When a charged ebonite rod is brought close to the knob of a neutral metal-leaf electroscope, electrons on the electroscope will move as far away as possible from the negative rod. This results in an **induced charge separation** on the electroscope since the positive charges cannot move. (If the charged rod is removed the electrons will return to their original distribution.) If you **ground** the electroscope by touching it, some electrons are induced to flow through your finger, leaving the electroscope. When you remove your finger from the electroscope, the device is left with a deficit of electrons, resulting in a net positive charge (**Figure 10(a)**).

A positively charged rod held near the knob of an electroscope will also result in an induced charge separation (**Figure 10(b)**). If grounded, electrons will move through your finger onto the electroscope. When you remove your finger from the electroscope, the device is left with an excess of electrons, resulting in a net negative charge. From this example we can conclude that an object that is charged by induction has the opposite charge to that of the charging rod.

induced charge separation: distribution of charge that results from a change in the position of electrons in an object

ground: a connection of an object to Earth through a conductor

Activity 12.1.1

Charging by Induction

In this activity, you will observe how an object is charged by induction.

Procedure

1. Place two short metal rods on top of two glass beakers so they are touching each other, as shown in **Figure 11**.

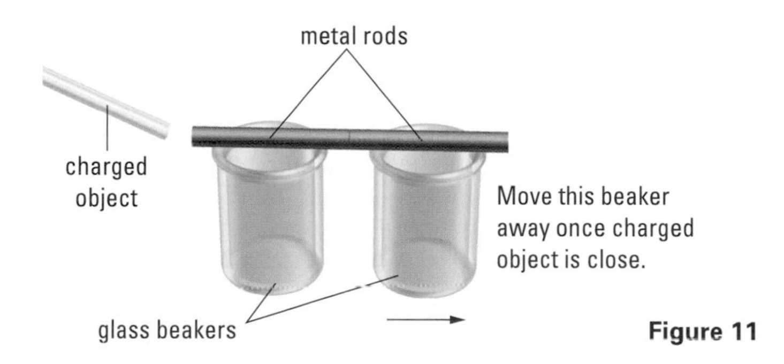

Figure 11

2. Bring a charged object close to one of the metal rods and then pull the other metal rod away by touching the glass beaker only. Put down the charged object and test each object for a charge with an electroscope.
3. Put the two metal rods back together by touching only the beakers and test them again for a charge.

Analysis

(a) What happens to the electroscope when it is brought close to the metal rods after separating them? Explain why this occurs by discussing the movement of electrons.

(b) What happens to the electroscope after the metal rods are put back together? Explain.

(c) Draw a diagram of the charged rod near the two metal rods before they are separated, showing the positions of the positive and negative charges.

(a)

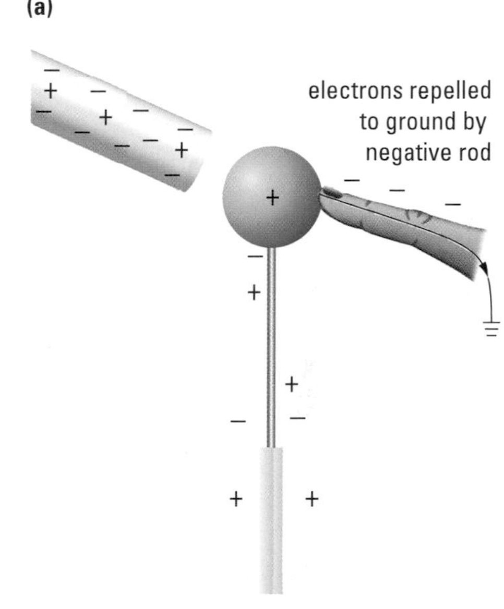

grounded electroscope in the presence of a negatively charged rod

(b)

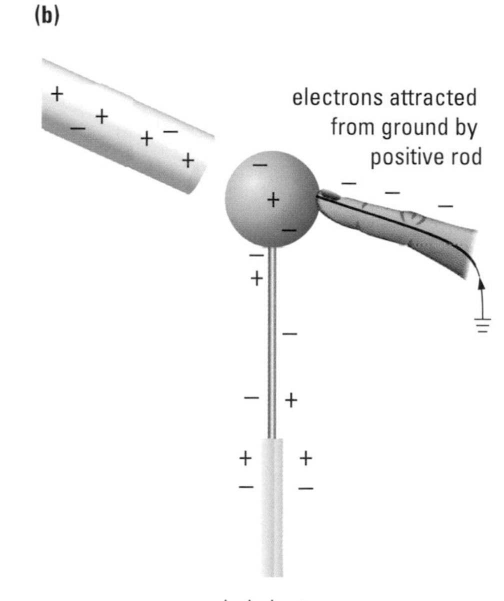

grounded electroscope in the presence of a positively charged rod

Figure 10
Notice that the leaves of the electroscope fall, indicating a neutral condition of the leaves while the finger is in position.

A summary of the methods of charging objects is shown in **Table 2**.

Table 2 Summary of Methods of Charging Objects

Method of charging	Friction	Contact	Induction and grounding
Initial charges on objects	both objects neutral	one neutral and the other charged	one neutral and the other charged
Steps	rub two objects together	touch charged object to neutral object	bring charged object close to neutral grounded object remove ground and then charged object
Final charges on objects	oppositely charged	similarly charged	oppositely charged

Practice

Understanding Concepts

1. If an ebonite rod is rubbed with fur, the rod becomes negatively charged. What is the source of the charge?
2. If the knob of a positively charged electroscope is approached slowly by a negatively charged rod, what happens to the leaves of the electroscope? Why?
3. Explain fully what happens when a positively charged rod touches the knob of a neutral metal-leaf electroscope.
4. Why are metallic fibres used in the pile of some carpets?
5. (a) Define electrostatic induction.
   (b) Given a glass rod, silk, and two metal spheres mounted on insulating stands, describe how to charge the spheres oppositely by electrostatic induction.
6. A negatively charged rod is brought near a neutral sphere supported by an insulating stand. What type of charge would you expect to find on the side of the sphere nearest the rod and on the other side if the sphere is (a) a conductor and (b) an insulator? Explain.

Controlling Static Charges in a Clean Room

How does a worker in a clean room use knowledge about the electrical nature of matter to prevent electric discharges from damaging sensitive devices? Keep in mind that people are prime generators of static charge and even just walking around can develop a significant charge. When the worker touches a sensitive piece of equipment, the charge will pass into it, causing damage. To prevent this from happening many precautions are taken.

Workers often wear wrist straps that are connected to a ground cord. Special clothing and shoes are worn to inhibit the build-up of charge and to help dissipate any charge that might develop. Special floors are installed, floor mats are used, and just about everything is grounded, including chairs, desks, and even the devices being constructed.

For those items that cannot be grounded, mostly insulators like plastic, the charge is dissipated by the air. Devices are installed that create charged particles in the air. These charged particles are attracted to the opposite charges that develop on objects and help to neutralize them.

SUMMARY Electrostatics

- The fundamental laws of electric charges state that opposite electric charges attract each other; similar electric charges repel each other; and charged objects attract some neutral objects.
- The net charge in solids is due to an excess or deficit of electrons.
- A conductor is a solid in which charge flows freely; an insulator is a solid that hinders the flow of charge.
- Solids can be charged by friction, contact, induction, or grounding.

Section 12.1 Questions

Understanding Concepts

1. Two rods, one brass and the other plastic, have been charged by contact at one end by an ebonite rod rubbed with fur while being supported by an insulator. Compare the distribution of electric charge on the two rods.
2. Sometimes when you are taking off a wool sweater by pulling it over your head, you find that your hair is attracted to the sweater. Explain why this is so by referring to the movement of electrons.
3. State the type of charge that would develop on each material when each pair of materials is rubbed together.
   (a) wool and ebonite
   (b) glass and paraffin wax
   (c) polyethylene and fur
4. A suspended polyethylene (plastic) strip is given a negative charge. What can you conclude for certain if you bring another object close to the polyethylene strip and the two objects
   (a) repel each other?
   (b) attract each other?
5. **Figure 12** shows an arrangement for demonstrating one of Newton's laws of motion. The three-armed device is called a pinwheel.
   (a) Decide the direction in which the pinwheel will rotate when the generator is turned on, and explain why.
   (b) Which one of Newton's laws does this action illustrate?

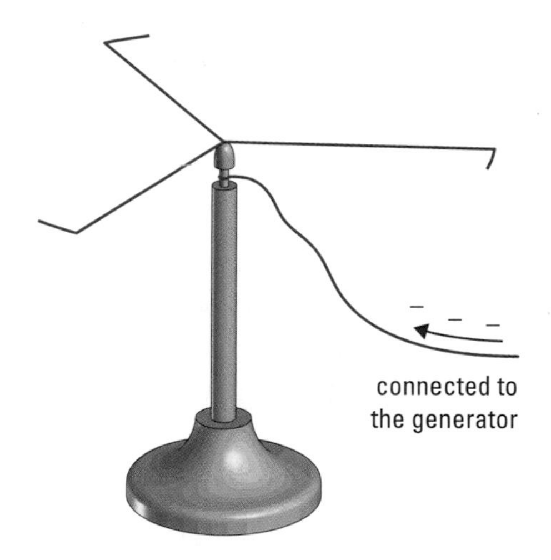

Figure 12
For question 5

6. In the 16th century, William Gilbert (1540–1603), an English physicist, discovered several insulators that would accept an electric charge, but no conductors. Give possible reasons why he did not discover that conductors will also accept charges.
7. The following observations are made of the interactions between various combinations of four pith balls, A, B, C, and D. The force between C and D is a repulsion. A attracts both B and D. If A is repelled by an ebonite rod that has been rubbed with fur, what are the possible charges on each pith ball?

Applying Inquiry Skills

8. Several dangerous explosions have occurred when powerful jets of water were used to wash ocean-going oil tankers. What was the likely cause of these explosions? What could be done to prevent them?

(continued)

9. A pith-ball electroscope may be in a charged or a neutral state. Describe the steps you would take to determine whether it is charged or neutral. If the electroscope is charged, how would you determine what charge it possesses?

Making Connections

10. While metal car bodies are being spray-painted at the factory, they are grounded. Explain the advantages of this. (*Hint:* As the spray leaves the nozzle at high speeds, it becomes charged.)
11. When clothes made of different materials tumble in a warm clothes dryer they have a tendency to cling to one another. Coated strips or spray may be purchased to prevent or remove the "static cling." Research how "anti-static cling" formulations work. Follow the links for Nelson Physics 11, 12.1.

GO TO www.science.nelson.com

12. Electrostatic printing, commonly known as xerography or photocopying, uses the principles of static electricity to print on paper or other materials. Research and explain the steps in the process where static electricity plays a role. Follow the links for Nelson Physics 11, 12.1.

GO TO www.science.nelson.com

Reflecting

13. Would you feel comfortable working in a high-tech clean room knowing that you would have to be extremely careful about static charge? List some advantages and disadvantages of such a job.

12.2 Electric Fields and Electric Charge

What causes electric force? How is electric charge measured? The study of charges will be far more productive if we address these questions.

Electric Fields

Electric charges exert forces that can attract and repel each other even when they are not in direct contact. What causes the force? We don't see anything between the charges that could be responsible for it. Yet this kind of force is already familiar to you. The force of gravity was explained in terms of a gravitational field of force—when a mass is placed in the gravitational field of another mass, the first mass experiences a force of attraction toward the second mass.

It is reasonable to assume that the forces between charged objects may also be due to a field of force. If this is true, then every charged object creates an **electric field** of force in the space around it (**Figure 1**) and any other charged object in that field will experience a force of electrical attraction or repulsion.

electric field: the space around a charged object where forces of attraction or repulsion act on other objects

The electric field is represented by drawing a series of field lines around the charged object. Field lines show the direction of the electric force on a small positive test charge placed at each and every point in the field. It is customary to use a positive charge as a test charge. The relative distance between adjacent field lines at a given point is an indication of the strength of the electric field at that point. Some typical electric fields are shown in **Figure 2**.

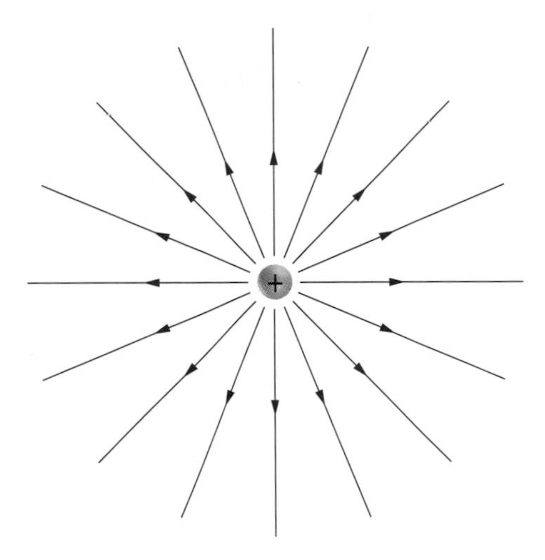

Figure 1
Electric field around a single positive charge

(a)

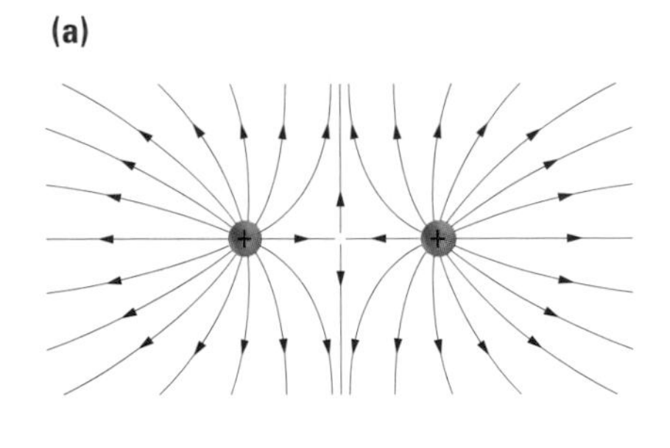

(b)

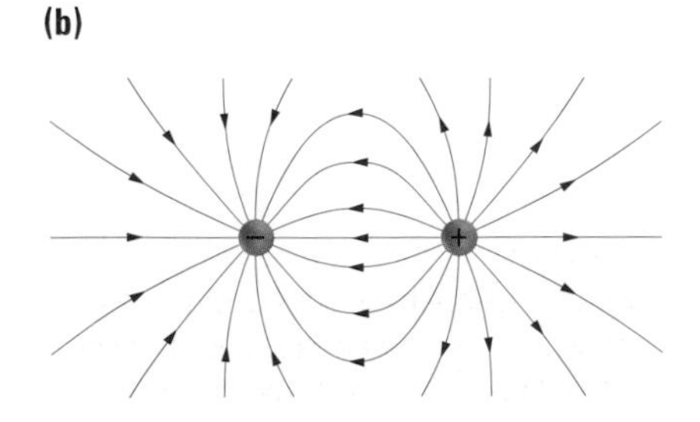

Figure 2
Some typical electric fields. Notice that these field lines are directed away from the positive spheres and toward the negative spheres.
(a) Like charges
(b) Opposite charges

Measuring Electric Charge

A quantitative analysis of the factors affecting electric force was first performed by the French physicist Charles Augustin de Coulomb (1736–1806). By performing experiments similar to that shown in **Figure 3**, Coulomb found that the magnitude of the force between two charged objects is directly proportional to the product of the charges and inversely proportional to the square of the distance between them. This famous relationship, called **Coulomb's law**, was of immense importance to our understanding of electric forces. In his honour, the unit of electric charge is called the **coulomb** (C).

To give you an idea of the magnitude of a coulomb: 1 C of electric charge is approximately the amount that would pass through a 100 W light bulb in 1 s, operating at 100 V. The relationship discovered by Coulomb can be written as:

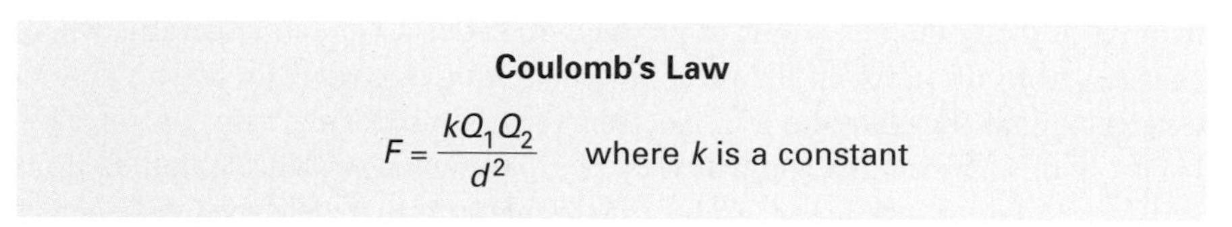

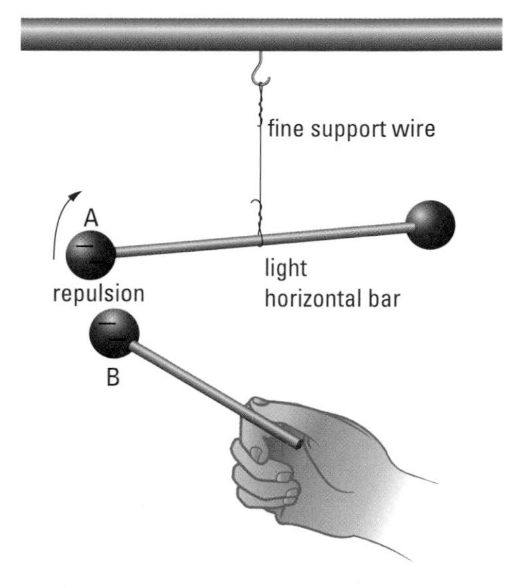

Figure 3
Coulomb devised an experiment to measure the force of repulsion between two charged objects, with one object suspended from a bar by a vertical wire. When charged object B was brought close to charged object A, it repelled A and caused the wire to twist a measurable amount.

Early in the 20th century Robert A. Millikan (1868–1953), an American physicist, performed a series of experiments proving the existence of a smallest unit of electric charge; all other electric charges are simple multiples of this smallest charge. He reasoned that this elementary charge (e) is the charge on a single electron.

Oil drops were sprayed into the space between two parallel metal plates (**Figure 4**). A light was shone on the oil drops, and they were observed through a telescope. A power supply was connected to the plates so that an electric force would act on the oil drops between the plates. An upward electric force was exerted on those drops whose charge was the same sign as the lower plate's. By adjusting the amount of charge on the plates, it was possible to isolate a single oil drop and balance it so that the downward gravitational force and the upward electrical force were equal.

Using measurements related to the "balancing field," and the speed with which the oil drops fell when the field was removed, Millikan was able to calculate the amount of electric charge on the oil drop, in coulombs.

Coulomb's law: the magnitude of the force between two charged objects is directly proportional to the product of the charges and inversely proportional to the square of the distance between them

coulomb: (C) the SI unit of electric charge

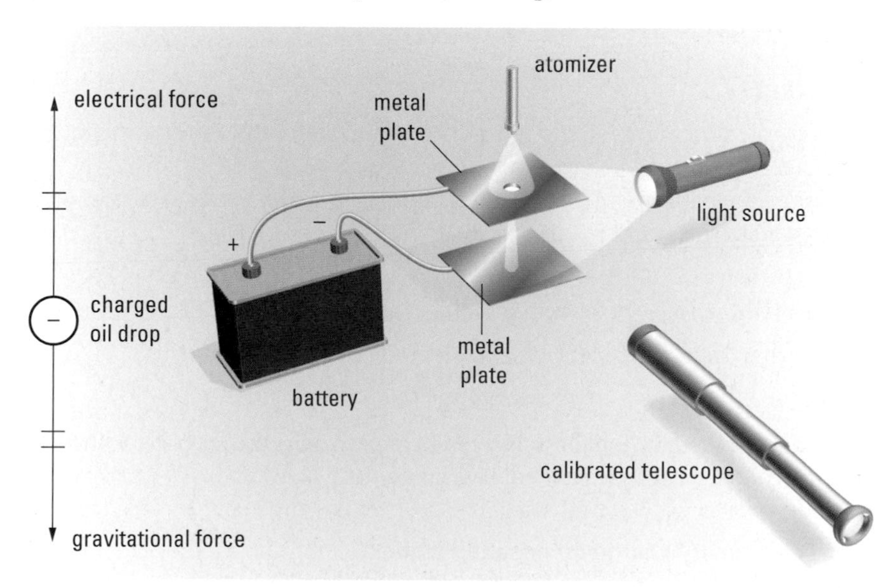

Figure 4
Millikan assumed that when tiny oil drops are sprayed from an atomizer, they become charged by friction—some acquiring an excess of a few electrons, while others have a deficit. Although there was no way of knowing how many extra electrons there were on an oil drop, or how many were missing, Millikan was able to devise a technique for measuring the total amount of charge on each individual drop.

By repeating this procedure many times, using the same oil drop with different amounts of charge on it, and using different oil drops, Millikan was able to compile a long list of values for the amount of charge on an oil drop. But how was he able to determine the value of the charge on an electron from this list of values for the total charge on a drop?

Lab Exercise 12.2.1

Investigating Data from Millikan's Oil Drop Experiment

Figure 5
Robert A. Millikan (1868–1953) was awarded the Nobel Prize for Physics in 1923.

By 1909, Robert Millikan (**Figure 5**) was able to determine the charge on an electron by studying the behaviour of charged oil drops. Using an apparatus where charged drops of oil fell in the presence of a strong electric field, he was able to determine that the charge on an electron was a fundamental constant of electricity. In this lab exercise, you will analyze experimental evidence obtained from Millikan's oil drop experiment and search for patterns that yield the fundamental charge on an electron.

Observations

Note: The values listed below represent the charges calculated on 12 oil drops, a very small portion of the data collected by Millikan.

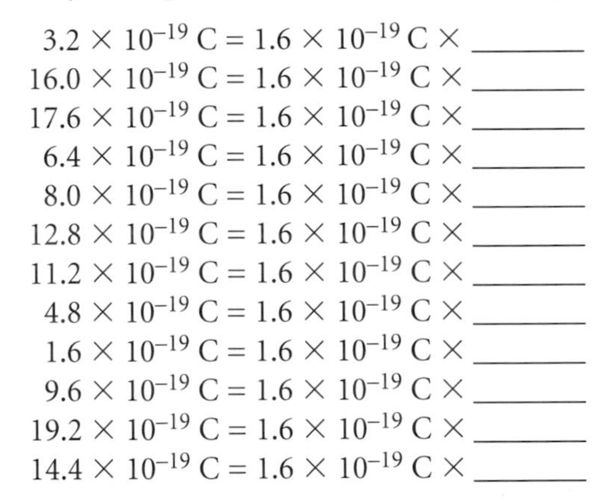

3.2×10^{-19} C = 1.6×10^{-19} C × ______
16.0×10^{-19} C = 1.6×10^{-19} C × ______
17.6×10^{-19} C = 1.6×10^{-19} C × ______
6.4×10^{-19} C = 1.6×10^{-19} C × ______
8.0×10^{-19} C = 1.6×10^{-19} C × ______
12.8×10^{-19} C = 1.6×10^{-19} C × ______
11.2×10^{-19} C = 1.6×10^{-19} C × ______
4.8×10^{-19} C = 1.6×10^{-19} C × ______
1.6×10^{-19} C = 1.6×10^{-19} C × ______
9.6×10^{-19} C = 1.6×10^{-19} C × ______
19.2×10^{-19} C = 1.6×10^{-19} C × ______
14.4×10^{-19} C = 1.6×10^{-19} C × ______

Analysis

(a) Copy the observations into your notebook. See if you can spot the patterns that Millikan did. (Use a calculator if necessary.)
(b) List all the patterns you can find. In your own words, try to describe what these patterns might mean.

DID YOU KNOW?

Nobel Prizes for physics, chemistry, and other fields have been awarded almost annually since 1901, according to the terms of the will of Alfred Bernard Nobel (1833–96), the Swedish industrialist who invented dynamite. The awards are made by the Swedish Royal Academy of Sciences. Each prize has a cash value, which increases from year to year.

Two observations were evident to Millikan when he analyzed his oil drop data:

1. The smallest value for the charge on an oil drop is 1.6×10^{-19} C.
2. All the other values are whole-numbered multiples of 1.6×10^{-19} C.

Millikan called the smallest unit of charge, which is the absolute value of the charge on an electron, the elementary charge (e).

The elementary charge (e) has magnitude

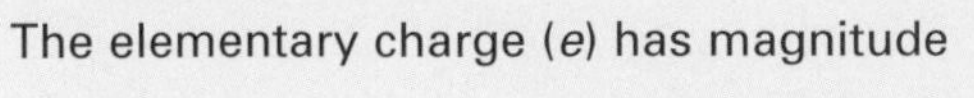

$$e = 1.60 \times 10^{-19} \text{ C}$$

Since an electron is a negative elementary charge and a proton is a positive elementary charge, we can conclude that the charge on one electron is $-e = -1.6 \times 10^{-19}$ C, and the charge on one proton is $e = 1.6 \times 10^{-19}$ C.

Also, if the elementary charge is 1.60×10^{-19} C, then it must take $\frac{1}{1.60 \times 10^{-19}}$ or 6.24×10^{18} electrons to make up 1 C of charge. For the present, we will use this as the value of a coulomb: 1 C = 6.24×10^{18} *e*.

Using this value for the elementary charge, we can devise an equation to make an important calculation. If a charged object has an excess or deficit of *N* electrons, each with a charge *e* (the elementary charge), then the total charge, *Q*, on the object, measured in coulombs, is given by

$$Q = Ne$$

Sample Problem

How many electrons have been removed from a positively charged pith-ball electroscope if it has a charge of 7.5×10^{-11} C?

Solution

$Q = 7.5 \times 10^{-11}$ C

$e = 1.6 \times 10^{-19}$ C

$N = ?$

$$N = \frac{Q}{e}$$

$$= \frac{7.5 \times 10^{-11}\ \text{C}}{1.6 \times 10^{-9}\ \text{C}}$$

$$N = 4.7 \times 10^{8} \text{ electrons}$$

The number of electrons removed was 4.7×10^{8} electrons.

Practice

Understanding Concepts

1. What is the charge in coulombs on an object that has
   (a) an excess of 6.25×10^{19} electrons?
   (b) a deficiency of 1.0×10^{8} electrons?
2. A polyethylene strip has a charge of -5.2×10^{-7} C. What is the excess number of electrons on the strip?

Answers

1. (a) -1.0×10^{1} C
   (b) $+1.6 \times 10^{-11}$ C
2. 3.3×10^{12}

SUMMARY Electric Fields and Electric Charge

- Every charged object creates an electric field of force in the space around it; any other charged object in that field will experience a force of electrical attraction or repulsion.
- Electric charge is measured in units called coulombs (C).
- Millikan's oil drop experiment proved the existence of a smallest unit of electric charge, which he called the elementary charge (*e*).
- The total charge, *Q*, on an object, measured in coulombs, is given by the equation $Q = Ne$.

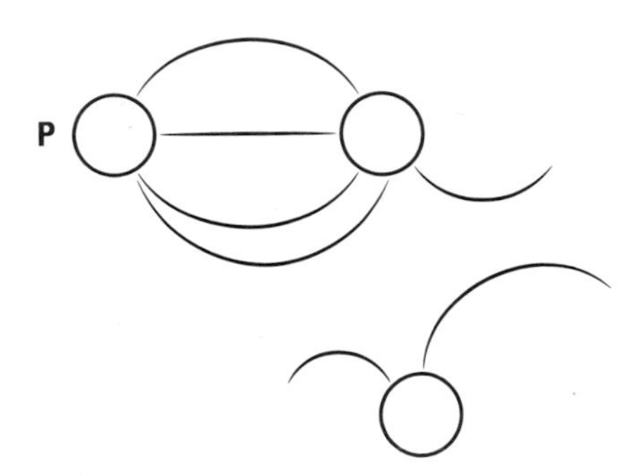

Figure 6

Section 12.2 Questions

Understanding Concepts

1. Assuming the object marked with a P in **Figure 6** is positively charged and the lines represent the electric field, determine the signs of the charges on the other objects and place arrows on the lines in the correct directions.
2. In a lightning bolt, it is estimated that a charge of 22.0 C is transferred from a cloud to Earth. How many electrons make up the lightning bolt?
3. A metal-leaf electroscope is given a negative charge of 1.2 μC by induction and grounding. How many electrons move through your finger when you touch the knob of the electroscope?
4. An ebonite rod with an excess of 6.4×10^8 electrons shares its charge equally with a pith ball when they touch. What is the charge on the pith ball, in coulombs?
5. In Millikan's experiment, oil drops that have similar charges to that on the bottom plate can become balanced so that they will float. Draw a free-body diagram of the oil drop and explain why this must be so.
6. Draw a straight horizontal line about 5 cm long in your notebook to represent a positively charged wire. Draw the electric field lines above and below the wire. (*Hint:* The direction of the electric field line is determined by the direction of the force experienced by a small positive test charge near the wire.)
7. Draw two straight horizontal lines about 5 cm long and parallel to each other, separated by about 2 cm. If the top line represents a positively charged plate and the bottom one a negatively charged plate, draw the electric field between the plates.

Applying Inquiry Skills

8. After completing an experiment similar to Millikan's, a student claims that the charge on the oil drop is 3.8×10^{-19} C. Is the measurement reasonable or is it suspect? Explain your reasoning.
9. Discuss how the mass of an individual marble could be calculated given several bags containing different numbers of identical marbles. You would not be allowed to see into the bag or handle it in any way. Assume the bags have very small masses and that an electronic balance is available. If the materials are available, try your method to see if it works.
10. Many people believe that electric fields can be used to treat human diseases. There are claims that exposure to low frequency electric fields is effective in treating conditions such as osteoporosis, arthritis, muscle pain, cancer, and AIDS. Research these controversial claims and determine whether they are credible. Follow the links for Nelson Physics 11, 12.2.

www.science.nelson.com

Reflecting

11. When Millikan was performing his famous oil drop experiment he was unaware that there existed a smallest electric charge. Initially he might have been discouraged since the measurements of the charge on the oil drop seemed to vary with no discernible pattern. What kinds of characteristics must a scientist (like Millikan) have to make such a discovery? Do you think you possess these characteristics?

DID YOU KNOW ?

Electric Fields in the Ocean

Sharks are sensitive to the small electric field surrounding possible prey, such as fish. Experiments have shown that a shark will attack an artificial electric field and ignore a piece of food. Artificial electric fields were used at the 2000 Olympics in Australia to keep sharks away from triathalon competitors swimming in Sydney harbour.

12.3 Electric Current

We know that objects can develop charges and that these charges can move from one object to another and be distributed over a conductor. When electric charges move from one place to another, we say they constitute an electric current.

In metals, these moving charges are electrons, which have a negative charge. Consider a cylindrical wire (**Figure 1**) of known cross-sectional area, with a total charge Q (in coulombs) flowing through the area A in a time Δt (in seconds). Then the electric current I through the wire is

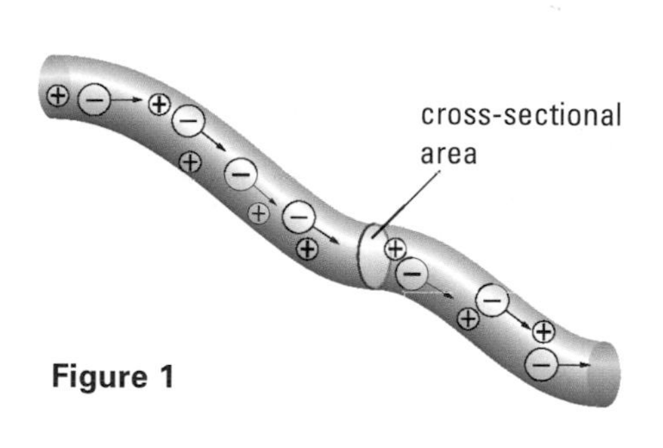

Figure 1

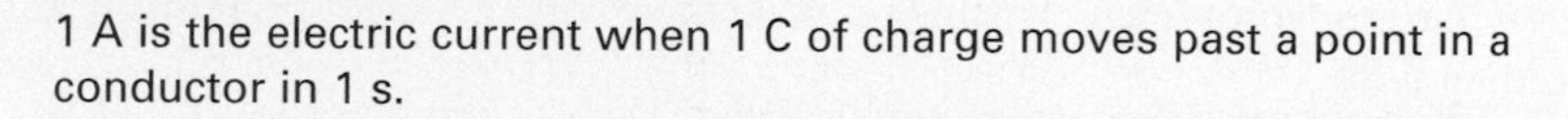

$$I = \frac{Q}{\Delta t}$$

Electric current is measured in units called **amperes** (A), after the French physicist André Marie Ampère (1775–1836) (**Figure 2**). The definition of an ampere depends upon a knowledge of the magnetic force acting between parallel current-carrying conductors, but, in terms of the amount of the charge that is flowing and the time,

ampere: (A) the SI unit of electric current, 1 A = 1 C/s

1 A is the electric current when 1 C of charge moves past a point in a conductor in 1 s.

$$1\text{ A} = 1\text{ C/s}$$

The charge (Q) involved in the calculation of current is the magnitude of the net charge that passes through the cross-section. It is calculated by multiplying the elementary charge (e) by the number of elementary charges (N) that pass through the cross-section:

$$Q = Ne$$

Figure 2
André Marie Ampère (1775–1836) devised a rule relating the direction of electron flow and the direction of the N-pole of a magnetic compass. He showed and explained that two parallel current-carrying wires either attracted or repelled each other. To explain the magnetic properties of materials such as iron, Ampère proposed the existence of tiny electrical currents within the atom, long before the existence of the electron was even suggested. Although his contemporaries regarded his theories with skepticism, Ampère was honoured by the naming of the unit of electric current, the ampere, after him.

Sample Problem 1

Calculate the current in an electric toaster if it takes 9.0×10^2 C of charge to toast two slices of bread in 1.5 min.

Solution

$\Delta t = 1.5$ min

$\Delta t = 9.0 \times 10^1$s

$Q = 9.0 \times 10^2$ C

$I = ?$

$$I = \frac{Q}{t}$$

$$= \frac{9.0 \times 10^2\text{ C}}{9.0 \times 10^1\text{ s}}$$

$$I = 1.0 \times 10^1\text{ A}$$

The current in the toaster is 1.0×10^1 A.

DID YOU KNOW ?

Healing with Electric Current

Since the body's nervous system operates on electric impulses, electricity can be used to relieve pain as well as promote healing. The ancient Egyptians applied this principle when they used the electric charges generated by torpedo fish to ease pain. The Romans evidently used electric eels to treat headaches and arthritis. Today, physicians use electrical nerve stimulation to treat certain types of pain. A 9-V battery supplies a weak electric current across the patient's skin into the nerve cells beneath the skin's surface. The current stimulates the body's natural ability to fight pain.

Sample Problem 2

A light bulb with a current of 0.80 A is left burning for 25 min. How much electric charge passes through the filament of the bulb?

Solution

$\Delta t = 25$ min

$\Delta t = 1.5 \times 10^3$ s

$I = 0.80$ A

$Q = ?$

$$Q = I\Delta t$$
$$= (0.80\text{ A})(1.5 \times 10^3\text{ s})$$
$$Q = 1.2 \times 10^3\text{ C}$$

1.2×10^3 C passes through the filament of the bulb.

Sample Problem 3

A metal-leaf electroscope with 1.25×10^{10} excess positive charges is grounded and discharges completely in 0.50 s. Calculate the average current in the grounding wire.

Solution

$N = 1.25 \times 10^{10}$

$e = 1.6 \times 10^{-19}$ C

$\Delta t = 0.50$ s

$I = ?$

$$Q = Ne$$
$$= (1.25 \times 10^{10})(1.6 \times 10^{-19}\text{ C})$$
$$Q = 2.0 \times 10^{-9}\text{ C}$$

$$I = \frac{Q}{\Delta t}$$
$$= \frac{2.0 \times 10^{-9}\text{ C}}{0.50\text{ s}}$$
$$I = 4.0 \times 10^{-9}\text{ A}$$

The average current in the grounding wire is 4.0×10^{-9} A.

Practice

Understanding Concepts

1. A charge of 32 C passes a point in an electric circuit in 2.0 min. Calculate the current in the wire.
2. Determine the current in each case:
   (a) $Q = 4.2$ C, $\Delta t = 0.30$ s
   (b) A safety fuse in an electric circuit burns out when more than 68 C pass by in 8.0 s.
   (c) In one minute, 51 C of charge passes through a 1.0×10^2 W light bulb.
3. Write an equation for each of the following:
   (a) the charge in terms of the electric current and the time interval
   (b) the time interval in terms of the current and the charge

Answers

1. 0.27 A
2. (a) 14 A
   (b) 8.5 A
   (c) 0.85 A

4. Calculate the total charge in each situation.
   (a) $I = 15$ A, $\Delta t = 12$ s
   (b) A cell provides 0.15 A of current to a radio for 32 min.
   (c) A calculator draws 2.0×10^{-5} A of current from a solar cell for 26 s.
5. Determine the time interval in each case.
   (a) $Q = 8.6$ C, $I = 2.0$ A
   (b) A current of 22 A delivers 1.1 C of charge.
   (c) A charge of 75 μC is delivered by a current of 150 μA.

Answers

4. (a) 1.8×10^2 C
   (b) 2.9×10^2 C
   (c) 5.2×10^{-4} C
5. (a) 4.3 s
   (b) 5.0×10^{-2} s
   (c) 0.50 s

The Direction of Electric Current

Early in the 19th century, Benjamin Franklin made the assumption that there were two electrical states: one with more than the normal amount of electricity, which he called a positive charge, and one with less than the normal amount of electricity, which he called a negative charge.

Electric current was defined as the rate of movement of electrically charged particles past a point, so it was natural to assume that the charge moved from an area where there was an excess (positive charge) to an area where there was a deficit (negative charge). The direction of the electric current was defined as moving from the positive terminal to the negative terminal of the source of electric potential. This assumption about the direction of the electric current was called **conventional current** or just **electric current**. The circuit diagram in **Figure 3(a)** corresponds to the current direction, using the conventional current assumption.

conventional current or **electric current:** describes electric charges travelling through a conductor from the positive terminal to the negative terminal of the source of electric potential

After the conventional current assumption had become firmly entrenched in scientific literature, the electron was discovered. It soon became clear that what actually constituted an electric current in a metallic conductor (such as a wire) was a flow of negatively charged electrons from the negative terminal to the positive terminal of the source of electric potential. This is referred to as **electron flow**. The same circuit, showing electron flow, is shown in **Figure 3(b)**.

electron flow: a term used to indicate that the electric current in metals is due to the motion of electrons

In this book we will use the widely accepted *electric current* convention to avoid introducing negative signs into equations. Only a few changes in thinking are required to translate from "electric current" to "electron flow." These changes are summarized below.

1. In all circuit diagrams, the direction of electric current is indicated by arrows on the conductors. The direction of electron flow is opposite to the direction of the arrows.
2. Any references in the text to "electric current through a conductor from positive to negative" are equivalent to "electron flow through a conductor from negative to positive."

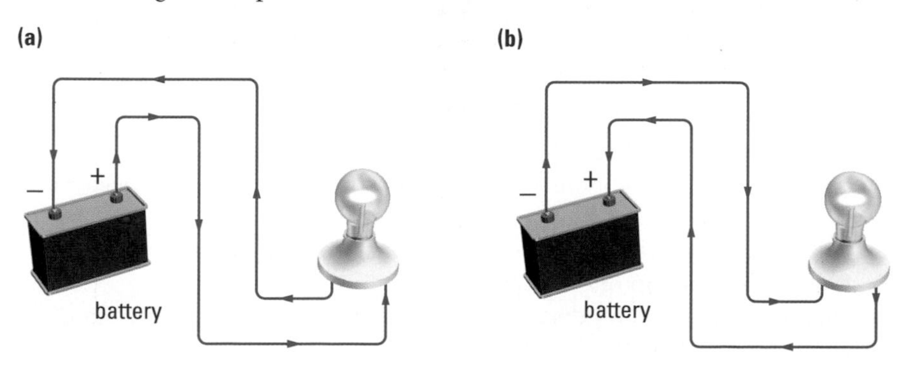

Figure 3
(a) Conventional current
(b) Electron flow

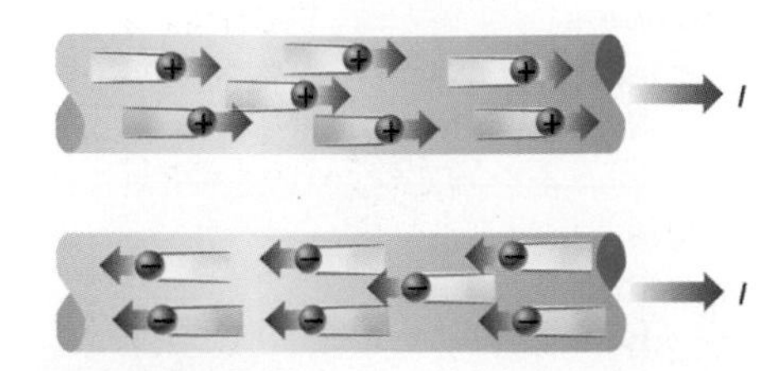

Figure 4
The flow of positive charge in one direction is the same as the flow of negative charge in the other direction.

Note that the above comments refer only to the flow of electrically charged particles in solids. In liquids and gases the charged particles can flow in either direction, and sometimes simultaneously. It might help to realize that positive charge flowing east along a wire is electrically equivalent, in every way, to negative charge flowing west (**Figure 4**).

The electric current supplied by a battery is significantly different from the electric current from a wall socket (**Figure 5**). Batteries supply **direct current** (DC) and wall sockets supply **alternating current** (AC). The direct current supplied by a battery is in a fixed single direction and it doesn't increase or decrease in magnitude, while the alternating current supplied by a wall socket periodically reverses direction in the circuit and the amount of current varies continuously.

An **ammeter** is a device that measures the amount of electric current in a circuit. It is connected directly into the path of the moving charges, as shown in **Figure 6**. This type of connection is called a series connection.

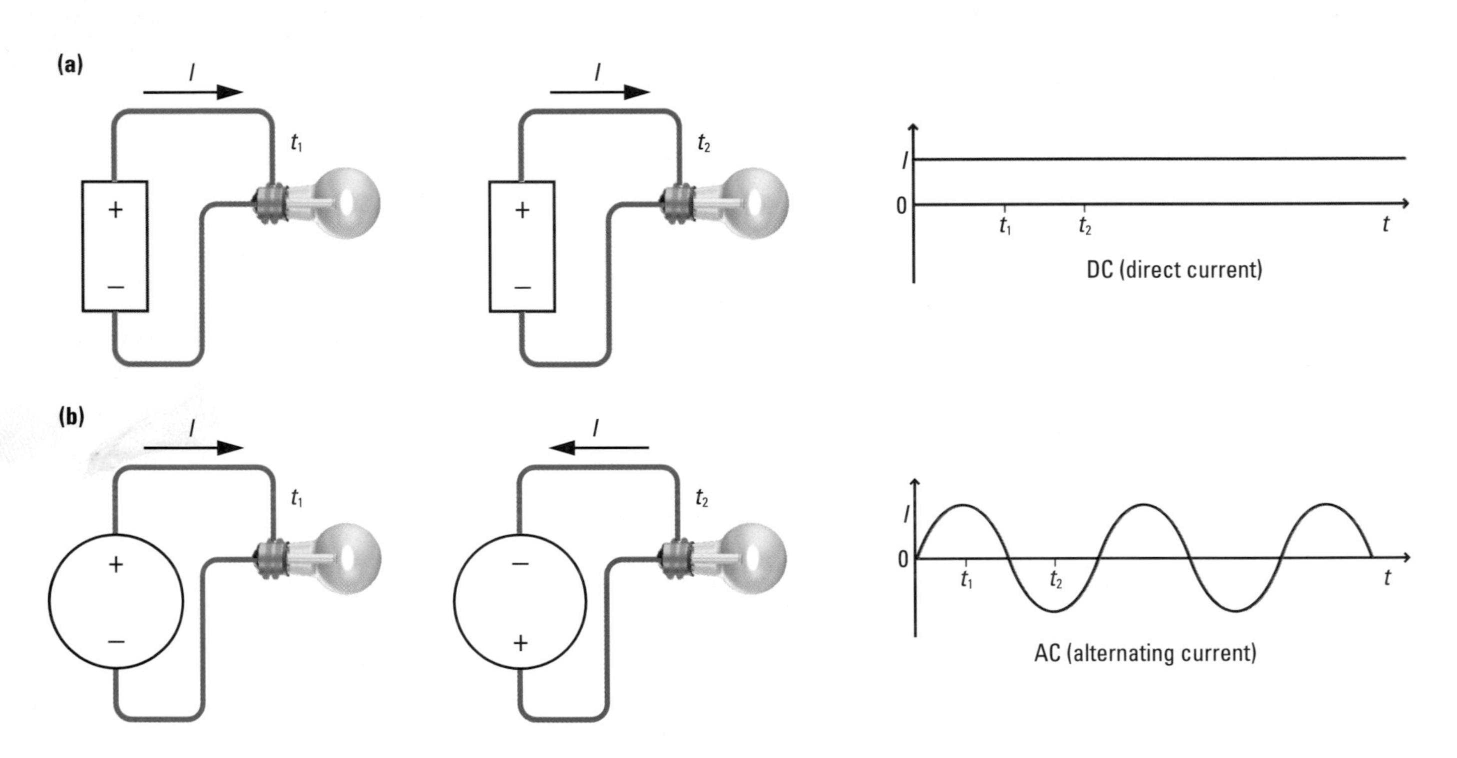

Figure 5
Comparing DC and AC currents
(a) DC current
(b) AC current

direct current: (DC) results when charges flow in a particular direction

alternating current: (AC) results when charges periodically reverse direction

ammeter: a device that measures the amount of electric current in a circuit

SUMMARY Electric Current

- When electric charges move from one place to another, this constitutes an electric current.
- The electric current, I, through a cross-section of cylindrical wire is represented by the equation $I = \frac{Q}{t}$.
- Electric current or conventional current describes electric charges travelling through a conductor from the positive terminal to the negative terminal of the source of electric potential.
- Electron flow is a term used to indicated that the electric current in metals is due to the motion of electrons.
- Direct current (DC) results when charges flow in a particular direction; alternating current (AC) results when charges periodically reverse direction.

Section 12.3 Questions

Understanding Concepts

1. How much electric current passes through a conductor when 12 C of charge passes a point in a conductor in 4.0 s?
2. What is the current in a light bulb when it takes 24 s for 18 C of charge to pass through its filament?
3. How much charge passes through the starting motor, if it takes 4.0 s to start a car and there is a current of 225 A during that time?
4. A small electric motor draws a current of 0.40 A. How long will it take for 8.0 C of charge to pass through it?
5. A student measured the current supplied by a particular cell over a period of 200 s and presented her results in the graph shown in **Figure 7**. Calculate the total charge delivered by the cell.
6. How many elementary charges pass through a light bulb in each second if the bulb has a current of 0.50 A in it?
7. Explain the difference between an electric current of 3.0 A and 2.0 A in terms of the definition of electric current.
8. Describe the difference between static electricity and current electricity.
9. (a) Compare and contrast AC and DC currents.
   (b) Give three examples of devices that would use each type of current.

Applying Inquiry Skills

10. If asked to measure the flow of water out of a tap, counting the number of water molecules that come out of the tap in a certain amount of time would be impractical because the number would be too large. It would be more convenient to calculate the water current in litres per second.
    (a) Using similar reasoning, explain why electric current is not measured in electrons per second.
    (b) List three ways that water current is similar to electric current.
    (c) Come up with another example of something that is similar to electric current.

Making Connections

11. (a) What is the difference between electron flow and electric current?
    (b) What are some of the disadvantages to using the convention that electric current consists of the flow of positive charge?

Reflecting

12. Electric current and electron flow are often confusing to students introduced to these concepts. How are you going to remember the difference between the two concepts? Why do you think people are sometimes reluctant to use the electric current convention at first?

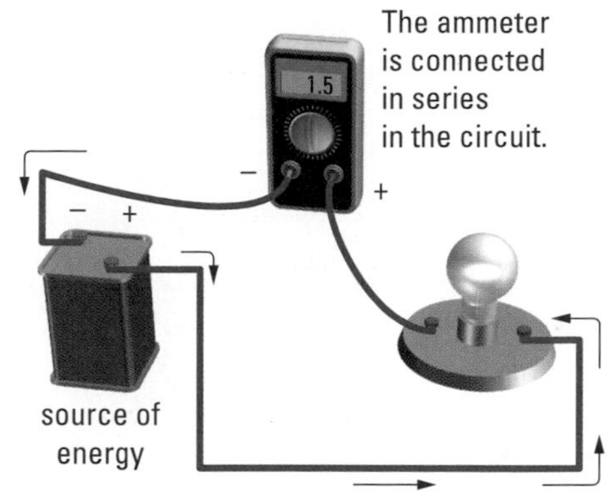

Figure 6
Measuring electric current with an ammeter. Notice that the positive terminal of the ammeter is connected to the positive terminal of the battery.

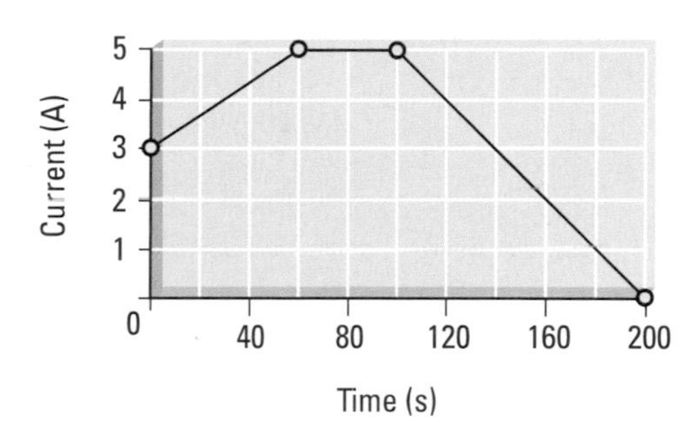

Figure 7
For question 5

Figure 1
The stove element has current passing through it and it heats up. What is taken from the current to cause the element to heat up?

12.4 Electric Potential Difference

Suppose that an electric stove element is connected to a battery and the element begins to heat up (**Figure 1**). If an ammeter is connected to the circuit it will measure a fairly large current. You might say that the element is "using electricity," but what does this mean? What does the element take from the current? To answer these questions, let's first look at a few analogies.

First, imagine a ball held some distance above the surface of Earth, as in **Figure 2(a)**. Work must have been done on the ball to overcome the gravitational field of Earth and lift it from the ground to its present position, increasing the ball's gravitational potential energy. If the ball is then released, the gravitational field will cause it to move back toward Earth, converting gravitational potential energy into kinetic energy as the ball falls.

Now, imagine a small positive charge held at rest a certain distance away from a negatively charged sphere (**Figure 2(b)**). The negative sphere is surrounded by a field of force, and the positive charge is pulled toward the sphere. Again, work must be done on the small positive charge to overcome this electric force and pull it away from the negative sphere. In this case, the small positive charge has an increase in electric potential energy as a result. If released, the positive charge will move back toward the negative sphere, in much the same way that the ball moved back toward Earth, thereby losing electric potential energy. Charged particles moving in the presence of an electric field and converting electric potential energy into some other form of energy constitute an electric current.

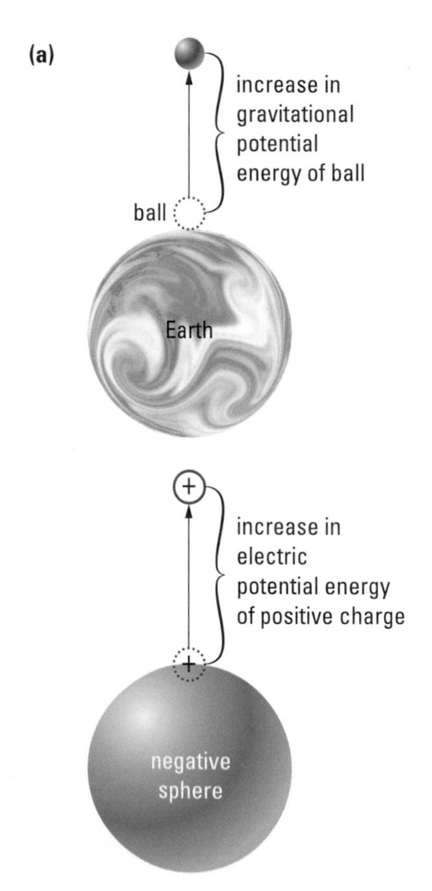

Figure 2

Similarly, a current flows in the stove element because an electric field is present that does work on the charges in the circuit, causing them to move. As current passes through the stove element, it experiences opposition to the flow, resulting in a loss of electric potential energy. The energy lost is transferred to the molecules and atoms of the conductor as the current moves through it. This causes the element to heat up and glow—electric potential energy has changed into heat and light energy.

Since the charge loses energy it also loses electric potential, resulting in an **electric potential difference** (V) between two points, A and B. This can be represented as

$$V = \frac{W}{Q}$$

where W is the amount of work that must be done to move a small positive charge, Q, from point A to point B (strictly speaking, this gives the electric potential of point B with respect to point A).

The SI unit for electric potential difference is the **volt** (V), named after Alessandro Volta (**Figure 3**).

1 V is the electric potential difference between two points if it takes 1 J of work per coulomb to move a positive charge from one point to the other.

1 V = 1 J/C

electric potential difference: (V) the amount of work required per unit charge to move a positive charge from one point to another in the presence of an electric field

volt: (V) the SI unit for electric potential difference; 1 V = 1 J/C

Because of the units in which it is measured, electric potential difference is often referred to as "voltage." A 12-V car battery is a battery that does 12 J of work on each coulomb of charge that flows through it. Electric potential difference between two points in a circuit is measured with a device called a **voltmeter**. To measure electric potential difference, a voltmeter is connected across the source and the bulb in the circuit (**Figure 4**). This type of connection is called a parallel connection.

The electrical energy lost or work done by a charge, Q, going through a potential difference, V, can be written

$$\Delta E = QV$$

Since it is often easier to measure the current and the time during which it lasts, we can use the equation

$$Q = I\Delta t$$

and, substituting in the first equation, we get

$$\Delta E = VI\Delta t$$

as an expression for the electrical energy lost by a current, I, through a potential difference, V, for a time interval, Δt.

Figure 3
Alessandro Volta (1745–1827). After a childhood showing little promise as a scholar, Volta forged ahead, becoming professor of physics at the high school in Como, Italy. His first major contribution was the discovery of the electrophorus, a device capable of producing and storing a large electrical charge. His reward was a professorship at the University of Pavia and membership in the Royal Society. Volta's most significant achievement was the invention of the first electric battery using bowls of salt solution connected by arcs of metal with copper at one end and tin or zinc at the other. For this invention, Napoleon awarded him the medal of the Legion of Honour and made him a Count.

voltmeter: a device that measures electric potential difference between two points in a circuit

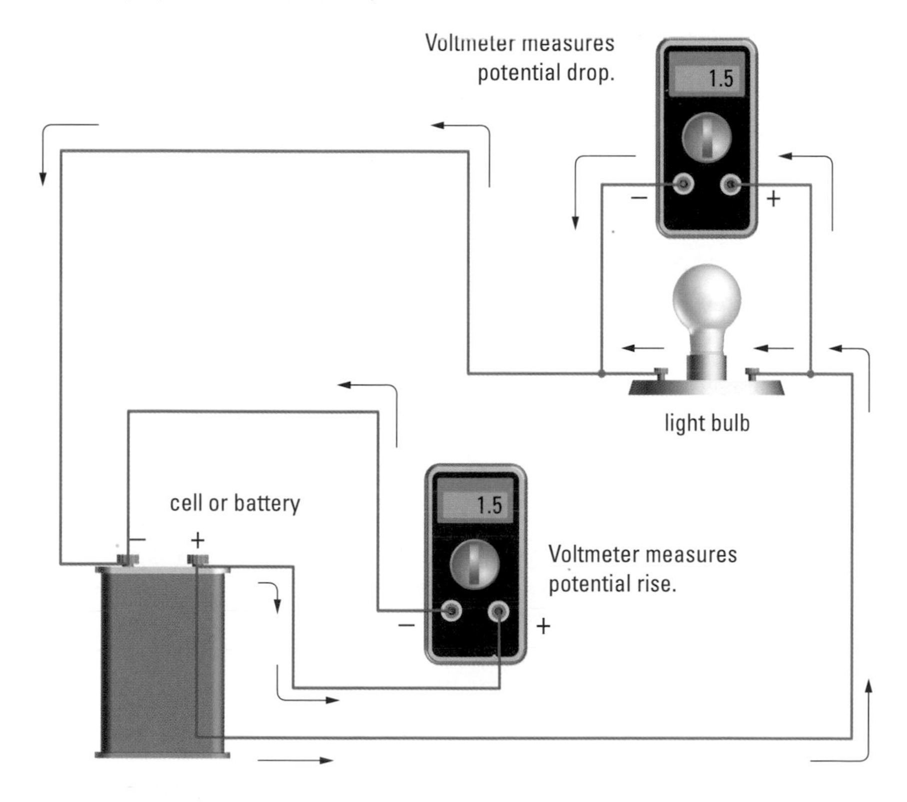

Figure 4
Measuring potential differences with a voltmeter connected in parallel

Sample Problem 1

A 12-V car battery supplies 1.0×10^3 C of charge to the starting motor. How much energy is used to start the car?

Solution

$V = 12$ V

$Q = 1.0 \times 10^3$ C

$\Delta E = ?$

$$\begin{aligned}\Delta E &= QV \\ &= (1.0 \times 10^3 \text{ C})(12 \text{ V}) \\ \Delta E &= 1.2 \times 10^4 \text{ J}\end{aligned}$$

The amount of energy used to start the car is 1.2×10^4 J.

Sample Problem 2

If a current of 10.0 A takes 3.0×10^2 s to boil a kettle of water requiring 3.6×10^5 J of energy, what is the potential difference (voltage) across the kettle?

Solution

$I = 10.0$ A

$\Delta t = 3.0 \times 10^2$ s

$\Delta E = 3.6 \times 10^5$ J

$V = ?$

$$\begin{aligned}V &= \frac{\Delta E}{I\Delta t} \\ &= \frac{3.6 \times 10^5 \text{ J}}{(10.0 \text{ A})(3.0 \times 10^2 \text{ s})} \\ V &= 1.2 \times 10^2 \text{ V}\end{aligned}$$

There is a potential difference of 1.2×10^2 V across the kettle.

DID YOU KNOW?

Neurons in the Human Body

An important part of the body's nervous system is the neuron, a nerve cell that can receive, interpret, or transmit electrical messages. An electric potential difference exists across the surface of every neuron because of a greater number of negative charges on the inside than on the outside. The potential difference is typically about 60 mV to 90 mV.

Practice

Understanding Concepts

1. A hair dryer has 1.2×10^3 C of charge passing through a point in its circuit when it is connected to a 120-V power supply. How much energy is used by the hair dryer?
2. Write an equation for each of the following:
   (a) charge in terms of energy used and voltage
   (b) voltage in terms of charge and energy used
3. A light bulb operating on an electric current of 0.83 A is used for 2.5 h at an electric potential difference of 120 V. How much energy is used by the light bulb?

Answers

1. 1.4×10^5 J
3. 9.0×10^5 J

4. Write an equation for each of the following:
   (a) electric current in terms of energy used, voltage, and time interval
   (b) voltage in terms of electric current, energy used, and time interval
   (c) time interval the device is used in terms of current, voltage, and energy used
5. A 120-V electric sander operating for 5.0 min uses 1.0×10^5 J of energy. Find the current through the sander.
6. An electric can opener used in a 120-V circuit operates at 2.2 A using 9.5×10^5 J of energy all year. How long was the can opener used for that year?

Answers

5. 2.8 A
6. 3.6×10^3 s

SUMMARY Electric Potential Difference

- Electric potential difference (V) is the amount of work required per unit charge to move a positive charge from one point to another in the presence of an electric field.
- The electric potential difference (V) between two points, A and B, is represented by the equation $V = \frac{W}{Q}$.
- Because of the units in which it is measured, electric potential difference is often referred to as "voltage."
- The electrical energy lost by a current, I, through a potential difference, V, for a time, Δt, is represented by the equation $\Delta E = VI\Delta t$.

Section 12.4 Questions

Understanding Concepts

1. What amount of energy does a kettle use to boil water if it has 810 C of charge passing through it with a potential difference of 120 V?
2. What is the potential difference across a refrigerator if 75 C of charge transfers 9.0×10^3 J of energy to the compressor motor?
3. An electric baseboard heater draws a current of 6.0 A and has a potential difference of 240 V. For how long must it remain on to use 2.2×10^5 J of electrical energy?
4. A flash of lightning transfers 2.0×10^9 J of electrical energy through a potential difference of 7.0×10^7 V between a cloud and the ground. Calculate the quantity of charge transferred in the lightning bolt.
5. Calculate the energy stored in a 9.0 V battery that can deliver a continuous current of 4.0 mA for 2.0×10^3 s.
6. If a charge of 0.30 C moves from one point to another in a conductor and, in doing so, releases 5.4 J of electrical energy, what is the potential difference between the two points?
7. Describe the significance of two points in a conductor that are at the same electric potential. How much work must be done to move a charge between the two points?

(continued)

8. How are electric potential energy and gravitational potential energy different and how are they similar?
9. Using the definition of a volt, compare a 6.0-V battery to a 12.0-V battery.
10. Compare and contrast the methods used to connect ammeters and voltmeters to electric circuits.

Applying Inquiry Skills

11. An electric current passes through a heating element on a stove. The element heats up and begins to glow. Discuss the energy transformations involved. How is the law of conservation of energy involved in these energy transformations?
12. Gravitational potential energy has been used as an analogy in understanding electric potential difference between a small positive charge and a negative sphere. Will the analogy be as useful if both charges have the same sign?

12.5 Kirchhoff's Laws for Electric Circuits

circuit: a path for electric current

load: any device in a circuit that transforms electrical potential energy into some other form of energy, causing an electric potential drop

Electrons possess electric potential energy that can be transformed into heat, light, and motion. For such transformations to occur, we need to connect a source of electric potential energy to one or more components by means of a **circuit**, which is a path for electric current. Any component or device in a circuit that transforms electric potential energy into some other form of energy, causing an electric potential drop is called a **load**.

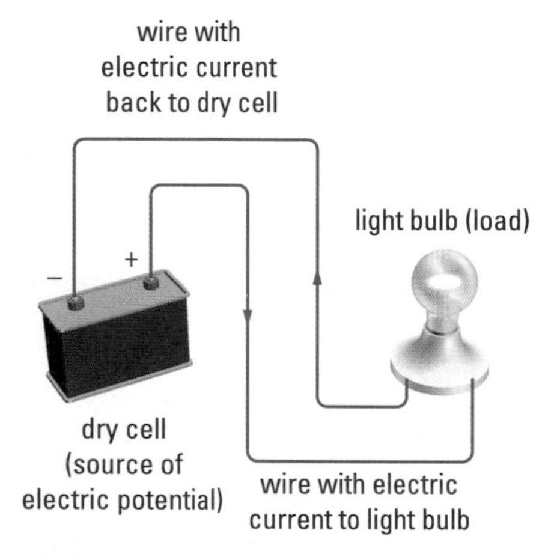

Figure 1
A simple electric circuit

The simple circuit in **Figure 1** provides a complete path for the electric current. The path can be traced from the positive terminal of the dry cell through the light bulb, and back to the negative terminal of the dry cell. As electric charges go through this simple circuit, they transfer the electric potential energy they acquired from the dry cell to the light bulb. In a sense, then, the charges act as carriers of energy from the source of electrical energy (the dry cell) to the converter of electrical energy (the light bulb).

Electric current can go through a circuit only if the circuit provides a complete path. Any break in the circuit will cause the electric current to cease. The circuit is then said to be an *open circuit*. If, by chance, two wires in a circuit touch, so that charge can pass from one wire to the other and return to the negative terminal of the source without passing through the load, the circuit is said to be a *short circuit*, or, simply, a "short."

Circuit Symbols

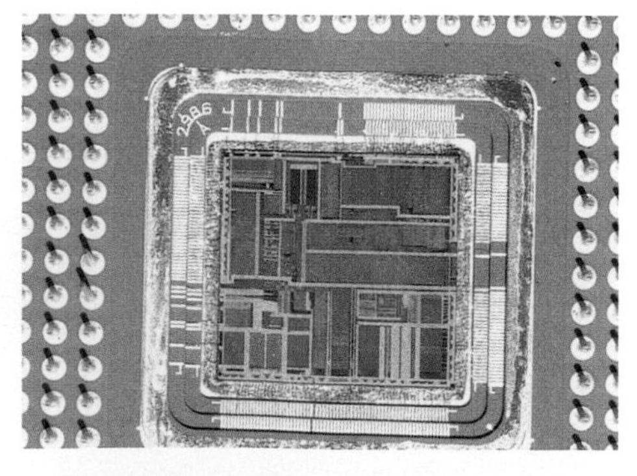

Figure 2
The Intel Pentium micro-processor contains thousands of electric circuits on a silicon chip the size of a fingernail and can process hundreds of millions of instructions per second.

elements: the components of an electric circuit

The various paths through a circuit can be complex and they can contain many different types of electrical devices and connectors (**Figure 2**). To simplify descriptions of these paths, circuit diagrams are drawn, using symbols that show exactly how each device is connected to the others. The components of an electric circuit are called **elements**, and the symbols most commonly used in such schematic diagrams are displayed in **Figure 3**.

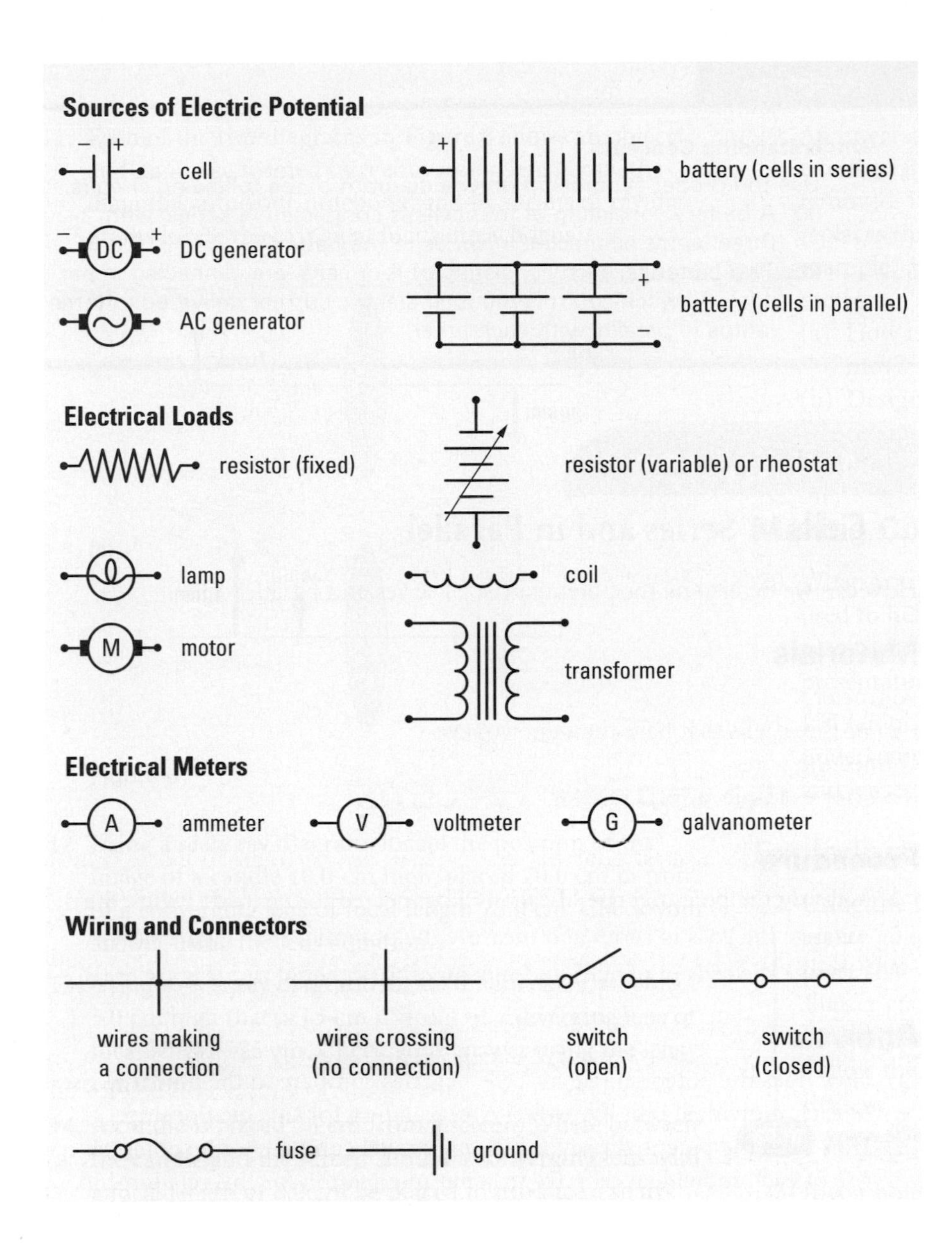

Figure 3
In the circuit diagrams in this chapter, a small dot on a conductor indicates a point in the circuit where a connection between two or more wires must be made.

series circuit: circuit in which charges have only one path to follow

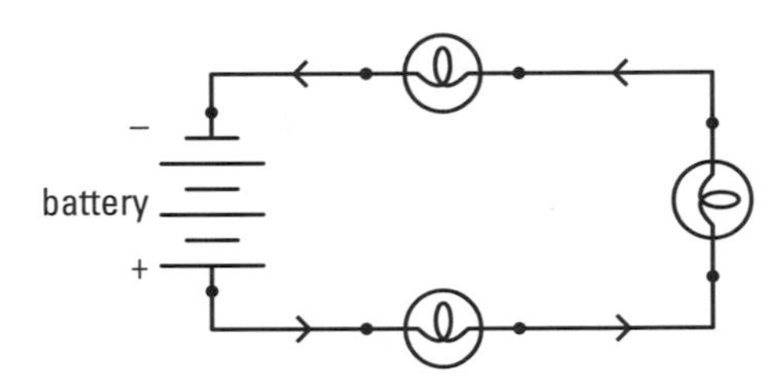

Figure 4
In a series circuit, a key word to remember is "and": the charges pass through one load and then the next and so on, as they return to the negative terminal of the source. The current is exactly the same at any point in a series circuit.

parallel circuit: circuit in which charges can move along more than one path

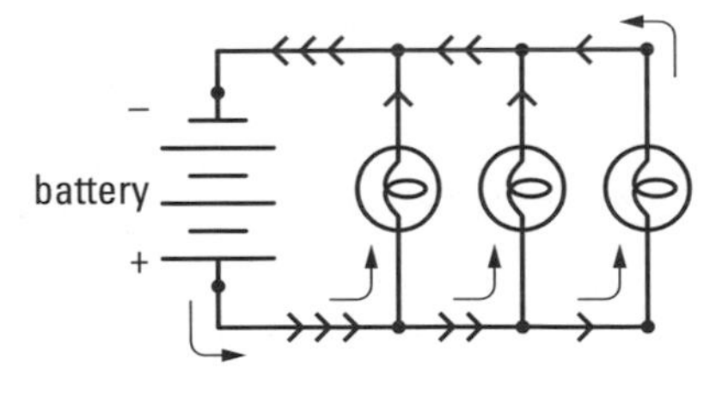

Figure 5
In a parallel circuit, a key word to remember is "or": charges pass through one load or another load, and so on, as they pass through the circuit. Notice that in this circuit, there are three different paths through the circuit. Each electron will take only one of the three paths.

Series Circuits

One simple way of joining several loads together is to connect them in a **series circuit** with a source of electric potential (**Figure 4**). In this type of connection, the charges have only one path to follow through the circuit, and as a result each charge must go through each load in turn: every charge that goes through a series circuit goes through each load in the circuit before returning to the source.

Parallel Circuits

In a **parallel circuit**, the charges can move along several paths through the circuit and, as a result, pass through any one of the several loads in the circuit. Every charge that goes through the parallel circuit goes through only one of the circuit's loads before returning to the source. An example of a parallel circuit is shown in **Figure 5**.

Practice

Understanding Concepts

1. Use the proper symbols to draw a diagram of the following circuits:
   (a) A battery consisting of four cells is connected in series, with three lamps connected in series to the battery.
   (b) Two batteries, each consisting of four cells, are connected in parallel. A switch controls the total electric current delivered to three lamps in parallel with each other.

Activity 12.5.1

Cells in Series and in Parallel

Do not "short" cells (connect one terminal directly to another).

How can we determine the potential rise of series and parallel cells?

Materials

voltmeter
two or three cells that have the same voltage
connecting wires
load (for example, a 25 Ω resistor)

Procedure

1. Measure the potential rise of each cell connected to the load (**Figure 6(a)**).
2. Arrange the cells in series and measure the potential rise (**Figure 6(b)**).
3. Arrange the cells in parallel and measure the potential rise (**Figure 6(c)**).

Analysis

(a) How does the potential rise for cells in series compare to the potential rise of each individual cell? Try to write an equation for this relationship.
(b) How does the potential rise for cells in parallel compare to the potential rise of each individual cell? Try to write an equation for this relationship.

(a)

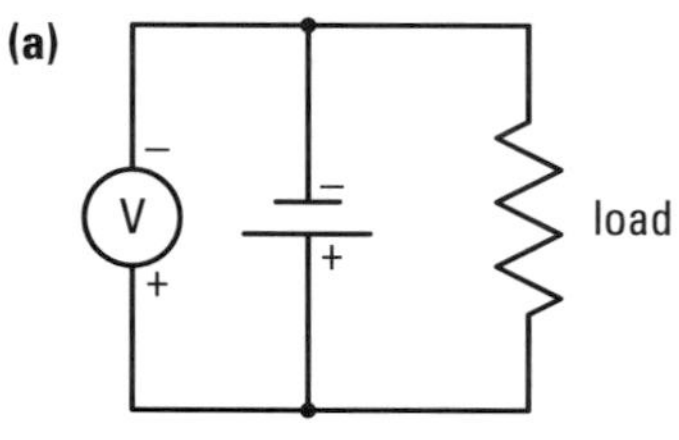

(b)

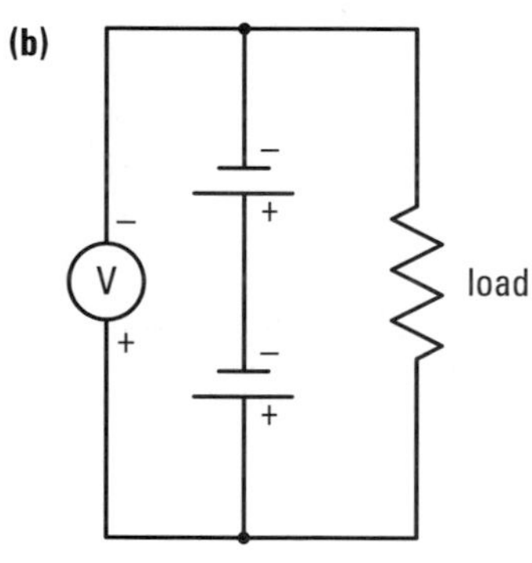

(c)

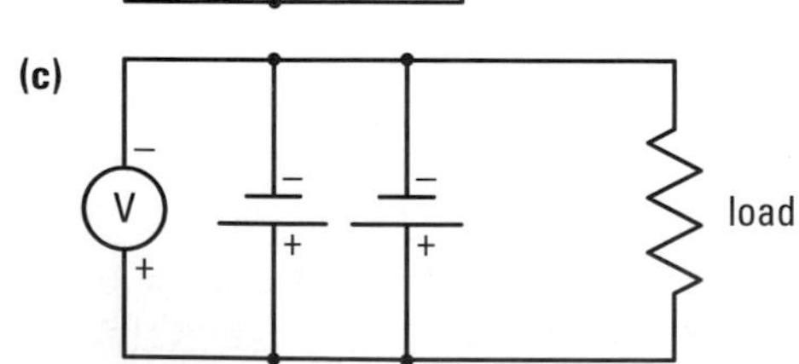

Figure 6
(a) Single cell
(b) Cells in series
(c) Cells in parallel

Kirchhoff's Laws for Electric Circuits

To understand how simple series and parallel circuits operate, we need to answer two basic questions:

1. When charges have several loads to pass through, what governs the amount of electric potential energy they will lose at each load?
2. When charges may follow several possible paths, what governs the number of charges that will take each path?

German physicist, Gustav Robert Kirchhoff (1824–1887) addressed these questions with two fundamental conservation laws for electric circuits: the **law of conservation of energy** and the **law of conservation of charge**.

law of conservation of energy: As electrons move through an electric circuit they gain energy in sources and lose energy in loads, but the total energy gained in one trip through a circuit is equal to the total energy lost.

Law of Conservation of Energy

As electrons move through an electric circuit they gain energy in sources and lose energy in loads, but the total energy gained in one trip through a circuit is equal to the total energy lost.

Law of Conservation of Charge

Electric charge is neither created nor lost in an electric circuit, nor does it accumulate at any point in the circuit.

law of conservation of charge: Electric charge is neither created nor lost in an electric circuit, nor does it accumulate at any point in the circuit.

By performing careful experiments, Kirchhoff was able to describe each of these conservation laws in terms of quantities easily measurable in electric circuits. They have since become known as **Kirchhoff's voltage law** (KVL) and **Kirchhoff's current law** (KCL).

Kirchhoff's Voltage Law (KVL)

Around any complete path through an electric circuit, the sum of the increases in electric potential is equal to the sum of the decreases in electric potential.

Kirchhoff's Current Law (KCL)

At any junction point in an electric circuit, the total electric current into the junction is equal to the total electric current out.

Kirchhoff's voltage law: (KVL) Around any complete path through an electric circuit, the sum of the increases in electric potential is equal to the sum of the decreases in electric potential.

Kirchhoff's current law: (KCL) At any junction point in an electric circuit, the total electric current into the junction is equal to the total electric current out.

These relationships are crucial to understanding the transfer of electrical energy in a circuit, and will provide the basis for electric circuit analysis in this chapter.

Sample Problem 1

Calculate the potential difference, V_2, in the circuit shown in **Figure 7**.

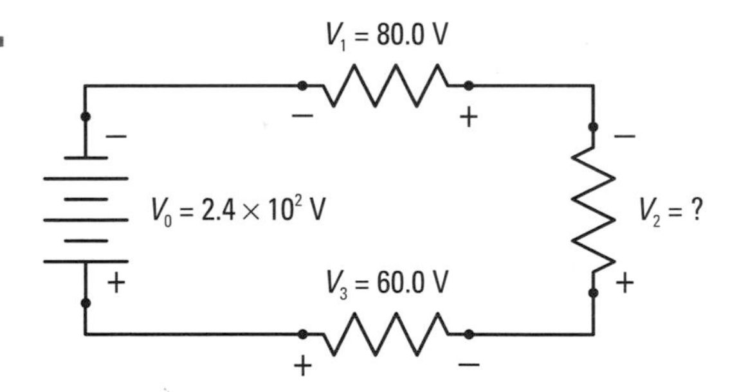

Figure 7
For Sample Problem 1

Solution

Applying KVL to the circuit,

$$V_0 = V_1 + V_2 + V_3$$
$$V_2 = V_0 - V_1 - V_3$$
$$= 2.4 \times 10^2 \text{ V} - 8.0 \times 10^1 \text{ V} - 6.0 \times 10^1 \text{ V}$$
$$V_2 = 1.0 \times 10^2 \text{ V}$$

The potential difference in this circuit is 1.0×10^2 V.

Sample Problem 2

Calculate the electric current, I_3, in the circuit shown in **Figure 8**.

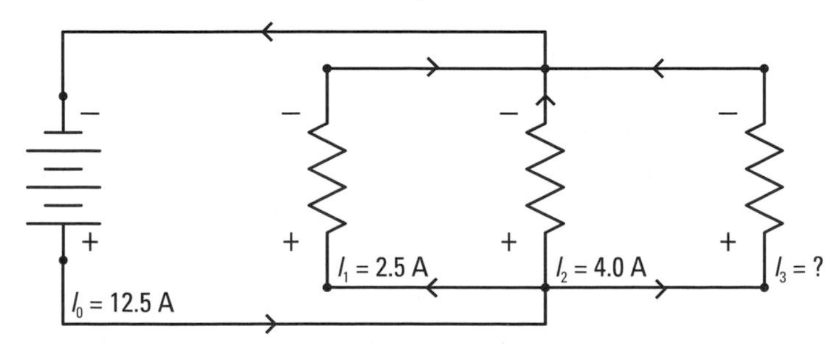

Figure 8

DID YOU KNOW?

The Drift Speed of Charges

It's easy to get the impression that charges "race" through a circuit from the positive terminal to the negative terminal of the source. In fact, they move relatively slowly, and it would take a long time for any one charge to make its way completely through any practical circuit. In a copper wire with a diameter of about 1 mm and an electric current of 1 A, the charges are drifting through the conductor with a speed of approximately 0.01 cm/s. However, the current starts and stops at essentially the same time in all parts of the circuit.

Solution

Applying KCL to the circuit,

$$I_0 = I_1 + I_2 + I_3$$
$$I_3 = I_0 - I_1 - I_2$$
$$= 12.5\text{ A} - 2.5\text{ A} - 4.0\text{ A}$$
$$I_3 = 6.0\text{ A}$$

The electric current in I_3 is 6.0 A.

Practice

Understanding Concepts

2. Find the missing electric potential differences for the circuits in **Figure 9**.
3. Find the missing electric currents and electric potential differences in the circuit in **Figure 10**.

Answers

2. (a) $V_0 = 6.0$ V
   (b) $V_1 = 2.0$ V
3. $V_1 = 6.0$ V, $V_2 = 6.0$ V, $I_2 = 0.7$ A, $I_3 = 1.5$ A

(a)

$V_3 = 1.0$ V

$V_0 = ?$

$V_2 = 3.0$ V

$V_1 = 2.0$ V

(b)

$V_3 = 3.0$ V

$V_0 = 10.0$ V

$V_2 = 5.0$ V

$V_1 = ?$

Figure 9
For question 2

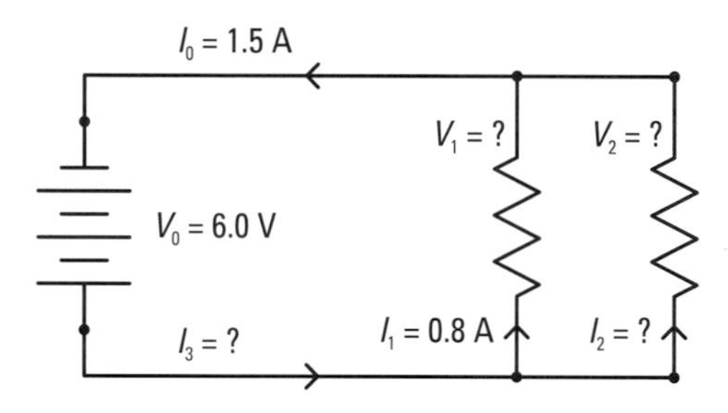

Figure 10

Here is a summary of the properties of simple series and parallel circuits:

- Series: the current is constant throughout the entire circuit and the total potential increase is equal to the sum of the potential decreases.
- Parallel: the total current equals the sum of the individual currents and the potential decrease is constant across all the loads.

Another way to remember this is that loads connected in series have the same current, while loads connected in parallel have the same potential difference.

SUMMARY Electric Circuits

- Sources of electric potential energy are connected to electric loads by means of a circuit.
- In a series circuit, the charges have only one path to follow, and as a result each charge goes through each load in turn.
- In a parallel circuit, charges can move along several paths, and as a result pass through any one of several loads in the circuit.
- Kirchhoff's voltage law (KVL) states that around any complete path through an electric circuit, the sum of the increases in electric potential is equal to the sum of the decreases in electric potential.
- Kirchhoff's current law (KCL) states that at any junction point in an electric circuit, the total electric current into the junction is equal to the total electric current out.

Section 12.5 Questions

Understanding Concepts

1. Find V_0 in the circuit shown in **Figure 11**.
2. Find I_0 in the circuit shown in **Figure 12**.

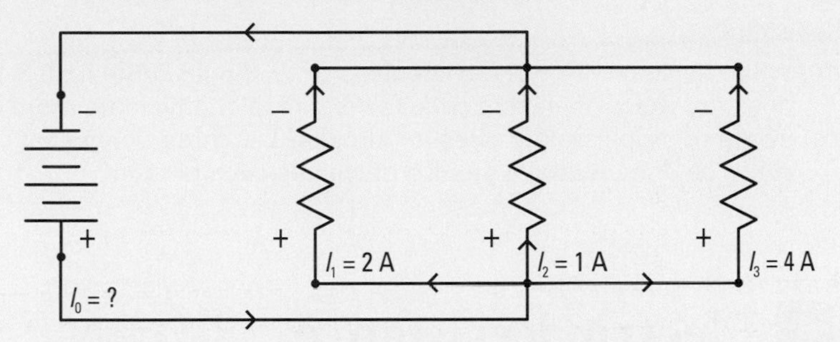

Figure 12

3. Calculate the total electric potential difference across three 6.0 V batteries connected (a) in series and (b) in parallel.
4. Find V_2, V_4, I_3, and I_4 in the circuit shown in **Figure 13**.

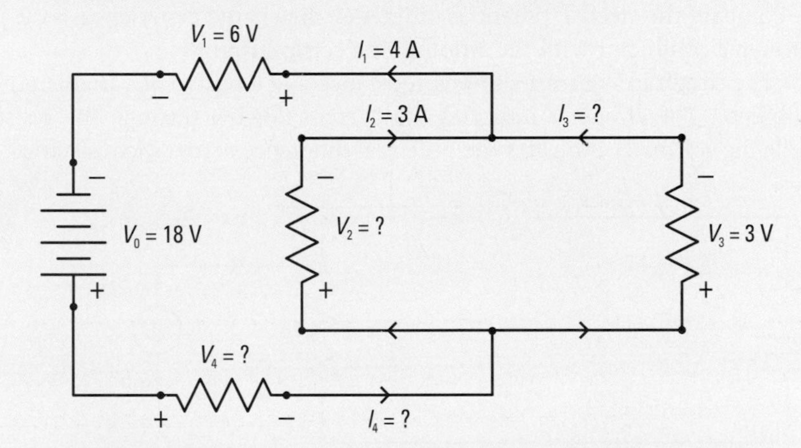

Figure 13

5. To construct a battery, cells can be connected in series or in parallel. If four 1.5-V cells are to be used to construct a battery, how will the batteries compare if they are all connected in series or all in parallel? Discuss in terms of the volts and battery life.
6. Draw a circuit diagram that includes an electric motor, a green light, a red light, and a switch. If the motor is on, the red light must also be on (to warn people of possible danger). If the motor is off, the green light should be on.

Applying Inquiry Skills

7. Design a simple circuit that could be used as a smoke detector or smoke alarm. (*Hint:* Smoke particles can be ionized readily and will conduct electricity across a narrow gap. You might try proving this by forcing smoke near a charged leaf electroscope.)

Making Connections

8. What are the advantages to connecting coloured lights in parallel rather than in series?

(continued)

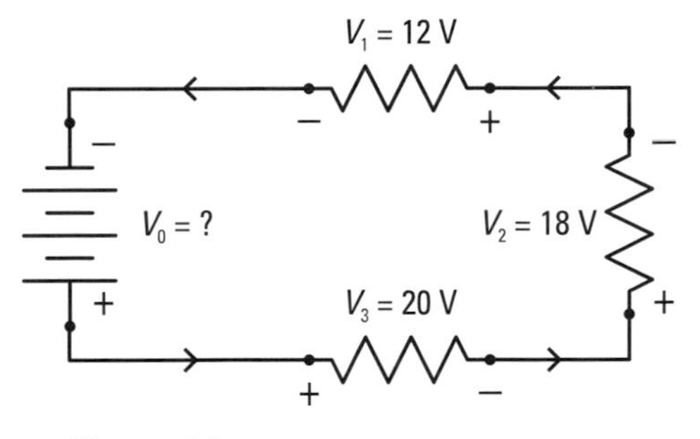

Figure 11
For question 1

9. Often a battery from a second vehicle is connected to the discharged battery of a vehicle that won't start in order to recharge or "boost" the battery. Should the batteries be connected in series or parallel? Explain your answer. What terminals should be connected?

Reflecting

10. When dealing with electric circuits, safety should come first. How does the study of electric circuits help people to become more aware of proper safety rules for circuits? List three things that could be done to make the use of electric circuits safer.

12.6 Electric Resistance

resistance: an opposition to the flow of charge, resulting in a loss of potential energy

When charges pass through a material or device, they experience an opposition or **resistance** to their flow, resulting in a loss of electric potential energy. To measure the amount of resistance that a quantity of moving charge encounters, we compare the electric potential difference the charge experiences as it passes through a conductor with the amount of electric current.

The circuit in **Figure 1** shows a resistance and a source of variable potential difference. The ammeter indicates the current flowing through the resistance, while the voltmeter indicates the potential difference across the resistance.

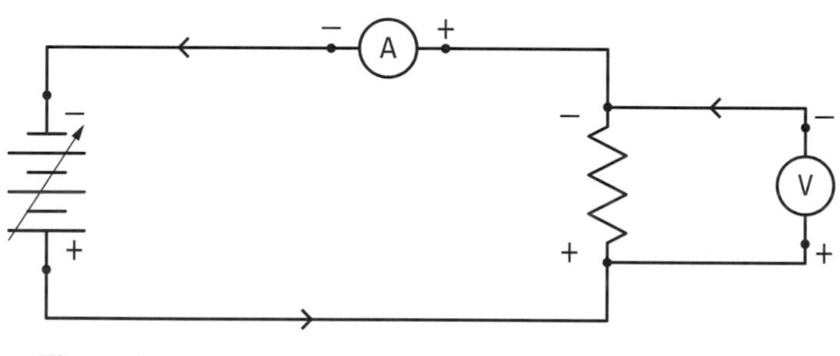

Figure 1

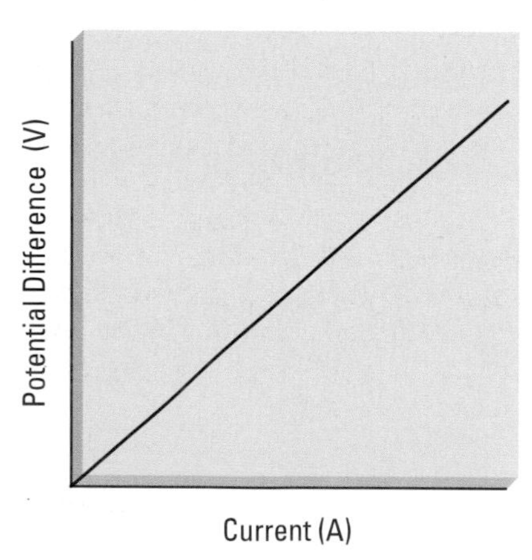

Figure 2
Mathematically, this graph is of the form $y = mx$, where m is the slope of the straight line.

By varying the potential difference of the source and making simultaneous measurements of current and potential difference, the graph in **Figure 2** can be obtained.

Ohm's Law

Ohm's law: The potential difference between any two points in a conductor varies directly as the current between the two points if the temperature remains constant.

The German physicist Georg Simon Ohm (1787–1854), shown in **Figure 3**, found that, for a given conductor, the ratio $\frac{V}{I}$ is a constant. From this constant ratio, he formulated what we now call **Ohm's law**:

Ohm's Law

The potential difference between any two points in a conductor varies directly as the current between the two points if the temperature remains constant.

This relationship can be written as

$$\frac{V}{I} = \text{constant}$$

Since the constant depends on the properties of the particular resistor being used, we give it the symbol R and call it resistance. Therefore, Ohm's law can be written as:

$$\frac{V}{I} = R$$

V is measured in volts, and I is measured in amperes, so R is measured in volts per ampere, and this new unit, the unit of electric resistance, is called the **ohm** (Ω).

1 Ω is the electric resistance of a conductor that has a current of 1 A through it when the potential difference across it is 1 V.

$1\ \Omega = 1\ \text{V/A}$

Figure 3
Georg Simon Ohm (1787–1854) became a high school teacher but longed to work at a university. To secure a university post, he began to do research in the area of electrical conduction. After his discovery of the law that bears his name, he received so much public criticism that he was forced to resign even his high school position. His fellow scientists felt that his discoveries were based too much on theory and lacked the experimental proof to make them acceptable. After years of living in poverty, he was finally recognized by the Royal Society, which conferred membership on him in 1842. He was also awarded a professorship at the University of Munich, where he spent the last five years of his life with his ambition realized.

ohm: (Ω) the SI unit for electric resistance

Sample Problem 1

What is the potential difference across a toaster of resistance 13.8 Ω when the current through it is 8.7 A?

Solution

$R = 13.8\ \Omega$

$I = 8.7\ \text{A}$

$$V = IR$$
$$= (8.7\ \text{A})\ (13.8\text{V})$$
$$V = 1.2 \times 10^2\ \text{V}$$

The potential difference across the toaster is 1.2×10^2 V.

Practice

Understanding Concepts

1. Calculate the value of the resistance in each case.
   (a) $V = 12$ V, $I = 0.25$ A
   (b) $V = 1.50$ V, $I = 30.0$ mA
   (c) $V = 2.4 \times 10^4$ V, $I = 6.0 \times 10^{-3}$ A
2. Calculate the maximum rating (in volts) of a battery used to operate a toy electric motor that has a resistance of 2.4 Ω and runs at top speed with a current of 2.5 A.
3. Write an equation for electric current in terms of electric potential and resistance.
4. Find the unknown quantities.
   (a) $R = 35\ \Omega$, $I = 0.45$ A, $V = ?$ (b) $R = 2.2$ kΩ, $I = 1.5$ A, $V = ?$
   (c) $V = 6.0$ V, $R = 18\ \Omega$, $I = ?$ (d) $V = 52$ mV, $R = 26\ \Omega$, $I = ?$
5. What current is drawn by a vacuum cleaner from a 115-V circuit having a resistance of 28 Ω?
6. A walkie-talkie receiver operates on a 9.0-V battery. If the receiver draws 3.0×10^2 mA of current, what is the resistance?

Answers

1. (a) 48 Ω
   (b) 50.0 Ω
   (c) $4.0 \times 10^6\ \Omega$
2. 6.0 V
4. (a) 16 V
   (b) 3.3×10^3 V
   (c) 0.33 A
   (d) 2.0×10^{-3} A
5. 4.1 A
6. $3.0 \times 10^1\ \Omega$

Resistance in Series

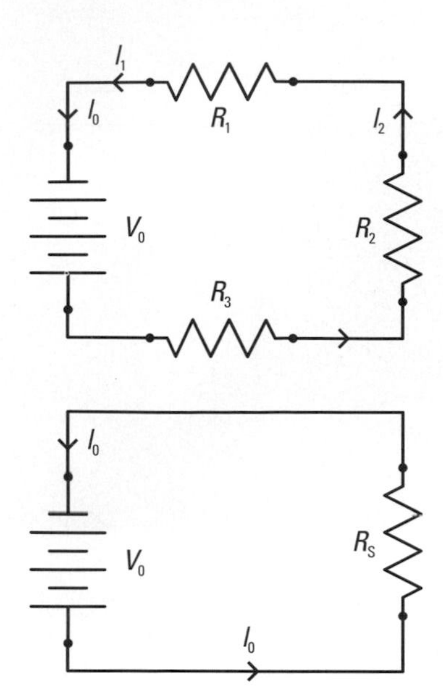

Figure 4

Figure 4 shows a circuit with one source of electric potential, V_0, and three resistors, R_1, R_2, and R_3, connected in series as indicated. We would like to find the value of the **equivalent resistor**, R_S, then the circuit would simply contain the source, V_0, and the resistor, R_S. The amount of current through the circuit from the source is I_0.

equivalent resistor: resistor that has the same current and potential difference as the resistors it replaces

Applying Kirchhoff's voltage law to the circuit,

$$(1)\ V_0 = V_1 + V_2 + V_3$$

Applying Ohm's law to each individual resistor,

$$(2)\ V_1 = I_1R_1 \qquad V_2 = I_2R_2 \qquad V_3 = I_3R_3 \qquad V_0 = I_0R_S$$

Substituting equation (2) into (1),

$$I_0R_S = I_1R_1 + I_2R_2 + I_3R_3$$

Applying Kirchhoff's current law to the circuit,

$$I_0 = I_1 = I_2 = I_3$$

Therefore, as an expression for the equivalent resistor of R_1, R_2, and R_3, in series, we get

$$R_S = R_1 + R_2 + R_3$$

And if the number of resistors connected in series is n, the equivalent resistor will be given by

$$R_S = R_1 + R_2 + ... + R_n$$

Sample Problem 2

What is the equivalent resistor in a series circuit containing a 16 Ω light bulb, a 27 Ω heater, and a 12 Ω motor?

Solution

$R_1 = 16\ \Omega$
$R_2 = 27\ \Omega$
$R_3 = 12\ \Omega$
$R_S = ?$

$$R_S = R_1 + R_2 + R_3$$
$$= 16\ \Omega + 27\ \Omega + 12\ \Omega$$
$$R_S = 55\ \Omega$$

The equivalent resistor is 55 Ω.

Practice

Understanding Concepts

7. Find the equivalent resistance in each of these cases.
   (a) A 12-Ω, a 25-Ω, and a 42-Ω resistor are connected in series.
   (b) Two 30-Ω light bulbs and two 20-Ω heating elements are connected in series.
   (c) Two strings of Christmas tree lights are connected in series; the first string has eight 4.0-Ω bulbs in series, and the second has twelve 3.0-Ω bulbs in series.
8. Find the value of the unknown resistance in each of these cases.
   (a) A 22-Ω, an 18-Ω, and an unknown resistor are connected in series to give an equivalent resistance of 64 Ω.
   (b) Two identical unknown bulbs are connected in series with a 48-Ω and a 64-Ω heater to produce an equivalent resistance of 150 Ω.
   (c) Each light in a series string of 24 identical bulbs has an equivalent resistance of 48 Ω.

Answers

7. (a) 79 Ω
   (b) 100 Ω
   (c) 68 Ω
8. (a) 24 Ω
   (b) 19 Ω
   (c) 2.0 Ω

Resistance in Parallel

We can use the same approach to find the equivalent resistance of several resistors connected in parallel (**Figure 5**). If we call the equivalent resistance of R_1, R_2, and R_3 connected in parallel R_P, then the circuit appears as shown.

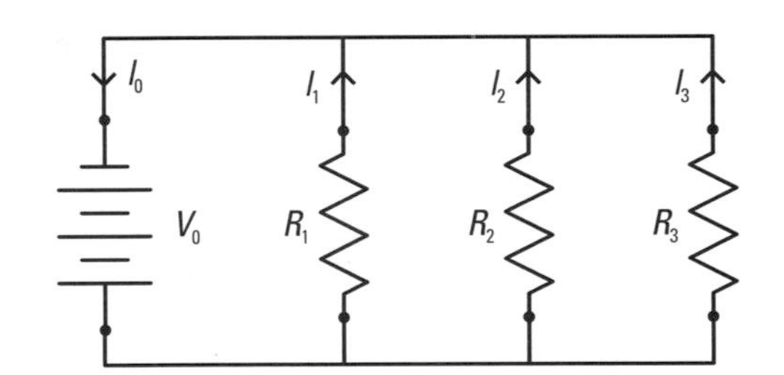

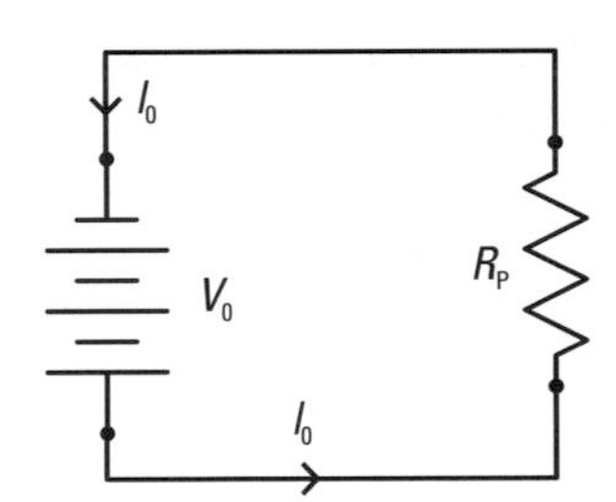

Figure 5

Applying Kirchhoff's current law to the circuit,

$$(1)\ I_0 = I_1 + I_2 + I_3$$

Applying Ohm's law to each individual resistance,

$$(2)\ I_1 = \frac{V_1}{R_1} \qquad I_2 = \frac{V_2}{R_2} \qquad I_3 = \frac{V_3}{R_3} \qquad I_0 = \frac{V_0}{R_P}$$

Substituting equation (2) into (1),

$$\frac{V_0}{R_P} = \frac{V_1}{R_1} + \frac{V_2}{R_2} + \frac{V_3}{R_3}$$

Applying Kirchhoff's voltage law to the circuit,

$$V_0 = V_1 = V_2 = V_3$$

Therefore, as an expression for the equivalent resistance of R_1, R_2, and R_3, in parallel, we get

$$\frac{1}{R_P} = \frac{1}{R_1} + \frac{1}{R_2} + \frac{1}{R_3}$$

And, if the number of resistors connected in parallel is n, the equivalent resistor will be given by

$$\frac{1}{R_P} = \frac{1}{R_1} + \frac{1}{R_2} + \ldots + \frac{1}{R_n}$$

Sample Problem 3

Find the equivalent resistor when a 4.0-Ω bulb and an 8.0-Ω bulb are connected in parallel.

Solution

$R_1 = 8.0\ \Omega$

$R_2 = 4.0\ \Omega$

$R_P = ?$

$$\frac{1}{R_P} = \frac{1}{R_1} + \frac{1}{R_2}$$

$$\frac{1}{R_P} = \frac{1}{8.0\ \Omega} + \frac{1}{4.0\ \Omega}$$

$$= \frac{1}{8.0\ \Omega} + \frac{2}{8.0\ \Omega}$$

$$\frac{1}{R_P} = \frac{3}{8.0\ \Omega}$$

$$R_P = \frac{8.0\ \Omega}{3}$$

$$R_P = 2.7\ \Omega$$

The equivalent resistor is 2.7 Ω.

Practice

Understanding Concepts

9. Find the equivalent resistance in each of these cases.
   (a) 16 Ω and 8.0 Ω connected in parallel
   (b) 22 Ω, 12 Ω, and 5.0 Ω connected in parallel
10. Calculate the equivalent resistance of two, three, four, and five 60-Ω bulbs in parallel. What is the simple relationship for the equivalent resistance of *n* equal resistances in parallel?

Answers

9. (a) 5.3 Ω
   (b) 3.0 Ω
10. 30 Ω; 20 Ω; 15 Ω; 12 Ω

Electric Circuit Analysis

With what you've learned about electricity you should be able to do a complete analysis of any simple series or parallel electric circuit containing resistances. However, because so many different electric circuits are possible, there is no standard approach to analyzing a circuit. The steps to take in each case will depend upon the information you have about the circuit, and what you want to find out.

Sample Problem 4

Find V_1, V_2, V_3, I_1, I_2, I_3, R_2 for the circuit in **Figure 6**.

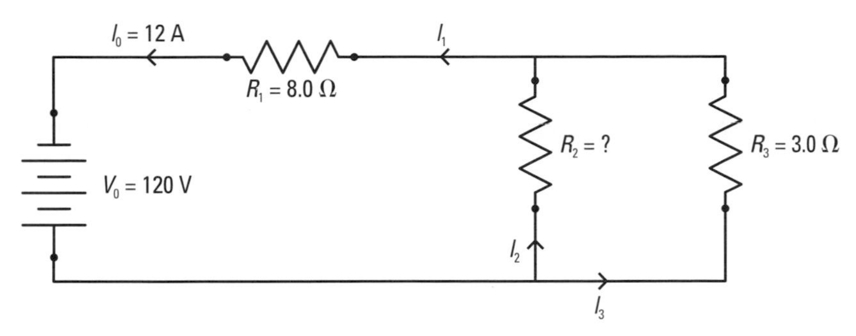

Figure 6

Solution

Using Kirchhoff's current law,

$$I_0 = I_1 = I_2 + I_3$$

Applying Ohm's law to the entire circuit

$$R_0 = \frac{V_0}{I_0}$$
$$= \frac{120 \text{ V}}{12 \text{ A}}$$
$$R_0 = 1.0 \times 10^1 \ \Omega$$

If the parallel pair of resistors R_2 and R_3 is, for the moment, thought of as one single resistor, R_P, then R_1 and R_P are connected in series, and their total resistance is given by

$$R_t = R_1 + R_P$$
$$R_p = R_t - R_1$$
$$= 1.0 \times 10^1 \ \Omega - 8.0 \ \Omega$$
$$R_P = 2.0 \ \Omega$$

Then, using the relationship for the equivalent resistance in parallel,

$$\frac{1}{R_P} = \frac{1}{R_2} + \frac{1}{R_3}$$
$$\frac{1}{R_2} = \frac{1}{R_P} - \frac{1}{R_3}$$
$$= \frac{1}{2.0 \ \Omega} - \frac{1}{3.0 \ \Omega}$$
$$= \frac{3}{6.0 \ \Omega} - \frac{2}{6.0 \ \Omega}$$
$$\frac{1}{R_2} = \frac{1}{6.0 \ \Omega}$$
$$R_2 = 6.0 \ \Omega$$

Using Ohm's law,

$$V_1 = I_1 R_1$$
$$= (12\text{ A})(8.0\text{ V})$$
$$V_1 = 96\text{ V}$$

Then, applying Kirchhoff's voltage law around each of the two paths through the circuit,

$$V_0 = V_1 + V_2 \qquad V_0 = V_1 + V_3$$
$$V_2 = V_0 - V_1 \qquad V_3 = V_0 - V_1$$
$$= 120\text{ V} - 96\text{ V} \qquad = 120\text{ V} - 96\text{ V}$$
$$V_2 = 24\text{ V} \qquad V_3 = 24\text{ V}$$

Finally,

$$I_2 = \frac{V_2}{R_2} \qquad I_3 = \frac{V_3}{R_3}$$
$$= \frac{24\text{ V}}{6.0\ \Omega} \qquad = \frac{24\text{ V}}{3.0\ \Omega}$$
$$I_2 = 4.0\text{ A} \qquad I_3 = 8.0\text{ A}$$

As a check, $I_0 = I_2 + I_3 = 4.0\text{ A} + 8.0\text{ A} = 12\text{ A}$, as given.

(a)

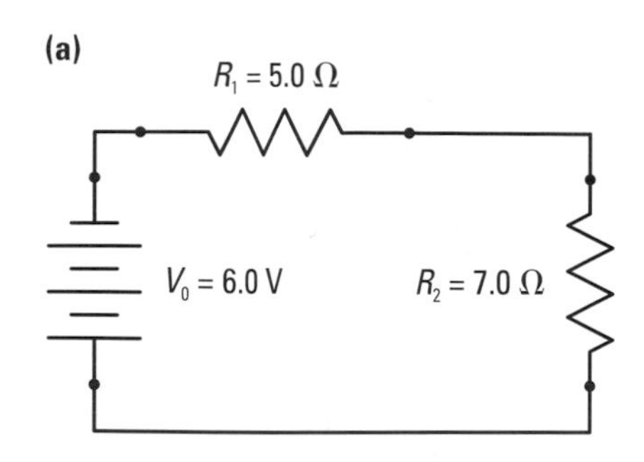

(b)

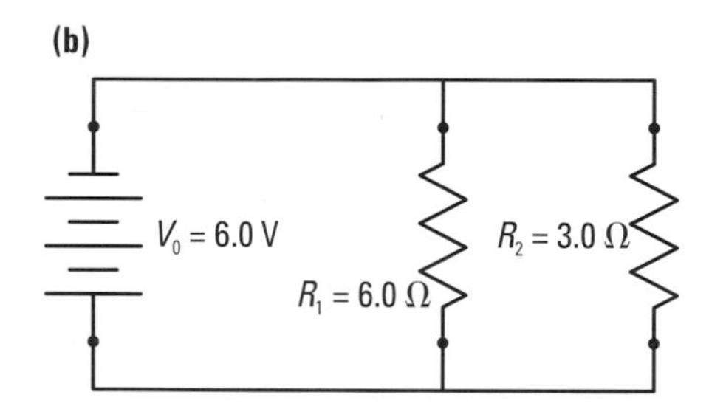

Figure 7
For question 11

Practice

Understanding Concepts

11. Complete each of the circuits in **Figure 7** by finding the unknown quantities indicated.
    (a) I_0, I_1, I_2, V_1, V_2
    (b) V_1, V_2, I_0, I_1, I_2

Answers

11. (a) $I_0 = 0.5$ A, $I_1 = 0.5$ A, $I_2 = 0.5$ A, $V_1 = 2.5$ V, $V_2 = 3.5$ V
    (b) $V_1 = 6.0$ V, $V_2 = 6.0$ V, $I_0 = 3.0$ A, $I_1 = 1.0$ A, $I_2 = 2.0$ A

SUMMARY Electric Resistance

- When charges pass through a material or device, they experience an opposition known as resistance.
- Ohm's law is represented by the equation $\frac{V}{I} = R$.
- The unit of electric resistance is called the ohm (Ω).
- If the number of resistors connected in series is n, the equivalent resistor is given by $R_S = R_1 + R_2 + ... + R_n$.
- If the number of resistors connected in parallel is n, the equivalent resistor is given by $\frac{1}{R_P} = \frac{1}{R_1} + \frac{1}{R_2} + \ldots + \frac{1}{R_n}$.

Section 12.6 Questions

Understanding Concepts

1. Find V_1, V_2, I_0, I_1, and R_2 for the circuit in **Figure 8**.
2. Find V_0, V_1, I_2, R_1, and R_2 for the circuit in **Figure 9**.

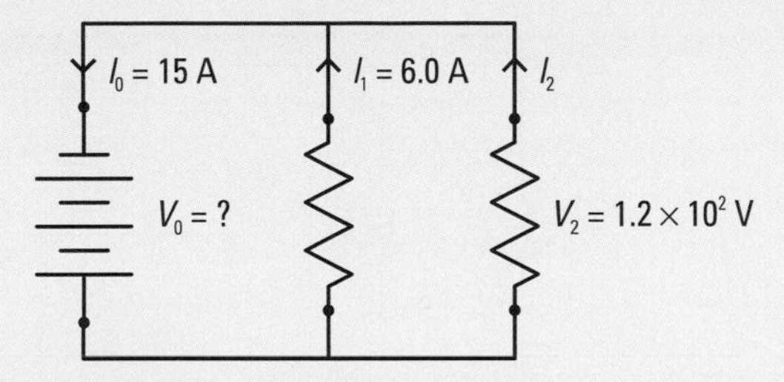

Figure 9

3. Find V_1, V_3, I_1, I_2, I_3, and R_3 for the circuit in **Figure 10**.

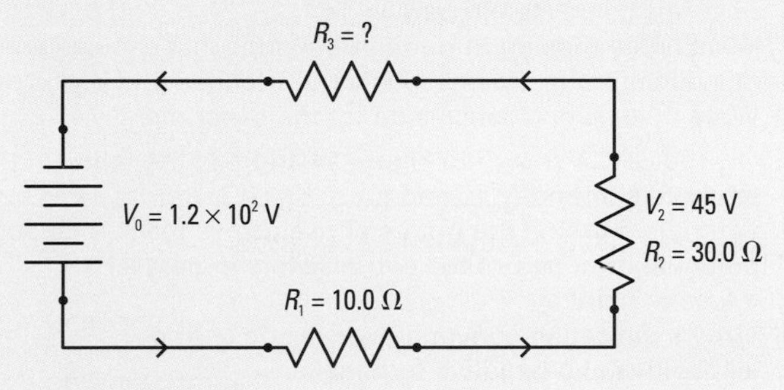

Figure 10

4. Find V_0, V_1, V_2, V_3, I_0, I_1, and I_2 for the circuit in **Figure 11**.

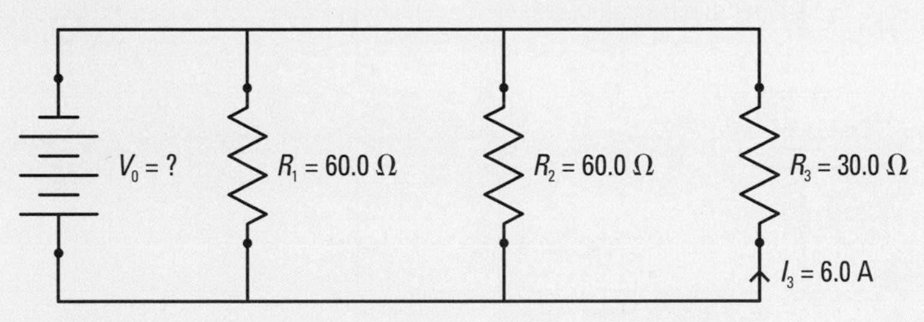

Figure 11

5. For each of the circuits in **Figure 12**, find the current through and the potential drop across each resistor.
6. What is the effect on the total resistance in a circuit when an extra resistor is added (a) in series and (b) in parallel?
7. Draw a graph of electric potential difference versus current for two different resistances, and indicate which has the greater resistance, and why.

(continued)

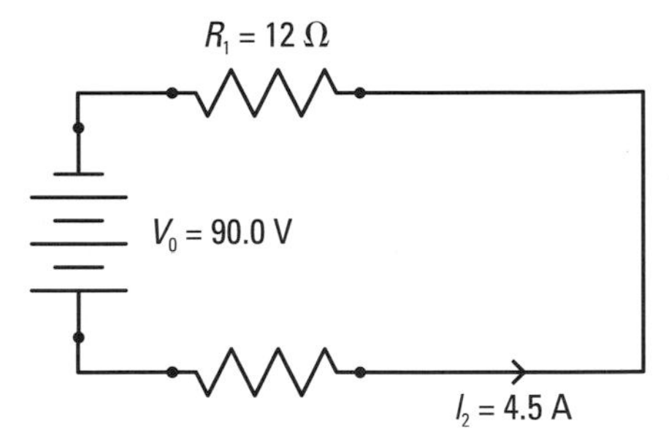

Figure 8
For question 1

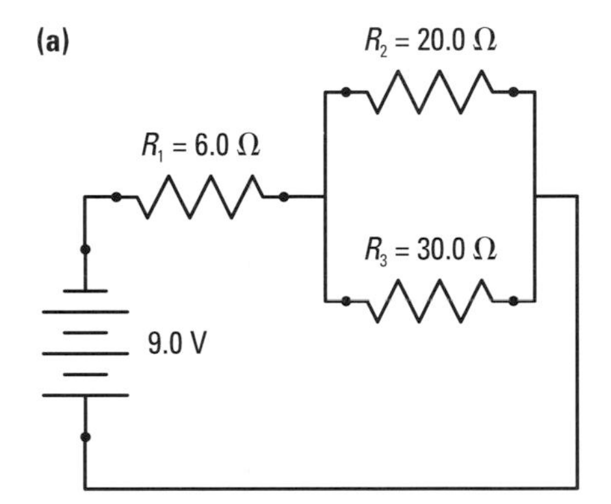

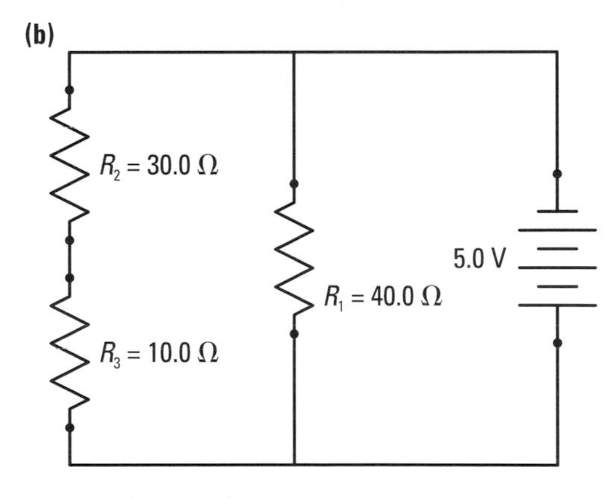

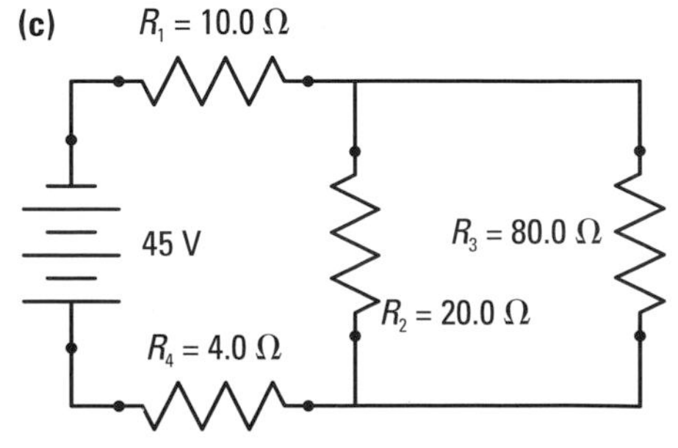

Figure 12
For question 5

Applying Inquiry Skills

8. A 12-V battery, an ammeter, a 5.0-A fuse (which will burn out if more than 5.0 A of current is in the circuit), and several 10.0-Ω lamps are used in an experiment to find the effect of connecting resistances in parallel (**Figure 13**).

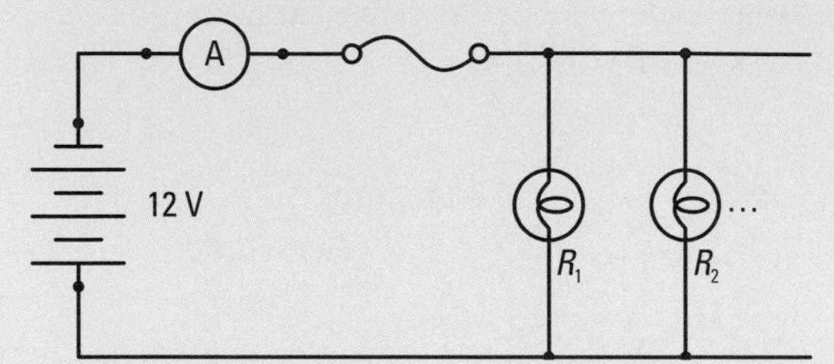

Figure 13

   (a) Determine the total resistance and current when the number of lamps connected in parallel is 1, 2, 3, 4, 5, and 6.
   (b) What is the maximum number of lamps that can be connected before the fuse becomes overloaded and burns out?
   (c) Write at least one conclusion for the experiment.

9. Redraw the circuit shown in **Figure 14** and add the following:
   (a) ammeters to find I_0, I_c, and I_e
   (b) voltmeters to find the potential rise across the source and the potential drop across the two resistors in parallel
   (c) a fuse at point a
   (d) arrows indicating conventional electric current
   (e) all positive and negative terminals

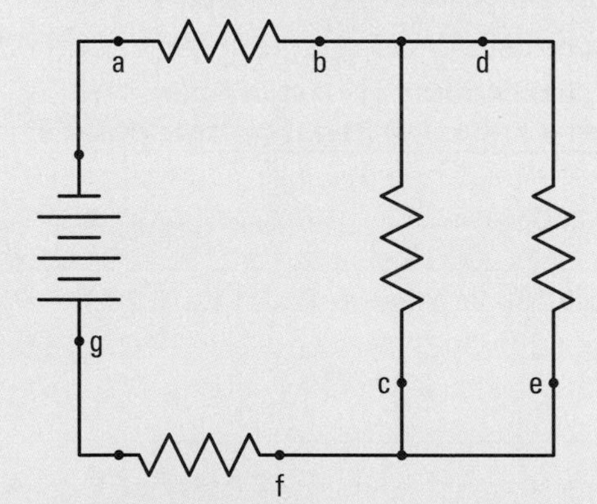

Figure 14

10. Parts of an electric circuit may heat up and start to smoke. What should you do if this occurs (a) in the classroom or (b) at home?

Making Connections

11. A dimmer switch is used in some household circuits to control the brightness of incandescent light bulbs. Describe how this type of switch operates. (A diagram will aid your explanation.)

12. The bulbs in an older string of Christmas tree lights are connected in series. They will all go out if one bulb is burnt out or removed. In a modern set, if one bulb burns out, the others stay lit. However, if a good bulb is removed or twisted, they all go out. Are modern sets connected in series or parallel?

12.7 Power in Electric Circuits

To predict the amount of energy used by an electrical device, such as a radio, stove, lights, or television, we first need to know the amount of time the device will be used. For some devices such as freezers, refrigerators, and clocks, we can predict the amount of time the device is in operation, and the amount of energy used can be calculated using the formula for energy from section 12.4:

$$\Delta E = VI\Delta t$$

(For most devices in the home, $V = 120$ V. For other devices, like stoves and dryers, $V = 240$ V.)

Why would you want to know how much energy an appliance would use in a year? Because energy costs money—electrical energy is not free. To calculate the cost for devices that are not used for predictable amounts of time, it is often more useful to know the rate at which the device uses energy, or the **power** (P). Power is the rate at which energy is used or supplied, and it does not depend on how long the device is on (**Figure 1**). Therefore, as covered in Chapter 4,

power: (P) the rate at which energy is used or supplied

$$P = \frac{\Delta E}{\Delta t}$$

If we substitute $\Delta E = VI\Delta t$, we get

$$P = \frac{VI\Delta t}{\Delta t}$$

$$P = VI$$

Figure 1
With many devices using a lot of electrical energy at the same time, the power rating is very high.

The equation $P = VI$ is an expression for the electric power dissipated by a current, I, through a potential difference, V.

Note that potential difference is measured in volts (joules per coulomb) and current is measured in amperes (coulombs per second), and the unit of electric power is the product of the units of potential difference and current, namely,

$$\begin{aligned}(\text{volt})(\text{ampere}) &= (\text{joule/coulomb})(\text{coulomb/second}) \\ &= \text{joule/second} \\ &= \text{watt}\end{aligned}$$

Therefore, electric power is measured in watts—the same unit that was used to measure mechanical power in Chapter 4. If the load has a resistance, R, then two different expressions can be derived by applying Ohm's law.

From Ohm's Law, $I = \frac{V}{R}$ and $V = IR$, we can write

$$\begin{aligned}P &= VI & \text{also} \quad P &= VI \\ &= (IR)I & &= V\left(\frac{V}{R}\right) \\ P &= I^2R & P &= \frac{V^2}{R}\end{aligned}$$

To summarize, for a load with a current, I, potential difference, V, and resistance, R, the power is given by

$$P = VI$$
$$P = I^2R$$
$$P = \frac{V^2}{R}$$

and, when SI units are used for I, V, and R, power is measured in watts.

Sample Problem 1

What is the current drawn by a 1.0×10^2-W light bulb operating at a potential difference of 120 V?

Solution

$P = 1.0 \times 10^2$ W

$V = 1.2 \times 10^2$ V

$I = ?$

$$P = VI$$
$$I = \frac{P}{V}$$
$$= \frac{1.0 \times 10^2 \text{ W}}{1.2 \times 10^2 \text{ V}}$$
$$I = 0.83 \text{ A}$$

The current drawn is 0.83 A.

Sample Problem 2

What is the resistance of a 6.0×10^2-W kettle that draws a current of 5.0 A?

Solution

$P = 6.0 \times 10^2$ W

$I = 5.0$ A

$R = ?$

$$P = I^2R$$
$$R = \frac{P}{I^2}$$
$$= \frac{6.0 \times 10^2 \text{ W}}{(5.0 \text{ A})^2}$$
$$R = 24\ \Omega$$

The resistance is 24 Ω.

Sample Problem 3

What power is dissipated by an electric frying pan that has a resistance of 12 Ω and operates at a potential difference of 120 V?

Solution

$R = 1.2 \times 10^1\ \Omega$

$V = 1.2 \times 10^2\ \text{V}$

$P = ?$

$$P = \frac{V^2}{R}$$

$$= \frac{(1.2 \times 10^2\ \text{V})^2}{1.2 \times 10^1\ \Omega}$$

$$P = 1.2 \times 10^3\ \text{W}$$

The power dissipated is 1.2×10^3 W, or 1.2 kW.

Note that the appliances discussed in the problems above, or in any other examples or questions, typically operate on AC and not DC. The values of V and I in all these problems are actually RMS values (the square root of the mean of the square of the instantaneous current or voltage). This concept will be covered in more advanced courses.

Practice

Understanding Concepts

1. A large refrigerator operates at a voltage of 120 V, drawing a current of 4.6 A. What is the power rating of the refrigerator?
2. Write an equation for each of the following:
   (a) voltage in terms of power and electric current
   (b) electric current in terms of power and voltage
   (c) electric resistance in terms of power and electric current
   (d) electric resistance in terms of power and voltage
3. Using a voltage of 1.2×10^2 V, what current do the following draw?
   (a) a 2.0×10^2-W light bulb (b) a 1200-W heater
4. Calculate the resistance of a 360-W hair dryer designed for a 120-V power supply.
5. The current through a device of resistance 22 Ω is 2.0 A. Find the power rating of the device.

Answers

1. 5.5×10^2 W
3. (a) 1.7 A
   (b) 1.0×10^1 A
4. 4.0×10^1 Ω
5. 88 W

The Cost of Electricity

To calculate the energy used in operating an electrical device, all we need to know is the power of the device and the amount of time it is used. However, the basic unit of energy, the joule, is a very small unit, and, for this reason, the electrical energy consumed is measured by means of a larger unit derived from the joule.

Energy used can be written as

$$\Delta E = P\Delta t$$

kilowatt hour: (kW·h) the energy dissipated in exactly 1 h by a load with a power of exactly 1 kW

Then, if power is measured in kilowatts (kW) and time is measured in hours (h), electrical energy can be measured in **kilowatt hours** (kW·h), which is the energy dissipated in 1 h by a load with a power of 1 kW.

It is often useful to calculate the relationship between the joule and the kilowatt hour, using the equation

$$\begin{aligned}\Delta E &= P\Delta t\\ &= (1.0\ \text{kW})\ (1.0\ \text{h})\\ &= 1.0\ \text{kW·h}\\ &= (1.0 \times 10^3\ \text{J/s})\ (3.6 \times 10^3\ \text{s})\\ \Delta E &= 3.6 \times 10^6\ \text{J}\end{aligned}$$

Therefore, $1.0\ \text{kW·h} = 3.6 \times 10^6\ \text{J} = 3.6\ \text{MJ}$.

Figure 2
Readings from electrical meters are used to calculate the amount of energy used.

If several appliances are operating simultaneously, you can calculate the total power usage by adding the amount of power dissipated by each of the appliances. The utility company installs a meter on each house to measure and keep track of the total energy usage (**Figure 2**). The total cost of electricity is obtained by multiplying the number of kilowatt hours of energy used by the rate (price per kilowatt hour). Generally, rates vary by region and by type of user (residential or commercial).

Sample Problem 4

Calculate the cost of operating a 4.0×10^2-W spotlight for 2.0 h a day for 30 d at a rate of 9.0¢/kW·h.

Solution

$P = 4.0 \times 10^2\ \text{W} = 0.40\ \text{kW}$
$\Delta t = (2.0\ \text{h/d})(30\ \text{d}) = 60\ \text{h}$
cost = ?

$$\begin{aligned}\Delta E &= P\Delta t\\ &= (0.40\ \text{kW})\ (60\ \text{h})\\ \Delta E &= 24\ \text{kW·h}\end{aligned}$$

cost = (24 kW·h) (9.0¢/kW·h)
cost = 216¢, or $2.16

Sample Problem 5

Find the cost of operating an oven for 3.0 h if it draws 25 A from a 240-V supply at a rate of 8.5¢/kW·h.

Solution

$\Delta t = 3.0\ \text{h}$
$I = 25\ \text{A}$
$V = 240\ \text{V}$

$$\begin{aligned}P &= IV\\ &= (25\ \text{A})(240\ \text{V})\\ P &= 6.0 \times 10^3\ \text{W, or } 6.0\ \text{kW}\end{aligned}$$

$$\Delta E = P\Delta t$$
$$= (6.0 \text{ kW})(3.0 \text{ h})$$
$$\Delta E - 18 \text{ kW·h}$$

$$\text{cost} = (18 \text{ kW·h})(8.5¢/\text{kW·h})$$
$$\text{cost} = 153¢, \text{ or } \$1.53$$

The Energuide label (**Figure 3**) for major electrical household appliances shows the annual energy rating in kilowatt hours, based on average use. The label is required by law on all appliances sold in Canada. Some common values of power ratings are shown in **Table 1**.

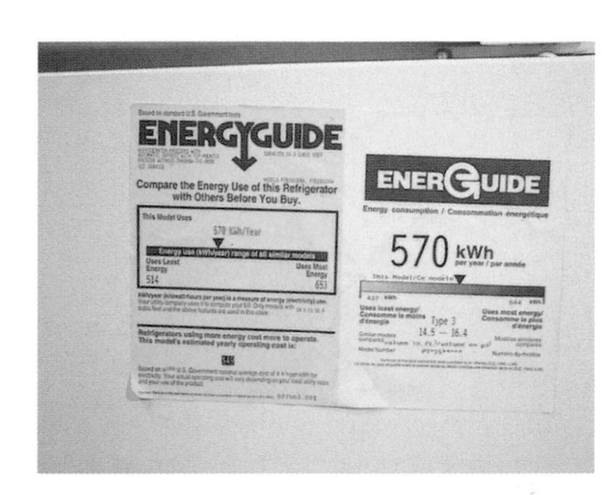

Figure 3
To keep the cost of energy low, buy the appliance with the lowest energy consumption rating capable of fulfilling your requirements.

Table 1 Power Rating of Electrical Appliances

Appliance	Power rating	Current (with 120 V supply)
iron	1200 W	10 A
stove	6 to 10 kW	*
refrigerator	200 W	1.7 A
toaster	1000 W	8.3 A
microwave oven	600 W	5.0 A
vacuum cleaner	500 W	4.2 A

* Stoves operate at 240 V and draw currents in the range of 25 A to 40 A.
Note: These power ratings refer to the electrical power used by the appliances, and not their power output, which, due to their relative efficiencies, may be considerably lower.

Practice

Understanding Concepts

6. If it costs 8.9¢/kW·h for electrical energy, how much would it cost to leave a 60.0-W light bulb on for
   (a) one hour? (b) one day? (c) one year?
7. A small television is connected to a 120-V power supply and draws a current of 1.2 A. How much would it cost to use the television for an average of 2.0 h a day for 30 days if the rate for electrical energy is 8.5¢/kW·h?
8. A student decides to help lower the electric bill by listening to the 120-W stereo for an average of 45 minutes a day instead of the usual hour. If the rate is 9.2¢/kW·h, how much money will be saved on the electric bill in a year?

Answers

6. (a) 0.53¢
   (b) 13¢
   (c) $46.78
7. 73¢
8. $1.01

SUMMARY Power in Electric Circuits

- The rate at which a device uses energy is known as power, represented by the equation $P = VI$.
- Electrical energy is measured in kilowatt hours (kW·h).
- The total cost of electricity can be calculated by multiplying the number of kilowatt hours of energy used by the rate (price per kilowatt hour).

Section 12.7 Questions

Understanding Concepts

1. Find the cost of operating an electric toaster for 3.0 h if it draws 5.0 A from a 120-V outlet. Electric energy costs 8.8¢/kW·h.
2. The blower motor on an oil furnace, rated at 250 W, comes on, for an average of 5.0 min at a time, a total of 48 times a day. What is the monthly (30 d) cost of operating the motor if electricity costs 9.4¢/kW·h?
3. The following appliances were operated for a 30-day month, in a 120-V circuit: a coffee percolator of resistance 15 Ω for 0.50 h/d, a 250-W electric drill for 2.0 h/d, and a toaster that draws 5.0 A for 15 min/d. Calculate the electric bill for these devices for the month at an average cost of 9.0¢/kW·h.
4. What is the potential difference across a 1250-W baseboard heater that draws 5.2 A?
5. If a 7.0×10^2-W toaster and an 1100-W kettle are plugged into the same 120-V outlet in parallel, what total current will they draw?
6. What is the maximum power that can be used in a circuit with a potential difference of 120 V and a maximum current of 20.0 A?
7. A portable heater is plugged into a 120-V outlet and draws a current of 8.0 A for 10.0 min. Calculate each of the following:
   (a) the quantity of electric charge that flows through the heater
   (b) the energy consumed by the heater
   (c) the power dissipated by the heater
8. A student starts working and decides to contribute to the family electric bill by paying for the energy he uses. He makes a table (**Table 2**) of the electric devices he uses, the power rating of the device, and the time of use for a month. If the cost of electric energy is 9.2¢/kW·h, how much does he owe his parents at the end of the month?
9. Given two resistors R_1 = 12 Ω and R_2 = 4.0 Ω connected in series to an 18.0-V battery, find (a) the total power and (b) the power dissipated by each resistor. Compare your answers for both parts.
10. Repeat question 9 with the two resistors connected in parallel.

Table 2

Electrical device	Power rating (kW)	Time used per month (h)
television	0.08	90
computer	0.2	45
VCR	0.04	8
stereo	0.03	80
washer	0.5	4
dryer	5.0	4
microwave	0.75	1

Applying Inquiry Skills

11. As an electrical engineer for Westinghouse, you are asked to design an immersion heater that can be placed into a glass of water to boil it for coffee.
    (a) Using a 12-V battery, bare high resistance wire, and a hollow glass tube, draw a diagram of a prototype immersion heater.
    (b) Calculate the energy used and the power rating of the heater if 250 g of water is heated from 15°C to 85°C in 240 s. (Assume 100% efficiency.)
    (c) Suggest one or two improvements you could make to your heater that would make it heat the water faster.

Making Connections

12. What are the advantages of connecting appliances in parallel in a household circuit?
13. Electric power companies charge a rate that decreases when the amount of energy consumed by a household exceeds a certain minimum quantity. Comment on the wisdom of this method.
14. In many households, the total power dissipated is higher in the middle of the summer and the middle of the winter. Explain why.

Chapter 12 Summary

Key Expectations

Throughout this chapter, you have had opportunities to

- define and describe the concepts and units related to electricity such as electric charge, electric current, electric potential difference, electron flow, energy, power, and kilowatt hour; (12.1–12.7)
- describe the two conventions used to denote the direction of movement of electric charge in an electric circuit (electric current and electron flow), and recognize electric current as the preferred convention; (12.3)
- compare direct current (DC) and alternating current (AC) in qualitative terms; (12.3)

Key Terms

electric charge
static electricity
atom
electron
proton
nucleus
elementary charge
neutron
negative ion
positive ion
fundamental laws of electric charges
conductor
insulator
electrostatic series
electroscope
induced charge separation
ground
electric field
Coulomb's law
coulomb
ampere
conventional current
electric current
electron flow
direct current
alternating current
ammeter
electric potential difference
volt
voltmeter
circuit
load
elements
series circuit
parallel circuit
law of conservation of energy
law of conservation of charge
Kirchhoff's voltage law
Kirchhoff's current law
resistance
Ohm's law
ohm
equivalent resistor
power
kilowatt hour

Make a Summary

Using the Key Expectations and the Key Terms, make a concept map to summarize what you have learned in this chapter and include word descriptions, units, defining equations, and graph relationships.

Reflect on your Learning

Revisit your answers to the Reflect on Your Learning questions at the beginning of this chapter.

- How has your thinking changed?
- What new questions do you have?

Chapter 12 Review

Understanding Concepts

1. (a) Explain the meaning of the term "elementary charge."
   (b) What is the charge on an electroscope with an excess of 3.0×10^{11} electrons?
   (c) An object has a positive charge of 0.30 C. How many electrons has the object lost?
2. Object 1, a conductor with a positive charge of 0.020 C, is brought into contact with an identical neutral object 2.
   (a) How many electrons move from one object to the other?
   (b) Which way do the electrons move?
3. Find V_2, V_4, V_5, I_1, and I_4 in the circuit shown in **Figure 1**.

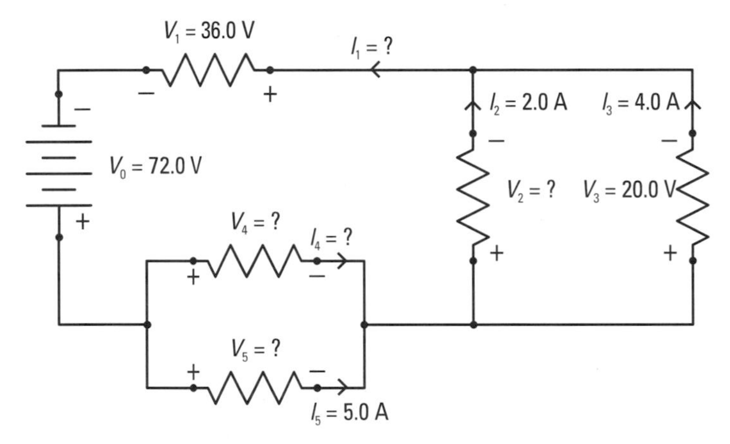

Figure 1

4. Find the charge that passes through a device if a constant electric current of 0.30 A flows in it for 5.0 min.
5. How long would it take to transfer 1.8 C of charge with a current of 9.5 mA?
6. It takes 15.4 s for 3.2 C of charge to pass through a resistor. What is the electric current through the resistor?
7. Find V_1, V_2, V_3, I_0, I_1, I_2, and I_3 for the circuit in **Figure 2**.

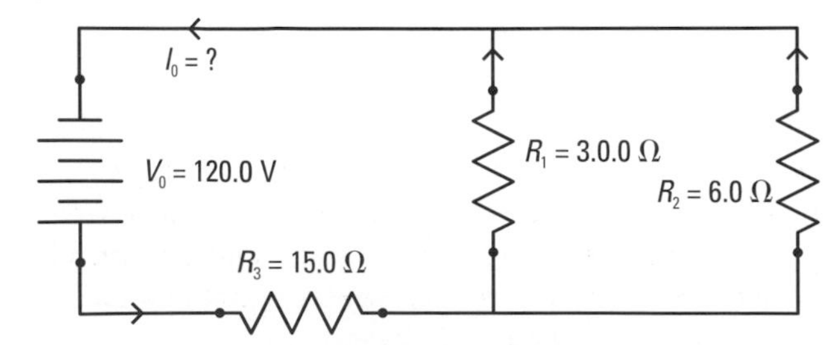

Figure 2

8. For the circuit shown in **Figure 3**, determine the value of R_3 and V_0.

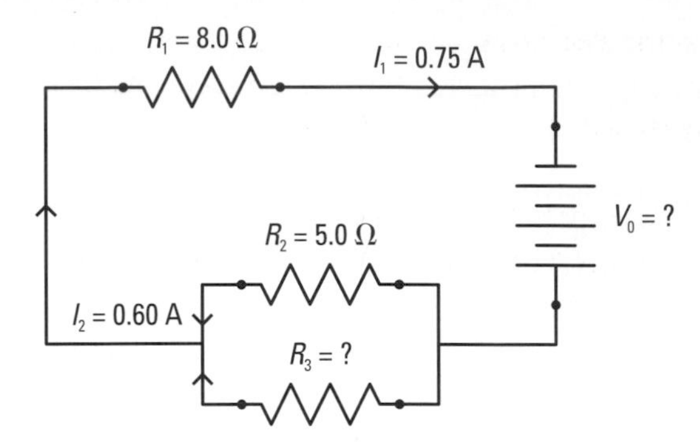

Figure 3

9. The circuit in **Figure 4(a)** was used to determine the resistance of R_2. The variable power supply gave several values for current and voltage on the meters pictured in the circuit. When these values were plotted, the graph in **Figure 4(b)** was generated. From this information, determine the value of R_2.

(a)

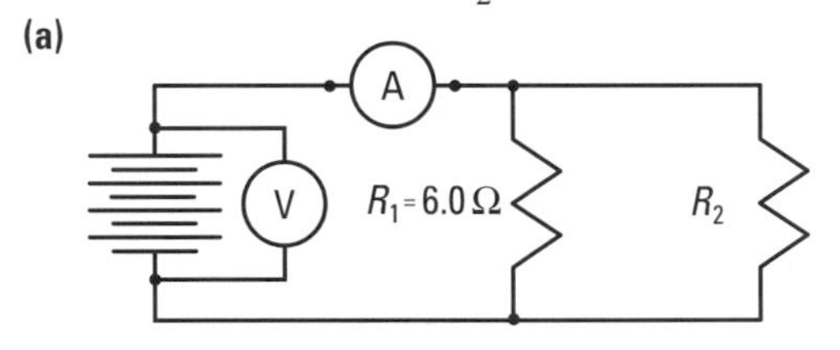

(b)

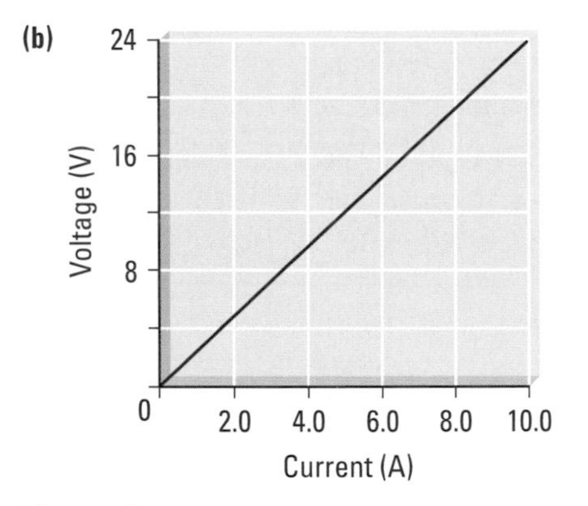

Figure 4

10. Find the potential difference between two points if 10.0 kJ of work is required to move 2.5 C of charge between the two points.
11. Research physicists use high electric potential differences to accelerate particles they are studying. How much energy is given to a proton as it accelerates through an electric potential difference of 4.5×10^7 V? The charge on a single proton is 1.6×10^{-19} C.
12. An electron is accelerated from rest through a potential difference of 1.5 MV. What is the final energy of the electron?
13. It takes 20.0 J of work to move a charge through a potential difference of 2.0×10^2 V. Find the charge.

14. When the electric potential difference across a series of lights is 12.0 V, a current of 0.35 A flows through it. The lights are left on for 0.50 h. What is the energy used by the lights?
15. A hair drier uses 10 800 J of energy in 45.0 s when the current through it is 2.0 A. What is the potential difference across the hair drier?
16. A bolt of lightning delivers 25 C of charge, releasing 1.2×10^9 J of energy in 30.0 ms.
    (a) What is the power associated with the bolt?
    (b) What is the current?
    (c) What is the potential difference across the bolt of lightning?
17. If you examine a diagram of the electric fields near a small point charge and between two parallel plates, you'll notice the field around the point charge is nonuniform and the field between the two plates is said to be uniform. Explain why this terminology is used for each field.
18. A worker in a clean room forgets to wear his wrist strap. He has no charge on him and works at his station for a while successfully. After a little more time he develops a charge, touches some sensitive equipment, and damages it. Explain how the charge might have developed.
19. The CN Tower in Toronto, Ontario, is one of the tallest buildings in the world. It is frequently struck by lightning and yet it is rarely damaged. Explain how this is possible.
20. What is the cost to a storekeeper of leaving a 40.0-W light burning near his safe over the weekend, for 60.0 h, if electricity costs 9.4¢/kW•h?

Applying Inquiry Skills

21. An experiment is performed to find the relationship between current and electric potential difference for two resistors, A and B. The data collected have been plotted in the graph shown in **Figure 5**. Study the graph and answer the questions that follow.

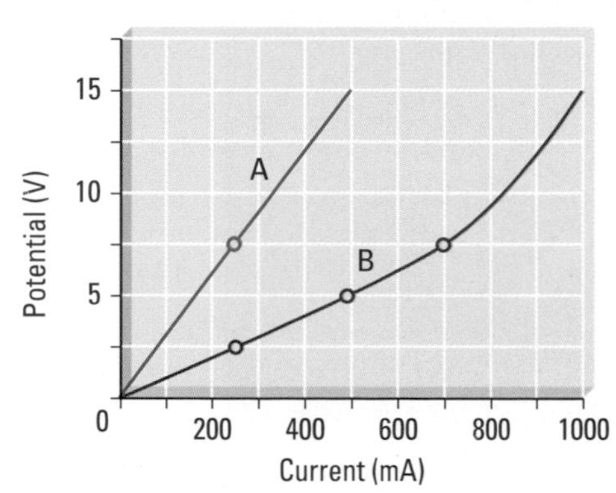

Figure 5

    (a) Find the resistance of each resistor for a current that does not exceed 5.00×10^2 mA.
    (b) What do you think happens to B when the current exceeds 5.00×10^2 mA?
22. Many common household appliances use an LED as an on/off indicator light. These are usually small red lights, but may be found in an assortment of colours, shapes, and sizes. **Table 1** shows the corresponding voltage and current values for an LED.

Table 1

Voltage (V)	0	1.0	2.0	3.0	4.0	5.0
Current (mA)	0	0.1	0.8	12.2	58.0	182.0

    (a) Plot a graph of potential difference versus current for the LED.
    (b) Does the LED obey Ohm's law?
    (c) Describe the change in resistance between the following voltage intervals:
        (i) 0 V and 2.0 V
        (ii) 4.0 V and 5.0 V

Making Connections

23. Make a table of the electrical devices you use in your home, the power rating of the devices, and an estimate of the amount of time you use them each month. Calculate the cost for operating the devices for a month. Use the rate in your area. Compare the cost of electricity with the benefits we gain by using these devices.
24. With the recent deregulation of electricity in some provinces, what are some of the implications at
    (a) the production level?
    (b) the distribution level?
    (c) the user level and local utilities?
25. Discuss the advantages and disadvantages of developing more uses of current electricity in our society.

Exploring

26. Research a device or a use for static electricity in industry or in the home. Discuss how it works, what it is used for, the environmental impacts, and any dangers associated with the device. Follow the links for Nelson Physics 11, Chapter 12 Review.

GO TO www.science.nelson.com

Chapter

13

Electromagnetism

In this chapter, you will be able to

- describe and understand the properties of magnetic fields around permanent magnets
- understand the domain theory of magnetism and be able to apply it to a variety of situations
- draw magnetic field lines around permanent magnets, conductors, and solenoids and understand factors that influence their strength
- understand the motor principle and apply it to a variety of situations and technologies
- research information about electromagnetism using a variety of technologies

How many magnets do you have in your home (**Figure 1**)? Many of us have magnets on the fridge to hold up papers. Magnets are also used to find a direction, as with a compass. Is that all? After more thought, you might remember that sometimes magnets are used on cabinet doors and electric can openers. These magnets are made of metal and are usually rectangular, U-shaped, or circular.

Did you know that magnets can be found in even more places such as fans, air conditioners, cars, computers, door bells, and many electronic devices? These magnets are made of coils of wire wrapped around a piece of iron or another metal.

With so many magnets used in the home, and even more used in industry and new technologies, it is essential to have an understanding of the properties of magnetism. In this chapter you will apply what you've learned about electricity to study the properties and applications of electromagnetism.

Reflect on your Learning

1. (a) Distinguish between a permanent magnet and an electromagnet.
   (b) What type of magnet is illustrated in **Figure 2**?
2. In **Figure 2**, how do you think the metal is held up by the crane? How can the metal be released? Would your answer be the same if just a paper clip were supported by the same crane?
3. What happens to the poles of a bar magnet when it is broken in two?

Throughout this chapter, note any changes in your ideas as you learn new concepts and develop your skills.

Figure 2

Figure 1
Some uses of magnets

Try This **Activity**

Making a Magnet

Use a magnet to change a steel paper clip into a magnet. Use a compass to test the results.

Materials: steel paper clips, compass, bar magnet

- Straighten out the paper clip. Bring the compass close to the paper clip and move it all the way around the paper clip. Note the effect on the compass needle.
- Drag the north end of the magnet in a sweeping motion along the paper clip several times as shown in **Figure 3(a)**. Make sure the sweeping motion is from left to right and that the magnet is lifted high above the paper clip when moving from right to left.
- Bring the compass close to the paper clip and move it all the way around the paper clip (**Figure 3(b)**). Note the effect on the compass needle.

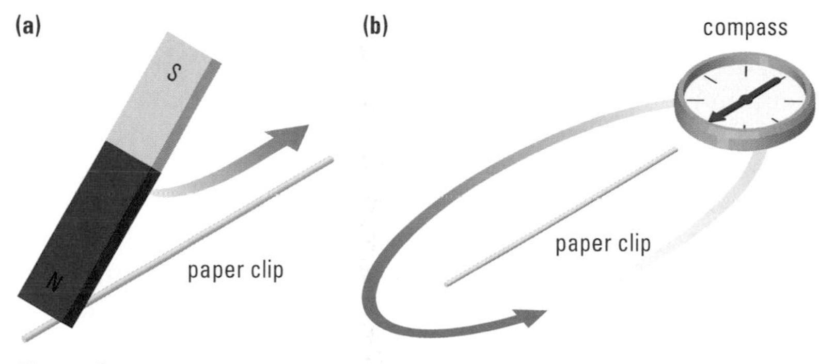

Figure 3

(a) Describe the behaviour of the compass needle when it was moved around the paper clip (i) before sweeping the magnet across it, and (ii) after sweeping the magnet across it.
(b) How can a compass be used to determine whether a metal object is a magnet?
(c) How would your observations change if the south end of the magnet was used in the same way? Try it again using another paper clip.

Figure 1
Lodestone consists mainly of iron oxide, a mineral that was first found near Magnesia, in Greece, hence the term "magnetism."

13.1 Magnetic Force and Fields

As early as 600 B.C., the Greeks discovered that a certain type of iron ore, later known as lodestone, or magnetite, was able to attract other small pieces of iron (**Figure 1**). Also, when pivoted and allowed to rotate freely, a piece of lodestone would come to rest in a north-south position. Because of this property, lodestone was widely used in navigation.

Today, lodestone is hardly ever used for its magnetic property. Artificial magnets are made from various alloys of iron, nickel, and cobalt.

When a magnet is dipped in iron filings, the filings are attracted to the magnet and accumulate most noticeably at the opposite ends of the magnet. We call these areas of concentrated magnetic force **poles**. When a magnet is allowed to rotate freely, one of the poles tends to "seek" the northerly direction on Earth and it is called the north-seeking pole, or **N-pole**; the other pole is called the south-seeking, or **S-pole**.

poles: areas of concentrated magnetic force

N-pole: the end of a magnet that seeks the northerly direction

S-pole: the end of a magnet that seeks the southerly direction

When the N-pole of one magnet is brought near the N-pole of another freely swinging magnet, the magnets repel each other, as shown in **Figure 2**. Similarly, two S-poles repel each other. On the other hand, N-poles and S-poles always attract each other. These observations lead to what is called the **law of magnetic poles**.

Law of Magnetic Poles

- Opposite magnetic poles attract.
- Similar magnetic poles repel.

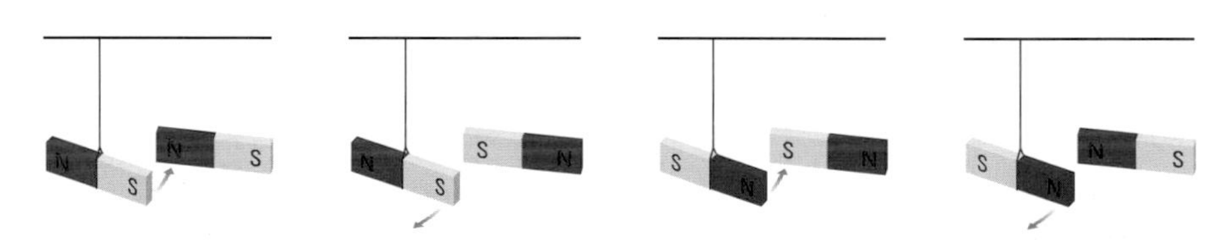

Figure 2
Testing the laws of magnetic poles

law of magnetic poles: Opposite magnetic poles attract. Similar magnetic poles repel.

magnetic field of force: the space around a magnet in which magnetic forces are exerted

When an N-pole and an S-pole are brought close to each other, they begin to attract even before they touch. This "action-at-a-distance" type of force is already familiar from your examination of gravitational force and electric force. The effect of those forces was described in terms of a "field of force" in the surrounding space. Similarly, we will consider the space around a magnet in which magnetic forces are exerted as a **magnetic field of force**.

To detect the presence of a magnetic field, we need a delicate instrument that is affected by magnetic forces. Small filings of iron respond to magnetic forces, but because their poles are not marked, we cannot determine which way they are pointing.

A magnetic field is represented by a series of lines around a magnet, showing the path the N-pole of a small test compass would take if it were allowed to move freely in the direction of the magnetic force. Then, at any point in the field, a magnetic field line indicates the direction in which the N-pole of the test compass would point.

Activity 13.1.1

Magnetic Fields

To observe the magnetic force at a given point in a magnetic field we can use a small test compass with clearly marked poles. To demonstrate the magnetic field around a magnet, iron filings are sprinkled around the magnet.

Materials

two bar magnets
iron filings
sheet of acetate (or paper)
small compass

Be careful not to get iron filings in your eyes. Wash your hands after handling iron filings.

Procedure

1. Cover the bar magnet with the sheet of acetate (or paper). Carefully sprinkle the iron filings onto the sheet of acetate around the bar magnet and look for any patterns formed by the iron filings, especially near the poles of the magnet. Sketch any patterns that you find.
2. Use the compass to indicate the direction of the field around the space above the bar magnet. Draw a three-dimensional picture of the magnetic field around the magnet.
3. Repeat the procedure for the following cases:
   (i) like poles close to each other
   (ii) opposite poles close to each other
   (iii) magnets side-by-side

Observations

(a) From what area of the magnets do field lines seem to originate? To what region do they seem to return? Which field lines, if any, leave the magnets but seem not to return?
(b) Do magnetic field lines ever cross each other? Could there be any magnetic lines of force in the regions of space between the field lines you have drawn? Check to see.
(c) What do you notice about the spacing of the field lines as you move away from the poles? What does this spacing indicate about the strength of the magnetic field?
(d) There is a theory of magnetism that states that every magnetic field line is a closed curve. However, the field lines we have drawn seem to start at the N-pole and return at the S-pole. For the magnetic field you have drawn for the bar magnet, choose several field lines and sketch the portion of each line as you believe it would exist within the magnet.

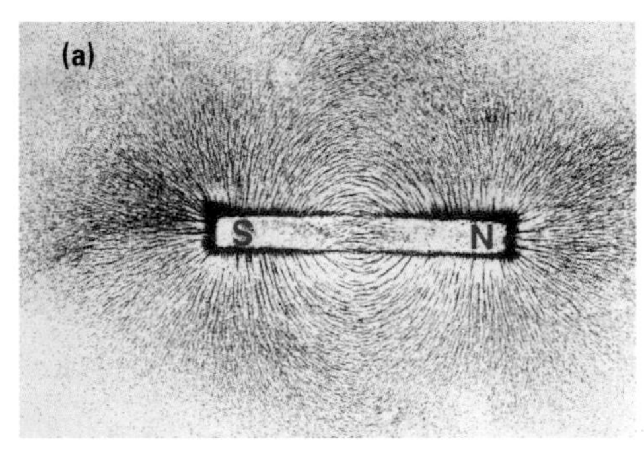

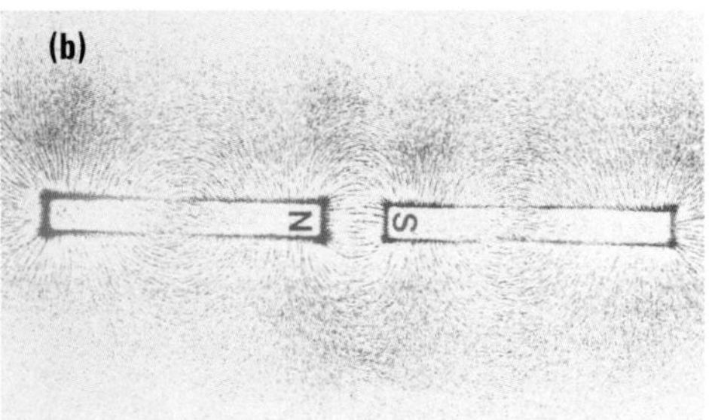

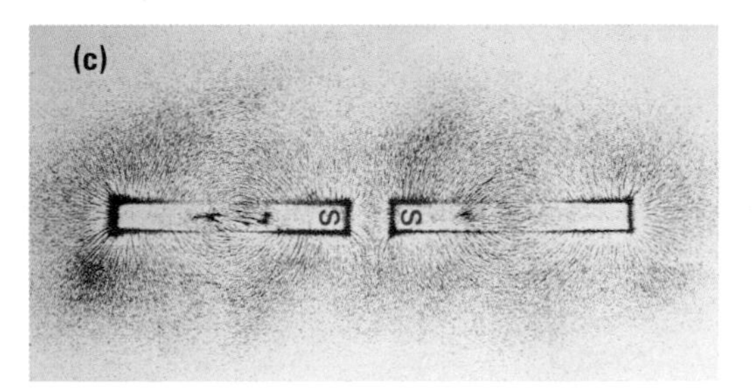

Figure 3
The nature of the magnetic fields
(a) Around a bar magnet
(b) Between a pair of opposite poles
(c) Between similar poles

Characteristics of Magnetic Field Lines

Figure 3 shows the magnetic fields around a single bar magnet, and around pairs of bar magnets close together.

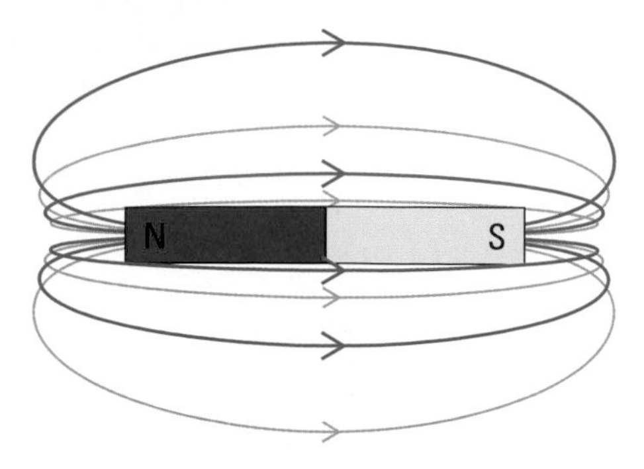

Figure 4
The three-dimensional magnetic field around a bar magnet. It is often easier to draw the fields as they appear in the horizontal; experiments are done that way, which may lead you to think that the magnetic fields are two-dimensional.

The characteristics of magnetic field lines are summarized below.

1. The spacing of the lines indicates the relative strength of the force. The closer together the lines are, the greater the force.
2. Outside a magnet, the lines are concentrated at the poles. They are closest within the magnet itself.
3. By convention, the lines proceed from S to N inside a magnet and from N to S outside a magnet, forming closed loops. (A plotting compass indicates these directions.)
4. The lines do not cross one another.

Note that the magnetic field around a bar magnet is three-dimensional in nature (**Figure 4**); it does not exist just in the horizontal plane.

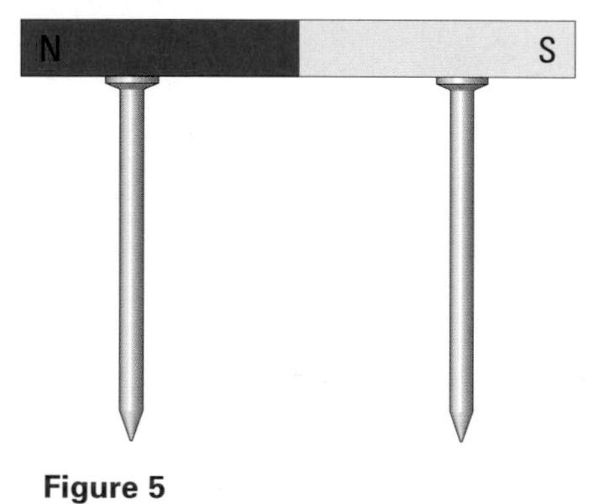

Figure 5
For question 5

Practice

Understanding Concepts

1. What name is given to the region in which a magnet influences other magnetic materials?
2. Is the magnetic pole area in the northern hemisphere an N-pole or an S-pole? Explain.

Applying Inquiry Skills

3. Draw a diagram showing the magnetic field lines around a U-shaped magnet. Include the directions of the field lines.
4. (a) Describe two methods that could be used to detect the presence of a magnetic field.
   (b) Is there a magnetic field present between the lines indicated by iron filings? Explain.
5. Two iron nails are held to a magnet, as shown in **Figure 5**. Predict what will happen when the nails are released. If possible, verify your prediction experimentally.
6. You are given two bars of steel: one is a perfectly good magnet and the other is unmagnetized. Using only these two bars, how can you determine which one is the magnet?

DID YOU KNOW?

Auroras

Earth's strong magnetic field helps to cause auroras, the dancing lights that spread across the sky above the north and south magnetic poles. Charged particles, such as electrons and protons, streaming from the Sun spiral in toward these poles, and collide with atoms in the upper atmosphere. These energized atoms then give off energy in the form of visible light with a variety of colours. Sometimes large solar storms cause the auroras to increase and result in the disruption of our communications systems for hours or even days. In the Northern Hemisphere, the auroras are called *aurora borealis*, or the Northern Lights.

SUMMARY Magnetic Force and Fields

- The law of magnetic fields states that opposite magnetic poles attract, and similar magnetic poles repel.
- The space around a magnet in which magnetic forces are exerted is known as the magnetic field of force.
- The magnetic field around a bar magnet is three-dimensional in nature.

Section 13.1 Questions

Understanding Concepts

1. Describe what would happen to a magnetic compass placed at
   (a) the magnetic north pole
   (b) the equator

2. In the diagrams in **Figure 6**, each circle represents a compass. Copy the diagrams into your notebook and show the direction of the needle in each compass.
3. Compare the law of magnetic poles with the law of electric charges.

Applying Inquiry Skills

4. Copy the diagrams in **Figure 7** into your notebook and draw the magnetic field lines between the ends of three bar magnets if
   (a) all the S-poles are close together
   (b) one N-pole and two S-poles are close together

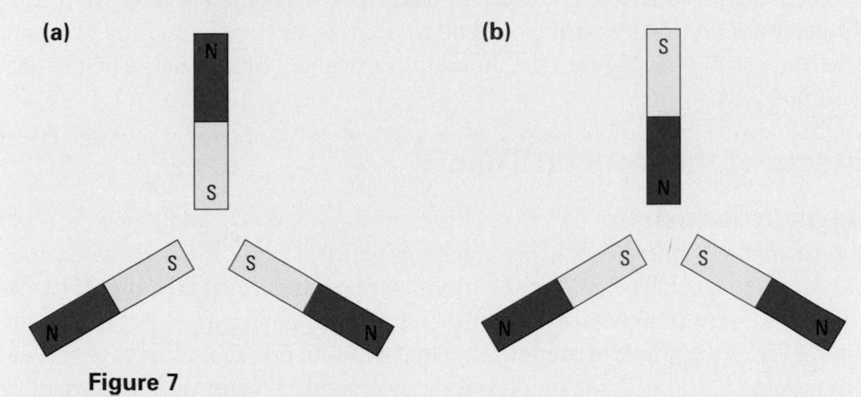

Figure 7

Making Connections

5. Biomagnetism is the study of the relationship between magnetism and living organisms. For example, research has found that pigeons have a magnetic sense that can detect magnetic fields. Research more about biomagnetism, including the magnetic sense of pigeons and other animals. Write a brief story summarizing what you discover. Follow the links for Nelson Physics 11, 13.1.

GO TO www.science.nelson.com

Reflecting

6. Often students are left with the impression that magnetic field lines exist around a bar magnet only in the horizontal; actually, there is a three-dimensional nature to the field.
   (a) Why do you think many students misunderstand the three-dimensional nature of magnetic field lines?
   (b) How would you demonstrate that the magnetic field around a bar magnet is three-dimensional?

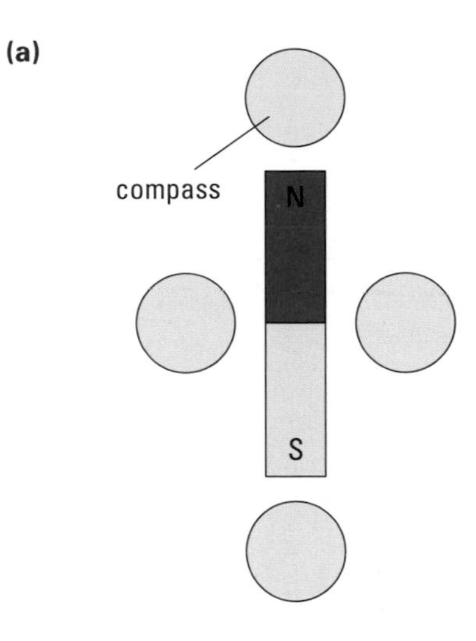

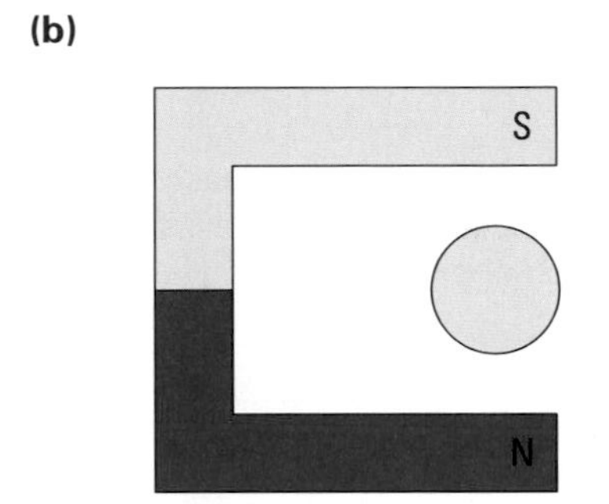

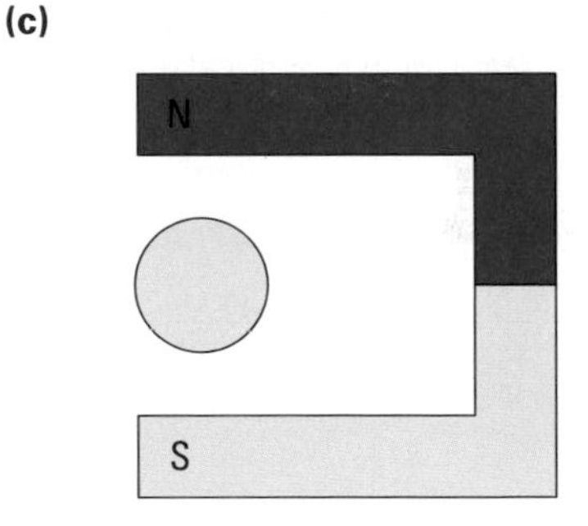

Figure 6
For question 2

13.2 Magnetic Materials

Small pieces of iron rubbed in one direction with lodestone become magnetized. Even bringing a piece of iron near a magnet causes the iron to be magnetized. Nickel and cobalt, and any alloy containing nickel, cobalt, or iron, behave in the same way. These substances are called **ferromagnetic**, and you can induce them to become magnetized by placing them in a magnetic field.

ferromagnetic: a substance that can become magnetized

Domain Theory of Magnetism

dipoles: atoms of ferromagnetic substances that act like tiny magnets

magnetic domain: formed by a group of dipoles lined up with their magnetic axes in the same direction

The atoms of ferromagnetic substances can be thought of as tiny magnets with N-poles and S-poles. These atomic magnets, or **dipoles**, interact with the nearest neighbouring dipoles and a group of them line up with their magnetic axes in the same direction to form a **magnetic domain**. An unmagnetized piece of iron contains millions of these domains, but they are pointing in random directions so that the piece of iron, as a whole, is not magnetized. **Figure 1(a)** represents this unmagnetized condition.

When an unmagnetized piece of iron is placed in a magnetic field (that is, near another magnet), the dipoles act like small compasses and rotate until they are aligned with the field. The piece of iron will then contain a large number of dipoles pointing north, causing one end to become an N-pole, and the other end to become an S-pole. **Figure 1(b)** shows the same piece of iron in the magnetized condition.

(a)

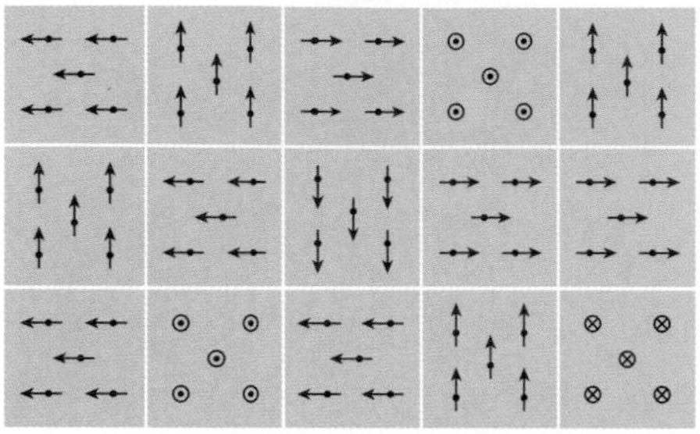

The atomic dipoles are lined up in each domain. The domains point in random directions. The magnetic material is unmagnetized.

(b)

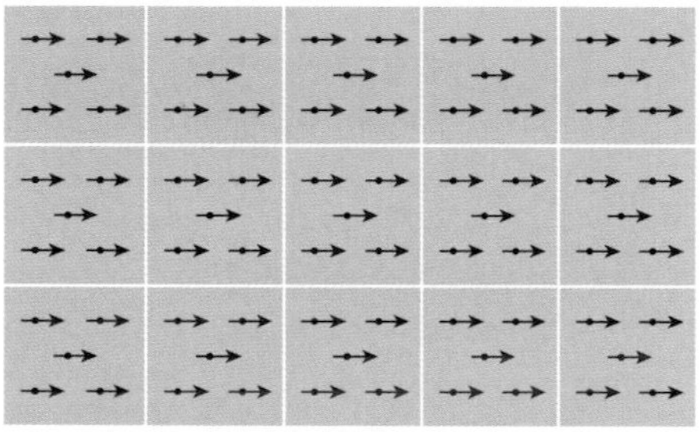

The atomic dipoles (not the domains) turn so that all domains point in the direction of the magnetizing field. The magnetic material is fully magnetized.

Figure 1
The actual boundaries between the magnetic domains are very irregular in shape and the domains are of varying size. These diagrams are greatly simplified for easier presentation.

Effects of the Domain Theory

Magnetic Induction

magnetic induction: the process of magnetizing an object from a distance

A permanent magnet brought near an iron nail will cause the nail to become a temporary magnet. The field of the permanent magnet causes the dipoles in the iron nail to align momentarily. This process of magnetizing an object from a distance is called **magnetic induction**. If magnetic induction is applied to a steel nail rather than an iron nail, the dipoles of the steel tend to retain their alignment for a longer time due to the carbon atoms in the steel. This causes the steel to act more like a permanent magnet.

Figure 2 illustrates the magnetic induction of an iron nail, which is shown with its own N- and S-poles. If the iron in the nail loses its magnetism as the nail moves away, it is called soft iron; if the iron retains its magnetism as the magnet moves away, it is referred to as hard iron. The terms "soft" and "hard" refer to magnetic induction and have nothing to do with how the iron feels to the touch.

Try This **Activity**

Magnetic Induction

Will a nail lose its magnetism when a magnet is removed?

- Place a bar magnet against a nail as shown in **Figure 2** and use the nail to lift several steel washers.
- Slowly pull the magnet away from the nail and observe what happens.
  (a) What happens to the washers as the magnet is pulled away?
  (b) In terms of magnetic induction, is the iron of the nail soft or hard? Explain.

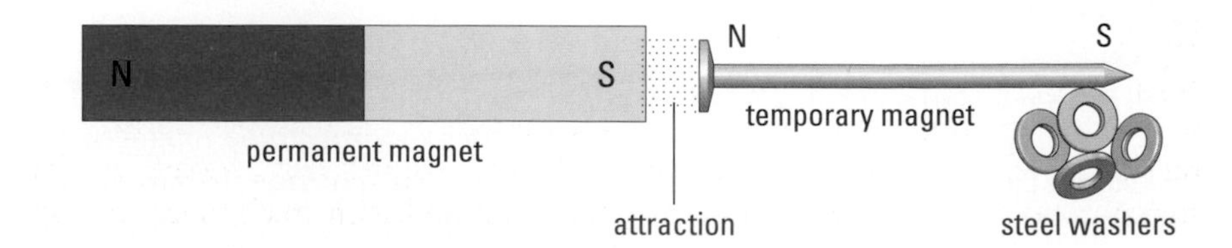

Figure 2
In the magnetic induction of an iron nail, the domains of the nail become aligned in the same direction as the domains in the permanent magnet.

Demagnetization

When a piece of iron becomes demagnetized, its aligned dipoles return to random directions. Dropping or heating an induced magnet will cause this to occur. Some materials, such as pure iron, revert to random alignment as soon as they are removed from the magnetizing field. Substances that can become instantly demagnetized are called soft ferromagnetic materials. Iron can be alloyed with certain materials, such as aluminum and silicon, that have the effect of keeping the dipoles aligned even when the magnetizing field is removed. These alloys are used to make permanent magnets and are referred to as hard ferromagnetic materials.

Reverse Magnetization

The bar magnets used in classrooms are made of hard ferromagnetic alloys, and they remain magnetized for a long time. The letter "N" is stamped on that end of the magnet to which all the N-poles of the aligned domains point. If a bar magnet is placed in a strong enough magnetic field of opposite polarity, its domains can turn and point in the opposite direction. In that case, the N-pole of the magnet is at the end marked "S." The magnet is reverse-magnetized. Small compass needles easily become reverse-magnetized.

Breaking a Bar Magnet

Breaking a bar magnet produces two pieces of iron whose dipole alignment is identical to the original piece (**Figure 3**). Both pieces will also be magnets, with N-poles and S-poles at opposite ends. Continued breaking will produce the same results, since the domains within the magnet remain aligned even when the magnet is broken.

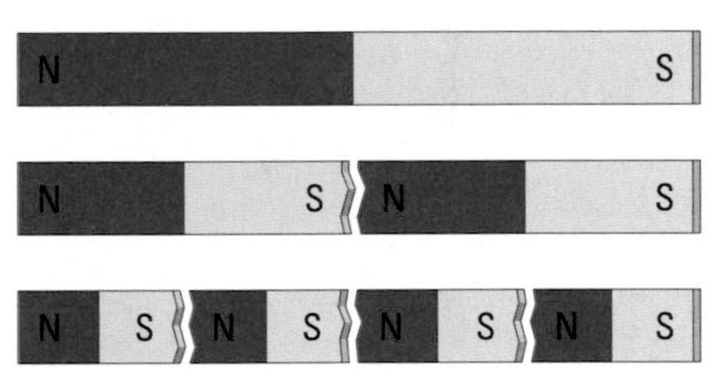

Figure 3
The effect of breaking a bar magnet

Magnetic Saturation

In most magnets many (but not all) of the dipoles are aligned in the same direction. The strength of a bar magnet can be increased only up to a certain point. The peak will occur when the maximum number of dipoles are aligned. The material is then said to have reached its **magnetic saturation.**

magnetic saturation: occurs when the maximum amount of dipoles in a material are aligned

Induced Magnetism by Earth

If a piece of iron is held in Earth's magnetic field and its atoms are agitated, either by heating or by mechanical vibration (that is, by hitting the iron with a hammer), its dipoles will align. This is most easily accomplished by holding the piece of iron pointing north and at the local angle of inclination, while tapping it with a hammer.

Steel columns and beams used in building construction are usually found to be magnetized. Steel hulls of ships and railroad tracks are also magnetized by Earth's magnetic field.

Keepers for Bar Magnets

A bar magnet will become demagnetized over time as the poles at its ends start to reverse the polarity of the atomic dipoles inside; this occurs because of random thermal motion of the atoms of the bar magnet. Bar magnets can be stored in pairs with their opposite poles adjacent and with small pieces of soft iron (called "keepers") across the ends, so that demagnetization does not occur. The keepers themselves become strong induced magnets and form closed loops of magnetic dipoles that prevent the poles from demagnetizing.

DID YOU **KNOW ?**

Magnetic Fluids

One modern application of magnetism is the use of magnetic fluids in loudspeakers. The fluid is placed between the voice coil of the loudspeaker and the permanent magnet surrounding it. Because the fluid is magnetic, it is held in place by the magnet. The fluid helps to distribute heat efficiently, so the power output can be increased, and it improves the frequency response of the speaker.

Practice

Understanding Concepts

1. What name is given to materials that are strongly attracted by a magnet? Name two such materials, other than iron and steel.
2. Describe how a screwdriver can be magnetized. What might happen if the magnetized screwdriver were heated or dropped? Explain.
3. Use the domain theory to explain the difference between soft iron and steel, and indicate which you would select for use as
   (a) a compass needle (b) keepers for a pair of bar magnets
4. Use the domain theory to explain each of the following statements:
   (a) When a magnet reaches magnetic saturation, it cannot become any stronger.
   (b) A magnet can be demagnetized if hammered repeatedly.
5. The needle of a small test compass is found to be facing south rather than north. Explain how you would use the N-pole of a magnet to magnetize the compass correctly.
6. A positively charged glass rod is hung from a string and balanced so that it is free to rotate. What will happen if a strong bar magnet is brought close to the glass rod?

SUMMARY Magnetic Materials

- Certain substances are ferromagnetic and can be induced to become magnetized by placing them in a magnetic field.
- When a ferromagnetic substance becomes magnetized, all its dipoles temporarily align in the same direction to form a magnetic domain.

Section 13.2 Questions

Understanding Concepts

1. Are all magnetic substances permanent magnets? Are all permanent magnets magnetic substances? Explain.
2. Your keys drop through a slot that is a few centimetres wide into a hole that is approximately 50 cm deep. How will you apply your knowledge of the concepts in this section to recover the keys? Explain the physics principles involved.
3. Does the result of breaking a magnet into several pieces support or refute the domain theory of magnetism? Why?

Applying Inquiry Skills

4. A tray filled with magnetic compasses is available for experimentation in your class.
   (a) How will you decide which compasses are magnetized correctly?
   (b) How can compasses become magnetized incorrectly?
   (c) Describe how you would get a faulty compass needle to face the correct direction.

Making Connections

5. The lodestone discovered in magnesia was a naturally magnetized substance. Describe the process whereby the magnetism likely occurred. (*Hint:* Volcanic eruptions are common geological events.)

13.3 Oersted's Discovery

For centuries, people believed that electricity and magnetism were somehow related, but no one could prove a connecting link between them. Then, in 1819, the Danish physicist Hans Christian Oersted (1777–1851) discovered the connection by accident while lecturing on electric circuits at the University of Copenhagen. Oersted noticed that a compass needle placed just below a wire carrying a current would take up a position nearly perpendicular to the wire while the current was flowing (**Figure 1**). When the direction of the current was reversed, the compass needle again set itself at right angles to the wire, but with its ends reversed. The effect lasted only while the current flowed. Much to his own surprise, Oersted had discovered the basic **principle of electromagnetism.**

Principle of Electromagnetism

Whenever an electric current moves through a conductor, a magnetic field is created in the region around the conductor.

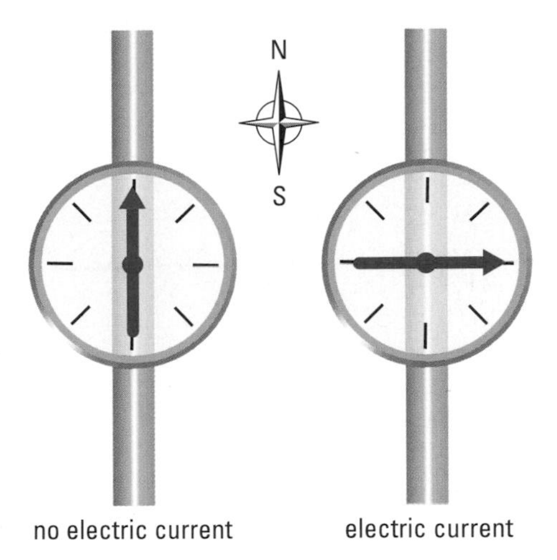

Figure 1
When there is no electric current, the compass needle points to the north. When there is a current, the needle turns so that it is perpendicular to the wire.

principle of electromagnetism: Whenever an electric current moves through a conductor, a magnetic field is created in the region around the conductor.

The Magnetic Field of a Straight Conductor

The magnetic field lines for a straight conductor are concentric circles around the conductor (**Figure 2**). As the distance from the conductor increases, the field gets weaker and the lines become more widely spaced. There are no poles; the field lines are continuous and give the direction of the plotting compass at every point.

Reversing the direction of electric current through the conductor causes the field lines to point in the opposite direction, though their pattern remains the same. To help you remember the relationship between the direction of the magnetic field lines and the direction of electric current there is the **right-hand rule for a conductor** (Figure 3).

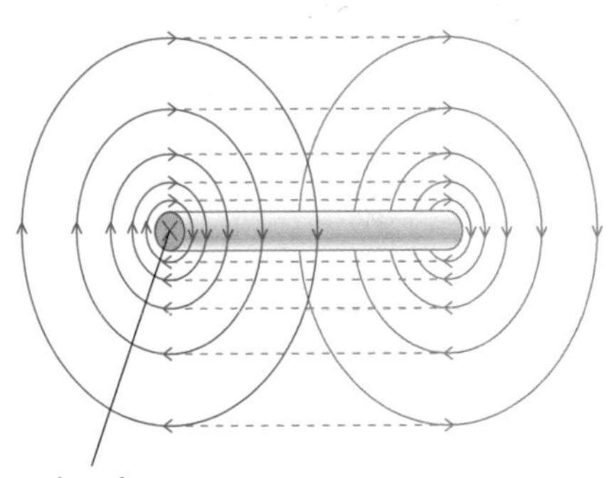

Figure 2
The field around a long straight current carrying conductor is three-dimensional in nature.

Right-Hand Rule for a Conductor

If a straight conductor is held in the right hand with the right thumb pointing in the direction of the electric current, the curled fingers will point in the direction of the magnetic field lines.

right-hand rule for a conductor: If a straight conductor is held in the right hand with the right thumb pointing in the direction of the electric current, the curled fingers will point in the direction of the magnetic field lines.

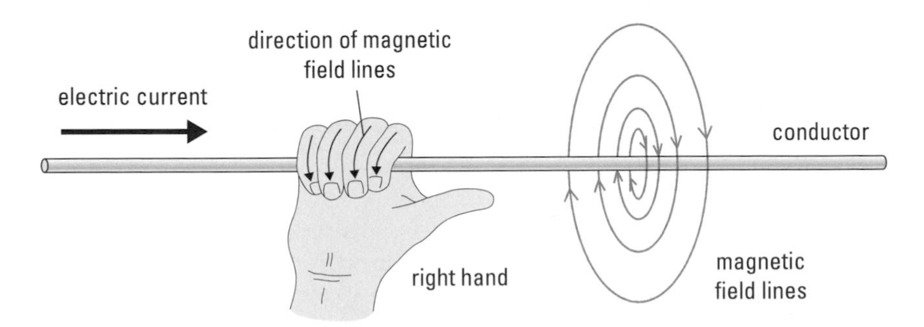

Figure 3
The right-hand rule for a straight conductor

(a)

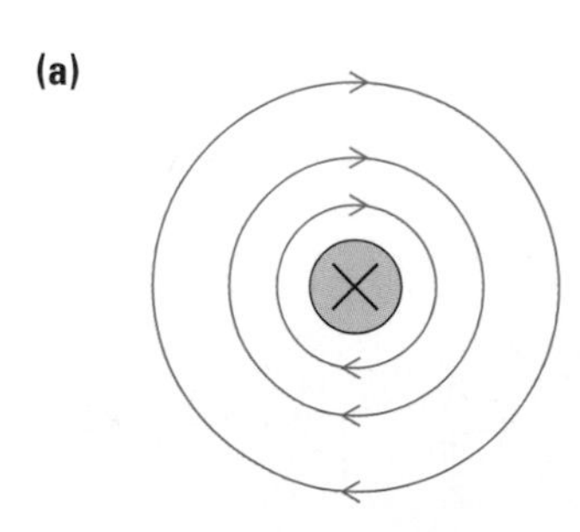

current into page

(b)

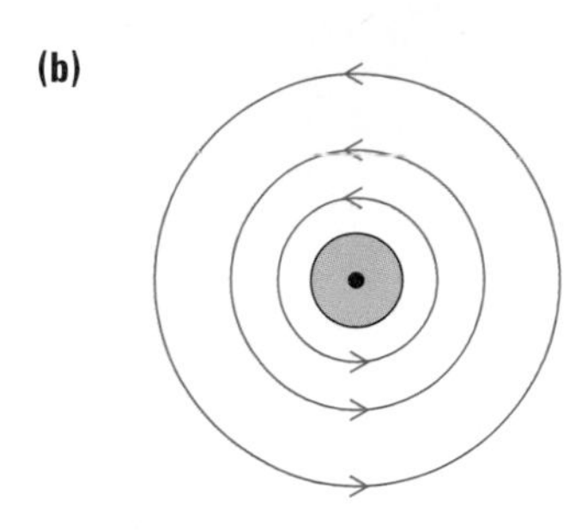

current out of page

Figure 4
Models of the magnetic field of a straight conductor
(a) Imagine the X as being the tail of an arrow moving away from you.
(b) Imagine the dot as being the tip of an arrow facing you.

Rather than drawing the conductor as a cylinder and using an arrow to indicate direction, it is often more convenient to use a two-dimensional picture, as in **Figure 4**. A circle is used to represent a cross-section of the conductor. A circle with an X in it represents an electric current moving into the page. A circle with a dot in it represents an electric current moving out of the page.

Activity 13.3.1

Magnetic Field of a Straight Conductor

Question

What are the characteristics of the magnetic field around a straight conductor?

Materials

20 cm of bare 12-gauge copper wire
piece of stiff cardboard, 15 cm × 15 cm
battery (6 V–12 V) or DC power supply
iron filings
connecting wires with alligator clips
four compasses

Procedure

1. Push the short piece of bare copper wire through the middle of the cardboard square and support the cardboard in a horizontal position, as shown in **Figure 5**.
2. Connect the upper end of the copper wire to either terminal of the battery, using a wire with an alligator clip. Connect another wire with a clip to the bottom of the copper wire, but do not connect it to the battery.
3. Lightly sprinkle iron filings on the piece of cardboard. Momentarily touch the loose wire to the other terminal of the battery, and tap the cardboard gently. Once the iron filings have assumed a pattern, disconnect the battery and sketch the pattern. Be sure to include the copper wire in your sketch. From the battery terminals used, determine the direction of the electric current and mark it in your sketch.
4. Place four plotting compasses on the cardboard, as shown in **Figure 5**. Connect the battery and note the directions in which the compasses point. Add these directions to your sketch of the iron filings.
5. Without moving the compasses, reverse the connections to the battery. Make another sketch. Show the direction in which the electric current is moving.

One of the wires from the conductor should be connected firmly to one of the terminals; the other wire should be touched momentarily to the other terminal. The resistance of the bare wire is very low. As a result, it draws a large current. This will cause the wire to get hot and the battery to discharge quickly if the terminals are connected for too long a time.

Observations

(a) Describe the shape of the magnetic field lines produced by the electric current in a straight conductor.
(b) Describe the spacing and clarity of the field lines farther away from the conductor. What does this indicate about the magnetic field in these regions?
(c) Compare the compass direction pattern obtained in step 4 with that obtained in step 5. Does the right-hand rule provide an adequate description of these patterns? Explain.

Be careful not to get iron filings in your eyes. Wash your hands after handling iron filings.

Practice

Understanding Concepts

1. **Figure 6** shows three current-carrying conductors with their magnetic fields. Copy the diagrams into your notebook and indicate the direction of electric current in each wire.
2. **Figure 7** shows three conductors with the direction of the electric current. Copy the diagrams into your notebook and draw magnetic field lines around each, indicating polarities where applicable.

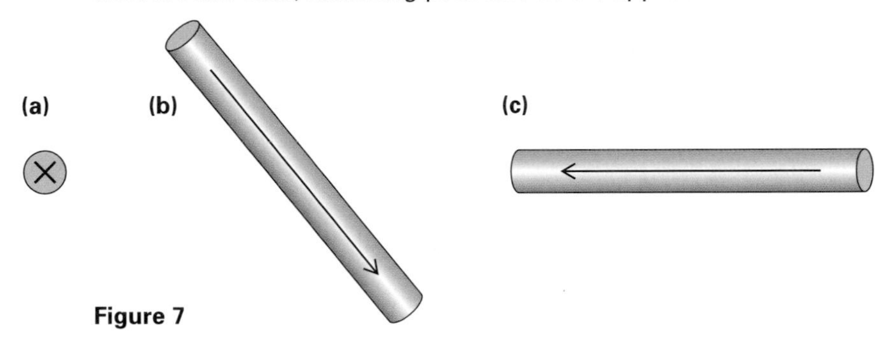

Figure 7

3. Choose the diagram from **Figure 8** that best illustrates the strength of the magnetic field surrounding a conductor. Explain your answer.

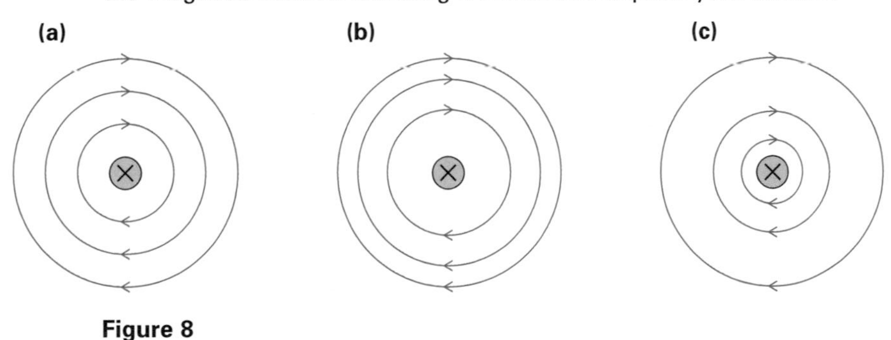

Figure 8

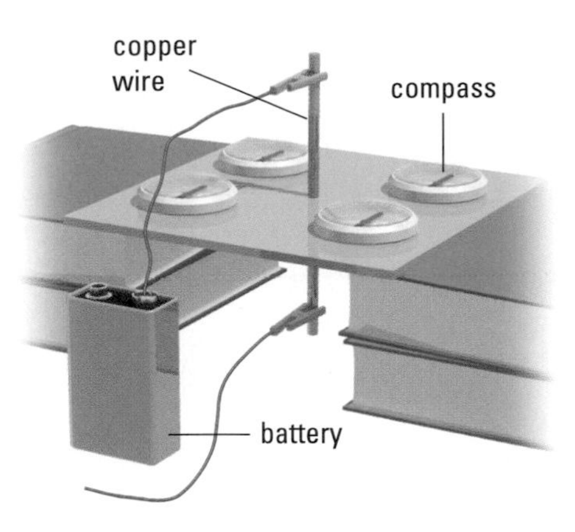

Figure 5
Setup for Activity 13.3.1

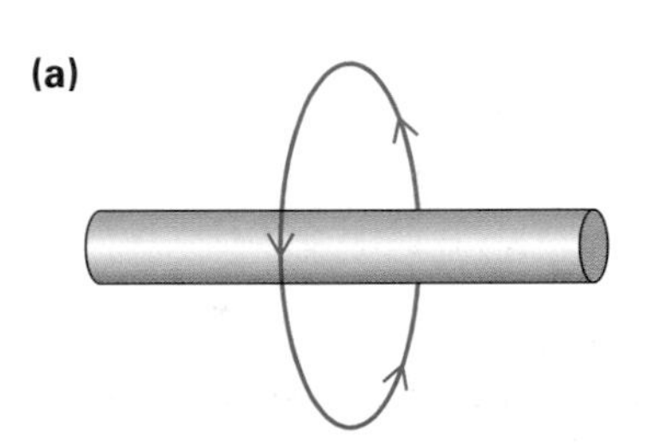

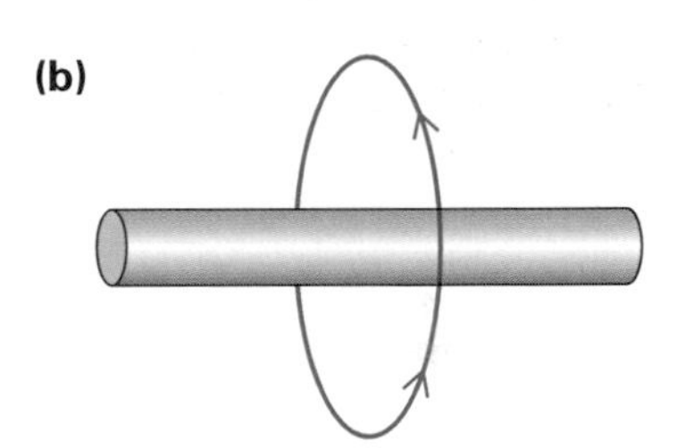

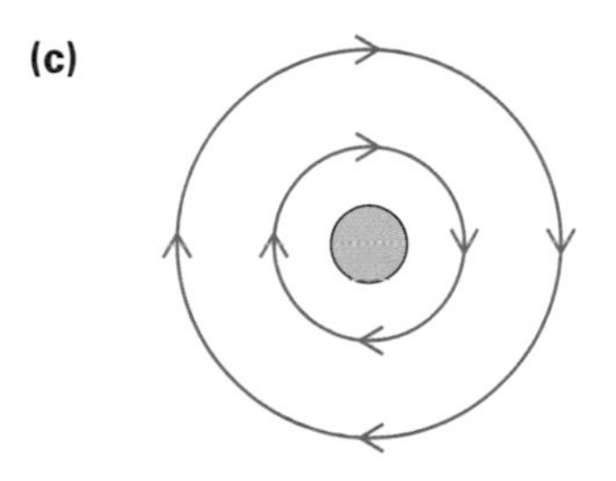

Figure 6
For question 1

SUMMARY Oersted's Discovery

- The principle of electromagnetism states that whenever an electric current moves through a conductor, a magnetic field is created in the region around the conductor.
- The magnetic field lines around a straight conductor are in concentric circles.
- The right-hand rule for a straight conductor states that if a conductor is held in the right hand with the right thumb pointing in the direction of the electric current, the curled fingers will point in the direction of the magnetic field lines.

Section 13.3 Questions

Understanding Concepts

1. An electric current is travelling southward in a straight, horizontal conductor. State the direction of the magnetic field
   (a) below the conductor (b) above the conductor
   (c) east of the conductor (d) west of the conductor
2. What is the direction of the magnetic field lines around a conductor with the electric current travelling (a) away from you, and (b) toward you?
3. The diagram in **Figure 9** represents two parallel current-carrying conductors. Determine whether the conductors attract or repel one another. Explain your reasoning.

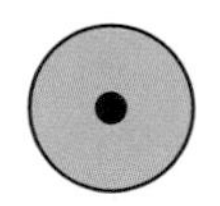

Figure 9
For question 3

Reflecting

4. Why do you think the connection between electricity and magnetism was not discovered until the early 19th century? (*Hint:* Research when magnetism, static electricity, and current electricity were discovered.)

13.4 The Magnetic Field of a Coil or Solenoid

Figure 1

In a junk yard, a crane lifts a pile of scrap metal to be compressed and eventually recycled (see **Figure 1**). How is the scrap metal held up by the crane? You might say by a magnet, but it couldn't be a permanent magnet; otherwise how would the metal be released? It is held by an **electromagnet**, a device that exerts a magnetic force using electricity.

The magnetic field around a straight conductor can be intensified by bending the wire into a loop, as illustrated in **Figure 2**. The loop can be thought of as a series of segments, each an arc of a circle, and each with its own magnetic field (**Figure 2(a)**). The field inside the loop is the sum of the fields of all the segments. Notice that the field lines are no longer circles but have become more like lopsided ovals (**Figure 2(b)**).

The magnetic field can be further intensified (**Figure 3**) by combining the effects of a large number of loops wound close together to form a coil, or **solenoid**. The field lines inside the coil are straight, almost equally spaced, and all point in the same direction. We call this a **uniform magnetic field**; the magnetic field is of the same strength and is acting in the same direction at all points.

If the direction of electric current through the coil is reversed, the direction of the field lines is also reversed but the magnetic field pattern, as indicated by a pattern of iron filings, looks the same as it did before. To help you remember the relationship between the direction of electric current through a coil and the direction of the coil's magnetic field, there is the **right-hand rule for a coil** (**Figure 4**).

Right-Hand Rule for a Coil

If a coil is grasped in the right hand with the curled fingers representing the direction of electric current, the thumb points in the direction of the magnetic field inside the coil.

electromagnet: object that exerts a magnetic force using electricity

solenoid: a coil of wire that acts like a magnet when an electric current passes through it

uniform magnetic field: the magnetic field is the same strength and acts in the same direction at all points

right-hand rule for a coil: If a coil is grasped in the right hand with the curled fingers representing the direction of electric current, the thumb points in the direction of the magnetic field inside the coil.

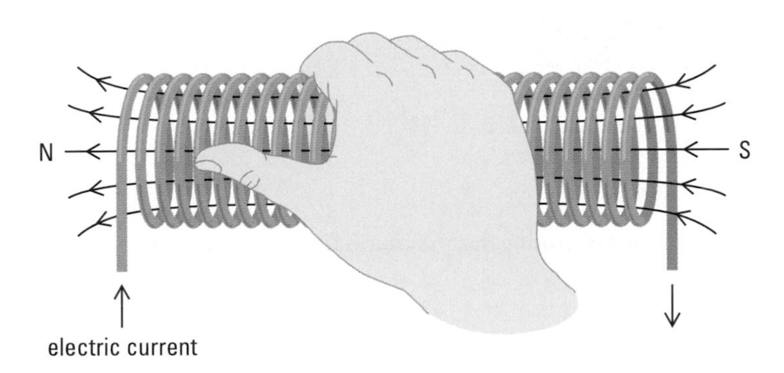

Figure 4
The right-hand rule for a coil

Figure 5 shows the similarity between the magnetic fields of a bar magnet (**Figure 5**(a)) and those of a coil (**Figure 5**(b)). Notice that the coil is wrapped around an iron core, which helps to increase the strength of the magnetic field.

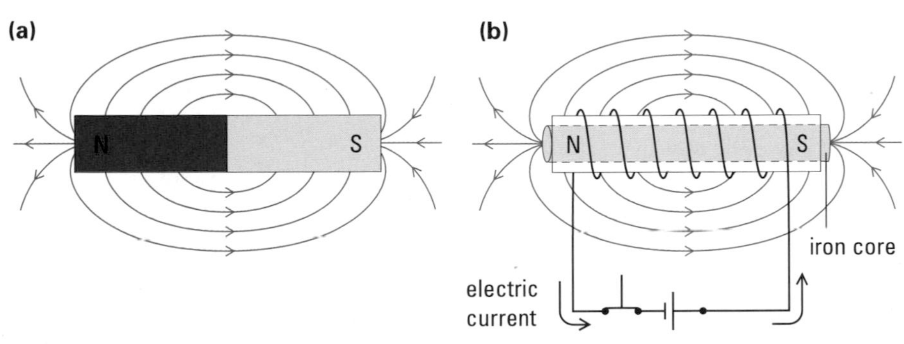

Figure 5
The ends of a coil can be thought of as an N-pole and an S-pole, and the field within the core resembles the field in the interior of a bar magnet, with its ferromagnetic dipoles aligned so that they all point in the same direction.

Practice

Understanding Concepts

1. Copy **Figure 6** into your notebook and indicate on each diagram the direction of the electric current, the magnetic field lines around the coil, and the north and south poles of the coil.

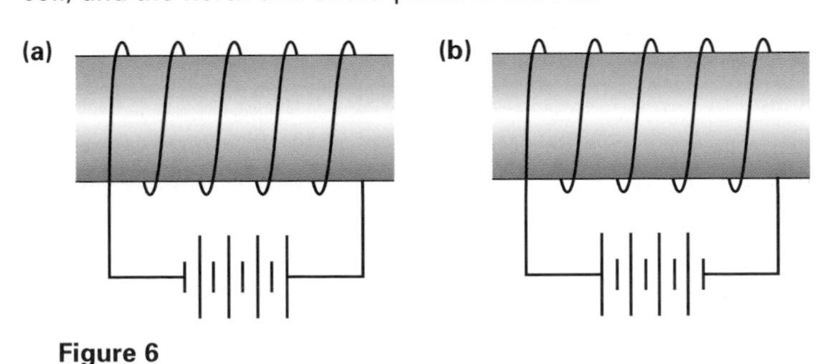

Figure 6

2. If two identical cardboard tubes are wound with wire in exactly the same way and placed end-to-end with current passing through them in the same direction, will the two tubes attract or repel one another? Explain.

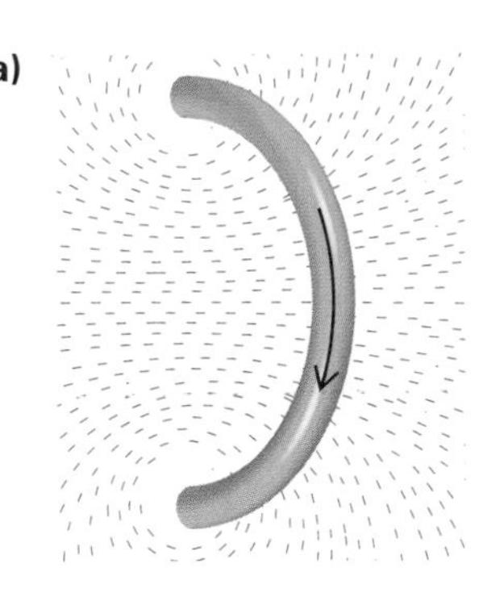

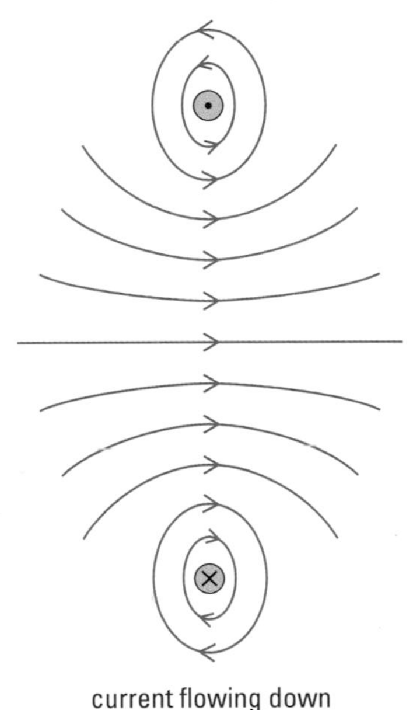

Figure 2

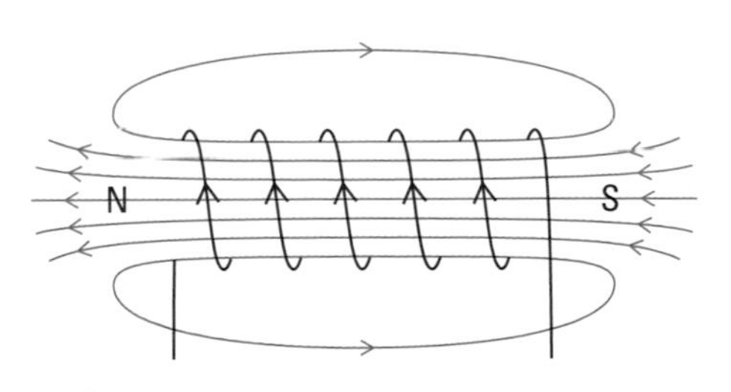

Figure 3
At both ends of the coil, the field is still strong but along the sides, the magnetic field is weaker, as the field lines begin to bend and spread out.

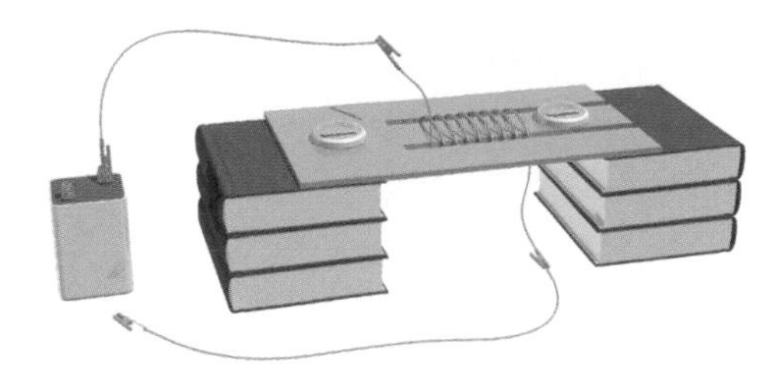

Figure 7
Setup for Activity 13.4.1

One of the wires from the conductor should be connected firmly to one of the terminals; the other wire should be touched momentarily to the other terminal. The resistance of the bare wire is very low. As a result, it draws a large current. This will cause the wire to get hot and the battery to discharge quickly if the terminals are connected for too long a time.

Be careful not to get iron filings in your eyes. Wash your hands after handling iron filings.

Activity 13.4.1

Magnetic Field of a Coil

Question

What are the characteristics of the magnetic field of a coil?

Materials

50 cm of bare 12-gauge copper wire
wooden dowel (about 15 cm long × 4 cm diameter)
stiff cardboard and scissors
battery (6 V-12 V) or DC power supply
connecting wires with alligator clips
iron filings
2 compasses

Procedure

1. Make a coil by winding the bare copper wire around the dowel as many times as possible. Spread the loops out so that they are about 0.5 cm apart. Remove the dowel.
2. Using scissors, cut a piece of cardboard to fit snugly into the core of the coil, as shown in **Figure 7**. Insert the cardboard into one end of the coil, and support the coil so that the cardboard is horizontal.
3. Connect one end of the coil to one of the terminals of the battery using a wire with an alligator clip. Connect another wire with a clip to the other end of the coil, but do not connect the wire to the battery.
4. Lightly sprinkle iron filings on the cardboard, both inside and outside the coil. Momentarily touch the loose wire to the other terminal of the battery, and tap the cardboard gently. Once the iron filings have assumed a pattern, disconnect the battery and sketch the pattern, including the coil. From the battery terminals you used, determine in which direction the electric current was moving through the coil, and mark the direction on your sketch.
5. Place the compasses on the cardboard, one at each end of the coil. Reconnect the battery and note the directions in which the compasses point. Add these directions to your sketch of the iron filings.
6. Without moving the compasses, reverse the connections to the battery. Make another sketch. Show the direction in which the electric current is moving.

Observations

(a) Describe the direction, shape, and spacing of the magnetic field lines in the core of the coil.
(b) What happens to these magnetic field lines at the ends of the coil?
(c) Describe the magnetic field in the region at either side of the coil.
(d) Compare the compass direction pattern obtained in step 5 to that obtained in step 6. Does the right-hand rule provide an adequate description of these patterns? Explain.

Factors Affecting the Magnetic Field of a Coil

A coil with an electric current flowing through it has a magnetic field that can be plotted in the same way as for a bar magnet; however, it can be much more useful than a bar magnet. A bar magnet is able to pick up small pieces of iron but it cannot release them, and its strength cannot be varied. A coil, on the other hand, has a magnetic field that can be turned on and off and altered in strength.

The strength of a magnetic field is related to the degree of concentration of its magnetic field lines. To increase the strength of the magnetic field in a coil, you must increase the number of magnetic field lines or bring them closer together. The magnetic field strength in a coil depends on the following factors.

Current in the Coil

Since the electric current flowing through the coil creates the magnetic field in the core of the coil, the more electric current there is, the greater the concentration of magnetic field lines in the core. In fact, in an air core coil (a coil with no material inside it), the magnetic field strength varies directly with the current in a coil: doubling the current doubles the magnetic field strength.

Number of Loops in the Coil

Each loop of wire produces its own magnetic field, and since the magnetic field of a coil is the sum of the magnetic fields of all its loops, the more loops that are wound in the coil, the stronger its magnetic field. Magnetic field strength varies directly as the number of loops per unit length in a coil: doubling the number of loops in a coil doubles the magnetic field strength.

Type of Core Material

The material that makes up the core of a coil can greatly affect the coil's magnetic field strength. For example, if a cylinder of iron—rather than air—is used as the core for a coil, the coil's magnetic field strength can be several thousand times stronger than with air. An aluminum core will have almost no effect on the strength.

The core material becomes an induced magnet, as its atomic dipoles align with the magnetic field of the coil. As a result, the core itself becomes an induced magnet and the magnetic field strength increases.

The factor by which a core material increases the magnetic field strength is called the material's **relative magnetic permeability** (K). The permeability is the ratio of the magnetic field strength in a material to the magnetic field strength that would exist in the same region if a vacuum replaced the material.

relative magnetic permeability: (K) the factor by which a core material increases the magnetic field strength that would exist in the same region if a vacuum replaced the material

$$K = \frac{\text{magnetic field strength in material}}{\text{magnetic field strength in vacuum}}$$

For example, a material with a relative magnetic permeability of 3 will make the field in a coil three times as strong as it would be in a vacuum.

The magnetic field strength of a coil varies directly as the permeability of its core material: doubling the relative magnetic permeability of the core doubles the magnetic field strength.

paramagnetic: material that magnetizes very slightly when placed in a coil and increases the field strength by a barely measurable amount

diamagnetic: materials that cause a very slight decrease in the magnetic field of a coil

Ferromagnetism, Paramagnetism, and Diamagnetism

Core materials are divided into three groups according to their relative magnetic permeability.

As you have learned, ferromagnetic materials become strong induced magnets when placed in a coil; that is, they have very high relative magnetic permeability. Iron, nickel, cobalt, and their alloys are ferromagnetic.

Paramagnetic materials magnetize very slightly when placed in a coil and increase the field strength by a barely measurable amount. As a result, they have a relative magnetic permeability only slightly greater than 1. Oxygen and aluminum are paramagnetic.

Diamagnetic materials cause a very slight decrease in the magnetic field of a coil and, as a result, their relative magnetic permeabilities are slightly less than 1. Copper, silver, and water are diamagnetic.

The crane in **Figure 1** of this section uses a huge electromagnet to lift the car. The electromagnet has a ferromagnetic core that passes through the coil and almost completely surrounds it. When the electromagnet is activated the core becomes completely magnetized. When the electromagnet is deactivated the core loses its magnetism and the car is released. (Even when the current is completely shut off, the magnet will not lose all of its strength because the core becomes a temporary magnet.)

DID YOU KNOW ?

Applying Paramagnetism and Diamagnetism

One of the uses of paramagnetism is to analyze the components of materials such as powders and glasses. This application is performed using "electron paramagnetic resonance" apparatus. An interesting use of diamagnetism is found in levitation devices, such as a frictionless bearing in which a diamagnetic material is used in conjunction with a ferromagnetic material.

Sample Problem

Calculate the effect on the strength of the magnetic field in a coil when each of the following separate changes is made:

(a) the current in the coil is increased from 2.0 A to 5.0 A;
(b) the number of loops in the coil is changed from 4400 to 1100; the length of the coil is unchanged;
(c) the core is changed from steel with a relative magnetic permeability of 3.0×10^3 to iron with a relative magnetic permeability of 8.0×10^3.

Solution

(a) Magnetic field strength varies directly with current.

$$\frac{\text{magnetic field after change}}{\text{magnetic field before change}} = \frac{5.0\ \text{A}}{2.0\ \text{A}}$$
$$= 2.5$$

The magnetic field strength is 2.5 times as great.

(b) Magnetic field strength varies directly with number of loops per unit length.

$$\frac{\text{magnetic field after change}}{\text{magnetic field before change}} = \frac{1100}{4400}$$
$$= 0.25$$

The magnetic field strength is 0.25 times as great.

(c) Magnetic field strength varies directly with permeability of core.

$$\frac{\text{magnetic field after change}}{\text{magnetic field before change}} = \frac{8.0 \times 10^3}{3.0 \times 10^3}$$
$$= 2.7$$

The magnetic field strength is 2.7 times greater.

Practice

Understanding Concepts

3. An electromagnet is able to exert a force of 1.5×10^2 N when lifting an object. The electromagnet has 1000 turns, a current of 1.5 A, and a material in the core with a relative magnetic permeability of 2.0×10^3. What force will the electromagnet exert if the following changes are made, each considered separately?
   (a) The current is increased to 6.0 A.
   (b) The number of turns in the coil is increased to 1400 without increasing the length of the coil.
   (c) All of the above changes are considered simultaneously.

switch: a device that is used to make or break an electric circuit

Answers

(a) 6.0×10^2 N

(b) 2.1×10^2 N

(c) 8.4×10^2 N

Applications of Electromagnetism

Many appliances, tools, vehicles, and machines use a current-carrying coil to create a magnetic field. In most cases, the magnetic field is used to cause another component to move by magnetic attraction. A few examples will illustrate how electromagnets are used.

Lifting Electromagnet

Large steel plates, girders, and pieces of scrap iron can be lifted and transported by a lifting electromagnet (**Figure 8**). A soft ferromagnetic core of high relative magnetic permeability is wound with a copper conductor. The ends of the coil are connected to a source of electric potential through a **switch**. Closing the switch causes an electric current in the coil, and the soft iron core becomes a very strong induced magnet.

When the switch is opened and the electric current stops, the soft iron core becomes demagnetized and releases its load. A U-shaped core is often used, with a coil wrapped around each leg of the device. If the coils are wound in opposite directions, the legs become oppositely magnetized and the lifting ability of the magnet is doubled.

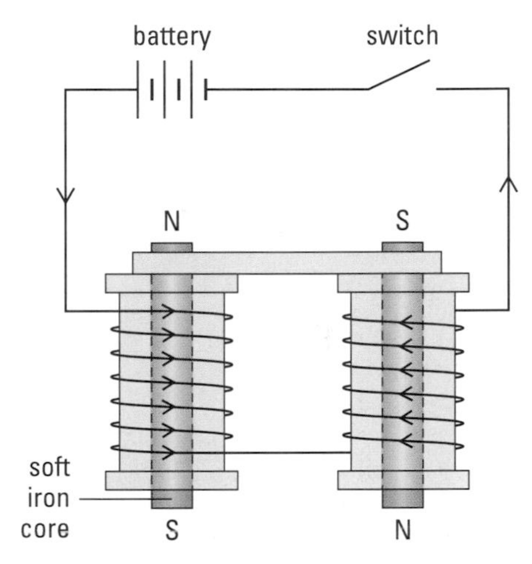

Figure 8
A lifting electromagnet

Electromagnetic Relay

A **relay** is a device in which a switch is closed by the action of an electromagnet (**Figure 9**). A relatively small current in the coil of the electromagnet can be used to switch on a large current without the circuits being electrically linked.

A pivoted bar of soft iron, called an **armature**, is held clear of the contact point by a light spring. There is no current in the left-hand circuit, and the lamp is off. When the switch is closed, there is a current in the right-hand circuit, and the soft iron U-shaped core becomes magnetized. The magnetized core attracts the armature and pulls it to the right until it touches the contact point, completing the circuit. Now there is a current in the left-hand circuit, and the lamp goes on.

When the switch is opened, the electric current drops to zero, the core becomes demagnetized, and the armature is released. When the spring pulls the armature away from the contact point, current drops to zero in the left-hand circuit and the lamp goes off.

If the contact points were on the opposite side of the armature from the electromagnet, the relay would operate in reverse. Closing the switch would then turn the left-hand circuit off, and vice versa.

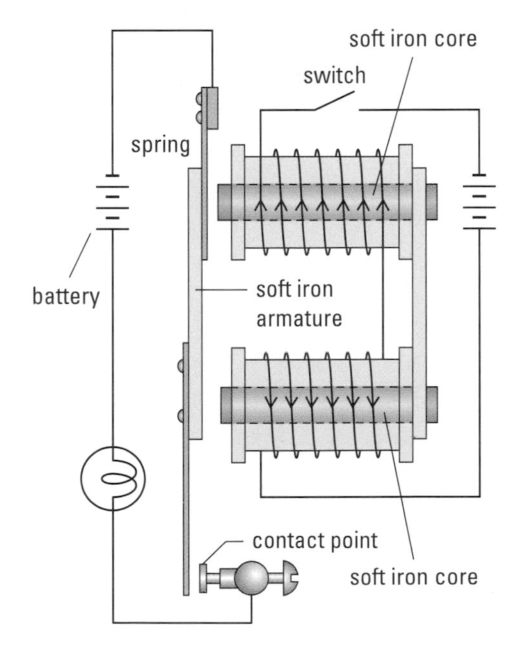

Figure 9
An electromagnetic relay

relay: a device in which a switch is closed by the action of an electromagnet

armature: a pivoted bar of soft iron

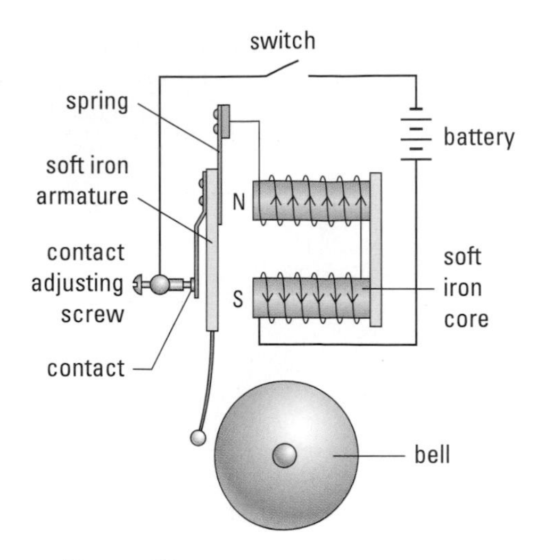

Figure 10
An electric bell

Electric Bell

In an electric bell, a small hammer is attached to the armature. The armature is vibrated back and forth several times a second, striking a metal bell. **Figure 10** shows the circuit that causes the armature to move.

When a button is pushed, the switch is closed. An electric current flows through the contacts and the spring to the coils, and the soft iron cores become magnetized. The cores attract the iron armature, which moves toward the electromagnet, causing the hammer to strike the bell. As the hammer strikes the bell, the movement of the armature opens the contacts. The electric current stops flowing to the coils and the soft iron cores become demagnetized, releasing the armature. A spring pulls the armature back to re-establish contact, thereby completing the circuit, and the entire cycle begins again. Small sparks, evidence of charge jumping across the gap, can be observed at the contact points as the circuit is alternately completed and broken.

Try This **Activity**

Testing an Electromagnet

Perform an activity in which you can use an electromagnet to lift iron objects, such as washers. The coil of the electromagnet can be connected to a variable DC power supply. Remember not to leave the current flowing for more than a few seconds at a time. Test the electromagnet both with and without the core, and with various currents and numbers of loops. Summarize your findings in tabular form.

Practice

Understanding Concepts

4. **Figure 11** shows coils of wire wound on cardboard cylinders. Copy the diagrams into your notebook, and on each diagram mark the direction of electric current, the direction of the field lines at each end of the coil, and the N-pole and S-pole of the coil.
5. A coil with an iron core is used as an electromagnet. With 500 loops and a current of 1.5 A, it can exert a lifting force of 30.0 N. What force will it be able to lift if the following changes are made?
   (a) The current is increased to 3.0 A.
   (b) The number of loops is increased to 750 without increasing the length of the coil.
   (c) Both the above changes are made together.
6. In both electromagnetic relays and doorbells, soft iron cores are used in the electromagnets. If a hard iron core is used, explain what would happen in
   (a) a relay that is used to turn a light on and off
   (b) an electric bell

Making Connections

7. Describe some commercial or industrial applications in which electromagnets are better than permanent magnets. Describe some applications in which permanent magnets are better than electromagnets.

Answers

5. (a) 60.0 N
   (b) 45.0 N
   (c) 90.0 N

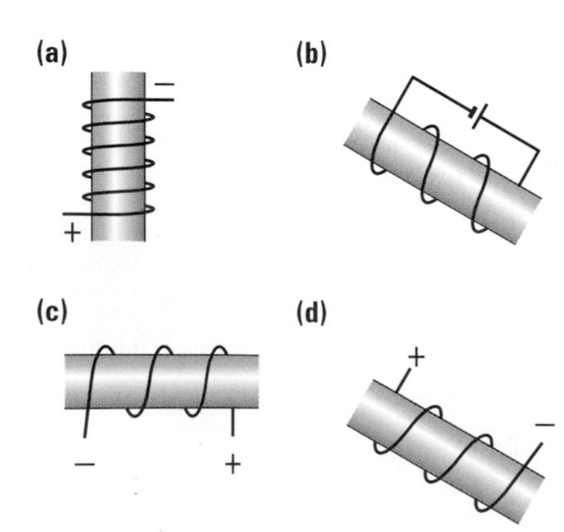

Figure 11
For question 4

SUMMARY Magnetic Field of a Coil or Solenoid

- The magnetic field around a coil is the sum of the magnetic fields of each of its segments.
- The right-hand rule for a coil states that if a coil is grasped in the right hand with the curled fingers representing the direction of electric current, the thumb points in the direction of the magnetic field inside the coil.
- The magnetic field strength of a coil depends on the current in the coil, the number of loops in the coil, and the type of core material.

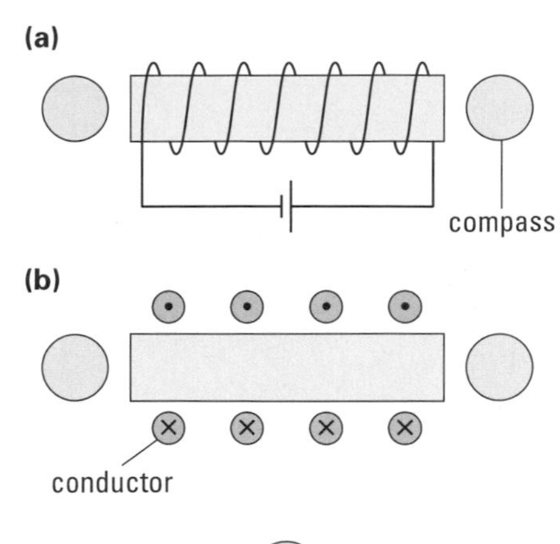

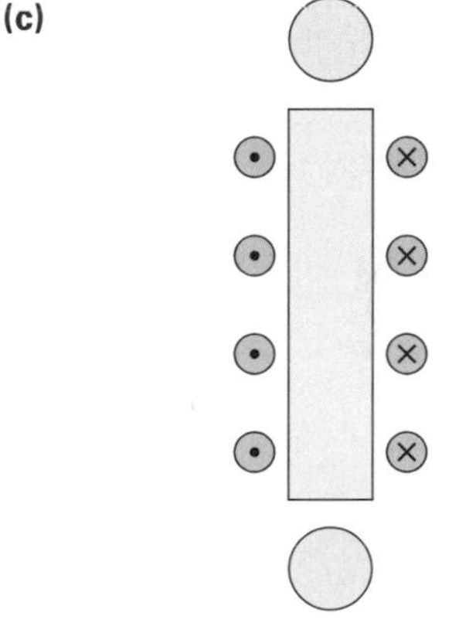

Figure 12
For question 1

Section 13.4 Questions

Understanding Concepts

1. Each empty circle in **Figure 12** represents a compass near one end of a coiled conductor. Redraw the diagrams, label the ends of the coils N or S, and show the direction of the compass needle in each case.
2. Two soft iron rods are placed inside a coiled conductor as shown in **Figure 13**. Determine whether the rods will attract or repel each other when the switch is closed.
3. Is insulation necessary for the conducting wires of a tightly-wound coil? Explain.
4. Determine the polarity of the poles of the electromagnet in **Figure 14**.
5. State the relationship between the strength of the magnetic field of a coil or solenoid and:
   (a) the amount of current in the coil
   (b) the number of loops in the coil
   (c) the type of core material in the coil
6. A coil of 600 turns, wound using 90.0 m of copper wire, is connected to a 6.0-V battery and is just able to support the weight of a toy truck. If 200 turns are removed from the coil, but the wire is uncoiled and left in the circuit, what battery voltage would be needed in order to support the truck?
7. Explain the series of events involved in the operation of a doorbell.
8. The circuit for an electric bell is just a slight variation of the circuit for an electromagnetic relay. How are the circuits the same? Describe how the circuits are different.

Making Connections

9. Describe how to construct a very strong lifting electromagnet.
10. Find out more about the uses of diamagnetic and paramagnetic materials. Write a brief report on what you discover. Follow the links for Nelson Physics 11, 13.4.

www.science.nelson.com

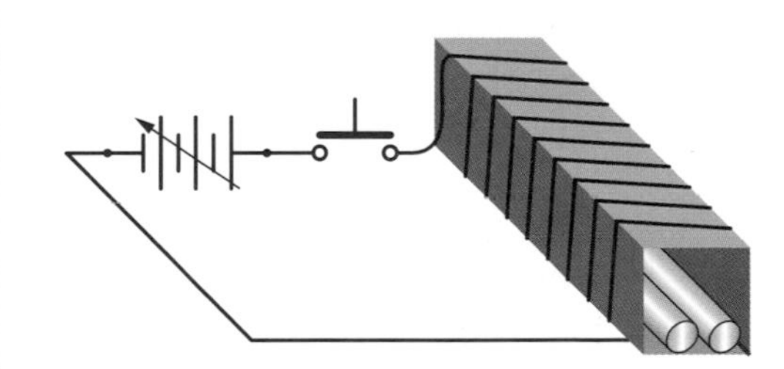

Figure 13
For question 2

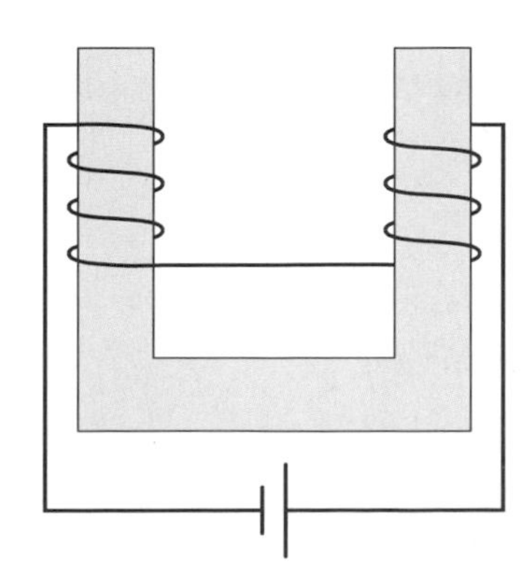

Figure 14
For question 4

13.5 Conductor in a Magnetic Field—The Motor Principle

Electric motors are used in cooling fans for computers, in refrigerators, in electric cars, in power tools such as electric drills, and even in many remote-controlled toys. Many subway systems use electric motors as a clean and silent way to power their trains. But what do electric motors have to do with magnetic fields?

In 1821, following Oersted's discovery of electromagnetism, English physicist Michael Faraday (1791–1867) set out to prove that, as a wire carrying electric current could cause a magnetized compass needle to move, so in reverse a magnet could cause a current-carrying wire to move. Suspending a piece of wire above a bowl of mercury in which he had fixed a magnet upright, Faraday connected the wire to a battery, and the wire began to rotate.

Faraday determined that the magnetic field of a permanent magnet can exert a force on the charges in a current-carrying conductor. **Figure 1** shows how the direction of this force is related to the magnetic field of the conductor and to the external magnetic field.

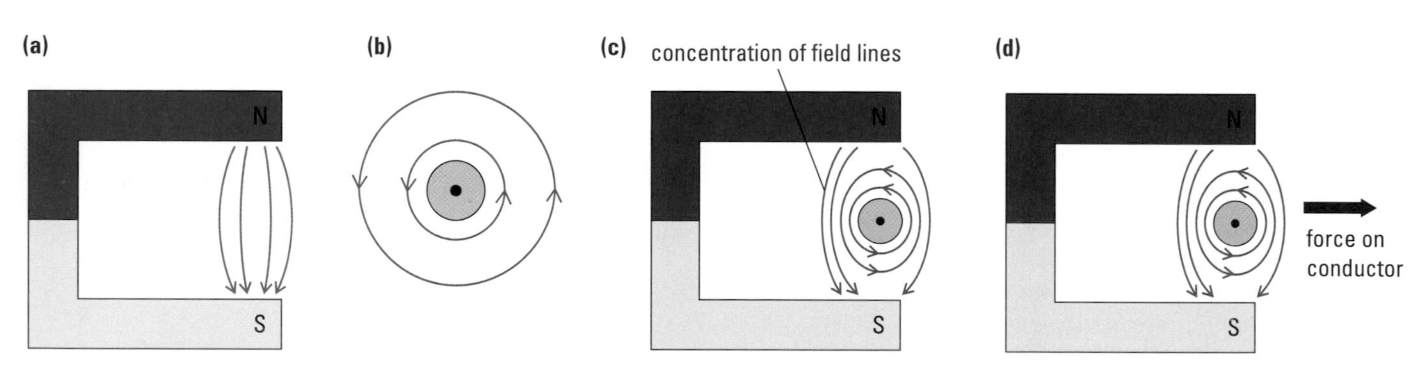

Figure 1
Determining the force on an electric conductor in a magnetic field
(a) Magnetic field of the permanent magnet
(b) Magnetic field of the current-carrying conductor
(c) Shape of the magnetic field when the fields in (a) and (b) are superimposed
(d) The direction of the force on the conductor is away from the region of concentrated field lines

To the left of the conductor, the field lines point in the same direction and tend to reinforce one another, producing a strong magnetic field. To the right, the fields are opposed and, as a result, tend to cancel one another, producing a weaker field. This difference in field strength results in a force to the right on the conductor. If either the external field or the direction of the electric current were reversed, the force would act in the opposite direction.

A more detailed investigation would show that the actual magnitude of the force depends on the magnitude of both the current and the magnetic field.

These effects are summarized in the **motor principle**.

motor principle: A current-carrying conductor that cuts across external magnetic field lines experiences a force perpendicular to both the magnetic field and the direction of electric current. The magnitude of this force depends on the magnitude of both the external field and the current, as well as the angle between the conductor and the magnetic field it cuts across.

Motor Principle

A current-carrying conductor that cuts across external magnetic field lines experiences a force perpendicular to both the magnetic field and the direction of electric current. The magnitude of this force depends on the magnitude of both the external field and the current, as well as the angle between the conductor and the magnetic field it cuts across.

The direction of the force on the conductor depends on the direction of electric current and the direction of the external magnetic field (**Figure 2**). The direction of the force can be determined using what is called the **right-hand rule for the motor principle.**

right-hand rule for the motor principle: If the fingers of the open right hand point in the direction of the external magnetic field, and the thumb represents the direction of electric current, the force on the conductor will be in the direction in which the right palm faces.

Right-Hand Rule for the Motor Principle

If the fingers of the open right hand point in the direction of the external magnetic field, and the thumb represents the direction of electric current, the force on the conductor will be in the direction in which the right palm faces.

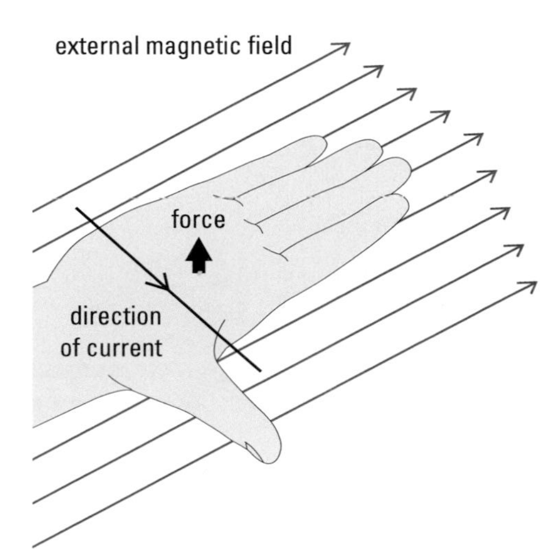

Figure 2
The right-hand rule for the motor principle

The motor principle is also used in the definition of the ampere. If two long wires are parallel to each other as in **Figure 3**, and electric current is in each, the wires will experience a force, as shown in **Figure 4**. One ampere is the current that, when flowing through each of two straight parallel wires placed 1 m apart in a vacuum, produces a force of 2.0×10^{-7} N between the wires for each 1 m of their length.

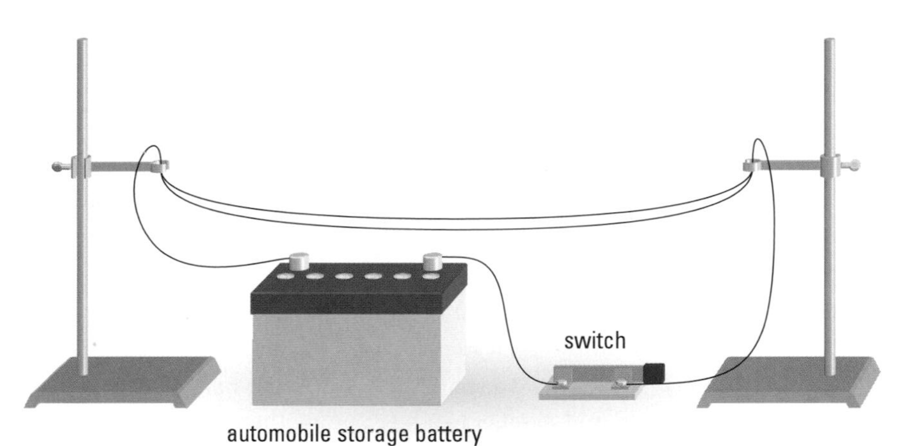

Figure 3

Figure 4
(a) If the electric currents in the parallel wires are in opposite directions, the force is repulsive.
(b) If the electric currents in the parallel wires are in the same direction, the force is attractive.

The motor principle allows us to introduce the SI definition of the coulomb, which states that 1 C is the amount of charge flowing past a point in 1 s when the current is 1 A.

$$1\ \text{C} = 1\ \text{A·s}$$

INQUIRY SKILLS

- ● Questioning
- ○ Hypothesizing
- ● Predicting
- ○ Planning
- ● Conducting
- ● Recording
- ● Analyzing
- ● Evaluating
- ● Communicating

Investigation 13.5.1

The Motor Principle

As its name suggests, the "motor principle" is the basis of operation of all electric motors, from the tiny ones used in toys to the massive ones used to propel electric commuter trains. To help you visualize how an electric motor actually works, it is wise to start by performing an investigation involving the motor principle.

Question

Under what conditions does a conductor experience an electromagnetic force?

Materials

insulated wire (fine)
utility stand, clamp, and metre stick
5-cm length of bare 12-gauge copper wire
pair of bar magnets
6-V battery or DC power supply

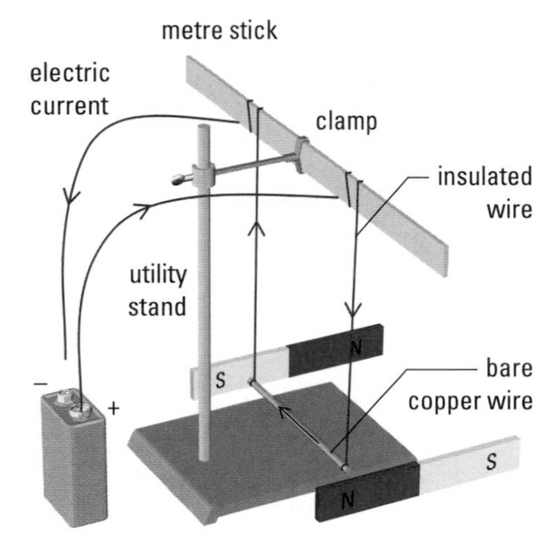

Figure 5
Setup for Investigation 13.5.1

Prediction

(a) Predict what will happen to the bare copper wire when it is placed between two bar magnets (i) perpendicular to the magnets, and (ii) parallel to the magnets when a current passes through the copper wire.

Procedure

1. Using the insulated wire, retort stand, clamp, metre stick, and bare copper wire, set up the apparatus as shown in **Figure 5**. Remove some insulation from the wire before attaching it to the bare copper wire.
2. Place the bar magnets so that the bare copper wire lies between opposite poles of the magnets, and parallel to a line joining them.
3. Connect the battery momentarily, noting any effect this has on the conductor.
4. Rotate the magnets by 90°, so that the conductor now lies between the poles but perpendicular to the line joining them with one magnet above the conductor and one below it.
5. Reconnect the battery and observe any effect this has on the conductor.
6. Reverse the poles of the magnets. What effect does this have on the conductor? Reverse the connections to the battery and note any effects.
7. Wind the wire into a rectangular coil of 15 turns as shown in **Figure 6** and suspend the coil between the poles of the magnets. Connect the battery and note the effect this has on the coil. Reverse the battery connections and repeat.

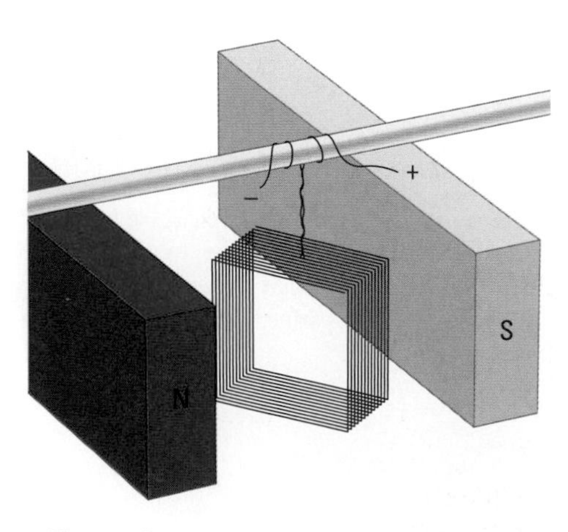

Figure 6

Analysis

(b) What happens to a current-carrying conductor when it is placed in a magnetic field so that it is
  (i) parallel to the magnetic field lines?
  (ii) perpendicular to the magnetic field lines?
(c) What factors affect the direction of the force on the conductor?
(d) What factors will affect the magnitude of this force?

(e) What happens to the rectangular coil when there is electric current in it? Do all four sides of the coil experience a force? Explain.
(f) What device does this simple coil and magnet simulate?

Evaluation

(g) How did your observations compare with your predictions?
(h) Describe the usefulness of the right-hand rule for the motor principle in predicting the direction of the force on the conductor in this investigation. Can you apply the same right-hand rule to determine the direction of the force on the suspended coil? Explain.

SUMMARY The Motor Principle

- The motor principle states that a current-carrying conductor that cuts across external magnetic field lines experiences a force perpendicular to both the magnetic field and the direction of electric current. The magnitude of this force depends on the magnitude of both the external field and the current, as well as the angle between the conductor and the magnetic field it cuts across.
- The right-hand rule for the motor principle states that if the fingers of the open right hand point in the direction of the external magnetic field, and the thumb represents the direction of electric current, the force on the conductor will be in the direction in which the right palm faces.

Section 13.5 Questions

Understanding Concepts

1. Copy each of the diagrams in **Figure 7** into your notes and then use them to do the following:
   (a) Draw the magnetic fields of the permanent magnet and the conductor.
   (b) Determine the direction of the force on the conductor.
2. Copy each of the diagrams in **Figure 8** into your notebook and then use them to do the following:
   (a) Draw the magnetic fields around each conductor.
   (b) Determine the direction of the force on each conductor.

(a)

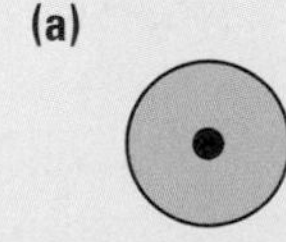

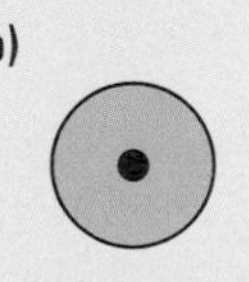

(b)

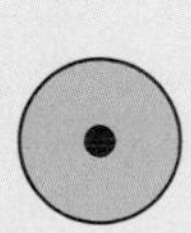

Figure 8

Applying Inquiry Skills

3. A student sets up a successful demonstration of the motor principle, but notices that the force on the conductor is very weak. What two changes could the student make to increase the force?

(continued)

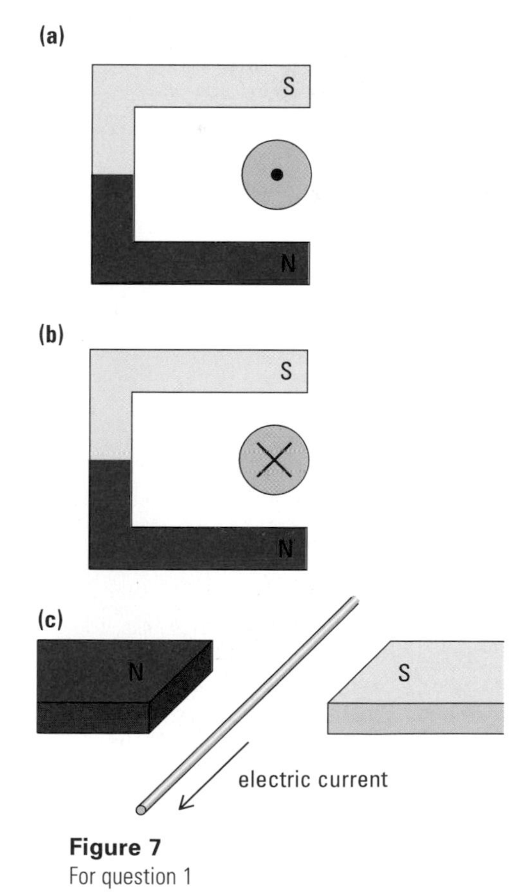

Figure 7
For question 1

4. When you look at the apparatus used to demonstrate the motor principle using a straight conductor (**Figure 5**), you can imagine that the suspended bare copper wire might act like a swing. What would you do to get the wire to swing back and forth with a regular period of vibration? (With your teacher's approval, you might be able to try your design.)

Reflecting

5. The explanation of the definition of the ampere is included in this section that discusses the motor principle. Explain why this is logical.

13.6 Applications of the Motor Principle

The motor principle refers to a force acting on a conductor carrying a current in a magnetic field. It is the most important principle used in the development of electric motors. However, the development of electric motors is not the only application of the motor principle. The motor principle has also been applied in the development of devices such as loudspeakers for stereos and in ammeters and voltmeters.

The Moving-Coil Loudspeaker

A loudspeaker reproduces sound waves by rapidly moving a paper or plastic sound cone back and forth in response to electrical signals from an amplifier. **Figure 1** shows side and front views of a magnetically driven speaker.

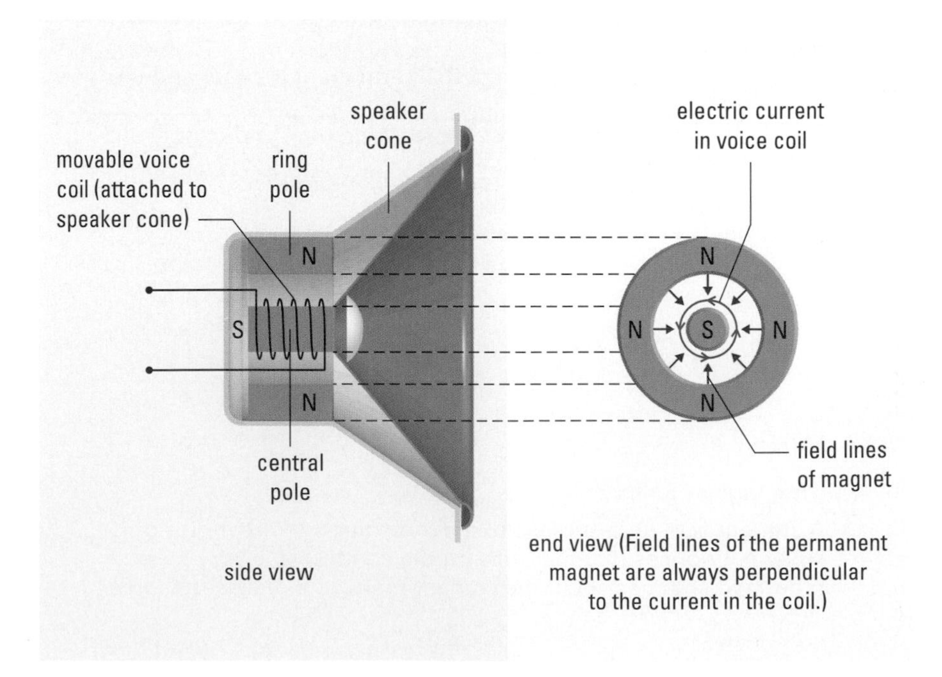

Figure 1
In a moving-coil loudspeaker, a movable coil is attached to the sound cone and placed over the central shaft of a tubular permanent magnet. The external magnetic field lines run radially from the outer tubular magnet to the central shaft. As a result, when electric current runs through the voice coil, it is in a magnetic field that is always perpendicular to it.

According to the motor principle, the movable coil will experience a force that is parallel to the axis of the coil, causing the sound cone to move. The magnitude and frequency of the force on the coil will depend on the amount and frequency of the current flowing through the voice coil. This will in part determine the loudness and frequency of the sound produced. The suspension mechanism holding the vibrating coil returns it to its original position when there is no current flowing through it.

galvanometer: device used to measure the magnitude and direction of small electric currents

The Moving-Coil Galvanometer

A **galvanometer** is a delicate device used to measure the magnitude and direction of small electric currents. As shown in **Figure 2**, a movable coil is wound around a light frame which surrounds a fixed iron core. The iron core increases the magnitude of the magnetic field, causing a larger force on the movable coil.

The coil is free to rotate when a current runs through it. The amount of rotation depends directly on the amount of current running through the coil. The direction of rotation depends on the direction of the electric current flowing through the coil. The amount of rotation is limited by an attached spring. The amount of current (or some other quantity) is then indicated by the attached needle and the calibrated scale.

According to the motor principle, there will be a force on the movable coil when an electric current is flowing through it (**Figure 3**). The N-pole of the coil will be attracted to the S-pole of the permanent magnet and repelled by the N-pole. This will cause the coil and the attached needle to turn. Using the motor principle, we can see that the front and back of the coil will not experience any force since they are parallel to the magnetic field lines, and the sides will experience opposite forces since the currents are opposite and perpendicular to the magnetic field lines. These opposite forces on the sides will cause the coil to turn.

If zero is marked at the centre of the scale, the galvanometer will be able to measure current flowing in either direction. An amount of current that causes the pointer to move completely across the scale is called the full-scale deflection current, and it is usually just a few milliamperes. A galvanometer must be protected from any current greater than its full-scale deflection current.

To protect the galvanometer, a device called a resistor is connected with it to limit the current passing through it. For a voltmeter (**Figure 4(a)**), the galvanometer is connected in series with a high resistance (*multiplier* resistance).

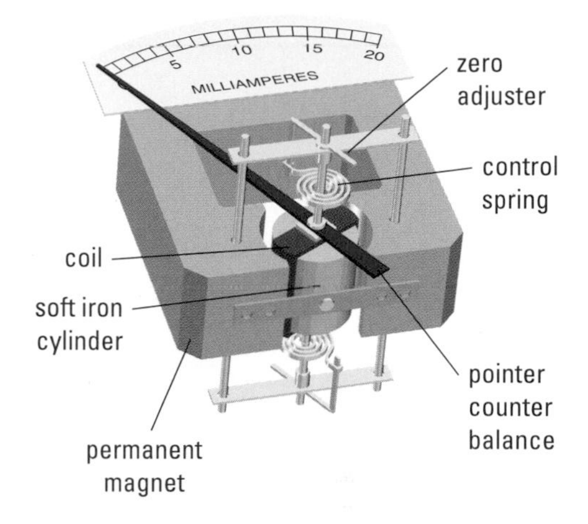

Figure 2
In a moving-coil galvanometer, the round iron core and the curved ends of the permanent magnet ensure that the magnetic field lines radiate through the core and stay perpendicular to the sides of the movable coil.

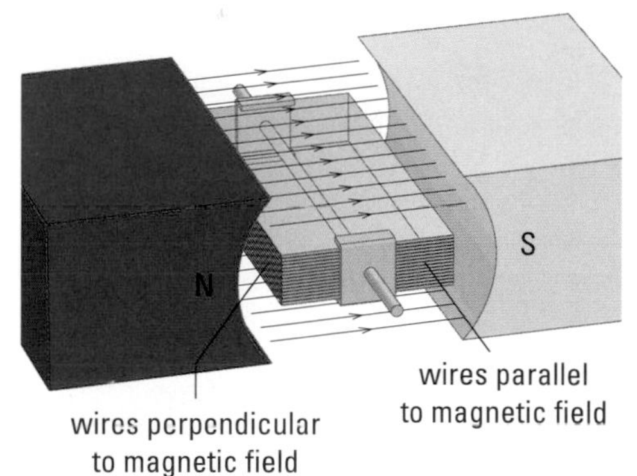

Figure 3
Wires perpendicular to the magnetic field experience a force causing the coil to turn. Wires parallel to the magnetic field experience no force.

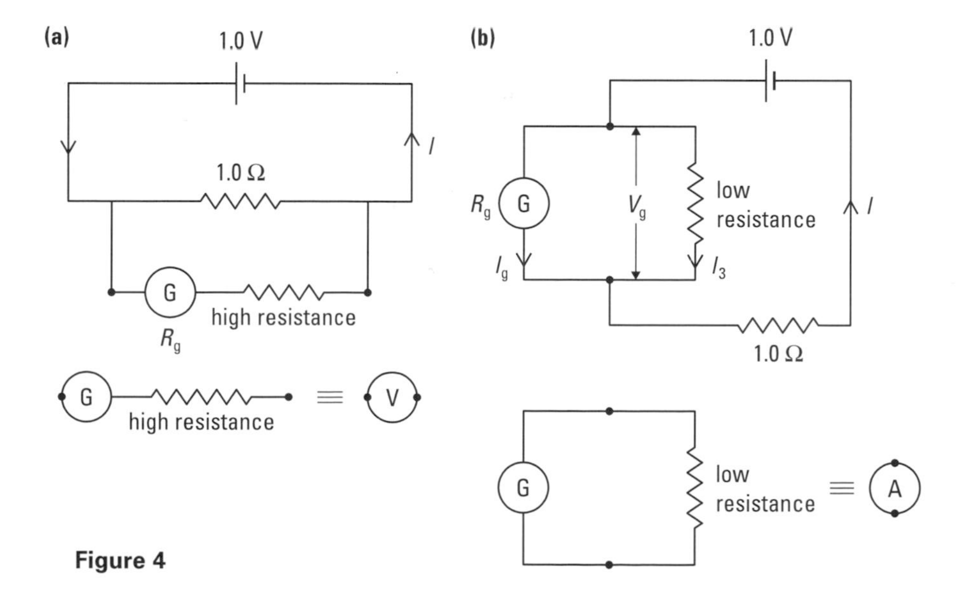

Figure 4

Since a voltmeter is connected in parallel, this will limit the amount of current flowing through the galvanometer. For an ammeter, a resistor of low resistance (a *shunt* resistance) is connected to the galvanometer in parallel (**Figure 4(b)**). The ammeter is connected in series, allowing it to measure large currents while allowing only a small current through the galvanometer.

The Electric Motor

As we saw with the moving-coil galvanometer, a current-carrying coil pivoted in a uniform magnetic field will begin to rotate. A closer examination would reveal that the coil will rotate only until it is at right angles to the field, and then it will stop. For the coil to continue to rotate, the direction of the force on it would have to change every half rotation. This could happen only by changing the direction of either the external magnetic field or the current flowing through the coil.

Figure 5 shows how it is possible to switch the direction of the current every half rotation.

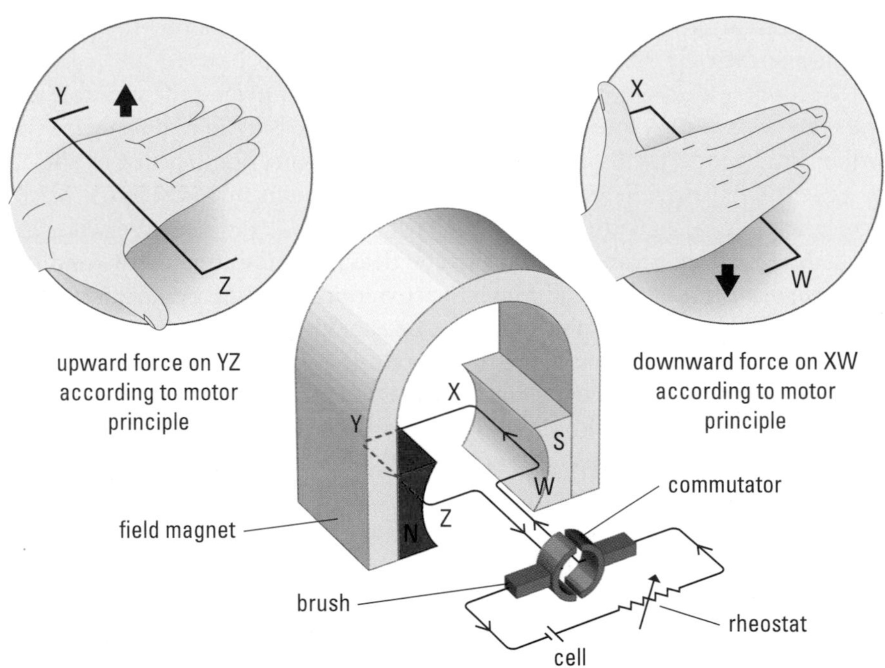

Figure 5
Basic design of an electric motor
In an electric motor, the ends of the coil are attached to a split copper ring, or commutator, that rotates with the coil. Continuous contact with the commutator is made by two stationary graphite brushes that push gently against the rotating commutator. The brushes are connected to a battery. Electric current enters the coil through one brush and leaves through the other. A **rheostat** is used to vary the current in the circuit.

rheostat: device in an electric circuit that can be adjusted to different resistances, changing the current in the circuit

According to the right-hand rule for the motor principle, when electric current flows through the circuit, side YZ of the coil experiences an upward force and side XW experiences a downward force, causing the coil to rotate in a clockwise direction, as illustrated. As the rotating coil reaches the vertical position, both brushes come opposite the gap between the commutator segments and no charge flows. However, the inertia of the coil keeps it rotating until the brushes make contact again, this time each with the other half of the ring. This causes the direction of the electric current through the coil to be reversed, so there is now a downward force on YZ, causing it to continue rotating in a clockwise direction. This switching procedure is repeated every half cycle as long as there is electric current in the brushes. Reversing the polarity of either the magnet or the battery will cause the coil to rotate in the opposite direction.

Figure 6 shows the relative positions of the armature, coil, graphite brushes, and commutator at four positions during one cycle of a DC motor, with an iron armature and an external source connected.

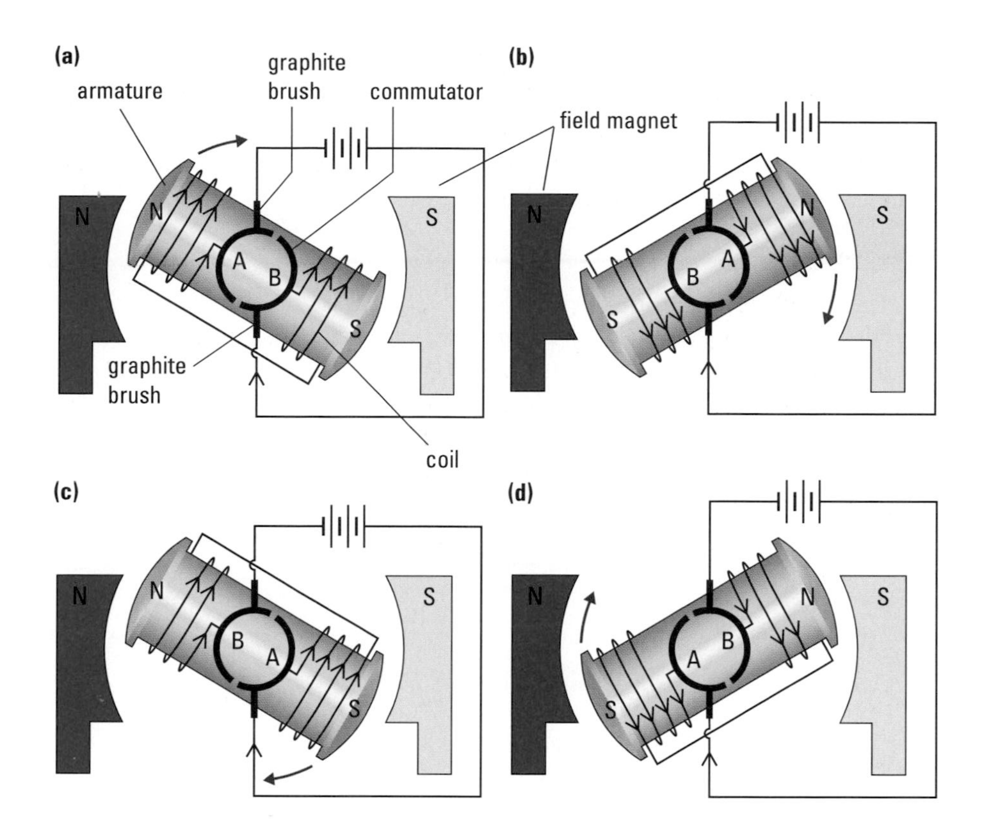

Figure 6
The design and operation of an electric motor

In **Figure 6**(a), electric current flows in through the bottom brush, into commutator segment B, and through the coil, eventually entering commutator segment A and leaving the motor through the top brush. End A of the armature becomes an N-pole, using the right-hand rule, and is repelled by the N-pole of the field magnet, causing it to move away and rotate clockwise.

In **Figure 6**(b), tracing the path of electric current through the motor verifies that end A remains an N-pole and is, therefore, attracted toward the S-pole of the field magnet.

In **Figure 6**(c), a significant change occurs. The top brush is now in contact with commutator segment B. Electric current continues to flow up through the coils, leaving by commutator segment B and the top brush. End A of the armature now becomes an S-pole and is repelled by the S-pole of the field magnet, causing the clockwise motion to continue.

Again, tracing the flow of electric current through the motor in **Figure 6**(d) confirms that end A of the armature remains an S-pole and is attracted toward the N-pole of the field magnet, completing one full rotation of the motor.

This simple electric motor is not very powerful or efficient. To increase its power an armature is used. The high relative magnetic permeability of the iron core and the large number of windings increases the magnetic field strength of the armature. These factors combine to produce a large force on the coil, causing it to rotate rapidly. A strong electromagnet is often used as the field magnet.

Practical electric motors have more coils connected to a multi-segmented commutator. Each coil is connected to two oppositely located commutator segments that allow current to flow through when the coil is perpendicular to the magnetic field. This will help to maximize the force on the armature, making a more powerful motor that doesn't require an initial push to get it going.

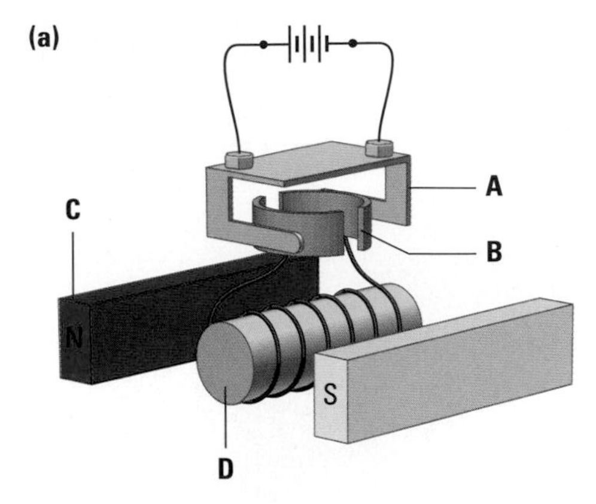

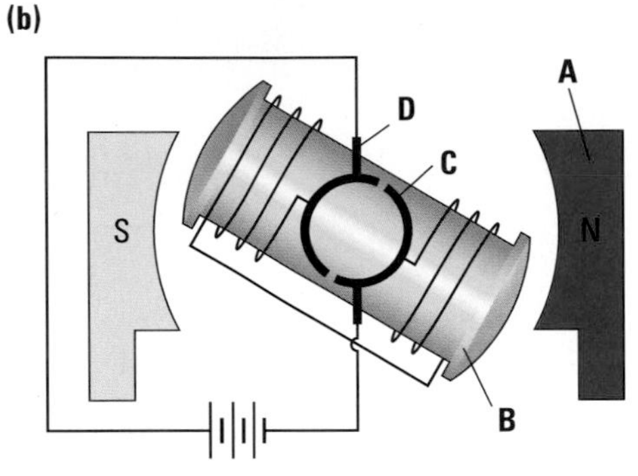

Figure 8
For question 3

The rate of rotation of such a motor is easily controlled by varying the current in the coils using a rheostat. Small motors controlled in this way are often used in battery-operated toys. Subway trains, street cars, and diesel electric locomotives use large-scale motors based on these principles.

Practice

Understanding Concepts

1. State the function of each of the following parts of a DC motor:
   (a) commutator (b) armature
   (c) brushes (d) field magnet
2. **Figure 7** represents a single loop in a DC electric motor. Determine the direction of the forces on each part shown and the rotation of the loop. Explain two ways to reverse the rotation of the loop.

Figure 7

3. For each of the diagrams in **Figure 8**,
   (a) name the parts of the motor that are labelled
   (b) determine which end of the coil is N
   (c) state in which direction the coil will spin
4. Examine **Figure 6** of this section, which shows the armature moving through one revolution. Explain in detail how the armature completes one revolution if the electric current is reversed.

Activity 13.6.1

Constructing a Simple DC Motor

How can a simple DC motor be constructed?

Materials

insulated wire
tape
pencil
pin
2 common iron nails (about 7 cm long)
6-V battery, or DC power supply
block of wood
cardboard

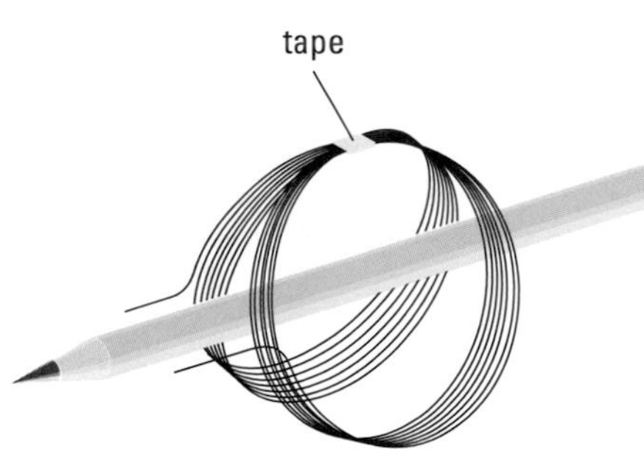

Figure 9

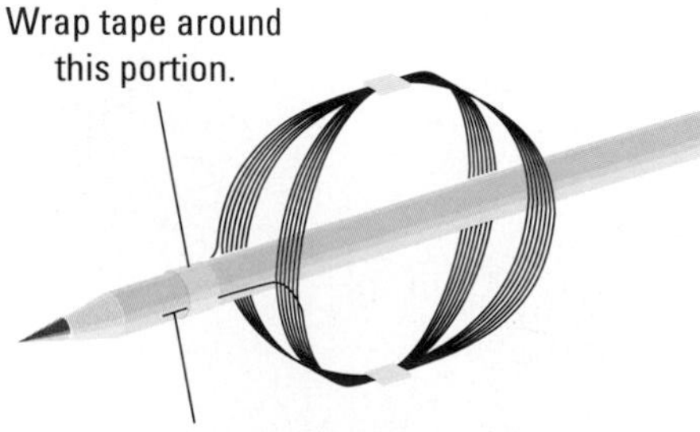

Figure 10

Procedure

1. Using something round and about 3 or 4 cm in diameter, make a coil consisting of about 20 turns of wire.
2. Remove the coil and tape the loops together on one side.
3. Split the coil windings in half, and position the coil straddling a sharp pencil as shown in **Figure 9**. When the coil is positioned so that the pencil will rotate like a well-balanced wheel on an axle, tape the coils on the side opposite that in step 2 (**Figure 10**).

4. Strip the insulation from the wires at the ends of the coil and position them along and on opposite sides of the pencil, as shown in **Figure 10**. Tape them in position with the bare wires exposed. This completes the armature for the motor.
5. Tightly wrap a double layer of coils around each nail, leaving the bottom 1 cm free of windings, and ensuring that the ends of the wires are at the pointed end of the nails.
6. Obtain a small wooden block about the same length as the pencil used for the armature. Prepare two armature support brackets using pieces of cardboard cut to 10 cm × 5 cm.
7. Mount all of the items as shown in **Figure 11**. The tops of the nails should be level with the pencil. A pin may be inserted in the eraser end of the pencil to act as a pivot. The windings on the nails have not been shown; they must be connected so the nails have the opposite polarity.
8. Complete the electrical connections as shown in **Figure 12**. When the battery is connected and the armature is given a small push, the motor should rotate. Minor adjustments to the alignment of the commutator or the balance of the armature may be necessary.

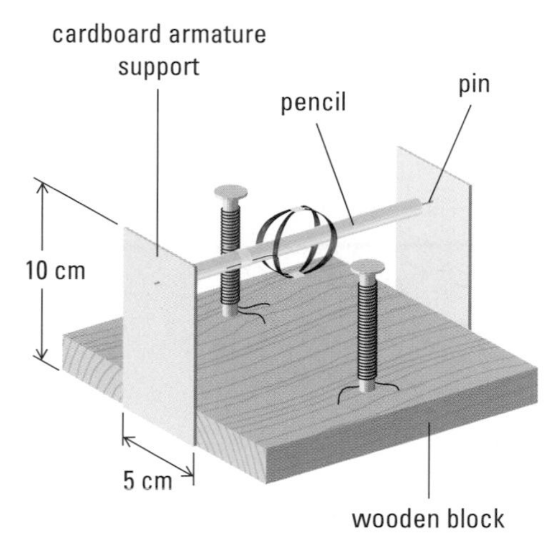

Figure 11

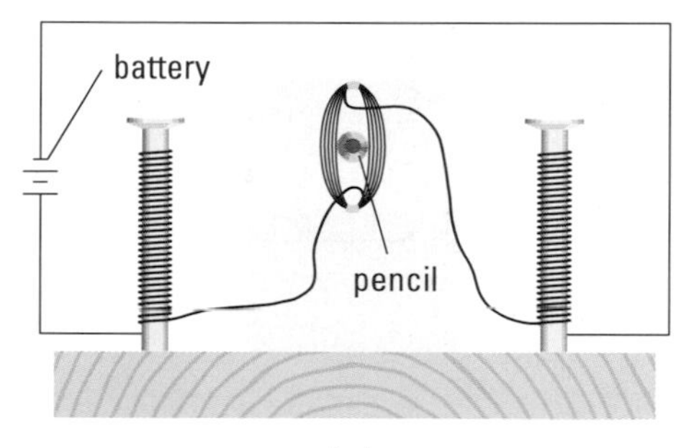

Figure 12

Analysis

(a) Explain in detail how the motor works. Make reference to the electric current in the motor and the magnetic fields involved. Use diagrams to help with your explanations.

(b) What improvements could be made to your motor to make it work better?

Case Study: Magnetic Resonance Imaging (MRI)

Magnetic Resonance Imaging (MRI) is a diagnostic technique that produces high-quality images of the inside of the human body. In 1977, Dr. Raymond V. Damadian, an American physician and scientist, developed the first MRI scanner as an improvement over X ray machines and other methods of diagnosing illnesses. Today, MRI is one of the most powerful tools doctors have for diagnosing illnesses; this life-saving technology enables them to see and study soft tissues of the body without the need for invasive exploratory surgery.

Magnetic Resonance Imaging: (MRI) diagnostic technique that produces high-quality images of soft tissue inside the body

The Basics of Magnetic Resonance Imaging

MRI machines are usually cube-shaped (**Figure 13**). A patient is set inside the machine and around her is a large electromagnet that generates an extremely powerful magnetic field. The magnetic field can be used to build a three-dimensional image of part of the patient's body or to visually "slice" through a portion, making a two-dimensional image.

The magnet is the most important part of the MRI machine. The magnet is so powerful that anything metallic will be drawn toward it. Objects with magnetic encoding, such as bank cards, must also be left outside the room or run the risk of being erased. People with certain medical conditions or equipment cannot go near an MRI machine.

Different MRI machines use three different types of magnets. The first are resistive magnets, which consist of a coil through which electric current is passed. The second are permanent magnets, which always maintain their magnetic field, and the third are powerful superconducting magnets. These superconducting

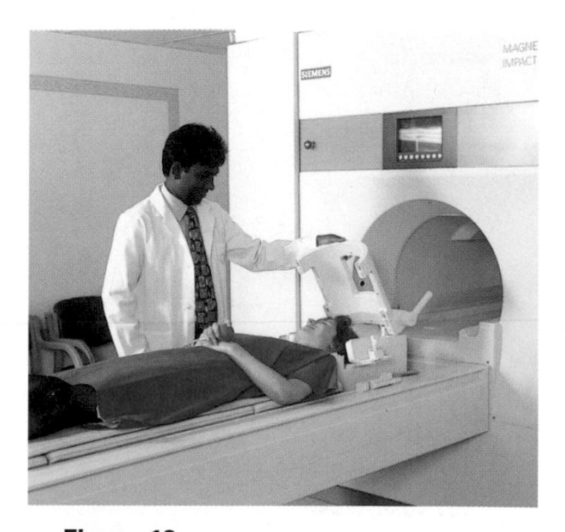

Figure 13
An MRI machine

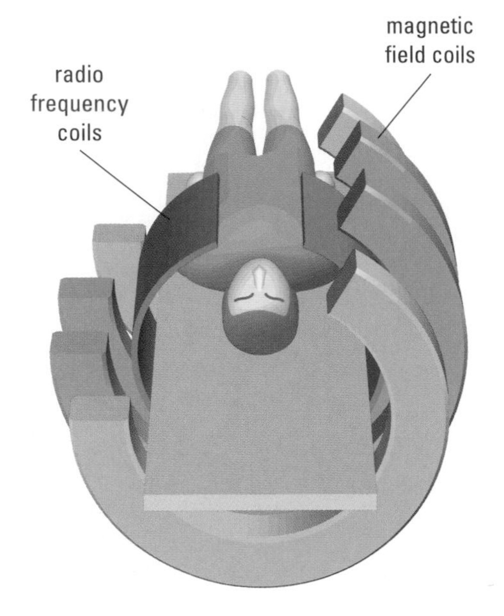

Figure 14

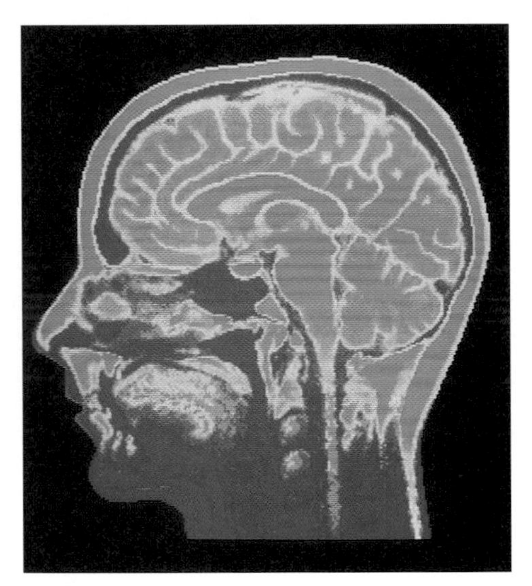

Figure 15
MRI image of a normal human brain

magnets are similar in design to resistive magnets, but use a material that when cooled down far enough, will allow electricity to pass through without resistance. As a result they are very efficient and far less expensive to use.

The superconductor is in the shape of a large coil. The large current in the superconductor makes a uniform magnetic field inside the coil. The patient is then moved inside the coil so that the area to be scanned is completely surrounded by the magnetic field (**Figure 14**).

How the MRI Looks Into You

The nuclei in atoms in the human body are just like the nuclei in anything else, and function under the same set of rules and properties. One of these nuclear properties is called spin, and this is the basis behind the idea of MRI.

The human body has a large number of hydrogen atoms, each of which has a single proton in its nucleus and a large magnetic moment, meaning that the spins of the protons tend to line up with any strong magnetic field. Once the magnetic field of the MRI machine has caused the hydrogen nuclei in the body to line up, a radio frequency pulse is directed at the person and causes the affected nuclei to spin in a different direction than before. When the pulse is stopped, the nuclei slowly return to their normal direction of spin. This gives off energy that can be detected, and the information is directed to a computer. The computer converts this data into images (**Figure 15**).

If the location of the problem is known, the radio frequency pulse can be localized to that area. Doctors can identify different types of problems by examining MRI images and comparing them with normal MRI images.

Sometimes a patient is injected with a certain kind of dye to help the doctor isolate different types of tissue under examination. This helps to identify abnormalities such as scar tissue and cancer.

Advantages and Disadvantages

MRI has some definite advantages over other types of imaging systems. It is good at detecting torn ligaments and muscles, abnormal growths, and infections, and is ideal for helping to diagnose tumours, cysts, multiple sclerosis, and conditions leading to strokes. It can image in any plane, allowing the doctor to see only what is needed. Unlike other imaging systems, the dyes do not use radiation.

However, MRI has some drawbacks as well. Some people cannot be scanned, such as those with pacemakers or those who experience claustrophobia. The machine is very noisy, which can cause headaches and discomfort. An MRI machine is expensive, and maintaining it is costly.

Practice

Understanding Concepts

5. Explain the function of the following in producing an MRI image:
   (a) the superconducting coils
   (b) the magnetic field
   (c) the radio pulse
   (d) the hydrogen atoms in the body

6. Discuss practical ways of increasing the strength of the magnetic field produced by an MRI machine. Why is putting a material inside the core of high relative magnetic permeability not an option?
7. A construction worker had an accident in the past and a small piece of metal was lodged in his eye and was never removed. Why would a doctor not allow him near an MRI machine?

Explore an Issue

Medical Funding

Today there is a great deal of interest in medical research and the treatment of illness. Much of the funding available goes to research. Research involves finding new and more effective ways to treat illness and detect it early, with the ultimate goal being to find a cure. Some of the funding goes to the treatment of the illness. These funds are used to buy new equipment (such as MRI machines), maintain this equipment, and pay doctors and other experts who make use of the equipment. One of the challenges faced by society today is deciding what percentage of the funds should be directed toward research and what percentage should go to treatment.

DECISION MAKING SKILLS

- ○ Define the Issue
- ● Identify Alternatives
- ● Research
- ● Analyze the Issue
- ● Defend the Proposition
- ○ Evaluate

Understanding the Issue

1. Why would it be beneficial to put a high percentage of the funding available into medical research?
2. Why would it be beneficial to put a high percentage of the funding available into medical treatment of illness?
3. List three possible impacts on society if funding was very low for
   (a) medical research
   (b) treatment of illness

Use the Internet or any other resource to investigate the funding models for medical research and treatment. Concentrate on a particular illness and investigate the current reseach and treatment techniques for this illness. Follow the links for Nelson Physics 11, 13.6.

www.science.nelson.com

Take a Stand

Working in small groups, write a short report on the research and treatment of a particular illness. Discuss how you would distribute the funds between each of these areas, giving reasons for the distribution.

SUMMARY Applications of the Motor Principle

- Applications of the motor principle include the moving-coil loudspeaker, the moving-coil galvanometer, and the electric motor.

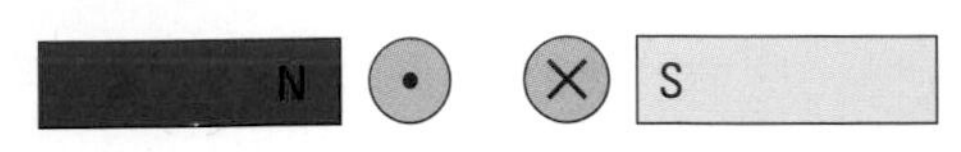

Figure 16
For question 3

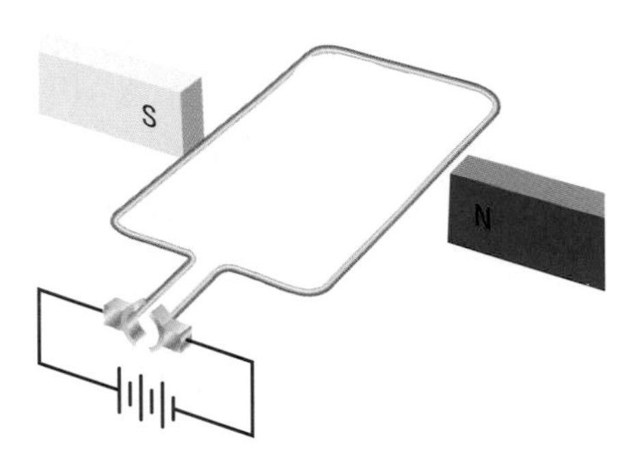

Figure 17
For question 4

Section 13.6 Questions

Understanding Concepts

1. Go back to **Figure 1** of this section, which shows a loudspeaker, and apply the right-hand rule for the motor principle to determine the direction of the force on the coil at the instant shown.
2. Both a voltmeter and an ammeter use a galvanometer as a basis of construction. How do these meters differ in their design and their connection in an electric circuit?
3. The conductors shown in **Figure 16** represent a loop in a magnetic field. Determine whether the force on the loop is clockwise or counterclockwise. Use a diagram to explain your answer.
4. For the instant shown in **Figure 17**, is the force on the loop clockwise or counterclockwise? Explain your reasoning.
5. Describe two possible ways of forcing the loop in **Figure 6** of this section to rotate counterclockwise.
6. (a) When would it be an advantage to use an electromagnet as a field magnet rather than a permanent field magnet? Under what conditions would you use a permanent magnet?
   (b) Why does the armature pass very close to the field magnet in a practical electric motor?
   (c) Why is it an advantage to have many coils and a multi-segmented commutator in a practical motor?

Applying Inquiry Skills

7. Use a demonstration electric motor to learn firsthand how it operates. Look at the armature windings carefully to determine the direction of the electric current when the motor is connected to the power supply. Predict the direction of the armature rotation, and then check your prediction. Write a short report explaining how the motor works.
8. One type of galvanometer or milliammeter has the zero at the centre of the scale, while another type has the zero at the extreme left side of the scale.
   (a) If you were using these galvanometers in the lab, how would the precautions differ for the two different types?
   (b) What other precautions would be wise to follow?

Making Connections

9. The motors described in this section are DC motors. However, AC motors are also commonly available. Find out, through research or looking at a model, how the construction and operation of an AC motor differs from that of a DC motor. With the aid of a diagram, describe the AC motor's operation.

Reflecting

10. If you know of anybody who has experienced an MRI scan, find out about the experience, and describe ways in which a patient could become mentally prepared for the scan.

Chapter 13 Summary

Key Expectations

Throughout this chapter, you have had opportunities to

- define and describe the concepts and units related to electricity and magnetism; (13.1–13.6)
- describe the properties, including the three-dimensional nature, of magnetic fields; (13.1)
- describe and illustrate the magnetic field produced by an electric current in a long straight conductor and in a solenoid; (13.3, 13.4, 13.5)
- analyze and predict, by applying the right-hand rule, the direction of the magnetic field produced when electric current flows through a long straight conductor and through a solenoid; (13.4, 13.5)
- state the motor principle, explain the factors that affect the force on a current-carrying conductor in a magnetic field, and, using the right-hand rule, illustrate the resulting motion of the conductor; (13.6)
- conduct an experiment to identify the properties of magnetic fields and describe the properties found; (13.1, 13.3, 13.4)
- interpret and illustrate, on the basis of experimental data, the magnetic field produced by a current flowing in a long straight conductor and in a coil; (13.3, 13.4, 13.5)
- analyze and describe the operation of industrial and domestic technological systems based on electromagnetic principles; (13.6)

Key Terms

poles
N-pole
S-pole
law of magnetic poles
magnetic field of force
ferromagnetic
dipoles
magnetic domain
magnetic induction
magnetic saturation
principle of electromagnetism
right-hand rule for a conductor
electromagnet
solenoid
uniform magnetic field
right-hand rule for a coil
relative magnetic permeability
paramagnetic
diamagnetic
switch
relay
armature
motor principle
right-hand rule for the motor principle
galvanometer
rheostat
Magnetic Resonance Imaging (MRI)

Make a Summary

Outline the different right-hand rules in this chapter and draw diagrams of the magnetic fields associated with them. Describe any factors that can affect the strength of the field, as well as examples of technological devices that apply any of the rules.

Reflect on your Learning

Revisit your answers to the Reflect on Your Learning questions at the beginning of this chapter.

- How has your thinking changed?
- What new questions do you have?

Chapter 13 Review

Understanding Concepts

1. Copy **Figure 1** into your notebook and finish drawing the rest of the field lines. Indicate the polarities of the magnets.

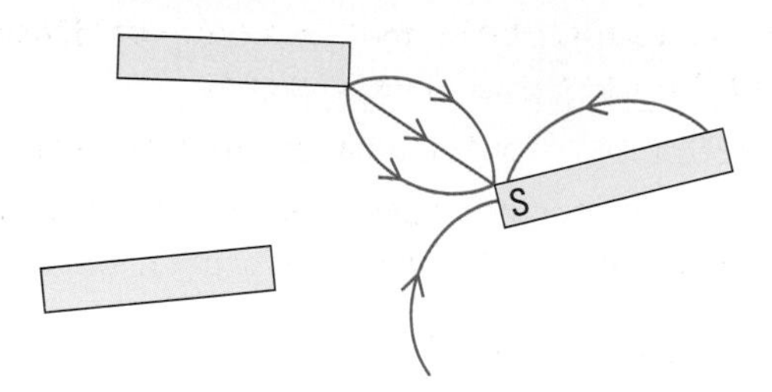

Figure 1

2. A straight east-west conductor is held near a plotting compass that is facing north. What direction of the electric current in the conductor allows the compass needle to remain stationary if the compass is
   (a) beneath the conductor?
   (b) above the conductor?

3. The diagram in **Figure 2** represents two parallel current-carrying conductors. Determine whether the conductors attract or repel each other. Explain your reasoning.

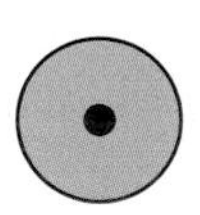
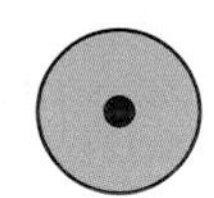

Figure 2

4. Each empty circle in **Figure 3** represents a plotting compass near a coiled conductor. Copy the diagrams, label the N- and S-poles of each coil, and indicate the direction of the needle of each compass.

(a)

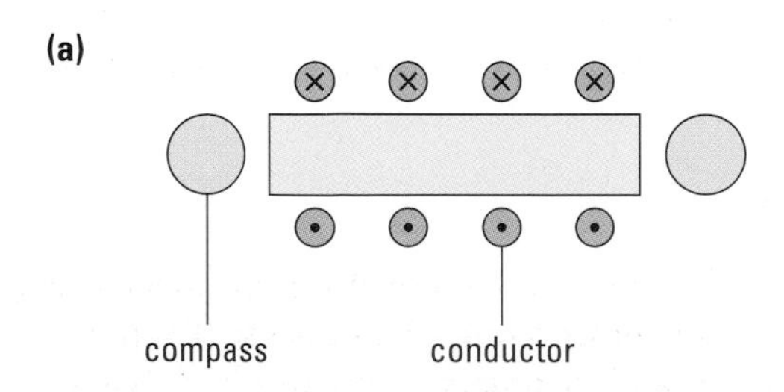

(b)

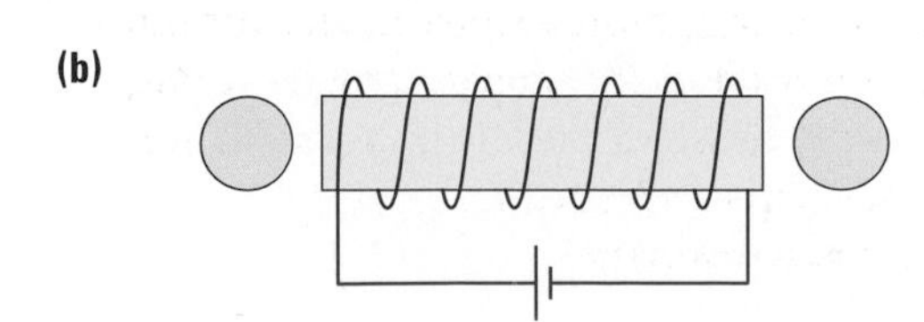

Figure 3

5. The circles in **Figure 4** represent a conductor in a coil. Determine which circles should have a dot and which an X.

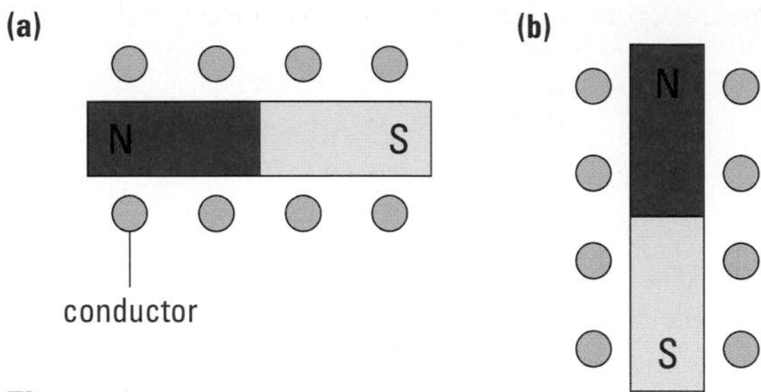

Figure 4

6. In **Figure 5**, a current-carrying conductor is in the magnetic field of a U-shaped magnet. With the aid of a diagram, determine the direction in which the conductor is forced.

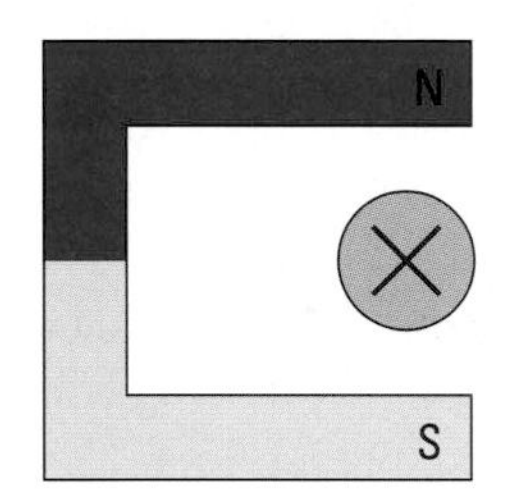

Figure 5

7. The diagram in **Figure 6** represents a single loop in a DC electric motor. Determine the direction of the force on the loop.

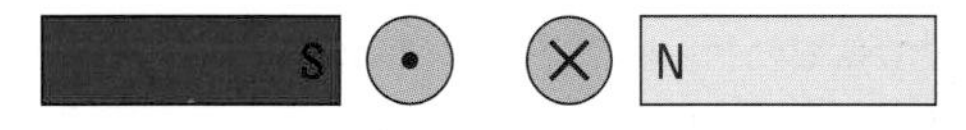

Figure 6

8. Two coils have the same length. One coil has 150 turns and a current of 2.0 A. The other coil has 300 turns and a current of 5.0 A. How will the strengths of the two coils compare?

9. **Figure 7** shows one design of an electric relay. Describe the sequence of events that occur when the switch in the left-hand circuit is closed.

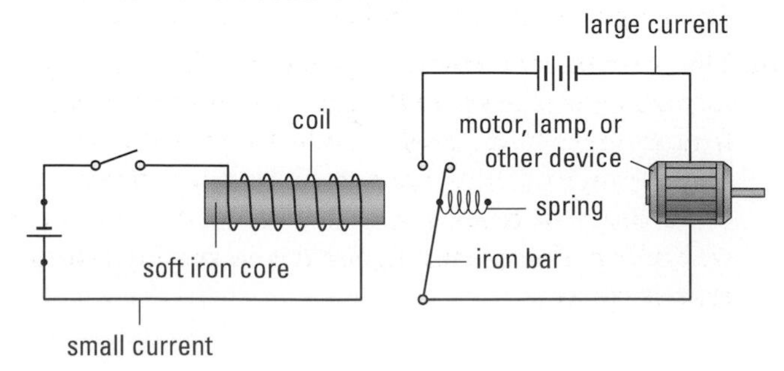

Figure 7

10. Each of the diagrams in **Figure 8** represents a simple motor. Copy each diagram and indicate direction of the electric current throughout the circuit, polarity of the armature, and the direction the armature will turn.

(a)

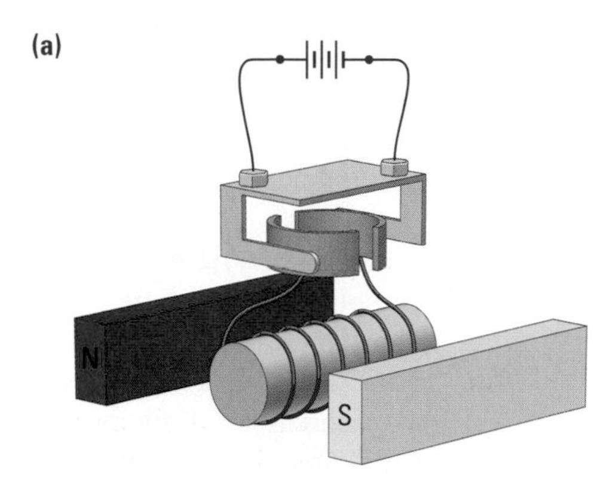

(b)

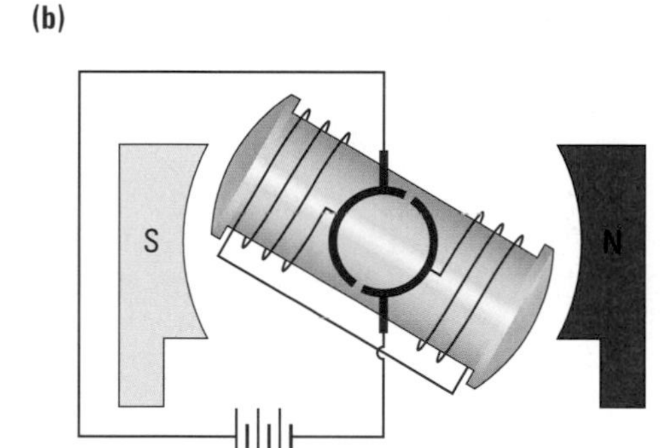

Figure 8

11. The domains of magnetic substances are composed of the atoms of those substances. Which particles within the atoms (electrons or protons) are likely responsible for the magnetic properties of domains? What experimental evidence supports your answer?
12. Do electromagnets have a maximum magnetic field strength that cannot be increased by increasing the number of loops or the current? Explain.
13. (a) Explain in detail how a galvanometer works.
    (b) Why must a voltmeter have a high resistance?
    (c) Why must an ammeter have a low resistance?
14. The strength of an electromagnet can be increased by increasing the current in the coil. How is the magnetic field around a long straight conductor related to the electric current in the conductor? Draw two diagrams of the magnetic fields around two different wires, one with a large electric current and the other with a small electric current.
15. (a) Use the domain theory to explain how a metal paper clip can become a magnet.
    (b) How can the paper clip be demagnetized? Use the domain theory to explain your method.
    (c) How can a paper clip be used as a compass? Does it matter if the paper clip is magnetized?
16. Explain in your own words how a loudspeaker works.
17. Compare a DC motor and a galvanometer with regard to how they are constructed and how they work.

Applying Inquiry Skills

18. (a) Sketch the magnetic field of a current-carrying coil, showing the direction of the field lines in the core and marking the magnetic polarities at each end of the coil.
    (b) Draw a sketch of an experimental arrangement that could be used to magnetize a bar made of iron so that it would have an S-pole at each end and an N-pole in the middle.
19. Design and carry out an experiment to discover whether magnetic field lines pass through paper, aluminum, iron, wood, or any other materials you wish to try. Discuss how you would design magnetic shielding to protect a sensitive electric instrument.
20. Design a device that would efficiently magnetize worn-out magnets.
21. How can you make a compass without using any magnetic substance? (*Hint:* Is copper magnetic?)

Making Connections

22. Many refrigerator magnets are flexible and appear to be made of rubber. How is this type of magnet made?
23. When repairing a broken magnetic tape for a tape recorder, it is better to use plastic scissors than steel scissors. Explain.

Exploring

24. Research other devices that use electromagnets, using the Web or any other source. Discuss uses for the device and how electromagnets are involved. Some possibilities are telephone earpieces, electromagnet relays, fusion reactors (torsional), maglev trains, and particle accelerators. Follow the links for Nelson Physics 11, Chapter 13 Review.

GO TO www.science.nelson.com

Chapter

14

In this chapter, you will be able to

- investigate and understand the principles of electromagnetic induction
- use and understand Lenz's law in various electromagnetic induction phenomena
- apply the principles of electromagnetic induction to the operation and use of electric generators
- investigate and solve problems involving transformers and the efficient distribution of electrical energy
- investigate careers in electricity and magnetism using a variety of resources
- analyze and understand the method of magnetic storage of information and investigate the various uses of this method of storage

Electromagnetic Induction

In Chapter 13 you learned that an electric current produces a magnetic field. It seems reasonable to expect that the reverse might also be true—that a magnetic field can induce an electric current. How can this be achieved? What properties of the magnetic field will induce an electric current? Is it sufficient to have a strong magnetic field near a conductor for an electric current to be produced? By the end of this chapter, you will be able to answer these questions. You will also understand how the principles of electromagnetic induction apply to huge electric generators like those shown in **Figure 1**.

André Marie Ampère conducted experiments and observed the effects of electromagnetic induction but it was not his emphasis at the time and so he made little of it. The French physicist François Arago (1786–1853) observed that a compass that was disturbed would settle down quickly if a copper base was beneath it (**Figure 2**). Also, if a copper base was rotated beneath a compass he found that the needle would rotate with the disk. These events seemed inexplicable to Arago at the time and it wasn't until Faraday's discovery that it was explained.

It is believed that the American physicist Joseph Henry (1797–1878) first discovered the principles of electromagnetic induction, although he did not publish his findings until after Michael Faraday announced his discoveries. As a result, Faraday has been credited for the discovery of the principles of electromagnetic induction even though Henry's experiments were somewhat more sophisticated and contained information on self-induction that Faraday missed.

Reflect on your Learning

1. Why is it reasonable to think that a magnetic field might induce an electric current?
2. Why do you think some scientists, like Ampère, sometimes make important observations but fail to investigate them in detail?
3. (a) Why is Joseph Henry not given credit for the explanation of electromagnetic induction? Do you think this is fair?
   (b) Give several reasons why the scientific community gives credit to the person who publishes findings first and not to the one who discovers them first.

Throughout this chapter, note any changes in your ideas as you learn new concepts and develop your skills.

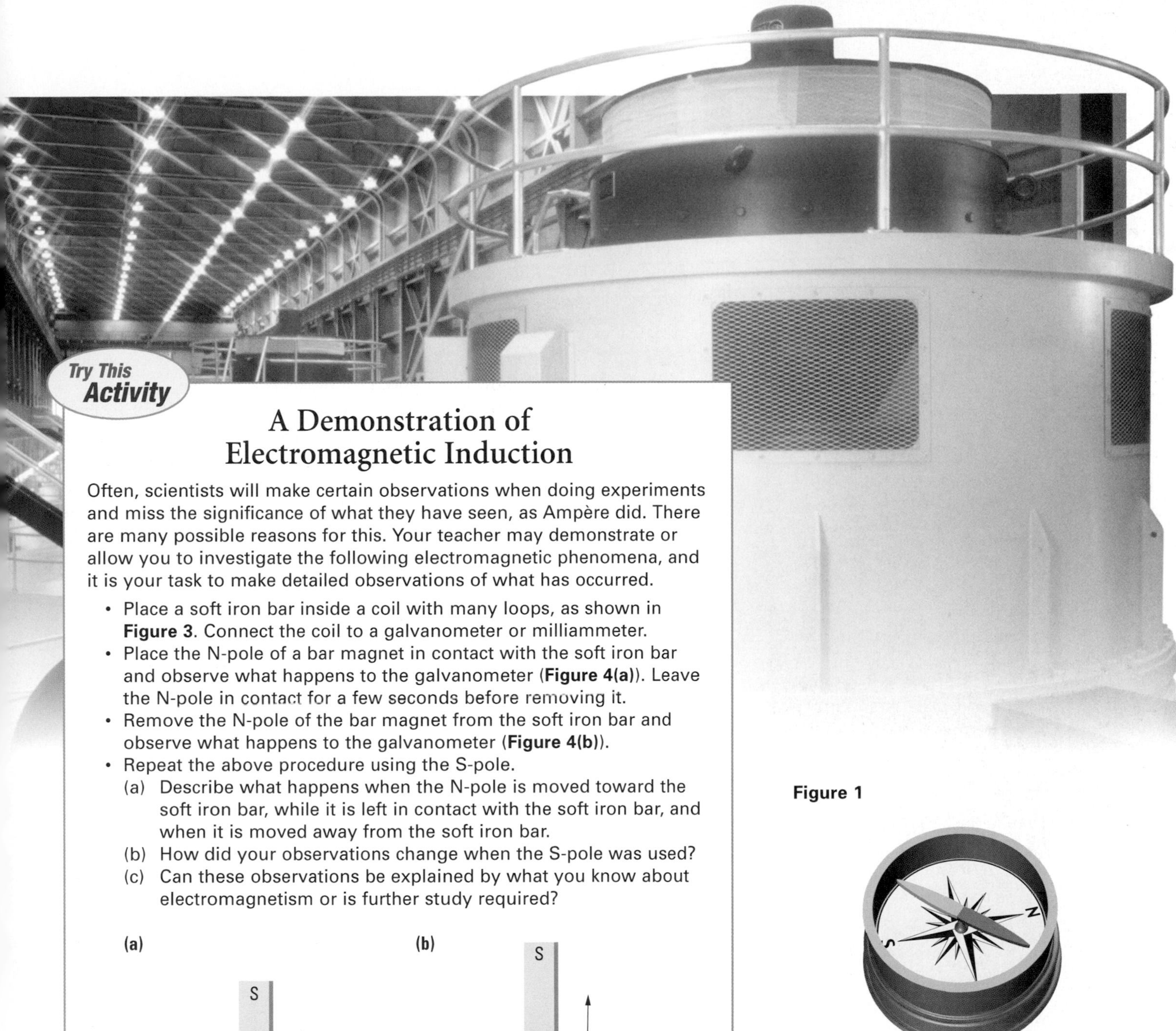

Figure 1

Figure 2
A compass needle will settle down more readily when it has a copper base. The compass needle will rotate with the copper disk if it is turned.

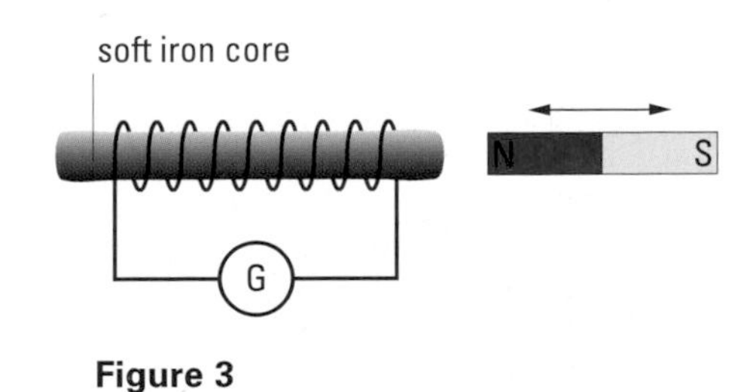

Figure 3

Try This **Activity**

A Demonstration of Electromagnetic Induction

Often, scientists will make certain observations when doing experiments and miss the significance of what they have seen, as Ampère did. There are many possible reasons for this. Your teacher may demonstrate or allow you to investigate the following electromagnetic phenomena, and it is your task to make detailed observations of what has occurred.

- Place a soft iron bar inside a coil with many loops, as shown in **Figure 3**. Connect the coil to a galvanometer or milliammeter.
- Place the N-pole of a bar magnet in contact with the soft iron bar and observe what happens to the galvanometer (**Figure 4(a)**). Leave the N-pole in contact for a few seconds before removing it.
- Remove the N-pole of the bar magnet from the soft iron bar and observe what happens to the galvanometer (**Figure 4(b)**).
- Repeat the above procedure using the S-pole.
  - (a) Describe what happens when the N-pole is moved toward the soft iron bar, while it is left in contact with the soft iron bar, and when it is moved away from the soft iron bar.
  - (b) How did your observations change when the S-pole was used?
  - (c) Can these observations be explained by what you know about electromagnetism or is further study required?

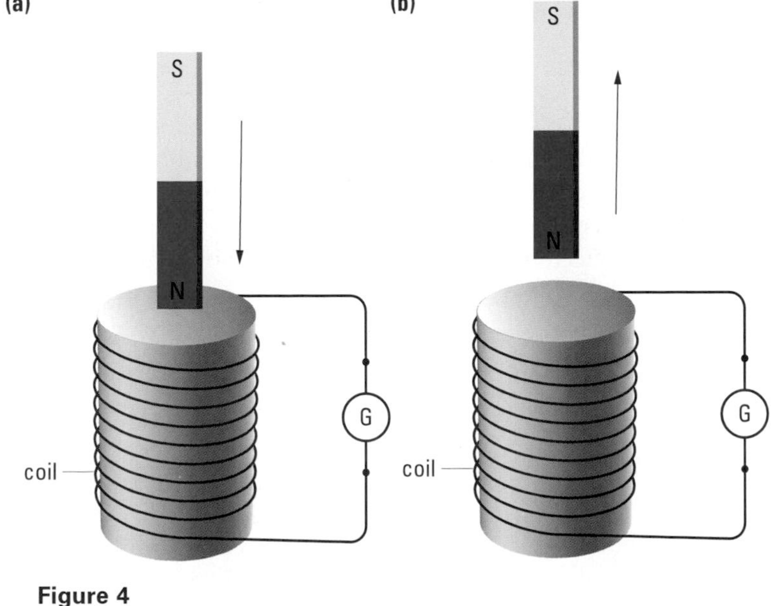

Figure 4

14.1 Faraday's Discovery

In 1819, when Oersted demonstrated the ability of a steady current to produce a steady magnetic field, scientists assumed that a steady magnetic field would produce a steady current. It seems logical and yet it doesn't work. It wasn't until 1831 that Michael Faraday (**Figure 1**) discovered the basic principle of electromagnetic induction. Faraday went on to develop the first electric generator, modern versions of which are now used to provide electricity to homes and industries all over the world.

INQUIRY SKILLS

- ○ Questioning
- ● Hypothesizing
- ● Predicting
- ○ Planning
- ● Conducting
- ○ Recording
- ● Analyzing
- ● Evaluating
- ● Communicating

Figure 1
Michael Faraday (1791–1867), one of ten children in a poor labourer's family in Surrey, England, became a bookbinder's apprentice. This exposure to books kindled tremendous interest in young Faraday, particularly in science. In addition to his work in electromagnetism, Faraday made great contributions in the fields of organic chemistry, electrolysis, optics, and field theory.

Investigation 14.1.1

Inducing Current in a Coiled Conductor

Question

Part A: How can a current be induced in a coil using a magnetic field?
Part B: What factors affect the magnitude of the induced current in a coil?

Materials

connecting wires
galvanometer or ammeter
horseshoe magnet
coil with a hollow core
pair of bar magnets
two similar solenoids of 300 turns and 600 turns

Hypothesis/Prediction

(a) State a hypothesis that will describe the effect a magnetic field will have on a coil of wire.
(b) Predict the effect on an electric current in a coil when
  (i) the number of turns in the coil increases and decreases
  (ii) the rate of change of the magnetic field increases or decreases
  (iii) the strength of the magnetic field changes.

Procedure

Part A

1. Connect the ends of a long piece of wire to the terminals of the galvanometer or ammeter, forming a single loop through the meter. Place the wire in the space between the poles of the horseshoe magnet (**Figure 2**) and note any effect this has on the galvanometer or ammeter.
2. Remove the wire from between the poles of the magnet and again note whether there is any electric current.
3. Move the wire back and forth and then up and down between the poles of the magnet. Try to determine exactly when electric current flows and when it does not.
4. Connect the ends of the coil to the galvanometer or ammeter. Plunge a bar magnet into the core of the coil. Allow it to remain there for a few seconds and then remove it. Note the effect of each of these actions on the galvanometer.
5. Repeat the procedure using the opposite pole of the bar magnet.

Part B

6. Connect the two coils and the galvanometer in series, as in **Figure 3**.
7. Using one of the bar magnets, insert its N-pole very slowly into the core of the 600-turn coil and measure the induced current. Repeat the procedure, this time moving the magnet more quickly into the coil. Follow the same procedure for a third time, moving the magnet as quickly as possible, and once again measure the induced current.
8. Insert the N-pole of the bar magnet steadily into the core of the 300-turn coil, measuring the induced current. Repeat this procedure moving the magnet at the same speed, this time using the 600-turn coil.
9. Holding the two bar magnets tightly together so that their N-poles are together, plunge the pair into the core of the 600-turn coil and measure the induced current. Repeat the procedure using a single bar magnet, and move the magnet at the same speed as you did when using the pair of magnets. Repeat again, using two magnets held side by side with opposite poles together and moving at the same speed.

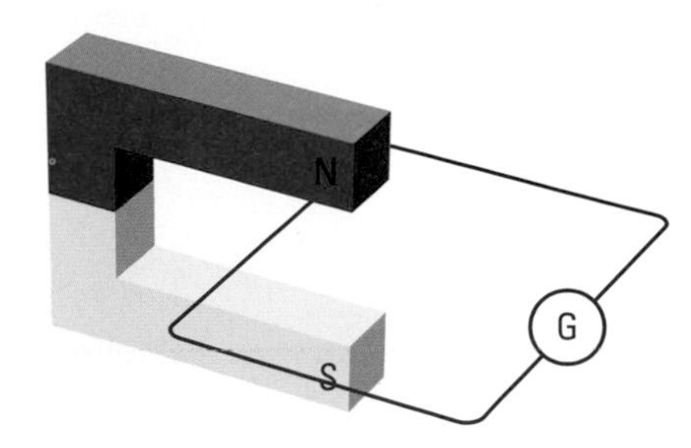

Figure 2

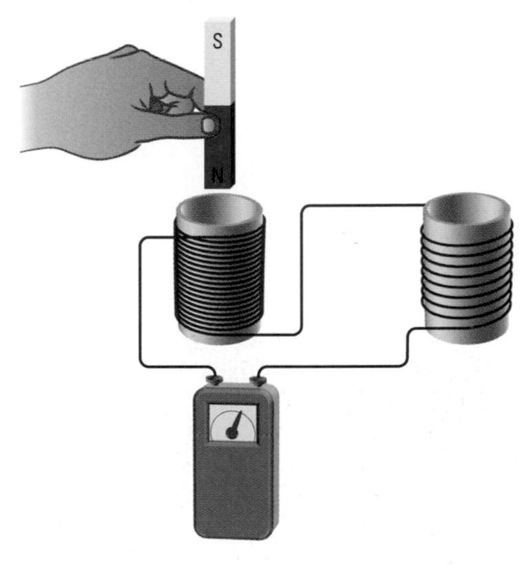

Figure 3

Analysis

Part A

(c) Describe different ways to produce an electric current in a coil using a magnetic field.

(d) Do the charges of an induced current always flow through the conductor in the same direction? What determines their direction?

(e) Oersted showed that moving charges cause a magnetic field. Make a similar statement about the cause of an induced current.

Part B

(f) List the three factors that affect the magnitude of the induced current in a coil.

(g) What set of conditions led to the maximum value of induced current?

(h) When a single bar magnet is inserted into the core of a 300-turn coil at a speed of 10 m/s, the induced current is measured and found to be 12 mA. What does the value of the induced current become if each of the following were used?

  (i) a magnet three times as strong
  (ii) a 900-turn coil of the same resistance
  (iii) a speed of 5 m/s for the magnet
  (iv) all the changes above, at once

Evaluation

(i) Describe any experimental errors and their sources. Suggest changes to the procedure that would help reduce errors.

Faraday's Law of Electromagnetic Induction

Faraday made his discovery by experimenting with conductors in the vicinity of magnetic fields. His investigations involved three situations:

- moving a wire through the jaws of a horseshoe magnet,
- plunging a bar magnet into and out of the core of a coil, and
- touching the iron core of a coil with a bar magnet, then removing the magnet.

In the first case, Faraday found that electric current only flowed while the conductor was cutting across the magnetic field. In the second case, electric current only began to flow when the bar magnet was moving into or out of the coil. And in the third situation, electric current was observed in the coil when the iron cylinder was being magnetized or demagnetized.

Faraday was able to combine these three conditions into one general statement, now known as the **law of electromagnetic induction.**

law of electromagnetic induction: An electric current is induced in a conductor whenever the magnetic field in the region of the conductor changes.

Law of Electromagnetic Induction

An electric current is induced in a conductor whenever the magnetic field in the region of the conductor changes.

Induction can be demonstrated with a device that Faraday constructed and used himself in his early studies of the induction effect. Known as Faraday's iron ring, it consists of a doughnut-shaped ring of soft iron with two separate coils of wire wound around it, as in **Figure 4.**

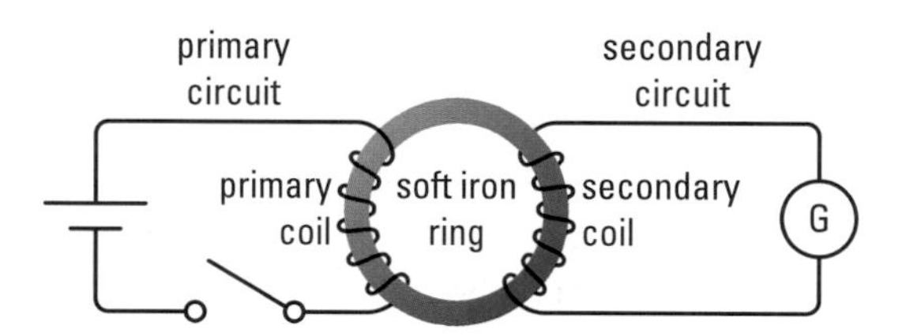

Figure 4
Faraday's iron ring

Faraday's Iron Ring

Under what conditions will a current be induced in the secondary coil of Faraday's iron ring apparatus?

- Use the Faraday's iron ring apparatus to investigate the induced current in the secondary coil.
- Turn on the primary circuit and observe the effect on the galvanometer or ammeter, wait for a few seconds, and then turn off the primary circuit, noting the effect on the galvanometer.
  (a) Under what conditions was a current observed in the secondary circuit?
  (b) Under what conditions was there no current in the secondary circuit?
  (c) How did the direction of the currents compare when the switch was being opened and closed?

The primary coil is connected through a switch to a voltage source. The secondary coil is connected directly to a galvanometer. When the switch is closed there is a current in the primary circuit, causing the entire iron ring to become magnetized. This sudden increase in magnetic field strength causes a current to be induced momentarily in the secondary coil. Once the current in the primary coil is steady and the magnetic field in the iron ring is established, the induced current no longer exists.

If the switch is then opened, the iron ring becomes demagnetized and the consequent decreasing magnetic field strength once again induces a momentary current in the secondary coil, this time in the opposite direction. This effect is known as **mutual induction**. In fact, the two coils do not have to be coupled with an iron ring; the iron ring merely acts to strengthen an effect that would be present in any case.

mutual induction: an effect that occurs whenever a changing current in one coil induces a current in a nearby coil

By observing what happens when a bar magnet is plunged into the core of a coil that is connected to a galvanometer, it is possible to conclude that the factors affecting the magnitude of the induced current are

- the number of turns on the induction coil
- the rate of change of the inducing magnetic field
- the strength of the inducing magnetic field

According to Ohm's law, the current through a circuit is proportional to the potential difference in the circuit when the resistance remains constant ($R = \frac{V}{I}$). The three factors just listed are really factors affecting the magnitude of the induced potential difference in the coil. The induced current that results from this potential difference is then given by Ohm's law as

$$I_{induced} = \frac{V_{induced}}{R}$$

Sample Problem

Figure 5(a) shows an experimental setup that can be used to show how a magnetic field produces an induced current. The three arrows, A, B, and C, show three mutually perpendicular ways in which the conductor can be moved in the magnetic field. Along which of these directions must the conductor be moved to induce a current?

Solution

If the conductor is moved vertically along direction A (**Figure 5(b)**), parallel to the magnetic field lines, no current will be produced since the magnetic field is not changing relative to the conductor. The same can be said for horizontal motion in the direction of B (**Figure 5(c)**)as long as the sides of the circuit do not enter the field lines. If the conductor is moved horizontally in the direction of C (**Figure 5(d)**), it must cut across magnetic field lines. Relative to the conductor, the magnetic field is changing, and current flows in the conductor. Notice in this case the circuit is moved instead of the magnet and an induced current is still produced.

(a)

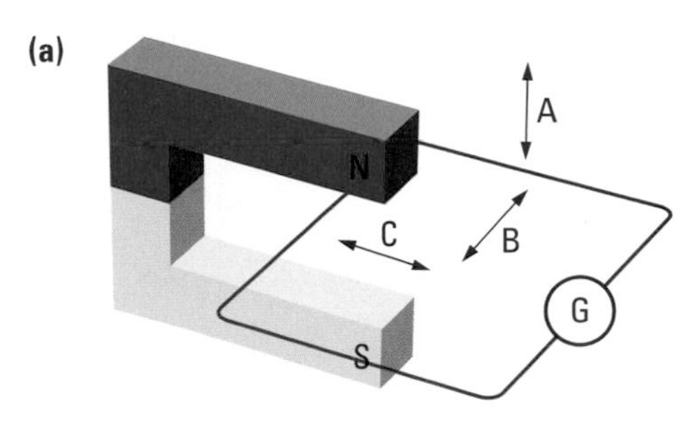

(b) moved parallel to A

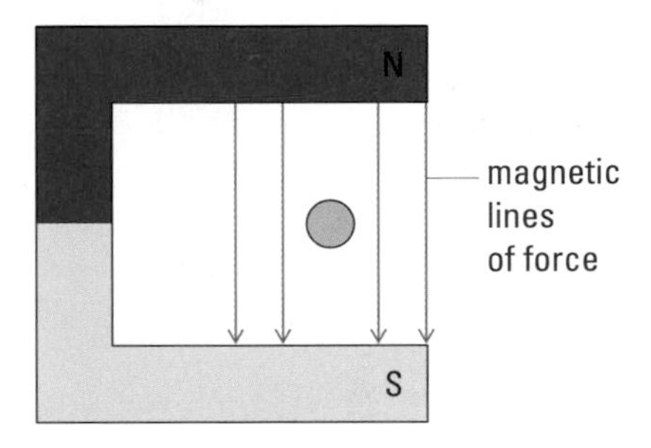

(c) moved parallel to B

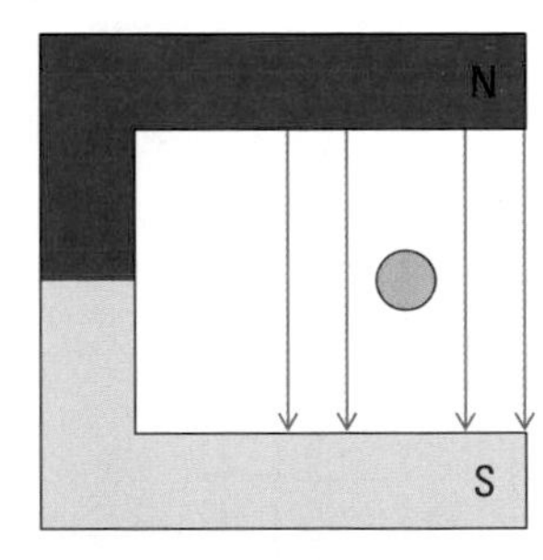

(d) moved parallel to C

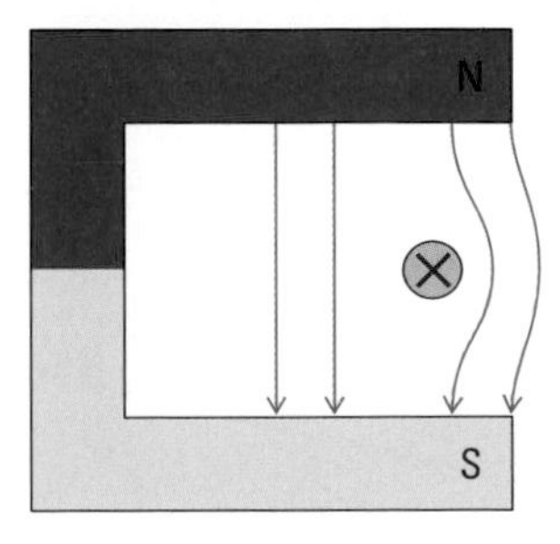

Figure 5

Practice

Understanding Concepts

1. An astronaut orbiting Earth in a space shuttle measures a small current in the gold ring she is wearing. There are no magnets in the vicinity of the astronaut's hands. What might be causing the current?

SUMMARY Faraday's Discovery

- The law of electromagnetic induction states that if the magnetic field inside a coil changes amount or direction, an electric current is induced in the coil.
- Mutual induction occurs whenever a changing current in one coil induces a current in a nearby coil.
- The magnitude of an induced current in a coil is affected by the number of turns on the induction coil, the rate of change of the inducing magnetic field, and the strength of the inducing magnetic field.

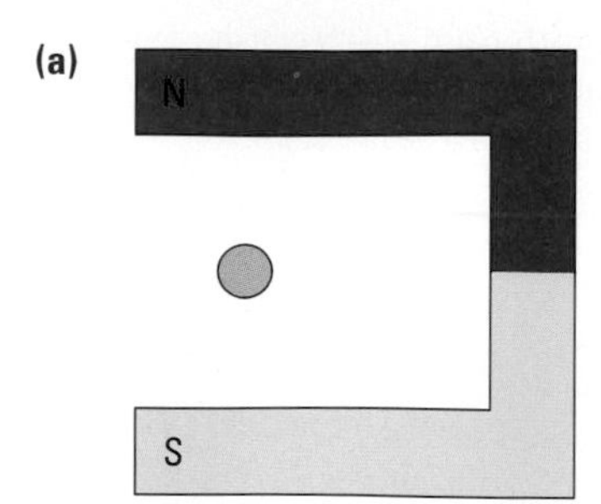

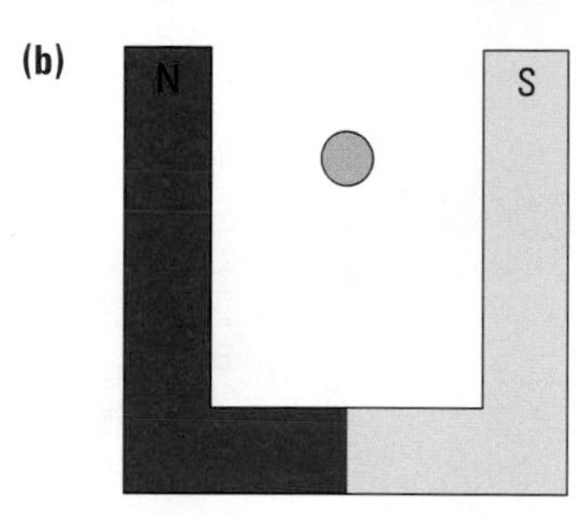

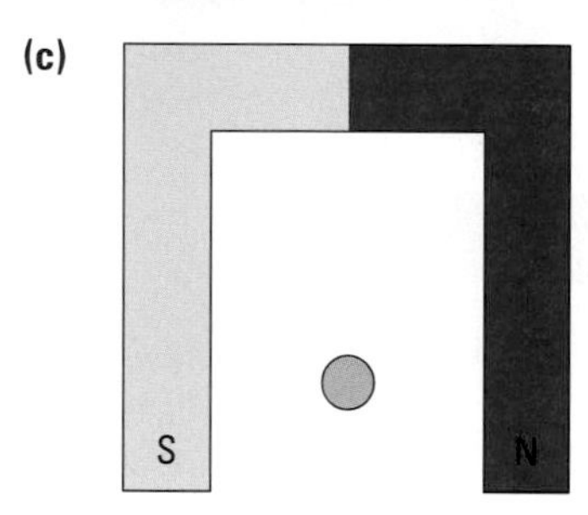

Figure 6
For question 2

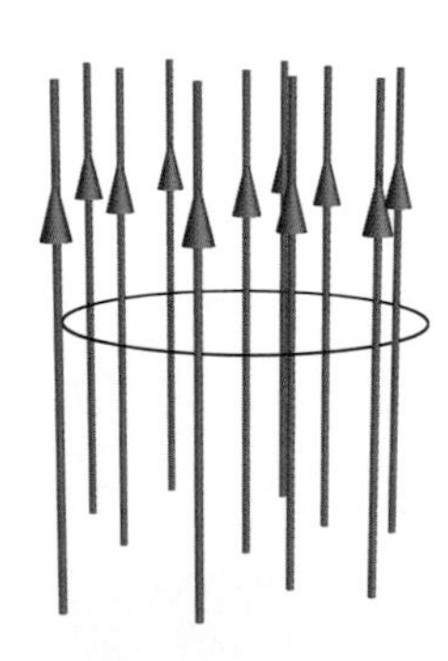

Figure 8
For question 5

Section 14.1 Questions

Understanding Concepts

1. What condition is necessary for a magnetic field to produce a current in a coil?
2. Each circle in **Figure 6** represents a conductor (which is part of a loop of wire not shown). In each case, state two ways of inducing a current in the conductor.
3. A bar magnet is used to induce a current in a coiled conductor. State the effect on the current in each of the following circumstances:
   (a) the number of coil windings is increased
   (b) the speed of the magnet relative to the coil is increased
   (c) the strength of the magnet is increased
   (d) the magnet and coil move at the same speed in the same direction
4. Two identical coils are placed side by side (**Figure 7**), and one is connected to a battery through a knife switch, while the other is connected to a galvanometer.

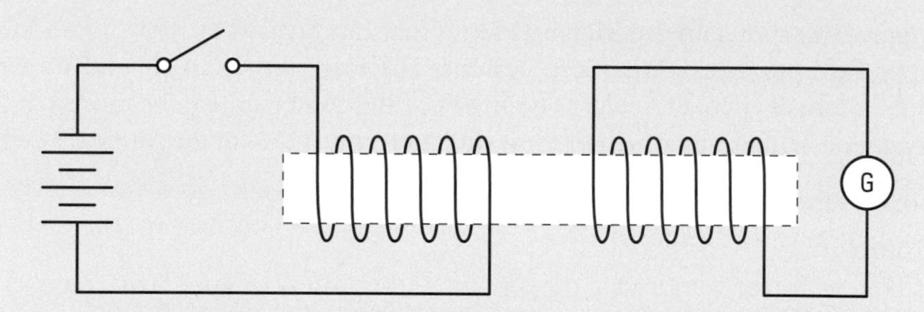

Figure 7

   (a) Describe the effect on the other coil of opening and closing the knife switch.
   (b) Would the same effect occur if one of the coils had been rotated by 90°, putting it at right angles to the other coil?
   (c) How would the insertion of a bar of iron into the two coils, as shown by the dotted line, change the effect of opening and closing the switch?
5. **Figure 8** shows a loop of wire inside a uniform magnetic field. Which of the following will produce an induced current in the loop? Explain.
   (a) the wire is moved up or down
   (b) the wire is moved horizontally out of the magnetic field
   (c) the loop is rotated in the magnetic field
   (d) the loop is compressed, decreasing its area
6. A closed circuit is placed near a nonuniform magnetic field of a bar magnet. Why would it be difficult to move the circuit without producing an induced current?

14.2 Direction of Induced Current

Faraday's discovery of induced currents opened the door to readily accessible and cheap sources of current. This current could be used to power electric motors and other devices, saving time and money for many people and industries. However, there were still some questions to be answered. What determines the direction of the induced current? Where does the energy associated with the induced current come from?

Activity 14.2.1

The Direction of Induced Current

Previously you explored ways of inducing an electric current in a coiled conductor, as well as factors affecting the size of the current. You may have wondered why the current was first in one direction and then in the other. In this activity you will discover the answer.

Question

What is the direction of the induced current in a coiled conductor?

Materials

coiled conductor
galvanometer or ammeter
magnet
connecting wires

Procedure

1. Examine the coil carefully to determine the direction in which the wire is wound. Connect the coil to the galvanometer or ammeter.
2. Plunge the N-pole of the magnet downward into the coil, and determine
   - the direction of the current through the galvanometer or ammeter
   - when viewed from above, the direction (clockwise or counterclockwise) of the electric current in the coil
   - whether the top of the coil becomes N or S when the N-pole of the magnet is pushed toward it (Apply the right-hand rule for a coiled conductor.)
   - whether the magnetic field of the coil helps or hinders the motion of the magnet

   Discuss your observations with your teacher before you proceed.
3. Determine each of the following when the N-pole of the magnet is being withdrawn from the coil:
   - the direction in which the galvanometer needle swings
   - the direction of the electric current in the coil
   - the polarity of the top of the coil
   - whether the motion of the magnet is helped or hindered
4. Repeat step 3, forcing the S-pole of the magnet first into and then out of the coil.

Analysis

(a) When a magnetic field is in motion near a conductor, an electric current is induced in the conductor. The current moves in a specific direction, causing the conductor to set up or induce its own magnetic field. How does the induced magnetic field interact with the magnetic field that caused it? (*Hint:* Does the induced field assist or oppose the action of the other field?)

(b) Relate your observations to the expression "Nature does not give something for nothing."

induced field: the magnetic field produced by the induced current

inducing field: the magnetic field that causes the induced current

Lenz's Law of Induced Current

Let's examine the case where the N-pole of a bar magnet is pushed into a coil. When the N-pole of a bar magnet enters a coil, a galvanometer will indicate an induced current through the coil. When the N-pole is removed, the galvanometer will indicate a current in the opposite direction. Using the S-pole of the bar magnet causes induced currents in directions opposite to the above. Evidently there is some simple relationship between the action of the inducing field and the direction of the induced current.

A few years after Faraday's discovery of induction, a German physicist working in Russia, Heinrich Lenz (1804–1865), applied the law of conservation of energy and succeeded in stating this relationship.

Lenz (pronounced "Lents") reasoned that when a current is induced through a conductor, the induced current itself sets up a magnetic field. This magnetic field, which we will call the **induced field**, then interacts with the **inducing field**, either attracting it or repelling it. In determining which of these interactions is the more likely one, it will help to consider an example.

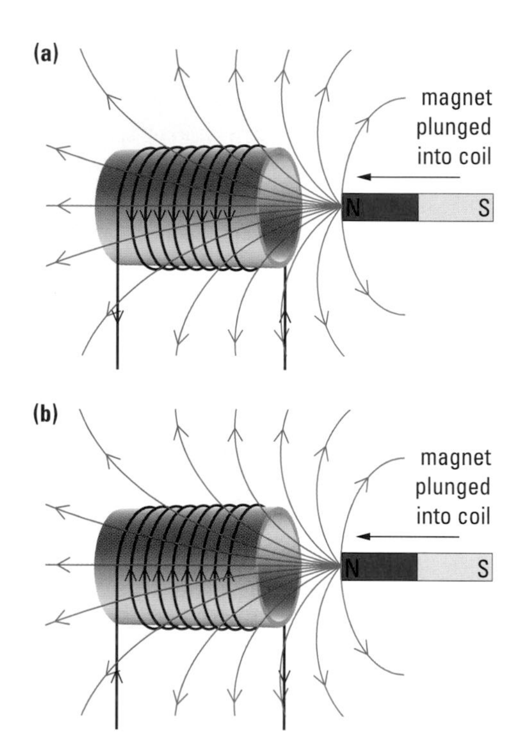

Figure 1
There are only two possible directions for the induced current in the coil as the N-pole is pushed forward.

If the N-pole of a bar magnet is inserted into the core of a coil, there will be an induced electric current through the coil. This current is either down (**Figure 1(a)**) or up (**Figure 1(b)**) across the front of the coil. It can take no other path.

Lenz began by assuming that the induced electric current is up, as in **Figure 2**. This current would produce an induced magnetic field in the coil, and, according to the right-hand rule, the right end of the coil would become an S-pole. The S-pole would then attract the N-pole of the bar magnet, pulling it into the coil. It would no longer be necessary to move the N-pole of the inducing magnet into the coil; it would be attracted into the coil by the S-pole of the induced field.

Lenz reasoned that this situation was impossible. The induction coil would have to produce electrical energy (in the form of an induced current and potential difference) by itself, without using any other form of energy. It would be a perpetual motion machine since the attraction would cause the magnet to accelerate, increasing its kinetic energy. The induced current would increase because the magnet is moving faster, which in turn would produce a stronger induced magnetic field, causing the magnet to speed up and so on and so on. The law of conservation of energy does not permit such a process to occur since in this case the total amount of energy of the system would have to increase. Energy can only be converted from one form to another; it cannot be created or destroyed.

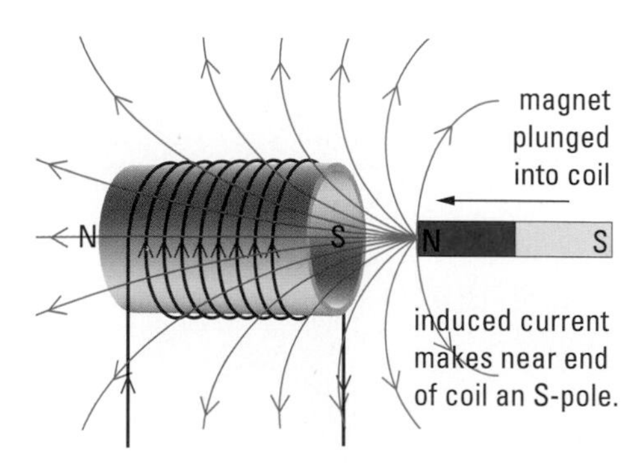

Figure 2
Violating the law of conservation of energy

But where was the flaw? It must be in the original assumption about the direction of the induced current. The electric current in the example being considered must be moving down across the front of the coil rather than up, as assumed.

Then the induced field has an N-pole at the right end, and this N-pole opposes the inward motion of the inducing magnet (**Figure 3**). That is, the inducing magnet must be pushed into the coil against the force of the induced field. The work done in moving the bar magnet into the coil against this opposing force is transformed into electrical energy in the induction coil.

In 1834, Lenz formulated this reasoning into a law for determining the direction of the induced current, now known as **Lenz's law**.

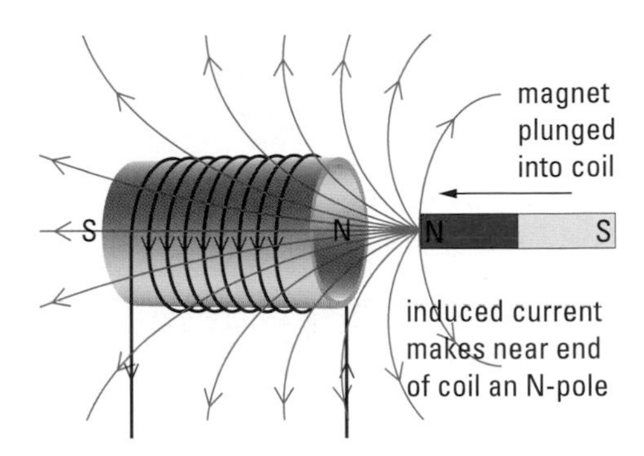

Figure 3
Obeying the law of conservation of energy

Lenz's Law

For a current induced in a coil by a changing magnetic field, the electric current is in such a direction that its own magnetic field opposes the change that produced it.

It should be noted that Lenz's law is really just a rule to determine the direction of the induced current and contains nothing that wasn't already implied by Faraday. Also, the changing magnetic field really produces an electric field, and the induced current is just a way that it can be detected.

Lenz's law: For a current induced in a coil by a changing magnetic field, the electric current is in such a direction that its own magnetic field opposes the change that produced it.

Sample Problem 1

Determine the direction of the electric current for the case in **Figure 4**.

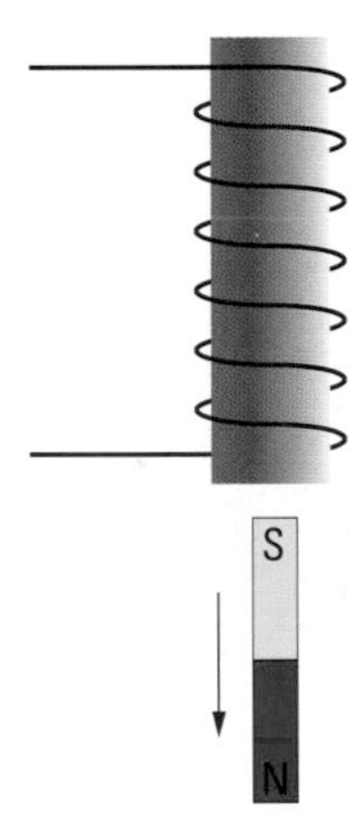

Figure 4
For Sample Problem 1

Solution

The solution of this problem requires two steps in the reasoning process.

(a) The lower end of the coil must become the N-pole of the induced field in order for it to oppose the removal of the S-pole of the inducing magnet (Lenz's law), as shown in **Figure 5**.

(b) Applying the right-hand rule, we see that the electric current must be flowing to the left, across the front of the coil.

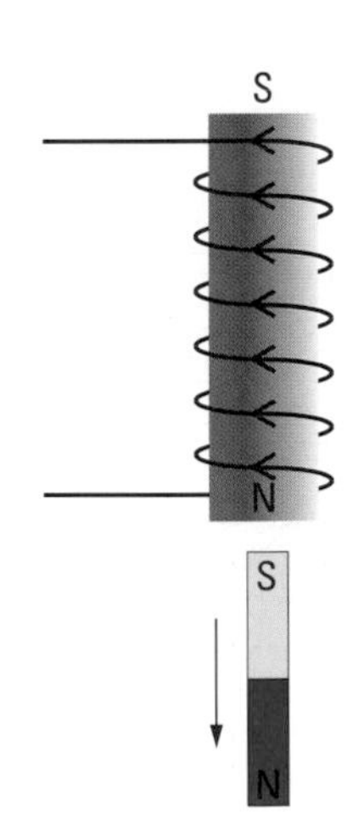

Figure 5

Sample Problem 2

Determine the pole of the bar magnet that is being inserted into the induction coil in **Figure 6**.

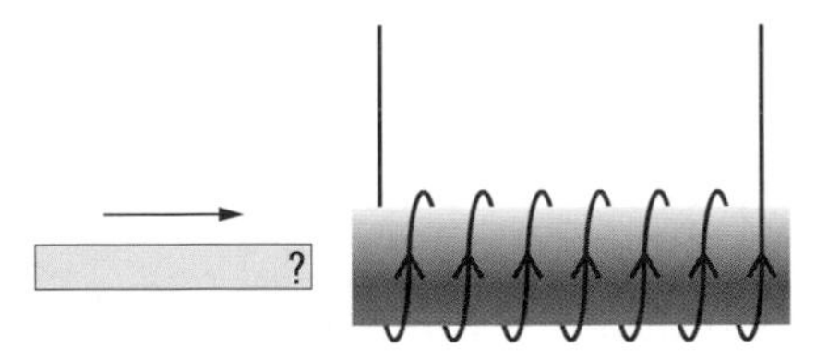

Figure 6

Solution

(a) Applying the right-hand rule, we see that the left end of the induction coil becomes an N-pole, as in **Figure 7**.

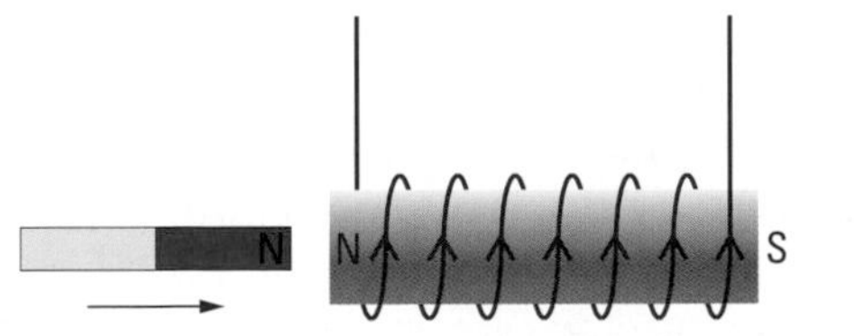

Figure 7

(b) According to Lenz's law, the pole of the bar magnet that would be opposed by this N-pole as it enters the coil is an N-pole itself.

Practice

Understanding Concepts

1. A bar magnet is held vertically above a vertical coiled conductor. For each situation, determine which end of the coil becomes the N-pole, and state whether the direction of the electric current in the coil is clockwise or counterclockwise when viewed from above the coil.
   (a) The N-pole of the magnet is pulled out of the coil.
   (b) The S-pole of the magnet is pushed toward the coil.
   (c) The S-pole of the magnet is pulled out of the coil.

How a Television Tube (Cathode Ray Tube) Works

The force that a magnetic field exerts on a current-carrying wire is a result of the interaction between the magnetic field and the individual charges flowing in the wire. The charged particles do not have to be moving in a wire to experience the force. In fact, electrons moving in a vacuum may be deflected by a magnetic field. This is what happens in a picture tube, or cathode ray tube (CRT).

In a typical cathode ray tube (**Figure 8**), a tiny filament heats a cathode, which is the source of electrons that accelerate toward the positively charged anode, which is cylindrically shaped to allow the electrons to pass through. Between the cathode and the anode is a grid of very fine wire called the control electrode. By making the control electrode more or less negative, the number of electrons passing through it can be increased or decreased. The electrons next pass through one or more focusing electrodes that concentrate the electrons in a fine beam. This collection of electrodes is called the *electron gun*.

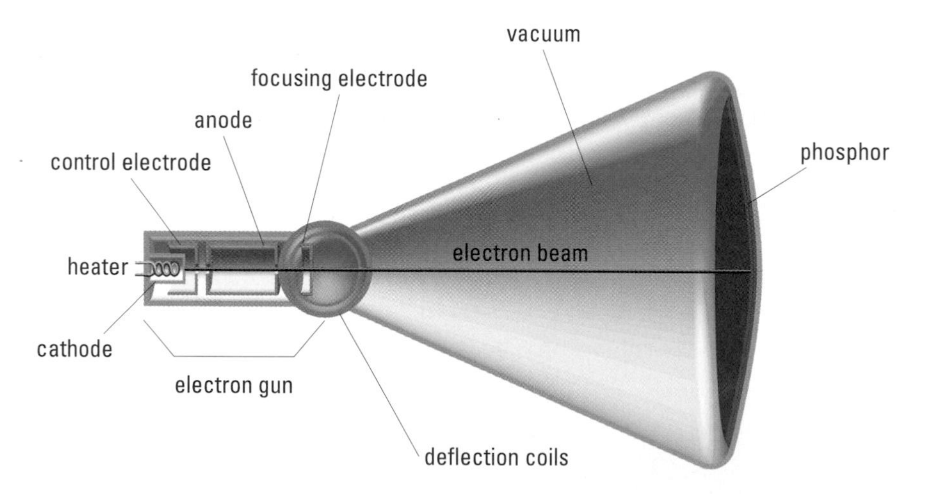

Figure 8
A black and white television picture tube

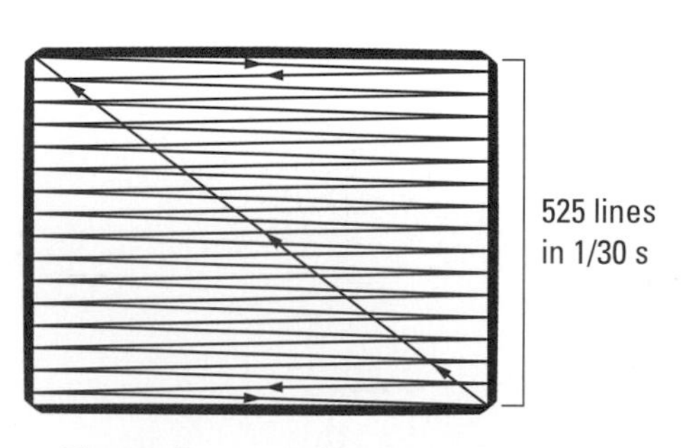

Figure 9
The electron beam paints one horizontal line on the screen at a time from left to right, then moves down slightly when moving from right to left before painting the next horizontal line.

The electron beam is deflected by pairs of magnetic coils or by charged plates. By varying the current in these coils or the charge on the plates, the electron beam can be directed to any point on the screen. The back of the screen is coated with a thin coat of phosphors that give off light when struck by electrons. The colour of light given off depends on the type of phosphor used.

An electronic circuit in the set generates currents at the proper frequency to "scan" the entire screen 30 times per second (**Figure 9**). While one set of magnetic deflection coils moves the beam back and forth across the screen 525 times, the other set of coils moves the beam down the screen, tracing out the pattern illustrated in **Figure 9**. Actually the beam moves down the screen twice to produce one complete picture, the first time scanning every other line, and the second time filling in the missing lines.

As the dot of light created by the electron beam moves across the screen, the brightness is controlled by the control electrode. The potential difference on the control electrode is determined by the signal received from the television station. Every $\frac{1}{30}$ s, a complete new picture is scanned out on the screen.

In a colour television, there are three electron guns, one each for red, blue, and green (**Figure 10**). Each gun produces an electron beam that scans the whole screen. The screen is covered with about 600 000 phosphor bars, 200 000 for each colour. Just behind the screen is a shadow mask, an opaque screen with 200 000 tiny slots in it that are very accurately aligned with the three electron beams and the phosphor bars. The green beam hits only the green bars, and so on.

The television station must provide information to produce a red picture, a green picture, and a blue picture every $\frac{1}{30}$ s and another, separate signal for the sound. Your eye combines the three images to produce a full-colour picture.

Both computer monitors and digital televisions work in a similar way but have higher resolutions and more sophisticated electronic hardware.

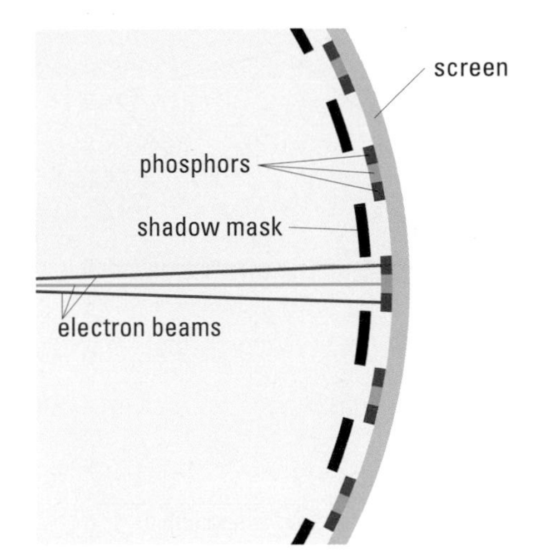

Figure 10

Practice

Understanding Concepts

2. If the control grid in a cathode ray tube becomes more positive, will the number of electrons in the electron beam increase, decrease, or remain the same? Explain.

Making Connections

3. How do the colours related to the different electron guns in a colour TV compare with the colours to which the cones in the human eye are sensitive? (The human eye was discussed in Chapter 10.)

SUMMARY Direction of Induced Current

- Lenz's law states that for a current induced in a coil by a changing magnetic field, the electric current is in such a direction that its own magnetic field opposes the change that produced it.

Section 14.2 Questions

Understanding Concepts

1. Copy the illustrations from **Figure 11** into your notebook, and for each show
   (a) the polarity of the induced magnetic field
   (b) the direction of the induced electric current
2. Conductors are being pushed to the left in each of the illustrations in **Figure 12**. For each illustration, use the following instructions and questions to determine the direction of the induced current in the conductor:
   (a) Copy the illustrations into your notebook and draw the magnetic lines of the magnet.

(continued)

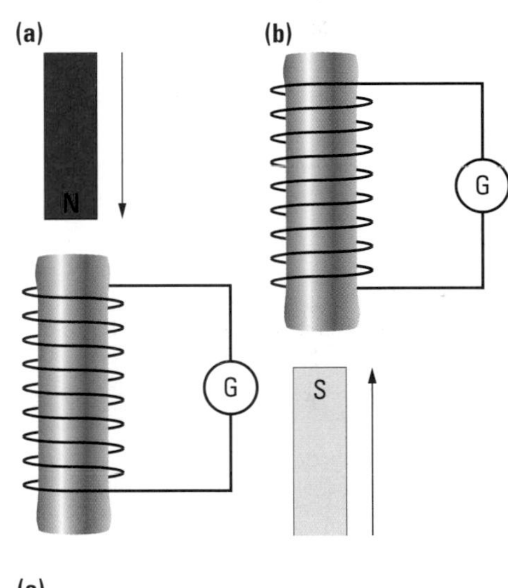

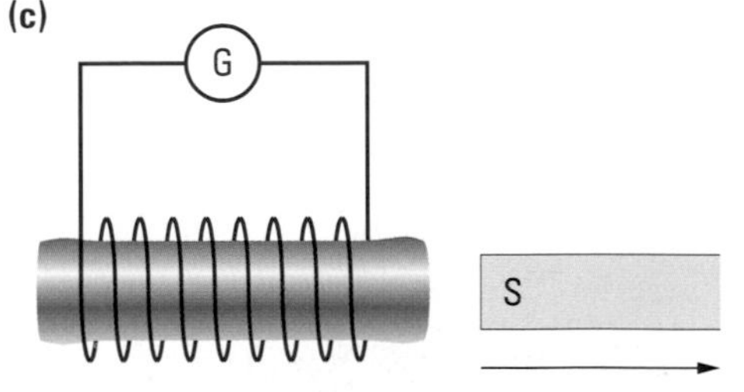

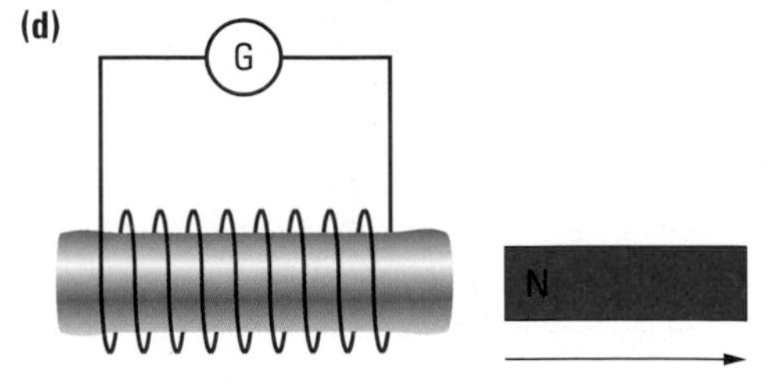

Figure 11
For question 1

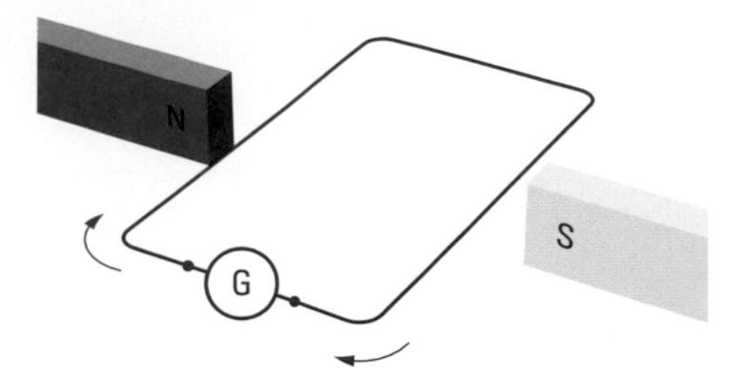

Figure 13
For question 3

electric generator: a device that converts the mechanical energy of motion into electrical energy

(b) According to Lenz's law, is the induced magnetic field (resulting from the induced current) clockwise or counterclockwise around the conductor?
(c) Apply the right-hand rule for straight conductors to determine the direction of the induced electric current.

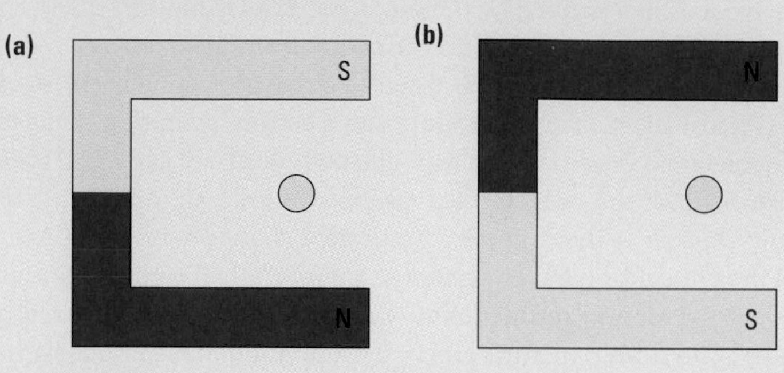

Figure 12

3. You are given the direction of the induced electric current in the loop of the circuit shown in **Figure 13**. Use Lenz's law to decide on the direction of the inducing motion. Explain your reasoning.
4. Comment on the statement "Lenz's law is a direct consequence of the law of conservation of energy."

14.3 Electric Generators: AC and DC

The **electric generator** provides the electricity that runs many devices we commonly use, such as computers, televisions, and even the electrical systems in automobiles. Without electric generators, our supply of current would have to be obtained from more expensive sources like voltaic cells, and many of the things that we have today would not be economically feasible.

The electric generator is a device that converts mechanical energy into electrical energy by rotational motion from a continuous source of energy such as wind, falling water, tides, or expanding steam.

The construction of a generator is identical to that of an electric motor, although its function is just the opposite. The main components of a generator are field magnets, an armature, brushes, and either slip rings (for an AC generator) or a split-ring commutator (for a DC generator). These components are shown for demonstration AC and DC generators in **Figure 1**.

To explain the operation of AC and DC generators, a single-loop coil will be used. Therefore, it is necessary to determine the direction of the induced current in a single conductor. **Figure 2(a)** shows an upward force being exerted on a conductor in a magnetic field. The conductor is cutting across the field lines, so the magnetic field is changing. Thus, according to Faraday's law, current is induced in the conductor. The direction of the current is such that the induced magnetic field repels or opposes the magnetic field that produced it. Therefore, the induced magnetic field must be clockwise, and the current must be flowing into the page (this is indicated by an x in the centre of the conductor). In **Figure 2(b)** the conductor is moving downward, thus, the current is moving out of the page (this is indicated by a dot in the centre of the conductor). **Figures 2(a)** and **2(b)** are combined in **Figure 2(c)**. **Figure 2(d)** shows the entire loop, with directions indicated.

(a)

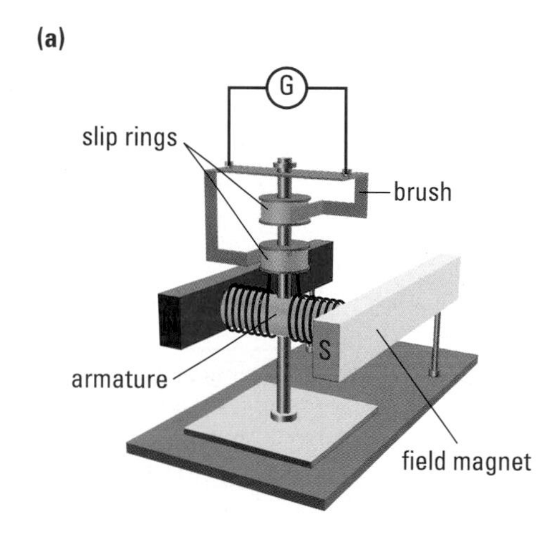

(b)

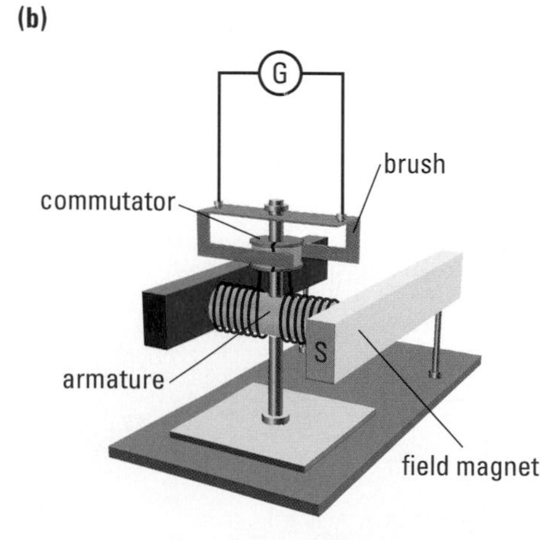

Figure 1
Demonstration generators
(a) AC generator
(b) DC generator

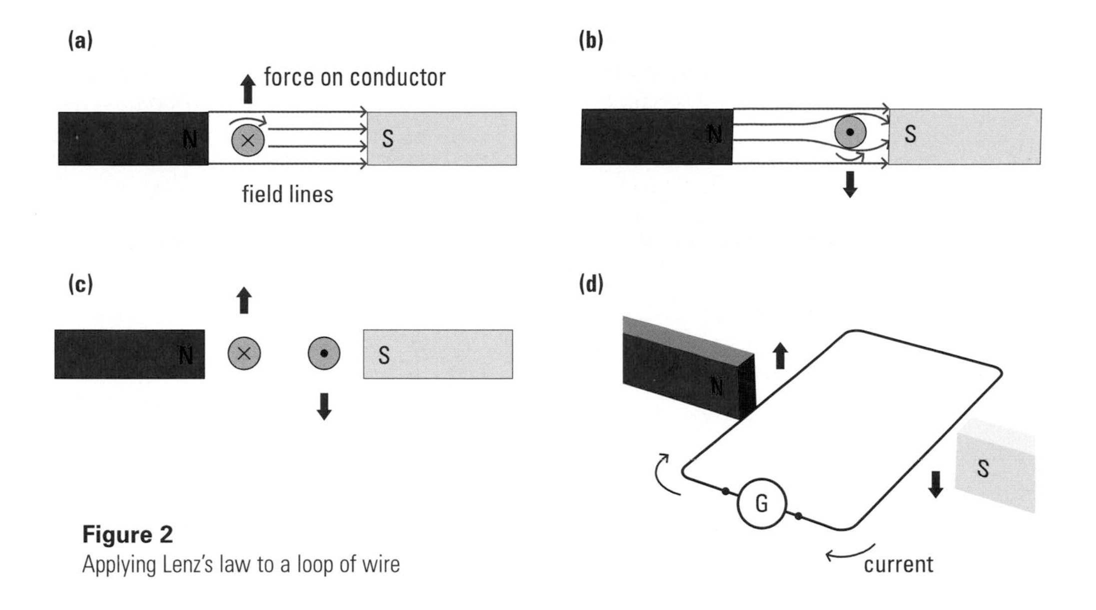

Figure 2
Applying Lenz's law to a loop of wire

The AC Electric Generator

To study the operation of an AC generator, refer to **Figure 3**. **Figure 3(a)** shows a simple AC generator with a coil of only one loop connected to the slip rings (R_1 and R_2), which rotate with the loop. The brushes (B_1 and B_2) touching the rings are stationary. The loop is being forced to spin clockwise by an external source of energy. In the position shown, the conductor is cutting across the field lines at a maximum rate, so the induced current is at a maximum. The direction of the current is found by applying Lenz's law; it is labelled in the figure.

As the loop rotates clockwise, the number of field lines crossed by the conductor drops gradually until it becomes zero when the loop is vertical. At this point, no current flows, as shown in **Figure 3(b)**. When the loop rotates further, the current begins to flow again, this time in the opposite direction. The current reaches its second maximum value in the horizontal position, **Figure 3(c)**. The current becomes zero in **Figure 3(d)**, and the process begins again.

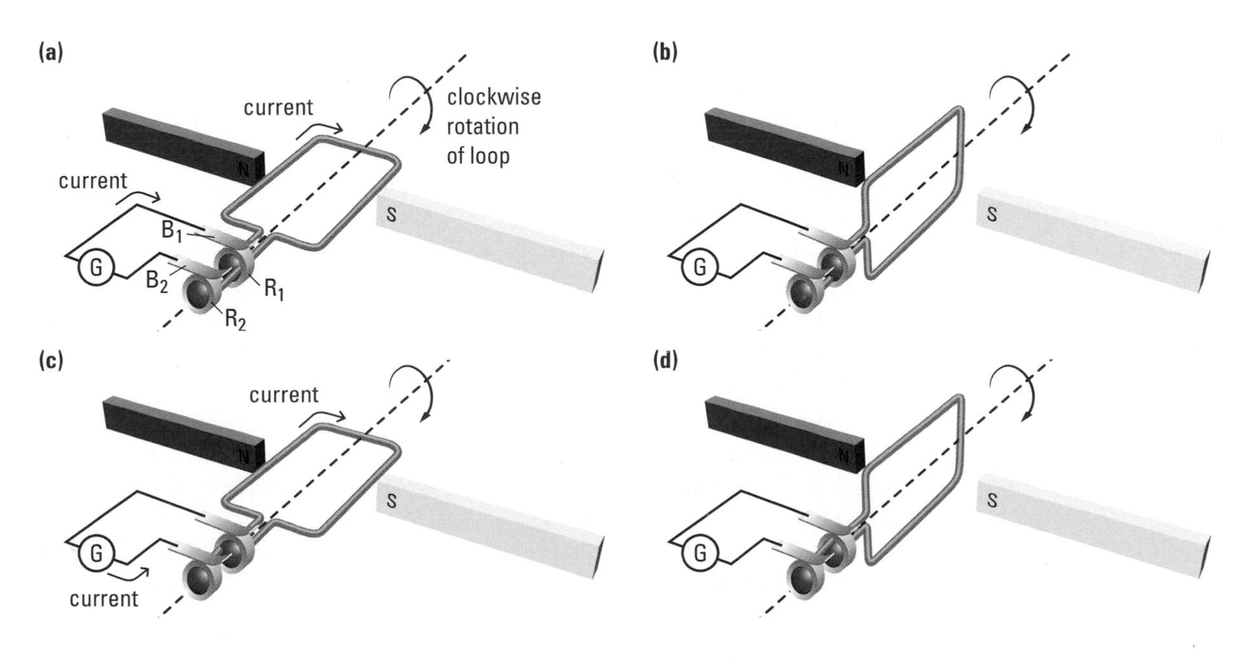

Figure 3
A single-loop AC generator
(a) The current is maximum and is leaving the loop at B_2.
(b) The current has dropped to zero.
(c) The current is maximum and is leaving the loop at B_1. Therefore, the current here is opposite in direction to the current in (a).
(d) Again the current has dropped to zero.

Because the current changes gradually, the output to the galvanometer is not constant. **Figure 4** shows the output current that corresponds to the AC generator described. The points labelled A, B, C, and D in **Figure 4** correspond to diagrams (a), (b), (c), and (d) in **Figure 3**.

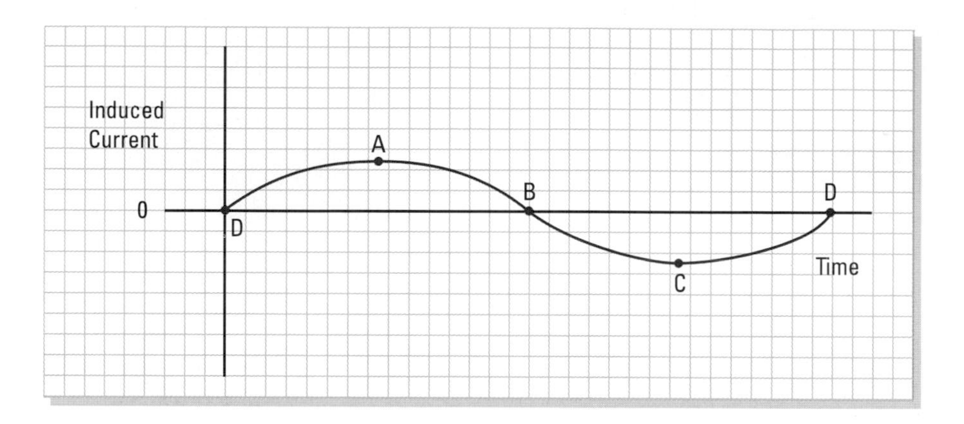

Figure 4
Output current of a simple AC generator

If the single-loop generator rotates at a certain frequency, the electrons in the conductor move back and forth at the same frequency. This produces an alternating current (AC). Commercial generators have several sets of coils, each with a great number of windings, and they often use electromagnets rather than permanent magnets for the field. Such generators produce a current much larger than the current produced by a demonstration generator. In North America, the frequency of AC generation is 60 Hz.

AC Generator Demonstration

In this activity you will compare the difficulty of turning the generator when it is not connected to a circuit to when the circuit is closed.

- Use a generator that can be cranked by hand. You can do this by connecting the terminals on the generator with wire.
- Insert a small light bulb if available.
  (a) In which case is there a greater resistance to motion?
  (b) Use what you know about electromagnetism to explain your observations.

The DC Electric Generator

If the slip rings of an AC generator are replaced by a split-ring commutator, the generator is able to produce direct current (DC). **Figure 5** shows the operation of a simple DC generator in which the external force on the loop is clockwise. In **Figure 5(a)**, the current is leaving the loop through R_2 and B_2. In **Figure 5(b)**, the current stops flowing at the same instant as the splits in the commutator reach the brushes, and the current becomes zero. In **Figure 5(c)**, the current is leaving the loop through R_1 and B_2, so it travels in the same direction in the external circuit. In **Figure 5(d)**, the current once again drops to zero.

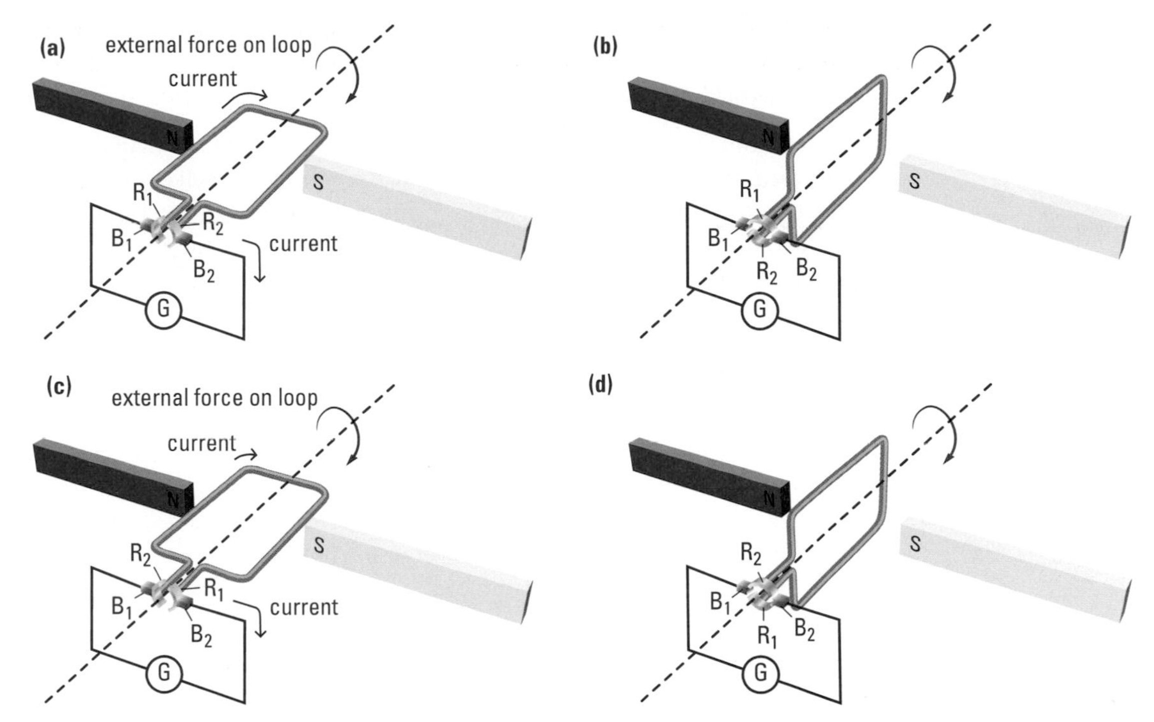

Figure 5
A single-loop DC generator
(a) The current is maximum and is leaving the loop through R_2 and B_2.
(b) The current is zero. R_1 is about to touch B_2, and R_2 is about to touch B_1.
(c) The current is maximum again and is leaving the loop through R_1 and B_2. Therefore, the current is in the same direction as the current in (a).
(d) Again the current is zero, and the cycle is about to start over.

The current produced by this simple DC generator is not the same as the constant DC current from a chemical source such as a battery. The electric generator produces a pulsating current in one direction, as illustrated in the graph in **Figure 6**. The points labelled A, B, C, and D on the graph correspond to (a), (b), (c), and (d) in **Figure 5**.

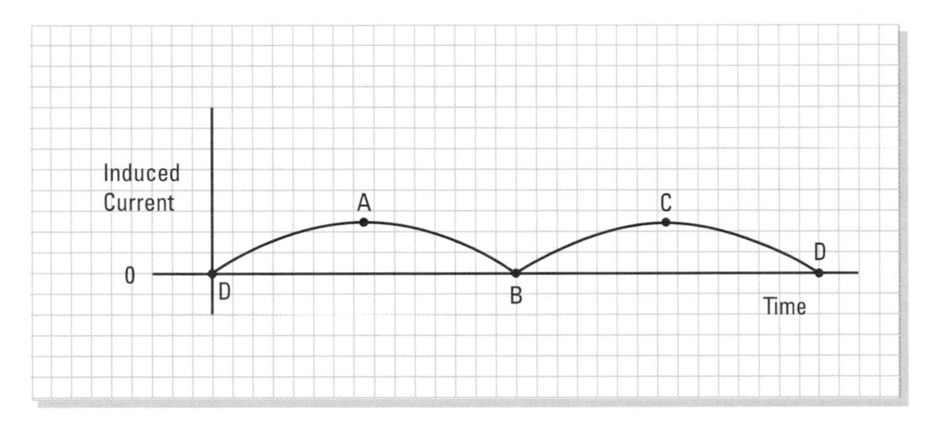

Figure 6
Output current of a simple DC generator

Commercial DC generators have coils with numerous windings and iron cores, and they also have more than two commutator segments. In this way, the output current is kept very nearly constant, and its value is much greater than in a demonstration model.

Maximizing Output from AC and DC Generators

For an electric generator to be economically viable, it must be able to produce as much electrical energy as possible in the most efficient way. Several factors can be used to help accomplish this:

- increase the number of coils and the number of windings in each coil
- place a soft iron core, called an armature, in the centre of each coil to increase the strength of the inducing field

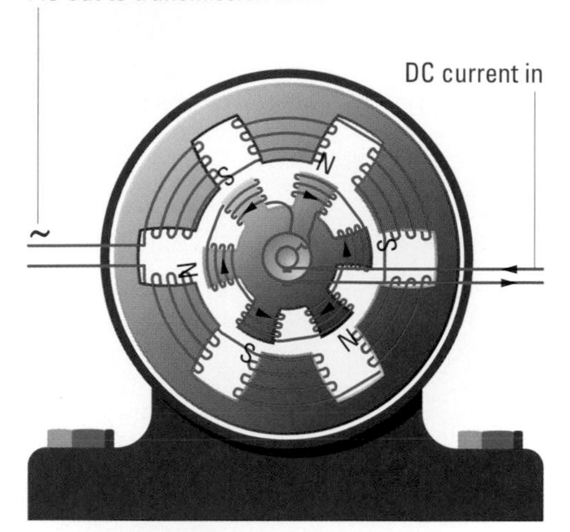

Figure 7
Cross-section of a typical AC generator

- Keep the relative motion between the armature and field magnet as high as possible.
- Make the field magnet as strong as possible.

An electromagnet is used to increase the strength of the field magnet. The current to run the field magnet is produced by the generator itself.

In keeping the relative motion between the coil and field magnet as high as possible, a lot of energy is lost to sparking and deterioration of the slip rings and brushes. This is resolved by rotating the field magnet on a shaft, which helps to keep the currents lower and reduce sparking and deterioration (**Figure 7**).

Electric Generating Stations

Huge AC generating stations supply much of the demand for electrical energy in our society. At the same time, the generators at these stations require vast amounts of energy to sustain their operation. One source of energy is the kinetic energy of falling water (**Figure 8**). Another source is thermal energy, which comes from chemical potential energy. The thermal energy for boiling water in this type of generating system can be obtained from coal, natural gas, oil, or nuclear reactions.

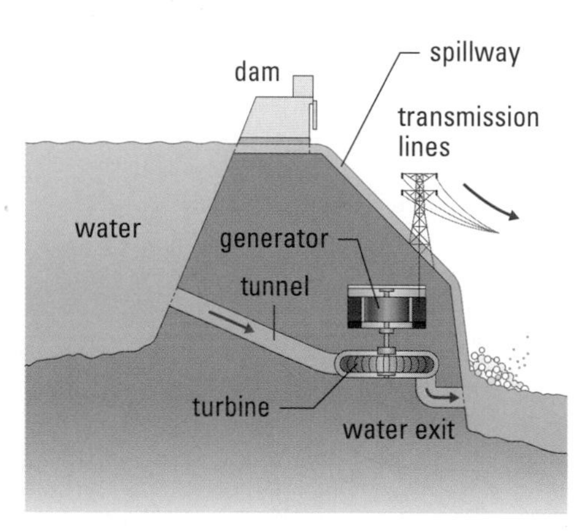

Figure 8
Hydroelectric generating plant. At the top of the dam, the water has gravitational potential energy. The falling water gains kinetic energy that causes huge turbines to spin. The generator changes the mechanical energy of spinning into electrical energy. (See Chapter 5 for more information about generating electricity.)

SUMMARY Electric Generators: AC and DC

- An electric generator is a device that converts mechanical energy into electrical energy using the rotational motion from a continuous source of energy.
- An AC generator provides current from an outside circuit connected by means of slip rings and brushes.
- A DC generator provides current from an outside circuit connected by means of a commutator.

Section 14.3 Questions

Understanding Concepts

1. (a) What is the major structural difference between an AC generator and a DC generator?
   (b) How does the current flowing through the armature coil of a DC generator differ from the current flowing through the armature coil of an AC generator?
   (c) In comparison with the number of separate armature coils in a DC generator, how many segments should the generator's commutator have?
2. (a) Sketch a simple AC generator, labelling four key parts, and describe the function of each.
   (b) Show how Lenz's law and the motor principle may be used together to determine the direction of the induced current at any point in the rotation.
   (c) What features in the design of a practical form of AC generator have an effect on the potential difference the generator will produce?
   (d) Draw a graph of induced potential difference versus time for a model AC generator, rotating three times per second and producing a peak potential difference of 2.5 V. Label the time axis at each point where the line crosses it.

3. (a) Explain the differences between a DC motor and an AC generator.
   (b) Explain the difference between a DC motor and a DC generator.
4. (a) What is the difference between AC and DC current?
   (b) Describe the difference between the current from a voltaic cell and the current from an AC generator.

Making Connections

5. (a) The original generators that began producing electrical energy at Niagara Falls over 100 years ago generated AC with a frequency of 25 Hz. This frequency was fine for operating electric motors, but it was not good for electric lights. Explain why, taking into consideration the properties of the human eye. (Information in Chapter 10 may help you answer this question.)
   (b) In many countries in the world, electricity is generated at a frequency of 50 Hz. Would the problem with electric lights be noticeable at that frequency? Explain your answer.

14.4 The Transformer

The typical electrical outlet in a North American home is 120 V AC. Some outlets, used for large appliances, are 240 V AC. If manufacturers produce devices that are required to run at some other voltage, it seems that they would have a problem. However, we have already studied a device that can solve this problem for us.

Our study of Faraday's iron ring has shown us that when the primary circuit switch is suddenly closed, a current will be momentarily produced in the secondary circuit. The same thing will occur when the switch in the primary circuit is suddenly opened. The changing current in the primary circuit will produce a varying magnetic field in the ring, which will pass through the secondary circuit and produce a current in the secondary circuit according to Lenz's law. When the switch is closed and the current is constant, no induced current will be produced in the secondary circuit since the inducing magnetic field is constant.

It is not necessary to turn the switch on and off to produce a current in the secondary circuit. All we need is a varying magnetic field in the ring, which means we need a varying or alternating current. The current produced by an AC generator is perfect for this task. An alternating current periodically reverses direction, providing a means for the current to be turned on and off without manually operating a switch.

transformer: device that consists of a core of soft iron with two separate coils of wire used to change the voltage

Investigation 14.4.1

Transformers (Teacher Demonstration)

INQUIRY SKILLS

- ○ Questioning
- ○ Hypothesizing
- ● Predicting
- ○ Planning
- ● Conducting
- ● Recording
- ● Analyzing
- ● Evaluating
- ● Communicating

Another name for Faraday's iron ring is a **transformer** because it has been found that the electric potential difference induced in the secondary circuit can be changed (or transformed) by changing the number of windings around the ring in either of the circuits.

Question

How are the input and output voltages in a transformer related to the number of windings on the primary and secondary coils?

Materials

lab transformer set (iron core and several coils with different number of turns)
variac (AC variable voltage power source)
two AC multi-range voltmeters
6 V or 12 V battery
DC voltmeter
nichrome wire or long finishing nail
water
hollow dish-shaped ring of metal with a wooden handle

Prediction

(a) Predict what will happen to the output voltage in the secondary coil when the primary coil has
  (i) more turns than the secondary coil
  (ii) fewer turns than the secondary coil

Procedure

Laboratory transformers should not be handled by students. Treat transformers with due respect. Always disconnect the power sources before changing connections.

1. Set up the transformer as shown in **Figure 1**, using the coil with the fewest turns as the primary and the coil with the next fewest turns as the secondary.

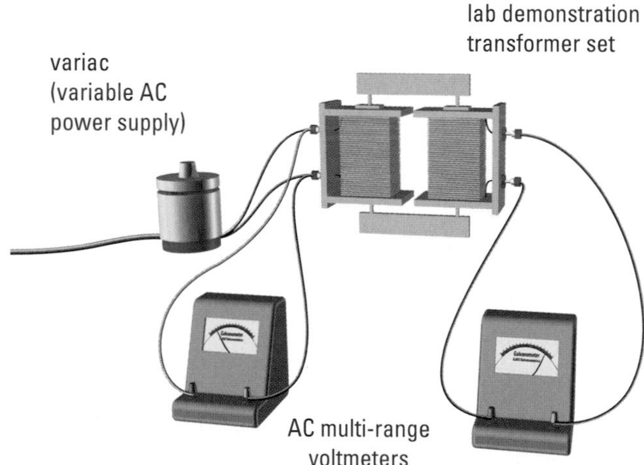

Figure 1
Setup for Investigation 14.4.1

Turn the variac off each time before changing coils.

2. Turn on the variac, and increase the voltage applied to the primary coil from zero to a maximum in several steps. Note both the primary voltage and the secondary voltage at each step.
3. With the primary voltage at a moderate value, substitute each of the remaining coils for the secondary coil, noting the secondary voltage each time.
4. Repeat the entire procedure, starting with the coil with the most turns as the primary.
5. Summarize your observations in tabular form using these headings:

N_1	V_1	N_2	V_2	$\frac{N_1}{N_2}$	$\frac{V_1}{V_2}$

6. Connect the ends of the primary coil to the terminals of the battery, and connect the DC voltmeter across the secondary coil. Note the reading on the voltmeter. Bring a nail or a small compass near the iron core of the transformer and determine whether the core is magnetized.
7. **YOUR TEACHER WILL PERFORM THE FOLLOWING STEPS:**
   - Using the coil with the most turns as the primary, and the coil with the fewest turns as the secondary, connect a piece of nichrome wire or a nail across the terminals of the secondary coil.
   - Connect the variac to the primary coil and increase it gradually to the maximum, observing the wire or nail connected to the secondary coil.
   - Remove the secondary coil and in its place use a hollow dish-shaped ring of metal with a wooden handle. Fill the ring with water, adjust the variac to maximum level, and, holding the ring by its wooden handle, observe the water.

Analysis

(b) What conditions led to a secondary voltage that was greater than the primary voltage? less than the primary voltage?
(c) What relationship exists between the ratio of windings and the ratio of voltages of the two coils? Express this relationship mathematically.
(d) What happened to the wire or the nail during the demonstration? Why? What happened to the water?
(e) During the investigation you may have noticed some humming and vibration. If you heard such sounds, what do you think caused them?
(f) What was the reading on the DC voltmeter connected across the secondary coil when the battery was connected to the primary? Was the iron core magnetized in that case? Explain why no potential difference is induced in the secondary coil when the primary coil is connected to a DC source.

Evaluation

(g) Evaluate your predictions.
(h) Describe the sources of error in the investigation and evaluate their effect on the results. Describe one or two improvements to the experimental design.

Step-Up and Step-Down Transformers

How does Faraday's iron ring help the manufacturer who needs a different voltage for a device? If the windings in the primary and secondary circuits are not the same, the voltage induced in the secondary circuit will be different from the voltage of the primary circuit. If the electric potential difference in the primary circuit is 120 V, the primary circuit had 60 windings around the ring, and the secondary circuit has 20 windings around the ring, the secondary circuit will have a potential of 40 V. The number of windings around the ring has decreased by a factor of three, so the electric potential difference has also decreased by a factor of three. Since the electric potential difference has been decreased, the device is called a **step-down transformer** (Figure 2).

Similarly, if the primary coil is the same as above but the secondary coil is changed so that it has 240 windings around the ring, the secondary potential difference will be 480 V. The number of windings has increased by a factor of four, so the potential difference has also increased by a factor of four. Since the electric potential difference has been increased, the device is called a **step-up transformer** (Figure 3).

step-down transformer: a transformer with fewer windings on the secondary coil, resulting in decreased voltage

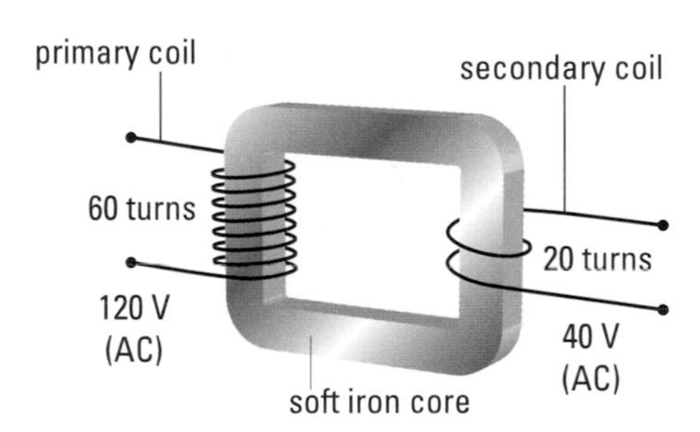

Figure 2
A step-down transformer

step-up transformer: a transformer with more windings on the secondary coil, resulting in increased voltage

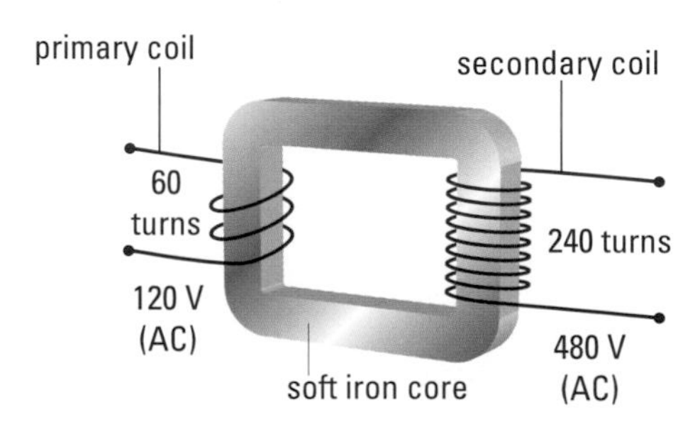

Figure 3
A step-up transformer

No charge flows from one circuit to the other through the ring. The change in potential difference is due to the changing magnetic field in the ring of the transformer and can be solved using the equation:

$$\frac{\text{secondary potential difference}}{\text{primary potential difference}} = \frac{\text{secondary windings}}{\text{primary windings}}$$

or

$$\frac{V_s}{V_p} = \frac{N_s}{N_p}$$

where V_s is the electric potential difference induced in the secondary circuit
V_p is the electric potential difference applied to the primary circuit
N_s is the number of windings in the secondary circuit
N_p is the number of windings in the primary circuit

To design a transformer that will lose as little energy as possible, the following steps are taken:

1. The coils are constructed so that they will have a small resistance. (Copper is often used.)
2. A core is used that is easily magnetized and demagnetized and properly shaped to transmit the magnetic field lines from the primary coil to the secondary coil.

Transformer coils are usually made not from a solid piece of soft iron but from many thin sheets of iron insulated from each other and attached together, much like plywood. This has the effect of reducing power losses in the core due to the presence of **eddy currents** that flow in the iron.

eddy currents: induced currents that form closed loops within a conductor

Demonstration of Eddy Currents

Using a copper tube and two similar cylindrical pieces of metal (one is a strong magnet and the other is unmagnetized), drop both into the tube one at a time and note any differences.

(a) Why does it take one of the objects longer to fall through the tube than the other?

These steps can reduce power losses to a negligible level and so we can assume that the transformer is 100% efficient, or ideal. For an ideal transformer,

power input (primary circuit) = power output (secondary circuit)

$$P_p = P_s$$

then

$$V_p I_p = V_s I_s \quad \text{since } P = VI$$

or

$$\frac{I_p}{I_s} = \frac{V_s}{V_p}$$

where I_p is the current in the primary circuit
I_s is the current in the secondary circuit

This equation brings up a very important point. First, the transformer may appear to be giving something for nothing. It seems like the electric potential difference can be increased at no cost. This is not true. If the electric potential difference increases by a factor of two, the electric current must decrease by a factor of two, since the power remains constant. As you might expect, transformers obey the law of conservation of energy.

Sample Problem

A door chime designed to operate at 8.0 V is connected to a 120-V power supply through a transformer. In the secondary coil the number of windings is 100 and the current is 1.8 A. Find

(a) the number of windings in the primary coil
(b) the current in the primary coil

Solution

(a) $V_s = 8.0$ V
$V_P = 120$ V
$N_s = 100$
$N_P = ?$

$$\frac{N_s}{N_P} = \frac{V_s}{V_P}, \text{ then}$$

$$\frac{N_P}{N_s} = \frac{V_P}{V_s}$$

$$N_P = \left(\frac{V_P}{V_s}\right)(N_s)$$

$$N_P = \left(\frac{120 \text{ V}}{8.0 \text{ V}}\right)(1.0 \times 10^2)$$

$$N_P = 1500$$

There are 1.5×10^3 windings in the primary coil.

(b) If the transformer is ideal, then
$I_s = 1.8$ A
$I_P = ?$

$$\frac{I_P}{I_s} = \frac{V_s}{V_P}$$

$$I_P = \left(\frac{V_s}{V_P}\right)(I_s)$$

$$I_P = \left(\frac{8.0 \text{ V}}{120 \text{ V}}\right)(1.8 \text{ A})$$

$$I_P = 0.12 \text{ A}$$

The current in the primary coil is 0.12 A.

What is being calculated here is actually the RMS electric current, since the current in both the primary and secondary coils is an alternating current. Many devices other than door chimes use transformers—televisions, some toys, high-intensity lamps and fluorescent lights, radios, battery chargers, and stereos, to name a few.

Answers

3. (a) 1.2×10^2 V
4. (a) 0.32 A
   (b) 48 V

Practice

Understanding Concepts

1. How many moving mechanical components are there in a transformer?
2. Draw a sketch of a transformer, labelling the three main parts. State the function of each part, and show which laws of electromagnetism and electromagnetic induction govern its operation.
3. An electric doorbell uses a transformer to obtain 6.0 V. The primary coil has 840 turns and the secondary coil has 42 turns.
   (a) What is the primary potential difference?
   (b) What type of transformer is this?
4. An ideal transformer has 120 primary turns and 300 secondary turns. If the current in the primary coil is 0.80 A and the electric potential difference in the secondary coil is 120 V, find
   (a) the current in the secondary coil
   (b) the electric potential difference across the primary coil
5. A step-up transformer could be called a step-down transformer if a variable other than potential difference were considered. What is that different variable? Explain your answer.

SUMMARY The Transformer

- A transformer is a device that can change the electric potential difference in a secondary circuit by varying the current in the primary circuit.
- The transformer equation states

$$\frac{\text{secondary potential difference}}{\text{primary potential difference}} = \frac{\text{secondary windings}}{\text{primary windings}}, \quad \text{or} \quad \frac{V_s}{V_p} = \frac{N_s}{N_p}.$$

Section 14.4 Questions

Understanding Concepts

For the following questions, assume that all transformers are ideal.

1. A transformer has 60 primary turns and 300 secondary turns. It is designed to supply a compressor motor requiring a current of 2.0 A at a potential difference of 5.5×10^2 V. What are the current and potential differences in the primary coil?
2. A transformer that is used to supply power to a toy train is plugged into a 1.2×10^2 V outlet and has various connections on its secondary coil to provide potential differences of 16.0 V, 12.0 V, 8.0 V, and 6.0 V to the train. If the primary coil has 1320 turns, how many turns must there be on the secondary coil at the position of each of the various connections?
3. A mercury vapour lamp operates at 3.0×10^2 V and has a resistance of 40.0 Ω. A transformer supplies the energy required, from a 1.2×10^2 V power line. Calculate
   (a) the power used by the transformer
   (b) the primary current
   (c) the ratio of primary turns to secondary turns in the transformer
4. For each of the transformers shown in **Figure 4**, state whether it is a step-up or step-down transformer, and calculate the unknown quantity.

(a)

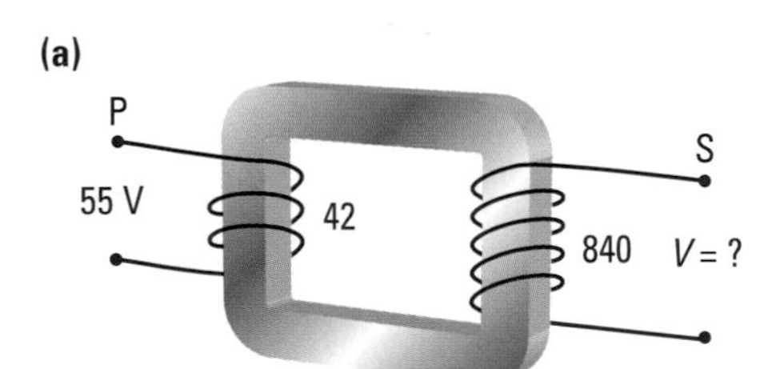

(b)

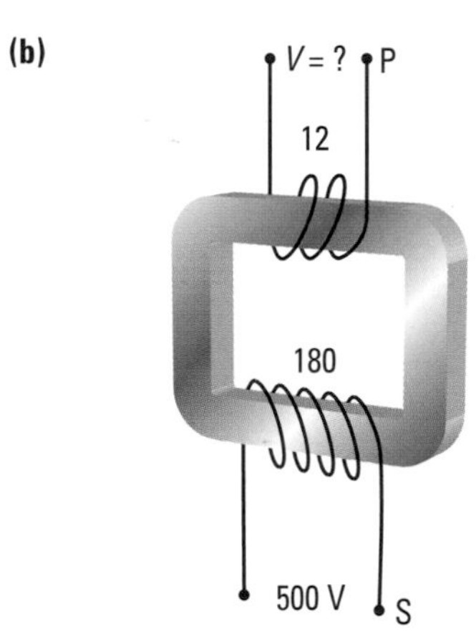

(c)

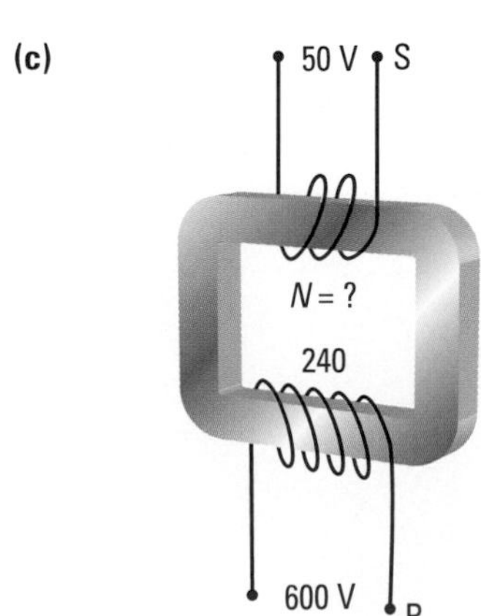

Figure 4
For question 4

5. (a) A battery is connected to the primary circuit in a transformer. Describe the current produced in the secondary circuit. Explain your reasoning.
   (b) Why will an AC current in the primary circuit produce an induced current in the secondary circuit when a constant DC current will not?
   (c) A practical DC generator produces a relatively ripple-free electric current. How effective would this current be if used to induce a current in the secondary circuit of a transformer?
6. A step-up transformer is used to increase the potential difference in a secondary circuit. What does it do to the current in the secondary circuit? Explain.

Making Connections

7. Sometimes a local electrical blackout occurs because a transformer burns out.
   (a) Where do you think the energy comes from in an electric transformer to cause its temperature to become high enough to burn it out?
   (b) Research to find out how efficient transformers are. Compare that efficiency with that of the "ideal" transformer described in this section.
   (c) Transformers are cooled in a variety of ways, some of which involve dangerous chemical substances. Find out through research some of the common methods of cooling transformers. Describe any controversial or dangerous methods you discover.

14.5 Distribution of Electrical Energy

With either AC generators or DC generators a tremendous amount of electrical power can be produced at the electric generating station. How can the electric power be delivered to the consumers in a safe, efficient, and useful way? Which of the two generators is better suited to the task? Both types of generators are equally easy to design and operate. Why is the AC generator the only one used for large-scale electrical energy supply?

Any type of generator needs an external force to turn the **turbine.** A hydroelectric generating station must be built near fast-flowing or falling water, which is often in remote areas. If fossil fuels or nuclear power is used, a large body of water is usually needed for cooling purposes, and the station should ideally be placed away from densely populated areas because of pollution. This means that the electrical power is usually produced far from the places where it is needed. This can be a problem, since some of the electrical energy is lost due to the resistance of the long lines connecting the generating station to the urban area. Most of the electrical energy lost is converted to heat in the transmission line. How can these losses be minimized?

turbine: machine in which the kinetic energy of a moving fluid is converted to rotary mechanical power

Since $P = VI$, it seems that we have only two variables that can be manipulated. Two obvious choices are to transmit the electric power at a low electric potential difference and a high current, or a high electric potential difference and a low current. Examine the following sample problem to see which choice is preferable.

Sample Problem

(a) Low electric potential difference and high current:
If 1.0×10^5 W of electric power is transmitted through a long cable with a resistance of 1.0 Ω, find the percentage of electric power lost for an electric potential difference of 1.0 kV.

(b) High electric potential difference and low current:
If 1.0×10^5 W of electric power is transmitted through a long cable with a resistance of 1.0 Ω, find the percentage of electric power lost for an electric potential difference of 1.0×10^5 V.

Solution

(a) $V = 1.0 \text{ kV} = 1.0 \times 10^3 \text{ V}$

$P = 1.0 \times 10^5 \text{ W}$

$R = 1.0\ \Omega$

$I = ?$

$$I = \frac{P}{V}$$

$$= \frac{1.0 \times 10^5 \text{ W}}{1.0 \times 10^3 \text{ V}}$$

$$I = 1.0 \times 10^2 \text{ A}$$

This is a very large current.

The power lost in the cable to heat would be

$$P = I^2R$$

$$= (1.0 \times 10^2 \text{ A})^2 (1.0\ \Omega)$$

$$P = 1.0 \times 10^4 \text{ W}$$

This is a loss of 10% of the original power in the transmission lines. A low potential difference and a high current is not very efficient.

(b) $P = 1.0 \times 10^5 \text{ W}$

$I = ?$

$$I = \frac{P}{V}$$

$$= \frac{1.0 \times 10^5 \text{ W}}{1.0 \times 10^5 \text{ V}}$$

$$I = 1.0 \text{ A}$$

This is a much smaller current.

The power lost in the cable to heat would be

$$P = I^2R$$

$$= (1.0 \text{ A})^2(1.0\ \Omega)$$

$$P = 1.0 \text{ W}$$

This is a loss of only 0.0010% of the original power. Therefore, having a high potential difference and a low current is much more efficient.

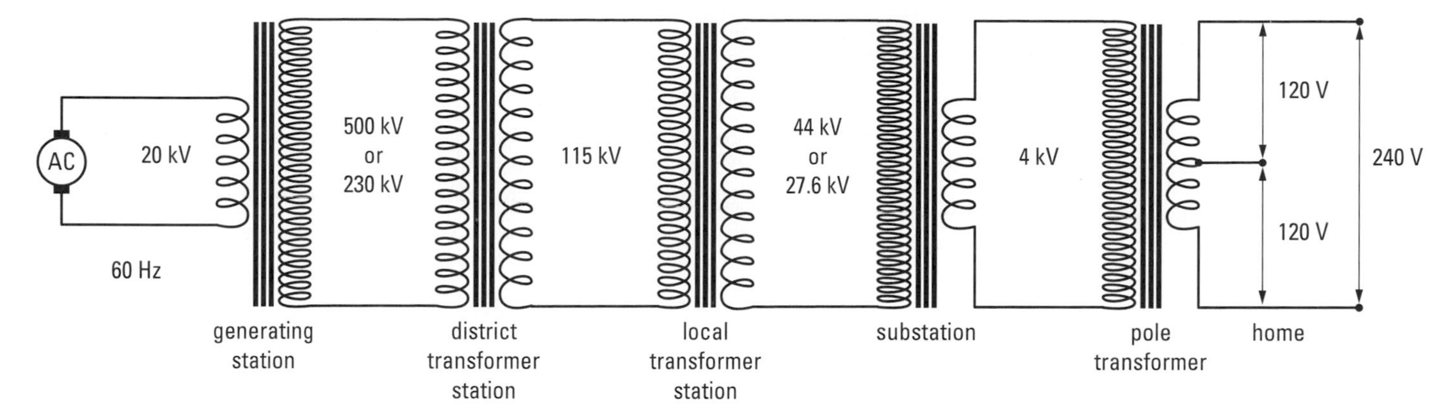

Figure 1
Distributing electrical energy

In the Sample Problem, the equation $P = VI$ is used to calculate the total current since the line is connected to an unknown resistance (all the light bulbs, televisions, refrigerators, and any other devices in the area). To determine how much of the power is dissipated as heat by the line itself, we ignore all these devices and use just the resistance of the line itself. Since the current through the line is known, the equation $P = I^2R$ is used.

To be efficient, electrical power must be transmitted at high potential differences. Such high potential differences can cause an **arc**; however, proper insulation of the wires can easily solve this since insulators do not conduct electricity. Another precaution is to place the wires up high to prevent arcing to the ground. The transmission of power is also more economical at low currents, since thinner, less expensive wires can be used in the transmission cables.

arc: light produced by air molecules when a current jumps a gap in an electric circuit

Another problem is that most generators produce electric potential differences of about 10 kV. To be more efficient, the potential difference should be much higher. Yet to be practical the consumer needs potential differences much lower, at 120 V to 240 V. A device is needed to increase the electric potential difference and decrease the current for transmission, and then decrease the electric potential difference and increase the current for use by the consumer. A transformer performs both of these functions very efficiently. Large networks of transmission lines connected to transformers are used all over the world to deliver electricity from generating stations to consumers (**Figure 1**).

In modern power distribution systems, electricity is generated at a potential of up to 20 kV and is immediately stepped up to 230 kV, 500 kV, 765 kV, or even more by transformers near the generating stations (**Figure 2(a)**). This very high-voltage power is then sent (with relatively little loss) along transmission lines whose rows of towers are a familiar sight in the countryside (**Figure 2(b)**).

Large district transformer stations are located along the transmission lines near cities, large towns, and industrial complexes. At these stations the power is stepped down to 115 kV, and it is transmitted at that potential difference to local transformer stations. The local transformer station in each small town or municipality further reduces the potential difference to either 44 kV or 27.6 kV. From there, the power is distributed to transformer substations in each neighbourhood, where it is further stepped down to about 4 kV. Wires carry this power along residential streets to the last transformer in the chain, the familiar hydro-pole transformer, which reduces the potential to 240 V. In some modern subdivisions, with underground wiring, pole transformers have been replaced by underground transformers buried beneath the lawns of every fourth or fifth house.

Three wires lead from the pole transformer into each residence—one from each end of the secondary coil, and one from its centre. In this way, both 240-V and 120-V potentials are made available for use in the home.

(a)

(b)

Figure 2
(a) Transformers near a generating station
(b) Power transmission lines

Answers

1. 6.4×10^{-5} %
2. (a) 2.0×10^2 A, 20 kW
   (b) 2.0 A, 2.0 W

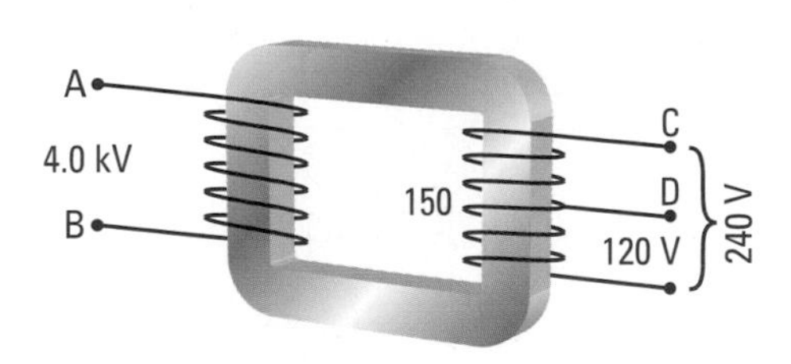

Figure 3
For question 3

Practice

Understanding Concepts

1. An electric potential difference of 5.0×10^2 kV is used to distribute 2.0×10^2 kW of electric power through a long cable with a resistance of 0.80 Ω. Find the percentage of electric power lost in the lines.
2. For each of the two transmission lines described below, determine the current and the power loss in the line.
   (a) power = 2.0×10^2 kW, resistance = 0.50 Ω, potential difference = 1.0 kV
   (b) power = 2.0×10^2 kW, resistance = 0.50 Ω, potential difference = 1.0×10^2 kV
3. **Figure 3** shows the final transformer in a network that delivers electrical energy to a home. Assume that there are 150 windings from C to D. How many windings are there from A to B?

Case Study: Magnetic Information Storage

Magnetic information storage is perhaps the most important element of computer technology. Without magnetic storage, most computers would lose a large part of the information once the power is turned off, so a hard disk is vital to the operation of the modern computer. Hard disks, or hard drives (**Figure 4(a)**), are a computer's long-term memory. A hard disk stores digital information in a relatively permanent form and allows the computer quick access to that information.

Hard disks were developed in the 1950s and have been steadily improving and changing over time. While the first hard disks could store only a megabyte or two of information, modern ones can store several gigabytes of computer files. They are known as "hard" disks to distinguish them from the external "floppy" disks. Floppy disks (**Figure 4(b)**) are made from plastic and are very thin, while hard disks are typically composed of aluminum or glass polished like a mirror.

Figure 4
(a) A hard disk drive
(b) A floppy disk

The Basics of Magnetic Information Storage

There are several advantages in using magnetic storage. It can be easily erased and rewritten, which gives the medium tremendous flexibility. Also the data can be stored for years and will usually last until it is outdated.

Each disk has a fixed amount of space for data storage on it. The information is stored on a thin film across the surface of the disk, which can be magnetized. The head of the drive never actually touches the surface of the hard disk, unlike a floppy drive. Modern heads can move to the edge of the disk and back to the hub upwards of 50 times per second! This allows for tremendous data transfer rates. The information on the hard disk can be accessed almost instantly, which has obvious advantages. Modern hard disks spin at speeds approaching 300 km/h at the outer edge! The head inside the hard drive can write and read information to and from the hard disk.

How the Computer Gets the Information

The computer receives and delivers data from and to the hard disk using two principles of electromagnetism. First, by passing an electric current through a coil a magnetic field is produced, and second, by applying a magnetic field to a coil an electric current will begin to flow. The magnetic field on the hard drive produces current as it spins. This electric current passes into the computer,

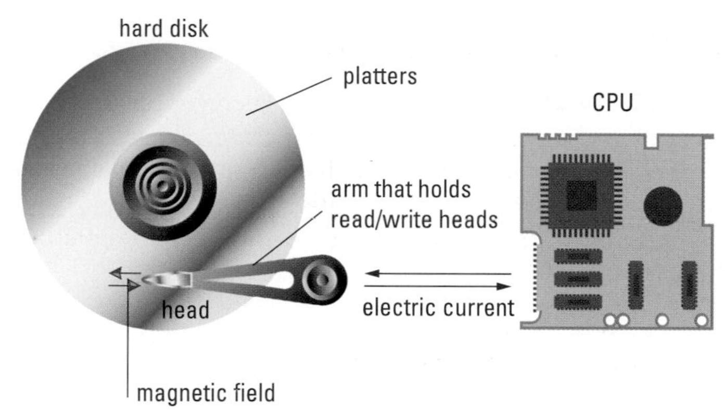

Figure 5
A computer hard disk system

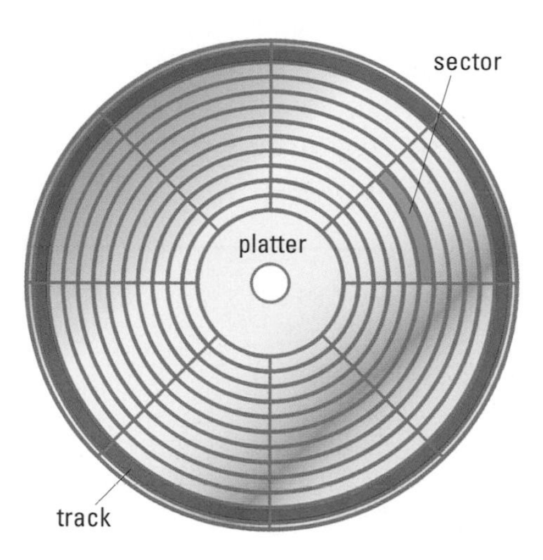

Figure 6
Tracks are concentric circles, and sectors are pie-shaped wedges on a track.

which interprets it as information. The reverse can also occur, when the computer passes out an electric current, which is then turned into a magnetic field, which is then stored on the hard disk as information (**Figure 5**).

The heads of a hard drive are small electromagnets that perform the conversion of electrical current to magnetic fields and back, allowing them to both read from and write to the hard disk.

The information on the hard disk is organized into bytes, which are in turn collected into sectors and tracks (**Figure 6**). The bytes form files that form the programs the computer can run. The head reads this information and passes it on to the central processing unit (CPU). The information comes in the form of electric current, which is represented digitally as ones and zeros. This is the computer's language, also known as binary code.

Weaknesses of Magnetic Storage

For all its strengths, magnetic information storage also has some weaknesses. First, it is susceptible to outside magnetic fields. These may corrupt or erase the data stored on the hard disk. It is important to keep any magnetic storage device away from large magnetic fields. Second, the data are difficult to preserve over time. The information on the disk may become corrupted. Also, it is possible for the disk to become damaged, thereby losing all stored information. It is because of this that many people back up their information, either on a second magnetic storage device, an optical one, or by producing a hard copy where possible.

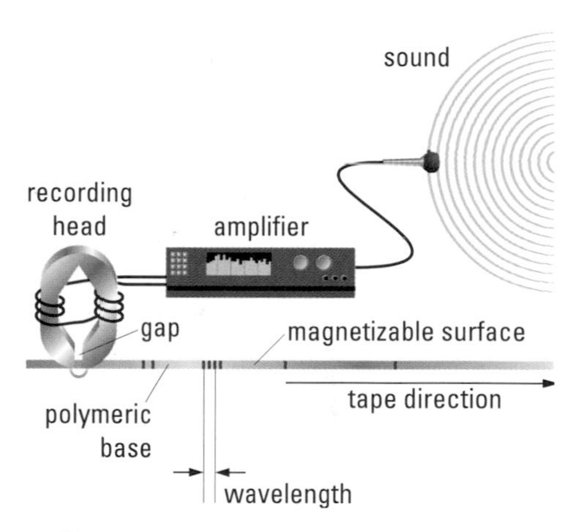

Figure 7
Audio or videotape system

Other Uses for Magnetic Information Storage

Magnetic storage is used for many other purposes, such as video and audio tape and the magnetic strip on credit cards. The principle, similar to that used in computers, involves moving a tape past a magnetic head (**Figure 7**). The tape consists of a thin plastic base coated with an even thinner layer of magnetic iron oxide particles in a polymer binder. As the tape moves past the the gap in the recording head, the field goes through the tape, aligning the dipoles in its magnetic oxide coating. The tape becomes a magnetic copy of the current in the recording head. When the current stops, so does the field, ending the magnetization.

If the tape is moved past the playback head, the moving magnetic field induces an electric current in the coil that is an exact copy of the information stored on the tape (**Figure 8**). The electric signal may then be amplified and converted into useful information.

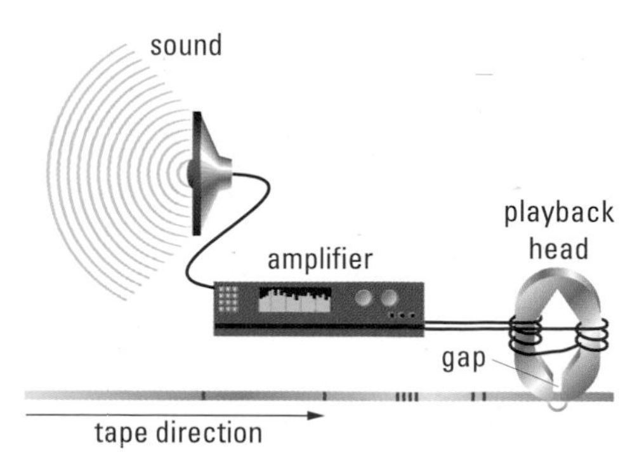

Figure 8
The playback head is identical to the recording head and is often the same head.

SUMMARY Distribution of Electrical Energy

- AC generators are used for large-scale electrical energy supply.
- Transformers allow the tremendous amount of electrical power produced at electric generating stations to be transformed into safe and efficient power for use by consumers.

Explore an ***Issue***

Computers in Today's Society

DECISION MAKING SKILLS

- ○ Define the Issue
- ● Identify Alternatives
- ● Research
- ● Analyze the Issue
- ○ Defend the Proposition
- ● Evaluate

There are many pressing issues involving computers and how they are used. In today's society, the advancements made in computers are both astounding and continuously changing. Many of today's computer companies are among the richest companies in the world, and as a result carry vast responsibilities. Information on almost anything is readily available over the Internet. Many wonderful computer-related careers are available, but the technology changes very quickly.

Take a Stand

Working in small groups, choose one of the issues listed below and discuss it as a group.

Understanding the Issues

1. For the consumer it is often frustrating that a recently purchased computer needs to be upgraded or replaced sooner than the buyer anticipated.
   (a) List three benefits in having rapid advancements in computer hardware.
   (b) List three disadvantages of these rapid advancements.
2. Should any one computer company be allowed to have a monopoly on any one essential part or program required for computers? List some advantages and disadvantages for consumers and companies for and against monopolies in the computer industry.
3. List some advantages and disadvantages to having so much information available over the Internet.
4. What characteristics are required for a person interested in a long-term career in computers? Would you consider a career in computers to be an excellent choice? Why or why not?

Forming an Opinion

Use the Internet or any other resources to further investigate rapid changes involving computers. Concentrate on one of the issues above. Write a short report on the issue explaining the particular aspect of the issue you are investigating. Provide any concrete examples involved. Follow the links for Nelson Physics 11, 14.5.

GO TO www.science.nelson.com

Section 14.5 Questions

Understanding Concepts

1. The power distributed at 5.0×10^2 V on a set of transmission lines is 5.0×10^2 kW, and the resistance of the lines is 1.0 Ω. Determine the following:
   (a) the current in the lines
   (b) the power loss if the potential difference is
      (i) 5.0×10^2 V
      (ii) 5.0×10^2 kV
   (c) the energy loss per hour
2. Why is it an advantage to distribute electrical power at a high potential difference and low current?
3. What is the process by which a computer gets data from a magnetic storage device?
4. (a) What are the differences between a hard disk and a floppy disk?
   (b) How is magnetic storage on a tape similar to storage on a hard disk?
5. How reliable is magnetic information storage? In what ways is it inferior to other forms of information storage? In what ways is it superior?
6. Assume that each of the transformers in **Figure 1** of this section is 96% efficient. What percentage of the original power would be left for the consumer? Why must transformers be as efficient as possible?

Making Connections

7. List some advantages to having access to both 120 V and 240 V in the home.
8. Theoretically, one transformer to step up the voltage for transmission and one to step down the voltage for consumers are the only transformers required. Why must there be so many transformers in a large-scale distribution of electrical energy?
9. Give some economic and societal reasons for and against distributing electrical power at both a low current for efficiency and a low potential difference (which would require no transformers).
10. Why is magnetic information storage important to the operation of the modern computer?

CAREER

Careers in Electricity and Magnetism

There are many different types of careers that involve the study of electricity and magnetism. Have a look at the careers described on this page and find out more about one of them or another career in electricity and magnetism that interests you.

Electronics Engineering Technologist
This is an 18-month community college diploma program that requires grade 12 math, physics, and English to gain entry. These technologists help design satellite systems and computer network support for telecommunications companies. They work with oscilloscopes, function generators, and electronic testing equipment.

Telecommunications Systems Technician
A telecommunications systems technician is qualified to install, plan, program, repair, or troubleshoot a telecommunications system such as a PBX (private branch exchange system) or an electronic key system. These technicians use software, hand tools, and network analysis instruments. They usually work for telephone companies, wireless companies, and information technology providers and manufacturers. To be hired for the job, they must complete a two- or three-year program at a community college.

Electrical Engineer
After obtaining a four-year bachelor's degree, electrical engineers must work under a professional engineer for two years to gain work experience. They must also pass a written examination in engineering law and ethics to become a member of the Association of Professional Engineers. Electrical engineers work in the private sector or for research centres, carrying out a wide variety of duties. For example, in aviation, they repair and overhaul generators and they also repair and design electrical circuits and systems, using computers and aircraft manuals.

Practice

Making Connections

1. Identify several careers that require knowledge about electricity and magnetism. Select a career you are interested in from the list you made or from the careers described above. Imagine that you have been employed in your chosen career for five years and that you are applying to work on a new project of interest.
   (a) Describe the project. It should be related to some of the new things you learned in this unit. Explain how the concepts from this unit are applied in the project.
   (b) Create a résumé listing your credentials and explaining why you are qualified to work on the project. Include in your résumé
   - your educational background—what university degree or diploma program you graduated with, which educational institute you attended, post-graduate training (if any);
   - your duties and skills in previous jobs; and
   - your salary expectations.

   Follow the links for Nelson Physics 11, 14.5.

GO TO www.science.nelson.com

Chapter 14 Summary

Key Expectations

Throughout this chapter, you have had opportunities to

- Define and describe the concepts and units related to electromagnetic induction; (14.1–14.7)
- Analyze and describe electromagnetic induction in qualitative terms and apply Lenz's law to explain, predict, and illustrate the direction of the electric current induced by a changing magnetic field, using the right-hand rule; (14.2, 14.3, 14.4)
- Compare direct current (DC) and alternating current (AC) in qualitative terms, and explain the importance of alternating current in the transmission of electrical energy; (14.3, 14.4)
- Explain, in terms of the interaction of electricity and magnetism, and analyze in quantitative terms, the operation of step-up and step-down transformers; (14.4, 14.5)
- Conduct an experiment to identify the factors that affect the magnitude and direction of the electric current induced by a changing magnetic field; (14.1–14.7)
- Construct, test, and refine a prototype of a device that operates using the principles of electromagnetism; (14.1–14.7)
- Analyze and describe the operation of industrial and domestic technologies based on principles related to magnetic fields such as cathode ray (TV) tubes, generators, and magnetic information storage; (14.2–14.7)
- Describe the historical development of technologies related to magnetic fields; (14.1–14.7)
- Identify and describe science and technology-based careers related to electricity and magnetism; (career feature)

Key Terms

law of electromagnetic induction
mutual induction
induced field
inducing field
Lenz's law
electric generator
transformer
step-down transformer
step-up transformer
eddy currents
turbine
arc

Make a Summary

Using the Key Expectations and Key Terms, make a concept map to summarize what you have learned in this chapter. Concentrate on the factors that affect magnetic induction and how they are used in generators and the distribution of electrical energy.

Reflect on your Learning

Revisit your answers to the Reflect on Your Learning questions at the beginning of this chapter.

- How has your thinking changed?
- What new questions do you have?

Chapter 14 Review

Understanding Concepts

1. (a) Explain what is meant by electromagnetic induction, and describe how an electric current or potential difference can be induced in a conductor.
   (b) State three factors that affect the magnitude of the potential difference induced in a coil.
   (c) Where does the electrical energy of the induced current and potential difference come from? What other form of energy is produced?
2. Sketch each of the illustrations in **Figure 1** into your notebook and use Lenz's law to complete them.

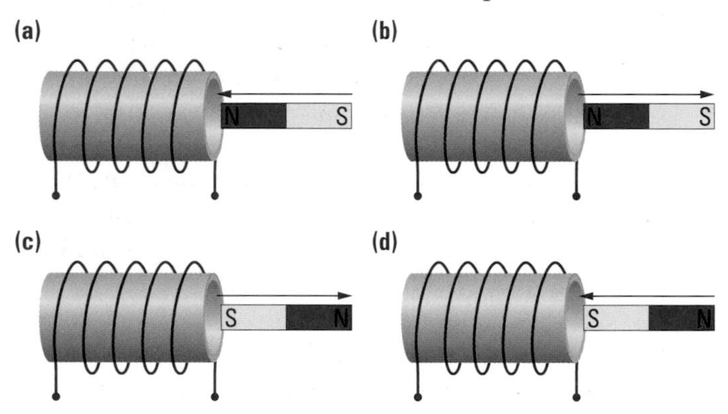

Figure 1

3. A bar magnet inserted completely into a coil of 5.0×10^2 turns produces an induced potential difference of 1.5 V. Determine the potential difference induced when each of the following changes is made, considered separately:
   (a) A 250-turn coil is used.
   (b) The bar magnet is moved twice as quickly.
   (c) Three identical magnets of the same polarity and strength are inserted at once, side by side.
4. State the effect on the size of the induced current from a generator, given the following:
   (a) The number of loops of the coil is increased.
   (b) The rate at which the loops cut across the field lines increases.
   (c) The field strength increases.
   (d) An iron core is used in the coil rather than an air core.
5. A step-up transformer is designed to operate from a 12-V AC supply and to deliver energy at 240 V. If the secondary winding is connected to a 60.0-W, 240-V lamp, determine
   (a) the ratio of the number of primary windings to the number of secondary windings of the transformer
   (b) the primary current
6. The transformer used to operate a model electric train has 1125 turns on its primary coil and 75 turns on its secondary coil. If it is plugged into a 120-V circuit, what potential difference does the model train receive?
7. A magnet induces a current of 8.0 mA in a 5.0×10^2-turn coil connected to a galvanometer. The total resistance of the coil and galvanometer is $1.0 \times 10^2\ \Omega$. The coil is then replaced by a 1500-turn coil, making the total resistance in the circuit 150 Ω. What current would then be induced by the same magnet moving at the same speed?
8. A generating station produces 15 000 kW of electric power. Assuming 100% efficiency of transmission, calculate the current flowing through a 5.0×10^5 V transmission line leading from the generating station.
9. **Figure 2** shows one coil partially inserted into another coil and connected in series with a resistance, a switch, and a DC source. Predict what will be observed on the galvanometer when the switch
   (a) is suddenly closed
   (b) remains closed
   (c) is suddenly opened

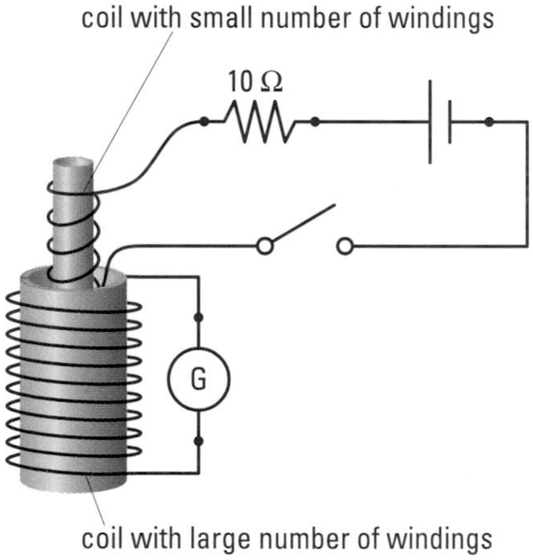

coil with small number of windings
10 Ω
G
coil with large number of windings

Figure 2

10. The N-pole of a magnet is pushed into a solenoid connected to a galvanometer and the needle moves to the right. The magnet continues to move into the coil and out the other side. As the magnet passes out the other side, which way will the needle on the galvanometer move? Draw an illustration and explain, using Lenz's law.
11. What properties of the induced current produced by an AC generator will change if the armature is rotated faster? Justify your answer using the principles of electromagnetic induction.

12. What is the source of the charges that vibrate back and forth in the electric circuits in your home? What is their source of energy?
13. What conditions minimize power loss in electric transmission lines?
14. How can a DC generator be converted into an AC generator?

Applying Inquiry Skills

15. If you were to use equipment similar to that shown in **Figure 3**, design an experiment that can be used to investigate the relationship between the ratios of the windings in each coil to the electric potential differences across each coil (and the electric current through each coil if possible.) Note: Laboratory transformers should not be handled by students.

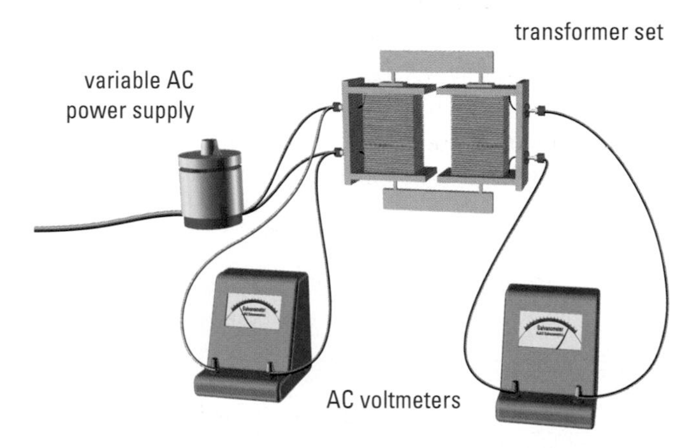

Figure 3

16. A solenoid with an iron core is connected to a circuit with a galvanometer. A strong magnet is suddenly brought into contact with the core, left there for a while, and then suddenly removed.
    (a) Discuss the induced current in the circuit.
    (b) Sketch a graph of induced current versus time and indicate any significant events on the graph.
    (c) Justify the graph using Faraday's discoveries and Lenz's law.
17. Perhaps you have seen somebody walking slowly along a beach searching for coins and other objects with a metal detector. How would you apply the principle of electromagnetic induction to construct such a device? (*Hint:* Think about how eddy currents are produced.) If you can't think of a design, perform research to get some ideas. Follow the links for Nelson Physics 11, Chapter 14 Review.

GO TO www.science.nelson.com

Making Connections

18. In AC transmission lines, electrons accelerate back and forth at a frequency of 60.0 Hz.
    (a) Research to find out how this acceleration relates to the production of 60.0-Hz electromagnetic radiation.
    (b) Calculate the wavelength of these electromagnetic waves.
    (c) Is the 60.0-Hz electromagnetic radiation from AC transmission lines or other electric sources considered to be dangerous? Do all scientists agree on this issue?
19. The principle of electromagnetic induction is applied in an early-warning system in regions of Australia where underground saltwater accumulations threaten to ruin farms. Irrigation and rain runoff following deforestation have caused underground water levels to rise, carrying salt toward the surface. In the detection system, an airplane fitted with a transmitter loop (the "primary circuit") flies low over the ground. An antenna trailing behind the plane receives signals from beneath the ground.
    (a) Where is the "secondary circuit" located?
    (b) Based on your experience in science or chemistry classes, how does the electrical conductivity in salt water compare with that in fresh water?
    (c) What is the relationship between the size of the induced current in the "secondary circuit" and the concentration of the salt?
    (d) What is the relationship between the timing of the return signal and the depth of the saltwater beneath the surface?
    (e) To find out more about this technology, follow the links for Nelson Physics 11, Chapter 14 Review. What are some of the limitations of this detection system?

GO TO www.science.nelson.com

Exploring

20. Using the Internet and other resources, research commercial generators and discuss the modifications made to make them more practical than those discussed in this section. Follow the links for Nelson Physics 11, Chapter 14 Review.

GO TO www.science.nelson.com

21. Trace the development of magnetic information storage from tapes to the modern hard drive. Use the Internet and other resources to research and outline a recent advance in the field. Follow the links for Nelson Physics 11, Chapter 14 Review.

GO TO www.science.nelson.com

Technology Trade Show

One good reason to study the physics of electricity and magnetism is to increase your understanding of the world around you. Another important reason is to apply this scientific knowledge to develop new technologies or to improve upon existing technologies.

A new technology or device is developed by a typical process: a technology company sees a need for a certain product (device) based on market research; the company considers introducing a new device for the market; a project team is set up which includes scientists, engineers, technicians, and technologists to work on research and design, as well as people on the business side for budgeting, sales, and marketing; once the team is assembled, research is done on how to build and market the device; when funding is approved, the design, production, and testing can then follow; refinements are made based on test results; and the device is manufactured and marketed.

One way to effectively introduce a new product to the consumer is to take advantage of a high tech trade show where displays and demonstrations of the device are made available to potential customers (**Figure 1**). It is not enough to just have a product that works—the presentation must be appealing and informative, demonstrating the product's practical value to the customers. Keeping in mind that there is always competition at these trade shows, the presenters must be well-informed and must present their well-made, reasonably priced device in an interesting and lively manner.

Figure 1
A trade show is a popular occasion for marketing new products.

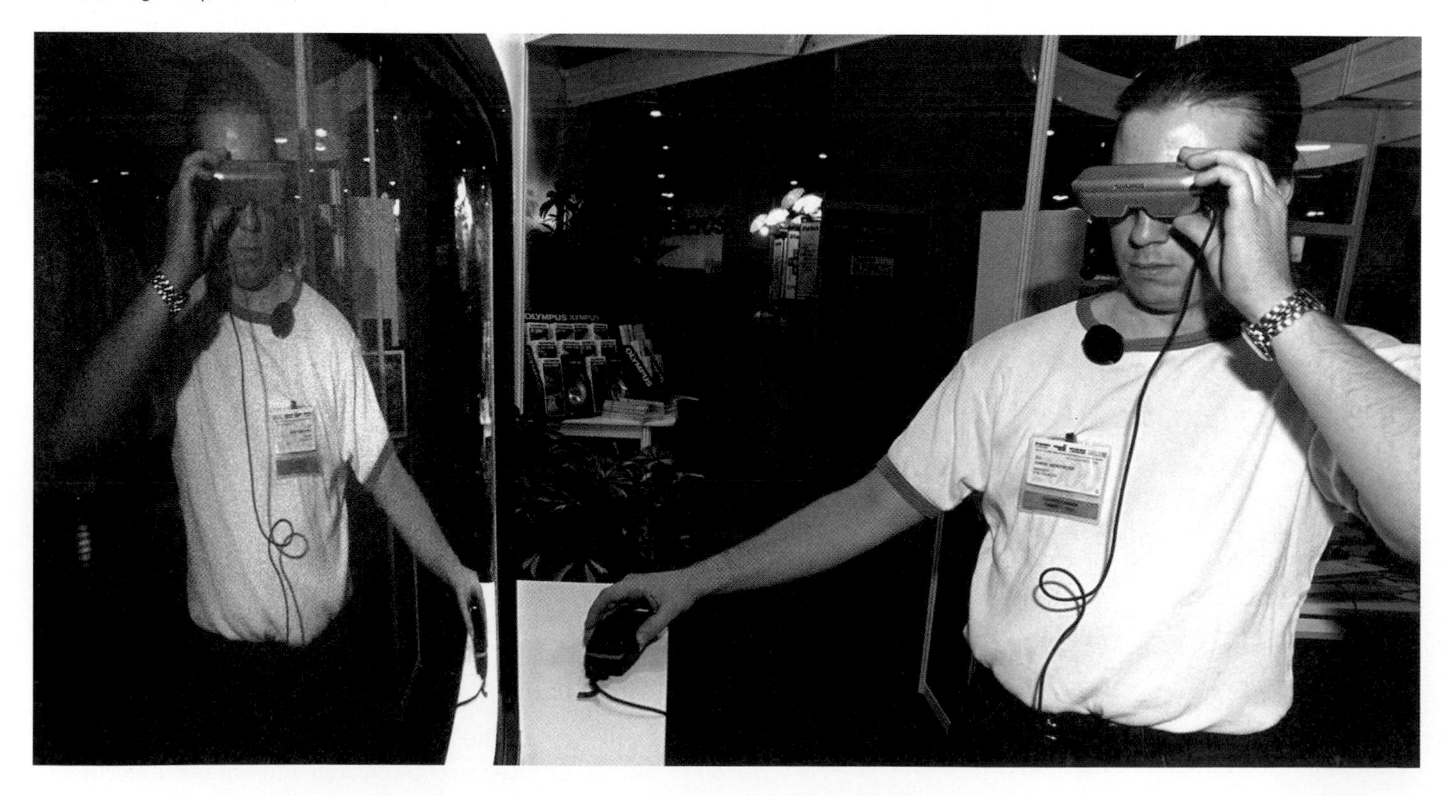

The Task

As you can see there is a lot involved in designing, producing, and marketing a device. It requires hard work, creativity, ingenuity, and teamwork. Your task is to research a prototype of a device that can be constructed with materials available in your classroom, construct the device that you have designed, test the device, refine your product, and prepare the presentation that you will use to sell it.

You may choose an electric bell, loudspeaker, ammeter, electric motor, electric generator, or any other device approved by your teacher for your task. Pride of workmanship can mean the difference between a well-made, useful device and one that does not work at all. Therefore, careful construction and patience are required for producing the device. Even the best devices on the market today can be improved upon. It is possible that even at a late stage of production, some further research and study may be required. However, this stage is also very close to the point where the product must be presented to the public.

In this Performance Task, you will be expected to design and construct the device, test the prototype in the presence of your teacher, and refine the device based on your test results. You will be expected to analyze the physics principles involved in using the device to explain how the device works, and answer questions that others may have during your presentation. As you work on the task, keep notes of your design, safety considerations, tests, and analyses, and design a plan for a creative and informative display to sell your product that you can submit with the device.

Analysis

Your analysis should include answers to the questions below.

(a) How does your device apply physics principles, laws, and theories?

(b) How has the basic design of this type of device evolved from historical times to the present?

(c) What are the main uses of the device, and who are the primary users?

(d) Select two models currently on the market that are similar to your device and compare their features and prices.

(e) What careers are related to the production or use of the device?

(f) What innovations related to your choice would you like to have implemented in the future?

(g) List problems that you encountered during the construction of the device and describe how you solved the problems, including any insights gained. (You will get ideas as you test and modify the device.)

Evaluation

(h) Evaluate the contributions of the device to domestic situations and industry.

(i) Evaluate the process you used in working on this task. If you were to begin this task again, what would you modify to ensure a better process and a better product?

Assessment

Your completed task will be assessed according to the following criteria:

Process

- Choose an appropriate electrical device to improve your skills.
- Draw up detailed plans of the technological design, tests, and modifications.
- Choose and safely use appropriate tools, equipment, materials, and computer software.
- Carry out the construction, tests, and modification of the device.
- Analyze the task (as described in Analysis).
- Evaluate the task (as described in Evaluation).

Product

- Demonstrate an understanding of the related physics principles, laws, and theories.
- Use terms, symbols, equations, and SI metric units correctly.
- Submit with a working model of the final product, a plan for a creative and informative display to sell your product.
- Prepare and submit a task portfolio with your notes on design, safety considerations, tests, analyses, and evaluation.

Understanding Concepts

1. When an ebonite rod and and a piece of fur are rubbed together, each object develops a charge.
   (a) What type of charge does each object develop?
   (b) How do the sizes of the charges compare?
   (c) What is the source of the charge?
2. (a) What is the charge on an object with a deficit of 4.0×10^{12} electrons?
   (b) What is the charge on an object with an excess of 6.2×10^{11} electrons?
3. One conductor has a charge of -2.0×10^{-8} C and another, identical conductor has a charge of 3.0×10^{-8} C.
   (a) If the two conductors are brought into contact, what will be the charge on each conductor?
   (b) How many electrons must move, and which way will they go?
4. A light bulb with a current of 0.70 A is left on for 2.50 h. How much electric charge passes through the filament of the bulb?
5. Sketch on the same diagram a graph of electric potential difference (from 0 V to 10 V) versus electric current (from 0 A to 2.0 A) for each of the following resistors:
   (a) a 5.0-Ω resistor
   (b) a 10.0-Ω resistor
   (c) a non-ohmic resistor
6. A particular 9.0-V battery can supply a current of 6.00 mA for 40.0 min. Another 12.0-V battery can supply a current of 7.00 mA for 35.0 min. Which battery has more electric potential energy?
7. A motor receives 6.0×10^3 J of energy from 50.0 C of charge. What is the potential difference across the motor?
8. A small light bulb with a resistance of 10.5 Ω is connected to a 9.0-V battery for 30.0 s.
   (a) What is the current through the bulb?
   (b) What power is dissipated by the bulb?
   (c) How much energy is used by the bulb?
   (d) How much charge is transferred through the bulb?
   (e) How many electrons pass through the bulb?
9. The following appliances are operated in a 120.0-V circuit for a thirty-day month:
   - six 100.0-W light bulbs for 8.0 h/d
   - a 10.0-Ω kettle for 10.0 min/d
   - a bread machine drawing on 6.0 A for 1.0 h/d

   Calculate the electric bill for the month at an average cost of 8.5¢/kW•h.
10. For each of the electric circuits in **Figure 1**, find
    (a) the electric current in each resistor
    (b) the electric potential difference across each resistor
    (c) the total energy dissipated by each circuit if left on for 30.0 min

(a)

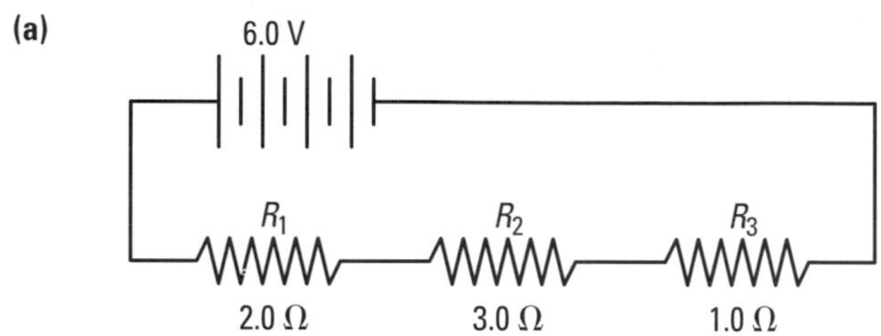

(b)

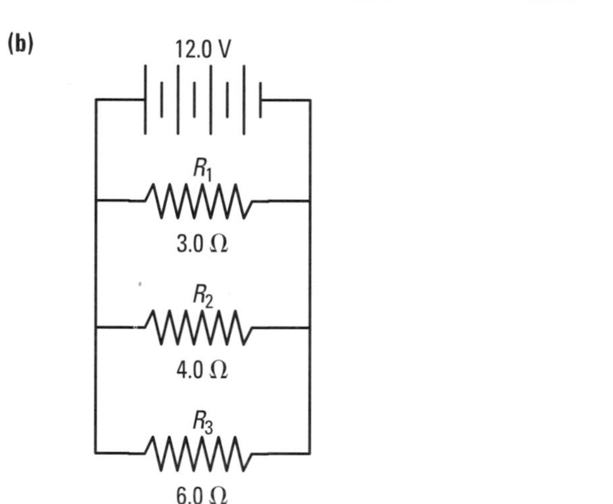

(c)

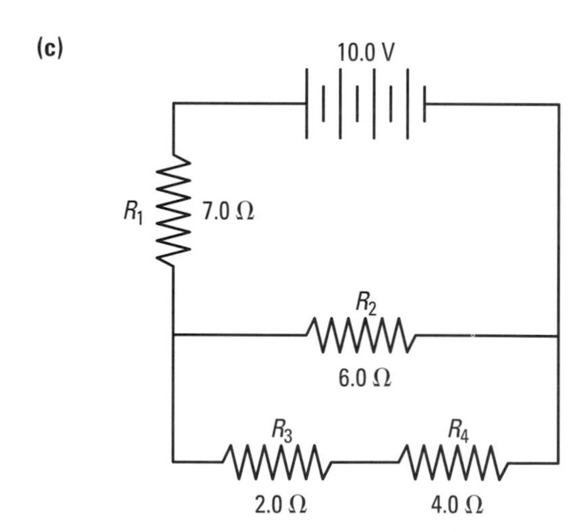

Figure 1

11. A student performs the following experiments while investigating the properties of ohmic resistors and records some of the data in the table shown (**Table 1**). Row 1 was found when only R_1 was in the circuit. Row 2 was found when R_1 and R_2 were in the circuit in series. Row 3 is for R_1 and R_2 in parallel. Complete the table by performing the following steps:
    (a) Calculate R_1 using row 1.
    (b) Calculate R_2 using row 2. (*Hint:* Find R_1 first.)
    (c) Complete row 3.

Table 1

Resistors	Voltage across resistors	Current in resistors
R_1	6.0 V	0.60 A
R_1 and R_2 in series	6.0 V	0.15 A
R_1 and R_2 in parallel	6.0 V	

12. Copy **Figure 2** into your notebook. Draw the magnetic lines of force for each of the diagrams.
    (a) two bar magnets
    (b) a current-carrying conductor
    (c) current through a solenoid

(a)
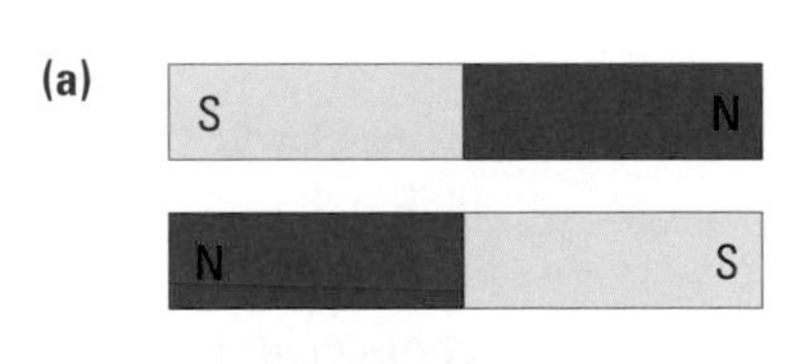

(b)
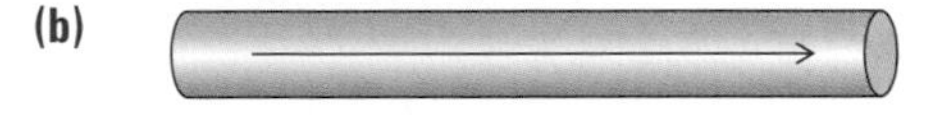

(c)
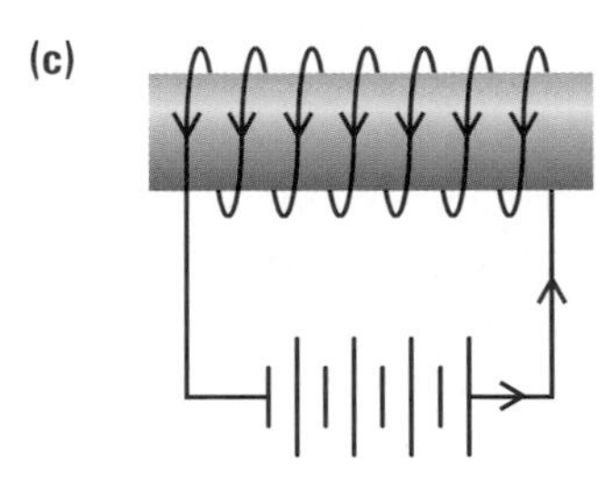

Figure 2

13. Copy each of the following diagrams in **Figure 3** into your notebook, and indicate the magnetic field and the direction of the force on each current-carrying conductor.
    (a) two parallel wires with currents
    (b) a conductor near a U-shaped magnet

(a)

(b)
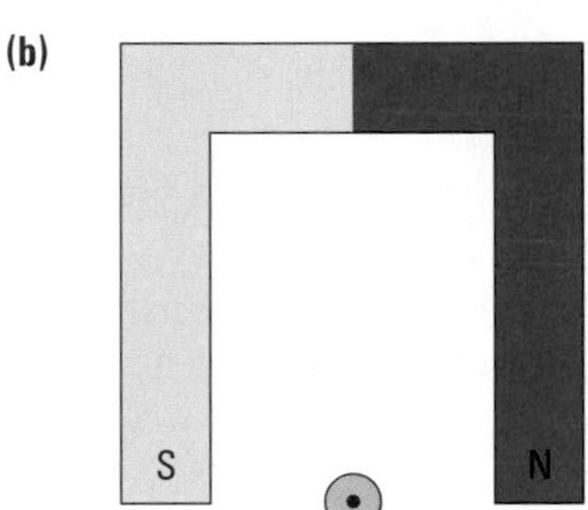

Figure 3

14. Copy each of the following diagrams in **Figure 4** into your notebook, and indicate the direction of the induced current.
    (a) a magnet being pulled out of a solenoid
    (b) a magnet being pushed into a solenoid
    (c) a conductor being pushed down through a magnetic field

(a)
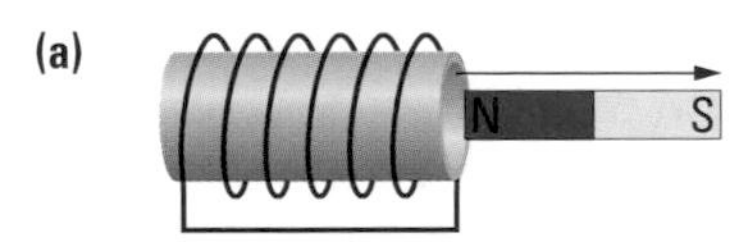

(b)

(c)
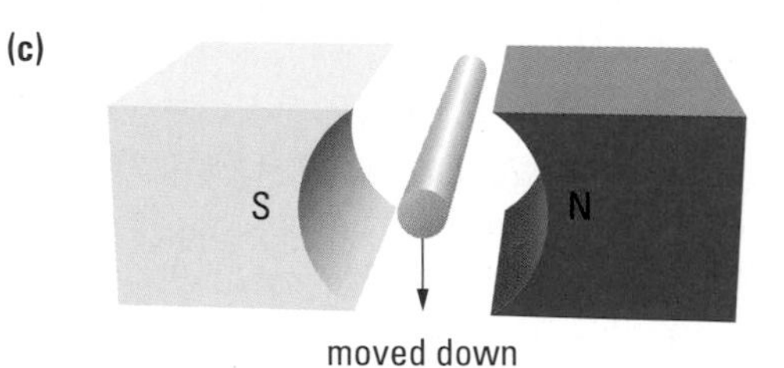

Figure 4

15. When a magnet is pushed into a coil, a resistance to the push can be detected. Why is the resistance greater for a coil of the same length but with more loops?

16. A bar magnet is pushed at a certain speed into a coil with 1000 turns, which then produces an induced electric potential difference of 4.0 V. Determine the electric potential difference produced when the following changes are made and considered separately:
    (a) The magnet is pulled out of the coil at the same speed.
    (b) The magnet is pushed at the same speed into a different coil with 700 turns.
    (c) The magnet is pushed half as fast into the 1000-turn coil.
    (d) Another bar magnet with twice the strength is pushed at the original speed into the 1000-turn coil.
17. For each of the following transformers in **Figure 5**, state whether it is a step-up or step-down transformer and calculate the
    (a) electric potential difference in the secondary circuit
    (b) number of windings in the secondary circuit

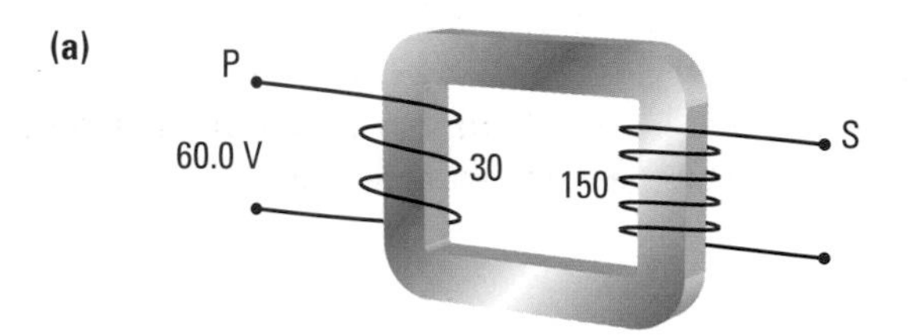

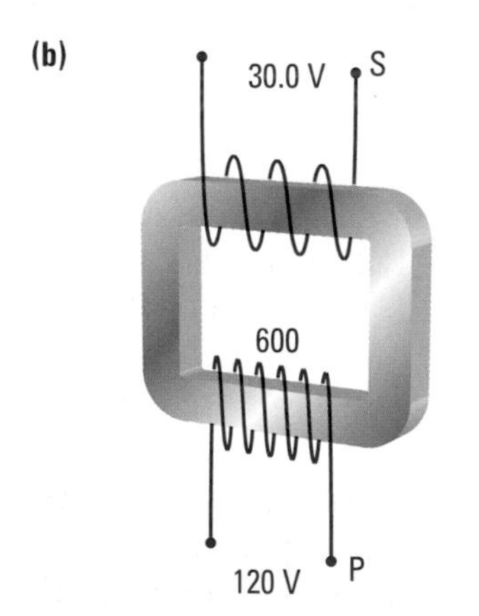

Figure 5

18. An electric device operates at 15.0 V. The device is connected to a transformer that is plugged into a 120-V energy supply and has 800 windings in the primary coil. How many windings are in the secondary coil?
19. Faraday's iron ring is also called a transformer because it transforms something from the primary circuit to the secondary circuit. What does it transform—electric current, electric potential difference, or energy? Explain your answer.
20. Transmission lines are used to distribute 400.0 kW of power at a potential difference of 500.0 kV. The power lines have a resistance of 1.2 Ω. Calculate the
    (a) current in the transmission lines
    (b) power loss in the transmission lines
    (c) fraction of the original power lost
21. When the top of a metal-leaf electroscope is touched by a charged object the leaves, which normally hang straight down, move apart.
    (a) Why do the metal leaves move apart? Draw a diagram to explain.
    (b) Does the charged object have to touch the top of the electroscope to cause the leaves to move apart? Draw a diagram to explain.
    (c) How can you use a metal-leaf electroscope to determine the relative amounts of charge on charged objects?
22. When charging by friction between two objects, the magnitude of the charge developed on each object is the same but the charges on the two objects are opposite in sign. Explain both of these observations.
23. To develop a charge on a conductor by providing it with an excess of electrons, the electrons spread out over the outer surface of the conductor. Explain why this happens.
24. Does a 12.0-V battery necessarily have twice the stored electrical potential energy that a 6.0-V battery has? Explain.
25. Several identical light bulbs are connected in parallel to a battery.
    (a) What will happen to the brightness of the bulbs if more of these identical lights are connected in parallel to the battery?
    (b) What will happen to the brightness of the bulbs if
        (i) a higher-resistance bulb is connected in parallel?
        (ii) a lower-resistance bulb is connected in parallel?
    (c) If a lot of charge flows in a battery, a significant amount of heating can occur, causing the internal resistance of the battery to increase. This will lower the external voltage supplied to the rest of the circuit. If too many bulbs are connected in parallel to a battery, what will happen to the brightness of bulbs? Explain.
26. Use the domain theory of magnetism to explain each of these statements.
    (a) When a magnet is being magnetized it reaches a point, called saturation, where it cannot become any stronger.

(b) A magnet can be demagnetized by being hammered repeatedly.

27. Is insulation necessary for the conducting wires of a coil? Explain.
28. A strong magnet and a weak magnet attract each other. Which magnet attracts the other with the greater amount of force? Explain.
29. Why is it important that the current in the armature of a motor periodically changes direction?
30. Someone makes the following statement: "If a step-up transformer increases electric potential difference, it must also increase the energy." Dispute this assumption.
31. In a step-up transformer the electric potential difference is increased and the electric current is decreased in the secondary circuit. Ohm's law implies that an increased electric potential difference will produce an increased electric current. Does Ohm's law contradict the properties of transformers? Explain.
32. The armature of a generator is more difficult to turn when it is connected to a circuit, causing an electric current in the circuit. Explain why this is so.
33. A manufacturer has made an electric bell, but after testing it decides that it is not loud enough. Describe different ways to make the electric bell ring louder. (Assume that the only way to make the electric bell ring louder is for the hammer to hit the bell faster.)
34. Consider the following idea:

    An electric motor is used to run a generator. The generator produces electricity at a particular voltage that is then stepped up with a transformer. The higher voltage produced is used to run the electric motor, which keeps the generator operating. The extra voltage can be used to run other devices. Since no energy is added to the motor to keep it running and there is extra energy to run other electric devices, this machine is called a perpetual motion machine. The machine continues to produce more electrical energy than it consumes.

    Comment on how effectively this perpetual motion machine will work.

Applying Inquiry Skills

35. Draw a circuit diagram that includes an electric motor, a green light, a red light, and a switch. If the motor is on, the red light must also be on (to warn people of possible danger). If the motor is off, the green light should be on.
36. Explain how to charge an object positively using only a negatively charged object. Draw a series of diagrams to explain your method. Explain how you would test the object for a positive charge. Test your method to see if it works.
37. A fellow student has constructed an electric motor and asks you for some advice on how to improve it. Give suggestions that might help with the following problems:
    (a) The motor does not work at all.
    (b) The motor works, but the armature turns very slowly.
    (c) The motor works, but the armature turns only after it is given a push to get started.
38. Rear window defrosters in cars are made of several strips of wire that heat up when connected to an electric potential difference. Assume that there are ten of these wires, each with a resistance of 5.0 Ω, and that they are connected to 12.0-V energy supply. Predict how the wires should be connected to deliver the greatest amount of energy to the window per unit time. Prove your prediction by calculating the power dissipated in the circuit when the wires are connected
    (a) in series
    (b) in parallel

39 Suggest a technique for making a magnet in the form of a flexible rubber strip.

Making Connections

40. Design a modern distribution system for the delivery of electrical energy. Your distribution system should include an electric generating station (located far from the consumers), several transformer stations, and pole transformers. Explain the function of each part of the distribution system. Provide details about the electric potential differences in each part of the system, and explain why you have chosen to step up or step down the potential differences. Include all your information on a neatly labelled diagram.

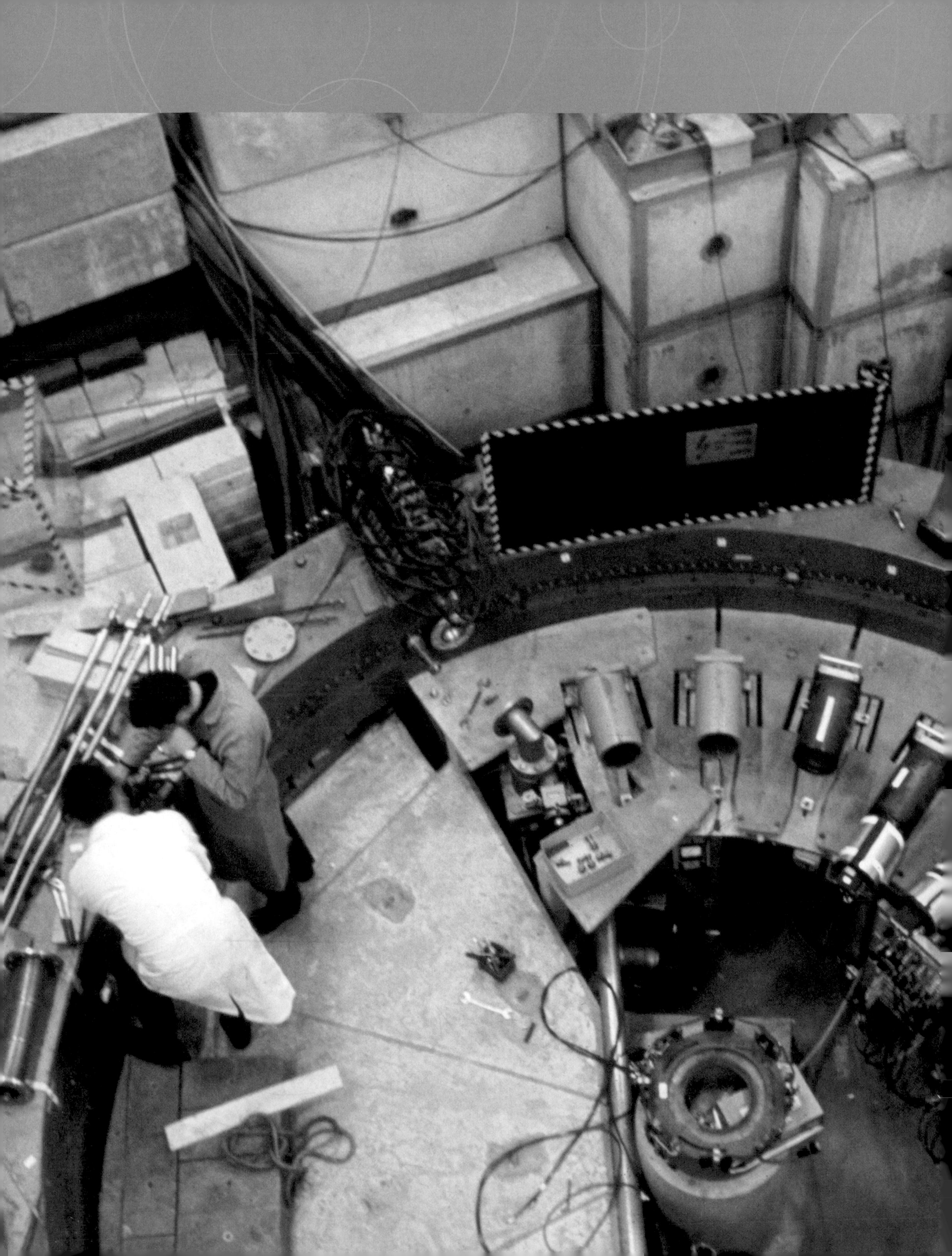

Contents

Scientific Inquiry

Planning an Investigation

In our attempts to further our understanding of the natural world, we encounter questions, mysteries, or events that are not readily explainable. We can use controlled experiments, correlational studies, or observational studies to attempt to answer these questions or explain the events. The methods used in scientific inquiry depend, to a large degree, on the purpose of the inquiry.

Controlled Experiments

A controlled experiment is an example of scientific inquiry in which an independent variable is purposefully and steadily changed to determine its effect on a second dependent variable. All other variables are controlled or kept constant. Controlled experiments are performed when the purpose of the inquiry is to create, test, or use a scientific concept.

The common components of controlled experiments are outlined in the flow chart below. *Even though the sequence is presented as linear, there are normally many cycles through the steps during the actual experiment.*

Process Description

Choose a topic that interests you. Determine whether you are going to create, test, or use a scientific concept and whether you are going to carry out a given procedure or develop a new experimental design. Indicate your decision in a statement of the purpose.

Your question forms the basis for your investigation. Controlled experiments are about relationships, so the question could be about the effects on variable A when variable B is changed. The question may also be about what causes the change in variable A. In this case, you might speculate about possible variables and determine which variable causes the change.

A hypothesis is a tentative explanation. You must be able to test your hypothesis, which can range in certainty from an educated guess to a concept that is widely accepted in the scientific community. A prediction is based upon a hypothesis or a more established scientific explanation, such as a theory. In the prediction you state what outcome you expect from your experiment.

The design of a controlled experiment identifies how you plan to manipulate the independent variable, measure the response of the dependent variable, and control all the other variables.

Stating the purpose → **Asking the question** → **Hypothesizing/predicting** → **Designing the investigation**

Example: Acceleration of Different Masses

When objects fall freely, it seems logical that objects of different masses would experience different accelerations.

The purpose of this investigation is to determine if the mass of a falling object affects its acceleration while falling.

What effect does the mass of a free-falling object have on the value of its acceleration?

Our hypothesis is that an object with a greater mass will experience a greater acceleration because it is pulled down to Earth with a greater force. Our prediction is that the greater the mass, the greater will be the acceleration of the object when it falls.

A ticker-tape timer of known frequency and a measuring tape will be used to measure the time and distance for falling balls of similar size but different masses. The size of the ball is a controlled variable. The ball's mass is an independent variable. The acceleration of the ball is a dependent variable.

Table 1 Time and Displacement

Ball	Ball mass (g)	Trial	Number of timer intervals	Total displacement (m [down])
1	37	1		
		2		
		3		
2	55	1		
		2		
		3		
3	89	1		
		2		
		3		

Table 2 Calculated Accelerations

Ball mass (g)	Trial	Total time (s)	Total displacement (m [down])	Acceleration (m/s² [down])	Average acceleration (m/s² [down])
37	1				
	2				
	3				
55	1				
	2				
	3				
89	1				
	2				
	3				

There are many ways to gather and record your observations during your investigation. It is helpful to plan ahead and think about what data you will need and how best to record them. This helps to clarify your thinking about the question posed at the beginning, the variables, the number of trials, the procedure, the materials, and your skills. It will also help you organize your evidence for easier analysis.

Gathering, recording, and organizing observations

We will perform three trials for each of the different balls. The number of timer intervals will be counted and the displacement will be measured. The data will be recorded in a table similar to **Table 1**. See the observations section of the Lab Report in Appendix A4 for a completed table.

After thoroughly analyzing your observations, you may have sufficient and appropriate evidence to enable you to answer the question posed at the beginning of the investigation.

Analyzing the observations

Using the number of timer intervals, we will determine the time of each fall using

$$\Delta t = \text{no. of intervals} \times \frac{1\ \text{s}}{60\ \text{intervals}}$$

We can then calculate the acceleration using $\Delta \vec{d} = \frac{1}{2}\vec{g}(\Delta t)^2$.

We will then find the average acceleration for each ball.

The results of all calculations will be presented in a table similar to **Table 2.** See the Analysis section of the Lab Report in Appendix A4 for an analysis of this example.

At this stage of the investigation, you will evaluate the processes that you followed to plan and perform the investigation. Evaluating the processes includes evaluating the materials, design, procedure, and your skills. You will also evaluate the outcome of the investigation, which involves evaluating the hypothesis (i.e., whether the evidence supports the hypothesis or not). You must identify and take into account any sources of error and uncertainty in your measurements. Compare the answer created in the hypothesis/prediction with the answer generated by analyzing the evidence. Is the hypothesis acceptable or not?

Evaluating the evidence and the hypothesis

After calculating the average acceleration for each of the three balls of different masses, we will be able to determine whether our hypothesis and prediction are supported by the evidence.

However, we must identify and take into account any sources of error and uncertainty in our measurements of time and displacement. We should also consider any problems with our procedure that may have contributed to errors.

We must decide whether the evidence supports or refutes our hypothesis.

In preparing your report, your objectives should be to describe your design and procedure accurately, and to report your observations accurately and honestly.

Reporting on the investigation

For the format of a typical lab report, see the sample lab report in Appendix A4.

Correlational Studies

When the purpose of scientific inquiry is to test a suspected relationship (hypothesis) between two different variables, but a controlled experiment is not possible, a correlational inquiry is conducted. In a correlational study, the investigator tries to determine whether one variable is affecting another without purposefully changing or controlling any of the variables. Instead, variables are allowed to change naturally. It is often difficult to isolate cause and effect in correlational studies. A correlational inquiry requires very large sample numbers and many replications to increase the certainty of the results.

The flow chart below outlines the components/processes that are important in designing a correlational study. The investigator can conduct the study without doing experiments or fieldwork, for example, by using databases prepared by other researchers to find relationships between two or more variables. The investigator can also make his or her own observations and measurements through fieldwork, interviews, and surveys.

Even though the sequence is presented as linear, there are normally many cycles through the steps during the actual study.

Process Description

Stating the purpose → **Asking the question** → **Hypothesizing/predicting** → **Designing the investigation**

Stating the purpose: **Choose a topic that interests you. Determine whether you are going to replicate or revise a previous study, or create a new one. Indicate your decision in a statement of the purpose.**

Asking the question: **In planning a correlational study, it is important to pose a question about a possible statistical relationship between variable A and variable B.**

Hypothesizing/predicting: **A hypothesis or prediction would not be useful. Correlational studies are not intended to establish cause–and–effect relationships.**

Designing the investigation: **The design of a correlational study identifies how you will gather data on the variables under study and also identifies potential sources. There are two possible sources—observation made by the investigator and existing data.**

Example: Speeding and Automobile Accidents

Stating the purpose: Speed kills! That's the message we see in advertisements and promotions that encourage us to slow down. Do the statistics actually support this claim? The purpose of this investigation is to attempt to determine if there is a relationship between speed and automobile accidents.

Asking the question: What is the relationship between the speed at which a driver is driving and the relative risk of being involved in an accident?

Hypothesizing/predicting: Does not apply.

Designing the investigation: We will gather data from police and insurance company records and from the department/ministry responsible for motor vehicle registration. These records will show the number of accidents in which the initial speed was 25 km/h below, to 25 km/h above the speed limit (in 5-km/h increments). We will attempt to select reports of accidents in which other factors (e.g., driving conditions, time of day, etc.) are the same.

Table 3 Accidents at Different Speeds

Pre-crash speed (km/h above/below speed limit, SL)	–25	–20	–15	–10	–5	0 (SL)	5	10	15	20	25
Accidents per 1000 vehicles											

Gathering, recording, and organizing observations

There are many ways to gather and record your observations during your investigation. It is helpful to plan ahead and think about what data you will need and how best to record them. This is an important step because it helps to clarify your thinking about the question posed at the beginning, the variables, the number of trials, the procedure, the materials, and your skills. It will also help you organize your observations for easier analysis.

We will record the data in a table similar to **Table 3**.

Analyzing the observations

After thoroughly analyzing your observations, you may have sufficient and appropriate evidence to enable you to answer the question posed at the beginning of the investigation.

By comparing the number of accidents at speeds above and below the speed limit with the number at the speed limit, we will be able to determine the relative risk of being involved in an accident.

To calculate the relative risk we can use

$$RR = \left(\frac{A_{110}}{N - A_{110}}\right) \div \left(\frac{A_{SL}}{N - A_{SL}}\right)$$

where RR is the relative risk, N = total number of vehicles (1000), and A represents the number of accidents at a particular speed.

Using this equation, the risk factor at the speed limit is 1.

Evaluating the evidence

At this stage of the investigation, you will evaluate the processes used to plan and perform the investigation. Evaluating the processes includes evaluating the materials, design, the procedure, and your skills.

The analysis of the data will provide us with the relative risk, not the absolute risk. The absolute risk cannot be determined; there is obviously some risk while driving at the speed limit. We will see whether the risk increases as the speed increases above the speed limit.

We will need to judge whether the evidence is appropriate and whether it reveals a valid correlation between the two variables.

This type of study could be extended to determine if there is any difference for different age groups. It could also lead to controlled studies that could help determine the reasons for an identified correlation.

Reporting on the investigation

In preparing your report, your objectives should be to describe your design and procedure accurately and to report your observations accurately and honestly.

For the format of a typical lab report, see the sample lab report in Appendix A4.

Observational Studies

Often the purpose of inquiry is simply to study a natural phenomenon with the intention of gaining scientifically significant information to answer a question. Observational studies involve observing a subject or phenomenon in an unobtrusive or unstructured manner, often with no specific hypothesis. A hypothesis to describe or explain the observations may, however, be generated over time, and modified as new information is collected.

The flow chart below summarizes the stages and processes of scientific inquiry through observational studies.

Even though the sequence is presented as linear, there are normally many cycles through the steps during the actual study.

	Stating the purpose	Asking the question	Hypothesizing/ predicting	Designing the investigation
Process Description	**Choose a topic that interests you. Determine whether you are going to replicate or revise a previous study, or create a new one. Indicate your decision in a statement of the purpose.**	**In planning an observational study, it is important to pose a general question about the natural world. You may or may not follow the question with the creation of a hypothesis.**	**A hypothesis is a tentative explanation. In an observational study, a hypothesis can be formed after observations have been made and information has been gathered on a topic. A hypothesis may be created in the analysis.**	**The design of an observational study describes how you will make observations relevant to the question.**
Example: School Noise Levels	We often ignore or grow accustomed to the noise that surrounds us during our daily life. The purpose of this study is to determine the noise levels at various locations around the school and during different school functions.	What is the noise level (in decibels) at common locations in school and during common school activities?	A hypothesis or prediction is not warranted in this study. This investigation is attempting to determine the noise levels that students are exposed to during their normal activities at school. This is an information-gathering study which will help us determine whether a noise problem exists. This knowledge will help us identify the sources of the noise. Since hearing loss from noise is normally gradual and painless, this knowledge may be used to raise awareness and encourage individuals to take corrective or protective measures.	This study involves the measurement of the noise level at various locations during normal everyday activities. No attempt is made to regulate or control any situation where noise is present. Measurements will be taken with a sound meter that can measure accurately to the nearest decibel. Measurements will be taken hourly during the school day (from one hour before classes start to one hour after classes end) at the following locations for a week (five school days): classrooms, science lab, gymnasium, industrial arts shop, cafeteria, music room, hallways, staffroom, bus (measurements only during transport times).

Table 4 Noise Levels in the Gymnasium

Day	Time	Noise level (dB)	Description of activity
Day 1	1st hour		
	2nd hour		
	3rd hour		
	4th hour		
	5th hour		
	6th hour		
	7th hour		

There are many ways to gather and record your observations during your investigation. During your observational study, you should quantify your observations where possible. All observations should be as objective and unambiguous as possible. Consider ways to organize your information for easier analysis.

Gathering, recording, and organizing observations

The measurements of noise levels will be recorded in tables similar to **Table 4.**

After thoroughly analyzing your observations, you may have sufficient and appropriate evidence to enable you to answer the question posed at the beginning of the investigation. You may also have enough observations to form a hypothesis.

Analyzing the observations

Analysis of the data will identify the kinds of activities that produce the loudest noise levels. We will enter the data into a database or spreadsheet and use it to generate a number of graphs. For each location we will plot a line graph showing the average noise level for each hour of the day. The daily and weekly average noise levels for each of the locations will be calculated and plotted on bar graphs.

At this stage of the investigation, you will evaluate the processes used to plan and perform the investigation. Evaluating the processes includes evaluating the materials, design, procedure, and your skills. The results of most such investigations will suggest further studies, perhaps correlational studies or controlled experiments, to explore tentative hypotheses you may have developed.

Evaluating the evidence and the hypothesis

Since we did not state a hypothesis or make a prediction we will not need to evaluate these. We will, however, need to decide whether our measurements are representative of a typical school/classroom situation.

The information generated by this study may have significant benefit to students and teachers. Any locations where noise levels are consistently above the acceptable standards will be identified so that individuals may take corrective or protective measures.

In preparing your report, your objectives should be to describe your design and procedure accurately, and to report your observations accurately and honestly.

Reporting on the investigation

For the format of a typical lab report, see the sample lab report in Appendix A4.

Decision Making

Modern life is filled with environmental and social issues that have scientific and technological dimensions. An issue is defined as a problem that has at least two possible solutions rather than a single answer. There can be many positions, generally determined by the values that an individual or a society holds, on a single issue. Which solution is "best" is a matter of opinion; ideally, the solution that is implemented is the one that is most appropriate for society as a whole.

The common processes involved in the decision-making process are outlined in the graphic below.

Even though the sequence is presented as linear, you may go through several cycles before deciding you are ready to defend a decision.

Process Description

Defining the issue

The first step in understanding an issue is to explain why it is an issue, describe the problems associated with the issue, and identify the individuals or groups, called stakeholders (Table 1), involved in the issue. You could brainstorm the following questions to research the issue: Who? What? Where? When? Why? How? Develop background information on the issue by clarifying facts and concepts, and identifying relevant attributes, features, or characteristics of the problem.

Identifying alternatives/positions

Examine the issue and think of as many alternative solutions as you can. At this point it does not matter if the solutions seem unrealistic. To analyze the alternatives, you should examine the issue from a variety of perspectives. Stakeholders may bring different viewpoints to an issue and these may influence their position on the issue. Brainstorm or hypothesize how different stakeholders would feel about your alternatives. Perspectives that stakeholders may adopt while approaching an issue are listed in Table 2.

Researching the issue

Formulate a research question that helps to limit, narrow, or define the issue. Then develop a plan to identify and find reliable and relevant sources of information. Outline the stages of your information search: gathering, sorting, evaluating, selecting, and integrating relevant information. You may consider using a flow chart, concept map, or other graphic organizer to outline the stages of your information search. Gather information from many sources, including newspapers, magazines, scientific journals, the Internet, and the library.

Example: Cellular Phone Use

Defining the issue

In the past few years there has been a tremendous surge in the number of cellular phone users around the world. An alarm bell has been sounded regarding the health risks associated with the use of cell phones. Claims have been made that the electromagnetic field (EMF) that surrounds the phone causes damage to human cells and can cause cancer. The existing scientific evidence is conflicting and inconclusive.

However, as of August 2000 the U.S. cellular telephone industry is required to publish information on the amount of radiation that enters users' heads when they use various wireless phones. Canadian authorities are studying these results carefully.

As a result of the public concerns, a number of companies have developed shielding devices that can be attached to the phone to eliminate or significantly reduce the amount of radiation that enters the user's head.

The issue is the health risk associated with using cellular phones. In this debate, there are basically two positions: you either accept the claims that cell phones do indeed pose a health risk or you consider the absence of scientific proof of risk as evidence that cell phones are safe. (See the Decision-Making section of the CD for a list of potential groups or stakeholders who may be positively or negatively affected by the issue.) Develop background information on the issue by clarifying facts and concepts, and identifying relevant attributes, features, or characteristics of the problem.

Identifying alternatives/positions

One possible solution for those concerned with the risk of cell phones is to have government legislate all manufacturers to install a shield on all appliances that emit electromagnetic radiation.

Think about how different stakeholders might feel about the alternative. For example, consumers may wish to forgo the convenience of a mobile telephone and revert back to regular phones. What would be their perspective? What would be the perspective of a cell phone manufacturer? A manufacturer of an EMF shield? A consumer advocate? A research scientist? An economist? A politician? (See the Decision-Making section of the CD for a description of the possible perspectives on an issue.)

Remember that one person could have more than one perspective. It is even possible that two people, looking at an issue from the same perspective, might disagree about the issue. For example, scientists might disagree about the health risk associated with cell phone use.

Researching the issue

Begin your search for reliable and relevant sources of information about the issue with a question such as "What does the existing research say about the health risk associated with cell phone use?" Outline the stages of your information search: gathering, sorting, evaluating, selecting, and integrating relevant information on the issue. Gather information from a number of different sources including newspapers, magazines, scientific journals, the Internet, and the library.

In this stage, you will analyze the issue and clarify where you stand. First, you should establish criteria for evaluating your information to determine its relevance and significance. You can then evaluate your sources, determine what assumptions may have been made, and assess whether you have enough information to make your decision.

Once the issue has been analyzed, you can begin to evaluate the alternative solutions. You may decide to carry out a risk-benefit analysis—a tool that enables you to look at each possible result of a proposed action and helps you make a decision.

There are five steps that must be completed to effectively analyze the issue:

1. **Establish criteria for determining the relevance and significance of the data you have gathered.**
2. **Evaluate the sources of information.**
3. **Identify and determine what assumptions have been made. Challenge unsupported evidence.**
4. **Determine any causal, sequential, or structural relationships associated with the issue.**
5. **Evaluate the alternative solutions, possibly by conducting a risk-benefit analysis.**

After analyzing your information, you can answer your research question and take an informed position on the issue. You should be able to defend your solution in an appropriate format—debate, class discussion, speech, position paper, multimedia presentation (e.g., computer slide show), brochure, poster, video.

Your position on the issue must be justified using supporting information that you have researched. You should be able to defend your position to people with different perspectives. Ask yourself the following questions:

- **Do I have supporting evidence from a variety of sources?**
- **Can I state my position clearly?**
- **Can I show why this issue is relevant and important to society?**
- **Do I have solid arguments (with solid evidence) supporting my position?**
- **Have I considered arguments against my position, and identified their faults?**
- **Have I analyzed the strong and weak points of each perspective?**

The final phase of decision making includes evaluating the decision itself and the process used to reach the decision. After you have made a decision, carefully examine the thinking that led to your decision.

Some questions to guide your evaluation include:

- **What was my initial perspective on the issue? How has my perspective changed since I first began to explore the issue?**
- **How did we make our decision? What process did we use? What steps did we follow?**
- **In what ways does our decision resolve the issue?**
- **What are the likely short- and long-term effects of the decision?**
- **To what extent am I satisfied with the final decision?**
- **What reasons would I give to explain our decision?**
- **If we had to make this decision again, what would I do differently?**

→ Analyzing the issue → Defending the decision → Evaluating the process

The issue of cell phone radiation has arisen as a consequence of the desire of people to maintain communication while away from their homes or offices. A secondary consequence may be that the demand for, and therefore the price of, regular home-based phones will increase. A related issue is that taxpayers' money is being used to conduct research into the safety of cell phones, an item considered by many to be a luxury.

Table 3 shows a risk-benefit analysis of cellular phone use, based on the analysis of viewpoints of those directly involved (**Table 1**), and general considerations for issue analysis (**Table 2**).

By reviewing research from a variety of sources, and performing a risk-benefit analysis using the information gathered, we have decided that there is minimal risk to human health as a result of short-term exposure to the EMF produced by cellular phones. More research is required on long-term exposure. Several correlational studies seem to indicate that there is a relationship between cell phone use and increased incidence of cancer, but such studies can't establish that cell phone use causes cancer. Cell phones are very useful communication tools, and their use should not be banned or restricted until evidence clearly shows that they cause cancer.

Every effort was made to obtain the most current information on cell phones and their health effects on humans. We reviewed studies that focused on the relationship between cell phone use and the incidence of cancer from a variety of reputable sources including universities, government agencies, cell phone manufacturers, and several independent research organizations. In addition to conducting a panel discussion among experts in the field, we carried out a risk-benefit analysis to help arrive at our decision. We realize that this decision will not satisfy all stakeholders, but represents the best solution given the evidence at our disposal. This decision will not drastically change cell phone use in the short term, but will stimulate further debate, discussion, and research.

Table 1 Potential Stakeholders in the Cellular Phone Debate

Stakeholder	Viewpoint (perspectives)
consumer	1. Cell phones should pose no risk to users. 2. Some risk is acceptable. 3. The truth about cell phone risk should be available so that users can make an informed decision.
scientist	1. Research shows a positive correlation between cell phone use and health problems. 2. Research shows no correlation between cell phone use and health problems.
doctor	The health care system is already overburdened. Err on the side of caution and ban cell phone use until zero risk is guaranteed.
cell phone manufacturer	Radiation from cell phones is far below government standards. Cell phones save lives.
radiation shield manufacturer	Users can be economically protected from cell phone radiation by installing a shielding device.
politician	Cell phone production and use is a significant contributor to the economy.
consumer advocate	Any level of risk is unacceptable.

Table 2 Perspectives on an Issue

Issue	Considerations
cultural	customs and practices of a particular group
ecological	an interaction among organisms and their natural habitat
economic	the production, distribution, and consumption of wealth
educational	the effects on learning
emotional	feelings and emotions
environmental	the effects on physical surroundings
aesthetic	artistic, tasteful, beautiful
moral/ethical	what is good/bad, right/wrong
legal	the rights and responsibilities of individuals and groups
spiritual	the effects on personal beliefs
political	the effects on the aims of a political group or party
scientific	logical or research-based
social	the effects on human relationships, the community, or society
technological	machines and industrial processes

A Risk-Benefit Analysis Model

- Research as many different aspects of the proposal as possible. Look at it from different perspectives. Collect as much evidence as you possibly can, including reasonable projections. Classify every individual potential result as being either a benefit or a risk.
- Quantify the size of the potential benefit or risk (perhaps as a dollar figure, or the number of lives affected, or on a scale of 1 to 5).
- Estimate the probability (percentage) of that event occurring.
- By multiplying the size of a benefit (or risk) by the probability of it happening, you can calculate a "probability value" of each potential result.
- Total the probability values of all the potential risks and all the potential benefits.
- Compare the sums to help you decide whether to proceed with the proposed action.

Table 3 Risk-Benefit Analysis of Cellular Phone Use

Risks				**Benefits**			
Possible result	*Cost of result (scale 1 to 5)*	*Probability of result occurring (%)*	*Cost × probability*	*Possible result*	*Benefit of result (scale 1 to 5)*	*Probability of result occurring (%)*	*Benefit × probability*
Cell phone use increases the risk of cancer.	very serious 5	inconclusive (50%)	250	Cell phones help save lives.	great 5	very likely (90%)	450
Cell phone radiation produces unknown effects on human cells.	serious 4	somewhat likely (70%)	280	Cell phone radiation may have beneficial effects on human health.	high 4	somewhat unlikely (30%)	120
Health care costs will increase because of problems from cell phone use.	serious 4	somewhat likely (60%)	240	Cell phones help people become more productive.	high 4	likely (80%)	320
Total risk value			**770**	**Total benefit value**			**890**

Technological Problem Solving

There is a difference between science and technology. The goal of science is to understand the natural world. The goal of technological problem solving is to develop or revise a product or a process in response to a human need. The product or process must fulfill its function but, in contrast to scientific problem solving, it is not essential to understand why or how it works. Technological solutions are evaluated based on such criteria as simplicity, reliability, efficiency, cost, and ecological and political ramifications.

Even though the sequence presented in the graphic below is linear, there are normally many cycles through the steps in any problem-solving attempt.

Process Description

Defining the problem

This process involves recognizing and identifying the need for a technological solution. You need to clearly state the question(s) that you want to investigate to solve the problem, and the criteria you will use as guidelines and to evaluate your solution. In any design, some criteria may be more important than others. For example, if the product solution measures accurately and is economical, but is not safe, then it is clearly unacceptable.

Identifying possible solutions

Use your prior knowledge and experience to propose possible solutions. Creativity is also important in suggesting novel solutions.

You should generate as many ideas as possible about the functioning of your solution and about potential designs. During brainstorming, the goal is to generate many ideas without judging them. They can be evaluated and accepted or rejected later.

To visualize the possible solutions it is helpful to draw sketches. Sketches are often better than verbal descriptions to communicate an idea.

Planning

Planning is the heart of the entire process. Your plan will outline your processes, identify potential sources of information and materials, define your resource parameters, and establish evaluation criteria.

Seven types of resources are generally used in developing technological solutions to problems—people, information, materials, tools, energy, capital, and time.

Example: Optical Instruments

Defining the problem

For example, What is the location of the site to be observed (e.g., Is it on land or under water)? How big (long) must the instrument be? What are the limits on the cost of construction materials? Will we need a light source? How much money do we have available? ... and so on.

In this case, you are asked to design and construct an optical instrument to meet the following criteria:

- a total reach of 1 m
- able to reach around two 45° turns and a 90° turn (as illustrated in **Figure 1**)
- incorporates a light source at the far viewing end
- moveable mirrors and/or lenses
- costs $50 or less
- reusable
- no negative impact on the viewing area

In any design, some criteria may be more important than others. For example, if a design is the right size, allows adequate observation, and can be reused easily but is likely to damage the observation area, then it is clearly unacceptable.

Identifying possible solutions

See sketches of two possible designs in **Figure 2**.

Design 1 shows a series of short sections joined together by pin hinges. The sections can be manoeuvred by strings running through the loops at the top of each section.

Design 2 shows a flexible bar, similar to a vertical blind bar, with suspended mirrors that can be moved along the bar. The angle of the mirrors can be controlled by strings run through loops on the bar.

Figure 1
Sketch of obscure area

Planning

People: The human resources required to solve this problem include you and your partner.
Information: You are already very familiar with mirrors. You will need to fully understand the law of reflection and be able to calculate angles of incidence and reflection.
Materials: As well as the limitations imposed by your proposed solution, cost, availability, and time, you are restricted to materials that can be obtained in school or at home. Materials to consider include mirrors, lenses, glue, wire, light bulb, wood or other casing material, etc.
Tools: Your proposed design could require the use of common hand tools such as a saw, pliers, shears, and glue guns, but your design should not require any specialized tools or machines.
Energy: The only energy requirement for this problem is a source of electrical energy for the light.
Capital: The capital resources must be minimal, otherwise it may be more efficient to purchase a commercial model. Limit your design to a cost of $50 or less.
Time: You should be able to construct your optical instrument within 60 minutes (not including designing and testing).

The solution will be evaluated on how well it meets the design criteria established earlier.

In this phase, you will construct and test your prototype using systematic trial and error. Try to manipulate only one variable at a time. Use failures to inform the decisions you make before your next trial. You may also complete a cost-benefit analysis on the prototype

To help you decide on the best solution, you can rate each potential solution on each of the design criteria using a five-point rating scale, with 1 being poor, 2 fair, 3 good, 4 very good, and 5 excellent. You can then compare your proposed solutions by totalling the scores.

Once you have made the choice among the possible solutions, you need to produce and test a prototype. While making the prototype you may need to experiment with the characteristics of different components. A model, on a smaller scale, might help you decide whether the product will be functional. The test of your prototype should answer three basic questions:

- **Does the prototype solve the problem?**
- **Does it satisfy the design criteria?**
- **Are there any unanticipated problems with the design?**

If these questions cannot be answered satisfactorily, you may have to modify the design or select another potential solution.

Constructing/testing solutions

Table 1 illustrates the rating for two different telescope designs. Note that although Design 1 came out with the highest rating, one factor (impact on viewing area) suggests we should go with Design 2. This is what is referred to as a trade-off. We have to compromise on other criteria such as reusability in order to ensure that the instrument does not damage or interfere with the viewing area. By reviewing or evaluating product and processes to this point, we may be able to modify Design 2 to optimize its performance on the other criteria.

Table 1 Design Analysis

Criterion	Design 1	Design 2
reach	4	3
navigate turns	4	3
light source	4	4
movable mirrors/lenses	4	3
cost	5	4
reusability	5	3
impact on viewing area	1	4
Total score	**27**	**24**

In presenting your solution, you will communicate your solution, identify potential applications, and put your solution to use.

Once the prototype has been produced and tested, the best presentation of the solution is a demonstration of its use—a test under actual conditions. This demonstration can also serve as a further test of the design. Any feedback should be considered for future redesign. Remember that no solution should be considered the absolute final solution.

Presenting the preferred solution

The optical device was presented to the class through a demonstration and a video.

The technological problem-solving process is cyclical. At this stage, evaluating your solution and the process you used to arrive at your solution may lead to a revision of the solution.

Evaluation is not restricted to the final step; however, it is important to evaluate the final product using the criteria established earlier, and to evaluate the processes used while arriving at the solution. Consider the following questions:

- **To what degree does the final product meet the design criteria?**
- **Did you have to make any compromises in the design? If so, are there ways to minimize the effects of the compromises?**
- **Did you exceed any of the resource parameters?**
- **Are there other possible solutions that deserve future consideration?**
- **How did your group work as a team?**

Evaluating the solution and process

The class was given an opportunity to provide feedback by filling out a survey questionnaire. The designers conducted a self and peer evaluation of the design and construction processes and produced a report.

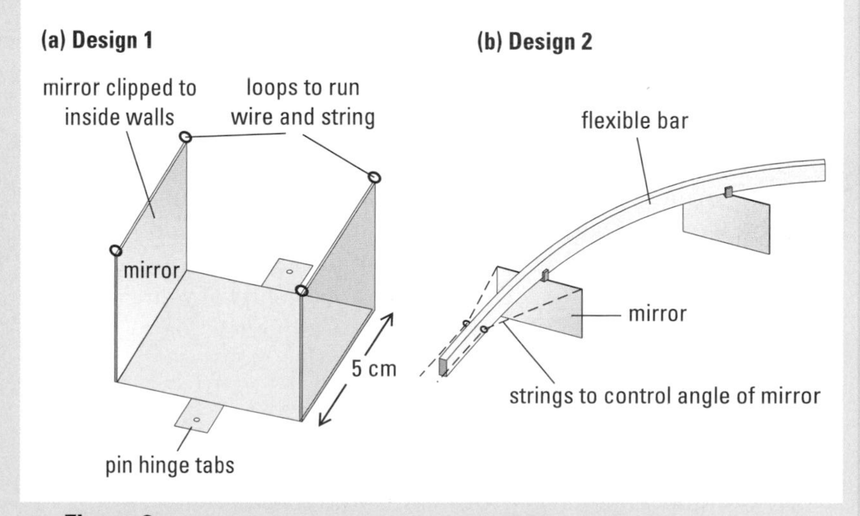

Figure 2

Lab Reports

When carrying out investigations, it is important that scientists keep records of their plans and results, and share their findings. In order to have their investigations repeated (replicated) and accepted by the scientific community, scientists generally share their work by publishing papers in which details of their design, materials, procedure, evidence, analysis, and evaluation are given.

Lab reports are prepared after an investigation is completed. To ensure that you can accurately describe the investigation, it is important to keep thorough and accurate records of your activities as you carry out the investigation.

Investigators use a similar format in their final reports or lab books, although the headings and order may vary. Your lab book or report should reflect the type of scientific inquiry that you used in the investigation and should be based on the following headings, as appropriate. (See **Figure 1** for a sample lab report.)

Title

At the beginning of your report, write the number and title of your investigation. In this course the title is usually given, but if you are designing your own investigation, create a title that suggests what the investigation is about. Include the date the investigation was conducted and the names of all lab partners (if you worked as a team).

Purpose

State the purpose of the investigation. Why are you doing this investigation?

Question

This is the question that you attempted to answer in the investigation. If it is appropriate to do so, state the question in terms of independent and dependent variables.

Hypothesis/Prediction

Based on your reasoning or on a concept that you have studied, formulate an explanation of what should happen (a hypothesis). From your hypothesis you may make a prediction, a statement of what you expect to observe, before carrying out the investigation. Depending on the nature of your investigation, you may or may not have a hypothesis or a prediction.

Design

This is a brief general overview (one to three sentences) of what was done. If your investigation involved independent, dependent, and controlled variables, list them. Identify any control or control group that was used in the investigation.

Materials

This is a detailed list of all materials used, including sizes and quantities where appropriate. Be sure to include safety equipment such as goggles, lab apron, latex gloves, and tongs, where needed. Draw a diagram to show any complicated setup of apparatus.

Procedure

Describe, in detailed, numbered, step-by-step format, the procedure you followed in carrying out your investigation. Include steps to clean up and dispose of waste.

Observations

This includes all qualitative and quantitative observations that you made. Be as precise as appropriate when describing quantitative observations, include any unexpected observations, and present your information in a form that is easily understood. If you have only a few observations, this could be a list; for controlled experiments and for many observations, a table will be more appropriate.

Analysis

Interpret your observations and present the evidence in the form of tables, graphs, or illustrations, each with a title. Include any calculations, the results of which can be shown in a table. Make statements about any patterns or trends you observed. Conclude the analysis with a statement based only on the evidence you have gathered, answering the question that initiated the investigation.

Evaluation

The evaluation is your judgment about the quality of evidence obtained and about the validity of the prediction and hypothesis (if present). This section can be divided into two parts:

- Did your observations provide reliable and valid evidence to enable you to answer the question? Are you confident enough in the evidence to use it to evaluate any prediction and/or hypothesis you made?
- Was the prediction you made before the investigation supported or falsified by the evidence? Based on your evaluation of the evidence or prediction, is the hypothesis supported or should it be rejected?

Investigation 2.5 – Acceleration of Different Masses

Conducted: February 2, 2001

By: Jenny Wilson and Sharif Khan

Purpose

The purpose of this investigation was to determine whether the mass of a falling object determines its acceleration while falling.

Question

What effect does the mass of a free-falling object have on the value of its acceleration?

Hypothesis/Prediction

It seems logical that an object with a greater mass will fall faster because it is pulled down to Earth at a greater rate. Our prediction was that the greater the mass, the greater the acceleration of the object when it falls.

Design

A ticker-tape timer of known frequency and a measuring tape were used to measure the time and distance for falling balls of similar size but different masses. The size of ball was the controlled variable, the ball's mass was the independent variable, and the acceleration of the ball was the dependent variable.

Materials

3 balls of similar size with different masses
ticker-tape timer
measuring tape

Procedure

1. A safe area to set up the apparatus was selected.
2. An observation table similar to **Table 1** was prepared.
3. The ticker-tape timer was clamped to a stand placed near the edge of a lab bench. The timer was set to 60 dots per second.
4. A piece of ticker tape slightly shorter than the distance the ball would fall was obtained so that the tape passed completely through the timer before the ball hit the floor.
5. The ticker tape was inserted through the timer and the end of the ticker tape was attached to the first ball with a small piece of masking tape.
6. The ball was kept steady, just below the timer, by holding the upper end of the piece of ticker tape. The tape was held directly above the timer so the tape would move freely through the timer.
7. The timer was started and the ball was immediately released.

Figure 1
Sample lab report

8. The timer was turned off when the tape passed completely through the timer.
9. The tape was checked to make sure that the dots were clear and that no dots were missing. If the tape wasn't usable, the procedure was repeated to get a good tape.
10. The number of dots were counted and the distance between the start and the last dot was measured.
11. Steps 4 to 10 were repeated twice more.
12. The observations were recorded in the observation table.
13. Steps 4 to 12 were repeated with two other balls of different masses.
14. All equipment was put away and all waste was recycled or disposed of.

Observations

The measurements we made when we carried out the procedure described above are recorded in **Table 1**.

Table 1 Time and Displacement

Ball	Ball mass (g)	Trial	Number of timer intervals	Total displacement (m [down])
1	37	1	25	0.860
		2	24	0.792
		3	26	0.923
2	55	1	25	0.855
		2	26	0.925
		3	25	0.860
3	89	1	25	0.858
		2	26	0.926
		3	26	0.924

Analysis

Using our observations we calculated the time of each fall, and we could then calculate the acceleration. The calculations below show how we figured out the time and how we calculated the acceleration of each fall and then the average acceleration. The results of all calculations are presented in **Table 2**.

Total Time

The total time for the 37-g ball, Trial 1, is calculated as follows:

$$\Delta t = 25 \text{ intervals} \times \frac{1 \text{ s}}{60 \text{ intervals}}$$

$$\Delta t = 0.417 \text{ s}$$

Acceleration

Assuming an initial velocity of 0 m/s and down to be positive, the acceleration for the 37-g ball (Trial 1) is calculated as follows:

$$\Delta\vec{d} = \frac{1}{2}\vec{g}\,(\Delta t)^2$$

Since $\vec{v}_i = 0$, then $\Delta\vec{d} = \frac{1}{2}\vec{a}_g\,(\Delta t)^2$, or $\vec{g} = \frac{2\Delta\vec{d}}{(\Delta t)^2}$

$$\vec{g} = \frac{2\Delta\vec{d}}{(\Delta t)^2}$$

$$= \frac{2(+0.860\text{ m})}{(0.417\text{ s})^2}$$

$$\vec{g} = +9.91\text{ m/s}^2$$

The acceleration for the 37-g ball in Trial 1 was 9.91 m/s^2 [down].

Average Acceleration

The average acceleration for all three trials of the 37-g ball is calculated as follows:

$$\vec{a}_{av} = \frac{(9.91 + 9.90 + 9.83)\text{ m/s}^2\text{ [down]}}{3}$$

$$\vec{a}_{av} = 9.88\text{ m/s}^2\text{ [down]}$$

The average acceleration for the 37-g ball is 9.88 m/s^2 [down].

According to the evidence in **Table 2**, the mass of the ball apparently has no significant effect on the value of the acceleration due to gravity.

Table 2 Calculated Accelerations

Ball mass (g)	Trial	Number of timer intervals	Total time (s)	Total displacement (m [down])	Acceleration (m/s^2 [down])	Average acceleration (m/s^2 [down])
	1	25	0.417	0.860	9.91	
37	2	24	0.400	0.792	9.90	9.88
	3	26	0.433	0.923	9.83	
	1	25	0.417	0.855	9.85	
55	2	26	0.433	0.925	9.87	9.88
	3	25	0.417	0.860	9.91	
	1	25	0.417	0.858	9.88	
89	2	26	0.433	0.926	9.86	9.86
	3	26	0.433	0.924	9.84	

Evaluation

The design of the investigation is adequate because the question was clearly answered and there are no obvious flaws. The materials, skills, and procedure were all satisfactory because they produced sufficient and reliable evidence. We consider the evidence reliable because all of the trials produced results that were very close.

The accepted value for the acceleration of gravity is 9.81 m/s^2. We calculated the percentage error as follows:

$$\% \text{ error} = \frac{|\text{experimental} - \text{accepted}|}{\text{accepted}} \times 100\%$$

$$= \frac{|(9.88 - 9.81)\ \text{m/s}^2|}{9.81\ \text{m/s}^2} \times 100\%$$

$$\% \text{ error} = 0.7\%$$

The percentage error between the experimental value of acceleration due to gravity for the 37-g ball and the accepted value is 0.7%. The percentage error for the 55-g ball was also 0.7% and for the 89-g ball, 0.5%. These low percentage errors indicate that the mass of the ball has no significant effect on the value of its acceleration due to gravity.

Some possible sources of error and uncertainty include the dots made by the timer. We assume that the timer frequency (60 dots per second) is constant. The dot marked as dot 1 is not really the precise starting point of the fall, because the tape must move enough to provide a clear dot from which to measure. Measurement uncertainties and some friction are also present.

The prediction is not supported because of the very similar values for the acceleration due to gravity for balls with very different masses. The small differences obtained for different balls can easily be accounted for by the sources of error and uncertainty.

The hypothesis that a greater mass with its greater weight is pulled down to Earth at a greater rate is clearly not supported by the evidence obtained in this investigation.

Math Skills

Scientific Notation

It is difficult to work with very large or very small numbers when they are written in common decimal notation. Usually it is possible to accommodate such numbers by changing the SI prefix so that the number falls between 0.1 and 1000; for example, 237 000 000 mm can be expressed as 237 km, and 0.000 000 895 kg can be expressed as 0.895 mg. However, this prefix change is not always possible, either because an appropriate prefix does not exist or because it is essential to use a particular unit of measurement. In these cases, the best method of dealing with very large and very small numbers is to write them using scientific notation. Scientific notation expresses a number by writing it in the form $a \times 10^n$, where $1 < |a| < 10$ and the digits in the coefficient a are all significant. **Table 1** shows situations where scientific notation would be used.

Table 1 Examples of Scientific Notation

Expression	Common decimal notation	Scientific notation
124.5 million kilometres	124 500 000 km	1.245×10^8 km
154 thousand picometres	154 000 pm	1.54×10^5 pm
602 sextillion/mol	602 000 000 000 000 000 000 000/mol	6.02×10^{23}/mol

To multiply numbers in scientific notation, multiply the coefficients and add the exponents; the answer is expressed in scientific notation. Note that when writing a number in scientific notation, the coefficient should be between 1 and 10 and should be rounded to the same certainty (number of significant digits) as the measurement with the least certainty (fewest number of significant digits). Look at the following examples:

$$(4.73 \times 10^5 \text{ m})(5.82 \times 10^7 \text{ m}) = 27.5 \times 10^{12} \text{ m}^2 = 2.75 \times 10^{13} \text{ m}^2$$

$$(3.9 \times 10^4 \text{ N}) \div (5.3 \times 10^{-3} \text{ m}) = 0.74 \times 10^7 \text{ N/m} = 7.4 \times 10^6 \text{ N/m}$$

On many calculators, scientific notation is entered using a special key, labelled EXP or EE. This key includes "× 10" from the scientific notation; you need to enter only the exponent.

For example, to enter

7.5×10^4 press 7.5 EXP 4

3.6×10^{-3} press 3.6 EXP +/–3

Uncertainty in Measurements

Two types of quantities are used in science: exact values and measurements. Exact values include defined quantities (1 m = 100 cm) and counted values (5 cars in a parking lot). Measurements, however, are not exact because there is always some uncertainty or error.

There are two types of measurement error. *Random error* results when an estimate is made to obtain the last significant digit for any measurement. The size of the random error is determined by the precision of the measuring instrument. For example, when measuring length, it is necessary to estimate between the marks on the measuring tape. If these marks are 1 cm apart, the random error will be greater and the precision will be less than if the marks are 1 mm apart.

Systematic error is associated with an inherent problem with the measuring system, such as the presence of an interfering substance, incorrect calibration, or room conditions. For example, if the balance is not zeroed at the beginning, all measurements will have a systematic error; using a slightly worn metre stick will also introduce error.

The precision of measurements depends on the gradations of the measuring device. *Precision* is the place value of the last measurable digit. For example, a measurement of 12.74 cm is more precise than one of 127.4 cm because the first value was measured to hundredths of a centimetre whereas the latter was measured to tenths of a centimetre.

When adding or subtracting measurements of different precision, the answer is rounded to the same precision as the least precise measurement. For example, using a calculator,

$$11.7 \text{ cm} + 3.29 \text{ cm} + 0.542 \text{ cm} = 15.532 \text{ cm}$$

The answer must be rounded to 15.5 cm because the first measurement limits the precision to a tenth of a centimetre.

No matter how precise a measurement is, it still may not be accurate. Accuracy refers to how close a value is to its accepted value. The *percentage error* is the absolute value of the difference between experimental and accepted values expressed as a percentage of the accepted value.

$$\% \text{ error} = \frac{|\text{experimental value} - \text{accepted value}|}{\text{accepted value}} \times 100\%$$

The percentage difference is the difference between a value determined by experiment and its predicted value. The *percentage difference* is calculated as

$$\% \text{ difference} = \frac{|\text{experimental value} - \text{predicted value}|}{\text{predicted value}} \times 100\%$$

(a)

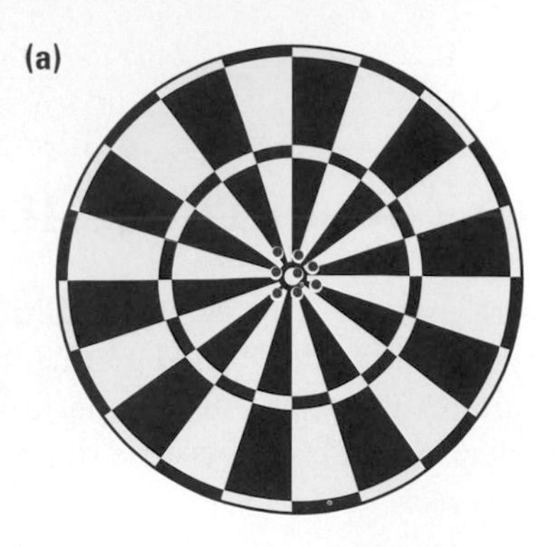

(b)

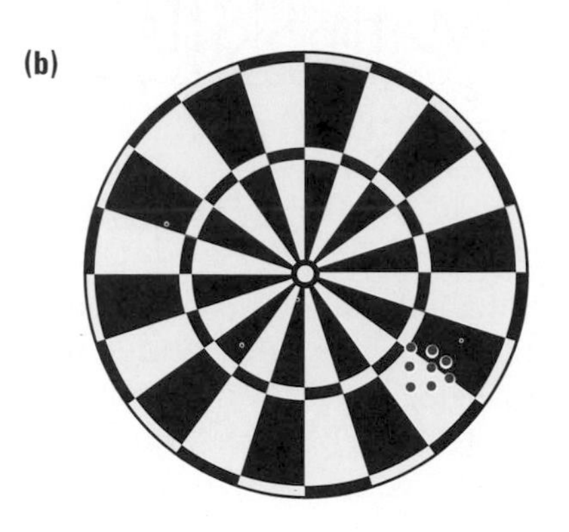

(c)

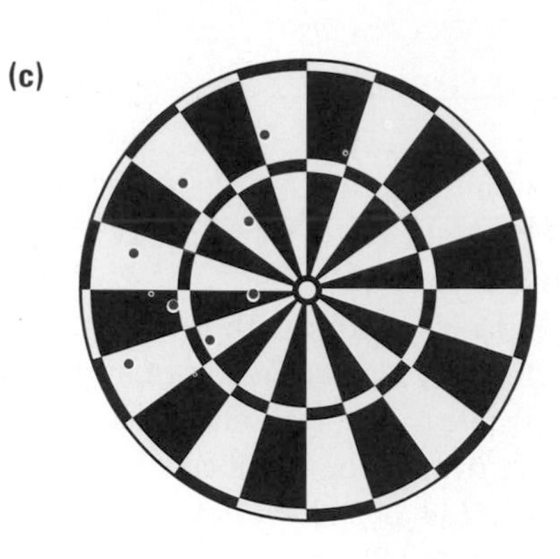

Figure 1
The positions of the darts in each of these figures are analogous to measured or calculated results in a laboratory setting.
The results in **(a)** are precise and accurate, in **(b)** they are precise but not accurate, and in **(c)** they are neither precise nor accurate.

Figure 1 shows an analogy between precision and accuracy, and the positions of darts thrown at a dartboard.

How certain you are about a measurement depends on two factors: the precision of the instrument used and the size of the measured quantity. More precise instruments give more certain values. For example, a mass measurement of 13 g is less precise than a measurement of 12.76 g; you are more certain about the second measurement than the first. Certainty also depends on the measurement. For example, consider the measurements 0.4 cm and 15.9 cm; both have the same precision. However, if the measuring instrument is precise to ± 0.1 cm, the first measurement is 0.4 ± 0.1 cm (0.3 cm or 0.5 cm) or an error of ± 25%, whereas the second measurement could be 15.9 ± 0.1 cm (15.8 cm or 16.0 cm) for an error of ± 0.6%. For both factors—the precision of the instrument used and the value of the measured quantity—the more digits there are in a measurement, the more certain you are about the measurement.

Significant Digits

The certainty of any measurement is communicated by the number of significant digits in the measurement. In a measured or calculated value, significant digits are the digits that are certain plus one estimated (uncertain) digit. Significant digits include all digits correctly reported from a measurement.

Follow these rules to decide whether a digit is significant:

1. If a decimal point is present, zeros to the left of the first non-zero digit (leading zeros) are not significant.
2. If a decimal point is not present, zeros to the right of the last non-zero digit (trailing zeros) are not significant.
3. All other digits are significant.
4. When a measurement is written in scientific notation, all digits in the coefficient are significant.
5. Counted and defined values have infinite significant digits.

Table 2 shows some examples of significant digits.

Table 2 Certainty in Significant Digits

Measurement	Number of significant digits
32.07 m	4
0.0041 g	2
5×10^5 kg	1
6400 s	2
204.0 cm	4
10.0 kJ	3
100 people (counted)	infinite

An answer obtained by multiplying and/or dividing measurements is rounded to the same number of significant digits as the measurement with the fewest number of significant digits. For example, if we use a calculator to solve the following equation:

$$(77.8 \text{ km/h})(0.8967 \text{ h}) = 69.76326 \text{ km}$$

However, the certainty of the answer is limited to three significant digits, so the answer is rounded up to 69.8 km.

Rounding Off

Use these rules when rounding answers to calculations:

1. When the first digit discarded is less than five, the last digit retained should not be changed.
   3.141 326 rounded to 4 digits is 3.141
2. When the first digit discarded is greater than five, or if it is a five followed by at least one digit other than zero, the last digit retained is increased by 1 unit.
   2.221 372 rounded to 5 digits is 2.2214
   4.168 501 rounded to 4 digits is 4.169
3. When the first digit discarded is five followed by only zeros, the last digit retained is increased by 1 if it is odd, but not changed if it is even.
   2.35 rounded to 2 digits is 2.4
   2.45 rounded to 2 digits is 2.4
   –6.35 rounded to 2 digits is –6.4

have

$- 2(AB)(BC) \cos B$

$+ (53.2\text{ cm})^2 - 2(47.3\text{ cm})(53.2\text{ cm}) \cos 115°$

$2830\text{ cm}^2 - (5033\text{ cm}^2)(\cos 115°)$

$(5033\text{ cm}^2)(-0.4226)$

...nd Graphs

...e written in the form $Ax + By = C$ is ...-degree equation in two variables. ... equation is rearranged as $y = mx + b$, ... variable (on the y-axis) and x is the ... the x-axis). This equation is known ...m because m is the slope of the line ... y-intercept.

... encountered in many areas of sci-... quation for the velocity ($\vec{v}$) of an ...t) is given by the linear equation ...e initial velocity and $\vec{a}$ is the accel-...s equation to the slope-intercept ...s $\vec{v} = \vec{a}\Delta t + \vec{v}_i$, where $\vec{v}$ represents ...sents the x variable.

...tion for the velocity of an object ... [E] and acceleration 5.0 m/s² [E].

...Δt

...an be calculated using the fol-...

... are any two points on the line.

...We can apply the sine ratio to determine the length of the two unknown sides of the triangle shown in **Figure 3**.

$$\sin R = \frac{\text{opp}}{\text{hyp}} = \frac{PQ}{PR}$$

$$\sin 54° = \frac{PQ}{2.3\text{ cm}}$$

...culator to find sin 54°, make sure it is in ... enter sin 54. This should produce the ...

... we can use the Pythagorean theorem or ...o to angle P.

...R / ...cm

...5878)(2.3 cm)

...4 cm (to two significant digits)

... trigonometric ratios that are frequently used ...g with right triangles are the cosine and tan-...bbreviated *cos* and *tan* respectively. They are

$$= \frac{\text{adj}}{\text{hyp}} \qquad \text{tangent } \theta = \frac{\text{opp}}{\text{adj}}$$

For the triangle shown in **Figure 4**

$$\cos B = \frac{BC}{AB} \qquad \tan B = \frac{AC}{BC}$$

$$\cos A = \frac{AC}{AB} \qquad \tan A = \frac{BC}{AC}$$

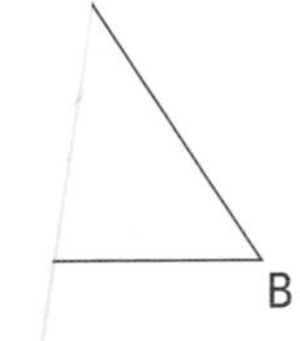

...ure 4
...t triangle

...rigonometry can also be used for triangles other than ... triangles. The sine law and the cosine law can be useful ... dealing with problems involving vectors.

The Sine Law

This law states that for any given triangle, *the ratio of the sine of an angle to the length of the opposite side is constant.* Thus, for **Figure 5**

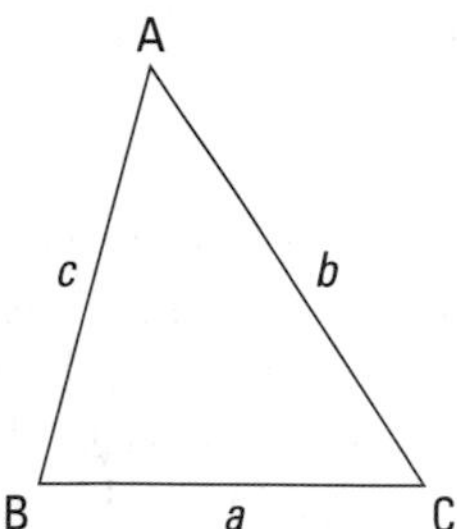

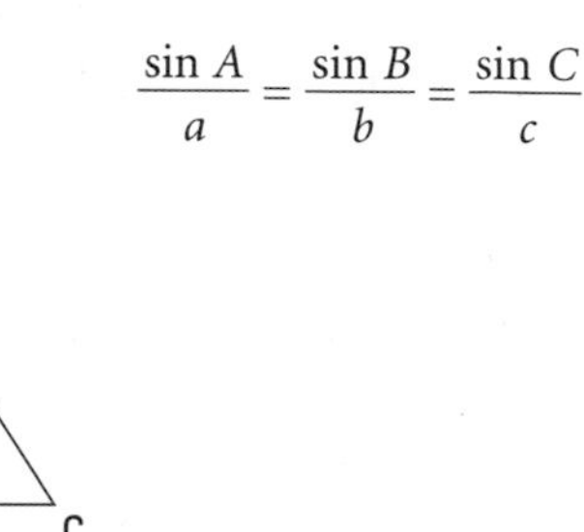

$$\frac{\sin A}{a} = \frac{\sin B}{b} = \frac{\sin C}{c}$$

Figure 5
Scalene triangle

The Cosine Law

This law states that for any given triangle, *the square of the length of any side is equal to the sum of the squares of the lengths of the other two sides, minus twice the product of the lengths of these two sides and the cosine of the angle between them (the included angle).* Thus, for **Figure 5**

$$a^2 = b^2 + c^2 - 2bc \cos A$$
$$b^2 = a^2 + c^2 - 2ac \cos B$$
$$c^2 = b^2 + a^2 - 2ab \cos C$$

Note that the first part of the cosine law is the Pythagorean theorem; the last factor simply adjusts for the fact that the angle is not a right angle. If the angle is a right angle, the cosine is zero and the term disappears, leaving the Pythagorean theorem.

Example

Look at **Figure 6** and calculate the length of side AC.

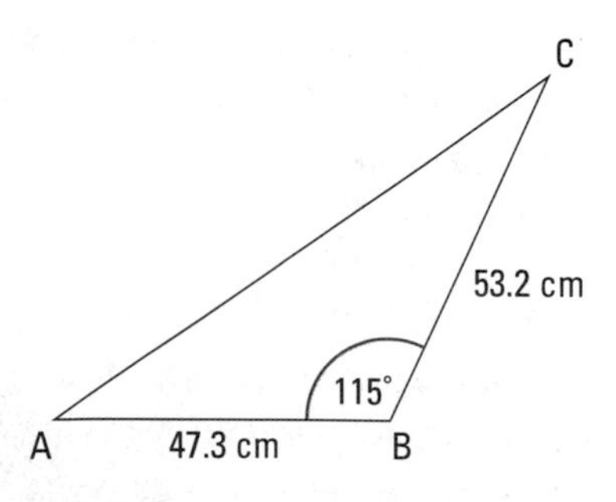

Figure 6

By the cosine law w

$$AC^2 = AB^2 + BC$$
$$= (47.3\ \text{cm})^2$$
$$= 2237\ \text{cm}^2 +$$
$$= 5067\ \text{cm}^2 -$$
$$AC^2 = 7194\ \text{cm}^2$$
$$AC = 84.8\ \text{cm}$$

Equations a

Linear Equations

Any equation that can b
called a *linear*, or firs
However, most often the
where y is the dependen
independent variable (o
as the slope-intercept fo
on the graph and b is the

Linear equations are
ence. For example, the e
object at a given time (
$\vec{v} = \vec{v}_i + \vec{a}\Delta t$, where $\vec{v}_i$ is t
eration. If we change th
form ($y = mx + b$), it rea
the y variable and Δt repre

Example

Plot the graph of the equa
with initial velocity 20.0 m/

$$\vec{v} = 20.0\ \text{m/s} + (5.0\ \text{m/s}^2)$$

The slope of the line
lowing equation:

$$m = \frac{\text{rise}}{\text{run}} = \frac{(y_2 - y_1)}{(x_2 - x_1)}$$

where y_1 and x_1, and y_2 and x

From **Figure 7**, we can choose two points (1, 25) and (5.5, 47.5); note that one is a data point and one is not. We can now calculate the slope.

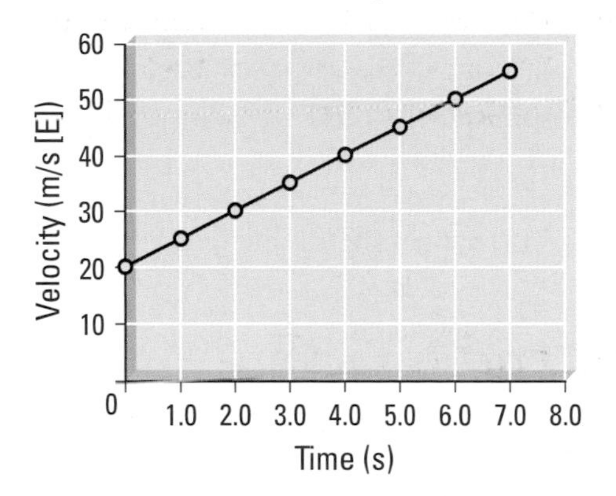

Figure 7
Velocity-time graph

$$m = \frac{v_2 - v_1}{t_2 - t_1}$$
$$= \frac{(47.5 - 25.0) \text{ m/s [E]}}{(5.5 - 1.0) \text{ s}}$$
$$= \frac{22.5 \text{ m/s [E]}}{4.5 \text{ s}}$$
$$m = 5.0 \text{ m/s}^2 \text{ [E]}$$

The slope of the line is 5.0 m/s^2 [E]. This is a positive value, indicating a positive slope. A negative slope (a line sloped the other way) would have a negative value.

If the value of one of the variables is known, the other value can be read from the graph or obtained by solving the equation using algebraic skills. For example, if $\Delta t = 2.0$ s you can see on the graph that the corresponding y coordinate is 30.0 m/s. Solving algebraically we get the same result:

$$\vec{v} = \vec{v}_i + \vec{a}\Delta t$$
$$= 20.0 \text{ m/s} + 5.0 \text{ m/s}^2 \text{ (2.0 s)}$$
$$= 20.0 \text{ m/s} + 10.0 \text{ m/s}$$
$$\vec{v} = 30.0 \text{ m/s}$$

Variation Equations

When y varies directly as x, written as $y \propto x$, it means that $y = kx$, where k is the constant of variation. When y varies inversely as x, written as $y \propto \frac{1}{x}$, it means that $y = \frac{k}{x}$ or $xy = k$.

In many situations there is a combination of direct and inverse variation, commonly referred to as joint variation. Problems dealing with joint variation are solved by substituting the values of the variables from a known experiment to calculate k and then using the value of k to determine the missing variables in another experiment.

Example

The electrical resistance of a wire (R) varies directly as its length and inversely as the square of its diameter. An investigation determines that 50.0 m of wire of diameter 3.0 mm has a resistance of 8.0 Ω. Determine the constant of variation for this type of wire. Without doing another investigation, determine the resistance of 40.0 m of the same type of wire if the diameter is 4.0 mm. The variation equation is

$$R = k\frac{l}{d^2}$$
$$8.0\ \Omega = k\frac{50.0 \text{ m}}{(3.0 \text{ mm})^2}$$
$$k = \frac{(8.0\ \Omega)(3.0 \text{ mm})^2}{50.0 \text{ m}}$$
$$= \left(\frac{1.44\ \Omega \cdot \text{mm}^2}{\text{m}}\right)\left(\frac{1 \text{ m}}{1000 \text{ mm}}\right)$$
$$k = 0.0014\ \Omega \cdot \text{mm}$$

Use this value of k to find R:

$$R = 0.0014\ \Omega \cdot \text{mm} \times \frac{40.0 \text{ m}}{(4.0 \text{ mm})^2} \times \left(\frac{1000 \text{ mm}}{1 \text{ m}}\right)$$
$$= 0.0014\ \Omega \cdot \text{mm} \times \left(\frac{4.00 \times 10^4}{16 \text{ mm}^2}\right)$$
$$= \frac{57.6\ \Omega \cdot \text{mm}^2}{16 \text{ mm}^2}$$
$$R = 3.6\ \Omega$$

Logarithms

Any positive number N can be expressed as a power of some base b where $b > 1$. Some obvious examples are

$16 = 2^4$	base 2, exponent 4
$25 = 5^2$	base 5, exponent 2
$27 = 3^3$	base 3, exponent 3
$0.001 = 10^{-3}$	base 10, exponent –3

In each example, the exponent is an integer. However, exponents may be any real number, not just an integer. If you use the x^y button on your calculator, you can experiment to get a better understanding of this concept.

The most common base is base 10. Some examples for base 10 are

$10^{0.5} = 3.162$
$10^{1.3} = 19.95$
$10^{-2.7} = 0.001995$

By definition, the exponent to which a base b must be raised to produce a given number N is called the *logarithm* of N to base b (abbreviated as $\log_b$). When the value of the base is not written it is assumed to be base 10. Logarithms to base 10 are called *common logarithms*. We can express the previous examples as logarithms:

$\log 3.162 = 0.5$
$\log 19.95 = 1.3$
$\log 0.001995 = -2.7$

Another base that is used extensively for logarithms is the base e (approximately 2.7183). Logarithms to base e are called *natural logarithms* (abbreviated as ln).

Most measurement scales are linear in nature. For example, a speed of 80 km/h is twice as fast as a speed of 40 km/h and four times as fast as a speed of 20 km/h. However, there are several examples in science where the range of values of the variable being measured is so great that it is more convenient to use a logarithmic scale to base 10.

One example of this is the scale for measuring the intensity level of sound. For example, a sound with an intensity level of 20 dB is 100 times (10^2) as loud as a sound with an intensity level of 0 dB and 40 dB is 10 000 (10^4) times more intense than a sound of 0 dB. Other situations that use logarithmic scales are the acidity of a solution (the pH scale) and the intensity of earthquakes (the Richter scale).

Logarithmic Graphs

Quite often graphing the results from experiments shows a logarithmic progression. For example, the series 1, 2, 3, 4, 5, 6 is a linear progression, whereas the series 10, 100, 1000, 10 000, 100 000 is a logarithmic progression. This means that the values increase exponentially. **Figure 8** is a graph of type $y = \log x$ to illustrate the sound intensity level scale where sound intensity level in decibels (dB) is equal to the logarithm of intensity in watts per metre squared (W/m^2).

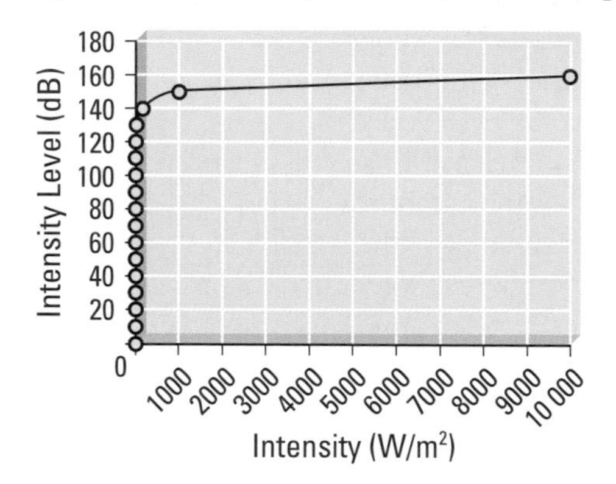

Figure 8
Graph of sound intensity level versus intensity

Notice that the scales on both axes are linear so that we can see very little of the detail on the x-axis. Where the data range on one axis is extremely large and/or does not follow a linear progression, it is more convenient to change the scale (usually on the x-axis) so that we can "see" more detail of the entire range of values. Semi-log graph paper can be used to construct such graphs. If the scale on the x-axis is changed to a logarithmic scale, the graph of sound intensity level versus sound intensity on semi-log graph paper is shown in **Figure 9**.

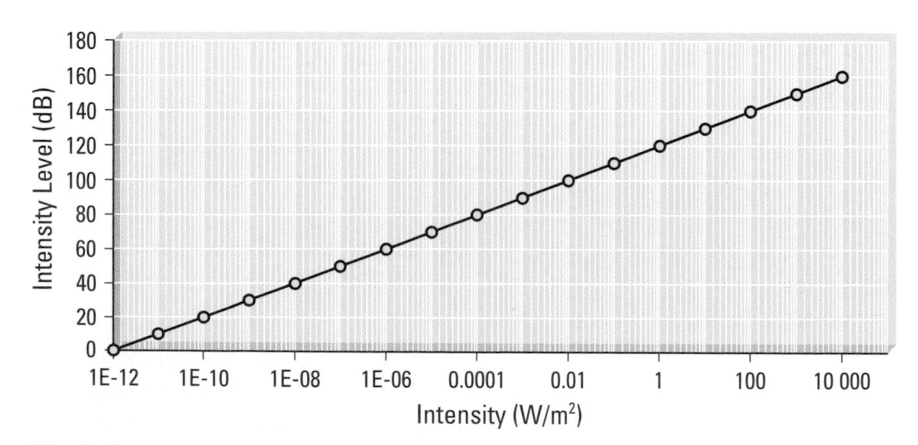

Figure 9
Semi-log graph of sound intensity level versus intensity

Dimensional Analysis

Dimensional analysis is a useful tool to determine whether an equation has been written correctly and to convert units. As is the case with many topics in physics, dimensional analysis can be "easy" or "hard" depending on the treatment we give it.

"Dimension" is a term that refers to quantities that we can measure in our universe. Three common dimensions are mass (m), length (l), and time (t). Note that the units of these dimensions are all *base units*—kilogram (kg), metre (m), and second (s). In dimensional analysis, all units are expressed as base units.

After a while, dimensional analysis becomes second nature. Suppose, for example, that after you solve an equation in which time Δt is the unknown, the final line in your solution is $\Delta t = 2.1$ kg. You know that something has gone seriously wrong on the right-hand side of the equation. It might be that care was not taken in cancelling certain units or that the equation was written incorrectly. For example, we can use dimensional analysis to determine if the following expression is valid:

$$\Delta d = v_i + \frac{1}{2}a\Delta t^2$$

One way to check is to insert the appropriate units. The usual technique when working with units is to put them in square brackets and to ignore numbers like the $\frac{1}{2}$ in the expression. The square brackets indicate that we are dealing with units only. The expression becomes

$$[\mathrm{m}] = \left[\frac{\mathrm{m}}{\mathrm{s}}\right] + \left[\frac{\mathrm{m}}{\mathrm{s}^2}\right][\mathrm{s}^2]$$

$$[\mathrm{m}] = \left[\frac{\mathrm{m}}{\mathrm{s}}\right] + [\mathrm{m}]$$

The expression is not valid because the units on the right-hand side of the equation do not equal the units on the left-hand side. The correct expression is

$$\Delta d = v_i\Delta t + \frac{1}{2}a\Delta t^2$$

You can check it out yourself by inserting the units in square brackets. If you wish, you can use the actual dimensions of length $[l]$ and time $[t]$ instead of substituting units. The dimensional analysis of the equation is

$$[l] = \left[\frac{l}{t}\right][t] + \left[\frac{l}{t^2}\right][t^2]$$

$$[l] = [l] + [l]$$

$$[l] = [l]$$

Remember that because we are dealing only with dimensions, there is no need to say $2l$ on the right-hand side.

You can also use dimensional analysis to change from one unit to another. For example, to convert 95 km/hr to m/s, kilometres must be changed to metres and hours to seconds. It helps to realize that 1 km = 1000 m and 1 hr = 3600 s. These two equivalencies allow the following two terms to be written:

$$\frac{1000\ \mathrm{km}}{1\ \mathrm{km}} = 1 \quad \text{and} \quad \frac{1\ \mathrm{hr}}{3600\ \mathrm{s}} = 1$$

Of course, the numerators and denominators could be switched (i.e., 1 km and 3600 s could be in the numerators) and the ratios would still be 1. However, as you will see, it is convenient to keep the ratios as they are for cancelling purposes. Because multiplying by 1 does not change the value of anything, we can write the following expression and cancel the units:

$$\frac{95\ \mathrm{km}}{\mathrm{hr}} = \frac{95\ \cancel{\mathrm{km}}}{\cancel{\mathrm{hr}}} \times \frac{1000\ \mathrm{m}}{1\ \cancel{\mathrm{km}}} \times \frac{1\ \cancel{\mathrm{hr}}}{3600\ \mathrm{s}}$$

Therefore,

$$\frac{95\ \mathrm{km}}{1\ \mathrm{hr}} = \frac{95\ 000\ \mathrm{m}}{3600\ \mathrm{s}}$$

$$= 26.4\ \mathrm{m/s},\ \text{or}\ 26\ \mathrm{m/s}\ \text{(to two significant digits)}$$

Example

What will be the magnitude of the acceleration of a 2100-g object that experiences a net force of magnitude 38.2 N?

First convert grams to kilograms:

$$\text{mass} = (2100\ \mathrm{g})\left(\frac{1\ \mathrm{kg}}{1000\ \mathrm{g}}\right)$$

$$\text{mass} = 2.1\ \mathrm{kg}$$

From Newton's second law,

$$F_{net} = ma$$

$$a = \frac{F_{net}}{m}$$

$$= \frac{38.2\ \mathrm{N}}{2.1\ \mathrm{kg}}$$

$$a = 18\ \mathrm{N/kg}$$

It is somewhat bothersome to leave acceleration with the units N/kg, so we will use dimensional analysis to change the units:

$$F_{net} = ma$$

$$[\mathrm{N}] = [\mathrm{kg}][\mathrm{m/s}^2]$$

$$a = \frac{18\ \mathrm{N}}{\mathrm{kg}} = \frac{18\ [\mathrm{kg}][\mathrm{m/s}^2]}{[\mathrm{kg}]}$$

$$a = 18\ \mathrm{m/s}^2$$

Safety Conventions and Symbols

Although every effort is undertaken to make the science experience a safe one, inherent risks are associated with some scientific investigations. These risks are generally associated with the materials and equipment used, and the disregard of safety instructions that accompany investigations and activities. However, there may also be risks associated with the location of the investigation, whether in the science laboratory, at home, or outdoors. Most of these risks pose no more danger than one would normally experience in everyday life. With an awareness of the possible hazards, knowledge of the rules, appropriate behaviour, and a little common sense, these risks can be practically eliminated.

Remember, you share the responsibility not only for your own safety, but also for the safety of those around you. Always alert the teacher in case of an accident.

In this text, equipment and procedures that are potentially hazardous are accompanied by the symbol and safety instructions.

WHMIS Symbols and HHPS

The Workplace Hazardous Materials Information System (WHMIS) provides workers and students with complete and accurate information regarding hazardous products. All chemical products supplied to schools, businesses, and industries must contain standardized labels and be accompanied by Material Safety Data Sheets (MSDS) providing detailed information about the product. Clear and standardized labelling is an important component of WHMIS (**Table 1**). These labels must be present on the product's original container or be added to other containers if the product is transferred.

The *Canadian Hazardous Products Act* requires manufacturers of consumer products containing chemicals to include a symbol specifying both the nature of the primary hazard and the degree of this hazard. In addition, any secondary hazards, first aid treatment, storage, and disposal must be noted. Household Hazardous Product Symbols (HHPS) are used to show the hazard and the degree of the hazard by the type of border surrounding the illustration (**Figure 1**).

	CORROSIVE This material can burn your skin and eyes. If you swallow it, it will damage your throat and stomach.
	FLAMMABLE This product or the gas (or vapour) from it can catch fire quickly. Keep this product away from heat, flames, and sparks.
	EXPLOSIVE Container will explode if it is heated or if a hole is punched in it. Metal or plastic can fly out and hurt your eyes and other parts of your body.
	POISON If you swallow or lick this product, you could become very sick or die. Some products with this symbol on the label can hurt you even if you breathe (or inhale) them.

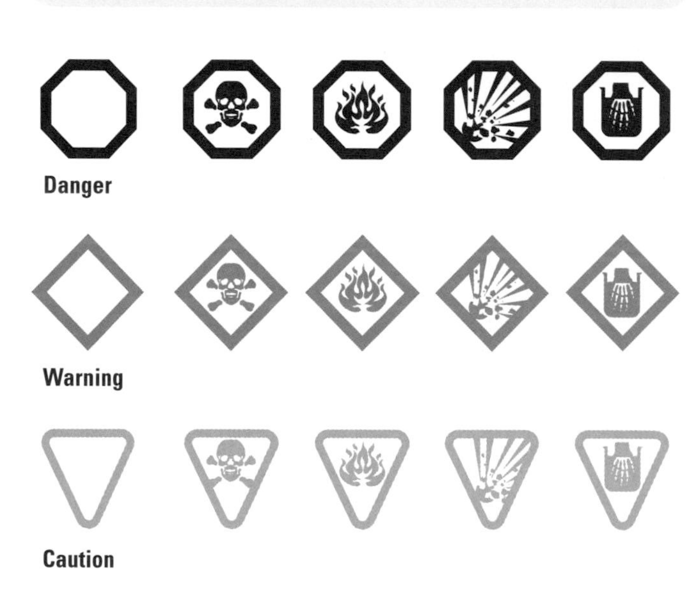

Figure 1
Hazardous household product symbols

Table 1 Workplace Hazardous Materials Information System (WHMIS)

Class and type of compounds	WHMIS symbol	Risks	Precautions
Class A: Compressed Gas Material that is normally gaseous and kept in a pressurized container		• could explode due to pressure • could explode if heated or dropped • possible hazard from both the force of explosion and the release of contents	• ensure container is always secured • store in designated areas • do not drop or allow to fall
Class B: Flammable and Combustible Materials Materials that will continue to burn after being exposed to a flame or other ignition source		• may ignite spontaneously • may release flammable products if allowed to degrade or when exposed to water	• store in properly designated areas • work in well-ventilated areas • avoid heating • avoid sparks and flames • ensure that electrical sources are safe
Class C: Oxidizing Materials Materials that can cause other materials to burn or support combustion		• can cause skin or eye burns • increase fire and explosion hazards • may cause combustibles to explode or react violently	• store away from combustibles • wear body, hand, face, and eye protection • store in proper container that will not rust or oxidize
Class D: Toxic Materials Immediate and Severe Poisons and potentially fatal materials that cause immediate and severe harm		• may be fatal if ingested or inhaled • may be absorbed through the skin • small volumes have a toxic effect	• avoid breathing dust or vapours • avoid contact with skin or eyes • wear protective clothing, and face and eye protection • work in well-ventilated areas and wear breathing protection
Class D: Toxic Materials Long Term Concealed Materials that have a harmful effect after repeated exposures or over a long period		• may cause death or permanent injury • may cause birth defects or sterility • may cause cancer • may be sensitizers causing allergies	• wear appropriate personal protection • work in a well-ventilated area • store in appropriate designated areas • avoid direct contact • use hand, body, face, and eye protection • ensure respiratory and body protection is appropriate for the specific hazard
Class D: Biohazardous Infectious Materials Infectious agents or a biological toxin causing a serious disease or death		• may cause anaphylactic shock • includes viruses, yeasts, moulds, bacteria, and parasites that affect humans • includes fluids containing toxic products • includes cellular components	• special training is required to handle materials • work in designated biological areas with appropriate engineering controls • avoid forming aerosols • avoid breathing vapours • avoid contamination of people and/or area • store in special designated areas
Class E: Corrosive Materials Materials that react with metals and living tissue		• eye and skin irritation on exposure • severe burns/tissue damage on longer exposure • lung damage if inhaled • may cause blindness if contacts eyes • environmental damage from fumes	• wear body, hand, face, and eye protection • use breathing apparatus • ensure protective equipment is appropriate • work in a well-ventilated area • avoid all direct body contact • use appropriate storage containers and ensure proper non-venting closures
Class F: Dangerously Reactive Materials Materials that may have unexpected reactions		• may react with water • may be chemically unstable • may explode if exposed to shock or heat • may release toxic or flammable vapours • may vigorously polymerize • may burn unexpectedly	• handle with care avoiding vibration, shocks, and sudden temperature changes • store in appropriate containers • ensure storage containers are sealed • store and work in designated areas

Safety in the Laboratory

General Safety Rules

Safety in the laboratory is an attitude and a habit more than it is a set of rules. It is easier to prevent accidents than to deal with the consequences of an accident. Most of the following rules are common sense.

- Do not enter a laboratory unless a teacher or other supervisor is present, or you have permission to do so.
- Familiarize yourself with your school's safety regulations.
- Make your teacher aware of any allergies or other health problems you may have.
- Wear eye protection, lab aprons or coats, and gloves when appropriate.
- Wear closed shoes (not sandals) when working in the laboratory.
- Place your books and bags away from the work area. Keep your work area clear of all materials except those that you will use in the investigation.
- Do not chew gum, eat, or drink in the laboratory. Food should not be stored in refrigerators in laboratories.
- Know the location of MSDS information, exits, and all safety equipment, such as the fire blanket, fire extinguisher, and eyewash station.
- Use stands, clamps, and holders to secure any potentially dangerous or fragile equipment that could be tipped over.
- Avoid sudden or rapid motion in the laboratory that may interfere with someone carrying or working with chemicals or using sharp instruments.
- Never engage in horseplay or practical jokes in the laboratory.
- Ask for assistance when you are not sure how to do a procedural step.
- When heating a test tube over a laboratory burner, use a test-tube holder and a spurt cap. Holding the test tube at an angle, facing away from you and others, gently move the test tube backwards and forwards through the flame.
- Never attempt any unauthorized experiments.
- Never work in a crowded area or alone in the laboratory.
- Clean up all spills, even spills of water, immediately.
- Always wash your hands with soap and water before or after you leave the laboratory. Definitely wash your hands before you touch any food.
- Do not forget safety procedures when you leave the laboratory. Accidents can also occur outdoors, at home, or at work.

Eye and Face Safety

- Always wear approved eye protection in a laboratory, no matter how simple or safe the task appears to be. Keep the safety glasses over your eyes, not on top of your head. For certain experiments, full-face protection may be necessary.
- Never look directly into the opening of flasks or test tubes.
- If, in spite of all precautions, you get a solution in your eye, quickly use the eyewash or nearest running water. Continue to rinse the eye with water for at least 15 minutes. This is a very long time—have someone time you. Unless you have a plumbed eyewash system, you will also need assistance in refilling the eyewash container. Have another student inform your teacher of the accident. The injured eye should be examined by a doctor.
- If you must wear contact lenses in the laboratory, be extra careful; whether or not you wear contact lenses, do not touch your eyes without first washing your hands. If you do wear contact lenses, make sure that your teacher is aware of it. Carry your lens case and a pair of glasses with you.
- If a piece of glass or other foreign object enters your eye, seek immediate medical attention.
- Do not stare directly at any bright source of light (e.g., a burning magnesium ribbon, lasers, or the Sun). You will not feel any pain if your retina is being damaged by intense radiation. You cannot rely on the sensation of pain to protect you.
- Be careful when working with lasers; be aware that a reflected laser beam can act like a direct beam on the eye.

Handling Glassware Safely

- Never use glassware that is cracked or chipped. Give such glassware to your teacher or dispose of it as directed. Do not put the item back into circulation.
- Never pick up broken glassware with your fingers. Use a broom and dustpan.
- Do not put broken glassware into garbage containers. Dispose of glass fragments in special containers marked "Broken Glass."
- Heat glassware only if it is approved for heating. Check with your teacher before heating any glassware.
- If you cut yourself, inform your teacher immediately.
- If you need to insert glass tubing or a thermometer into a rubber stopper, get a cork borer of a suitable size. Insert the borer in the hole of the rubber stopper starting from the small end of the stopper. Once the borer is pushed all the way through the hole, insert the tubing or thermometer through the borer. Ease the borer out of the hole, leaving the tubing or thermometer inside. To remove the tubing or thermometer from the stopper, push the borer from the small end

through the stopper until it shows from the other end. Ease the tubing or thermometer out of the borer.
- Protect your hands with heavy gloves or several layers of cloth before inserting glass into rubber stoppers.
- Be very careful when cleaning glassware. There is an increased risk of breakage from dropping when the glassware is wet and slippery.

Using Sharp Instruments Safely

- Make sure your instruments are sharp. Dull cutting instruments require more pressure than sharp instruments and are therefore much more likely to slip.
- Select the appropriate instrument for the task. Never use a knife when scissors would work best.
- Always cut away from yourself and others.
- If you cut yourself, inform your teacher immediately and get appropriate first aid.
- Be careful when working with wire cutters or wood saws. Use a cutting board where needed.

Heat and Fire Safety

- In a laboratory where burners or hot plates are being used, never pick up a glass object without first checking the temperature by lightly and quickly touching the item, or by placing your hand near, but not touching, it. Glass items that have been heated stay hot for a long time, but do not appear to be hot. Metal items such as ring stands and hot plates can also cause burns; take care when touching them.
- Do not use a laboratory burner near wooden shelves, flammable liquids, or any other item that is combustible.
- Before using a laboratory burner, make sure that long hair is always tied back. Do not wear loose clothing (wide long sleeves should be tied back or rolled up).
- Never look down the barrel of a laboratory burner.
- Always pick up a burner by the base, never by the barrel.
- Never leave a lighted Bunsen burner unattended.
- If you burn yourself, *immediately* run cold water gently over the burned area or immerse the burned area in cold water and inform your teacher.
- Make sure that heating equipment, such as a burner, hot plate, or electrical equipment is secure on the bench and clamped in place when necessary.
- Always assume that hot plates and electric heaters are hot and use protective gloves when handling.
- Keep a clear workplace when performing experiments with heat.
- Remember to include a "cooling" time in your experiment plan; do not put away hot equipment.
- Very small fires in a container may be extinguished by covering the container with a wet paper towel or ceramic square.
- For larger fires, inform the teacher and follow the teacher's instructions for using fire extinguishers, blankets, alarms, and for evacuation. Do not attempt to deal with a fire by yourself.
- If anyone's clothes or hair catch fire, tell the person to drop to the floor and roll. Then use a fire blanket to help smother the flames.

Electrical Safety

- Water or wet hands should never be used near electrical equipment.
- Do not operate electrical equipment near running water or a large container of water.
- Check the condition of electrical equipment. Do not use if wires or plugs are damaged.
- Make sure that electrical cords are not placed where someone could trip over them.
- When unplugging equipment, remove the plug gently from the socket. Do not pull on the cord.
- When using variable power supplies, start at low voltage and increase slowly.

Waste Disposal

Waste disposal at school, at home, or at work is a societal issue. Some laboratory waste can be washed down the drain or, if it is in solid form, placed in ordinary garbage containers. However, some waste must be treated more carefully. It is your responsibility to follow procedures and dispose of waste in the safest possible manner according to the teacher's instructions.

First Aid

The following guidelines apply if an injury, such as a burn, cut, chemical spill, ingestion, inhalation, or splash in eyes, is to yourself or to one of your classmates.

- If an injury occurs, inform your teacher immediately.
- Know the location of the first aid kit, fire blanket, eyewash station, and shower, and be familiar with the contents/operation.
- If you have ingested or inhaled a hazardous substance, inform your teacher immediately. The MSDS will give information about the first aid requirements for the substance in question. Contact the Poison Control Centre in your area.
- If the injury is from a burn, immediately immerse the affected area in cold water. This will reduce the temperature and prevent further tissue damage.
- If a classmate's injury has rendered him/her unconscious, notify the teacher immediately. The teacher will perform CPR if necessary. Do not administer CPR unless under specific instructions from the teacher. You can assist by keeping the person warm and reassured.

Reference

SI Units

Throughout *Nelson Physics 11* and in this reference section, we have attempted to be consistent in the presentation and usage of units. As far as possible, the text uses the International System of Units (SI). However, some other units have been included because of their practical importance, wide usage, or use in specialized fields.

The most recent *Canadian Metric Practice Guide* (CAN/CSA-Z234.1-89) was published in 1989 and was reaffirmed in 1995 by the Canadian Standards Association.

Numerical Prefixes

Powers and Subpowers of Ten		
Prefix	*Power*	*Symbol*
deca	10^1	da
hecto	10^2	h
kilo	10^3	k
mega	10^6	M
giga	10^9	G
tera	10^{12}	T
peta	10^{15}	P
exa	10^{18}	E
deci	10^{-1}	d
centi	10^{-2}	c
milli	10^{-3}	m
micro	10^{-6}	μ
nano	10^{-9}	n
pico	10^{-12}	p
femto	10^{-15}	f
atto	10^{-18}	a

Some Examples of Prefix Use

2000 metres	$= 2 \times 10^3$ metres	= 2 **kilo**metres	or 2 km
0.27 metres	$= 27 \times 10^{-2}$ metres	= 27 **centi**metres	or 27 cm
3 000 000 000 hertz	$= 3 \times 10^9$ hertz	= 3 **giga**hertz	or 3 gHz

Common Multiples

Multiple	Prefix
$\frac{1}{2}$	hemi-
1	mono-
$1\frac{1}{2}$	sesqui-
2	bi-, di-
$2\frac{1}{2}$	hemipenta-
3	tri-
4	tetra-
5	penta-
6	hexa-
7	hepta-
8	octa-
9	nona-
10	deca-

SI Base Units

Quantity	Symbol	Unit name	Abbreviation
amount of substance	n	mole	mol
electric current	I	ampere	A
length	L, l, h, d, w	metre	m
luminous intensity	I_v	candela	cd
mass	m	kilogram	kg
Celsius temperature	t	Celsius	°C
thermodynamic temperature	T	kelvin	K
time	t	second	s

REFERENCE

The Greek Alphabet

Α	α	alpha
Β	β	beta
Γ	γ	gamma
Δ	δ	delta
Ε	ε	epsilon
Ζ	ζ	zeta
Η	η	eta
Θ	θ	theta
Ι	ι	iota
Κ	κ	kappa
Λ	λ	lambda
Μ	μ	mu
Ν	ν	nu
Ξ	ξ	xi
Ο	ο	omicron
Π	π	pi
Ρ	ρ	rho
Σ	σ	sigma
Τ	τ	tau
Υ	υ	upsilon
Φ	ϕ	phi
Χ	χ	chi
Ψ	ψ	psi
Ω	ω	omega

Some SI Derived Units

Quantity	**Symbol**	**Unit**	**Unit symbol**	**SI base unit**
acceleration	$\vec{a}$	metre per second per second	m/s^2	m/s^2
area	A	square metre	m^2	m^2
density	ρ, D	kilogram per cubic metre	kg/m^3	kg/m^3
displacement	$\Delta\vec{d}$	metre	m	m
electric charge	Q, q	coulomb	C	A·s
electric field	E	volt per metre	V/m	$kg \cdot m/A \cdot s^3$
electric field intensity	E	newton per coulomb (Tesla)	N/C, T	$kg/A \cdot s^2$
electric potential	V	volt	V	$kg \cdot m^2/A \cdot s^3$
electric resistance	R	ohm	Ω	$kg \cdot m^2/A^2 \cdot s^3$
energy	E, E_k, E_p	joule	J	$kg \cdot m^2/s^2$
force	F	newton	N	$kg \cdot m/s^2$
frequency	f	hertz	Hz	s^{-1}
heat	Q	joule	J	$kg \cdot m^2/s^2$
magnetic field	B	weber per square metre (Tesla)	T	$kg/A \cdot s^2$
magnetic flux	Φ	weber	Wb	$kg \cdot m^2/A \cdot s^2$
momentum	P, p	kilogram metre per second	kg·m/s	kg·m/s
period	T	second	s	s
power	P	watt	W	$kg \cdot m^2/s^3$
pressure	P, p	newton per square metre	N/m^2	$kg/m \cdot s^2$
speed	v	metre per second	m/s	m/s
velocity	$\vec{v}$	metre per second	m/s	m/s
volume	V	cubic metre	m^3	m^3
wavelength	λ	metre	m	m
weight	W, w	newton	N	N, $kg \cdot m/s^2$
work	W	joule	J	$kg \cdot m^2/s^2$

Physical Constants

Quantity	**Symbol**	**Approximate value**
speed of light in a vacuum	c	3.00×10^8 m/s
gravitational constant	G	6.67×10^{-11} $N \cdot m^2/kg^2$
Coulomb's constant	k	9.00×10^9 $N \cdot m^2/C^2$
charge on electron	$-e$	-1.60×10^{-19} C
charge on proton	e	1.60×10^{-19} C
electron mass	m_e	9.11×10^{-31} kg
proton mass	m_p	1.673×10^{-27} kg
neutron mass	m_n	1.675×10^{-27} kg
atomic mass unit	u	1.660×10^{-27} kg
Planck's constant	h	6.63×10^{-34} J·s

Terrestrial and Lunar Data

Quantity	**Approximate value**
mass of Earth	5.98×10^{24} kg
radius of Earth at equator (mean)	6.38×10^6 m
radius of Earth's orbit (mean)	1.50×10^{11} m
acceleration due to gravity (g) on Earth at equator (sea level) on Earth at poles (sea level) on the Moon	 9.7804 m/s^2 9.8322 m/s^2 1.62 m/s^2
Earth-Moon distance (mean)	3.84×10^8 m
mass of Moon	7.35×10^{22} kg
radius of the Moon (mean)	1.74×10^6 m
radius of the Moon's orbit (mean)	3.84×10^8 m
Earth-Sun distance (mean)	1.50×10^{11} m
standard atmospheric pressure	1 atm or 1.013×10^5 Pa
length of Earth year	3.16×10^7 s

Answers

This section includes numerical answers to questions that require calculation in Chapter Reviews and Unit Reviews.

Chapter 1 Review, pp. 49–51

2. 3.76×10^8 m
3. 112 s
4. (b) 4.3 km/h
   (c) 2.0 km/h [40°S of E]
6. (b) 2.5 m/s [W]; 0.0 m/s; 7.5 m/s [E]
   (d) 0.0 m
7. (a) 6.2 m/s [38°E of S]
   (b) 1.7×10^2 s
   (c) 1.1×10^3 m [38°E of S]
9. (a) 9.8 m/s^2 [↓]
   (b) 9.8 m/s^2 [↓]
   (c) 9.8 m/s^2 [↓]
12. 0.44 m/s^2; 0.30 m/s^2
13. (a) 7.5 m/s [W]; 15 m/s [W]; 7.5 m/s [W]
    (b) 7.5 m/s^2 [W]; 0.0 m/s^2 [W]; 15 m/s^2 [E]
14. (a) 14 m/s [↑]
    (b) 9.6 m
15. 3.8×10^3 m/s^2 [fwd]
16. 7.3 m
17. (a) 45 s
    (b) 75 s
    (c) 9.0×10^2 m
20. (a) −0.81%
    (b) 1.3%
26. (a) 33 m
    (b) 56 m

Chapter 2 Review, pp. 80–81

3. 0 N [E]
4. (a) 5.4 N [fwd]
   (b) 4.5 m/s^2 [fwd]
7. (a) 1.0×10^2 N [fwd]
   (b) 1.0×10^2 N
8. 6.2×10^{-5} N
9. 3.0 m/s^2
10. (a) 8.8×10^{15} m/s^2
    (b) 1.9×10^7 m/s
11. (a) 8.0×10^3 N [N]
12. (a) 3.3 m/s^2 [fwd]
    (b) 2.9 m/s^2 [fwd]
14. (a) 8.0 m/s^2 [fwd]
    (b) 0.16 m
15. (a) 1.6×10^3 kg
    (b) 2.0×10^3 N [W]
    (c) 1.2 m/s^2 [W]
16. (a) 10 N [W]
    (b) 0 N
    (c) 7.0×10^2 N [↑]
18. 2.0%
21. 2.0 N [bkwd]

Chapter 3 Review, pp. 110–111

2. (b) 9.8 N/kg [↓]
3. (a) 4.9 N
   (b) 4.9 N
   (c) 0.0 N
4. (a) 47 kg
6. (a) 9.8×10^5 N [↓]
   (b) 2.4×10^3 N [↓]
   (c) 2.7×10^2 N [↓]
   (d) 1.8×10^{-3} N [↓]
7. 8.0×10^5 N
8. 9.1 N/kg [↓]
9. 5.3×10^2 N [↓]
10. (a) 2.3×10^5 N [↓]
    (b) 2.3×10^5 m/s^2 [↓]
11. $3.0r_E$
12. (b) 3.46×10^8 m from Earth's centre
15. (a) 74 N [fwd]
    (b) 0.22 m/s^2 [fwd]
    (c) 1.5×10^2 N [fwd]
16. (a) 8.2 m/s^2 [E]
    (b) 5.4×10^2 N [E]
    (c) 0.84

Unit 1 Review, pp. 114–117

2. 10 d 16 h 22 min
3. 4.0 km
5. 38.7 m
6. 24 min
7. (a) 23 km [S]
   (b) 51 km [27° S of E]
   (c) 25 km/h [E]
12. (a) 2.4 m/s [W]
    (b) 2.4 m/s [E]
14. (b) 5.0×10^3 N [E]
15. 5.5 m/s [↑]
16. (b) 1.0×10^1 m/s
17. (a) 0.57 s
    (b) 1.6 m
18. (a) 3.9 m/s^2 [fwd]
    (b) 63 m [fwd]
23. 2.6×10^3 N
24. (a) 0.33
25. 17 N
26. (b) 2.8 N [E]
    (c) 0.59 m/s^2 [E]
    (d) 0.59 m/s^2 [E]
    (f) 1.8 m/s^2 [W]
27. (a) 2.1×10^2 N
    (b) 0.22 N/kg
29. (a) −7.5%
    (b) 9.2%
34. (b) 4.8 N [E]
36. (a) 3.2 m

Chapter 4 Review, pp. 158–159

3. 2.5×10^4 J
5. (a) 2.6×10^3 N
   (b) 5.4×10^4 m/s^2
   (c) 7.0×10^1 m/s
6. 2.1 kg
7. 3.0 m
8. (a) 9.7×10^2 J
9. 4 times; 9 times
10. (a) 7.5×10^2 J
    (b) 3.0×10^3 J
    (c) 2.2×10^3 J
11. 1.7 kg
12. (a) 26 m/s
    (b) 34 m
14. (a) 2.3×10^5 J
    (b) 9.4×10^5 J
    (c) 9.4×10^6 J
    (d) 1.6×10^2 W
18. 4.8×10^5 J
19. (a) 10.9°C
    (b) 1.05°C
20. 79°C
21. 8.4×10^2 J/(kg·°C)
22. 0.096 kg

23. 37°C

24. 0.13°C

25. 2.5×10^2 W

26. 3.6×10^5 J

29. (a) 91%

30. (a) 23 kW
    (b) 29 kW

34. (a) 9.9×10^{11} J
    (b) $29 000

Chapter 5 Review, pp. 182–183

2. (a) 2
   (b) 4
   (c) 32

3. 1.4%/a

7. (a) 77%

12. (b) 1×10^6 drops
    (c) 100% full

15. (a) 3.7×10^5 MJ
    (b) 1.1×10^{13} MJ
    (c) 0.50%
    (d) 3.1%
    (e) 6.2 : 1.0

16. (a) 7.8×10^{10} m^3; 7.8×10^{13} kg
    (b) 1.5×10^{15} J
    (c) 35 GW

17. about 20 a

Unit 2 Review, pp. 186–189

4. 68 m

6. (b) 8.9×10^3 J
   (c) 1.4

8. 0.045 kg

9. 43 m

10. 9.6 times greater

11. 5.5×10^9 J

12. 13 m/s

13. 1.12×10^5 kg

18. (a) 1.1×10^4 J
    (b) 15 m/s
    (c) 13 m/s

19. (a) 4.0 m/s
    (b) 5.3 m/s

20. (a) 2.3 MJ
    (b) 0.24 kg

25. (a) 2.3×10^4 J
    (b) 4.0×10^5 J

26. 2.0°C

27. 5.00×10^2 W

28. 3.00×10^2 s

29. 8.6 MJ

30. 13 m/s

33. (a) 2.5×10^9 W
    (b) 2.5×10^9 J/s; 9.0×10^{12} J/h; 2.2×10^{14} J/d; 7.9×10^{16} J/a
    (c) 5.3×10^{14} J
    (d) 5.0°C

Chapter 6 Review, pp. 234–235

2. 3.5 cm

3. 0.76 Hz; 1.3 s

4. (b) 0.46 s
   (c) 2.2 Hz

6. 1.5×10^3 m/s

7. (a) 0.93 Hz
   (b) 1.1 s

8. 0.34 Hz

9. 4.8 m/s

10. (a) 2.5 m/s
    (b) 40 s
    (c) 1.5×10^2 m

11. 3.8 s

12. 15

15. (a) 36 cm
    (b) 3.6 m/s

Chapter 7 Review, pp. 274–275

2. (a) 300
   (b) 300
   (c) 300
   (d) 300

4. (a) 346 m/s
   (b) 338 m/s
   (c) 350 m/s

5. (a) 17 m
   (b) 1.7×10^{-2} m

6. (a) 341 m/s
   (b) 308 m/s

7. 1.0×10^3 m

8. 2.5×10^{-3} s

9. 14 s

10. (a) 340 m/s
    (b) 14°C

11. 1.7×10^2 m

12. 5.2×10^3 s

13. 57

14. 2.3 s

15. 1.5×10^3 m/s, 3.5×10^2 m/s

16. 10^4

17. 50 dB

18. 3.3×10^3 m

19. (a) 11.8 s
    (b) 2.72 s

20. 24

21. 297 Hz and 303 Hz

22. 2 Hz, 6 Hz, 7 Hz, 9 Hz, 13 Hz, 15 Hz

23. 303 Hz; 299 Hz and 301 Hz

24. 397.5 Hz

25. 377 Hz

26. (a) 1.3×10^3 km/h [W], 2.1×10^3 km/h [E]
    (b) 3.4×10^3 km/h

27. 5.7×10^3 s or 1.6 h

28. 926 Hz, 786 Hz

30. 6.0×10^7

Chapter 8 Review, pp. 312–313

3. (a) 5 : 4 high
   (b) 41 : 40 low

4. (a) 3200 Hz
   (b) 400 Hz

7. 480 Hz

8. (a) [illegible]
   (b) 400 Hz

9. (a) 600 Hz
   (b) 100 Hz

11. 1200 Hz

12. 4 Hz

13. (a) 143 Hz
    (b) 430 Hz
    (c) 717 Hz

14. (a) 1.00 m
    (b) 338 Hz

15. 0.66 m

16. 338 m/s

17. (a) 660 Hz, 990 Hz
    (b) 990 Hz, 1650 Hz

18. 0.54 m

24. 12 dB

Unit 3 Review, pp. 316–319

1. (a) 2.5 s, 3.0 s
   (c) 5
2. 0.4 s
3. 905 Hz
4. 3.4 m to 2.8 m
8. 0.70 m
12. 345 m/s, 319 m/s
13. 426 Hz
14. 0.39 s
15. 1.8×10^2 m
16. 198 km/h
17. 3.6×10^2 m/s
18. 3.4×10^2 m/s
21. 246 Hz
22. 14 Hz
23. (a) 2.33 m
    (b) 3.5×10^2 m/s
24. 261 Hz
25. 256 Hz, 1024 Hz
27. (a) 12 Hz
    (b) 14 m
29. (a) 17 cm
    (b) 50 cm, 84 cm
30. 340 m/s
31. 0.20 m, 0.59 m
32. (a) 340 Hz, 680 Hz
    (b) 170 Hz, 340 Hz
33. (a) 0.38 m
    (b) 226 Hz
34. 6.7 cm/s
36. (a) 1.4×10^3 m
    (b) 0.16 s
37. (a) 2.0×10^{-5} s to 2.0×10^{-3} s
    (b) 3.0×10^{-2} m to 3.0 m
38. (a) 3.3×10^{-5} s
    (b) 1.2×10^{-2} m
    (c) 8.0 m
41. 6.5
42. 160×, 10×
44. (b) 9.4 cm
49. 85 cm/s
55. (c) 1.4×10^3 km

Chapter 9 Review, pp. 352–353

7. 2.42 (diamond)
8. 1.96×10^8 m/s; 1.99×10^8 m/s
9. 10.7°
10. (a) 0.0°
    (b) 19.5°
    (c) 35.3°
11. 34.2°
12. (a) 36.5°
    (b) 1.56°

Chapter 10 Review, pp. 382–383

12. -1.0×10^2 cm
13. –9.4 cm
14. 12 cm or 24 cm
15. –10.0 cm; 110.0 cm mark (10.0 cm off end of metre stick)
16. 270 cm
20. (a) 15.8 cm
    (b) virtual
    (c) upright
    (d) 0.42 cm
21. (a) 30.0 cm
    (b) real
    (c) inverted
    (d) –20.0 cm
22. 91 cm
23. 40 cm
25. –10.0 d
26. (b) –3.97
    (c) 14.7
    (d) –58.4
32. 31°
35. (a) 50.0°
    (b) 1.53
    (c) 1.96×10^8 m/s
    (e) 40.8°

Chapter 11 Review, pp. 410–411

7. 5.7×10^{-5} m
11. 0.10 m
14. 400×
15. 75×
17. (b) –40.0×
    (c) 10.0×
    (d) –400×
21. (b) *f*/5.6

Unit 4 Review, pp. 414–417

1. 7.5×10^{14} Hz
4. 6.0 m
6. 1.20
8. 2.17×10^8 m/s
9. 49°
16. (a) $d_i = -56$ mm; $h_i = 3.4$ mm
    (b) $d_i = -28$ cm; $h_i = 9.3$ cm
17. 44 cm
18. 42 cm
19. 18 mm; –30 cm; virtual

Chapter 12 Review, pp. 468–469

1. (b) -4.8×10^{-8} C
   (c) 1.9×10^{18}
2. 6.3×10^{16}
3. 20 V; 16 V; 16 V; 6 A; 1 A
4. 9.0×10^1 C
5. 3.2 min
6. 0.21 A
7. 30 V; 30 V; 90 V; 6A; 1 A; 5 A; 6 A
8. 20 Ω; 9.0 V
9. 4 Ω
10. 4.0×10^3 V
11. 7.2×10^{-12} J
12. 2.4×10^{-13} J
13. 1.0×10^{-1} C
14. 7.6×10^3 J
15. 120 V
16. (a) 4.0×10^{10} W
    (b) 8.3×10^2 A
    (c) 4.8×10^7 V
20. 23¢
21. (a) A: 30 Ω; B: 10 Ω

Chapter 13 Review, pp. 504–505

8. 1 : 5

Chapter 14 Review, pp. 538–539

3. (a) 0.75 V
   (b) 3.0 V
   (c) 4.5 V
5. (a) 1 : 20
   (b) 5.0 A
6. 8.0 V
7. 16 mA
8. 3.0×10^1 A

Unit 5 Review, pp. 542–545

2. (a) 6.4×10^{-7} C
   (b) -9.9×10^{-8} C
3. (a) 5.0×10^{-9} C
   (b) -9.9×10^{11}
4. 6.3×10^3 C
7. 120 V
8. (a) 0.86 A
   (b) 7.7 W
   (c) 2.3×10^2 J
   (d) 26 C
   (e) 1.6×10^{20}
9. $15
10. Diagram (a)
    (a) $I_1 = I_2 = I_3 = 1.0$ A
    (b) $V_1 = 2.0$ V; $V_2 = 3.0$ V; $V_3 = 1.0$ V
    (c) 1.1×10^4 J

    Diagram (b)
    (a) $I_1 = 4.0$ A; $I_2 = 3.0$ A; $I_3 = 2.0$ A
    (b) $V_1 = V_2 = V_3 = 12.0$ V
    (c) 1.9×10^5 J

    Diagram (c)
    (a) $I_1 = 1.0$ A; $I_2 = I_3 = I_4 = 0.50$ A
    (b) $V_1 = 7.0$ V; $V_2 = 3.0$ V; $V_3 = 1.0$ V; $V_4 = 2.0$ V
    (c) 1.8×10^4 J
11. (a) 10 Ω
    (b) 30 Ω
    (c) 0.80 A
16. (a) –4.0 V
    (b) 2.8 V
    (c) 2.0 V
    (d) 8.0 V
17. (a) 300 V
    (b) 150
18. 100
20. (a) 0.80 A
    (b) 0.77 W
    (c) 1.9×10^{-6}

Glossary

A

accelerated motion: nonuniform motion that involves change in an object's speed or direction or both

acceleration: rate of change of velocity

acceleration due to gravity: the vector quantity 9.8 m/s^2 [down], represented by the symbol $\vec{g}$

acoustics: the total effect of sound produced in an enclosed space

active solar heating: the process of absorbing the Sun's energy and converting it into other forms of energy

alternating current: (AC) results when charges periodically reverse direction

ammeter: a device that measures the amount of electric current in a circuit

ampere: (A) the SI unit of electric current, 1 A = 1 C/s

amplitude: distance from the equilibrium position to maximum displacement

angle of incidence: (θ_i) the angle between the incident ray and the normal

angle of reflection: (θ_r) the angle between the reflected ray and the normal

angle of refraction: (θ_R) the angle between the refracted ray and the normal

antinode: point midway between the nodes where maximum constructive interference occurs

arc: light produced by air molecules when a current jumps a gap in an electric circuit

armature: a pivoted bar of soft iron

astigmatism: eye defect in which the cornea or the lens of the eye is not perfectly spherical

atmosphere: the air in a specific place that can be used as a source of heat

atom: sub-microscopic particle of which all matter is made

average acceleration: change of velocity divided by the time interval for that change

average speed: total distance of travel divided by total time of travel

average velocity: change of position divided by the time interval for that change

B

base unit: unit from which other units are derived or made up

beat frequency: the number of beats heard per second in hertz

beats: periodic changes in sound intensity caused by interference between two nearly identical sound waves

biomass energy: the chemical potential energy stored in plants and animal wastes

bitumen: a mixture of hydrocarbons and other substances that occurs naturally or is obtained by distillation from coal or petroleum

cataract: opaque, cloudy area that develops in a normally clear eye lens

charge coupled device: (CCD) a light-sensitive device used to store data

chromatic aberration: coloured fringes around objects viewed through lenses

circuit: a path for electric current

closed air column: column closed at one end and open at the other

coefficient of friction: ratio of the magnitude of friction to the magnitude of the normal force

coefficient of kinetic friction: ratio of the magnitude of kinetic friction to the magnitude of the normal force

coefficient of static friction: ratio of the magnitude of the maximum static friction to the magnitude of the normal force

cogeneration: the process of producing electricity and using the resulting thermal energy for heat

compression: region in a longitudinal wave where the particles are closer together than normal

conduction: the process of transferring heat through a material by the collision of atoms

conductor: solid in which charge flows freely

consonance: combinations of sounds of specific frequencies that are pleasing to the ear; the frequencies are often in a simple ratio

constructive interference: occurs when waves build each other up, resulting in the medium having a larger amplitude

convection: the process of transferring heat by a circulating path of fluid particles

conventional current or **electric current:** describes electric charges travelling through a conductor from the positive terminal to the negative terminal of the source of electric potential

converging lens: lens that causes parallel light rays to come together so that they cross at a single focal point

coulomb: (C) the SI unit of electric charge

Coulomb's Law: The magnitude of the force between two charged objects is directly proportional to the product of the charges and inversely proportional to the square of the distance between them.

crest or **positive pulse:** high section of a wave

critical angle: (θ_c) the angle in an optically denser medium at which total internal reflection occurs; at this angle the angle of refraction in the less dense medium is 90°

cycle: one complete vibration or oscillation

D

decibel: (dB) unit used to measure sound intensity level ($0 \text{ dB} = 10^{-12} \text{ W/m}^2$)

derived unit: unit that can be stated in terms of the seven base units

destructive interference: occurs when waves diminish one another and the amplitude of the medium is less than it would have been for either of the interfering waves acting alone

diamagnetic: materials that cause a very slight decrease in the magnetic field of a coil

diffraction: the bending effect on a wave's direction as it passes through an opening or by an obstacle

digital camera: camera that uses digital information instead of film to capture images

dipoles: atoms of ferromagnetic substances that act like tiny magnets

direct current: (DC) results when charges flow in a particular direction

dispersion: the spreading of white light into a spectrum of colours

displacement: change in position of an object in a given direction

dissonance: combinations of sounds of specific frequencies that have a harsh effect; the frequencies are not in a simple ratio

diverging lens: lens that causes parallel light rays to spread apart so that they appear to emerge from the virtual focal point

Doppler effect: when a source of sound approaches an observer, the observed frequency increases; when the source moves away from an observer, the observed frequency decreases

doubling time: the time required for an amount to double

dynamics: the study of the causes of motion

echo: reflected sound waves

echolocation: location of objects through the analysis of reflected sound.

eddy currents: induced currents that form closed loops within a conductor

efficiency: the ratio of the useful energy provided by a device to the energy required to operate the device

electric charge: a basic property of matter described as negative or positive

electric field: the space around a charged object where forces of attraction or repulsion act on other objects

electric generator: a device that converts the mechanical energy of motion into electrical energy

electric potential difference: (V) the amount of work required per unit charge to move a positive charge from one point to another in the presence of an electric field

electromagnet: object that exerts a magnetic force using electricity

electromagnetic force: force caused by electric charges

electron: negatively charged particle which moves around the nucleus of an atom

electron flow: a term used to indicate that the electric current in metals is due to the motion of electrons

electroscope: device that is used to detect the presence of an electric charge and to determine the charge's "sign" (that is, whether it is positive or negative)

electrostatic series: chart that shows a substance's tendency to gain or lose electrons

elementary charge: (e) electric charge of magnitude equal to the charge on a proton and an electron

elements: the components of an electric circuit

energy: the capacity to do work

energy resource: raw material obtained from nature that can be used to do work

energy transformation: the change from one form of energy to another

energy transformation technology or **energy converter:** a system that converts energy from some source into a usable form

equivalent resistor: resistor that has the same current and potential difference as the resistors it replaces

ferromagnetic: a substance that can become magnetized

first law of motion: If the net force acting on an object is zero, the object will maintain its state of rest or constant velocity.

fixed-end reflection: reflection from a rigid obstacle when a pulse is inverted

focal length: (f) the distance between the principal focus and the optical centre, measured along the principal axis

focal plane: the plane, perpendicular to the principal axis, on which all focal points lie

focal point: (waves) a specific place where straight waves are reflected to; (optics) the position where parallel incident rays meet, or appear to come from, after they pass through a lens

force: a push or a pull

force field: space surrounding an object in which the object exerts a force on other objects placed in the space

frame of reference: coordinate system relative to which motion can be observed

free-body diagram: (FBD) drawing in which only the object being analyzed is drawn, with arrows showing all the forces acting on the object

free-end reflection: reflection where the new medium is free to move and there is no inversion

frequency: (f) the number of cycles per second $f = \frac{\text{number of cycles}}{\text{total time}}$

friction: force between objects in contact and parallel to contact surfaces

***f*-stop number:** the ratio of the focal length of the lens to the diameter of the aperture; $f\text{-stop} = \frac{f}{d}$

fuel cell: device that changes chemical potential energy directly into electrical energy

fundamental forces: forces are classified into four categories—gravitational, electromagnetic, strong nuclear, and weak nuclear

fundamental frequency: (f_0) the lowest natural frequency

fundamental laws of electric charges: Opposite charges attract each other. Similar electric charges repel each other. Charged objects attract some neutral objects.

fundamental mode: simplest mode of vibration producing the lowest frequency

G

galvanometer: device used to measure the magnitude and direction of small electric currents

general equation of Snell's law: $n_1 \sin \theta_1 = n_2 \sin \theta_2$

geothermal energy: thermal energy or heat taken from beneath Earth's surface

glaucoma: eye defect in which the optic nerve is damaged

gravitational field strength: the amount of force per unit mass acting on objects in the gravitational field

gravitational force or **force of gravity:** force of attraction between all objects

gravitational potential energy: the energy possessed by an object because of its position relative to a lower position

ground: a connection of an object to Earth through a conductor

H

harmonics: whole-number multiples of the fundamental frequency

heat: the transfer of energy from a warmer body or region to a cooler one

heat pump: a device that uses evaporation and condensation to heat a home in winter and cool it in summer

hertz: 1 Hz = 1 cycle/s or 1 Hz = 1 s^{-1}, since cycle is a counted quantity, not a measured unit

hydraulic energy: energy generated by harnessing the potential energy of water

hydrocarbons: compounds that contain only carbon and hydrogen

hyperopia: eye defect in which nearby images are not seen clearly; also called farsightedness

I

image point: point at which light from an object point converges

incident ray: the approaching ray of light

index of refraction: (n) the ratio of the speed of light in a vacuum (c) to the speed of light in a given material (v); $n = \frac{c}{v}$

induced charge separation: distribution of charge that results from a change in the position of electrons in an object

induced field: the magnetic field produced by the induced current

inducing field: the magnetic field that causes the induced current

inertia: the property of matter that causes a body to resist changes in its state of motion

infrasonic: any sound with a frequency lower than the threshold of human hearing (approximately 20 Hz)

in phase: objects are vibrating in phase if they have the same period and pass through the rest position at the same time.

instantaneous acceleration: acceleration at a particular instant

instantaneous speed: speed at a particular instant

instantaneous velocity: velocity that occurs at a particular instant

insulator: solid that hinders the flow of charge

J

joule: (J) the SI unit for work

K

kilowatt hour: (kW•h) the energy dissipated in exactly 1 h by a load with a power of exactly 1 kW

kinematics: the study of motion

kinetic energy: the energy possessed by an object due to its motion

kinetic friction: the force that acts against an object's motion in a direction opposite to the direction of motion

Kirchhoff's current law: (KCL) At any junction point in an electric circuit, the total electric current into the junction is equal to the total electric current out.

Kirchhoff's voltage law: (KVL) Around any complete path through an electric circuit, the sum of the increases in electric potential is equal to the sum of the decreases in electric potential.

L

lateral displacement: sideways shifting of the path of a refracted ray

law of conservation of charge: Electric charge is neither created nor lost in an electric circuit, nor does it accumulate at any point in the circuit.

law of conservation of energy: When energy changes from one form to another, no energy is lost; (electricity) as electrons move through an electric circuit they gain energy in sources and lose energy in loads, but the total energy gained in one trip through a circuit is equal to the total energy lost.

law of electromagnetic induction: An electric current is induced in a conductor whenever the magnetic field in the region of the conductor changes.

law of magnetic poles: opposite magnetic poles attract; similar magnetic poles repel

law of universal gravitation: The force of gravitational attraction between any two objects is directly proportional to the product of the masses of the objects, and inversely proportional to the square of the distance between their centres.

laws of reflection: The angle of incidence is equal to the angle of reflection. The incident ray, the reflected ray, and the normal all lie in the same plane.

laws of refraction: The ratio of sin θ_i to sin θ_R is a constant (Snell's law). The incident ray and the refracted ray are on opposite sides of the normal at the point of incidence, and all three are in the same plane.

lens: transparent device with at least one curved surface that changes the direction of light passing through it

lens camera: light-proof box inside which a converging lens directs light onto a film to capture a real image

Lenz's law: For a current induced in a coil by a changing magnetic field, the electric current is in such a direction that its own magnetic field opposes the change that produced it.

light: a form of energy that is visible to the human eye

linear propagation: one of the properties of light; describes light as travelling in straight lines

load: any device in a circuit that transforms electrical potential energy into some other form of energy, causing an electric potential drop

local consumption: generating energy locally to avoid the transfer of energy over long distances using transmission lines

longitudinal vibration: occurs when an object vibrates parallel to its axis

longitudinal wave: particles vibrate parallel to the direction of motion of the wave

Mach number: the ratio of the speed of an object to the speed of sound in air; Mach number $= \frac{\text{speed of object}}{\text{speed of sound}}$

magnetic domain: formed by a group of dipoles lined up with their magnetic axes in the same direction

magnetic field of force: the space around a magnet in which magnetic forces are exerted

magnetic induction: the process of magnetizing an object from a distance

Magnetic Resonance Imaging: (MRI) diagnostic technique that produces high-quality images of soft tissue inside the body

magnetic saturation: occurs when the maximum amount of dipoles in a material are aligned

mass: the quantity of matter in an object

mechanical energy: the sum of the gravitational potential energy and the kinetic energy

microscope: instrument that produces enlarged images of objects too small to be seen by the unaided eye

motor principle: A current-carrying conductor that cuts across external magnetic field lines experiences a force perpendicular to both the magnetic field and the direction of electric current. The magnitude of this force depends on the magnitude of both the external field and the current, as well as the angle between the conductor and the magnetic field it cuts across.

music: sound that originates from a source with one or more constant frequencies

mutual induction: an effect that occurs whenever a changing current in one coil induces a current in a nearby coil

myopia: eye defect in which distant objects are not seen clearly; also called nearsightedness

negative ion: an atom that has at least one extra electron and is negatively charged

neutron: a neutral particle found in the nucleus

newton: (N) the SI unit of force

node or **nodal point:** (*N*) point that remains at rest

noise: sound that originates from a source where the frequencies are not constant

non-renewable energy resource: an energy resource that does not renew itself in the normal human lifespan

nonuniform motion: movement that involves change in speed or direction or both

normal: (waves) a straight line drawn perpendicular to a barrier struck by a wave; (optics) the line drawn at right angles to the reflecting surface at the point of incidence

normal force: force perpendicular to the surfaces of the objects in contact

N-pole: the end of a magnet that seeks the northerly direction

nuclear fission: process in which the nucleus of each atom splits and releases a relatively large amount of energy

nuclear fusion: the process in which the nuclei of the atoms of light elements join together at extremely high temperatures to become larger nuclei

nucleus: the central region of an atom, where protons and neutrons are found

O

objective: in an optical instrument, the lens closest to the object

object point: a point representing a source of rays, whether by production or reflection

octave: sounds that differ in frequency by a factor of two

ocular: in an optical instrument, the lens the eye looks into

ohm: (Ω) the SI unit for electric resistance

Ohm's law: The potential difference between any two points in a conductor varies directly as the current between the two points if the temperature remains constant.

open air column: column open at both ends

optical axis: (OA) a vertical line through the optical centre

optical centre: (O) the geometric centre of all lenses

out of phase: objects are vibrating out of phase if they do not have the same period or if they have the same period but they do not pass through the rest position at the same time.

overtones: the resulting modes of vibration when a string vibrates in more than one segment emitting more than one frequency

parallel circuit: circuit in which charges can move along more than one path

paramagnetic: material that magnetizes very slightly when placed in a coil and increases the field strength by a barely measurable amount

partial reflection: some of the energy is transmitted into the new medium and some is reflected back into the original medium

partial reflection and refraction: a beam of light splitting into two rays; one is reflected; the other is refracted

passive solar heating: the process of designing and building a structure to take best advantage of the Sun's energy at all times of the year

period: (T) the time required for one cycle $T = \frac{\text{total time}}{\text{number of cycles}}$

periodic motion: motion that occurs when the vibration, or oscillation, of an object is repeated in equal time intervals

periodic waves: originate from periodic vibrations where the motions are continuous and are repeated in the same time intervals

pitch: the perception of the highness or lowness of a sound; depends primarily on the frequency of the sound

pixel: image-forming unit

point of incidence: the point at which the incident ray strikes the reflecting surface

poles: areas of concentrated magnetic force

position: the distance and direction of an object from a reference point

positive ion: an atom that has lost at least one electron and is positively charged

power: (P) the rate of doing work or transforming energy, or the rate at which energy is used or supplied

presbyopia: eye defect in which the lens in the eye loses elasticity; a type of farsightedness

primary rainbow: the rainbow of visible spectral colours that results from a single internal reflection in rain drops

principal axis: (PA) a horizontal line drawn through the optical centre

principal focus: (F) the point on the principal axis through which a group of rays parallel to the principal axis is refracted

principle of electromagnetism: Whenever an electric current moves through a conductor, a magnetic field is created around the conductor.

principle of heat exchange: When heat is transferred from one body to another, the amount of heat lost by the hot body equals the amount of heat gained by the cold body.

principle of superposition: At any point the resulting amplitude of two interfering waves is the algebraic sum of the displacements of the individual waves.

prism: a transparent solid that has at least two plane refracting surfaces

proton: positively charged particle found in the nucleus of an atom

pulse: wave that consists of a single disturbance

pure tone: a sound that consists of one frequency

quality of a musical note: a property that depends on the number and relative intensity of harmonics that make up the sound

radiation: the process in which energy is transferred by means of electromagnetic waves

rarefaction: region in a longitudinal wave where the particles are farther apart than normal

ray: the path taken by light energy

reference level: the level to which an object may fall

reflected ray: the ray of light reflected from the reflecting surface

reflecting telescope: telescope that uses a parabolic mirror to focus light

refracting telescope: telescope that uses two converging lenses to focus light

refraction: (waves) the bending effect on a wave's direction that occurs when the wave enters a different medium at an angle; (optics) the change in direction of light as it passes from one medium into another of differing density

relative magnetic permeability: (K) the factor by which a core material increases the magnetic field strength that would exist in the same region if a vacuum replaced the material

relative velocity: velocity of a body relative to a particular frame of reference

relay: a device in which a switch is closed by the action of an electromagnet

renewable energy resource: an energy resource that renews itself in the normal human lifespan

resistance: an opposition to the flow of charge, resulting in a loss of potential energy

resonance or **mechanical resonance:** the transfer of energy from one object to another having the same natural frequency

resonator: an object, usually a hollow chamber, that vibrates in resonance with a source of sound

resultant displacement: vector sum of the individual displacements

resultant force or **net force:** the vector sum of all the forces acting on an object

reverberation time: the time required for the intensity of the sound to drop to 10^{-6} (one millionth) of its original value or until the sound is inaudible

rheostat: device in an electric circuit that can be adjusted to different resistances, changing the current in the circuit

right-hand rule for a coil: If a coil is grasped in the right hand with the curled fingers representing the direction of electric current, the thumb points in the direction of the magnetic field inside the coil.

right-hand rule for a conductor: If a straight conductor is held in the right hand with the right thumb pointing in the direction of the electric current, the curled fingers will point in the direction of the magnetic field lines.

right-hand rule for the motor principle: If the fingers of the open right hand point in the direction of the external magnetic field, and the thumb represents the direction of electric current, the force on the conductor will be in the direction in which the right palm faces.

S

scalar quantity: quantity that has magnitude, but no direction

second law of motion: if the net external force on an object is not zero, the object accelerates in the direction of the net force, with magnitude of acceleration proportional to the magnitude of the net force and inversely proportional to the object's mass

semiconductor: material used to transmit and amplify electronic signals

series circuit: circuit in which charges have only one path to follow parallel circuit: circuit in which charges can move along more than one path

single-lens reflex (SLR) camera: camera that uses a single lens and a movable mirror

Snell's law: The ratio of the sine of the angle of incidence to the sine of the angle of refraction is a constant.

solar energy: radiant energy from the Sun

solenoid: a coil of wire that acts like a magnet when an electric current passes through it

sonar: system that uses transmitted and reflected underwater sound waves to locate objects or measure the distance to the bottom of the water body

sonic boom: an explosive sound that radiates from an aircraft travelling at supersonic speeds

sound: a form of energy, produced by rapidly vibrating objects, that can be heard by the human ear

sound barrier: a high-pressure region produced as an airplane approaches a speed of Mach 1

sound intensity: power of a sound per unit area in watts per metre squared

specific heat capacity: (c) a measure of the amount of energy needed to raise the temperature of 1.0 kg of a substance by 1.0°C

speed of light: (c) the speed of light in a vacuum with a value of 299 792 458 m/s, or 3.00×10^8 m/s

S-pole: the end of a magnet that seeks the southerly direction

standing wave: created when waves travelling in opposite directions have the same amplitude and wavelength

starting friction: the amount of force that must be overcome to start a stationary object moving

static electricity: a build up of stationary electric charge on a substance

static friction: the force that tends to prevent a stationary object from starting to move

step-down transformer: a transformer with fewer windings on the secondary coil, resulting in decreased voltage

step-up transformer: a transformer with more windings on the secondary coil, resulting in increased voltage

strong nuclear force: force that holds protons and neutrons together in the nucleus of an atom

subsonic speed: speed less than the speed of sound in air

supercrest: occurs when a crest meets a crest

supersonic speed: speed greater than the speed of sound in air

supertrough: occurs when a trough meets a trough

switch: a device that is used to make or break an electric circuit

sympathetic vibration: the response to a vibration with the same natural frequency

system diagram: sketch of all the objects involved in a situation

tangent: a straight line that touches a curve at a single point and has the same slope as the curve at that point

tangent technique: a method of determining velocity on a position-time graph by drawing a line tangent to the curve and calculating the slope

temperature: a measure of the average kinetic energy of the atoms or molecules of a substance

tension: force exerted by string, ropes, fibres, and cables

terminal speed: maximum speed of a falling object at which point the speed remains constant and there is no further acceleration

terrestrial telescope: telescope that is similar to a refractor, but with the addition of a third lens, the erector lens, to obtain upright images

thermal energy: the total kinetic energy and potential energy of the atoms or molecules of a substance

thin lens equation: the quantitative relationship for thin lenses; $\frac{1}{d_o} + \frac{1}{d_i} = \frac{1}{f}$

third law of motion: for every action force, there is a reaction force equal in magnitude, but opposite in direction

tidal energy: energy generated by harnessing the gravitational forces of the Moon and the Sun that act on Earth

torsional vibration: occurs when an object twists around its axis

total internal reflection: the reflection of light in an optically denser medium; it occurs when the angle of incidence in the more dense medium is equal to or greater than a certain critical angle

transformer: device that consists of a core of soft iron with two separate coils of wire used to change the voltage

transverse vibration: occurs when an object vibrates perpendicular to its axis

transverse wave: particles in the medium move at right angles to the direction in which the wave travels

trough or **negative pulse:** low section of a wave

turbine: machine in which the kinetic energy of a moving fluid is converted to rotary mechanical power

U

ultrasonic: any sound with a frequency above the range of human hearing (approximately 20 000 Hz)

uniformly accelerated motion: motion that occurs when an object travelling in a straight line changes its speed uniformly with time

uniform magnetic field: the magnetic field is the same strength and acts in the same direction at all points

uniform motion: movement at a constant speed in a straight line

universal wave equation: $v = f\lambda$

V

vector quantity: quantity that has both magnitude and direction

velocity: the rate of change of position

virtual image point: point from which rays from an object point appear to diverge

volt: (V) the SI unit for electric potential difference; 1 V = 1 J/C

voltmeter: a device that measures electric potential difference between two points in a circuit

W

watt: (W) the SI unit for power

wave: a transfer of energy over a distance, in the form of a disturbance

wavefront: the leading edge of a wave

wave interference: occurs when two or more waves act simultaneously on the same particles of a medium

wavelength: (λ) distance between successive wave particles that are in phase

wave ray: a straight line drawn perpendicular to a wavefront that indicates the direction of transmission

weak nuclear force: force responsible for interactions involving elementary particles such as protons and neutrons

weight: the force of gravity on an object

wind energy: energy generated by harnessing the kinetic energy of wind

work: the energy transferred to an object by an applied force over a measured distance

Index

Q

R

S

T

Credits

Contents: Page vi: First Light; Page vii: Top: © Ron Stroud/Masterfile, Bottom: © Rich Fischer/Masterfile; Page viii: Stuart Westmorland/ Tony Stone Images; Page ix: G. Locke/First Light

Unit 1 Opener, Page x: First Light; Page 1: CP Picture Archive (Peter Cosgrove); Chapter 1 Opener, Pages 4-5: Photofest; Page 7: Top: National Research Council Canada, Bottom: © Duomo/CORBIS/ Magma; Page 8: © Bettman/CORBIS/Magma; Page 9: Top: Dr. H.E. Edgerton/Palm Press, Bottom: Photo courtesy of Boreal Laboratories 1-800-387-9393; Page 12: CP Picture Archive (Tony Bock); Page 18: © Larry Mulvehill/Photo Researchers; Page 37: Paramount Canada's Wonderland; Page 40: Top: Courtesy Department of National Defence, Bottom: NASA; Page 41: Top: © Richard Olivier/CORBIS/ Magma, Bottom: CP Picture Archive (James Forsyth); Page 45: Dave Starrett

Chapter 2 Opener, Page 52: Al Hirsch; Page 53: Both: Dave Starret; Page 54: © Chris Mattison; Frank Lane Picture Agency/CORBIS/ Magma; Page 55: © Michael Syamashita/CORBIS/Magma; Page 56: James Gritz/PhotoDisc; Page 57: © Bettman/CORBIS/Magma; Page 62: Top: © Hulton-Deutsch Collection/CORBIS/Magma, Bottom: Ryan McVay/PhotoDisc; Page 63: Ryan McVay/PhotoDisc; Page 64: Tim Wright/CORBIS/Magma; Page 66: CORBIS/Magma; Page 67: CP Picture Archive (Bruce Bunstead); Page 68: Dave Starrett; Page 73: Corel; Page 74: NASA; Page 77: NASA; Page 78: Top: Michael Newman/Photo Edit, Middle: © Steve Chenn/CORBIS/Magma, Bottom: CORBIS/Magma

Chapter 3 Opener, Page 83: © AFP/CORBIS/Magma; Page 84: Top: CP Picture Archive (Chuck Stoody), Middle: © Dave G. Houser/CORBIS/Magma; Page 88: NASA; Page 90: Top: CORBIS/Magma, Middle: NASA; Page 94: Top: © The Purcell Team/CORBIS/Magma, Bottom: NASA; Page 96: © Jens Meyer/Associated Press; Page 97: Right: Inga Spence/Visual Unlimited, Bottom: © Dean Conger/CORBIS/Magma; Page 98: Al Hirsch; Page 100: © Jan Hinsch/SPL/Photo Researchers; Page 106: Tom Kinsbergen/Science Photo Library; Page 112: Top: Corel, Bottom Left: PhotoDisc, Bottom Right: CORBIS/Magma; Page 112: NASA; Page 117: Paramount Canada's Wonderland

Unit 2 Opener, Pages 118-119: © Ron Stroud/Masterfile; Page 119: Courtesy of Edwin Lim; Chapter 4 Opener, Page 123: Right: CP Picture Archive (Peter Power), Inset: Dave Starrett, CP Picture Archive (Peter Power); Page 125: Top Left: Nelson Photo, Top Right: PhotoDisc, Bottom Left: CP Picture Archive (Frank Gunn), Bottom Right: CP Picture Archive (Tom Hanson); 127: © CORBIS/Magma; Page 130: CP Picture Archive (Adrian Wyld); Page 131: Top: © Lester Lefkowitz/Stock Market/First Light, Bottom: Jeff Greenberg/Visuals Unlimited; Page 133: Photo Researchers; Page 134: Top: A.J. Copley/ Visuals Unlimited, Middle: © Roger Ressmeyer/CORBIS/ Magma; Page 135: Jeremy Woodhouse/PhotoDisc; Page 145: CP Picture Archive (Jacques Boissinot); Page 148: © Tony Freeman/Photo Edit; Page 149: © James L. Amos/CORBIS/Magma; © Richard Cummins/ CORBIS/Magma; Page 154: © Bettman/CORBIS/Magma; Page 156: © CORBIS/Magma

Chapter 5 Opener, Page 157: © Paul A. Souders/CORBIS/Magma; Page 162: Superstock; Page 166: © Barrett & MacKay Photography; Page 171: Visuals Unlimited; Page 173: Superstock; Page 174: Environment Canada; Page 175: Top: Courtesy of Nova Scotia Power Inc.; Bottom: Al Hirsch; Page 178: Reproduced with the permission of Her Majesty the Queen in Right of Canada, 2001; Page 180: Left: © Dwayne Newton/Photo Edit, Middle: www.comstock.com, Right: Dave Thompson/Life File/PhotoDisc; Page 183: Al Hirsch, Bottom: CP Picture Archive (Jonathan Hayward); Page 186: Al Hirsch; Page 189: Paramount Canada's Wonderland

Unit 3 Opener: © Rick Fischer/Masterfile; Page 191: Courtesy of Dr. John Ford; Page 193: Courtesy of Bridgeview School; Chapter 6 Opener, Page 195: Corel, Inset (both): Dave Starrett; Page 199: Larry Stepanowicz/Visuals Unlimited; Page 212: Educational Development Center Inc.; Page 213: Educational Development Center Inc.; Page 215: Top: Larry Stepanowicz/Visuals Unlimited, Bottom: © Richard Megma/Fundamental Photographs; Page 217: Top: © Fundamental Photographs, Bottom: © Richard Megma/Fundamental Photographs; Page 224: a: Associated Press, b: Associates Press, c: CP Picture Archive (Associated Press), d: Associated Press; Page 229: © Richard Megma/Fundamental Photographs; Page 231: John D. Cunningham/Visuals Unlimited

Chapter 7 Opener, Pages 236-237: Comstock; Page 238: CP Picture Archive/Ian McAlpine; Page 239: Allan Pappe/PhotoDisc; Page 247: © CORBIS/Magma: Page 252: CORBIS/Magma; Page 253: a: Jane Shemitt/Science Photo Library, b: Jane Shemitt/Science Photo Library, c: Science VU/Visuals Unlimited; Page 254: Ontario Science Centre; Page 255: Top: Courtesy of The Centre In The Square Kitchener, Bottom: CP Picture Archive (Ho); Page 256: Ian Cartwright/PhotoDisc; Page 257: Top: © Tony Roberts/CORBIS/Magma, Bottom: Dept. of Clinical Radiology/Salisbury District Hospital/Science Photo Library; Page 260: Dick Hemingway; Page 263: Photo Courtesy of Sony; Page 270: CP Picture Archive (J. Conrad Williams Pool)

Chapter 8 Opener, Page 277: Courtesy of Roy Thomson Hall; Page 285: Educational Development Center Inc., Page 293: C Squared Studios/PhotoDisc; Page 294: Top: © Bob Krist/CORBIS/Magma; Bottom: CMCD/PhotoDisc; Page 295: © Kelly-Monney Photography/CORBIS/Magma; Page 297: C Squared Studios/PhotoDisc; Page 301: Top: C Squared Studios/PhotoDisc, Bottom: CORBIS/Magma; Page 305: Corel; Page 308: Top: © Roger Ressmeyer/CORBIS/Magma, © Elio Ciol/CORBIS/Magma; Page 310: Left: © AFP/CORBIS/Magma, Middle: First Light, Right: © Michael Newman/Photo Edit; Page 314: Top: Duo Trapeze act from the "O™" show by Cirque du Soleil®, Photo: Veronique Vial, Joan Francois Grafton, Costumes: Dominique Lemieux, © Cirque du Soleil; Bottom: © Jan Butschofsky-Houser/CORBIS/Magma

Unit Opener 4: Stuart Westmorland/Tony Stone Images; Page 321: Courtesy of Melanie Campbell; Page 322: Top: CORBIS/Magma, Bottom: Dave Starrett; Page 323: Dave Starrett; Chapter 9 Opener, Page 324: Courtesy ESO; Page 325: Courtesy ESO; Page 327: Both: David Parker/Science Photo Library; Page 331: © Richard Megma/Fundamental Photographs; Page 332: © Richard Megma/Fundamental Photographs; Page 345: © Spencer Grant/ Photo Researchers; Page 347: Top: Carl Tape, Bottom: © Charles O'Rear/ CORBIS/Magma; Page 348: Top: Corel,

Bottom: Photograph by Dr. Roy Bishop, Acadia University, Nova Scotia

Chapter 10 Opener, Page 355: Left: Mary Evans Picture Library, Middle: © Richard Cummins/CORBIS/Magma, Right: Chris Mills; Page 372: Jay M. Pasachoff/Visuals Unlimited; Page 373: Top: www.comstock.ca , Bottom: Superstock; Page 374: Graham Davis; Page 378: CP Picture Archive/David Lucas

Chapter 11 Opener, Pages 384-385: Bud Nielsen/Visuals Unlimited, Inset: Mary Evans Picture Library; Page 388: Top a, b, c: from "Seeing The Light: Optics in Nature, Photograph, Color, Vision and Holograph" by Falk, Brill and Stock, © 1986. Reprinted by permission of John Wiley & Sons, Inc., Bottom: Dave Starrett; Page 392: © Cindy Charles/Photo Edit; Page 404: IMAXCorporation, Page 405: Right: Photo Courtesy of Boreal Laboratories 1-800-387-9393, Bottom: www.comstock.ca; Page 408: Left: © Michael Newman/Photo Edit, Middle: Associated Press (Al Behrman), Right: © A. Ramey/Photo Edit; Page 412: © Paul. A. Souders/CORBIS/Magma; Page 413: Left: Siede Preis/PhotoDisc, Top Middle: CORBIS/Magma, Lower Middle: www.comstock.ca; Right: CORBIS/Magma

Unit 5 Opener, Pages 418-419: G. Locke/First Light; Page 419: Courtesy of Mehrdad Moallem; Chapter 12 Opener, Page 422: CP Picture Archive (Adrian Wyld); Page 423: Top: CP Picture Archive (Ryan Remiorz), Bottom: Fundamental Photographs; Page 424: © Richard T. Nowitz/CORBIS/Magma; Page 425: From "Physics: Algebra/Trig" by Eugene Hecht. Photo by Eugene Hecht; Page 434: © Bettman/CORBIS/Magma; Page 437: © Leonard de Selva/CORBIS/Magma; Page 442: Nelson Photo; Page 443: © Bettman/CORBIS/Magma; Page 446: Dan McCoy/Rainbow; Page 453: Hulton Getty/Archive Photos; Page 461: Corel; Page 464: Dick Hemingway; Page 465: Dick Hemingway

Page 470: © Spencer Grant/Photo Edit; Chapter 13 Opener, Page 471: Left: Corel, Right: www.comstock.ca; Bottom: Steve Cole/PhotoDisc; Page 472: Françoise Sauze/Science Photo Library; Page 473: All: Visuals Unlimited; Page 482: A. Bartel/Publiphoto/Science Photo Library; Page 499: Science Photo Library; Page 500: Mehau Kulyk/Science Photo Library

Chapter 14 Opener, Page 507: © Larry Lee/First Light; Page 508: © Hulton-Deutsch Collection/CORBIS/Magma; Page 531: Both: Dick Hemingway; Page 532: a: www.comstock.ca; b: David Chasey/PhotoDisc; Page 536: Left: © Roger Ressmeyer/CORBIS/Magma, Middle: © Mark Richards/Photo Edit, Right: © Jim Maguire/Image Network Inc.; Page 540: CP Picture Archive (Tannis Toohey)

Appendix Opener, Page 546: © Marc Garanger/CORBIS/Magma

Illustrators:
Cynthia Watada
Dave McKay
Deborah Crowle
Irma Ikonen
Linda Neale
Peter Papayanakis